John Fernandes

Jean-Marie Saudubray

Georges van den Berghe

John H. Walter (Editors)

Inborn Metabolic Diseases

Fourth, Revised Edition

John Fernandes
Jean-Marie Saudubray
Georges van den Berghe
John H. Walter (Editors)

Inborn Metabolic Diseases

Diagnosis and Treatment

Fourth, Revised Edition

With 65 Figures and 63 Tables

 Springer

John Fernandes
Department of Pediatrics
University Hospital Groningen
Private address: Burgemeester Weertslaan 31
8162 DP Epe, The Netherlands

Jean-Marie Saudubray
Unité de Métabolisme
Département de Pédiatrie
Hôpital Necker Enfants Malades
149 Rue de Sèvres
75043 Paris Cedex 15, France

Georges van den Berghe
Metabolic Research Group
Christian de Duve Institute of Cellular Pathology
University of Louvain Medical School
Avenue Hippocrate 75-39
1200 Brussels, Belgium

John H. Walter
Willink Biochemical Genetics Unit
Royal Manchester Children's Hospital
Hospital Road, Pendlebury
Manchester M27 4HA, Great Britain

Library of Congress Control Number: 2006928830

ISBN-10   3-540-28783-3 Springer Medizin Verlag Heidelberg
ISBN-13   978-3-540-28783-4 Springer Medizin Verlag Heidelberg

Springer Medizin Verlag
springer.com
© Springer Medizin Verlag Heidelberg 2000, 2006
Printed in Germany

Editor: Renate Scheddin
Project Manager: Meike Seeker
SPIN 10989814
Cover design: deblik Berlin
Typesetting: Fotosatz-Service Köhler GmbH, Würzburg, Germany
Printing and Binding: Stürtz GmbH, Würzburg, Germany

Printed on acid-free paper   2126 – 5 4 3 2 1 0

Preface

Since the publication of the first edition sixteen years ago, Inborn Metabolic Diseases – Diagnosis and Treatment has become a classic textbook, indispensable for those involved in the care of children and adults with inborn errors of metabolism, including pediatricians, biochemists, dieticians, neurologists, internists, geneticists, psychologists, nurses, and social workers. This new 4th edition has been extensively revised. An additional clinician, John Walter, has joined the three other editors, there is a new chapter on neonatal screening, including tandem MS/MS, and several new disorders have been included, for example defects involving the pentose phosphate pathway (polyol metabolism) and disorders of glucose transport. However, the focus of the book remains clinical, describing symptoms and signs at presentation, how to come to a diagnosis and methods for treatment.

As with the previous edition, the book can be used in two main ways.

If the diagnosis is not known the reader should first refer to Chapter 1. This chapter, which includes a number of algorithms and tables, lists the clinical findings under four main headings: the neonatal period and early infancy; acute presentation in late infancy and beyond; chronic and progressive disease; and specific organ involvement. In addition a list of important symptoms or signs can be found at the end of the chapter which then refers either to the text, a table, a figure, an algorithm, a list of disorders, or a combination of these. In this edition, the chapter incorporates many new findings, particularly in neuroradiology and neurophysiology, and emphasizes those disorders for which treatment is available.

If the diagnosis is already suspected, or is indicated from reference to Chapter 1, the reader can go directly to the relevant chapter to obtain more specific information. In order to simplify this process each of these chapters is presented in a uniform format.

For more detailed information, particularly with respect to pathophysiology and genetics, we recommend the eighth edition of the Molecular and Metabolic Bases of Inherited Disease, edited by Charles R. Scriver et al (Mc Graw-Hill, 2001). As in the previous edition, we also advocate referral to centres specialized in the diagnosis and treatment of inherited metabolic disorders. For countries of the European Union such a list is compiled by the Society for the Study of Inborn Errors of Metabolism (SSIEM), for the United States and Canada, Japan and Australia by the American, Japanese and Australian Societies of Inherited Metabolic Diseases (SIMD, JIMD, and AIMD, respectively).

The editors welcome new authors of old and new chapters and pay tribute to those authors who, though not participating this time, laid the framework for this book.

John Fernandes
Jean-Marie Saudubray
Georges van den Berghe
John H. Walter
Spring 2006

Contents

Contents

Contents

V Vitamin-Responsive Disorders

VI Neurotransmitter and Small Peptide Disorders

VII Disorders of Lipid and Bile Acid Metabolism

VIII Disorders of Nucleic Acid and Heme Metabolism

List of Contributors

Akman, Hasan O.
Department of Neurology,
Columbia University College
of Physicians & Surgeons,
630 West 168th Street, P & S 4-443,
New York, NY 10032, USA

Anderson, Karl E.
Department of Preventive Medicine and
Community Health,
The University of Texas Medical Branch,
700 Harborside Drive,
Galveston, TX 77555-11 09, USA

Andria, Generoso
Department of Pediatrics,
Federico II University,
Via Sergio Pansini 5,
80131 Naples, Italy

Aubourg, Patrick
Hôpital Saint-Vincent de Paul,
82 Avenue Denfert-Rochereau,
75674 Paris Cedex 14, France

Baumgartner, Matthias R.
Division of Metabolism and
Molecular Pediatrics,
University Children's Hospital,
Steinwiesstr. 75,
8032 Zürich, Switzerland

Bennett, Michael J.
Metabolic Diagnostic Laboratory,
The Children's Hospital of Philadelphia,
34th Street & Civic Center Boulevard,
Philadelphia, PA 19104, USA

Berry, Gerard T.
Division of Pediatrics,
Thomas Jefferson University,
1025 Walnut Street, Suite 102 College,
Philadelphia, PA 19107-5083, USA

Broyer, Michel
Unité de Néphrologie,
Département de Pédiatrie,
Hôpital Necker Enfants Malades,
149 Rue de Sèvres,
75743 Paris Cedex 15, France

Burgard, Peter
Department of General Pediatrics,
Universitäts-Kinderklinik,
Im Neuenheimer Feld 150,
69120 Heidelberg, Germany

Chakrapani, Anupam
Department of Clinical Inherited
Metabolic Disorders,
Birmingham Children's Hospital,
Steelhouse Lane,
Birmingham B4 6NH, UK

Charpentier, Christiane
Département de Biochimie,
Hôpital Necker Enfants Malades,
149 Rue de Sèvres,
75743 Paris Cedex 15, France

Clayton, Peter T.
Biochemistry Unit,
Institute of Child Health,
Great Ormond Street Hospital
for Children,
30 Guilford Street,
London WC1N 1EH, UK

Cochat, Pierre
Unité de Néphrologie Pédiatrique,
Hôpital Edouard Herriot &
Université Claude Bernard,
5, Place d'Arsonval,
69437 Lyon Cedex 03, France

de Lonlay, Pascale
Unité de Métabolisme,
Département de Pédiatrie,
Hôpital Necker-Enfants Malades,
149 Rue de Sèvres,
75743 Paris Cedex 15, France

De Meirleir, Linda J.
Department of Pediatrics,
Akademisch Ziekenhuis,
Vrije Universiteit Brussel,
Laarbeeklaan 101,
1090 Brussels, Belgium

Desguerre, Isabelle
Unité de Neurologie,
Département de Pédiatrie,
Hôpital Necker Enfants Malades,
149 Rue de Sèvres,
75743 Paris Cedex 15, France

DiMauro, Salvatore
Department of Neurology,
Columbia University College of
Physicians & Surgeons,
630 West 168th Street, P & S 4-443,
New York, NY 10032, USA

Dulac, Olivier
Unité de Neurologie,
Département de Pédiatrie,
Hôpital Necker Enfants Malades,
149 Rue de Sèvres,
75743 Paris Cedex 15, France

Egger, Norman G.
Department of Medicine,
Mayo Clinic,
200 First Street SW,
Rochester, MN 55905, USA

Fowler, Brian
University Children's Hospital,
Römergasse 8,
4005 Basel, Switzerland

Gitzelmann, Richard
Division of Metabolism and
Molecular Pediatrics,
University Children's Hospital,
Steinwiesstr. 75,
8032 Zürich, Switzerland

Hoffmann, Georg F.
Department of General Pediatrics,
Universitäts-Kinderklinik,
Im Neuenheimer Feld 150,
69120 Heidelberg, Germany

Holme, Elisabeth
Department of Clinical Chemistry,
Sahlgrenska University Hospital,
41345 Gothenburg, Sweden

Houwen, Roderick H.J.
Department of Pediatric
Gastroenterology,
Wilhelmina Children's Hospital,
University Medical Center Utrecht,
Lundlaan 6,
3584 EA Utrecht, The Netherlands

Huber, Jan
Department of Pathology,
Wilhelmina Children's Hospital,
University Medical Center Utrecht,
Lundlaan 6,
3584 EA Utrecht, The Netherlands

Jaeken, Jaak
Centre for Metabolic Diseases,
Department of Pediatrics,
University Hospital Gasthuisberg,
Herestraat 49,
3000 Leuven, Belgium

Jakobs, Cornelis
Metabolic Unit,
Department of Clinical Chemistry,
Vrije Universiteit Medical Center,
De Boelelaan 1117,
1081 HV Amsterdam, The Netherlands

Klepper, Jörg
University Children's Hospital,
Hufelandstr. 55,
45122 Essen, Germany

Kwiterovich Jr., Peter O.
Lipid Research-Atherosclerosis Division,
Department of Pediatrics,
The Johns Hopkins Hospital,
550 North Broadway, Suite 312,
Baltimore, MD 21205, USA

Larsson, Agne
Department of Pediatrics,
Karolinska University Hospital
Huddinge,
141-86 Stockholm, Sweden

Lee, Chul
Department of Preventive Medicine and
Community Health,
The University of Texas Medical Branch,
700 Harborside Drive,
Galveston, TX 77555-1109, USA

Lee, Philip J.
Charles Dent Metabolic Unit,
Post Box 92,
National Hospital for Neurology &
Neurosurgery,
Queens Square,
London WC1N 3BG, UK

Leonard, James V.
Biochemistry, Endocrinology
and Metabolic Unit,
Institute of Child Health,
Great Ormond Street Hospital
for Children,
30 Guilford Street,
London WC1N 1EH, UK

Lissens, Willy
Center for Medical Genetics,
Akademisch Ziekenhuis,
Vrije Universiteit Brussel,
Laarbeeklaan 101,
1090 Brussels, Belgium

Marie, Sandrine
Laboratory for Inherited Metabolic
Diseases,
Saint-Luc University Hospital,
University of Louvain Medical School,
Avenue Hippocrate 10,
1200 Brussels, Belgium

Mayatepek, Ertan
University Children's Hospital,
Moorenstr. 5,
40225 Düsseldorf, Germany

Morris, Andrew A.M.
Willink Biochemical Genetics Unit,
Royal Manchester Children's Hospital,
Hospital Road, Pendlebury,
Manchester M27 4HA, UK

Munnich, Arnold
Département de Génétique Médicale,
Hôpital Necker-Enfants Malades,
149 rue de Sèvres,
75743 Paris Cedex 15, France

Näntö-Salonen, Kirsti
Department of Pediatrics,
University of Turku,
Klinamyllynkatu 4-8,
20520 Turku, Finland

Ogier de Baulny, Hélène
Service de Neurologie et Maladies
Métaboliques,
Hôpital Robert Debré,
48 Boulevard Sérurier,
75019 Paris, France

Poll-The, Bwee Tien
Department of Pediatrics,
Academic Medical Center,
University of Amsterdam,
Meibergdreef 9,
1105 AZ Amsterdam, The Netherlands

Prietsch, Viola
Klinik für Kinder- und Jugendmedizin,
Städtisches Klinikum Karlsruhe,
Moltkestr. 90,
76133 Karlsruhe, Germany

Rake, Jan Peter
Beatrix Children's Hospital Groningen,
Hanzeplein 1,
9700 RB Groningen, The Netherlands

Ristoff, Ellinor
Department of Pediatrics,
Karolinska University Hospital
Huddinge,
141 86 Stockholm, Sweden

Rodriguez-Oquendo, Annabelle
Department of Medicine,
Johns Hopkins Bayview Medical Center,
4940 Eastern Avenue,
Baltimore, MD 21224, USA

Rolland, Marie-Odile
Inborn Errors of Metabolism
Biochemistry Centre,
Hôpital Debrousse,
29, rue Soeur Bouvier,
69322 Lyon Cedex 05, France

Rosenblatt, David S.
Division of Medical Genetics,
McGill University Health Centre,
Montreal General Hospital,
1650 Cedar Avenue, Room L3.319
Montreal, Quebec H3G 1A4, Canada

Salomons, Gajja S.
Metabolic Unit,
Department of Clinical Chemistry,
Vrije Universiteit Medical Center
De Boelelaan 1117,
1081 HV Amsterdam, The Netherlands

Santer, René
University Children's Hospital,
Martinistr. 52,
20246 Hamburg, Germany

Saudubray, Jean-Marie
Unité de Métabolisme,
Département de Pédiatrie,
Hôpital Necker Enfants Malades,
149 Rue de Sèvres,
75043 Paris Cedex 15, France

Sebastio, Gianfranco
Department of Pediatrics,
Federico II University,
Via Sergio Pansini 5,
80131 Naples, Italy

Sedel, Frédéric
Fédération de Neurologie,
Hôpital Pitié-Salpétrière,
47-83 Boulevard de l'Hôpital,
75651 Paris Cedex 13, France

Segal, Stanton
J. Stokes Jr Research Institute,
Children's Hospital of Philadelphia,
34th & Civic Centre Boulevard,
Philadelphia, PA 19104, USA

Shih, Vivian E.
Amino Acid Disorder Laboratory,
Pediatrics and Neurology Services,
Massachusetts General Hospital,
Building 149, 13th Street,
Boston, MA 02129, USA

Simell, Olli G.
Department of Pediatrics,
University of Turku,
Klinamyllynkatu 4-8,
20520 Turku, Finland

Smit, G. Peter A.
Beatrix Children's Hospital Groningen,
Hanzeplein 1,
9700 RB Groningen, The Netherlands

Stanley, Charles A.
Endocrine Division,
The Children's Hospital of Philadelphia,
34th Street & Civic Center Boulevard,
Philadelphia, PA 19104, USA

Steinmann, Beat
Division of Metabolism and Molecular
Pediatrics,
University Children's Hospital,
Steinwiesstr. 75,
8032 Zürich, Switzerland

Stöckler-Ipsiroglu, Sylvia
Division of Biochemical Diseases,
British Columbia Children's Hospital,
University of British Columbia,
4480 Oak Street,
Vancouver BC, V6H 3V4, Canada

Suormala, Terttu
Metabolic Unit,
University Children's Hospital,
Römergasse 8,
4005 Basel, Switzerland

Touati, Guy
Unité de Métabolisme,
Département de Pédiatrie,
Hôpital Necker Enfants Malades,
149 Rue de Sèvres,
75043 Paris Cedex 15, France

Van Coster, Rudy
Department of Pediatrics,
Ghent University Hospital,
De Pintelaan 185,
9000 Ghent, Belgium

van den Berghe, Georges
Metabolic Research Group,
Christian de Duve Institute
of Cellular Pathology,
University of Louvain Medical School,
Avenue Hippocrate 75-39,
1200 Brussels, Belgium

Vanier, Marie-Thérèse
Laboratoire Gillet-Mérieux, Bat 3B,
Centre Hospitalier Lyon-Sud,
69310 Pierre-Bénite, France

Verhoeven, Nanda M.
Metabolic Unit,
Department of Clinical Chemistry,
Vrije Universiteit Medical Center,
De Boelelaan 1117,
1081 HV Amsterdam, The Netherlands

Vincent, M.-Françoise
Laboratory for Inherited Metabolic
Diseases,
Saint-Luc University Hospital,
University of Louvain Medical School,
Avenue Hippocrate 10,
1200 Brussels, Belgium

Walker, Valerie
Department of Chemical Pathology,
Southampton General Hospital,
Tremona Road,
Southampton SO16 6YD, UK

Walter, John H.
Willink Biochemical Genetics Unit,
Royal Manchester Children's Hospital,
Hospital Road, Pendlebury
Manchester M27 4HA, UK

Wanders, Ronald J.A.
Laboratory for Genetic Metabolic
Diseases (F0-224),
Academic Medical Center,
University of Amsterdam,
Meibergdreef 9,
1105 AZ Amsterdam, The Netherlands

Waterham, Hans R.
Laboratory for Genetic Metabolic
Diseases (F0-224),
Academic Medical Center,
University of Amsterdam,
Meibergdreef 9,
1105 AZ Amsterdam, The Netherlands

Wendel, Udo
University Children's Hospital,
Moorenstr. 5,
40225 Düsseldorf, Germany

Wevers, Ron A.
Institute of Neurology,
University Medical Centre Nijmegen,
Reinier Postlaan 4,
6525 GC Nijmegen, The Netherlands

Wilcken, Bridget
The Children's Hospital at Westmead,
Locked Bag 4001,
Westmead, NSW 2145, Australia

Wraith, J. Ed
Willink Biochemical Genetics Unit,
Royal Manchester Children's Hospital,
Hospital Road, Pendlebury,
Manchester M27 4HA, UK

I Diagnosis and Treatment: General Principles

1   A Clinical Approach to Inherited Metabolic Diseases

Jean-Marie Saudubray, Isabelle Desguerre, Frédéric Sedel, Christiane Charpentier

Introduction

Inborn errors of metabolism (IEM) are individually rare, but collectively numerous. The recent application of tandem mass spectrometry (tandem MS) to newborn screening and prenatal diagnosis has enabled pre-symptomatic diagnosis for some IEM. However, for most, neonatal screening tests are either too slow, expensive or unreliable and, as a consequence, a simple method of clinical screening is mandatory before initi-ating sophisticated biochemical investigations. The clinical diagnosis of IEM relies upon a limited number of principles:

- To consider IEM in parallel with other more com-mon conditions; for example, sepsis or anoxic-ischemic encephalopathy in neonates, and intoxi-cation, encephalitis and brain tumors in older patients.
- To be aware of symptoms that persist and remain unexplained after the initial treatment and the usual investigations have been performed.
- To suspect that any neonatal death may possibly be due to an IEM, particularly those that have been attributed to sepsis.
- To carefully review all autopsy findings.
- Not to confuse a symptom (such as peripheral neuropathy, retinitis pigmentosa, cardiomyopathy, etc.) or a syndrome (such as Reye syndrome, Leigh syndrome, sudden infant death, etc.) with etiology.
- To remember that an IEM can present at any age, from fetal life to old age.
- To know that although most genetic metabolic errors are hereditary and transmitted as recessive disorders, the majority of individual cases appear sporadic because of the small size of sibships in developed countries.
- To initially consider inborn errors which are amena-ble to treatment (mainly those that cause intoxica-tion).
- In the acute, emergency situation, to undertake only those few investigations that are able to diag-nose treatable IEM.
- To obtain help from specialized centers.

Based mainly upon personal experience over 40 years, this chapter gives an overview of clinical clues to the diagnosis of inborn errors of metabolism in pediatrics and adulthood. In the following pages, inborn errors amenable to treatment are printed in bold.

**Do not miss a treatable disorder
First take care of the patient (emergency treatment) and then the family (genetic counselling)**

1.1 Classification of Inborn Errors of Metabolism

1.1.1 Pathophysiology

From a pathophysiological perspective, metabolic disorders can be divided into the following three diagnostically useful groups.

Group 1: Disorders which give rise to intoxication. This group includes inborn errors of intermediary metabolism that lead to an acute or progressive intoxication from the accumulation of toxic compounds proximal to the meta-bolic block. In this group are the inborn errors of amino acid catabolism (phenylketonuria, maple syrup urine dis-ease, homocystinuria, tyrosinemia etc.), most organic aci-durias (methylmalonic, propionic, isovaleric etc.), con-genital urea cycle defects, sugar intolerances (galactosemia, hereditary fructose intolerance), metal intoxication (Wilson, Menkes, hemochromatosis), and porphyrias. All the con-ditions in this group share clinical similarities: they do not interfere with the embryo-fetal development and they present with a symptom-free interval and clinical signs of »intoxication«, which may be acute (vomiting, coma, liver failure, thromboembolic complications etc.) or chronic (failure to thrive, developmental delay, ectopia lentis, car-diomyopathy etc.). Circumstances that can provoke acute metabolic attacks include catabolism, fever, intercurrent illness and food intake. Clinical expression is often both late in onset and intermittent. The diagnosis is straightforward and most commonly relies on plasma and urine amino acid, organic acid and acylcarnitine chromatography. Most of these disorders are treatable and require the emergency removal of the toxin by special diets, extra-corporeal proce-dures, or »cleansing« drugs (carnitine, sodium benzoate, penicillamine, etc.).

Although the pathophysiology is somewhat different the inborn errors of neurotransmitter synthesis and cata-bolism (monoamines, GABA and glycine) and the inborn errors of amino acid synthesis (serine, glutamine, and proline/ornithine) can also be included in this group since they share many characteristics: they are inborn errors of intermediary metabolism, their diagnosis relies on plasma, urine, and CSF investigations (amino acid, organic acid analyses etc.), and some are amenable to treatment even when the disorder starts in utero, for example 3-phos-phoglycerate dehydrogenase deficiency [1].

Group 2: Disorders involving energy metabolism. These consists of inborn errors of intermediary metabolism with symptoms due at least partly to a deficiency in energy pro-duction or utilization within liver, myocardium, muscle, brain or other tissues. This group can be divided into mito-chondrial and cytoplasmic energy defects. Mitochondrial defects are the most severe and are generally untreatable.

They encompass the congenital lactic acidemias (defects of pyruvate transporter, pyruvate carboxylase, pyruvate dehydrogenase, and the Krebs cycle), mitochondrial respiratory chain disorders and the fatty acid oxidation and ketone body defects. Only the latter are partly treatable. Cytoplasmic energy defects are generally less severe. They include disorders of glycolysis, glycogen metabolism and gluconeogenesis, hyperinsulinism (all treatable disorders), the more recently described disorders of creatine metabolism (partly treatable), and the new inborn errors of the pentose phosphate pathway (untreatable). Common symptoms in this group include hypoglycemia, hyperlactatemia, hepatomegaly, severe generalized hypotonia, myopathy, cardiomyopathy, failure to thrive, cardiac failure, circulatory collapse, sudden unexpected death in infancy, and brain involvement. Some of the mitochondrial disorders and pentose phosphate pathway defects can interfere with the embryo-fetal development and give rise to dysmorphism, dysplasia and malformations [2]. Diagnosis is difficult and relies on function tests, enzymatic analyses requiring biopsies or cell culture, and on molecular analyses.

Group 3: Disorders involving complex molecules. This group involves cellular organelles and includes diseases that disturb the synthesis or the catabolism of complex molecules. Symptoms are permanent, progressive, independent of intercurrent events and unrelated to food intake. All lysosomal storage disorders, peroxisomal disorders, disorders of intracellular trafficking and processing such as alpha-1-antitrypsin, congenital disorders of glycosylation (CDG), and inborn errors of cholesterol synthesis belong to this group. Almost none are treatable acutely; however enzyme replacement therapy is now available for several lysosomal disorders.

1.1.2   Clinical Presentation

Besides newborn screening in the general population (as for phenylketonuria) or in at-risk families, there are four groups of clinical circumstances in which physicians are faced with the possibility of a metabolic disorder:
- Early symptoms in the antenatal and neonatal period
- Later-onset acute and recurrent attacks of symptoms such as coma, ataxia, vomiting, and acidosis
- Chronic and progressive generalised symptoms which can be mainly gastrointestinal (chronic vomiting, failure to thrive), muscular or neurological (developmental delay, neurological deterioration)
- Specific and permanent organ presentations suggestive of an inborn error of metabolism, such as cardiomyopathy, hepatomegaly, lens dislocation etc.

These four categories of clinical conditions are presented in the following sections. For the first two categories, which often present as emergencies and are treatable, the clinical presentations, the metabolic abnormalities and the laboratory tests required for a tentative diagnosis, are described in detail. Chronic progressive generalised symptoms and signs which raise suspicion of an IEM are listed in tables that take into account system and organ involvement, leading symptoms and other signs, and age of onset. Specific organ presentations are listed in alphabetical order. For each presentation, the diagnostic possibilities listed encompass not only inborn errors of metabolism, but also diverse inherited syndromes which mimic and are possibly related to inborn errors, and a number of non-inherited disorders which should be considered in the differential diagnosis. All treatable disorders are printed in **bold**.

1.2   Acute Symptoms in the Neonatal Period and Early Infancy (<1 Year)

1.2.1   Clinical Presentation

The neonate has a limited repertoire of responses to severe illness [2–4]. IEM may present with non-specific symptoms such as respiratory distress, hypotonia, poor sucking reflex, vomiting, diarrhea, dehydration, lethargy, seizures; problems which can easily be attributed to infection or some other common cause. Where a previously affected sibling has died, this may have been falsely attributed to sepsis, heart failure, or intraventricular hemorrhage, and it is important to critically review clinical records and autopsy reports when they are available.

In group 1 disorders (IEM that give rise to intoxication), an extremely suggestive clinical picture is that of a baby, born at full-term after a normal pregnancy and delivery, who, after an initial entirely symptom-free period, relentlessly deteriorates for no apparent reason and does not respond to symptomatic therapy. The interval between birth and clinical symptoms may range from hours to weeks, depending on the nature of the metabolic block and the environment. Investigations, routinely performed in sick neonates, including a chest X-ray, CSF examination, bacteriologic studies, and cerebral ultrasound, yield normal results. This unexpected and »mysterious« deterioration after a normal initial period is the most important indication for this group of IEM. Careful re-evaluation of the child's condition is then warranted. In this context signs previously interpreted as non-specific manifestations of neonatal hypoxia, infection, or other common diagnoses take on a new significance. In energy deficiencies (group 2 disorders), clinical presentation is often less evocative and displays variable severity. A clinical algorithm for screening for treatable IEM in neonates is presented in ◻ Table 1.1.

A careful reappraisal of the child is warranted for the following:

▢ Table 1.1. The »sick« neonate: an algorithm for screening for treatable inborn errors of metabolism

Premature / Low birth weight →	Full term neonates appropriately grown for gestational age →			
• Not suggestive but possibility of fortuitous association	**Inborn metabolic diseases** →	»Traumatic« »Accidental« hypoxia Intracranial injury →	Infection →	Major electrolyte disturbances:
• If head circumference is small, think of maternal PKU				

»Traumatic« »Accidental« hypoxia Intracranial injury →	Infection →	Major electrolyte disturbances: ■ Hypo/hypercalcemia ■ Hypo/hypernatremia ■ Hypo/hyperkaliemia →	Isolated/multiple malformations Polymalformation syndromes →
Clinical history Chest X-ray Cranial ultrasound →	»Septic screen« Antibiotics →	»Simple metabolic screen« Hormonal, renal investigation →	Radiologic investigation Echography Genetic advice →

First think of treatable disorders
Emergency treatment must be undertaken in parallel with investigations (Chap. 3)

Neurological deterioration (intoxication) →	Predominant seizures →	Jaundice Liver failure →	Cardiac failure Heart beat disorders →	Persistent hypoglycemia →
MSUD MMA PA IVA MCD UCD →	**B6-responsive seizures** **PNPO** (pyridoxal-P) **MCD** (biotin) **folinic acid resp. seizures** 3PGD (serine) GLUT1 (ketotic diet) →	Galactosemia HFI Tyrosinemia CDG type Ib (with diarrhea) Bile acid synthesis defects (with cholestasis) LCHAD →	FAO defects →	Glycogenosis PHHI FAO defects

CDG, congenital disorders of glycosylation; FAO, fatty acid oxidation; HFI, hereditary fructose intolerance; IVA, isovaleric acidemia; LCHAD, 3-hydroxy long chain acyl-CoA dehydrogenase deficiency; MCD, multiple carboxylase deficiency; MMA, methylmalonic aciduria; MSUD, maple syrup urine disease; PA, propionic acidemia; PHHI, primary hyperinsulinemic hypoglycemia of infancy; PKU, phenylketonuria; PNPO, pyridox(am)ine-5'-phosphate oxidase deficiency; 3PGD, 3-phosphoglycerate dehydrogenase deficiency; UCD, urea cycle defects; **bold face,** treatable disorders.

Neurological Deterioration (Coma, Lethargy)

Most inborn errors that result in intoxication or energy deficiency are brought to a doctor's attention because of neurological deterioration. With intoxication, the initial symptom-free interval varies in duration depending on the condition. Typically, the first reported sign is poor sucking and feeding, after which the child sinks into an unexplained coma despite supportive measures. At a more advanced state, neurovegetative problems with respiratory abnormalities, hiccups, apneas, bradycardia, and hypothermia can appear. In the comatose state, characteristic changes in muscle tone and involuntary movements appear. In **maple syrup urine disease (MSUD)** generalized hypertonic episodes with opisthotonus are frequent, and boxing or pedalling movements as well as slow limb elevations, spontaneously or upon stimulation, are observed. Conversely, most non metabolic causes of coma are associated with hypotonia, so that the presence of »normal« peripheral muscle tone in a comatose child reflects a relative hypertonia. Another neurological pattern observed in **organic acidurias** is axial hypotonia and limb hypertonia with large amplitude tremors and myoclonic jerks which are often mistaken for convulsions. An abnormal urine and body odor is present in some diseases in which volatile metabolites accumulate; the most important examples are the maple syrup odor of **MSUD** and the sweaty feet odor of **isovaleric acidemia (IVA)** and **type II glutaric acidemia**. If any of the preceding signs or symptoms are present, metabolic disorders should be given a high diagnostic priority.

In energy deficiencies, the clinical presentation is less evocative and displays a more variable severity. In many conditions, there is no symptom-free interval. The most frequent findings are a severe generalized hypotonia, rapidly progressive neurological deterioration, and possible dysmorphism, or malformations. However, in contrast to the intoxication group, lethargy and coma are rarely initial signs. Hyperlactatemia with or without metabolic acidosis is very frequent. Cardiac and hepatic involvement are commonly associated (▶ below).

Only a few lysosomal storage disorders present in the neonatal period with neurological deterioration. By contrast, most peroxisomal biogenesis defects present at birth with dysmorphism and severe neurological dysfunction.

Seizures

Five treatable disorders can present in the neonatal period predominantly with intractable seizures: **pyridoxine responsive seizures**, **pyridox(am)ine-5'-phosphate oxidase deficiency**, **folinic acid responsive seizures**, **3-phosphoglycerate dehydrogenase deficiency** responsive to serine supplementation and persistent **hyperinsulinemic hypoglycemia**.

Biotin responsive **holocarboxylase synthetase deficiency** can also rarely present predominantly with neonatal seizures. **GLUT1 deficiency** (brain glucose transporter),

responsive to hyperketotic diet, and biotin responsive **biotinidase deficiency** can also present in the first months of life as an epileptic encephalopathy.

Many other non treatable inherited disorders can present in the neonatal period with severe epilepsy: non ketotic hyperglycinemia (NKH), D-2-hydroxyglutaric aciduria, and mitochondrial glutamate transporter defect (all three presenting with myoclonic epilepsy and a burst-suppression EEG pattern), peroxisomal biogenesis defects, respiratory chain disorders, sulfite oxidase deficiency and Menkes disease. In all these conditions, epilepsy is severe, with an early onset, and can present with spasms, myoclonus, partial or generalized tonic/clonic crises.

Hypotonia

Severe hypotonia is a common symptom in sick neonates. It is generally observed in non metabolic inherited diseases (mainly in severe fetal neuromuscular disorders). Only a few inborn errors of metabolism present with isolated hypotonia in the neonatal period and only very few are treatable. Discounting disorders in which hypotonia is part of a constellation of abnormalities, including, for example, major bone changes, dysmorphism, malformations, or visceral symptoms, the most severe metabolic hypotonias are observed in hereditary hyperlactatemia, respiratory chain disorders, urea cycle defects, NKH, sulfite oxidase (SO) deficiency, peroxisomal disorders, and trifunctional enzyme deficiency. Central hypotonia is associated with lethargy, coma, seizures, and neurological symptoms in NKH, SO deficiency, and peroxisomal disorders, and with the characteristic metabolic changes in congenital lactic acidosis and **urea cycle disorders** (hyperammonemia). Severe forms of Pompe disease (alpha-glucosidase deficiency) can initially mimic respiratory chain disorders or trifunctional enzyme deficiency when generalized hypotonia is associated with cardiomyopathy. However, Pompe disease does not strictly start in the neonatal period. Finally, one of the most frequent causes of neonatal hypotonia is Prader-Willi syndrome, where central hypotonia is apparently an isolated symptom at birth which can mimick the hypotonia cystinuria syndrome [4a].

The three neurological presentations are summarized in ◘ Table 1.2.

Hepatic Presentation and Hydrops Fetalis

Three main clinical groups of hepatic symptoms can be identified:

- Hepatomegaly with hypoglycemia and seizures suggest **glycogenosis type I or III**, **gluconeogenesis defects**, or **severe hyperinsulinism**.
- Liver failure (jaundice, coagulopathy, hepatocellular necrosis with elevated transaminases, and hypoglycemia with ascites and edema) suggests **hereditary fructose intolerance** (now very rare since infant formulas are fructose free), **galactosemia**, **tyrosinemia type I** (after

◻ **Table 1.2.** Neurological presentation

Predominant clinical symptom	Main clinical findings	Biological abnormalities	Most likely diagnoses (disorder or enzyme deficiency)
Neurological deterioration (mostly metabolic and **treatable**)	Lethargy, coma, hiccups Poor sucking, hypothermia Hypotonia, hypertonia Abnormal movements Large amplitude tremor Myoclonic jerks »Burst suppression« Abnormal odor	Ketosis, acidosis Hyperlactatemia Leuconeutropenia Thrombopenia Hyperammonemia Characteristic changes of AAC or OAC	**MSUD (odor)** **MMA, PA, IVA (odor)** **MCD** **Urea cycle defects** **GA type II (odor)**
Seizures (sometimes metabolic, sometimes treatable)	Isolated Generalized	Metabolic ketoacidosis Organic acid profile None None Hypocalcemia Hypomagnesemia Severe hypoglycemia	**MCD** **Pyridoxine responsive seizures** **Folinic acid responsive seizures** **Congenital magnesium malabsorption** **PHHI**
	Generalized Hypsarythmia Severe microcephaly	Low serine (plasma/CSF)	**3PGD**
	Severe hypotonia Myoclonic jerks Burst suppression EEG	Low HVA, 5HIAA in CSF, Vanillactic (urine) Hyperglycinemia None S-sulfocysteine (AAC)	**PNPO** (pyridoxal phosphate responsive seizures) NKH Glutamate transporter Sulfite oxidase
	Facial dysmorphism Malformations Severe hypotonia	VLCFA, phytanic/acid plasmalogens Glycosylated transferrin Sterols in plasma	Peroxisomal defects CDG Cholesterol biosynthesis defects
Severe hypotonia (rarely metabolic)	Isolated	None	Prader Willi syndrome Hypotonia cystinuria [4a]
Not treatable	Fetal distress Hydramnios Arthrogryposis Respiratory failure	None	Severe fetal neuromuscular diseases Steinert Myasthenia Congenital myopathy Sensitivo-motor neuropathy
	Predominant dysmorphism Malformations	VLCFA, phytanic acid plasmalogens Sterols in plasma Tubulopathy Glycosylated transferrin Aberrant protein O-glycosylation Chromosome analyses	Peroxisomal defects Cholesterol defects Lowe syndrome CDG Polymalformative syndromes with muscular dystrophy (Walker Warburg, muscle-eye-brain, etc.) Chromosomal abnormalities
	Cataract Tubulopathy	Hyperlactatemia Enzyme/DNA analyses	Lowe syndrome Respiratory chain
	Cardiomyopathy Macroglossia	Vacuolated lymphocytes Hyperlactatemia Acylcarnitines	Pompe disease Respiratory chain Trifunctional enzyme

AAC, amino acid chromatography; *CDG*, congenital disorders of glycosylation; *GA*, glutaric aciduria; *HVA*, homovanillic acid; *IVA*, isovaleric acidemia; *MCD*, multiple carboxylase deficiency; *MMA*, methylmalonic aciduria; *MSUD*, maple syrup urine disease; *NKH*, non ketotic hyperglycinemia; *OAC*, organic acid chromatography; *PA*, propionic acidemia; *PHHI*, primary hyperinsulinemic hypoglycemia of infancy; *PNPO*, pyridox(am)ine-5′-phosphate oxidase; *VLCFA*, very long chain fatty acids; *3PGD*, 3-phosphoglycerate dehydrogenase; *5HIAA*, 5-hydroxyindoleaceatic acid; **bold face**, treatable disorders.

3 weeks), neonatal hemochromatosis, respiratory chain disorders, and transaldolase deficiency, a disorder of the pentose phosphate pathway which can present with hydrops fetalis. Another disorder described in 15 newborns from Finland displays severe fetal growth retardation, lactic acidosis, failure to thrive, hyperaminoaciduria, very high serum ferritin, hemosiderosis of the liver and early death (GRACILE syndrome). The etiology of this syndrome has been recently identified [5, 6].

- Cholestatic jaundice with failure to thrive is a predominant finding in alpha-1-antitrypsin deficiency, Byler disease, **inborn errors of bile acid metabolism**, peroxisomal disorders, Niemann-Pick type C disease, CDG and citrin deficiency [7, 8].

   With the exception of **long-chain 3-hydroxyacyl-CoA dehydrogenase (LCHAD)** deficiency which can present early in infancy (but not strictly in the neonatal period) as cholestatic jaundice, liver failure and hepatic fibrosis, hepatic presentations of inherited fatty acid oxidation disorders and urea cycle defects consist of acute steatosis or Reye syndrome with normal bilirubin, slightly prolonged prothrombin time, and moderate elevation of transaminases rather than true liver failure. One must emphasize that there are frequent difficulties in investigating patients with severe hepatic failure. At an advanced state, many non specific abnormalities secondary to liver damage can be present. Mellituria (galactosuria, glycosuria, fructosuria), hyperammonemia, hyperlactatemia, hypoglycemia after a short fast, hypertyrosinemia (>200 µmol/l), and hypermethioninemia (sometimes higher than 500 µmol/l) are encountered in all cases of advanced hepatocellular disease.

Cardiac Presentation

Some metabolic disorders can present predominantly with cardiac disease. Cardiac failure and a dilated hypertrophic cardiomyopathy, most often associated with hypotonia, muscle weakness, and failure to thrive, suggests fatty acid oxidation disorders, respiratory chain disorders or Pompe disease. Methylglutaconic aciduria is found in Barth syndrome and ketoglutarate excretion in ketoglutarate dehydrogenase deficiency. Several observations suggest that some respiratory chain disorders are tissue specific and are only expressed in the myocardium. CDG type Ia can sometimes present in infancy with cardiac failure due to pericardial effusions, cardiac tamponnade, and cardiomyopathy. Many defects of long-chain fatty acid oxidation can present with cardiomyopathy and/or arrhythmias and conduction defects (auriculoventricular block, bundle branch blocks, ventricular tachycardia) which may lead to cardiac arrest [9, 10].

1.2.2   Metabolic Derangements and Diagnostic Tests

Initial Approach and Protocol for Investigation

As soon as there is clinical suspicion of an IEM, general supportive measures and laboratory investigations should be undertaken concurrently (◘ Table 1.3). Abnormal urine odors can be detected on a drying filter paper or by opening a container of urine which has been closed at room temperature for a few minutes. Although serum ketone bodies reach 0.5–1 mmol/l in early neonatal life, acetonuria, if observed in a newborn, is always abnormal and an important sign of a metabolic disease. The dinitrophenylhydrazine (DNPH) test screens for the presence of alpha-keto acids as occur in MSUD. However, it has now largely been abandoned because of its poor specificity and because amino acid chromatography has become much more readily available. Hypocalcemia and elevated or reduced blood glucose are frequently present in metabolic diseases and the physician should be wary of attributing marked neurological dysfunction purely to these findings.

The metabolic acidosis of organic acidurias is usually accompanied by an elevated anion gap. Urine pH should be below 5; otherwise, renal acidosis is a consideration. A normal blood pH does not exclude hyperlactatemia, as neutrality is usually maintained until serum levels reach 5 mmol/l and more. Ammonia and lactic acid should be determined systematically in newborns at risk. An elevated ammonia level in itself can induce respiratory alkalosis; hyperammonemia with ketoacidosis suggests an underlying organic acidemia (OA), but an isolated hyperammonemia can occur. Elevated lactic acid levels in the absence of infection or tissue hypoxia are a significant finding. Moderate elevations (3–6 mmol/l) are often observed in organic acidemias and in the hyperammonaemias; levels greater than 10 mmol/l are frequent in hypoxia. In most anoxic lactic acidoses, ketosis is absent. Whenever it is possible, the measurement of lactate (L), pyruvate (P), 3-hydroxybutyrate (3OHB), and acetoacetate (AcAc) on a plasma sample immediately deproteinized at the bedside, allows an assessment of the cytoplasmic and mitochondrial redox states through the measurement of L/P and 3OHB/AcAc ratios, respectively. PA, MMA and IVA can induce granulocytopenia and thrombocytopenia, which may be mistaken for sepsis. Transaldolase deficiency and early onset forms of mevalonate kinase deficiency present with severe recurrent hemolytic anemia.

The storage of adequate amounts of plasma, urine, blood on filter paper, and CSF, is an important element in reaching a diagnosis. The utilization of these precious samples should be carefully planned after taking advice from specialists in IEM.

Identification of Five Major Types of Metabolic Distress

Once the above clinical and laboratory data have been collected, specific therapeutic recommendations can be made (▶ Chap. 4). This process is completed within 2–4h and often precludes waiting long periods for the results of sophisticated diagnostic investigations. On the basis of this evaluation, most patients can be classified into one of five types (◻ Table 1.4). The experienced clinician will, of course, have to carefully interpret the metabolic data, particularly in relation to time of collection and ongoing treatment. At the same time it is important to collect all the biologic data listed in ◻ Table 1.3. Some very significant symptoms (such as metabolic acidosis and especially ketosis) can be moderate and transient, largely depending on the symptomatic therapy. Conversely, at an advanced stage, many non-specific abnormalities (such as respiratory acidosis, severe hyperlactatemia, secondary hyperammonemia) can disturb the original metabolic profile. This applies particularly to IEM with a rapid fatal course such as **urea cycle disorders**, in which the initial characteristic presentation of hyperammonemia with respiratory alkalosis shifts rapidly to a rather non-specific picture of acidosis and hyperlactatemia.

In our experience, types I and II (**MSUD, organic acidurias**), type IVa (**urea cycle defects** and **fatty acid oxidation disorders**), nonketotic hyperglycinemia and respiratory chain disorders, encompass more than 80% of the newborn infants with inborn errors of intermediary metabolism.

1.3 Later Onset Acute and Recurrent Attacks (Late Infancy and Beyond)

1.3.1 Clinical Presentation

In about 50% of the patients with inborn errors of intermediary metabolism, disease onset is later. The symptom-free period is often longer than 1 year and may extend into late childhood, adolescence, or even adulthood. Each attack can follow a rapid course ending either in spontaneous improvement or unexplained death, despite supportive measures in the intensive care unit. Between attacks the patient may appear normal. Onset of acute disease may be precipitated by an intercurrent event or may occur without overt cause. Excessive protein intake, prolonged fasting,

◻ **Table 1.3.** Protocol for emergency investigations

	Immediate investigations	Storage of samples
Urine	Smell (distinctive odor) Look (distinctive color) Acetone (Acetest, Ames) Reducing substances (Clinitest, Ames) Keto acids (DNPH) pH (pHstix Merck) Sulfitest (Merck) Electrolytes (Na, K), urea, creatinine Uric acid	Urine collection: collect fresh sample and put it in the refrigerator Freezing: freeze samples collected before and after treatment at −20°C, and collect an aliquot 24 h after treatment. Do not use them without having expert metabolic advice Metabolic investigations: OAC, AAC, orotic acid, porphyrins
Blood	Blood cell count Electrolytes (search for anion gap) Glucose, calcium Blood gases (pH, pCO_2, HCO_3^-, pO_2) Uric acid Prothrombin time Transaminases (and other liver tests) Ammonia Lactate, pyruvate 3-hydroxybutyrate, acetoacetate Free fatty acids	Plasma (5 ml) heparinized at −20°C Blood on filter paper: 2 spots (as »Guthrie« test) Whole blood (10-15 ml) collected on EDTA and frozen (for molecular biology studies) Major metabolic investigations: total homocysteine, AAC, acylcarnitines (tandem MS), OAC, porphyrins, neurotransmitters (HPLC, tandem MS)
Miscellaneous	Lumbar puncture Chest X-ray Cardiac echography, ECG Cerebral ultrasound, EEG	Skin biopsy (fibroblast culture) CSF (1 ml), frozen (neurotransmitters, AA) Postmortem: liver, muscle biopsies (Chap. 3)

AA, amino acid; *AAC*, amino acid chromatography; *CSF*, cerebrospinal fluid; *DNPH*, dinitrophenylhydrazine; *FCG*, electrocardiogram; *EDTA*, ethylenediaminetetra-acetic acid; *EEG*, electroencephalogram; *MS*, mass spectrometry; *HPLC*, high performance liquid chromatography; *OAC*, organic acid chromatography.

Table 1.4. Classification of inborn errors revealed in the neonatal period and early in infancy

Types	Clinical type	Acidosis/ Ketosis	Other signs	Most usual diagnosis (disorder or enzyme deficiency)	Elective methods of investigation
I	**Neurological deterioration,** »Intoxication« type Slow movements Hypertonia	Acidosis – DNPH +++ Acetest –/±	NH₃ N or ↑ ± Lactate N Blood count N Glucose N Calcium N	**MSUD** (distinctive odor)	Aminoacid chromatography (plasma, urine) Blood spot for tandem MS-MS
II	**Neurological deterioration,** »Intoxication« type Fast movements Dehydration	Acidosis ++ Acetest ++ DNPH –/± Ketoacidosis	NH₃ ↑ +/++ Lactate N or ↑ ± Blood count: leucopenia, thrombopenia Glucose N or ↑ + Calcium N or ↓ +	**Organic acidurias (MMA, PA, IVA, MCD) Ketolysis defects**	OAC by GLCMS (urine, plasma) Carnitine (plasma) Carnitine esters by tandem MS (urine, plasma) Blood spot for tandem MS-MS
	Neurological deterioration, »energy deficiency« type, with liver or cardiac symptoms	Acidosis ++/± Acetest – DNPH – No ketosis	NH₃ ↑ +/++ Lactate ↑ ±/++ Blood count N Glucose ↓ +/++ Calcium N or ↓ + Hypoketotic hypoglycemia	**Fatty acid oxidation and ketogenesis defects** (GA II, CPT II, CAT, VLCAD, MCKAT, HMG-CoA lyase)	Idem above Loading test Fasting test Fatty acid oxidation studies on lymphocytes or fibroblasts
III	**Neurological deterioration,** »energy deficiency« type, Tachypnea Hypotonia Lactic acidosis (may be well tolerated)	Acidosis +++/+ Acetest ++/– Lactate +++/+ Lactic acidosis	NH₃ N or ↑ ± Blood count: anemia or N Glucose N or ± Calcium N	Congenital lactic acidoses (pyruvate carrier, PC, **PDH**, Krebs cycle, respiratory chain) **MCD**	Plasma redox states ratios (L:P, 3OHB:AcAc) OAC (urine), AAC (plasma) Polarographic studies Enzyme assays (muscle, lymphocytes, fibroblasts)
IV a)	**Neurological deterioration,** »intoxication« type, Moderate hepatocellular disturbances Hypotonia, seizures, coma	Acidosis – (alkalosis) Acetest –/+ DNPH –	NH₃ ↑ +/+++ Lactate N or ↑ + Blood count N Glucose N Calcium N	**Urea cycle defects Triple H syndrome Fatty acid oxidation defects** (GA II, CPT II, VLCAD, LCHAD, CAT) **PA, MMA, IVA**	AAC, OAC (plasma, urine) Orotic acid (urine) Liver or intestine enzyme studies (CPS, OTC)
b)	**Neurological deterioration,** Seizures Myoclonic jerks Severe hypotonia	Acidosis – Acetest – DNPH – No major metabolic disturbance	NH₃ N Lactate N or ↑ + Blood count N Glucose N	NKH, SO plus XO **3PGD B6-responsive seizures PNPO, neurotransmitter defects** Peroxisomal defects Trifunctional enzyme Respiratory chain CDG Cholesterol synthesis defects	AAC (plasma, CSF) AAC (plasma, CSF) OAC OAC, neurotransmitters (plasma, urine, CSF) VLCFA, phytanic acid in plasma Acylcarnitine profile, OAC Lactate (plasma) Glycosylated transferrin (plasma) Sterols (plasma)

□ Table 1.4 (continued)

Types	Clinical type	Acidosis/ Ketosis	Other signs	Most usual diagnosis (disorder or enzyme deficiency)	Elective methods of investigation
V a)	**Recurrent hypo- glycemia with hepatomegaly**	Acidosis ++/– Acetest +/–	Lactate $\uparrow$ +/++ NH_3 $\uparrow$ +/– Intractable hypo- glycemia	**Glycogenosis type I** (acetest –) **Glycogenosis type III** (acetest +) **FBPase** **FAO defects** **PHHI**	Fasting test, Loading test DNA analyses, enzyme studies (liver, lymphocytes, fibroblasts) Organic acids, acylcarnitine Insulin plasma levels
b)	Hepatomegaly Jaundice Liver failure Hepatocellular necrosis	Acidosis +/– Acetest +/–	NH_3 N or $\uparrow$ + Lactate / +/++ Glucose N or $\downarrow$ ++	**HFI** **Galactosemia** **Tyrosinemia type I** Neonatal hemochromatosis Respiratory chain defects TALDO	DNA analyses, enzyme studies Succinyl acetone Iron-ferritin in salivary glands Organic acids, enzyme/DNA analyses Polyols (HPLC)
c)	**Hepatomegaly Cholestatic Jaundice** ± Failure to thrive ± Chronic diarrhea ± osteoporosis ± rickets	Acidosis – Ketosis –	NH_3 N Lactate N Glucose N	Alpha-1-antitrypsin Inborn errors of **bile acid metabolism** Peroxisomal defects CDG Niemann-Pick type C **LCHAD** Mevalonic aciduria Cholesterol meta- bolism **Cerebrotendinous xanthomatosis** Citrin	Protein electrophoresis Bile acids (plasma, urine, bile by tandem MS) VLCFA, phytanic & pipecolic acid Glycosylated transferrin Fibroblasts studies OAC, acylcarnitine profile OAC Plasma sterols Plasma sterols AAC (citrulline can be normal)
d)	**Hepatosplenomegaly »Storage« signs** (coarse facies, ascites, hydrops fetalis, macro- glossia, bone changes, cherry red spot, vacuolated lympho- cytes) ± Failure to thrive ± Chronic diarrhea ± Hemolytic anemia	Acidosis – Acetest – Ketosis – DNPH –	NH_3 N Lactate N or $\uparrow$ Glucose N Hepatic signs ±/++	**Congenital erythro- poietic porphyria** GM1 gangliosidosis ISSD (sialidosis type II) I-cell disease Niemann-Pick type IA MPS VII Galactosialidosis CDG Mevalonic aciduria TALDO	Porphyrins Oligosaccharides, sialic acid Mucopolysaccharides Enzyme studies (lymphocytes, fibroblasts) Glycosylated transferrin OAC Polyols (HPLC)

N, normal (normal values = NH_3 < 80 µM; lactate < 1.5 mM; glucose 3.5-5.5 mM); ±, slight; +, moderate; ++, marked; +++, significant/mas-
sive; $\uparrow$ elevated; $\downarrow$ decreased; –, absent (acidosis) or negative (acetest, dinitrophenylhydrazine, DNPH).
L, lactate; *P*, pyruvate; *3OHB*, 3-hydroxybutyrate; *AcAc*, acetoacetate; *GLCMS*, gas liquid chromatography mass spectrometry; *HPLC*, high
performance liquid chromatography; *VLCFA*, very-long-chain fatty acids.
AAC, amino acid chromatography; *CAT*, carnitine acylcarnitine translocase; *CDG*, congenital disorders of glycosylation; *CPS*, carbamyl
phosphate synthetase; *CPT II*, carnitine palmitoyltransferase II; *FAO*, fatty acid oxidation; *FBPase*, fructose-1,6-bisphosphatase; *GA II*, glutaric aciduria type II; *HFI*, hereditary fructose intolerance; *HMG-CoA*, 3-OH-3-methylglutaryl coenzyme A; *ISSD*, infantile sialic acid storage
disease; *IVA*, isovaleric acidemia; *LCHAD*, 3-OH long-chain acyl CoA dehydrogenase; *MCD*, multiple carboxylase; *MCKAT*, medium-chain
3-ketoacylCoA A thiolase; *MMA*, methylmalonic acidemia; *MPS VII*, mucopolysaccharidosis type VII; *MSUD*, maple syrup urine disease;
NKH, nonketotic hyperglycinemia; *OAC*, organic acid chromatography; *OTC*, ornithine transcarbamylase; *PA*, propionic acidemia; *PC*, pyruvate carboxylase; *PDH*, pyruvate dehydrogenase; *PNPO*, pyridox(am)ine-5′-phosphate oxidase; *SO*, sulfite oxidase; *TALDO*, transaldolase;
VLCAD, very-long-chain acyl CoA dehydrogenase; *XO*, xanthine oxidase; *3PGD*, 3-phosphoglycerate dehydrogenase; **bold face,** treatable
disorders.

■ **Table 1.5.** Diagnostic approach to recurrent attacks of coma and vomiting with lethargy

Clinical Presentation	Metabolic derangements and other important signs		Most frequent diagnosis (disorder or enzyme deficiency)	Differential diagnosis
Metabolic coma (without focal neurological signs)	Acidosis (metabolic) pH < 7.20 HCO₃⁻ < 10 mmol/l pCO₂ < 25 mmHg	Ketosis + (acetest ++)	Respiratory chain defects **MCD**, PC **MMA, PA, IVA, GA I, MSUD*** **Ketolysis defects** **Gluconeogenesis defects**	Diabetes Intoxication Encephalitis
		Ketosis −	**PDH, Ketogenesis defects** **FAO, FBPase**, EPEMA	
	Hyperammonemia NH₃ > 100 µmol/l Respiratory alkalosis pH > 7.45 pCO₂ < 25 mmHg	Normal glucose	**Urea cycle defects*** **Triple H syndrome** **LPI**	Reye syndrome Encephalitis Intoxication
		Hypoglycemia	**FAO (MCAD*)** **HMG-CoA lyase**	
	Hypoglycemia (< 2 mmol/l)	Acidosis +	**Gluconeogenesis defects** **MSUD** **HMG-CoA lyase** **FAO**	Drugs and toxins Ketotic hypoglycemia Adrenal insufficiency GH deficiency Hypopituitarism
	Hyperlactatemia (> 4 mmol/l)	Normal glucose	PC, **MCD**, Krebs cycle defects Respiratory chain* **PDH*** (without ketosis) EPEMA syndrome	
		Hypoglycemia	**Gluconeogenesis defects** (ketosis variable) **FAO** (moderate hyperlactatemia, no ketosis)	
Neurological coma (with focal signs, seizures, or intracranial hypertension)	Biological signs are very variable, can be absent or moderate; ▶ »Metabolic coma«	Cerebral edema Hemiplegia Extrapyramidal signs	**MSUD, OTC** **MSUD, OTC, MMA, PA**, PGK **GA I, Wilson disease*** **Homocystinuria***	Cerebral tumor Migraine Encephalitis
		Caudate nucleus and putamen necrosis	**BBGD**	
		Stroke-like	**UCD, MMA, PA, IVA*** Respiratory chain (MELAS*) **Homocystinurias*** CDG **Thiamine responsive megaloblastic anemia** **Fabry disease*** (rarely presenting) **Acid maltase*** (rare)	Moya Moya syndrome Vascular hemiplegia Cerebral thrombophlebitis Cerebral tumor
	Abnormal coagulation, Hemolytic anemia	Thromboembolic accidents	AT III, Protein C,S **Homocystinurias*** Sickle cell anemia CDG, PGK	

◘ Table 1.5 (continued)

Clinical Presentation	Metabolic derangements and other important signs		Most frequent diagnosis (disorder or enzyme deficiency)	Differential diagnosis
Hepatic coma (hepatomegaly cytolysis or liver failure) Reye syndrome	Normal bilirubin Slight elevation of transaminases	Steatosis and Fibrosis	**FAO, UCD**	Reye syndrome Hepatitis Intoxication
	Hyperlactatemia	Liver failure	Respiratory chain defects	
	Hemolytic jaundice	Cirrhosis Chronic hepatic dysfunction	**Wilson disease***	
	Hypoglycemia	Exudative entero-pathy	Hepatic fibrosis with enteropathy (**CDG Ib**)	

AT III, antithrombin III; *BBGD*, biotin-responsive basal ganglia disease; *CDG*, congenital disorders of glycosylation; *EPEMA*, ence-phalopathy, petechiae, ethylmalonic aciduria syndrome; *FAO*, fatty acid oxidation; *FBPase*, fructose-1,6-bisphosphatase; *GA*, glutaric aciduria; *GH*, growth hormone; *HMG-CoA*, 3-hydroxy-3-methylglutaryl coenzyme A; *IVA*, isovaleric acidemia; *LPI*, lysinuric protein intolerance; *MCAD*, medium chain acyl-CoA dehydrogenase; *MCD*, multiple carboxylase deficiency; *MELAS*, mitochondrial ence-phalopathy with lactic acidosis and stroke-like episodes; *MMA*, methylmalonic acidemia; *MSUD*, maple syrup urine disease; *OTC*, ornithine transcarbamylase; *PA*, propionic acidemia; *PC*, pyruvate carboxylase; *PDH*, pyruvate dehydrogenase; *PGK*, phosphoglycerate kinase; *UCD*, urea cycle disorders; **bold face**, treatable disorders; * presenting or predominant feature reported in adults.

prolonged exercise, and all conditions that enhance protein catabolism, may exacerbate such decompensations. The diagnostic approach to recurrent attacks of coma, vomiting, ataxia, and psychiatric disturbances, dehydration, Reye and SIDS are presented in ◘ Tables 1.5 to 1.9. Other acute manifestations are listed in Sect. 1.5.

Coma, Strokes and Attacks of Vomiting with Lethargy (◘ Table 1.5)

Acute encephalopathy is a common problem in patients (children and adults) with IEM. All types of coma can be indicative of an IEM, including those presenting with focal neurological signs. Neither the age at onset, the accompanying clinical signs (hepatic, gastrointestinal, neurological, psychiatric etc.), the mode of evolution (improvement, sequelae, death), nor the routine laboratory data, allow an inborn error of metabolism to be ruled out a priori. Two categories can be distinguished:

1. *Metabolic coma without focal neurological signs*
   The main varieties of metabolic comas may all be observed in these late-onset, acute diseases: coma with predominant metabolic acidosis, coma with predominant hyperammonemia, coma with predominant hypoglycemia, and combinations of these three major abnormalities. A rather confusing finding in some organic acidurias and ketolytic defects is ketoacidosis with hyperglycemia and glycosuria that mimics diabetic coma. The diagnostic approach to these metabolic derangements is developed below (▶ Sect. 1.3.2).

2. *Neurological coma with focal signs, seizures, severe intracranial hypertension, strokes or stroke-like episodes*
   Although most recurrent metabolic comas are not accompanied by neurological signs other than encephalopathy, some patients with **organic acidemias** and **urea cycle defects** present with focal neurological signs or cerebral edema. These patients can be mistakenly diagnosed as having a cerebrovascular accident or cerebral tumor. In these disorders, stopping the protein intake, delivering glucose at high infusion rate and giving »cleansing drugs« (carnitine, sodium benzoate, etc.) can be life saving. Another treatable condition is **biotin-responsive basal ganglia disease** which presents in childhood with a subacute encephalopathic picture of undefined origin including confusion, vomiting, and a vague history of febrile illness [11, 12].

   All severe forms of **homocystinuria** (total homocysteine >100 µmol/l) can cause an acute cerebrovascular accident from late childhood to adulthood. These include **cystathionine-β-synthase deficiency** (usually B6-responsive in the late onset presentations), the severe **MTHFR** defects (folate responsive) and **CblC, CblD** defects (hydroxocobalamin responsive). Patients with **MMA** may, after first presenting with metabolic decompensation, have acute extrapyramidal and corticospinal tract involvement as a result of bilateral destruction of the globus pallidus with variable involvement of the internal capsule. Cerebellar hemorrhage has also been observed in IVA, PA, and MMA.

EPEMA syndrome, the molecular mechanism of which has been recently identified [13], starts in general early in infancy and is characterized by the association of progressive encephalopathy with mental retardation, pyramidal signs, and bilateral lesions in the striatum, resembling Leigh syndrome, relapsing petechiae, orthostatic acrocyanosis, and recurrent attacks of metabolic decompensation with lactic acidosis without ketosis, during which there is an exacerbation of ethylmalonic and methylsuccinic excretions.

Two patients with 3-hydroxyisobutyric aciduria presenting with recurrent episodes of vomiting and keto-acidotic coma have been described [14]. Patients with mitochondrial DNA mutations have presented with cyclical vomiting associated with intermittent lactic acidosis [15, 16]. GA type I frequently presents with an encephalopathic episode, mimicking encephalitis, in association with an intercurrent gastrointestinal or viral infection. Mitochondrial encephalopathy with lactic acidosis and stroke-like episodes (MELAS) syndrome is another important diagnostic consideration in such late-onset and recurrent comas. Early episodic central nervous system problems, possibly associated with liver insufficiency or cardiac failure, have been the initial findings in some cases of CDG. **Wilson disease** can rarely present with an acute episode of encephalopathy with extrapyramidal signs.

In summary, all these disorders should be considered in the differential diagnosis of strokes or stroke-like episodes. Vaguely defined and/or undocumented diagnoses such as encephalitis, basilar migraine, intoxication, poisoning, or cerebral thrombophlebitis should therefore be questioned, particularly when even moderate ketoacidosis, hyperlactatemia, or hyperammonemia is present. In fact, these apparent initial acute manifestations are frequently preceded by other premonitory symptoms, which may be unrecognized or misinterpreted. Such symptoms include acute ataxia, persistent anorexia, chronic vomiting, failure to thrive, hypotonia, and progressive developmental delay – all symptoms that are often observed in **urea cycle disorders**, respiratory chain defects, and **organic acidurias**. Late onset forms of **PDH** can present in childhood with recurrent attacks of ataxia, sometimes described by the patient as recurrent episodes of pain or muscular weakness (due to dystonia or to peripheral neuropathy).

Certain features or symptoms are characteristic of particular disorders. For example, macrocephaly is a frequent finding in **glutaric aciduria type I**; unexplained episodes of dehydration may occur in **organic acidurias**; and hepatomegaly at the time of coma is an important although inconsistent finding in **fructose-1,6-bisphosphatase deficiency**. Severe hematologic manifestations and recurrent infections are common in **IVA, PA, and MMA**. Macrocytic

anemia may be an important clue indicating a **cobalamin or folate disorder**.

When coma is associated with hepatic dysfunction, Reye syndrome secondary to disorders of **fatty acid oxidation** or the **urea cycle** should be considered. Hepatic coma with liver failure and hyperlactatemia can be the presenting sign of respiratory chain disorders. Finally, hepatic coma with cirrhosis, chronic hepatic dysfunction, hemolytic jaundice, and various neurological signs (psychiatric, extrapyramidal) is a classic, but underdiagnosed manifestation of **Wilson disease**.

Recurrent Attacks of Ataxia (◘ Table 1.6)

Intermittent acute ataxia and disturbed behavior can be the presenting signs of late-onset **MSUD** and **organic acidurias**, where they are associated with ketoacidosis and sometimes with hyperglycemia which can mimic diabetic ketoacidosis. Late onset **ornithine transcarbamylase (OTC) deficiency** and **argininosuccinate synthetase (ASS)** deficiency can present with recurrent attacks of ataxia. Acute ataxia associated with peripheral neuropathy is a frequent presenting sign of **pyruvate dehydrogenase (PDH)** deficiency; moderate hyperlactatemia with a normal L/P ratio supports this diagnosis. Hartnup disease (the molecular mechanism of which has been recently identified) is a classical but very rare cause of acute recurrent ataxia.

Acute Psychiatric Symptoms (◘ Table 1.7)

Late-onset forms of congenital hyperammonemia, mainly partial **OTC** deficiency, can present late in childhood or in adolescence with psychiatric symptoms. Because hyperammonemia and liver dysfunction can be mild even at the time of acute attacks, these intermittent late-onset forms of **urea cycle disorders** can easily be misdiagnosed as hysteria, schizophrenia, or alcohol or drug intoxication. **Acute intermittent porphyria** and **hereditary coproporphyria** present classically with recurrent attacks of vomiting, abdominal pain, neuropathy, and psychiatric symptoms. Finally, patients with **homocysteine remethylation defects** may present with schizophrenia-like, folate-responsive episodes. In view of **these possible diagnoses, it is justified to systematically measure ammonia, porphyrins, plasma homocysteine and copper in every patient presenting with unexplained acute psychiatric symptoms**. Episodes of acute psychosis also occur in the newly described autosomal dominant disorder neuroferritinopathy which is associated with low serum ferritin [17, 18].

Dehydration (◘ Table 1.8)

In pediatrics, dehydration is a common consequence of diarrhea caused by a variety of enteral or parenteral acute infections. However, these common infectious diseases can occasionally trigger acute decompensation of an IEM. Moreover, aside from dehydration due to gastrointestinal losses, some IEM can present as recurrent attacks of de-

☐ **Table 1.6.** Diagnostic approach to recurrent attacks of ataxia/± lethargy

Clinical presentation	Metabolic derangements or other important signs	Additional symptoms	Most frequent diagnosis (disorder or enzyme deficiency)	Differential diagnosis
Acute ataxia	Ketoacidosis Characteristic AAC and OAC profiles	Distinctive odor Neutropenia Thrombopenia Hyperglycemia	**Late onset MSUD MMA, PA, IVA**	Diabetes
	Hyperammonemia (sometimes slight elevation) AAC, orotic acid	Respiratory alkalosis Hepatomegaly	**Urea cycle defects (OTC, ASA)**	Intoxication Encephalitis Brain tumor
	Hyperlactatemia (sometimes very moderate and only in post-prandial state)	Normal L/P ratio No ketosis Peripheral neuropathy	**PDH**	Migraine Cerebellitis (varicella) Polymyoclonia Acetazolamide responsive ataxia Acute exacerbation in chronic ataxias
		High L/P ratio Ketosis Cutaneous signs	**MCD** Respiratory chain defects	
	AAC (neutral AA in urines)	Skin rashes, pellagra, sun intolerance	Hartnup disease	

AAC, amino acid chromatography; *ASA*, arginosuccinic aciduria; *IVA*, isovaleric acidemia; *L*, lactate; *LPI*, lysinuric protein intolerance; *MCD*, multiple carboxylase deficiency; *MMA*, methymalonic acidemia; *MSUD*, maple syrup urine disease; *OAC*, organic acid chromatography; *OTC*, ornithine transcarbamylase; *P*, pyruvate; *PA*, propionic acidemia; *PDH*, pyruvate dehydrogenase; **bold face**, treatable disorders.

☐ **Table 1.7.** Diagnostic approach to recurrent attacks of psychiatric symptoms

Clinical presentation	Metabolic derangements or other important findings	Additional symptoms	Most frequent diagnosis (disorder or enzyme deficiency)	Differential diagnosis
Psychiatric symptoms (hallucinations, delirium, dizziness, aggressiveness, anxiety, schizophrenic-like behaviour, agitation)	Hyperammonemia (sometimes moderate) AAC, orotic acid	Slight liver dysfunction Vomiting Failure to thrive	**Urea cycle defects (OTC, ASA, arginase) LPI**	
	Ketoacidosis AAC, OAC	Ataxia, neutropenia	**Organic acid defects, MSUD**	
	Port-wine urine Porphyrins in plasma/urine	Abdominal pain All kinds of neuropathy Vomiting	Acute intermittent **porphyria** **Hereditary coproporphyria**	Hysteria
	Homocystinuria (total homocysteine > 100 µM)	Stroke, seizures Myelopathy	**Methylene tetrahydrofolate reductase**	Schizophrenia
	AAC (neutral AA in urines)	Skin rashes, pellagra	Hartnup disease	
	Low serum ferritin	Dystonia, Parkinsonism, Pallidal necrosis	Neuroferritinopathy [17,18]	Hallervorden-Spatz
	Low serum copper acanthocytosis		**Wilson disease**	
	Foam cells in bone marrow	Vertical ophthalmoplegia	Niemann Pick type C	
	None	Epilepsy, retinitis pigmentosa	Ceroid lipofuscinosis	

AAC, amino acid chromatography; *ASA*, arginosuccinic aciduria; *MSUD*, maple syrup urine disease; *LPI*, lysinuric protein intolerance; *OAC*, organic acid chromatography; *OTC*, ornithine transcarbamylase; **bold face**, treatable disorders.

□ Table 1.8. Attacks of dehydration

Leading symptoms	Other signs	Age at onset	Diagnosis (disorder or enzyme deficiency)
Severe diarrhea: »gastrointestinal causes«	Severe watery acidic diarrhea Glycosuria	Neonatal	**Glucose galactose malabsorption** Lactase
	Hydramnios, no meconium Severe watery nonacidic diarrhea Metabolic alkalosis Low K$^+$, Cl$^-$	Congenital	**Congenital chloride diarrhea**
	Severe watery diarrhea	After weaning or when sucrose or starch dextrins are added to the diet	**Sucrase isomaltase**
	Anorexia, failure to thrive Weight loss (before cutaneous lesions and alopecia)	2-4 Weeks or after weaning	**Acrodermatitis enteropathica**
Ketoacidosis: »organic acidurias«	Polyuria Tachypnea Hyperglycemia Glycosuria	Infancy to early childhood	Diabetic coma **MMA, PA, IVA** **3-Ketothiolase** Hydroxyisobutyric aciduria
Failure to thrive, anorexia, poor feeding, polydipsia, polyuria: renal tubular dysfunction	Photophobia Renal Fanconi syndrome	Infancy 3-6 months	**Cystinosis**
	Hypernatremia, vomiting Psychomotor retardation Spasticity	Neonatal to first month	**Nephrogenic diabetes insipidus** (X-linked)
	Hyperchloremia Metabolic acidosis Alkaline urine pH	Early in infancy	**RTA type I (distal)** **RTA type II (proximal)** **RTA type IV**
	Hypoglycemia Hepatic glycogenosis Fanconi syndrome	Early in infancy	**Fanconi-Bickel syndrome** (GLUT2 mutation)
	Pulmonary infections Chronic diarrhea Salty sweet	Infancy to early childhood	**Cystic fibrosis**
Salt-losing syndrome: »adrenal dysfunctions»	Severe hyponatremia Ambiguous genitalia	End of first week of life	**Congenital adrenal hyperplasia**
	Unambiguous genitalia	End of first week	**Hypoaldosteronism**
	Ambiguous genitalia	Infancy to early childhood	**Congenital adrenal hypoplasia** **Congenital adrenal hyperplasia, late-onset forms**
	Unambiguous genitalia		Hypo & pseudohypoaldosteronism
	Hypoketotic hypoglycemia		**FAO defects (CPT I and II)**

CPT, carnitine palmitoyl transferase; *FAO*, fatty acid oxidation; *GLUT*, glucose transporter; *IVA*, isovaleric acidemia; *MMA*, methylmalonic aciduria; *PA*, propionic aciduria; *RTA*, renal tubular acidosis; **bold face**, treatable disorders.

hydration secondary to polyuria, hyperventilation or excessive sweating. The main accompanying findings (severe diarrhea, salt wasting, ketoacidosis, failure to thrive, Fanconi syndrome) can be used to classify dehydration due to IEM as shown in □ Table 1.8.

Reye Syndrome, Sudden Unexpected Death in Infancy (SUDI) and Near-miss (□ Table 1.9)

Within the last decade, an increasing number of IEM have been described that produce episodes fitting the criteria originally used to define Reye syndrome. There is now con-

◻ Table 1.9. Diseases in which Reye syndrome, sudden unexpected death in infancy (SUDI), and near-miss have been reported

Disorder or Enzyme deficiency	Incidence of the syndrome
Disorders of ureagenesis	Frequent
Partial OTC	Frequent
Partial carbamylphosphate synthetase	Rare
Partial argininosuccinic acid synthetase	Rare
Lysinuric protein intolerance	Rare
Triple H syndrome	Very rare
Disorders of mitochondrial fatty acid oxidation & ketogenesis	Frequent in neonates
Carnitine transport defect	Rare
CPT I	Rare
CPT II	Rare
Translocase	Rare
VLCAD / LCAD	Very rare
LHCAD, trifunctional protein	Rare
MCAD	Rare
Multiple acyl-CoA dehydrogenase	Frequent
3-Hydroxymethylglutaryl-CoA lyase	Rare
Unknown (3-ketoacyl-CoA thiolase?)	Very rare
Organic acidurias	Frequent
3-methylcrotonyl-CoA carboxylase	Very rare
Glutaric aciduria type I	Rare
Respiratory chain defects	Rare
Others (MMA, PA, IVA, biotinidase …)	Rare
Carbohydrate metabolism	Very rare
Hereditary fructose intolerance	
Glycogenosis type I	
Fructose-1,6-bisphosphatase	
Metabolic diseases causing Leigh and Leigh-like syndromes (PDH, PC, respiratory chain defects …)	Very rare
Peroxisomal defects	Very rare

CPT, carnitine palmitoyl transferase; *IVA*, isovaleric acidemia; *LCAD*, long chain acyl-CoA dehydrogenase; *LCHAD*, 3-hydroxy-long chain acyl-CoA dehydrogenase; *MCAD*, medium chain acyl-CoA dehydrogenase; *MMA*, methylmalonic acidemia; *OTC*, ornithine transcarbamylase; *PA*, propionic acidemia; *VLCAD*, very long chain acyl-CoA dehydrogenase; **bold face**, treatable disorders.

siderable evidence that many of the disorders responsible for Reye syndrome were misdiagnosed in the past because of inadequate investigations for IEM. Another important reason for this underestimation is the necessity of collecting blood and urine specimens for metabolic investigations at an appropriate time in relation to the illness since most disorders affecting the mitochondrial pathway, **urea cycle** and **FAO disorders** may produce only intermittent abnormalities. In addition, in contrast to the usual belief, a normal or non specific urinary organic acid and acylcarnitine pattern, even at the time of an acute attack, does not exclude an inherited **FAO disorder**.

True SUDI due to an IEM is, however, a rare event despite the large number of publications on the topic and despite the fact that at least 31 metabolic defects are possible causes. This assertion is not true in the first week of life in which **unexpected death or near-miss is a priori due to a fatty acid oxidation disorder and the investigation of which is mandatory**.

1.3.2 Metabolic Derangements and Diagnostic Tests

Initial Approach, Protocol of Investigation

The initial approach to the late-onset acute forms of inherited metabolic disorders, as with the approach to acute neonatal distress, is based on the proper use of a few screening tests. As with neonates, the laboratory data listed in ◻ Table 1.3 must be collected during the acute attack and both before and after treatment.

Main Metabolic Presentations
Metabolic Acidosis (◻ Fig. 1.1)

Metabolic acidosis is a very common finding in pediatrics. It can be observed in a large variety of acquired conditions, including infections, severe catabolic states, tissue anoxia, severe dehydration, and intoxication, all of which should be ruled out. However, these can also trigger an acute decompensation of an unrecognized IEM. The presence or absence of ketonuria associated with metabolic acidosis is the major clinical clue to the diagnosis.

When metabolic acidosis is not associated with ketosis, **PDH deficiency**, **fatty acid oxidation disorders**, and some **disorders of gluconeogenesis** should be considered, particularly when there is moderate to severe hyperlactatemia. All these disorders except **PDH deficiency** have concomitant fasting hypoglycemia. Although **fructose-1,6-bisphosphatase** deficiency is classically considered to give rise to ketoacidosis, some patients have had low concentrations of ketone bodies during hypoglycemia. When metabolic acidosis occurs with a normal anion gap and without hyperlactatemia or hypoglycemia, the most frequent cause is renal tubular acidosis (RTA) type I or II. Pyroglutamic aciduria also can present early in life with permanent, isolated metabolic acidosis, which can be mistaken for RTA type II.

A number of IEM cause a metabolic acidosis with an associated ketosis. The range of serum ketone body concentration varies with age and nutritional state (▶ Chap. 2). Insulin-dependent **diabetes**, inborn errors of **branched-chain amino acid metabolism**, congenital lactic acidoses such as **multiple carboxylase** and PC deficiencies, inherited defects in enzymes of **gluconeogenesis** and of glycogen synthesis (**glycogen synthase**), and **ketolytic defects** are the main groups of metabolic disorders. The glucose level which can be high, normal, or low is the first parameter to be considered to classify these disorders.

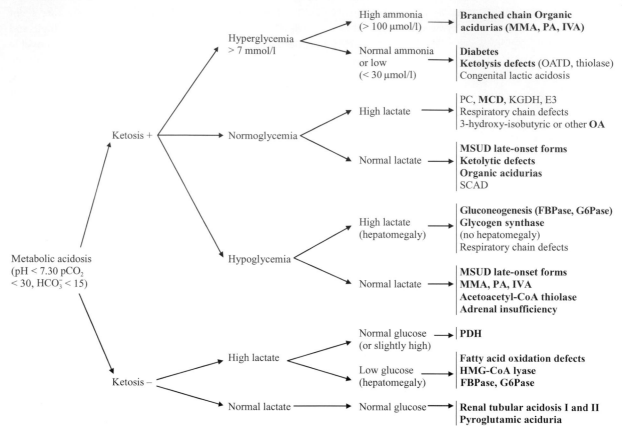

◘ Fig. 1.1. Metabolic acidosis. *E3*, lipoamide oxidoreductase; *FBPase*, fructose-1,6-bisphosphatase; *G6Pase*, glucose-6-phosphatase; *GS*, glycogen synthase; *HMG-CoA*, 3-hydroxy-3-methylglutaryl coenzyme A; *IVA*, isovaleric acidemia; *KGDH*, alpha-ketoglutarate dehydrogenase; *MCD*, multiple carboxylase deficiency; *MMA*, methylmalonic aciduria; *MSUD*, maple syrup urine disease; *OA*, organic aciduria; *OATD*, oxoacid-CoA transferase; *PA*, propionic acidemia; *PC*, pyruvate carboxylase; *PDH*, pyruvate dehydrogenase; *SCAD*, short chain acyl-CoA dehydrogenase; **bold face**, treatable disorders

In the case of hyperglycemia, the classic diagnosis is diabetic ketoacidosis. However, organic acidurias such as **propionic, methylmalonic, or isovaleric acidemia and ketolytic defects** can also be associated with hyperglycemia and glycosuria, mimicking diabetes. The distinction between the different disorders is also based on ammonia and lactate levels, which are generally increased in organic acidemias and normal or low in ketolytic defects.

In the case of hypoglycemia, the first group of disorders to be considered is the **gluconeogenesis defects** and the **glycogenoses**. The main findings suggestive of this group are hepatomegaly and hyperlactatemia, although they are not constant. When there is no significant hepatomegaly, late-onset forms of **MSUD** and **organic acidurias** and **glycogen synthase** (**GS**) deficiency should be considered. A classic differential diagnosis is **adrenal insufficiency**, which can cause a ketoacidotic attack with hypoglycemia.

If the glucose level is normal, congenital lactic acidosis must be considered in addition to the disorders discussed above. According to this schematic approach to inherited ketoacidotic states, a simplistic diagnosis of fasting ketoacidosis or ketotic hypoglycemia should be questioned when there is a concomitant severe metabolic acidosis.

Ketosis (◘ Fig. 1.2)

While ketonuria should always be considered abnormal in neonates, it is a physiological result of catabolism in late infancy, childhood, and even adolescence. However as a general rule, hyperketosis at a level that produces metabolic acidosis is not physiological. Ketosis which is not associated with acidosis, hyperlactatemia, or hypoglycemia, is likely to be a normal physiological reflection of the nutritional state (fasting, catabolism, vomiting, medium-chain triglyceride-enriched or other ketogenic diets). Of interest are ketolytic defects (**succinyl-CoA transferase** and **3-ketothiolase** deficiencies) that can present with persistent moderate ketonuria occurring mainly in the fed state at the end of the day.

Significant fasting ketonuria without acidosis is often observed in **glycogenosis type III** in childhood (with marked hepatomegaly) and in the very rare **GS** defect in infancy (with normal liver size). In both disorders, there is fasting hypoglycemia, and postprandial hyperlactatemia and hyperglycemia.

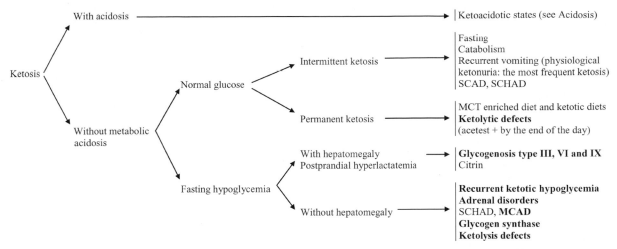

Fig. 1.2. Ketosis (▶ also Fig. 1.1). *MCAD,* medium chain acyl co-enzyme A dehydrogenase; *MCT,* medium chain triglycerides; *SCAD,* short-chain acyl-coenzyme A dehydrogenase; *SCHAD,* hydroxy short-chain acyl-coenzyme A dehydrogenase; **bold face,** treatable disorders

Ketosis without acidosis is observed in ketotic hypoglycemias of childhood (a frequent condition) and in association with hypoglycemias due to **adrenal insufficiency**. **Absence of ketonuria in hypoglycemic states**, as well as in fasting and catabolic circumstances induced by vomiting, anorexia, or intercurrent infections, **is an important observation, suggesting an inherited disorder of fatty acid oxidation or ketogenesis disorder** and can also be observed in hyperinsulinemic states at any age and in growth hormone deficiency in infancy. However, short-chain 3-hydroxy acyl-CoA dehydrogenase (SCHAD), SCAD, and **MCAD** deficiencies can present as recurrent attacks of ketotic hypoglycemia as these enzymes are both sufficiently far down the β-oxidation pathway to be able to generate some ketones from long chain fatty acids [19].

Hyperlactatemia (☐ Table 1.10)

Lactate and pyruvate are normal metabolites. Their plasma levels reflect the equilibrium between their cytoplasmic production from glycolysis and their mitochondrial consumption by different tissues. The blood levels of lactate and pyruvate and the L/P ratio reflect the redox state of the cells.

Blood lactate accumulates in circulatory collapse, in hypoxic insult, and in other conditions involving failure of cellular respiration. These conditions must be excluded before an inborn error of lactate-pyruvate oxidation is sought. Persistent hyperlactatemias can also result from many acquired conditions, such as diarrhea, persistent infections (mainly of the urinary tract), hyperventilation, and hepatic failure. Ketosis is absent in most hyperlactatemias secondary to tissue hypoxia, while it is a nearly constant finding in most inborn errors of metabolism (except in **PDH deficiency, glycogenosis type I** and **FAO disorders**). On the other hand, the level of lactate is not discriminating; some acquired disorders are associated with very high levels, whereas it is only moderately raised in some inborn errors

of lactate-pyruvate metabolism. Nutritional state also influences the levels of lactate and pyruvate.

Once the **organic acidurias, urea cycle defects** (mainly **citrullinemia**), and **fatty acid oxidation defects** that cause secondary hyperlactatemia have been excluded as possible diagnoses, four types of inherited disorders remain to be considered: disorders of liver glycogen metabolism, disorders of liver gluconeogenesis, abnormalities of lactate-pyruvate oxidation, PDH deficiency, Krebs cycle defects, and deficient activity in one of the components of the respiratory chain. The diagnosis of hyperlactatemias is largely based upon two metabolic criteria:

— *Time of occurrence of lactic acidosis relative to feeding:* in disorders of gluconeogenesis (**fructose-1,6-bisphosphatase** and **glucose-6-phosphatase deficiencies**), hyperlactatemia reaches its maximum level (up to 15 mM) when the patient is fasting and hypoglycemic. By contrast, in **glycogenosis types III and VI** and in **glycogen synthase deficiency**, hyperlactatemia is observed only in the postprandial period in patients on a carbohydrate-rich diet. Here, hyperlactatemia never exceeds 7 mM. In pyruvate carboxylase deficiency, hyperlactatemia is present in both the fed and the fasted state, but tends to decrease with a short fast. In disorders of **PDH**, alpha-ketoglutarate dehydrogenase, and respiratory chain function, maximum lactate levels are observed in the fed state (although all hyperlactatemias exceeding 7 mM appear more or less permanent). In these disorders, there is a real risk of missing a moderate (although significant) hyperlactatemia when the level is checked only before breakfast after an overnight fast (as it is usual for laboratory determinations).

— *Determinations of L/P and ketone bodies ratios before and after meals.* These ratios indirectly reflect cytoplasmic (L/P) and mitochondrial (3OHB/AcAc) redox potential states. They must be measured in carefully collected

◻ **Table 1.10.** Diagnostic approach to hyperlactatemias

Time of occurence	Main clinical signs	Redox potential states	Diagnosis (disorder or enzyme deficiency)
Only after feeding (or exacerbated after feeding)	Hepatomegaly Fasting ketotic hypoglycemia	Not diagnostic	**Glycogenosis type III** **Glycogen synthase**
	Neurological signs Encephalomyopathy	Normal L/P ratio, no ketosis	**PDH**, pyruvate carrier
		L/P high, 3OHB/AcAc low Postprandial hyperketosis	PC (high citrulline, low glutamine) **MCD** α-KDH (isolated or E3)
		L/P high, 3OHB/AcAc high Postprandial ketosis	Respiratory chain defects (3-methyl-glutaconic aciduria, Krebs cycle intermediates)
		L/P high, No ketosis	Respiratory chain
Only after fasting (or exacerbated after fasting)	Prominent hepatomegaly Hypoglycemia	Not diagnostic	**Glycogenosis type I** **Fructose-1,6-bisphosphatase** (ketosis inconstant)
	Moderate or no hepatomegaly Hypoketotic hypoglycemia	Not diagnostic	**FAO (cardiac, muscle symptoms)** **Fructose-1,6-bisphosphatase**
Permanent	Moderate hyperlactatemia Recurrent attacks of ketoacidosis	Not diagnostic	**Organic acidurias** (MMA, PA, IVA) MAMEL syndrome [27]
	Predominant hyperammonemia	Not diagnostic	**Urea cycle defects** (in neonates)
	Predominant hypoglycemia Hepatomegaly	Not diagnostic	**Glycogenosis type I** **Fructose-1,6-bisphosphatase**
	Neurological signs, encephalomyopathy, important hyperlactatemia (>10 mM)	Highly diagnostic (► above »after feeding«)	Congenital lactic acidemias (► above »after feeding«)

3OHB, 3-hydroxybutyrate; *AcAc*, acetoacetate; *IVA*, isovaleric acidemia; *KDH*, ketoglutarate dehydrogenase; *L*, lactate; *MAMEL*, methylmalonic aciduria, mitochondrial encephalopathy Leigh-like; *MMA*, methylmalonic acidemia; *MCD*, multiple carboxylase deficiency; *P*, pyruvate; *PA* propionic acidemia; *PC*, pyruvate carboxylase; *PDH*, pyruvate dehydrogenase; **bold face**, treatable disorders.

blood samples (► Chap. 2). Three abnormal hyperlactatemia/pyruvicemia profiles are nearly pathognomonic of an inborn error of lactate-pyruvate metabolism:

When hyperpyruvicemia is associated with a normal or low L/P ratio (<10) without hyperketonemia, **PDH deficiency** or pyruvate transporter defect are highly probable, regardless of the lactate level.

When the L/P ratio is very high (>30) and is associated with postprandial hyperketonemia and with a normal or low 3OHB/AcAc ratio (<1.5), a diagnosis of pyruvate carboxylase (PC) deficiency (isolated or secondary to **biotinidase** or **holocarboxylase synthetase deficiency**) or alpha-ketoglutarate dehydrogenase deficiency is virtually certain. In PC deficiency, there is a very characteristic amino acid profile with hyperammonemia, high citrulline and low glutamine.

When both L/P and 3OHB/AcAc ratios are elevated and associated with a significant postprandial hyperketonemia, respiratory chain disorders should be suspected.

All other situations, especially when the L/P ratio is high without hyperketonemia, are compatible with respiratory chain disorders, but all acquired anoxic conditions should also be ruled out (► above).

Hypoglycemia (◻ Table 1.11)

Our approach to hypoglycemia is based on four major clinical criteria: liver size, characteristic timing of hypoglycemia (unpredictable, only postprandial or only after fasting), association with lactic acidosis, and association with hyperketosis or hypoketosis. Other clinical findings of interest are hepatic failure, vascular hypotension, dehydration, short stature, neonatal body size (head circumference, weight & height), and evidence of encephalopathy, myopathy, or cardiomyopathy. Based on the liver size, hypoglycemias can be classified into two major groups:

— *Hypoglycemia with permanent hepatomegaly:* Hypoglycemia associated with permanent hepatomegaly is usually due to an inborn error of metabolism. However, all conditions, both acquired or inherited, that are as-

◻ **Table 1.11.** Hypoglycemia: general approach

Leading symptoms	Other signs	Age at onset	Diagnosis (disorder or enzyme deficiency)
With permanent hepatomegaly			
Permanent short-fast hypoglycemia	Severe liver failure Hepatic necrosis	Neonatal to early infancy	**Galactosemia** **Hereditary fructose intolerance** **Tyrosinemia type I** Neonatal hemochromatosis Respiratory chain defects Other severe hepatic failure
Fibrosis, Cirrhosis	Postprandial hypoglycemia (triggered by fructose), Vomiting	Neonatal to early infancy	**Hereditary fructose intolerance** Glycerol intolerance
	Mental retardation Hypermethioninemia Hepatic failure induced by methionine	Early in infancy	Glycogenosis type IV SAH hydrolase Respiratory chain defects
	Exudative enteropathy Cholangitis attacks Short fast hypoglycemia	Early in infancy	CDG Ib
Isolated Hepatomegaly	Fasting hypoglycemia and Lactic acidosis, Ketosis	Infancy	**G6Pase** **FBPase**
	Protuberant abdomen Fasting hypoglycemia and ketosis Postprandial hyperlactatemia	Infancy	**Glycogenosis type III and VI**
Hypotonia	Abnormal glycosylated transferrin	Infancy	CDG
Failure to thrive Chronic diarrhea	Fanconi-like tubulopathy Postprandial hyperglycemia	Infancy	**Fanconi-Bickel syndrome** (GLUT II mutations)
Without permanent hepatomegaly			
With ketoacidosis	Recurrent attacks Hyperlactatemia	Infancy to childhood	**Organic acidurias** Late-onset **MSUD** **Ketolysis defects** Glycerol kinase **FBPase** SCHAD, **MCAD** Respiratory chain defects
	Dehydration, collapse Hyponatremia	Neonatal to childhood	**Adrenal insufficiency** (central or peripheral)
Acidosis without ketosis	Moderate hyperlactatemia Reye syndrome (with muscle/cardiac symptoms)	Neonatal to infancy	**HMG-CoA lyase** (frequent) **HMG-CoA synthase** (rare) **FAO defects** (frequent) Reye syndrome (idiopathic)
Ketosis without acidosis	Fasting hypoglycemia Low lactate levels Small size for age Macrocephaly	1–6 years	**Recurrent ketotic hypoglycaemia** Adrenal insufficiency SCHAD, **MCAD,** **Glycogen synthase** **Ketolysis defects**
Without acidosis or ketosis	Unpredictable and postprandial hypoglycemia reactive to glucagon	Neonatal to childhood	**Hyperinsulinisms** **Cortisol deficiency** CDG **Factitious or induced illness**
	Short stature, short-fast hypoglycemia	Infancy	**GH deficiency & related disorders**
	Long-fast hypoglycemia Reye syndrome Moderate hepatomegaly Transient cytolysis	Neonatal to infancy	**FAO defects** (frequent) **HMG-CoA lyase** (rare) **FBPase** (rare) **HMG-CoA synthase** (rare)

CDG, congenital disorders of glycosylation; *FAO* fatty acid oxidation disorders; *FBPase*, fructose-1,6-bisphosphatase; *GH*, growth hormone; *G6Pase*, glucose-6-phosphatase; *HMG-CoA*, 3-hydroxy-3-methylglutaryl coenzyme A; *MSUD*, maple syrup urine disease; A; *SAH*, S-adenosyl-homocysteine hydrolase; **bold face**, treatable disorders.

sociated with severe liver failure, can give rise to severe hypoglycemia, which appears after 2-3h of fasting and is associated with moderate lactic acidosis and no ketosis. When hepatomegaly is the most prominent feature without liver insufficiency, gluconeogenesis defects (**glucose-6-phosphatase deficiency, fructose-1,6-bisphosphatase deficiency**) and **glycogenosis type III**, are the most likely diagnoses. Disorders presenting with hepatic fibrosis and cirrhosis, such as hereditary **tyrosinemia type I**, also can give rise to hypoglycemia. The late-onset form of **hereditary fructose intolerance** rarely, if ever, presents with isolated postprandial hypoglycemic attacks. S-adenosylhomocysteine hydrolase deficiency presents with fasting hypoglycemia and hepatocellular insufficiency, often triggered by high protein or methionine ingestion, and is associated with hepatic fibrosis, mental retardation, and marked hypermethioninemia. Respiratory chain disorders can present with hepatic failure and hypoglycemia. **CDG type Ib** (**phosphomannose isomerase** deficiency) with hepatic fibrosis and exudative enteropathy can cause hypoglycemia early in infancy [20, 21].

— *Hypoglycemia without permanent hepatomegaly:* It is important to determine the timing of hypoglycemia and to look for metabolic acidosis and ketosis when the patient is hypoglycemic. Most episodes of hypoglycemia, due to IEM that are not accompanied by permanent hepatomegaly, appear after at least 8h of fasting. This is particularly true for inherited **fatty acid oxidation disorders** except in the neonatal period. Conversely, unpredictable postprandial or hypoglycaemia occurring after a very short fast (2–6h) is mostly due to **hyperinsulinism** and **growth hormone deficiency** or related disorders. When ketoacidosis is present at the time of hypoglycemia, **organic acidurias**, **ketolytic defects**, **late-onset MSUD**, and glycerol kinase deficiencies should be considered. Here, hypoglycemia is very rarely the initial metabolic abnormality. **Adrenal insufficiencies** must be considered in the differential diagnosis, especially when vascular hypotension, dehydration, and hyponatremia are present. Severe hypoglycemia with metabolic acidosis and absence of ketosis, in the context of Reye syndrome, suggests **HMG-CoA lyase deficiency**, **HMG-CoA synthase deficiency** or **fatty acid oxidation disorders**. Fasting hypoglycemia with ketosis occurring mainly in the morning and in the absence of metabolic acidosis suggests recurrent functional ketotic hypoglycemia, which presents mostly in late infancy or childhood in those who were small for gestational age or with macrocephaly. All types of **adrenal insufficiencies** (peripheral or central) can share this presentation. SCHAD and **MCAD** deficiency can on occasions present as recurrent attacks of ketotic hypoglycemia [19] as can **glycogen synthase deficiency**. However, in our experience,

this pattern is rarely associated with inborn errors of metabolism.

Hypoketotic hypoglycemias encompass several groups of disorders including **hyperinsulinemic states**, **growth hormone deficiency**, inborn errors of **fatty acid oxidation**, and **ketogenesis defects** (▶ »Ketosis« above).

Hyperammonemia

The diagnostic approach to hyperammonemia is developed in ▶ Chap. 20.

1.4 Chronic and Progressive General Symptoms/Signs

As already stated, many acute presentations of inherited disorders that are apparently of delayed onset are preceded by insidious premonitory symptoms and which may have been ignored or misinterpreted. These signs fall schematically into three categories according to whether there is gastrointestinal, muscle or neurological involvement.

1.4.1 Gastrointestinal Symptoms

Gastrointestinal findings (GI) (anorexia, failure to thrive, osteoporosis, chronic vomiting) occur in a wide variety of inborn errors of metabolism. Unfortunately, their cause often remains unrecognized, thus delaying the correct diagnosis. Persistent anorexia, feeding difficulties, chronic vomiting, failure to thrive, frequent infections, osteopenia, and generalized hypotonia in association with chronic diarrhea are the presenting symptoms and signs in a number of constitutional and acquired diseases in pediatrics. They are easily misdiagnosed as cow's milk protein intolerance, celiac disease, chronic ear, nose, and throat infections, late-onset chronic pyloric stenosis etc. Congenital immuno-deficiencies are also frequently considered, although only a few present early in infancy with this clinical picture.

From a pathophysiological point of view, it is possible to define two groups of inborn errors of metabolism presenting with chronic diarrhea and failure to thrive:
— Disorders of the intestinal mucosa or the exocrine function of the pancreas with almost exclusive intestinal effects, for example congenital chloride diarrhea, glucose-galactose malabsorption, lactase and sucrase-isomaltase deficiencies, abetalipoproteinemia type II (Anderson disease), enterokinase deficiency, acrodermatitis enteropathica, and selective intestinal malabsorption of folate and vitamin B_{12}, the latter also causing systemic disease.
— Systemic disorders which also give rise to GI abnormalities.

□ Table 1.12. Chronic diarrhea, poor feeding, vomiting, failure to thrive

Leading symptoms	Other signs	Age of onset	Diagnosis (disorder or enzyme deficiency)
Severe watery diarrhea Attacks of dehydration	Nonacidic diarrhea, Hypochloremic alkalosis	Congenital to infancy	**Congenital chloride diarrhea**
	Acidic diarrhea, Reducing substances in stools	Neonatal	**Glucose galactose malabsorption Lactase**
	Acidic diarrhea, Reducing substances in stools after weaning	Neonatal to infancy	**Sucrase isomaltase**
	Skin lesions, alopecia	Neonatal or post weaning	**Acrodermatitis enteropathica**
Protein losing enteropathy	Cholangitis crisis Hypoglycemia	Infancy	**CDG type Ib and Ih**
Fat-soluble vitamins malabsorption Severe hypo-cholesterolemia Osteopenia Steatorrhea	Cholestatic jaundice	Neonatal to infancy	**Bile acid synthesis defects** Infantile Refsum
	Hepatomegaly, hypotonia, retinitis pigmentosa, deafness	Infancy	Infantile Refsum CDG type I
	Abdominal distension, ataxia, acanthocytosis, peripheral neuropathy, retinitis pigmentosa	Infancy	**ABL I and II** (no acanthocytes, no neurological signs in type II)
	Pancreatic insufficiency, neutropenia, pancytopenia	Early in infancy	Pearson syndrome Schwachman syndrome
Severe failure to thrive, anorexia, poor feeding, with predominant hepato-splenomegaly	Severe hypoglycemia, inflammatory bowel disease, neutropenia,	Neonatal to early infancy	**Glycogenosis type Ib** (no splenomegaly)
	Hypotonia, vacuolated lymphocytes, adrenal gland calcifications	Neonatal	Wolman disease
	Recurrent infections, inflammatory bowel disease,	Infancy	**Chronic granulomatosis** (X-linked)
	Megaloblastic anemia, neuropathy, homocystinuria, MMA	1–5 years	**Intrinsic factor**
	Leuconeutropenia, osteopenia, hyperammonemia, interstitial pneumonia,	Infancy	**Lysinuric protein intolerance**
	Recurrent fever, inflammatory bowel syndrome, hyper-IgD	Infancy	Mevalonate kinase
Severe failure to thrive, anorexia, poor feeding, with megaloblastic anemia	Oral lesion, neuropathy, infections, pancytopenia, homocystinuria, MMA	1-2 years	**TC II Intrinsic factor**
	Stomatitis, peripheral neuropathy, infections, intracranial calcifications	Infancy	**Congenital folate malabsorption**
	Severe pancytopenia, abnormal marrow precursors, lactic acidosis	Neonatal	Pearson syndrome
Severe failure to thrive, anorexia, poor feeding, no significant hepato-splenomegaly, no megaloblastic anemia	Severe hypoproteinemia, putrefaction diarrhea	Infancy	**Enterokinase**
	Diarrhea after weaning, cutaneous lesions (periorificial), low plasma zinc	Infancy	**Acrodermatitis enteropathica**
	Ketoacidotic attacks, vomiting	Infancy	**Organic acidurias (MMA, PA)** Mitochondrial DNA deletions
	Vomiting, lethargy, hypotonia, hyperammonemia	Infancy	**Urea cycle defects** (mainly **OTC**)
	Frequent infections, lymphopenia,	Infancy	**Adenosine deaminase**
	Developmental delay, relapsing petechiae, orthostatic acrocyanosis	Infancy	EPEMA syndrome [13]

ABL, abetalipoproteinemia; *CDG*, congenital disorders of glycosylation; *EPEMA*, encephalopathy, petechiae, and ethylmalonic aciduria; *MMA*, methylmalonic acidemia; *OTC*, ornithine transcarbamylase; *PA*, propionic acidemia; *TC*, transcobalamin; **bold face**, treatable disorders.

In clinical practice, these groups are sometimes very difficult to distinguish, because a number of specific intestinal disorders can give rise to various systemic clinical abnormalities and vice versa. This is summarized in ◘ Table 1.12.

1.4.2   Muscle Symptoms

Many inborn errors of metabolism can present with severe hypotonia, muscular weakness, and poor muscle mass. These include most of the late-onset forms of urea cycle defects and many organic acidurias. Severe neonatal generalized hypotonia, progressive myopathy with or without an associated nonobstructive idiopathic cardiomyopathy, can be the specific presenting findings in a number of inherited energy deficiencies; the most frequent conditions are mitochondrial respiratory chain disorders and other congenital hyperlactatemias, fatty acid oxidation defects, peroxisomal disorders, muscular glycogenolysis defects, alpha-glucosidase deficiency, and some other lysosomal disorders (▶ also Sect. 1.2.1 and 1.5.8). Hypotonia, generalized weakness, reduced muscle mass and developmental delay are also the presenting features of the Allan-Herndon-Dudley syndrome due to mutations in the monocarboxylate transporter 8 gene. This X-linked mental retardation syndrome involves the transport of triiodothyronine into neurones and disturbs blood levels of thyroid hormone [22].

1.4.3   Neurological Symptoms

Neurological symptoms are very frequent in inborn errors and encompass progressive psychomotor retardation, seizures, and a number of neurological abnormalities, in both the central and peripheral system, sensorineural defects and psychiatric symptoms.

A large number of inborn errors of intermediary metabolism present with an early and non-specific progressive developmental delay, poor feeding, hypotonia, some degree of ataxia, and frequent autistic features. The list has lengthened rapidly as new laboratory techniques have been applied. The relationship between clinical and biochemical abnormalities is not always firmly established. Many aminoacidopathies that were first described in the late 1950s and 1960s, when plasma and urine amino acid chromatography was systematically used in studying mentally retarded children, must now be questioned as definitely being the cause of neurological disease. This is the case for histidinemia, hyperlysinemia, hyperprolinemia, alpha-amino-adipic aciduria, saccharopinuria, Hartnup »disease« and the recently described acetyl amino aciduria due to amino acylase I deficiency [23a].

A similar picture is now emerging with organic acidurias and it is therefore important to link clinical symptoms and metabolic disturbances. Conversely, it becomes more and more difficult to screen patients on clinical grounds when the clinical symptoms consist only of rather non specific signs such as developmental delay, microcephaly, hypotonia or convulsions. Among the new categories of inborn errors of intermediary metabolism that can present with uninformative clinical manifestations are, for example, adenylosuccinase deficiency, dihydropyrimidine dehydrogenase deficiency, 4-hydroxybutyric aciduria, L-2- and D-2-hydroxyglutaric acidurias, late onset NKH and a number of other inborn errors (▶ bottom of ◘ Table 1.14). These disorders rarely, if ever, cause true development arrest; rather, they cause progressive subacute developmental delay. Conversely, there is still an important gap between neurological descriptions and biological investigations. Many well-known heritable neurological or polymalformative syndromes have not been considered from a pathophysiological perspective and should be submitted to a comprehensive biochemical evaluation. This is illustrated for example by the story of Canavan disease, in which N-acetylaspartic aciduria was only found in 1988, even though the clinical phenotype had been identified in 1949 and the procedure for identifying N-acetylaspartate in urine was available in 1972.

In the following pages, IEM are listed:

a. according to their age at onset and their association with particular abnormalities, both neurological and extraneurological (▶ Tables 1.13–1.17) and also

b. alphabetically, under specific associated neurological abnormalities that they can cause (▶ Sect. 1.4.4).

Of course, these two categorizations are complementary and inevitably involve redundancies. It is recommended that one looks at both.

Progressive Neurological and Mental Deterioration Related to Age (Overview)

◘ Tables 1.13 to 1.17 present a general approach to inborn errors of metabolism involving neurological and/or mental deterioration. Diseases are classified according to their age at onset, the presence or absence of associated extraneurological signs, and the neurological presentation itself; the last is based largely on the clinical classification of Lyon and Adams [24]. Inborn errors of metabolism with neurological signs presenting in the neonate (birth to 1 month; ◘ Table 1.2) and those presenting intermittently as acute attacks of coma, lethargy, ataxia, or acute psychiatric symptoms, were discussed earlier (◘ Tables 1.5 to 1.7).

Early Infancy (◘ Table 1.13)

Three general categories can be identified:

— **Category 1**: **Disorders Associated with Extraneurological Symptoms.** Visceral signs appear in lysosomal disorders. A cardiomyopathy (associated with early neurological dysfunction, failure to thrive, and hypo-

■ Table 1.13. Progressive neurological and mental deterioration with extraneurological symptoms (1 to 12 months) (▶ also Tables 1.2 and 1.4)

Leading symptoms	Other signs	Diagnosis (disorder or enzyme deficiency)
Visceral signs	Hepatosplenomegaly Storage signs, coarse facies	Landing, I-Cell disease Sialidosis type II, Niemann-Pick A Lactosyl ceramidosis
	Hepatosplenomegaly Opisthotonos, spasticity	Gaucher type II
	Hepatomegaly Retinitis pigmentosa	Peroxisomal defects CDG
Hair and cutaneous symptoms	Steely brittle hair	Menkes (X-linked) Trichothiodystrophy
	Trichorrexis nodosa	**Argininosuccinic aciduria**
	Ichthyosis, spastic paraplegia	Sjögren-Larsson syndrome **Serine deficiency syndrome**, CDG
	Alopecia, cutaneous rashes	**Biotinidase** Respiratory chain defects
	Peculiar fat pads on buttocks	CDG
	Cyanosis, hypertonicity	Cytochrome b-5 reductase
	Kernicterus, athetosis	Crigler-Najjar
	Acrocyanosis, petechiae	EPEMA syndrome [13]
Megabloblastic anemia	Failure to thrive, RP	**Folate and cobalamin defects UMP synthase**
Cardiac symptoms	Cardiomyopathy Heart failure, heart beat disorders	D-2-hydroxyglutaric acidemia Respiratory chain defects, CDG
Ocular symptoms	Cherry-red spot, hydrops fetalis	Landing, Galactosialidosis, Sialidosis type I
	Myoclonic jerks, macrocephaly	Tay-Sachs, Sandhoff
	Optic atrophy, macrocephaly	Canavan
	Nystagmus, dystonia, stridor	Pelizaeus-Merzbacher (X-linked)
	Retinitis pigmentosa	▶ Section 1.4.4
	Abnormal eye movements	Aromatic amino acid decarboxylase
	Strabism	CDG
	Supranuclear paralysis	Gaucher, Niemann-Pick type C

CDG, congenital disorders of glycosylation; *EPEMA*, encephalopathy, petechiae, and ethylmalonic aciduria; *RP*, retinitis pigmentosa; *UMP*, uridine monophosphate; **bold face**, treatable disorders.

tonia), sometimes responsible for cardiac failure, is suggestive of respiratory-chain disorders, D-2-hydroxyglutaric aciduria (with atrioventricular block), or CDG. Abnormal hair and cutaneous signs appear in Menkes disease, Sjögren-Larsson syndrome, **biotinidase deficiency**, and respiratory-chain disorders. Peculiar fat pads of the buttocks and thick and sticky skin (like tallow, peau d'orange), and inverted nipples are highly suggestive of CDG. A generalized cyanosis, unresponsive to oxygen, suggests methemoglobinemia, which is associated with severe hypertonicity in cytochrome-b5 reductase deficiency. Kernicterus and athetosis are complications of Criggler-Najjar syndrome. The recently described EPEMA syndrome is characterized by an orthostatic acrocyanosis, relapsing petechiae, pyramidal signs, mental retardation, and recurrent attacks of lactic acidosis. The presence of megaloblastic anemia suggests an inborn error of **folate and cobalamin (Cbl)** metabolism. Ocular abnormalities can be extremely helpful diagnostic signs, for example cherry-red spot, optic atrophy, nystagmus, abnormal eye movements, and retinitis pigmentosa.

- **Category 2: Disorders with Specific or Suggestive Neurological Signs.** Predominant extrapyramidal symptoms are associated with inborn errors of biopterin and aromatic-amino-acid metabolism, **pyridox(am)ine phosphate oxidase**, Lesch-Nyhan syndrome, cytochrome-b5 reductase deficiency, Criggler-Najjar syndrome, the early-onset form of GA type I, and **cerebral creatine deficiency**. Dystonia can also be observed as a subtle but presenting sign in X-linked Pelizaeus-Merzbacher syndrome. It can be also associated with psychomotor retardation, spastic paraplegia and ataxia in the **cerebral folate deficiency syndrome** [25].

    Macrocephaly with a startle response to sound, incessant crying, and irritability are frequent early signs in GM-2 gangliosidosis, Canavan disease, Alexander leukodystrophy, infantile Krabbe disease, and GA type I. Macrocephaly can be also an initial sign in L-2-hydroxyglutaric aciduria and in respiratory-chain disorders due to complex-I deficiency (association with hypertrophic cardiomyopathy).

    Recurrent attacks of neurological crisis associated with progressive neurological and mental deterioration suggest Leigh syndrome, which can present at any age from early in infancy to late childhood. Leigh syndrome is not a specific disorder but, rather, the clinical phenotype of any of several inborn errors of metabolism, some of which still remain to be identified. Recurrent stroke-like episodes often associated with anorexia, failure to thrive, and hypotonia can be presenting symptoms in **urea-cycle defects** (mostly **OTC deficiency**), late-onset **MSUD**, **organic acidurias**, GA type I, CDG and respiratory-chain disorders. Thromboembolic events can be the presenting sign of **classical homocystinuria** and CDG. Angelman syndrome sometimes displays a very suggestive picture, with early-onset encephalopathy, happy-puppet appearance, and epilepsy with a highly suggestive EEG pattern.

- **Category 3: Disorders with Non-specific Developmental Delay.** A large number of inborn errors present with non-specific early progressive developmental delay, poor feeding, hypotonia, some degree of ataxia, frequent autistic features, and seizures. Many IEM can masquerade as a cerebral palsy by presenting as a permanent impairment of movement or posture (Table 1.14). Consequently, it is mandatory to systematically screen such children for those IEM which can be at least partly treatable. In this context, late-onset **subacute**

forms of hyperammonemia (usually **OTC** deficiency in girls) can present with an apparently non-specific early encephalopathy and **inborn errors of neurotransmitter synthesis**, especially **dopa-responsive dystonia** due to cyclohydrolase deficiency, tyrosine hydroxylase deficiency, and aromatic-L-amino-acid decarboxylase deficiency, can masquerade as cerebral palsy. Recurrent attacks of seizures unresponsive to anticonvulsant drugs occurring in the first year of life is the presenting symptom of the **blood brain-barrier glucose-transporter (GLUT1)** defect, a disorder that is improved by a hyperketotic diet. The diagnosis relies on the finding of a low glucose level in the CSF while the simultaneous blood glucose level is normal. The new treatable **cerebral folate deficiency syndrome** [25] (improved by folinic acid) should be also systematically screened for.

Late Infancy to Early Childhood (1–5 years) (Table 1.15)

In this period, diagnosis becomes easier. Five general categories can be defined (Table 1.15):

- **Category 1: with visceral, craniovertebral, ocular, or other somatic abnormalities.** These symptoms associated with a slowing or regression of development, suggest mucopolysaccharidosis types I and II, mucolipidosis type III, oligosaccharidosis, Austin disease, Niemann-Pick disease type C, Gaucher disease type III, and lactosyl ceramidosis, all disorders which are usually easy to recognize. Mucolipidosis type IV, which causes major visual impairment by the end of the first year of life, sometimes associated with dystonia, presents with characteristic cytoplasmic membranous bodies in cells. In Sanfilippo syndrome, coarse facies and bone changes may be very subtle or absent. Peroxisomal disorders may present at this age, with progressive mental deterioration, retinitis pigmentosa, and deafness, and in a very similar manner to Usher syndrome type II. Pyrroline-5-carboxylate synthase deficiency presents with slowly progressive neurological and mental deterioration, severe hypotonia, joint laxity, and congenital cataracts.

- **Category 2: with progressive paraplegia and spasticity.** Progressive paraplegia and spasticity are characteristic of six IEM. Metachromatic leukodystrophy and neuroaxonal dystrophy present between 12 and 24 months of age with flaccid paraparesis, hypotonia, and weakness. CSF protein content and nerve conduction velocity are disturbed in the former but normal in the latter. Schindler disease is roughly similar to neuroaxonal dystrophy, though it is often associated with myoclonic jerks. **Arginase deficiency** is a rare disorder that presents early in infancy to childhood (2 months to 5 years) with progressive spastic diplegia, scissoring or tiptoe gait, and developmental arrest. A rapidly progressive flaccid paraparesis resembling subacute degeneration of the cord can be the presenting sign of inherited

▢ Table 1.14. Progressive neurological and mental deterioration (1 to 12 months)

Leading symptoms	Other signs	Diagnosis (disorder or enzyme deficiency)
With suggestive neurological signs		
Extrapyramidal signs	Major parkinsonism Abnormal neurotransmittors	**Inborn errors of biopterin metabolism** **Aromatic amino acid decarboxylase** **Tyrosine hydroxylase, PNPO**
	Choreoathetotis, self-mutilation	Lesch-Nyhan (X-linked)
	Bilateral athetosis, hypertonicity	Cytochrome b5 reductase
	Dystonia, stridor	Pelizaeus Merzbacher (X-linked)
	Kernicterus syndrome	**Criggler-Najjar**
	Acute-onset pseudoencephalitis	**Glutaric aciduria type I**
	Low cerebral creatine	**Creatine deficiency (GAMT)**
	Spastic paraplegia, ataxia, epilepsy	**Cerebral folate deficiency**
	Leigh syndrome	PDH, complex I
Painful pyramidal hypertonia	Opisthotonos	Krabbe, Gaucher III, Nieman-Pick type C
Early epilepsy infantile spasm	Spasticity	NKH, SO, **untreated MSUD & OA** **MCD**, Menkes
Macrocephaly, startle response to sound	Cherry red spot, myoclonic jerks	Tay Sachs, Sandhoff, Canavan, Alexander Vacuolizing leucoencephalopathy
Ocular symptoms	Optic atrophy, incessant crying	Krabbe (infantile)
	Dystonia, choreoathetosis	**GA I**, L-2-hydroxyglutaric aciduria
	Progressive irritability	Respiratory chain, peroxisomal defects
Recurrent attacks of neurological crisis (▶ also Sect. 1.3.1)	Failure to thrive, hyperventilation attacks	Leigh syndrome (PC, PDH, respiratory chain, MAMEL syndrome [27])
	Stroke-like episodes	**Urea cycle defects, MSUD, OA, GA I** CDG, respiratory chain
	Thromboembolic accidents	**Homocystinurias**, CDG
Without suggestive neurological signs		
Evidence of developmental arrest	Infantile spasms, hypsarrhythmia autistic features	**Untreated PKU, biopterin defects** Peroxisomal defects, Rett syndrome
Non specific symptoms, Apparently non-progressive disorder	Frequent autistic features Poor feeding, failure to thrive	**Hyperammonemia** (late-onset subacute) 4-OH-butyric, L-2-OH-, D-2-OH-glutaric acidurias
	Hypotonia, seizures	Mevalonic aciduria
	With diverse neurological findings simulating cerebral palsy	Adenylosuccinase, pyrimidine defects 3-methylglutaconic, fumarase **Other OA, creatine deficiency** **3-PGD**, 3-phosphoserine phosphatase **Homocystinurias**, Salla **Neurotransmittor defects**, Cerebral folate deficiency Angelman, **GLUT1**

CDG, congenital disorders of glycosylation; *GA*, glutaric aciduria; *GAMT*, guanidino acetate methyltransferase; *MAMEL*, methylmalonic aciduria, mitochondrial encephalopathy Leigh-like; *MCD*, multiple carboxylase deficiency; *MSUD*, maple syrup urine disease; *NKH,* non ketotic hyperglycinemia; *OA*, organic acidurias; *PC*, pyruvate carboxylase; *PDH*, pyruvate dehydrogenase; *3-PGD*, 3-phosphoglycerate dehydrogenase; *PNPO*, pyridox(am)ine phosphate oxidase; *SO,* sulfite oxidase; **bold face**, treatable disorders.

⬛ Table 1.15. Progressive neurological and mental deterioration (1 to 5 years)

Symptoms	Diagnosis (disorder or enzyme deficiency)
With visceral, craniovertebral, or other somatic abnormalities	
▪ Coarse facies, skeletal changes, hirsutism, corneal opacities	MPS I, MPS II, MPS III, MLP III
▪ Coarse facies, subtle bone changes, lens/corneal opacities, hepatosplenomegaly, vacuolated lymphocytes,	Mannosidosis (gingival hyperplasia)
	Fucosidosis (angiokeratoma)
	Aspartylglucosaminuria (joint laxity)
	Austin (ichthyosis)
▪ Hepatosplenomegaly, progressive dementia, myoclonic jerks	Niemann-Pick type C and related disorders (vertical supranuclear ophtalmoplegia)
▪ Splenomegaly + hepatomegaly, osseous lesions, (ataxia, myoclonus)	Gaucher type III (supranuclear ophtalmoplegia)
▪ Major visual impairment, blindness	Mucolipidosis type IV (corneal clouding)
▪ Retinitis pigmentosa, deafness	Peroxisomal defects, Usher type II
▪ Cataract, joint laxity, hypotonia	Pyrroline-5-carboxylase synthase
With paraplegia, hypotonia, or spasticity due to corticospinal tract involvement or to peripheral neuropathy	
▪ Flaccid paraparesis, pyramidal signs, hyperproteinorrhachia	Metachromatic leukodystrophy (abnormal NCV)
▪ Flaccid paraparesis, no change in CSF, optic atrophy	Neuro-axonal dystrophy Schindler (normal NCV)
▪ Progressive spastic diplegia, scissoring or »tiptoe« gait	**Arginase** (high arginine, high orotic)
	Cbl C (subacute cord degeneration)
	Triple H (recurrent attacks of hyper NH_3)
	Costeff syndrome (OPA3 gene mutation with 3-methylglutaconic aciduria)
With unsteady gait, uncoordinated movements due to cerebellar syndrome, sensory defects or myoclonia	
▪ Without disturbances of organic acid excretion	
– Ataxia, choreoathetosis, oculocephalic asynergia	Ataxia telangiectasia
– Ataxia, difficulty in walking, mental/speech deterioration	GM1 (spastic quadriparesis, pseudobulbar signs)
	BBGD (caudate nucleus, putamen necrosis)
– Ataxia, spinocerebellar degeneration, psychotic behavior	GM2 (Tay-Sachs, Sandhoff) (late infantile form)
– Ataxia, pyramidal signs, vision loss	Krabbe (late infantile, peripheral neuropathy)
– Ataxia, muscular atrophy, peripheral neuropathy	CDG, trifunctional enzyme, peroxisomal defects
– Seizures, myoclonic jerks, postictal coma, transient hemiplegia	Alpers (hepatic signs, hyperlactatemia)
▪ With disturbances of organic and amino acid excretion	
– Progressive ataxia, intention tremor, cerebellar atrophy	L-2-OH-glutaric (spongiform encephalopathy)
	Combined degeneration of the spinal cord
	Cobalamin defects (CblC, CblE, CblF, CblG)
– Ataxia, peripheral neuropathy, dystonia	**PDH** (moderate hyperlactactemia)
– Ataxia, weakness, RP, myoclonic epilepsy	Respiratory chain, MERFF, methylglutaconic
– Extrapyramidal signs	**Creatine deficiency** (GAMT)
– Ataxia, peripheral neuropathy, RP	**LCHAD** (organic acids, acylcarnitine)
– Acute attacks encephalitis-like, temporal lobe atrophy	GA I (dystonia, macrocephaly)
– Dystonia, athetosis, acute attacks	**MMA, PA, homocystinurias**
– Ataxia, dysarthria, optic atrophy, nystagmus	Ribose-5-phosphate isomerase (polyols)
With seizures and myoclonus, ataxia, frequent falling due to intention myoclonus or to the cerebellar ataxia	
▪ Rapid mental regression, myoclonic jerks, blindness	INCL (early-flattening EEG, CLN1 mutations)
▪ Akinetic myoclonic petit mal, RP, typical EEG pattern	LINCL (misdiagnosed with Lennox-Gastaut)
▪ Rapid regression, myoclonic seizures, spasticity	Schindler (optic atrophy, severe osteoporosis)
▪ Myoclonic epilepsy, volitional and intentional myoclonias, muscular weakness	MERFF, Niemann-Pick C, Gaucher III (ophtalmoplegia, hepatosplenomegaly)
▪ Seizures and myoclonic jerks, uncoordinated movements	Alpers (hepatic symptoms, hyperlactatemia)
Disorders with arrest or regression of psychic and perceptual functions as presenting symptom	
▪ Autistic behaviour, regression of high-level achievements, stereotyped movements of fingers	Rett syndrome (girls), sporadic (acquired microcephaly, secondary epilepsy)
▪ Regression of high-level achievements, loss of speech,	Sanfilippo (hirsutism, agitation)

BBGD, biotin-responsive basal ganglia disease; *Cbl*, cobalamin; *CoA*, coenzyme A; *CSF*, cerebrospinal fluid; *EEG*, electroencephalogram; *GAMT*, guanidino acetate methyltransferase; *Ig*, immunoglobulin; *INCL*, Infantile ceroid lipofuscinosis (CLN1 mutations); *LINCL*, late infantile ceroid lipofuscinosis (CLN2 mutations); *MERRF*, myoclonic epilepsy with ragged red fibers; *MLP*, mucolipidosis; *MPS*, mucopolysaccharidosis; *NCV*, nerve conduction velocity; *OPA*, optic atrophy; *PDH*, pyruvate dehydrogenase; *RP*, retinitis pigmentosa; **bold face**, treatable disorders.

Cbl-synthesis defects. Spastic paraparesis is an almost constant finding in the **triple H syndrome**.

- **Category 3: with unsteady gait and uncoordinated movements** (when standing, walking, sitting, reaching for objects, speaking, and swallowing). Several groups of disorders must be considered. A careful investigation of organic acid and amino acid metabolism is always mandatory, especially during episodes of metabolic stress.

  - Disorders without disturbances of urinary organic acid excretion and lactic acid metabolism are the late-onset forms of GM-1 and GM-2 gangliosidosis, late infantile Krabbe disease, ataxia telangiectasia, and CDG; each presents with signs that are sufficiently characteristic to warrant specific investigation. A severe early-onset encephalopathy with seizures and myoclonic jerks associated with hepatic disease is highly suggestive of Alpers syndrome due to respiratory-chain disorders. **Creatine deficiency** due to guanidinoacetate-methyltransferase deficiency can present in infancy, with an extrapyramidal disorder associated with epilepsy, neurological regression, and failure to thrive.

  - Disorders with disturbances of organic and amino acid metabolism are numerous. PDH deficiency presents frequently with peripheral neuropathy, intermittent ataxia, dystonia and slight or moderate hyperlactatemia (► Hyperlactatemias above). Several respiratory-chain disorders initially cause ataxia, intention tremor, dysarthria, epilepsy, myopathy, and (eventually) multiorgan failure. **LCHAD deficiency**, L-2-hydroxyglutaric aciduria, 3-methylglutaconic aciduria, **MMA**, and **PA** significantly disturb organic acid excretion, although sometimes only slightly and intermittently. In these disorders, the acylcarnitine profile determined (by tandem MS) from blood spots collected on dry filter paper can be very helpful in identifying characteristic abnormalities. GA type I can also present with a permanent unsteady gait due to choreoathetosis and with dystonia developing abruptly after an acute episode resembling encephalitis.

- **Category 4: with predominant epilepsy and myoclonus.** Predominant epilepsy and myoclonus result in ataxia and frequent falling and include two ceroid lipofuscinoses: Santavuori-Hagberg disease (CLN1) and Jansky-Bielchowski disease (CLN2), which is similar to Lennox-Gastaut syndrome (akinetic myoclonic petit mal). Late-onset forms of Niemann-Pick type C and Gaucher disease are easily suspected because of hepatosplenomegaly and supranuclear paralysis. Two other disorders must also be considered: myoclonic-epilepsy with ragged red fibers (MERRF) syndrome and Schindler disease, which is similar to neuroaxonal dystrophy.

- **Category 5: isolated developmental arrest or regression.** Only a few disorders present between 1 and 5 years of age with an isolated developmental arrest or regression of cognitive and perceptual abilities without other significant neurological or extraneurological signs. Sanfilippo disease is one, although regression of high-level achievements, loss of speech, and agitation usually begin later than 5 years of age. Although non-metabolic, Rett syndrome is another such disease; it should be considered when a girl, without a family history, presents between 1 and 2 years of age with autistic behavior, developmental regression, typical stereotyped hand movements, and microcephaly.

Late Childhood to Adolescence (5–15 years)
(◘ Table 1.16)

It is important to distinguish between conditions in which cognitive function is primarily affected and those disorders with more extensive neurological involvement with normal or subnormal intellectual functioning. According to Lyon and Adams [23], there are six clinical categories.

- **Category 1: with predominant extrapyramidal signs** (parkinsonian syndrome, dystonia, choreoathetosis).

- **Category 2: with severe neurological and mental deterioration and diffuse central nervous system involvement.** Category-2 patients have in common severe neurological dysfunction with bipyramidal paralysis, incoordination, seizures, visual failure, impaired school performance, and dementia. In association with splenomegaly or hepatomegaly, these signs suggest Niemann-Pick disease type C or Gaucher disease type III. When visceral signs are absent, they may indicate juvenile metachromatic leukodystrophy, X-linked adrenoleukodystrophy, Krabbe disease, juvenile GM-1 and GM-2 gangliosidoses, or respiratory-chain disorders. Peroxisomal biogenesis defects can also present in the second decade of life with peripheral neuropathy initially mimicking Charcot-Marie-Tooth type II disease, but which then evolves into a pyramidal syndrome, intellectual deterioration, dementia and, shortly thereafter, a neurovegetative state.

- **Category 3: with polymyoclonus and epilepsy.** The juvenile form of ceroid lipofuscinosis (Spielmeyer-Vogt or Batten disease due to CLN3 gene mutations), which presents with loss of sight, retinitis, ataxia, and (at an advanced stage) extrapyramidal signs, should be suspected with the onset of polymyoclonus and epilepsy. After puberty, Lafora disease should also be considered. Gaucher disease type III, late onset GM-2 gangliosidosis, Niemann-Pick disease type C, and respiratory-chain disorders can also begin with polymyoclonus as an early major sign.

- **Category 4: with predominant cerebellar ataxia.** Friedreich ataxia and other hereditary ataxias should be considered and are recognized on clinical and genetic

▫ Table 1.16. Progressive neurological and mental deterioration (5 to 15 years)

Symptoms	Diagnosis (disorder or enzyme deficiency)
With predominant extrapyramidal signs, parkinson syndrome, dystonia, choreoathetosis	
■ Torsion, dystonia, no mental retardation	**Dystonia musculorum deformans**
■ Dystonia on lower extremities, gait difficulties, normal IQ	**Segawa (GTP cyclohydrolase)**
	Tyrosine hydroxylase
■ Lens dislocation, marfanoid morphology	**Classic homocystinuria**
■ Generalized parkinsonian rigidity, scholastic failure	**Wilson**
■ Parkinsonism, reading/writing difficulties, alacrima, dysphagia	**Familial glucocorticoid deficiency** (with hypoglycemia)
■ Dysarthria, dysphagia, cogwhill rigidity	**Biotin responsive basal ganglia disease**
■ Walking difficulties, dystonic posture, mental regression	Panthotenate kinase (RP, acanthocytosis)
■ Orofacial dyskinesia	HARP syndrome (panthotenate kinase)
■ Acute psychosis, pallidal necrosis	Neuroferritinopathy [17,18]
With diffuse central nervous system disorders, seizures, visual failure, dementia	
■ With hepatosplenomegaly	Niemann Pick type C, Gaucher type III
■ Without visceral signs	Metachromatic leucodystrophy, X-ALD
	Peroxisomal biogenesis defects
	Krabbe, GM1 and GM2
	Leigh syndrome, respiratory chain defects
With polymyoclonia	
■ Intellectual deterioration, loss of sight, RP	JNCL (Batten, CLN3 mutations)
■ Proeminent seizures, myoclonic epilepsy, dementia	Gaucher type III (splenomegaly, osseous signs)
■ Cerebellar ataxia, cherry red spot	Late GM2 gangliosidosis (Sandhoff, Tay-Sachs)
■ Hepatomegaly, splenomegaly	Niemann-Pick type C
■ Myoclonic epilepsy, lactic acidosis	Respiratory chain defects (MERFF, etc.)
With predominant cerebellar ataxia	
■ Without significant mental deterioration	
– Dysarthria, pes cavus, cardiomyopathy	Friedreich ataxia
– Spinocerebellar degeneration	Other hereditary ataxias, Peroxisomal defects
– Chronic diarrhea, low cholesterol, acanthocytosis	**Abetalipoproteinemia**
– Retinitis pigmentosa, peripheral neuropathy	Refsum, peroxisomal defects, CDG
– Oculocephalic asynergia, conjunctival telangiectasias	Ataxia telangiectasia
■ With deterioration and dementia	**CTX**, Lafora, GM1, GM2, Gaucher, Niemann-Pick type C, Krabbe Metachromatic leukodystrophy, respiratory chain
With predominant polyneuropathy	
■ Acute attacks	**Porphyrias, tyrosinemia type I**
■ Progressive	
– With demyelination (low NCV)	Metachromatic leucodystrophy, Krabbe β-mannosidase, Refsum, peroxisomal biogenesis MNGIE syndrome
– Predominantly axonal (normal NCV)	**LCHAD, trifunctional enzyme, PDH, homocysteine remethylation defects, CTX**, peroxisomal biogenesis defects, α-methylacyl-CoA racemase, **serine deficiency**, P5C synthase, **ornithine amino transferase** Leigh syndrome, respiratory chain defects **Abetalipoproteinemia**
With psychiatric symptoms as the only presenting sign	
Behaviour disturbances, personality and character changes, mental regression, dementia, schizophrenia before any significant neurological or extraneurologic sign	**OTC, homocystinurias (CBS, MTHFR, CblC)** Sanfilippo, metachromatic leucodystrophy, Krabbe, Niemann-Pick C, X-ALD Leigh syndrome, JNCL (Batten), Hallervorden-Spatz (PK deficiency), **Wilson**, **CTX**, Huntington chorea (juvenile form) Neuroferritinopathy [17,18]

CBS, cystathionine β-synthase; *CDG*, congenital disorders of glycosylation; *CTX*, cerebrotendinous xanthomatosis; *HARP*, hypobetalipoproteinemia, acanthosis, retinitis pigmentosa, pallidal degeneration; *JNCL*, juvenile neuronal ceroid lipofuscinosis; *MTHFR*, methylene tetrahydrofolate reductase; *NCV*, nerve conduction velocity; *OTC*, ornithinetranscarbamylase; *RP*, retinitis pigmentosa; *X-ALD*, X-linked adrenoleukodystrophy; **bold face**, treatable disorders.

grounds. Abetalipoproteinemia and ataxia telangiectasia are usually suspected because of the associated extraneurological signs. Peroxisomal disorders, CDG, and Refsum disease (which can all present similarly to a peripheral neuropathy and retinitis pigmentosa) can be demonstrated by the analysis of plasma very-long-chain fatty acids, glycosylated transferrin profile, and plasma phytanic acid, respectively. Cerebellar ataxia in association with progressive mental deterioration, dementia, and epilepsy suggests Lafora disease, **cerebrotendinous xanthomatosis**, late-onset forms of gangliosidosis, Krabbe disease, Gaucher disease, Niemann-Pick disease type C, and metachromatic leukodystrophy. Respiratory-chain disorders also can present with a predominant ataxia.

- **Category 5: with predominant polyneuropathy. Porphyrias** and **tyrosinemia type I** can present with an acute attack of polyneuropathy mimicking Guillain-Barre syndrome. Many other disorders can present with a late-onset progressive polyneuropathy that can mimic hereditary ataxia, such as Charcot-Marie-Tooth disease. These include lysosomal diseases (Krabbe disease, metachromatic leukodystrophy, β-mannosidase), peroxisomal disorders (peroxin 7, other peroxisomal biogenesis defects, Refsum disease with demyelination and reduced nerve conduction velocities), defects of energy metabolism (Leigh syndrome, respiratory-chain disorders, PDH deficiency, LCHAD and trifunctional-enzyme deficiencies), abetalipoproteinemia, CDG (▶ also Sect. 1.4.3).

- **Category 6: with behavioral disturbances as the presenting signs.** Some inborn errors of metabolism can present between 5 and 15 years of age as psychiatric disorders. Behavioral disturbances (personality and character changes), loss of speech, scholastic failure, mental regression, dementia, psychosis, and schizophrenia-like syndrome are the most frequent symptoms. In addition **OTC deficiency** can present with episodes of abnormal behavior and character change until hyperammonemia and coma reveal the true situation (▶ Recurrent Attacks of Coma above). **Homocystinuria** due to methylenetetrahydrofolate reductase deficiency has presented as isolated schizophrenia. **Searching for these treatable disorders is always mandatory including CTX and Wilson disease.**

Onset in Adulthood (15–70 years)

As metabolic investigations become increasingly common in adult neurological practice, a rapidly increasing number of adults patients with IEM has been identified. When neurologists extend metabolic investigations to late-onset, progressive, neurological deterioration, currently considered to be degenerative, inflammatory, or of vascular origin, it is highly probable that many other disorders will be discovered. Some disorders truely start in adulthood; others

present in the pediatric age with symptoms that are not recognised, misdiagnosed or inadequately interpreted. Metabolic diagnosis in adults based on substrate measurements can be very difficult because metabolic changes can be very moderate or even absent. A recent reappraisal of the metabolic profile of the three infantile Refsum disease index cases described in 1986 [26] showed an almost total normalization of the initial characteristic biochemical abnormalities (personal unpublished observation). In ◻ Table 1.17 is the »state of the art« of adult metabolic neurology based upon the personal experience of the adult metabolic clinic of La Pitié Salpétrière hospital and on the literature analysis mostly composed of isolated case reports. There is no doubt that this will change significantly very soon. ◻ Table 1.17 does not describe the full phenotype of the listed disorders, but rather emphasizes the main presenting symptom.

1.4.4 Specific Associated Neurological Abnormalities

Cherry-red Spot
- Cytochrome C oxidase deficiency
- Galactosialidosis (neuraminidase deficiency)
- Gangliosidosis GM1 (Landing)
- Gangliosidosis GM2 (Sandhoff, Tay-Sachs)
- Nephrosialidosis
- Niemann-Pick type A, C and D
- Sialidosis type I

Deafness (Sensorineural)

Detectable in neonatal to early infancy:
- Acyl-CoA oxidase deficiency
- Alport syndrome
- Cockayne syndrome
- Encephalopathy with hyperkinurininuria
- Rhizomelic chondrodysplasia punctata
- Zellweger and variants

Detectable in late infancy to childhood:
- **Biotinidase deficiency (biotin responsive) (untreated or treated late)**
- Infantile Refsum disease (pseudo Usher syndrome)
- Mannosidosis (alpha)
- **Megaloblastic anemia, diabetes and deafness (B1-responsive)**
- Mitochondrial encephalomyopathy
- Mitochondrial encephalopathy with lactic acidosis and stroke-like episodes (MELAS), myoclonic epilepsy with ragged red fibers (MERFF), Kearns-Sayre syndrome
- Mucolipidosis type II (I cell disease)
- Mucopolysaccharidosis type I, II and IV
- Neutral lipid storage disorder
- PRPP synthetase superactivity
- Wolframm syndrome

□ Table 1.17. Progressive neurological and mental deterioration (adulthood)

Diseases	Major presenting signs							
(Disorder or enzyme deficiency)	Parkinson syndrome	Dystonia, chorea	EMG poly-neuropathy	Epilepsy	Psychiatric signs	Spastic paraparesia	Cerebellar ataxia	Multi-systemic
α-mannosidase						+	+	+
α-methylacyl-CoA race-mase			+	+		+		
β-mannosidosis			+		+			+
L-2-OH-glutaric aciduria	+	+		+		+		
Acerulo-plasminemia	+	+			+		+	+
AMN			+			+	+	
ANCL	+			+	+			
Arginase						+		
Biotin responsive encephalopathy		+		+				
Cerebral ALD		+ (chorea)		+	+	+		
Co-enzyme Q10							+	
CTX	+		+	+	+	+	+	+
Dopa responsive dystonia		+				+		
Fabry			+*					+
Gaucher III	+	+ (chorea)		+			+	+
GM1	+	+						
GM2		+	+*		+		+	
GTP cyclohydrolase I	+	+				+		
Hemochromatosis	+							
Homocysteine methylation defects			+	+	+	+		+
Homocystinuria	+	+ +	+		+			+
Huntington chorea	+	+		+	+		+	
Krabbe			+			+		
Lafora				+	+		+	
LCHAD			+					+
Leigh, MELAS, Leber	+	+ +	+	+			+	+
Lesch-Nyhan		+ +			+			
MAO-A					+			
Mevalonate kinase							+	+

◘ **Table 1.17** (continued)

Diseases	Major presenting signs							
(Disorder or enzyme deficiency)	Parkinson syndrome	Dystonia, chorea	EMG poly-neuropathy	Epilepsy	Psychiatric signs	Spastic paraparesia	Cerebellar ataxia	Multi-systemic
Mitochondrial defects	+	+	+	+			+	+
MLD		+	+		+	+	+	
MNGIE			+					+
MPS III					+			
Neuraminidase				+				
Neuro-ferritinopathy	+	+			+			
Niemann-Pick C	+	+		+	+		+	
NKH		+ (chorea)				+	+	
Oligosaccharidosis								+
PA		+ (chorea)						
PBD			+				+	+
PDH (E1)	+	+	+				+	
PK		+						
PKU	+				+	+		
Polyglucosan body disease	+		+			+		
Polymerase γ	+		+	+			+	
Porphyria			+*	+	+			
PTP synthase		+						
Refsum			+				+	+
Serine deficiency syndrome			+					
Sjögren-Larsson						+		
Tangier			+*					+
Triple H syndrome						+		
Tyrosine hydroxylase	+					+		
Urea cycle				+	+		+	
Wilson	+	+		+	+			

ALD, adrenoleukodystrophy; *AMN*, adrenomyeloneuropathy; *ANCL*, adult neuronal ceroid lipofuscinosis (CLN4 mutations); *CTX*, cerebro-tendinous xanthomatosis; *LCHAD*, 3-hydroxy long chain acyl-CoA dehydrogenase; *MAO*, mono amine oxidase; *MLD*, metachromatic leucodystrophy; *MNGIE*, mitochondrial neuro-gastro-intestinal encephalopathy; *MPS*, mucopolysaccharidosis; *NKH*, non ketotic hyperglycinemia; *PA*, propionic acidemia; *PBD*, Peroxisome biogenesis defects; *PDH*, pyruvate dehydrogenase; *PK*, panthotenate kinase; *PKU*, phenylketonuria; *PTP*, 6-pyruvoyl tetrahydropterin; * polyneuropathy affecting small sensitive fibers and autonomic nervous system; **bold face**, treatable disorders.

Detectable in late childhood to adolescence:
- Beta-mannosidosis
- MERFF, Kearns-Sayre syndromes
- Refsum disease (adult form)
- Usher syndrome type II

Leigh Syndrome
- **Biotinidase deficiency**
- EPEMA syndrome [13]
- Fumarase deficiency
- MAMEL syndrome [26]
- Pyruvate carboxylase deficiency
- **Pyruvate dehydrogenase deficiency**
- Respiratory chain disorders
- Sulfite oxidase deficiency
- 3-methylglutaconic aciduria

Macrocephaly
- Alexander
- Canavan (acetylaspartaturia)
- Gangliosidosis GM2 (Sandhoff, Tay-Sachs)
- **Glutaric aciduria type I**
- Krabbe (infantile form)
- L-2-hydroglutaric aciduria
- Respiratory chain disorders
- Vacuolizing encephalopathy

Microcephaly
Congenital:
- Infant born to untreated PKU mother
- Sulfite oxidase deficiency
- **3P-Glycerate phosphate dehydrogenase** (improved by serine)

Acquired:
- Rett syndrome
- Many untreated disorders in which microcephaly is a symptom of a non specific cerebral atrophy

Neuro-imaging (CT Scan, MRI, ¹H-MRS) Abnormalities
(**Bold face**, treatable disorders; *reported in adults as presenting or preponderant symptoms)
Calcifications on CT-scan:
- Aicardi-Goutières syndrome
- **Biopterin metabolism defects**
- Cockayne syndrome
- Congenital lactic acidemias
- **Folic acid metabolism defects**
- GM2 Gangliosidosis
- Kearns-Sayre
- Leigh syndrome
- MELAS syndrome
- Respiratory chain disorders
- 3-hydroxyisobutyric aciduria

White matter hyperintensity:
- With increased head circumference
    - Alexander (anterior)
    - Canavan
    - **Glutaric aciduria type I** (bi-temporal atrophy)
    - L-2-hydroxyglutaric aciduria
    - Mucopolysaccharidosis (with vacuoles)
    - Vacuolizing leucoencephalopathy

- With normal head circumference
  Predominantly periventricular white matter
    - Aicardi-Goutières syndrome (with calcifications)
    - CACH (vanishing white matter disease)
    - **Cerebrotendinous xanthomatosis***
    - Cockayne (with calcifications)
    - **Homocysteine remethylation defects***
    - **Glutaric aciduria type I***
    - Kearns-Sayre
    - L-2-hydroxyglutaric aciduria
    - Menkes
    - Metachromatic leucodystrophy*
    - Mitochondrial cytopathy
    - MNGIE (with supratentorial cortical atrophy)
    - Pelizaeus-Merzbacher (myelination arrest)
    - Peroxisomal biogenesis defects*, PEX-7
    - **PKU** (untreated, reversible)*
    - Polyglucosan body disease*
    - Ribose-5-phosphate isomerase* (arabitol, ribitol)
    - **X-ALD** (posterior)
    - 3-methylglutaryl-CoA lyase*
  Predominant pyramidal tracts
    - Adrenomyeloneuropathy*
    - **Cerebrotendinous xanthomatosis***
    - Krabbe disease*
    - Mitochondrial cytopathies*
  Affecting U fibers
    - Mitochondrial cytopathies*
    - Polyglucosan body disease*
    - Ribose-5-phosphate isomerase* (arabitol, ribitol)
    - L-2-hydroxyglutaric aciduria*

Basal ganglia/brain stem hyperintensities:
- **Biotin-responsive basal ganglia disease** [12] (bilateral necrosis of caudate nucleus and putamen)
- **Cerebrotendinous xanthomatosis***
- GM1 Gangliosidosis*
- Hypoceruleoplasminemia* (diffuse hypointensity)
- Infantile bilateral striatal necrosis [28]
- Leigh syndrome (putamen, caudate nuclei)
- L-2-hydroxyglutaric aciduria
- **Methylmalonic aciduria** (pallidum)
- Mitochondrial cytopathies*
- Neuroferritinopathy* (pallidum) [17, 18]
- PKAN* (Hallervorden-Spatz, HARP syndrome: hypointensity: tiger eye)

- **Pyruvate dehydrogenase deficiency***
- **Wernicke encephalopathy*** (thalami, brain stem)
- **Wilson disease***

Dentate nuclei of the cerebellum (hyperintensities):
- **Cerebrotendinous xanthomatosis***
- L-2-hydroxyglutaric aciduria
- Mitochondrial encephalopathy*
- Polyglucosan body disease*
- Semialdehyde succinate dehydrogenase*
- **Wilson disease***

Gyration abnormalities:
- CEDNIK (snare protein mutation) [29]
- Glutamine synthetase
- Congenital muscular dystrophy: DMC1-C (fukutin related protein), DMC1-D (LARGE protein)
- O-glycosylation disorders: muscle-eye-brain disease, Walker-Warburg syndrome, Fukuyama disease
- Peroxisomal disorders (Zellweger and others)

Corpus callosum agenesis:
- With gyration abnormalities (▶ above)
- **ACTH deficiency**
- Aicardi syndrome (with calcifications)
- Complex II mitochondrial cytopathies (with leucodystrophy)
- Non ketotic hyperglycinemia
- PDH (with basal ganglia abnormalities)
- 3-hydroxyisobutyric aciduria

Posterior fossa (and olivo-ponto-cerebellar):
- Hypoplasia
  - CDG
  - Congenital muscular dystrophies
  - Joubert syndrome
  - Mitochondrial cytopathies
  - Peroxisomal disorders
- Progressive atrophy
  - Ceroid lipofuscinosis*
  - GM1 Gangliosidosis (Landing)
  - L-2-hydroxyglutaric aciduria
  - Mevalonic aciduria (mevalonate kinase)
  - Neuroaxonal dystrophy (infantile)
  - Schindler
  - Smith-Lemli-Opitz
  - Succinyl semialdehyde dehydrogenase deficiency
  - 3-methylglutaconic aciduria

Stroke and stroke-like episodes:
- CDG
- **Homocystinurias***
- MELAS syndrome*

Nystagmus

With retinitis pigmentosa:
- Abetalipoproteinemia*
- Ceroid lipofuscinosis (CLN1, CLN2, CLN3*, CLN4*)
- **LCHAD***
- Mitochondrial cytopathies (Kearns-Sayre* etc.)
- Peroxisomal defects (infantile to childhood)
- Sjögren-Larsson (fatty acid alcohol oxido-reductase)
- All causes of severe retinitis pigmentosa

With optic atrophy:
- All causes of optic atrophy in adulthood*
- Canavan disease (early sign)
- Ceroid lipofuscinosis (CLN3*, CLN4*)
- Krabbe disease (infantile)
- Leber due to mitochondrial DNA deletions*
- Leigh syndrome (all causes)
- Metachromatic leucodystrophy*
- Mitochondrial cytopathies*
- Neuroaxonal dystrophy – Schindler (infantile)
- Pelizaeus-Merzbacher (presenting sign early in infancy)
- Peroxisomal biogenesis defects*
- Pyruvate dehydrogenase deficiency*
- Ribose-5-phosphate isomerase*
- Sulfite oxidase (infantile)
- X-ALD*
- 3-methylglutaconic aciduria

With corneal opacities, cataract:
- **Fabry disease***
- **Homocystinurias***
- Lowe syndrome (infancy)
- Mucopolysaccharidosis (childhood)
- **Wilson disease***

Opthalmoplegia, Ptosis, Eye Movements, Strabismus (▶ also Sect. 1.4.3)

Neonatal to early infancy (oculogyric crisis):
- Aromatic amino acid decarboxylase
- CDG Ia (with congenital strabismus)
- Cogan syndrome (ocular contraversion)
- **Pyridox(am)ine-5-phosphate oxidase**
- Tyrosine hydroxylase

Infancy to childhood:
- Ataxia telangiectasia (ocular contraversion, telangiectasia)
- Gaucher type III (horizontal supranuclear paralysis)
- Leigh syndrome (acute attacks of abnormal movements)
- Nicmann-Pick C and D (vertical supranuclear paralysis)
- Pyruvate dehydrogenase (acute attacks of abnormal movements)
- Respiratory chain (acute attacks of abnormal movements)

Adulthood:
- **Glutaric aciduria type I**
- GM2 gangliosidosis (abnormal eye movements)
- Mitochondrial cytopathies: Kearns-Sayre (abnormal movements)
- Niemann-Pick C, Gaucher III (▶ above)
- Non ketotic hyperglycinemia
- Pyruvate dehydrogenase (abnormal movements)
- **Wilson disease**

Peripheral Neuropathy – EMG, NCV Findings

Acute (recurrent attacks):
- **Porphyrias***
- **Tyrosinemia type I**

Chronic:
- Predominantly demyelination (low NCV)
  - Presenting or preponderant
    - Refsum disease (late childhood to adulthood)
    - X-ALD (childhood to adulthood): leucodystrophy
    - AMN (adulthood)
  - Accompanying symptom
    - Austin disease
    - β-mannosidosis
    - Farber lipogranulomatosis
    - **Homocysteine remethylation defects (MTHFR, CblC)**
    - Krabbe (leucodystrophy)
    - Metachromatic leucodystrophy (leucodystrophy)
    - MNGIE syndrome (leucodystrophy)
    - Refsum disease
    - **Tangier disease**
- Predominantly axonal (normal NCV)
  - Presenting or preponderant
    - Abetalipoproteinemia (childhood)
    - α-methylacyl-CoA racemase (adolescence to adulthood)
    - CDG type I (childhood)
    - GM2 gangliosidosis*
    - **LCHAD**, trifunctional (childhood to adolescence)
    - Peroxisomal biogenesis defects (late childhood to adult)
    - Polyglucosan body disease* (leucodystrophy)
    - Pyruvate dehydrogenase (childhood to adulthood)
    - **Vitamin E malabsorption** (tocopherol carrier)
  - Accompanying symptom
    - **Cerebrotendinous xanthomatosis*** (leucodystrophy)
    - Neuroaxonal dystrophy, Schindler (early childhood) (leucodystrophy)
    - **Ornithine amino transferase** (late complications)
    - P5C synthase (late childhood)

- **Porphyria***
- Pyroglutamic aciduria (late complication)
- Respiratory chain (early childhood to adolescence)
- **Serine deficiency syndrome** (adolescence)
- Triose phosphate isomerase
- Affecting small sensitive fibers and the autonomic nervous system
  - **Fabry disease*** (presenting sign)
  - GM2 gangliosidosis*
  - **Porphyria***
  - **Tangier disease***
- Affecting anterior horn
  - GM2 gangliosidosis, Krabbe disease
  - **Homocysteine remethylation defects (ClbC)**
  - Non ketotic hyperglycinemia
  - Panthotenate kinase (Hallervorden-Spatz) (basal ganglia)
  - Polyglucosan body disease*

Retinitis Pigmentosa

- Aceruleoplasminemia*
- Congenital disorders of glycosylation
- Ceroid lipofuscinosis: CLN1, CLN2; LCN3
- Cobalamin metabolism defects: CblC*
- Gyrate atrophy with ornithine aminotransferase deficiency
- Inborn errors of lipid metabolism:
  - **Abetalipoproteinemia**
  - Sjögren-Larsson syndrome
  - **Vitamin E malabsorption (tocopherol carrier)**,
  - **3-hydroxyacyl-CoA dehydrogenase**
- Panthothenate kinase* (Hallervorden-Spatz, HARP syndrome)
- Peroxisomal biogenesis defects:
  - α-methylacyl-CoA racemase*
  - Classical Refsum disease*
  - Isolated fatty acid oxidation defects
  - Peroxisomal biogenesis defects (Zellweger, NALD, Refsum and variant forms)
- Respiratory chain disorders:
  - Kearns-Sayre syndrome*
  - NARP
  - Other mitochondrial DNA deletions
- Recessive autosomal syndromes (Cockayne, Laurence-Moon-Biedl, Usher type II, Joubert, Senior-Loken etc.)
- »Primary retinitis pigmentosa« X-linked, autosomal recessive or dominant

Self Mutilation, Auto-aggression

- Lesch-Nyhan syndrome
- **Phenylketonuria** (untreated)
- **Tyrosinemia type I** (crisis)
- 3-methylglutaconic aciduria

1.5   Specific Organ Symptoms

A number of clinical or biological abnormalities can be associated with inherited inborn errors of metabolism. Some of these phenotypes are rare and very distinctive (e.g., lens dislocation and thromboembolic accidents in homocystinuria) whereas others are common and rather non-specific (e.g., hepatomegaly, seizures, mental retardation). The most important ones are listed below. The following diagnostic checklist presented is primarily based upon the authors' personal experience and, of course, is not exhaustive. It should be progressively extended by the personal experiences of all readers.

It is important to reemphasise the difference between a syndrome where the underlying pathophysiology has not been described and a disorder where the aetiology is known. Some well-known recessive syndromes (such as Joubert, Usher, Cockayne etc.) have been listed under inborn errors of metabolism, highlighting the need to perform extensive metabolic and genetic investigations. The demonstration of cholesterol synthesis defects in a number of malformative syndromes or the more recent demonstration of O-glycosylation defects in congenital muscular dystrophies illustrate this statement.

1.5.1   Cardiology

Arrhythmias, Conduction Defects (Heart Beat Disorders)

Primitive heart beat disorders:
- **Adrenal dysfunction** (hyperkalemia)
- AMP activated protein kinase (PRKAG2 mutations with cardiac glycogenosis and Wolf-Parkinson-White)
- Triose phosphate isomerase deficiency
- D-2-hydroxyglutaric aciduria (AV block)
- **Fatty acid oxidation disorders** (CPT II, carnitine translocase, LCAD, LCHAD, TF, VLCAD)
- **Hypoparathyroidism** (hypocalcemia)
- Kearns-Sayre syndrome (respiratory chain disorders)
- **Thiamine deficiency-dependent states**

With cardiac/multiorgan failure: ► below
With cardiomyopathy: ► below

Cardiac Failure, Collapse

With tamponade, multiorgan failure:
- Congenital disorders of glycosylation

With apparently primitive heart beat disorders: ► above
With cardiomyopathy: ► below

Cardiomyopathy
- AMP activated protein kinase (presenting sign) [30, 31]
- Barth syndrome
- Congenital disorders of glycosylation (with pericardial effusion, can be the presenting sign)
- Congenital muscle dystrophies
- D-2-hydroxyglutaric aciduria
- **Fabry disease**
- **Fatty acid oxidation disorders** (presenting sign)
- Friedreich ataxia (presenting sign)
- **Glycogenosis type III** and IV
- GM1 gangliosidosis
- Isobutyryl-CoA dehydrogenase
- **Methylmalonic aciduria (Cbl C)**, malonic aciduria
- Mucopolysaccharidosis
- Muscle glycogen synthase (presenting sign) [32]
- Pompe disease, Danon disease (presenting sign)
- **Propionic acidemia**
- Respiratory chain disorders (presenting sign)
- **Selenium deficiency**
- Steinert disease – myotonic dystrophy
- **Thiamine deficiency** (presenting sign)
- **Thiamine-responsive anemia**
- 3-methylglutaconic aciduria

1.5.2   Dermatology

Acrocyanosis (Orthostatic)
- EPEMA syndrome [13]

Alopecia

Age at onset: neonatal to infancy
- **Acrodermatitis enteropathica**
- **Biotin-responsive multiple carboxylase defects**
- **Calciferol metabolism defects** (vitamin-D-dependent rickets)
- **Congenital erythropoietic porphyria**
- Conradi-Hünermann syndrome
- Ehlers-Danlos type IV
- **Essential fatty acid deficiency**
- **Hepatoerythropoietic porphyria**
- Menkes disease (X-linked)
- **Methylmalonic and propionic acidurias**
- Netherton syndrome
- **Zinc deficiency**

Age at onset: adulthood
- **Porphyria cutanea tarda**
- Steinert

Angiokeratosis

- Aspartylglucosaminuria
- β-mannosidosis
- **Fabry disease** (presenting sign)
- Fucosidosis
- Galactosialidosis
- Kanzaki disease
- Schindler disease (adult form)

Brittle Hair

- **Argininosuccinic aciduria**
- **Citrullinemia**
- Menkes syndrome
- Pollitt's syndrome
- Trichothiodystrophy

Hyperkeratosis

- CEDNIK (neuro-cutaneous syndrome: keratosis on palms and soles) [29]
- Ichthyosis (▶ below)
- **Tyrosinemia type II** (keratosis on palms and soles)

Ichthyosis (with Congenital Erythrodermia)

- Austin disease
- CEDNIK (neuro-cutaneous syndrome: SNARE protein mutation) [29]
- Conradi-Hünermann syndrome (chondrodysplasia punctata X-linked)
- Multisystemic triglyceride storage disease
- Netherton syndrome
- **Refsum disease (adult form)**
- **Serine deficiency syndrome**
- Sjögren-Larsson syndrome
- Steroid sulfatase deficiency (X-linked)

Laxity (Dysmorphic Scarring, Easy Bruising)

Inborn errors of collagen:
- Cutis laxa
- Ehlers-Danlos syndrome (nine types)
- Occipital horn syndrome
- Pyrroline-5-carboxylate synthase

Nodules

- Congenital disorders of glycosylation
- Farber lipogranulomatosis

Pellagra

- **Hartnup disease**

Photosensitivity and Skin Rashes

Age at onset: neonatal to childhood
- **Congenital erythropoietic porphyria**
- **Erythrohepatic porphyria**

- **Erythropoietic protoporphyria**
- **Hartnup disease**
- Mevalonic aciduria (with fever and arthralgia)
- Respiratory chain disorders
- Xeroderma pigmentosa (nine varieties)

Age at onset: adulthood
- **Hereditary coproporphyria**
- **Porphyria variegata**
- **Porphyria cutanea tarda**

Pili Torti

- Menkes disease
- Netherton syndrome

Telangiectasias - Purpuras - Petechiae

- Ethylmalonic aciduria (EPEMA syndrome)
- Prolidase deficiency

Trichorrhexis Nodosa

- **Argininemia**
- **Argininosuccinic aciduria**
- **Lysinuric protein intolerance**
- Menkes disease
- Netherton syndrome

Ulceration (Skin Ulcers)

- Prolidase deficiency

Vesiculo-Bullous Skin Lesions

- **Acrodermatitis enteropathica**
- **Biotinidase deficiency** (biotin-responsive)
- **Holocarboxylase synthetase deficiency** (biotin-responsive)
- **Methylmalonic, propionic acidemias** (isoleucine deficiency)
- **Zinc deficiency**

Xanthoma

- Apo CII (eruptive)
- Apolipoprotein A1 (planar)
- **Familial dominant hypercholesterolemia:**
  - homozygote (childhood)
  - heterozygote (adulthood)
- Dysbetalipoproteinemia (hyperlipoproteinemia type III)
- Hepatic lipase
- Lipoprotein lipase (eruptive)
- Sitosterolemia (childhood)

1.5.3 Dysmorphism

Coarse Facies

Age at onset: present at birth
- Galactosialidosis (early infancy)
- I-cell disease
- Landing
- Sialidosis type II
- Sly (mucopolysaccharidosis (MPS) type VII) (rare)

Age at onset: early infancy
- Austin
- Fucosidosis type I
- Hurler (MPS type IH)
- Mannosidosis
- Maroteaux-Lamy (MPS type V)
- Salla disease
- Sialidosis type II
- Sly (MPS type VII)

Age at onset: childhood
- Aspartylglucosaminuria
- Hunter (MPS type II)
- Pseudo-Hurler polydystrophy
- Sanfilippo (MPS type III)

Dysplasia, Dysmorphic

Maternal metabolic disturbances (untreated pregnancy):
- **PKU** (dysmaturity, heart defect, microcephaly, specific face, hypotrophy)
- Alcohol (dysmorphic, hypotrophy)
- **Diabetes** (macrosomia)
- Drugs (dysmorphic, hypotrophy)
- **Vitamin deficiencies** (riboflavin)

Inborn errors affecting the fetus:
- Carnitine palmitoyl transferase II deficiency (renal cysts)
- D-2-hydroxyglutaric aciduria
- Glutaric aciduria type II (MADD) (renal cysts)
- Inborn errors of collagen
- **Hyperinsulinism** (macrosomia, dysmorphia)
- Hypoparathyroidism
- Hypophosphatasia
- Leprechaunism
- Lysosomal storage disorders (hydrops fetalis)
- Mevalonic aciduria (mevalonate kinase deficiency)
- Peroxisomal biogenesis defects (renal cysts, migration defects)
- Chondrodysplasia punctata
- Pyruvate dehydrogenase deficiency
- Respiratory chain defects
- Serine synthesis (microcephaly)
- Transaldolase deficiency (hydrops fetalis)

Malformations
- 3-OH-isobutyryl-CoA deacylase deficiency (limbs, vertebrae)
- Cholesterol synthesis defects:
  - Smith-Lemli-Opitz
  - Conradi-Hünermann-Happle syndrome
  - Desmosterolosis
  - Greenberg dysplasia
  - Antley-Bixler syndrome
  - Mevalonic aciduria
  - CHILD syndrome
  - Lathosterolosis
- Glutamine synthetase
- Non-ketotic hyperglycinemia
- O-glycosylation and related defects:
  - Walter-Warburg (POMT1)
  - Muscle-eye-brain disease (POMGMT),
  - Fukuyama (Fukutin)
  - DMC1-C (fukutin related protein)
  - DMC1-D (LARGE protein)

Intra-uterine Growth Retardation
- Fetal alcoholic syndrome
- Infants born to mothers with untreated phenylketonuria
- Cholesterol biosynthesis defects
- Lysosomal storage disorders
- Many non-metabolic polymalformative syndromes
- Peroxisomal disorders
- Respiratory chain disorders
- Transaldolase deficiency

1.5.4 Endocrinology

Diabetes (and Pseudodiabetes)
- Abnormal pro-insulin cleavage
- **Diabetes, deafness and thiamine responsive megaloblastic anemia**
- Diabetes type II: fatty acid oxidation disorders, Kir 6.2, glucokinase
- **Organic acidurias (methylmalonic, propionic, isovaleric acidemias, ketolysis defects)**
- Respiratory chain disorders – Wolfram syndrome

Hyperinsulinism
- **SUR1 and KIR6.2 mutations (potassium channel)**
- **Glucokinase overactivity**
- **Glutamate dehydrogenase overactivity**
- Short chain L-3-OH-acyl-CoA dehydrogenase
- Wiedemann-Beckwith syndrome

Hyperthyroidism
- Glutaric aciduria type I (glutaryl-CoA dehydrogenase deficiency)

Hypogonadism – Sterility

- CDG type I
- Galactosemia

Hypoparathyroidism

- **Long-chain 3-hydroxyacyl-CoA dehydrogenase deficiency**
- Respiratory chain disorders
- **Trifunctional enzyme deficiency**

Hypothyroidism

- Allan-Herndon-Dudley syndrome (monocarboxylate transporter 8) [22]

Salt-Losing Syndrome

- **Disorders of adrenal steroid metabolism**
- **Fatty acid oxidation disorders** (carnitine palmitoyl transferase II)
- Respiratory chain disorders (mitochondrial DNA deletions)

Sexual Ambiguity

- Congenital adrenal hyper- and hypoplasia
- Disorders of adrenal steroid metabolism

Short Stature – Growth Hormone Deficiency

- Respiratory chain disorders

1.5.5   Gastroenterology

Abdominal Pain (Recurrent)

With flatulence, diarrhea, lose stools:
- **Lactose malabsorption**
- **Congenital sucrase isomaltase deficiency**

With vomiting, lethargy, ketoacidosis:
- **Urea cycle defects** (OTC, ASA)
- **Organic acidurias** (MMA, PA, IVA)
- **Ketolysis defects**
- Respiratory chain disorders
- **Diabetes**

With neuropathy, psychiatric symptoms:
- MNGIE syndrome
- **OTC** (late onset)
- **Porphyrias**
- **Tyrosinemia type I**

With hepatomegaly and splenomegaly:
- Cholesterol ester storage disease
- Lipoprotein lipase deficiency
- **Lysinuric protein intolerance**
- Hemochromatosis
- Mevalonate kinase deficiency

With pain in extremities:
- **Fabry disease**
- δ-aminolevulinate dehydratase deficiency
- **Sickle cell anemia**

With hemolytic anemia:
- **Coproporphyria**
- Hereditary spherocytosis
- **Sickle cell anemia**
- Nocturnal paroxysmal hemoglobinuria

With Crohn disease (and pseudo-Crohn):
- **Glycogenosis type 1b**
- **Trifunctional enzyme deficiency**
- **Carnitine transporter** (OCTN2)

With inflammatory syndrome (fever rash, IC reactive protein):
- HyperIgD syndrome (mevalonate kinase deficiency)

Acute Pancreatitis

- Hyperlipoproteinemia type I and IV
- **Lysinuric protein intolerance**
- **Organic acidurias** (MMA, PA, IVA, MSUD)
- Respiratory chain disorders (Pearson, MELAS)

Chronic Diarrhea, Failure to Thrive, Osteoporosis
(▶ Table 1.12)

Hypocholesterolemia

- Abetalipoproteinemia type I and II
- Congenital disorders of glycosylation type I
- Infantile Refsum disease
- Mevalonic aciduria
- Peroxisomal disorders
- Smith-Lemli-Opitz syndrome
- **Tangier disease** (alpha-lipoprotein deficiency)

HELLP Syndrome (Baby Born to Mothers with)

- **Carnitine palmitoyl transferase I deficiency**
- **LCHAD deficiency** and other fatty acid β-oxidation disorders
- Respiratory chain defects

Intestinal Obstruction

- MNGIE syndrome (mitochondrial cytopathy)

1.5.6   Hematology

Acanthocytosis

- Abetalipoproteinemia
- Hallervorden-Spatz syndrome (panthothenate kinase)
- Inborn errors of cobalamin (**Cbl C**)
- Wolman disease

Anemias (Macrocytic)

- **Hereditary orotic aciduria**
- Inborn errors of cobalamin metabolism:
    - **Imerslund-Gräsbeck syndrome**
    - **Intrinsic factor deficiency**
    - **TC II deficiency**
    - **Cbl C, Cbl E, Cbl G**
    - **Methionine synthase deficiency**
- Inborn errors of folate metabolism:
    - Dihydrofolate reductase deficiency,
    - Glutamate formimino transferase deficiency
    - **Congenital folate malabsorption**
- Mevalonic aciduria
- Pearson syndrome (due to mitochondrial DNA deletion) (dyserythropoiesis)
- Respiratory chain disorders
- **Thiamine responsive megaloblastic anemia**

Anemias (Non-macrocytic, Hemolytic or Due to Combined Mechanisms)

- Abetalipoproteinemia
- **Carnitine transport defect**
- **Congenital erythropoietic porphyria**
- **Erythropoietic porphyria**
- **Erythropoietic protoporphyria**
- **Galactosemia**
- Hemochromatosis
- Lecithin cholesterol acyltransferase deficiency
- Mevalonic aciduria
- Pyroglutamic aciduria
- Red blood cells glycolysis defects
- Severe liver failure
- Transaldolase deficiency
- **Wilson disease**
- Wolman disease

Bleeding Tendency, Hemorragic Syndromes

- Gaucher disease
- **Glycogenosis type Ia and Ib**
- Inborn errors with severe liver failure
- Primitive disorders of homeostasis
- Severe thrombocytopenia

Pancytopenia – Thrombocytopenia – Leucopenia

- Aspartylglucosaminuria
- CDG IIf (CMP sialic acid transporter)
- **Gaucher type I and III**
- **Glycogenosis type Ib** (neutropenia)
- **Inborn errors of cobalamin metabolism**
- **Inborn errors of folate metabolism**
- Johansson-Blizzard syndrome
- **Lysinuric protein intolerance**
- **Organic acidurias** (methylmalonic, propionic, iso-valeric)
- Other conditions with large splenomegaly

- Pearson syndrome
- Respiratory chain disorders
- Schwachman syndrome
- Transaldolase deficiency

Vacuolated Lymphocytes

- Aspartylglucosaminuria
- Austin disease
- Ceroid lipofuscinosis
- I-cell disease (mucolipidosis type II)
- Landing disease (GM1)
- Mucopolysaccharidosis
- Niemann-Pick type Ia
- Pompe disease
- Sialidosis
- Wolman disease

Hyperleucocytosis (>100.000/mm^3)

- Leucocyte adhesion deficiency syndrome (CDG IIc: GDP fucose transporter 1)

Hemophagocytosis

- Gaucher disease
- **Lysinuric protein intolerance**
- Niemann-Pick

1.5.7 Hepatology

Cholestatic Jaundice

- α-1-antitrypsin deficiency
- **Arginase deficiency**
- Byler disease
- Congenital disorders of glycosylation
- **Cerebrotendinous xanthomatosis**
- Cholesterol synthesis defects (Smith-Lemli-Opitz)
- Citrin deficiency
- COG 7 deficiency
- Cystic fibrosis
- **Galactosemia**
- **Inborn errors of bile acid metabolism**
- **Long-chain 3-hydroxyacyl-CoA dehydrogenase**
- α-methylacyl-CoA racemase
- Mevalonic aciduria
- Niemann-Pick type C
- Peroxisomal disorders
- Transaldolase deficiency
- **Tyrosinemia type I**

Cirrhosis

- Alpers progressive infantile polydystrophy
- Alpha-1-antitrypsin deficiency
- **Arginase deficiency**
- Congenital disorders of glycosylation
- Cholesterol ester storage disease

- Cystic fibrosis
- **CDG Ib**
- **Galactosemia**
- **Gaucher disease**
- Glycogenosis type IV
- Hemochromatosis
- **Hereditary fructose intolerance**
- **Long-chain 3-hydroxyacyl-CoA dehydrogenase deficiency**
- Niemann-Pick disease
- Peroxisomal disorders
- S-adenosylhomocysteine hydrolase deficiency
- Sitosterolemia
- Transaldolase deficiency
- **Tyrosinemia type I**
- **Wilson disease**
- Wolman disease

Liver Failure (Ascites, Edema) (▶ also Reye Syndrome, Sect. 1.3.1)

Age at onset: congenital (hydrops fetalis)
- Barth hemoglobin
- CDG
- **Erythropoietic porphyria**
- Galactosialidosis
- GM1 gangliosidosis (Landing)
- Mevalonic aciduria
- Mucopolysaccharidosis type VII
- Niemann-Pick A and C
- Sialidosis type II
- Transaldolase deficiency

Age at onset: neonatal to early infancy
- **Fatty acid oxidation disorders**
- **Fructose-1,6-bisphosphatase deficiency**
- **Hereditary fructose intolerance**
- **Galactosemia**
- Mevalonic aciduria
- Mitochondrial DNA depletion (DOGK)
- **Neonatal hemochromatosis** (prevented by immuno-globulin to the pregnant mother)
- Respiratory chain disorders
- **Tyrosinemia type I** (after 3 weeks)

Age at onset: infancy
- Same defects as in neonatal period
- ACAD 9
- Alpha-1-antitrypsin deficiency
- Congenital disorders of glycosylation
- Cholesterol ester storage disease
- Cystic fibrosis
- Familial hepatic fibrosis with exudative enteropathy (**CDG Ib**)
- **Ketogenesis defects**
- Pyruvate carboxylase deficiency

- S-adenosylhomocysteine hydrolase deficiency
- **Urea cycle defects**
- Wolman disease

Age at onset: childhood to adolescence
- **Wilson disease**

1.5.8   Immune System

Inflammatory Syndrome
- Hyper-IgD syndrome
- Mevalonate kinase deficiency

Macrophage Activating Syndrome
- **Gaucher disease**
- **Lysinuric protein intolerance**
- Niemann-Pick disease
- **Propionic acidemia**

Severe Combined Immune Deficiency
- Adenosine deaminase deficiency
- Purine nucleoside phosphorylase deficiency
- **Hereditary orotic aciduria**

1.5.9   Myology

Exercise Intolerance, Myoglobinuria, Cramps, Muscle Pain, Elevated CK

Glycolytic defects (muscle glycogenosis):
- Phosphorylase deficiency (McArdle)
- Phosphofructokinase deficiency
- Phosphoglycerate kinase deficiency
- Phosphoglycerate mutase deficiency
- Lactate dehydrogenase deficiency
- **Glucose-6-phosphate dehydrogenase deficiency**
- Phosphorylase b kinase deficiency

Fatty acid oxidation defects:
- **Carnitine palmitoyl transferase II**
- **VLCAD, LCHAD, translocase, trifunctional**
- SCHAD (restricted to muscles), MCKAT
- Others undescribed (CPT I, SCAD ?)

Miscellaneous:
- **Channelopathies** (hyperkalemic paralysis)
- Duchenne and Becker muscular dystrophies
- Idiopathic familial recurrent myoglobinuria
- Lipoamide dehydrogenase deficiency
- Acid maltase (adult)
- Myoadenylate deaminase deficiency
- Respiratory chain disorders

Myopathy (Progressive)

- Adenylate deaminase deficiency
- **Fatty acid oxidation disorders**
- Glycogenosis type II (acid maltase deficiency)
- Glycogenosis type III
- Multisystemic triglyceride storage disease
- Respiratory chain disorders (Kearns-Sayre and others)
- Steinert disease

1.5.10   Nephrology

Hemolytic Uremic Syndrome

- **Inborn errors of cobalamin metabolism** (Cbl C, Cbl G)

Nephrolithiasis/Nephrocalcinosis

- APRT deficiency (2-8 dihydroxyadenine)
- **Cystinuria** (cystine)
- Hereditary hyperparathyroidism (calcium)
- Hereditary renal hypouricemia (uric acid)
- **Hyperoxaluria type I and II** (oxalic)
- Lesch-Nyhan (uric acid)
- Molybdenum cofactor deficiency (xanthine)
- PRPP synthase superactivity (uric acid)
- **Renal tubular acidosis type I**
- **Xanthine oxidase deficiency** (xanthine)
- Familial juvenile hyperuricemic nephropathy (uromodulin mutation)

Nephrotic Syndrome

- Respiratory chain disorders

Nephropathy (Tubulointerstitial)

- **Glycogenosis type I**
- **Methylmalonic aciduria**
- Respiratory chain disorders (pseudo Senior-Loken syndrome)

Polycystic Kidneys

- Congenital disorders of glycosylation
- CPT II deficiency
- Glutaric aciduria type II
- Zellweger syndrome

Tubulopathy

Fanconi syndrome:
- **Galactosemia – hereditary fructose intolerance**
- Respiratory chain disorders (complex IV or mito DNA deletion)
- **Tyrosinemia type I**
- Bickel-Fanconi syndrome: GLUT2 mutations
- Lowe syndrome (OCRL1 X-linked mutations)
- **Cystinosis**

Renal tubular acidosis:
- **Renal tubular acidosis type I** (distal)
- **Renal tubular acidosis type II** (proximal)
- Pyruvate carboxylase deficiency
- **Methylmalonic aciduria**
- **Glycogenosis type I**
- **Carnitine palmitoyl transferase I deficiency**
- Dent disease (CLCN5 mutations)

Urine – Abnormal Color

- Alkaptonuria (black)
- Indicanuria (blue)
- Myoglobinuria (red)
- Porphyria (red)

Urine – Abnormal Odor

- Dimethylglycine dehydrogenase (fish)
- 3-methyl-crotonylglycinuria (cat)
- Glutaric aciduria type II (sweaty feet)
- **Isovaleric acidemia** (sweaty feet)
- **MSUD** (maple syrup)
- **Phenylketonuria** (musty odor)
- Trimethylaminuria (fish)
- **Tyrosinemia type I** (boiled cabbage)

1.5.11   Neurology (▶ Sect. 1.4.2)

1.5.12   Ophthalmology (▶ also Cherry-red Spot, Ophthalmoplegia, and Retinitis Pigmentosa, Sect. 1.4.4)

Cataracts

Detectable at birth (congenital):
- Cockayne syndrome
- Lowe syndrome (OCRL1 X-linked mutation)
- Peroxisomal biogenesis defects (Zellweger and variants)
- Phosphoglycerate dehydrogenase deficiency
- Rhizomelic chondrodysplasia punctata
- Sorbitol dehydrogenase deficiency

Detectable in the newborn period (1st week to 1st month):
- **Galactosemias**
- Marginal maternal galactokinase deficiency
- Peripheral epimerase deficiency (homozygotes and heterozygotes)

Detectable in infancy (1st month to 1st year):
- Alpha-mannosidosis
- Galactitol or sorbitol accumulation of unknown origin
- Galactokinase deficiency
- Hypoglycemia (various origins)
- P5C synthase deficiency
- Respiratory chain disorders
- Sialidosis

Detectable in childhood (1 to 15 years):

- **Diabetes mellitus**
- Dominant cataract with high serum ferritin
- **Hypoparathyroidism**
- **Lysinuric protein intolerance**
- Mevalonic aciduria
- Neutral lipid storage disorders (unknown cause)
- Pseudo-hypoparathyroidism
- Sjögren-Larsson syndrome
- **Wilson disease**

Detectable in adulthood (> 15 years):

- Carriers for Lowe syndrome
- **Cerebrotendinous xanthomatosis**
- **Fabry disease**
- Glucose-6-phosphate dehydrogenase deficiency
- **Heterozygotes for GALT and galactokinase**
- **Homocystinurias**
- **Lactose malabsorbers**
- Mevalonate kinase
- Mitochondrial cytopathies
- **Ornithine aminotransferase deficiency**
- PEX 7
- Refsum disease
- Steinert dystrophy (cataract can be presenting sign)
- **Tangier disease**

Corneal Opacities (Clouding)

Visible in early infancy (3 to 12 months):

- **Tyrosinemia type II** (presenting sign)
- **Cystinosis** (presenting sign)
- Hurler, Sheie (MPS I)
- I-cell disease (mucolipidosis type II)
- Maroteaux-Lamy (MPS VI)
- Steroid sulfatase deficiency

Visible in late infancy to early childhood (1 to 6 years):

- Mucolipidosis type IV (presenting sign)
- Alpha-mannosidosis (late-onset form)
- Lecithin cholesterol acyltransferase deficiency
- Morquio (MPS IV)
- Pyroglutamic aciduria (presenting sign)
- **Tangier disease**

Visible in late childhood, adolescence to adulthood:

- **Fabry disease** (X-linked)
- Galactosialidosis (juvenile form)
- **Wilson disease** (green Kaiser Fleischer ring)

Ectopia Lentis (Dislocation of the Lens)

- **Classical homocystinuria** (downwards dislocation)
- Sulfite oxidase deficiency
- Marfan syndrome (upwards dislocation)
- Marchesani syndrome

Keratitis, Corneal Opacities

- **Tyrosinemia type II**
- **Fabry disease** (X-linked)

Microcornea

- Ehlers-Danlos type IV

1.5.13   Osteology

Osteopenia

- **Cerebrotendinous xanthomatosis**
- **CDG**
- **Glycogenosis type I**
- **Homocystinuria**
- I-cell disease (mucolipidosis type II)
- Infantile Refsum disease
- **Lysinuric protein intolerance**
- **All organic acidurias** (chronic forms)

Punctate Epiphyseal Calcifications

- Beta-glucuronidase deficiency
- Chondrodysplasia punctata rhizomelic type
- Conradi-Hünermann syndrome
- Familial resistance to thyroid hormone
- Peroxisomal disorders (Zellweger and variants)
- Spondyloenchondromatosis
- Warfarin embryopathy

Exostosis (Hereditary Multiple)

- O-glycosylation defects (EXT1-EXT2)

1.5.14   Pneumology

Hyperventilation Attacks

- Gazeous alkalosis
- **Hyperammonemias**
- Joubert syndrome
- Leigh syndrome (idiopathic or due to various inborn errors)
- Metabolic acidosis
- Rett syndrome (only girls)

Pneumopathy (Interstitial)

- **Gaucher disease**
- **Lysinuric protein intolerance**
- Niemann-Pick type B

Stridor

- **Biotinidase deficiency**
- Hypocalcemia
- Hypomagnesemia
- **MADD** (riboflavin responsive)
- Pelizaeus-Merzbacher

Pulmonary Hypertension
- **Glycogenosis type I**
- Non ketotic hyperglycinemia

1.5.15 Psychiatry (▶ Sect. 1.3.1 and 1.4.3)

1.5.16 Rheumatology

Arthritis – Joint Contractures – Bone Necrosis
- Alkaptonuria
- Familial Gout
- Farber disease
- **Gaucher type I**
- **Homocystinuria**
- I-cell disease, Mucolipidosis type III
- Lesch-Nyhan syndrome
- Mevalonic aciduria (recurrent crisis of arthralgia)
- Mucopolysaccharidosis type IS
- PRPP synthetase superactivity, HGPRT deficiency
- Uromoduline mutation (familial hyperuricemic nephropathy)

Bone Crisis
With bone changes (rickets):
- **Calciferol metabolism deficiency**
- **Hereditary hypophosphatemic rickets**

With hemolytic crises and abdominal pain:
- **Porphyrias**
- **Sickle cell anemia**
- **Tyrosinemia type I**

With progressive neurological signs:
- **Gaucher type III**
- Krabbe disease
- Metachromatic leucodystrophy

Apparently isolated (presenting symptom):
- **Fabry disease**
- Gaucher type I

1.5.17 Stomatology

Glossitis, Stomatitis
- **Clb F**
- **Folate malabsorption**
- **Intrinsic factor deficiency**
- **Transcobalamin II**

Macroglossia
- Beckwith-Wiedemann syndrome
- Congenital muscular dystrophies (DMC1C)

- Complex IV deficiency
- Pompe disease

Hypodontia
- Leucoencephalopathy with ataxia [33]

1.5.18 Vascular Symptoms

Raynaud Syndrome
- **Fabry disease**

Thromboembolic Accidents – Stroke-like Episodes (▶ also Sect. 1.3.1)
- CDG
- Ehlers-Danlos type IV
- **Fabry disease**
- **Homocystinuria** (all types)
- Menkes disease
- **Organic acidurias** (methylmalonic, propionic)
- Respiratory chain disorders (MELAS and others ...)
- **Urea cycle disorders** (OTC deficiency)

References

1. De Koning TJ, Klomp LW, van Oppen AC et al (2004) Prenatal and early postnatal treatment in 3-phosphoglycerate-dehydrogenase deficiency. Lancet 364: 2221-2222
2. Van Spronsen FJ, Smit GPA, Erwich JJHM (2005) Inherited metabolic diseases and pregnancy. BJOG: an International Journal of Obstetrics and Gynaecology 112: 2-11
3. Saudubray JM (2002) Inborn errors of metabolism. Semin Neonatol 7 (issue 1)
4. Saudubray JM, Nassogne MC, de Lonlay P, Touati G (2002) Clinical approach to inherited metabolic disorders in neonates: an overview. Semin Neonatol 7: 3-15
4a. Jaeken J, Martens K, François I et al (2006) Deletion of PREPL, a gene encoding a putative serine oligopeptidase, in patients with hypotonia cystinuria syndrome. Am J Hum Genet 78: 38–51
5. Fellman V, Rapola J, Pihko H, Varilo T, Raivio KO (1998) Iron-overload disease in infants involving fetal growth retardation, lactic acidosis, liver haemosiderosis, and aminoaciduria. Lancet 351: 490-493
6. Visapaa I, Fellman V, Vesa J et al (2002) GRACILE syndrome, a lethal metabolic disorder with iron overload, is caused by a point mutation in BCS1L. Am J Hum Genet 71: 863-876
7. Tazawa Y, Kobayashi K, Abukawa D et al (2004) Clinical heterogeneity of neonatal intrahepatic cholestasis caused by citrin deficiency: case reports from 16 patients. Mol Genet Metab 83: 213-219
8. Tazawa Y, Abukawa D, Sakamoto O et al (2005) A possible mechanism of neonatal intrahepatic cholestasis caused by citrin deficiency. Hepatol Res 31: 168-171
9. Saudubray JM, Martin D, De Lonlay P et al (1999) Recognition and management of fatty acid oxidation defects: a series of 107 patients. J Inherit Metab Dis 22: 488-502
10. Bonnet D, Martin D, De Lonlay P, Villain E, Jouvet P, Rabier D, Brivet M, Kachaner J, Saudubray JM (1999) Arrhythmias and conduction defects as a presenting symptom of fatty-acid oxidation disorders in children. Circulation 100: 2248-2253

11. Ozand PT, Gascon GG, Al Essa M et al (1998) Biotin-responsive basal ganglia disease: a novel entity. Brain 121:1267-1279
12. Zeng WQ, Al-Yamani E, Acierno JS et al (2005) Biotin-responsive basal ganglia disease maps to 2q36.3 and is due to mutations in the SLC19A3 gene. Am J Hum Genet 77:16-26
13. Tiranti V, d'Adamo P, Briem E et al (2004) Ethylmalonic encephalopathy is caused by mutations in ETHE1, a gene encoding a mitochondrial matrix protein. Am J Hum Genet 74: 239-252
14. Sasaki M, Kimura M, Sugai K, Hashimoto T, Yamaguchi S (1998) 3-Hydroxyisobutyric aciduria in two brothers. Pediatr Neurol 18: 253-255
15. Boles RG, Chun N, Senadheera D, Wong LJC (1997) Cyclic vomiting syndrome and mitochondrial DNA mutations. Lancet 350: 1299-1300
16. Chinnery PF, Turnbull DM (1997) Vomiting, anorexia, and mitochondrial DNA disease. Lancet 6351: 448
17. Curtis A, Fey C, Morris C et al (2001) Mutation in the gene encoding ferritin light polypeptide causes dominant adult-onset basal ganglia disease. Nat Genet 28: 350-354
18. Maciel P, Cruz VT, Constante M et al (2005) Neuroferritinopathy: missense mutation in FTL causing early-onset bilateral pallidal involvement. Neurology 65: 603-605
19. Bennett MJ, Weinberger MJ, Kobori JA, Rinaldo P, Burlina AB (1996) Mitochondrial short-chain L-3-hydroxyacyl-coenzyme A dehydrogenase deficiency: a new defect of fatty acid oxidation. Pediatr Res 39: 185-188
20. Pelletier VA, Galéano N, Brochu P, Morin CL, Weber AM, Roy CC (1986) Secretory diarrhea with protein-losing enteropathy, enterocolitis cystica superficialis, intestinal lymphangiectasia, and congenital hepatic fibrosis: a new syndrome. J Pediatr 108: 61-65
21. Vuillaumier-Barrot S, Le Bizec C, de Lonlay P et al (2002) Protein losing enteropathy-hepatic fibrosis syndrome in Saguenay-Lac St-Jean, Quebec is a congenital disorder of glycosylation type Ib. J Med Genet 39: 849-851
22. Schwartz CE, May MM, Carpenter NJ et al (2005) Allan-Herndon-Dudley syndrome and the monocarboxylate transporter 8 (MCD8) gene. Am J Hum Genet 77:41-53
23. Van Coster R, Lissens W, Giardina T et al (2005) An infant with large amounts of several N-acetylated amino acids in the urine has aminoacylase I deficiency. J Inherit Metab Dis 28(Suppl 1): 50
23a. Sass JO, Mohr V, Olbrich H et al (2006) Mutations in ACY1, the gene encoding aminocyclase 1, cause a novel inborn error of metabolism. Am J Hum Genet 78: 401–409
24. Lyon G, Adams RD, Kolodny EH (1996) Neurology of hereditary metabolic diseases of children, 2nd edn. McGraw-Hill, New York
25. Ramaekers VT, Rothenberg SP, Sequeira JM et al (2005) Autoantibodies to folate receptors in the cerebral folate deficiency syndrome. N Engl J Med 352: 1985-1991
26. Poll-The BT, Saudubray JM, Ogier H et al (1986) Infantile Refsum's disease: biochemical findings suggesting multiple peroxisomal dysfunction. J Inherit Metab Dis 9: 169-174
27. Deodato F, Brancati F, Valente EM et al (2005) MAMEL (Methylmalonic Aciduria Mitochondrial Encephalopathy Leigh-like): a new mitochondrial encephalopathy. J Inherit Metab Dis 28(Suppl 1): 130
28. Straussberg R, Shorer Z, Weitz R et al (2002) Familial infantile bilateral striatal necrosis: clinical features and response to biotin treatment. Neurology 59: 983-989
29. Sprecher E, Ishida-Yamamoto A, Mizrahi-Koren M et al (2005) A mutation in SNAP29, coding for a SNARE protein involved in intracellular trafficking, causes a novel neurocutaneous syndrome characterized by cerebral dysgenesis, neuropathy, ichthyosis, and palmoplantar keratoderma. Am J Hum Genet 77:242-251
30. Gollob MH (2003) Glycogen storage disease as a unifying mechanism of disease in the PRKAG2 cardiac syndrome. Biochem Soc Trans 31: 228-231
31. Arad M, Maron BJ, Gorham JM et al (2005) Glycogen storage diseases presenting as hypertrophic cardiomyopathy. N Engl J Med 352: 362-72
32. Holme E, Kollberg G, Oldfors A, Gilljam T, Tulinius M (2005) Muscular glycogen storage disease 0: a new disease entity in a child with hypertrophic cardiomyopathy and myopathy due to a homozygous stop mutation in the muscular glycogen synthase gene (GYS1). J Inherit Metab Dis 28(Suppl 1): 214
33. Wolf NI, Harting I, Boltshauser E et al (2005) Leucoencephalopathy with ataxia, hypodontia and hypomyelination. Neurology 64: 1461-1464

2 Newborn Screening for Inborn Errors of Metabolism

Bridget Wilcken

2.1 Introduction

Newborn screening was first applied to the detection of phenylketonuria (PKU) by a bacterial inhibition assay pioneered in 1961 by Guthrie, who was also responsible for the introduction of the use of a dried blood sample [1]. This was followed by further bacterial inhibition assays to detect other aminoacidopathies (maple syrup urine disease, homocystinuria, urea cycle disorders and so on) but only screening for PKU was widely adopted. In 1975 Dussault described screening for congenital hypothyroidism (CH) [2], and since then other disorders covered in some screening programmes have included congenital adrenal hyperplasia, the galactosaemias, cystic fibrosis, biotinidase deficiency, glucose-6-phosphate dehydrogenase deficiency and many others. The application of tandem mass spectrometry to newborn screening was first described in 1990 [3]. This new technology has greatly changed both newborn screening and the diagnosis of many inborn errors of metabolism.

2.2 General Aspects of Newborn Screening

2.2.1 Aims and Criteria

The initial aim of newborn screening was to identify infants with serious but treatable disorders, so as to facilitate interventions to prevent or ameliorate the clinical consequences of the disease. In recent years, with the advent of tandem mass-spectrometry which can detect many disorders at one time, and hence the ability for early detection of currently untreatable disorders (▶ below), there has been discussion about how the aims of screening might be widened to encompass a benefit to families, rather than individual babies.

The classic criteria for screening are those of Wilson and Jungner [4]. More recently the World Health Organisation has published guidelines [5] as has the United States [6], and the United Kingdom National Screening Committee has extended the Wilson and Jungner criteria [7]. In reality the criteria can be simplified and reduced to two main considerations which would justify screening for any specific disorder: there should be a benefit from neonatal detection, and the overall benefit should be reasonably balanced by the costs of all kinds: the financial costs (opportunity costs) and the cost of the harm, if any, to individuals by early detection of the disorder, or false assignment of a positive or negative result. It is important to remember that newborn screening covers the whole process from sampling to the appropriate referral of an affected baby for the start of treatment, and assessment of overall outcome.

2.2.2 Sensitivity, Specificity, and Positive Predictive Value

In assessing screening tests and understanding the screening process, some definitions are important:
Sensitivity: The proportion of subjects with the disorder in question detected by the test.
Specificity: The proportion of subjects without the disorder that have a negative test result.
False negative rate: The percentage of affected subjects not detected by the test.
False positive rate: The percentage of healthy subjects with a positive test result.
Positive predictive value: The chance that a positive result actually indicates an affected individual. Similarly, a *negative predictive value* is the chance that a negative result excludes the disorder. These values depend not only on the specificity or sensitivity of the test, but also on the frequency of the disorder.

The sensitivity of a test depends to a large extent on chosen cut-off values, and is a balancing act: the higher the sensitivity the lower the specificity. The sensitivity and specificity of course will vary according to decisions about cut-off points and what range of false positive and negative results can be tolerated in a particular programme.

2.2.3 Technical Aspects of Newborn Screening Tests

Blood-collection-paper Samples
Newborn screening tests are mainly carried out on blood spots dried on specially manufactured filter paper, usually obtained by heel-stick. Although various forms of venous blood sampling may also be used, filter paper blood samples will be considered here.

Methods
A wide variety of technologies can be applied to filter paper samples, including bacterial inhibition assays, chromatographic techniques, enzyme-linked immunosorbent assays (ELISA), fluorescent immunoassays (FIA), radioimmunoassays (RIA), and most recently electrospray ionisation tandem mass spectrometry (MSMS). Methods for newborn screening prior to MSMS screening are well-described by Therrell [8]. DNA analysis is also performed as a part of some screening tests. The methodology to some extent varies with the analyte of interest (for example, hormone analyses are usually immunoassays), and cost, sensitivity and specificity vary according to the method used. While quantitative results are obtained, the precision of tests using a paper sample is less than for a plasma or serum sample because of the matrix, collection process, haematocrit variations and so forth.

Timing of the Test

The timing of the test also has an important influence on the results. A newborn during the first 72 h is catabolic to some extent, and this is very useful for detecting disorders of intermediary metabolism. There is a paucity of data, except for PKU and CH screening, about results of screening in the first 24 h of life, and this is generally not recommended. Cut-off values adopted to indicate a positive test result for the different analytes will vary with the age of the child at screening, and the sensitivity of a test for detecting certain disorders may also vary according to age.

Cut-off Values

Determination of the cut-off point for each analyte is always a compromise between the aim for perfect sensitivity (detecting all the cases) and keeping the false negative rate as low as possible. It is important for the laboratory to establish age-dependent cut-off values, as these can vary greatly. In general, the cut-points for amino acids and free carnitine may be lower for babies aged, say, 48–72 h than for babies aged 7–10 days. The converse is true for acylacarnitines, where levels decrease with increasing age, and use of a 2–3 day cut-point at 7–10 days could lead to a case being missed. Physicians must bear in mind that no screening test ever gives a perfect performance, although some may come close. If clinical presentation suggests a disorder which is included in newborn screening, a test should be done, even if the screening test was negative.

2.2.4 Range of Possibilities from Early Detection

Newborn screening has opened new perspectives in preventive medicine. Babies with disorders of amino acid, organic acid, and fatty acid metabolism are now often detected in the newborn screening laboratory, rather than by the clinical metabolic service. Early detection provides 3 possibilities:

1. The disorder may present in the first days of life, before any newborn screening result is likely. Disorders in this category include neonatal presentations of urea cycle defects, organic acidaemias such as methylmalonic acidaemia, and less commonly, almost any of the fatty acid oxidation defects. Detection by newborn screening is unlikely to benefit directly most cases in this category. However, it seems appropriate to include these early-presenting disorders in the screening suite, as some may have delayed diagnosis, and on occasion a diagnosis may never be made – the baby having been thought to have died from sepsis.
2. The disorder may be later presenting, and an effective treatment can beneficially alter the natural history. Cases in this category include the less severe urea cycle disorders, most aminoacidopathies, such as homocys-

tinuria, some organic acidaemias, and most fatty acid oxidation disorder cases.

3. The disorder may be benign, or largely so, and most cases will have no benefit from early diagnosis. It is hard to know yet which cases will fit into this category. If that was clear, then the disorders could be removed from the screening suite, but newborn screening, if carefully and sensitively conducted, provides an excellent opportunity for elucidating the natural history of disorders which might or might not fall into this category. What is clear is that mild forms of several disorders will readily be detected by newborn screening, but may not need treatment. One example is mild citrullinaemia (argininosuccinate synthase deficiency) [9].

2.2.5 Tandem Mass Spectrometry

MSMS has so revolutionised newborn screening that a brief general description of the method is appropriate here.

Methods

MSMS is a method of measuring analytes by both mass and structure. After ionisation of the compounds, the first mass spectrometer selects ions of interest, sorted by weight. They pass into a collision cell, are dissociated to »signature« fragments, and then pass into a second mass spectrometer where ions are selected for detection. The two mass spectrometers are linked by computer. Only selected ions can pass through the assembly – the others are deflected. Two important points to note are, firstly, the ions to be analysed can be selected by the operator, using the multiple reaction monitoring (MRM) mode, and thus in most cases it is possible to select what disorders might be detected. Secondly, using both mass and structure gives high specificity, except in the case of isomers. The initial methods described for aminoacid and acylcarnitine analysis used derivatisation of samples to form either methyl or butyl esters [10]. Most newborn screening programmes use derivatisation but with modern instruments adequate sensitivity can be achieved without derivatising, simplifying sample preparation. For some analytes, elevations may indicate more than one disorder and the use of ratios of analytes improves sensitivity and specificity [11]. Cut-off values need to be determined for each laboratory, and actual values are not discussed here. Confirmation of a disorder always requires appropriate follow-up tests in a specialised biochemical genetics laboratory. These follow-up tests may include aminoacid analysis, organic acid analysis by gas chromatography/mass spectrometry, and plasma acylcarnitine profile by MSMS.

Range of Disorders Detectable

Current screening by MSMS involves the measurement of aminoacids and acylcarnitines, to detect selected disorders of aminoacid, organic acid, and fatty acid metabolism. It is

◻ **Table 2.1.** Reliability of detection by tandem mass spectrometry (when the sample is appropriate)

High	Probably high	Uncertain	Unreliable
Phenylketonuria	Many other organic acidaemias	LCHAD	Homocystinuria (CBS): B6-responsive cases probably not detected
Tyrosinaemia type II	VLCAD	CPT I	Fumaryl acetoacetase deficiency: requires succinyl acetone assay
MSUD classical	CACT/neonatal CPT II	SCAD	Glutaric aciduria I: low excretors may be missed
ASS; ASL			Carnitine transporter
PA, MMA, IVA neonatal			MSUD variant
MCAD	MADD		OTC, CPS, NAGS

MSUD, maple syrup urine disease; *ASS*, argininosuccinate synthase deficiency; *ASL*, argininosuccinate lyase deficiency; *PA*, propionic acidaemia; *MMA*, methylmalonic acidaemia; *IVA*, isovaleric acidaemia; *SCAD, MCAD, VLCAD, LCHAD, MADD*, short-chain, medium-chain, very long-chain, long-chain 3-hydroxy and multiple acyl-CoA dehydrogenase deficiencies; *CACT*, carnitine acylcarnitine transporter defect; *CPT I* and *II*, carnitine palmitoyl transferase deficiencies; *CBS*, cystathionine beta-synthase deficiency; *B6*, pyridoxine; *OTC, CPS, NAGS*, ornithine transcarbamylase, carbamyl phosphate synthetase and N-acetyl glutamate synthetase deficiencies.

still too early to be sure of the reliability of MSMS for the detection of the rarer disorders. ◻ Table 2.1 gives an estimate, based on current knowledge, of the reliability of detection by MSMS (when the sample is appropriate). The selection of disorders that should be included is the subject of intense discussion and some disagreement in various screening jurisdictions. Other factors which affect performance and interpretation include the flux through the metabolic pathway, depending on the degree of catabolism present.

Performance and Outcome

There are as yet no comprehensive studies of outcomes for MSMS screening, but these are on the way. A false positive rate of approximately 0.2–0.3% has been achieved by several programmes that have also attained an apparently high sensitivity [11, 12]. While it is clear which disorders selected analytes can indicate, what is not yet clear is the screening performance for each disorder – i.e. the sensitivity and specificity that can be achieved. Moreover, it has emerged, not surprisingly, that screening detects more cases than clinical diagnosis. This is particularly so for medium-chain acyl-CoA dehydrogenase deficiency, to a lesser extent other fatty acid oxidation disorders, and additionally, for some organic acidopathies, previously thought very rare [13, 14]. Some maternal disorders can also be detected by testing of the baby. Reported results from several comprehensive programmes are available [11, 12, 14].

2.3   Screening for Individual Inborn Errors of Metabolism

Well over 40 inborn errors of metabolism can now be detected by newborn screening, with varying degrees of cer-

tainty. This section will concentrate on inborn errors that are the province of the metabolic physician.

2.3.1 Phenylketonuria

Test Methods

Screening for PKU has been usual in most developed countries since the late 1960s. The initial test was the »Guthrie test«, a bacterial inhibition assay, and this is still widely used. Alternative methods include those using fluorimetry and calorimetry. More recently, PKU screening has been by MSMS where this is available.

Timing

Screening in the United States may take place after 24 h, but elsewhere 48 h plus is usual. As with other aminoacids, levels of phenylalanine rise steadily during the first days of life, and there is a theoretical risk of missing PKU if the test is conducted too early. However, testing by MSMS and measuring the ratio of phenylalanine to tyrosine overcomes this problem [11, 15].

Reliability

The bacterial inhibition assay is a robust test, and if it is applied to samples taken at 48 h or more, any false negative tests in subjects with classical PKU have almost always been due to errors of process (sample not taken, clerical error etc.) rather than biological variation or problems with the actual test procedure [16]. MSMS screening may be even more reliable, and overall, the sensitivity of all PKU tests is very high indeed. With experience, the specificity is also high. There are no established benchmarks.

Outcome

Patients obtaining good control of phenylalanine levels by 3–4 weeks and maintaining good average control have a good neuropsychological outcome. There are still minor deficits, and *maternal PKU* remains a potential problem. These aspects are reviewed in Chap. 17.

2.3.2 Galactosaemias

Galactose-1-phosphate uridyl transferase (GALT) deficiency, galactokinase deficiency and galactose epimerase deficiency can all be detected by newborn screening.

Test Methods

Methods used in screening are measures of metabolites, galactose and galactose-1-phosphate, or measures of enzyme activity, confined to the Beutler test for GALT. Most commonly nowadays, a metabolite assay is followed by confirmatory testing using a GALT assay and quantitative determination of galactose and galactose-1-phosphate.

Reliability

A combination of these methods will provide a precise diagnosis of transferase deficiency (but ▶ below), and an indication of galactokinase deficiency, in which there is elevation of galactose alone and a normal GALT activity. However, the differentiation of red-cell epimerase deficiency (a benign condition) and systemic epimerase deficiency, clinically similar to transferase deficiency, may not be clear. Additionally, moderate metabolite elevations and severely reduced, but not absent, GALT activity is seen in combined heterozygosity for a severe mutation in the transferase *GALT* gene, and a common »Duarte« mutation. The dif-

ferentiation is important, as a Duarte/galactosaemia double heterozygote needs no treatment.

Outcome

Despite early identification and treatment, the long-term outcome for transferase deficiency is not particularly good, with about half the children having early intellectual problems, and some evidence of ongoing deterioration in most (▶ Chap. 7). There is no evidence that pre-symptomatic treatment alters outcome, (although death may be avoided in a few) and because of this, not all developed countries screen for the galactosaemias. Treated galactokinase deficiency would be expected to have a good outcome, but is much rarer than transferase deficiency, and systemic epimerase deficiency is rarer still, and little is known of long-term effects of screening.

2.3.3 Aminoacidopathies

▣ Table 2.2 shows the amino acids most commonly analysed by tandem mass-spectrometry, the secondary markers in use, and the follow-up tests indicated.

Disorders of the Urea Cycle

Citrullinaemia and *argininosuccinic aciduria*, either severe or later-presenting, can be diagnosed with apparently high sensitivity by measuring citrulline. There are problems because of the recent description of mild, asymptomatic citrullinaemia. Detection of severe, early-presenting *argininaemia*, by measuring arginine, has also been described. *Carbamyl phosphate synthase* and *ornithine transcarbamylase (OTC) deficiencies* cannot be so easily detected. Low citrulline is an indicator, but a low cut-off for citrulline

▣ **Table 2.2.** The main amino acids analysed in newborn screening by tandem mass spectrometry. Also shown are the possible diagnoses, the secondary markers in use, and the initial follow-up tests indicated for confirmation

Amino acids	Possible diagnoses	Secondary markers	Initial follow-up tests
Arginine (ARG)	Arginase deficiency		PLAA, enzyme analysis
Citrulline (CIT)	ASS, ASL, Citrullinaemia type II	CIT/ARG ratio	PLAA, UAA, LFTs
Leucines (LEUs)	Maple syrup urine disease	LEUs/PHE	PLAA
Methionine (MET)	Cystathionine synthase deficiency, hypermethioninaemias	MET/PHE	PLAA, tHCY
Phenylalanine (PHE)	Phenylketonuria, hyperphenylalaninaemia, pterin disorders	PHE/TYR	PLAA, pterin load, urinary pterins, DHPR
Tyrosine (TYR)	Tyrosinaemias: fumaryl acetoacetase deficiency, tyrosine aminotransferase deficiency, tyrosinaemia III	Succinylacetone	UOAs, PLAA, LFTs etc.

ASS, argininosuccinate synthase deficiency (citrullinaemia type I); *ASL*, argininosuccinate lyase deficiency; *PLAA*, plasma amino acid analysis; *UAA*, urinary amino acid analysis; *UOAs*, urinary organic acid analysis; *LFTs*, liver function tests; *tHCY*, total homocysteine (plasma); *DHPR*, dihydropteridine reductase assay (dried blood spot).

overlaps with low citrulline seen in sick neonates in general. Diagnosis of OTC deficiency has been described by the detection of pyroglutamic acid, derived from glutamine, and a blood-spot method for detecting glutamine has been described [17]. Probably newborn screening is quite unreliable for these early disorders. A related disorder, *citrin deficiency* causing neonatal hepatitis, (citrullinaemia type II) could be detected by a disturbance of several amino acids, especially moderately elevated citrulline, and detection of one case by MSMS has been described. If ornithine is one of the analytes included, *hyperornithinaemia, hyperammonaemia, homocitrullinuria (HHH syndrome)* could theoretically be detected, but no cases have been reported and this analyte is often not included. Hyperornithinaemia due to *ornithine aminotransferase deficiency* probably does not cause elevated ornithine levels in early infancy, and would not be detected [18].

Other Aminoacidopathies

When screening for *phenylketonuria* by MSMS is the method used, some but not all programmes have found it useful to use a phenylalanine/tyrosine ratio for identifying positive results [15]. The *tyrosinaemias* present some problem. In *tyrosinaemia type 1 (fumaryl acetoacetase deficiency)* the blood tyrosine level in newborns is often not high, and there is considerable overlap with transient tyrosinaemia cases. Several cases of tyrosinaemia type 1 have been reported as missed by MSMS screening, and unless a separate assay of succinylacetone is performed as a back-up, it is likely that this disorder is usually not detectable by MSMS without an unacceptable false-positive rate [12]. *Tyrosinaemia type II* is readily detectable, and at least one case of tyrosinaemia type III has been found. *Maple syrup urine disease* is detected by assay of leucines and perhaps valine. As MSMS detection cannot distinguish among isotopes, the leucine peak encompasses isoleucine, alloisoleucine, and hydroxyproline in addition. Classical MSUD can readily be detected (although a positive result could also indicate the benign hyperhydroxyprolinaemia) but it is not clear that all variant cases can be distinguished, even by the use of the leucine/phenylalanine ratio. It goes without saying that a result indicating classic MSUD needs to be handled as an emergency. *Cystathionine synthase deficiency (homocystinuria)* is currently detected by an elevated methionine level. The likelihood is that this test, as with the bacterial inhibition assay for methionine formerly used, will miss most cases of pyridoxine-responsive homocystinuria. It is too early yet to be sure of the sensitivity for non-responsive cases, but it is likely to be good. A back-up assay of homocysteine, which cannot be done using the same method, would be ideal, but no experience of this approach has been reported.

Several other aminoacids can be measured simultaneously by MSMS without altering the method, including ornithine, serine, valine and glutamine, but most have not been reported as useful in newborn screening. We found that elevated alanine was not a useful discriminator for babies later diagnosed with mitochondrial respiratory chain disorders.

2.3.4 Organic Acid Disorders

Organic acids that form acylcarnitines can be detected by MSMS, and a large number of organic acid disorders have been so detected (◘ Table 2.3). The classic organic acid disorders, *methylmalonic (MMA), propionic (PA),* and *isovaleric acidaemias (IVA)* can readily be detected, although the baby will probably be symptomatic before newborn screening results are available. An elevation of propionylcarnitine (C_3) might indicate either PA, MMA, vitamin B_{12} deficiency secondary to maternal deficiency, or *cobalamin C defect* (methylmalonic aciduria with homocystinuria), or possibly *cobalamin D or F defects*. While severe neonatal onset MMA will have elevated C_3 levels, other defects can be more reliably detected by using the ratio of C_3 to acetylcarnitine. Of special importance is *glutaric aciduria type I* (glutaryl CoA dehydrogenase deficiency). The marker compound, glutarylcarnitine (C_5DC), which is also one of several markers for *glutaric aciduria type II*, may be only marginally elevated, and may not be detected, especially if the infant is sampled after the first week. Again, a ratio, this time with palmitoylcarnitine, is more discriminatory [11].

Newborn screening has uncovered an unexpectedly high frequency of cases of *3-methylcrotonyl CoA carboxylase deficiency* (MCCC), previously thought to be exceptionally rare, and asymptomatic cases of maternal MCCC are also detected regularly by newborn screening [19]. This disorder is one of several that might be benign in most instances. Other maternal disorders uncovered by abnormal results on neonatal screening include mild holocarboxylase synthase deficiency and vitamin B_{12} deficiency.

2.3.5 Fatty Acid Oxidation Disorders

Disorders of carnitine uptake, the carnitine cycle, and mitochondrial beta-oxidation can be detected by MSMS testing of acylcarnitines (◘ Table 2.3). For several disorders, newborn screening programmes have detected more cases than have historically presented clinically [12, 13]. While some of these subjects might never have experienced episodes of decompensation it is not possible at present to distinguish who is at most risk, and all have by definition a functional defect in oxidation rates. This is especially true of *medium-chain acyl-CoA dehydrogenase deficiency (MCAD),* the most frequently occurring fatty acid oxidation disorder, in which the detection rate is nearly doubled. MCAD is reliably indicated by elevated octanoylcarnitine

◨ Table 2.3. The major acylcarnitine species measured in newborn screening by tandem mass-spectrometry. Also shown are the possible diagnoses, the secondary markers in use, and the initial follow-up tests indicated for confirmation. Other acylcarnitine species are often routinely measured simultaneously with those listed, and give further support to possible diagnoses

Abbreviation	Acyl-carnitines	Possible diagnoses	Secondary markers	Initial follow-up tests
C_0	Carnitine	Low: carnitine transporter defect High: CPT I		PLAC, Fibroblast assay, uptake enzyme
C_2	Acetylcarnitine	No inborn error. (high: ketoacidosis, low: low carnitine status)		
C_3	Propionylcarnitine	Methylmalonic acidaemias (incl CblC) propionic acidaemia, vitamin B_{12} deficiency	C_3/C_2	UOAs, tHCY, enzyme assays, serum B12, infant and mother
C_3DC	Malonylcarnitine	Malonic aciduria		UOAs
C_4	Butyrylcarnitine	SCAD, MADD, isobutyrylCoA dehydrogenase		PLAC, UOAs
C_5	Isovalerylcarnitine, 3-methylbutyrylcarnitine	Isovaleric acidaemia, or 2-methylbutyryl CoA dehydrogenase deficiency		UOAs
C_6	Hexanoylcarnitine	MCAD, MADD		
C_5-OH	3-Hydroxyisovaleryl carnitine, or 2-methyl-3-hydroxy-butyrylcarnitine	Biotinidase, HMG-CoA lyase, β-ketothiolase 3-methylcrotonyl-CoA carboxylase, or multiple carboxylase deficiencies		UOAs, maternal UOAs
C_8	Octanoylcarnitine	MCAD, MADD	C_8/C_{10}	UOAs, DNA mutation
C_{10}	Decanoylcarnitine	MCAD, MADD		
C_5DC	Glutarylcarnitine	Glutaric acidaemia I		UOAs, enzyme assay
CH_3-C_5DC	3-Methyl-glutarylcarnitine	HMG-CoA lyase deficiency		
$C_{14:1}$	Tetradecenoylcarnitine	VLCAD, LCHAD		PLAC
C_{14}	Tetradecanoylcarnitine	LCHAD, VLCAD		PLAC
C_{16}	Hexadecanoylcarnitine	CPT II, CACT (low: CPT 1)	$C_{16}+C_{18}/C_2$	PLAC
C_{16}-OH	Hydroxyhexadecanoyl-carnitine	LCHAD		PLAC

CPT, carnitine palmitoyltransferase; *CblC*, cobalamin C defect; *DC*, dicarboxylic; *SCAD, MCAD, VLCAD, LCHAD, MADD,* short-chain, medium-chain, very long-chain, long-chain 3-hydroxy and multiple acyl-CoA dehydrogenase deficiencies; *CACT*, carnitine acylcarnitine transporter defect; *CPT I* and *II*, carnitine palmitoyl transferase deficiencies; *UOAs*, urinary organic acids; *tHCY*, plasma total homocysteine; B_{12}, vitamin B_{12}; *PLAC*, plasma acylcarnitine profile.

(C_8). The ratio of C_8 to C_{10} (decanoylcarnitine) is also used as a secondary marker. Often, mutational analysis to determine the presence of at least the common Northern European mutation, A985G, is performed as a secondary test. However, the allele frequency of this mutation is somewhat lower (around 75%) in screen-detected than in clinically presenting patients, and some other mutations found by screening have not been recorded in clinical presentations [20]. Elevated urinary acylglycines and an abnormal plasma acylcarnitine profile virtually confirm the diagnosis of MCAD. More definitive confirmation can be obtained by enzymatic or DNA analysis or more simply by acylcarnitine profiling in cultured skin fibroblasts. *Short-chain acyl-CoA*

dehydrogenase deficiency is found more commonly in screened patients than would have been expected from previous experience, and all cases reported so far have been asymptomatic. Other fatty acid oxidation disorders found by newborn screening are shown in ◨ Table 2.3. Amongst these, some cannot be reliably detected, but with a multiplex test this does not seem to be a reason not to screen. An example is the carnitine transporter defect. It is certain that some cases can be detected, but not necessarily all. Treatment is simple and as the disorder is potentially fatal, detection of even some of the cases seems justified.

2.3.6 Other Neonatal Screening Programmes

Other disorders often tested for in various combinations in newborn screening programmes, but of more importance for general paediatricians or those with other specialties, include the following:

Congenital Hypothyroidism

Screening for primary CH is universal in countries with well-developed health systems, and widely practised in less well-developed countries. Usually the primary test is measurement of thyroid stimulating hormone (TSH). False negative results may occur in cases with well compensated ectopic thyroid glands, or sometimes with dyshormonogenesis. Trial withdrawal of treatment at 2–3 years will identify cases that were transient and not needing life-long treatment.

Cystic Fibrosis

Screening and early diagnosis has produced clear-cut nutritional benefit and probable improvement in lung function. The test involves measurement of immunoreactive trypsin (trypsinogen) and usually some mutational analysis on the original blood spot. Some screening programmes have been functioning for almost 25 years, and screening is widely recommended.

Congenital Adrenal Hyperplasia

Measurement of 17-hydroxyprogesterone is the primary test for congenital adrenal hyperplasia (CAH). The test readily identifies cases, but the false positive rate is high, especially among pre-term babies, and strategies to reduce this are necessary. In the future, a secondary MSMS test may be the best option.

Biotinidase Deficiency

A specific enzyme assay on dried blood spots is often used to detect biotinidase deficiency. This is more sensitive than MSMS testing. The utility of using a separate test for this disorder has not been substantiated.

Glucose-6-phosphate Dehydrogenase Deficiency

An NAD/NADH-based enzyme assay can detect this disorder. The test is widely used in Asian countries. The usefulness in avoiding kernicterus and haemolytic crises is likely but has not been well demonstrated.

Haemoglobinopathies

Sickle-cell disease is included in areas where this is prevalent. There is now the possibility of a wider screen for haemoglobinopathies using MSMS.

References

1. Guthrie R, Susi A (1963) A simple phenylalanine method for detecting phenylketonuria in large populations of newborn infants. Pediatrics 32:338–343
2. Dussault JH, Coulombe P, Laberge C et al. (1975). Preliminary report on a mass screening program for neonatal hypothyroidism. J Pediatr 6:670–674
3. Millington DS, Kodo N, Norwood DL, Roe CR (1990) Tandem mass spectrometry: a new method for acylcarnitine profiling with potential for neonatal screening for inborn errors of metabolism. J Inherit Metab Dis 13:321–324
4. Wilson JMG, Jungner G (1968) Principles and practice of screening for disease. World Health Organization, Geneva
5. Proposed international guidelines on ethical issues in medical genetics and genetics services. WHO/HGN/GL/ETH/98 1 2000
6. Serving the family from birth to the medical home. A report from the Newborn Screening Task Force convened in Washington DC, May 10-11, 1999 (2000). Pediatrics 106(2 Pt 2):383–427
7. Green A, Pollitt RJ (1999) Population newborn screening for inherited metabolic disease: current UK perspectives. J Inherit Metab Dis 22:572–579
8. Therrell BL Jr (1993) Laboratory methods for neonatal screening. American Public Health Association, Washington, DC
9. Sander J, Janzen N, Sander S et al. (2003) Neonatal screening for citrullinaemia. Eur J Pediatr 162:417–420
10. Rashed MS, Rahbeeni Z, Ozand PT (1999) Application of electrospray tandem mass spectrometry to neonatal screening. Semin Perinatol 23:183–193
11. Zytkovicz TH, Fitzgerald EF, Marsden D et al. (2001) Tandem mass spectrometric analysis for amino, organic, and fatty acid disorders in newborn dried blood spots: a two-year summary from the New England Newborn Screening Program. Clin Chem 47:1945-1955
12. Wilcken B, Wiley V, Hammond J, Carpenter K (2003) Screening newborns for inborn errors of metabolism by tandem mass spectrometry. N Engl J Med 348:2304–2312
13. Schulze A, Lindner M, Kohlmuller D et al. (2003) Expanded newborn screening for inborn errors of metabolism by electrospray ionization-tandem mass spectrometry: results, outcome, and implications. Pediatrics 111:1399–1406
14. Chace DH, Kalas TA, Naylor EW (2003) Use of tandem mass spectrometry for multianalyte screening of dried blood specimens from newborns. Clin Chem 49:1797–1817
15. Chace DH, Millington DS, Terada N et al. (1993) Rapid diagnosis of phenylketonuria by quantitative analysis for phenylalanine and tyrosine in neonatal blood spots by tandem mass spectrometry. Clin Chem 39:66–71
16. Holzman , Slazyk WE, Cordero JF, Hannon WH (1986) Descriptive epidemiology of missed cases of phenylketonuria and congenital hypothyroidism. Pediatrics 78:553–558
17. Trinh MU, Blake J, Harrison JR et al. (2003) Quantification of glutamine in dried blood spots and plasma by tandem mass spectrometry for the biochemical diagnosis and monitoring of ornithine transcarbamylase deficiency. Clin Chem 49:681 684
18. Wang T, Lawler AM, Steel G et al. (1995) Mice lacking ornithine-delta-aminotransferase have paradoxical neonatal hypoornithinaemia and retinal degeneration. Nat Genet 11:185–190
19. Gibson KM, Bennett MJ, Naylor EW, Morton DH (1998) 3-Methylcrotonyl-coenzyme A carboxylase deficiency in Amish/Mennonite adults identified by detection of increased acylcarnitines in blood spots of their children. J Pediatr 132(3 Pt 1):519–523
20. Andresen BS, Dobrowolski SF, O‹Reilly L et al. (2001) Medium-chain acyl-CoA dehydrogenase (MCAD) mutations identified by MS/MS-based prospective screening of newborns differ from those observed in patients with clinical symptoms: identification and characterization of a new, prevalent mutation that results in mild MCAD deficiency. Am J Hum Genet 68:1408–1418

3 Diagnostic Procedures: Function Tests and Postmortem Protocol

Guy Touati, Jan Huber, Jean-Marie Saudubray

3.1 Introduction

Functional investigations are based on the measurement of intermediary metabolites in body fluids. They are most useful in disorders that give rise to toxicity or energy deficiency. The best functional test is elicited by nature itself during episodes that cause metabolic stress, including acute infection, inadvertent fasting, or consumption of a nutrient that induces a metabolic intolerance. If an inherited metabolic disease is suspected then blood, urine and cerebrospinal fluid should be collected for the appropriate investigations (▶ Chap. 1, ◘ Table 1.3). If no material is available or if the results are ambiguous, a provocative test that challenges a metabolic pathway may provide clues for a diagnosis and indicate which specific enzymatic or genetic analysis should be undertaken.

When performing a functional test, it is important to adhere to a strictly defined protocol in order to attain a maximum of interpretable diagnostic information and to minimize the risk of metabolic complications. Some of the tests described in this chapter are now used infrequently, since more simple direct assays of metabolites and DNA have reduced their diagnostic value. This applies to the galactose, fructose, or fat-loading tests. Other tests have fallen into total disuse and are not considered here. These include the glucagon test for the differentiation of glycogen-storage diseases and the phenylpropionate loading test for diagnosis of medium-chain acyl-coenzyme A dehydrogenase deficiency.

3.2 Functional Tests

3.2.1 Metabolic Profile over the Course of the Day

Indications

This first line of evaluation of intermediary metabolism may be performed following an initial or recurrent clinical incident associated with a disturbance of intermediary metabolism in which the aetiology is unknown. The investigation is used in the metabolic/endocrine investigation of hypoglycaemia, hyperlactatemia, hyperketosis or hypoketosis and, in these situations, should always be undertaken before any provocative test that may lead to metabolic decompensation. The metabolic profile is also used for monitoring treatment in many disorders.

Procedure

Blood samples from an indwelling venous catheter (kept patent with a saline infusion) are taken before and after meals, and once during the night, as outlined in ◘ Table 3.1.

◘ **Table 3.1.** Assessment of intermediary metabolism over the course of the day

Parameters in blood	Breakfast		Lunch		Dinner		Night
	Before	1 h after	Before	1 h after	Before	1 h after	04 h
Glucose[1]	X	X	X	X	X	X	X
Acid-base	X	X					
Lactate[2]	X	X	X	X	X	X	X
Pyruvate[2]	X	X	X	X	X	X	X
Free fatty acids	X	X	X	X	X	X	X
Ketone bodies	X	X	X	X	X	X	X
Ammonia	X	X	X	X	X	X	X
Amino acids	X						
Carnitine	X						
Acylcarnitines	X						
Hormones[3]	X	X	X	X	X	X	X
Urine 24 h collection[4]	Amino acids, organic acids, ketone bodies, urea, creatinine						

[1] Glucose should be determined immediately.
[2] Immediate deproteinization (with perchloric acid) at the bedside is the only way of ensuring that the results for calculating redox potential ratios can be accurately interpreted.
[3] Hormones (insulinemia, cortisol, growth hormone) are useful in the investigation of hypoglycaemia.
[4] Urine samples are collected both overnight and during the day and should be frozen immediately.

In order for the results to be interpreted reliably, the correct method of sampling and processing of blood and urine is specified in ◘ Table 3.5. It is important to record the conditions under which sampling is undertaken, for example local or general anoxia may influence the results for lactate, lactate/pyruvate ratio and ammonia measurements.

Interpretation

This investigation may show abnormalities in the metabolic and endocrine profiles throughout the day or specifically only during either the fasting or fed states. The data must be compared to age related reference values [1, 2]. All physiological (food refusal) or pathological conditions (malnutrition, cardiac, renal or liver failure) that may influence the results need to be taken in account.

1. In glycogenosis (GSD) type I and in disorders of gluconeogenesis, blood glucose and lactate move in opposite directions, with hypoglycaemia and hyperlactatemia more pronounced in the fasted than in the fed state. In GSD type III and VI, glucose and lactate evolve in parallel, with a moderate increase of glucose and lactate in the post-prandial state. Fasting hypoglycaemia with post-prandial hyperlactatemia is usual in glycogen synthase deficiency. Repeated assays are required for glucose and insulin in primary hyperinsulinism, as hyperinsulinemia is frequently erratic and difficult to prove. An insulin level >3 μU/ml with a glucose concentration lower than 2.8 mmol/l should be considered abnormal.

2. Plasma lactate may be continuously elevated but usually decreases during fasting in patients with pyruvate dehydrogenase (PDH) deficiency and increases in those with glycogenosis type I. Lactate may be normal, moderately raised or very high in mitochondrial respiratory chain (RC) disorders [1]. It may be difficult to distinguish a moderate elevation of lactate from a falsely raised level due to difficult sampling. However, the presence of a lactaturia with an elevation of alanine in blood is very suggestive of a true hyperlactatemia. Lactate measurement in cerebrospinal fluid (CSF) may also be of help in patients with neurological disorders.

3. Ketone bodies measurements are useful for the diagnosis of hyperketotic states, i.e. ketolysis defects or some RC disorders. The simultaneous measurement of blood glucose, free fatty acids and ketone bodies is necessary for the diagnostic and therapeutic evaluation of hypoketotic states, i.e disorders of fatty acid oxidation (FAO) or ketogenesis (Chap. 13 and 14); data must be interpreted with regard to age and length of fasting (▶ below Fasting Test and also Fig.1.2).

4. The lactate/pyruvate ratio (L/P) and the 3-hydroxybutyrate/acetoacetate ratio (3OHB/AcAc), reflect the redox states of the cytoplasm and the mitochondrion, respectively, and may provide additional information [2] as follows:

- L/P increased, 3OHB/AcAc normal or decreased: pyruvate carboxylase (PC) deficiency or 3-ketoglutarate dehydrogenase deficiency.
- L/P and 3OHB/AcAc both increased with persistent hyperlactatemia: RC disorders.
- L/P normal or low and 3OHB/AcAc normal, with varying hyperlactatemia: PDH deficiency, pyruvate carrier defect.

The usual metabolic abnormalities observed in lactic acidosis due to inborn errors of metabolism are summarized in ◘ Table 3.2 (▶ also ◘ Table 1.10).

3.2.2 Fasting Test

Indications

This test [3] has been used for the clarification of hypoglycaemia observed in disorders of gluconeogenesis, fatty acid oxidation and ketogenesis, ketolysis and in some endocrinopathies. However, as it can be a highly dangerous procedure, its indications are now restricted to unexplained hypoglycaemia when basal metabolic investigations (organic acids analysis, acylcarnitines profile and enzymatic studies) have ruled out a FAO disorder or as a means to assess fasting tolerance during the treatment of certain disorders.

Procedures

The fasting test should only be performed in a specialized metabolic unit and under close medical supervision. The results of the basal investigations should be known prior to planning the test. If permanent abnormalities exist, the diagnostic work-up should be changed accordingly. During the three days before the test the patient should be adequately fed and the energy intake appropriate for his age. No intercurrent infection or metabolic incident should have occurred during the preceding week.

Fasting tolerance differs considerably depending on the age of the patient and on the disorder. The recommended period of fasting is as follows: 12 h for children less than 6 months of age, 20 h for those 6–12 months, and 24 h from age one year onwards. The test should be planned to ensure that the final and most important period (during which complications may arise) takes place during the daytime, when the best facilities for close supervision are available.

An indwelling venous catheter with a saline drip is inserted at zero time. The patient is encouraged to drink plain water during fasting. ◘ Table 3.3 gives the time schedule for the laboratory investigations. The main metabolic monitors for continuing the test safely are glucose and HCO_3^- concentrations in blood. Blood for a complete metabolic and endocrine profile is collected at the start of the test and twice before the end. If glucose drops below

◻ **Table 3.2.** Main metabolic abnormalities in lactic acidosis due to inborn errors of metabolism (from [2])

Hyperlactatemia	Gluconeogenesis (G6Pase, FBPase deficiencies)	Glycogenosis type III, VI	PDH deficiency	PC deficiency Hyperlactatemia	KGDH deficiency	Fumarase deficiency	E3 deficiency	Respiratory chain defects
	Maximum during fasting and when hypoglycaemic	Only in fed state	Permanent, maximum in fed state; can be moderate	Permanent	Permanent	Moderate	Permanent	Permanent, maximum in fed state
L/P ratio	<15	<15	<10	>30	15–30	<15	15–30	>20
Ketone bodies	↑at fast or N	only at fast	absent	+	+	N	N	↑ + or N
3OHB/AcAc	N	N	N	↓↓	N or ↓	N	N	↑
Glucose	↓at fast	↓at fast	N	N or ↓	N	N	N	N or ↓
Ammonia	N	N	N	↑	N or ↑	N	N	N
Alanine	↑at fast	N	↑post-prandial	N or ↓	N	N	↑	↑
Glutamine	N	N	N	↓	↑	N	↑	↑
Proline	N	N	N or ↑	↑	N	N	↑	↑
BCAA	N	N	N	↓	N	N	↑	↑
Citrulline	N	N	N	↑	N	N	N	N or ↓
Organic acids in urine	lactate	lactate	lactate, pyruvate	lactate KB	αKG, lactate fumarate	fumarate	branched-chain ketoacids	N or lactate ± Krebs intermediates, methylglutaconic acid

It should be noted that all metabolic abnormalities are highly variable and that many patients affected with respiratory chain defects have no hyperlactatemia. *BCAA,* branched chain amino acids; *FBPase,* fructose-1,6-bisphosphatase; *G6Pase,* glucose-6-phosphatase; *KB,* ketone bodies; *αKG,* α-ketoglutarate; *KGDH,* α-ketoglutarate dehydrogenase; *3OHB/AcAc,* 3-hydroxybutyrate/acetoacetate; *PC,* pyruvate carboxylase; *PDH,* pyruvate dehydrogenase; *N,* normal; ↑ increased; ↓ decreased.

◻ Table 3.3. Fasting-test flow sheet. The duration of the test is adapted to the age of the patient or is determined by the length of time for the onset of spontaneous symptoms (▶ text). A complete sample is taken at the end of the fast if the test is stopped before 24 h

Time (h)	0	8	12	16	20	24
Blood						
Glucose	+	+	+	+	+	+
HCO_3^-	+	+	+	+	+	+
Lactate	+	+	+	+	+	+
3-Hydroxybutyrate	+	+	+		+	+
FFA	+	+	+	+	+	+
Carnitine	+		+		+	+
Acylcarnitines	+	+	+	+	+	+
Amino acids	+		+		+	+
Insulin	+		+		+	+
Cortisol	+					+
ACTH	+					+
Growth hormone	+					+

Urine	0–8		16–24			
Amino acids	+		+			
Organic acids	+		+			

ACTH, adrenocorticotrophic hormone; *FFA*, free fatty-acids.

3.2 mmol/l, glucose and HCO_3^- should then be determined at 30-min intervals. If glucose drops below 2.6 mmol/l and/or HCO_3^- drops below 15 mmol/l, or if neurological symptoms develop, the test should be stopped. At that time, blood is taken for the complete metabolic and endocrine profile. The urine is collected and kept on ice for each 8-h period of the fast and for a further 4-h period after the end. From each 8-h or 4-h collection, a sample of 10 ml should be frozen at –70°C for the determination of lactate, ketone bodies, amino acids and organic acids.

Interpretation

The interpretation of this investigation is difficult and the results must be compared with the normal values for the particular age (◻ Table 3.4).

Blood measurements: the tentative diagnoses are as follows:
- Hyperinsulinemia: glucose <2.8 mmol/l, insulin >3 mU/l, and FFA <0.6 mmol/l, simultaneously. Ketone bodies (KB) remain very low during the fast.
- Fatty-acid oxidation and ketogenesis defects: glucose <2.8 mmol/l, increased FFA and low KB with FFA/KB ratio >2 (normal <1) and glucose/total KB <4.
- Gluconeogenesis defect: glucose <2.8 mmol/l and lactate >3.0 mmol/l simultaneously.
- Ketolysis defect: ketone bodies are already high in the basal state and increase dramatically during the fast, with possible acidosis. Glucose x total KB >10.
- PDH defect: high lactate (L) and pyruvate (P) with normal L/P ratio, L and P decrease during the test.
- Defects of the citric-acid cycle and the respiratory chain: variable levels of lactate and KB. The fasting test is usually not informative for these disorders.
- Adrenal-cortex insufficiency: glucose <2.8 mmol/l and cortisol <250 nmol/l simultaneously. Adrenocorticotrophin hormone (ACTH) deficiency: ACTH <80 pg/l.
- Human growth hormone (hGH) deficiency: glucose <2.8 mmol/l and hGH <10 ng/ml simultaneously.

◻ Table 3.4. Metabolic profiles during fasting tests in children of different ages (from [3]). Normal blood values of hormones at the end of the fast or when the patient is hypoglycaemic, irrespective of age, are: insulin <3 mU/l at a glucose level of <2.8 mmol/l; cortisol >120 ng/ml; adrenocorticotrophic hormone (ACTH) <80 pg/ml; growth hormone >10 ng/ml

	Less than 12 months	1–7 years		7–15 years	
	20 h	20 h	24 h	20 h	24 h
Glucose (mM)	3.5–4.6	2.8–4.3	2.8–3.8	3.8–4.9	3.0–4.3
Lactate (mM)	0.9–1.8	0.5–1.7	0.7–1.6	0.6–0.9	0.4–0.9
FFA (mM)	0.6–1.3	0.9–2.6	1.1–2.8	0.6–1.3	1.0-1.8
KB (mM)	0.6–3.2	1.2–3.7	2.2–5.8	0.1–1.3	0.7–3.7
3OH-B (mM)	0.5–2.3	0.8–2.6	1.7–3.2	<0.1–0.8	0.5–1.3
3OH-B/AcAc	1.9–3.1	2.7–3.3	2.7–3.5	1.3–2.8	1.6–3.1
FFA/KB	0.3–1.4	0.4–1.5	0.4–0.9	0.7–4.6	0.5–2.0
Carnitine (free; μM)	15–26	16–27	11–18	24–46	18–30

AcAc, acetoacetate; *FFA*, free fatty-acids; *3OH-B*, 3-hydroxybutyrate; *KB*, ketone bodies.

Urine measurements: the safest approach is to compare the results of the last period with those of the first.

Complications

Hypoglycaemia, metabolic acidosis, cardiac dysrythmia, cardiomyopathy, organ failure. Fluids and medication must be immediately available in the patient's room.

3.2.3 Glucose Loading Test

Indications

This test is used for elucidation of hypoglycaemia or hyperlactatemia of unknown origin.

Procedures

It should follow a period of fasting of 3–8 h, depending on the patient's usual interval between meals. In the case of previously recorded hypoglycaemia, the test is started at a plasma glucose concentration between 3.3 mmol/l and 2.8 mmol/l. An indwelling venous catheter is inserted 30 min before the expected start of the test and kept patent with a saline drip. A glucose load (2 g/kg with a maximum of 50 g), as a 10% solution in water, is administered orally or through a nasogastric tube over 5–10 min. The blood is sampled from the indwelling venous catheter twice at zero time (just before glucose administration) and every 30 min thereafter until 3–4 h.

All blood samples are assayed for glucose, lactate, pyruvate, 3OHB and AcAc. A urine sample collected just before the test and a second aliquot from a sample collected during the 8 h after glucose administration are tested for lactate, ketone bodies and organic acids.

Interpretation

- Glucose: A short-lived increase followed by a precipitous decrease may be observed in some cases of hyperinsulinism.
- Lactate: A marked decrease from an elevated fasting level occurs in disorders of gluconeogenesis and glucose-6-phosphatase deficiency (GSD type I) [4]. An exaggerated increase from a normal fasting level occurs in other GSDs including glycogen-synthase deficiency. Lactate remains increased or increases even further after glucose administration in PDH deficiency and RC disorders [1, 5]. Any increase in lactate must be compared to control values. The L/P ratio, normally around 10:1, is usually increased in PC deficiency and in mitochondrial disorders and is normal or low in PDH deficiency and in mitochondrial pyruvate carrier defect.
- Ketone bodies: Ketone bodies may increase paradoxically in PC deficiency (with a low 3OHB/AcAc ratio) and in RC disorders (with a high 3OHB/AcAc ratio). Fasting ketone bodies are very low in hyperinsulinism.

Complications

In patients with PDH deficiency, a glucose load may precipitate lactic acidosis. The test should be stopped if plasma glucose drops below 2.6 mmol/l. The complete metabolic profile should be taken at that time.

3.2.4 Galactose Loading Test

Indications

This test has been used to screen for GSD (except in patients presumed to have GSD type I, for whom the test is, in effect, a continuation of the fasting state [4]). It should never be used in patients with galactosuria, since galactose administration is highly toxic in patients with galactosemias.

Procedures

The preparation is similar to that for the glucose loading test. Galactose (2 g/kg, with a maximum of 50 g) as a 10% solution in water is administered orally over 5–10 min. Blood is sampled twice at zero time and every 30 min until 3–4 h after the completion of galactose ingestion. All blood samples are assayed for glucose and lactate.

Interpretation

Serum lactate rises above 3.5 mmol/l and up to 10 mmol/l in GSD type III, VI and IX [4].

3.2.5 Fructose Loading Test

Indications

This test is used for the differentiation of disorders of fructose metabolism including fructose-1,6-bisphosphatase deficiency. Its use has becomes rare, now that hereditary fructose intolerance (HFI) can usually be diagnosed on the basis of molecular analysis.

Procedures

A diet devoid of fructose and sucrose is prescribed for 2 weeks before the test. Blood samples are collected at zero time (twice) and at 5, 10, 15, 30, 45, 60 and 90 min after fructose administration. Fructose 0.2 g/kg as a 10% solution in water is injected intravenously over 2 min. Blood samples are assayed for glucose, fructose, phosphate, magnesium and urate. The intravenous route is used as oral administration of fructose is accompanied by severe gastrointestinal side effects in HFI.

Interpretation

In HFI glucose and phosphate decrease within 10–20 min and magnesium and urate increase in parallel [6]. In fructose-1,6-bisphosphatase deficiency, these changes tend to be less pronounced.

Note of Caution

When starting the fructose test, the plasma glucose level should not be less than 3.3 mmol/l in view of a potential hypoglycaemic effect of fructose. The test is toxic in cases of impaired liver function.

3.2.6 Protein and Allopurinol Loading Test

Indications

These tests have been used for the detection of late-onset forms of ornithine transcarbamylase deficiency (OTC). They may be indicated in a patient with intermittent clinical signs suggesting of OTC deficiency when no samples have been taken during the acute episode and when basal investigations (ammonia levels, plasma amino acids, and urinary orotic acid) are normal. They may also be used to detect heterozygotes in a family of an affected patient [7]. This latter indication has now been superseded by molecular analysis.

Procedures

Three procedures have been used:
1. Acute protein-loading (1 g/kg) test.
2. Allopurinol loading test [8]: 100 mg in children <6 years, 200 mg in children aged 6 to 12 years, 300 mg in patients >12 years.
3. Combined protein + allopurinol loading test.

The patient should avoid caffeine, tea, chocolate or cola during the day before the test. After an overnight fast, an oral protein load (1 g/kg) is given. An indwelling venous catheter is inserted for the collection of blood samples at zero time and 1, 2 and 4 h after the protein load. Ammonia and amino acids are measured. If allopurinol is added, urine is collected in 5 fractions: before the test, 0–6, 6–12, 12–18 and 18–24 h after the challenge. Quantification of orotic acid and orotidine is performed on an aliquot from each collection using a specific method (HPLC).

A fourth procedure may be used in patients where there is a high suspicion of intermittent hyperammonaemia if these simple tests are negative. A protein-load is performed with a high protein diet: 5 g/kg/day for 5 days. This test has significant risks and should only be performed in a metabolic unit where blood ammonia can be measured rapidly (after each meal) and emergency procedures started if the level increases.

Interpretation

These tests are positive if the protein load leads to an increase in blood ammonia or if the protein and/or allopurinol load induces a high excretion of orotic acid. False positive results can occur due to a small increase in orotic acid excretion in normal subjects following allopurinol ingestion. More importantly, false negative tests are not uncommon in patients with partial OTC deficiency [9].

3.2.7 Fat Loading Test

Indications

The fat-loading test is used to investigate whether the in vivo production of ketone bodies from fatty-acid β-oxidation is intact. The additional measurement of certain metabolites in blood and urine enables the enzyme defects at various levels of the chain of fatty-acid β-oxidation to be characterised (Chap. 13). These metabolic markers include several acylcarnitines and organic acids [10]. The test may be contraindicated if a basal disturbance of organic acids or acylcarnitines profile has already been found.

Procedures

Following a normal diet for at least 3 days, the patient is fasted for 8–10 h, depending on his/her fasting tolerance. Sunflower oil or corn oil (2 g/kg; 2 ml/kg) is administered orally or via a nasogastric tube, which is then flushed with water. At zero time and after 1, 2, 3, 4, 5 and 6 h, blood from an indwelling venous catheter is sampled for measurement of triglycerides, FFA, 3OHB, AcAc, blood gases (anion gap) and acylcarnitines. The test is stopped earlier if the glucose level is ≤ 2.6 mmol/l. Urine is collected for assay of dicarboxylic acids in urine 12 h before and after fat loading.

Interpretation

High levels of FFA with steady, very low levels of ketone bodies and with blood glucose falling to hypoglycaemic levels is highly suggestive of an enzyme defect of fatty-acid β-oxidation. However, the test is unreliable in medium-chain and short-chain acyl-CoA-dehydrogenase deficiencies.

Side Effects

Muscle cramps and gastrointestinal symptoms such as abdominal pain, vomiting and diarrhea may occur.

3.2.8 Tetrahydrobiopterin Test

Indications

This test is performed in newborns with hyperphenylalaninemia in order to exclude tetrahydrobiopterin (BH_4) deficiency. BH_4 is the co-factor of phenylalanine hydroxylase (PAH) and its synthesis defects may lead to decreased activity of PAH. These defects represent 1–2% of all patients with hyperphenylalaninemia. This test is also indicated to identify BH_4-sensitivity in patients with primary PAH deficiency (▶ Chap. 17).

Procedures

For detection of BH_4 deficiency, the best procedure is to carry out this test immediately after the detection of hyperphenylalaninemia and before a phenylalanine-restricted diet is initiated. A *simple* BH_4-loading test may be used if the plasma phenylalanine level is more than 400 µmol/l. A *combined* BH_4 and phenylalanine loading is undertaken in newborns with plasma phenylalanine between 120 and 400 µmol/l. This additional phenylalanine is necessary to increase the blood phenylalanine sufficiently in order to detect any effect from the BH_4. Before the test, blood and urine should be sampled for the following assays:

- Plasma, or serum, for phenylalanine and tyrosine (blood sample 1).
- Blood spots on filter paper for dihydropteridine-reductase activity.
- Urine for pterins (urine sample 1).

Simple BH4 Test

BH_4 (20 mg/kg) dissolved in 20-30 ml water is administered in a bottle approximately 30 min before a normal bottle feed. Blood samples 2, 3 and 4 are collected at 4, 8 and 24 h, respectively, after BH_4 loading. Urine sample 2 is collected between 4 and 8 h after BH_4 loading. The plasma and urine samples should be protected against light and heat, and should be immediately frozen.

Combined Phenylalanine-BH$_4$ Test

L-phenylalanine (100 mg/kg) dissolved in fruit juice is administered 3 hours before the BH_4 load. Blood and urine are sampled at the same time intervals after BH_4 loading as used for the simple test.

Interpretation

In patients with defects of pterin metabolism, a marked drop of the blood phenylalanine concentration occurs within 4–8 h after BH_4-load [11].

A slow decrease in phenylalanine, reaching a minimum level at 24 h after the BH_4-load, is considered an indication of BH_4-sensitivity. This is mainly observed in patients with mild PKU or hyperphenylalaninemia. Criteria for the definition of BH_4-sensitivity are not well established and the frequency is highly variable according to the criteria used. For example, 87% of 31 patients with mild PKU and hyperphenylalaninemia were classified as responsive in one study [12] where a 30% fall in blood phenylalanine was considered significant. In another study, only 4% of 87 patients with hyperphenylalaninemia were considered responsive when a 50% fall was used to define a positive response [13].

3.2.9 Exercise Test

Indications

The exercise test is used to identify patients suspected of having a metabolic myopathy. Several methods exist:
1. Semi-ischemic forearm-exercise test [14].
2. Bicycle ergometer test [15].
3. Treadmill test.

The best exercise test for the widest age span is the treadmill test. The exercise test is also suitable for assessing the results of treatment (▶ also Chap. 6).

Procedures

The forearm and the bicycle test are only applicable in adults and older children who are able to squeeze the sphygmomanometer balloon or ride a bicycle. The treadmill test has the advantage that it can be used from the age at which the child is able to walk. All exercise tests should be carried out at a submaximal workload. This is a safeguard to prevent severe complications, such as rhabdomyolysis, myoglobinuric anuria and metabolic acidosis.

The original *forearm test* was an ischemic exercise test. Its successor, the semi-ischemic modification, is much safer. In this version, a sphygmomanometer cuff is inflated around the upper arm to the mean arterial pressure, but not above it. A hand manometer is squeezed by the patient for 2 min (or less) with submaximal exertion (to prevent muscle cramps).

In the *bicycle ergometer test*, the duration of the exercise and a submaximal workload associated with a pulse rate below 150 beats/min for adults or between 150 and 180 beats/min for children are adapted to the patient's condition.

In the *treadmill test*, the speed of the belt and its angle of inclination can be manipulated to a walking velocity of 3–5 km/h and a pulse rate of 150–180 beats/min. Exhaustion arises rapidly in those with myopathies due to defects of glycolysis and in defects of the citric-acid cycle and the respiratory chain. It occurs later in patients with defects of fatty-acid oxidation (after the exhaustion of energy from glycogen via aerobic and anaerobic glycolysis). The interpretation of the results of each exercise test should take this time sequence into consideration. In plasma and urine, the parameters to be compared before, during and after exercise are the following:
- Plasma: lactate, pyruvate, throughout the study. Acylcarnitines, ammonia, creatine kinase (CK) and potassium (K^+) at the start and the end of the test.
- Urine: lactate, organic acids.

Interpretation

Lactate normally rises during muscle contraction reflecting a disturbed equilibrium between its production from glycolysis and its expenditure in the citric-acid cycle. No increase in lactate reflects deficient glycolysis which can be caused by phosphorylase deficiency and other more rare muscle glycolysis defects (▶ Chap. 6). Abnormally high elevations of lactate can be found with mitochondriopathies and muscle AMP deaminase deficiency.

Ammonia normally rises owing to deamination of adenosine monophosphate (AMP) during muscle contraction. There is no increase in ammonia in muscle AMP deaminase deficiency (▶ Chap. 35).

Specific acylcarnitine accumulation can be observed in fatty acid oxidation disorders. An elevation of CK and K$^+$ reflects abnormal myolysis.

3.3   Postmortem Protocol

Since the first description of a post-mortem protocol by Kronick [16], some refinements have become available to enhance the diagnostic value of the original recommendations [17, 18]. In the protocol given below, the time schedule for proper preservation of specimens determines the sequence of the diagnostic procedures.

3.3.1   Cells and Tissues for Enzyme Assays

Liver (minimum 10–20 mg wet weight) and muscle (minimum 20–50 mg wet weight) biopsies are taken by needle puncture or, preferably, by open incision. The tissues are immediately frozen in small plastic cups in liquid nitrogen, followed by storage at –70°C. Part of the liver biopsy should be fixed for histological and electron-microscopic investigation prior to freezing (▶ »Autopsy« below). A total of 20 ml of blood is collected by peripheral or intracardiac puncture in a heparin-coated syringe; 10 ml is transferred to the laboratory for isolation of erythrocytes or white blood cells, and the biochemist is notified. At least 10 ml is conserved for chromosome analysis and DNA extraction (▶ below).

3.3.2   Cells and Tissues for Chromosome and DNA Investigations

Of the 10 ml of fresh heparinized blood collected, 1–2 ml is reserved for chromosome analysis; the remaining 8–9 ml can be used for DNA extraction. Alternatively, blood spots dried on filter paper (as in the Guthrie test) are useful for many investigations and should always be collected. These samples, and paraffin-embedded tissues, can also be used for DNA analysis after polymerase-chain-reaction (PCR) amplifications.

3.3.3   Skin Fibroblasts

At least two biopsies (diameter 3 mm) are taken under sterile conditions as early as possible; one from the forearm, one from the upper leg (fascia lata, ▶ above). Although a delay decreases the chance of successful fibroblast cultivation, fibroblasts may often be cultivated even from biopsies taken many hours after the death. A biopsy may also be taken from the pericardium in case of delayed autopsy. These samples are conserved in culture medium or, alterna-

tively, on sterile gauze wetted in sterile saline and sealed in a sterile tube for one night at room temperature.

3.3.4   Body Fluids for Chemical Investigations

Plasma from the centrifuged blood sample, urine (~10 ml), and cerebrospinal fluid (~4 ml) are immediately frozen at –20°C (◘ Table 3.5). If no urine can be obtained by suprapubic puncture or catheterization, the bladder may be filled with 20 ml of saline solution and diluted urine may be harvested. Alternatively, vitreous humor can also be collected (by intraocular puncture) and frozen. This liquid is comparable to blood plasma with respects to its solubility for organic acids. Recently, bile, readily available at autopsy, has been found to be useful material for the post-mortem assay of acylcarnitines [19].

Many biochemical parameters are impossible to interpret post-mortem due to rapid tissue lysis. These include lactate, ammonia, carnitine (total and free), and amino acids, all of which rapidly increase without any specific significance. In contrast, the acylcarnitine-ester profile, determined from dried blood spots or from bile, may be highly diagnostic for many disorders of fatty-acid oxidation and for organic acidurias.

3.3.5   Imaging

Photographs are made of the whole body and of specific dysmorphic anomalies, if present. Total-body radiographs in anterioposterior and lateral views are performed, as is ultrasound of the skull, thorax and abdomen. In those cases where autopsy is refused, much information may be obtained from post-mortem magnetic resonance imaging (MRI) [20].

3.3.6   Autopsy

The autopsy is important, particularly in undiagnosed patients, where it may give important clues to the underlying disorder. It should be as complete as possible and include the cranium, provided that the parents have given permission. The pathologist freezes fresh samples of liver, spleen, muscle, heart, kidney and brain and conserves important tissues for histology and electron microscopy in buffered formaldehyde (4%) and Karnofski fixative, respectively.

If a complete autopsy is refused, it is important to obtain permission to take photographs, X-rays, blood, urine and CSF samples, skin biopsies, and to do needle biopsies of liver and muscle. A kit containing all the material necessary for collecting and conserving speci-

---

▢ Table 3.5. Collection, processing and storage of blood, urine, and cerebrospinal fluid (CSF) for metabolic and endocrine investigation. The volumes of blood, urine, and CSF are subject to local practice, which must be taken into account

Blood
Hematology: 0.5 ml in EDTA tube
Blood gases: 0.5 ml on heparin-coated syringe (eject air bubble, cap syringe immediately)
Electrolytes, urea, creatinine, urate, total protein, liver function tests: 1–2 ml (centrifuge after clotting)
Glucose: 0.3 ml fluoride-heparin cup (dry heparin and fluoride salts, no solution)
Lactate/pyruvate and 3OHB/AcAc: 1 ml blood (no forcing), mix immediately with 0.5 ml perchloric acid (18% v/v, keep on ice, centrifuge under refrigeration, store supernatant at –20°C)
Ammonia: 0.5 ml in heparin-coated syringe on ice (eject air bubble, cap syringe immediately)
Amino acids: 1–2 ml in EDTA or heparin tube
Carnitine: 1–2 ml in EDTA tube on ice, centrifuge under refrigeration, store at –20°C
Free fatty acids: 0.3 ml in fluoride-heparin cup (dry heparin and fluoride salts, no solution)
Insulin: 1 ml in EDTA tube on ice, centrifuge under refrigeration, store at –20°C
Cortisol and ACTH: 1 ml in plastic, heparin-coated syringe (keep on ice, centrifuge under refrigeration in plastic tube, store at –20°C)
Growth hormone: 1 ml (centrifuge under refrigeration after clotting, store at –20°C)
Glucagon: 3 ml in heparin tube (centrifuge under refrigeration, store in plastic vial at –20°C)

Urine
pH, amino acids, organic acids, ketone bodies, lactate, reducing substances: 5 ml (at least), freeze at –20°C

Cerebrospinal fluid
Cells, protein, glucose: 0.5 ml in plastic tube
Lactate/pyruvate: 1 ml, add to 0.5 ml perchloric acid (18% v/v, keep on ice), centrifuge under refrigeration, store supernatant at –20°C
Amino acids: 0.5 ml in plastic tube
Culture: 1 ml in sterile tube

AcAc, acetoacetate; *ACTH*, adrenocorticotrophic hormone; *EDTA*, ethylenediaminetetraacetic acid; *3OHB*, 3-hydroxybutyrate.

---

mens is highly recommended as a means of ensuring that the post-mortem protocol is completed as fully and as quickly as possible.

References

1. Touati G, Rigal O, Lombes A et al. (1997) In vivo functional investigations of lactic acid in patients with respiratory chain disorders. Arch Dis Child 76:16–21
2. Poggi-Travert F, Martin D, Billette de Villemeur T et al. (1996) Metabolic intermediates in lactic acidosis: compounds, samples and interpretation. J Inherit Metab Dis 19:478–488
3. Bonnefont JP, Specola NB, Vassault A et al. (1990) The fasting test in paediatrics: application to the diagnosis of pathological hypo- and hyperketotic states. Eur J Pediatr 150:80–85
4. Fernandes J, Huying F, Van de Kamer JH (1969) A screening method for liver glycogen diseases. Arch Dis Child 44:311–317
5. Ching-Shiang Chi, Suk-Chun Mak, Wen-Jye Shian, Chao-Huei Chen (1992) Oral glucose lactate stimulation in mitochondrial diseases. Pediatr Neurol 8:445–449
6. Steinmann B, Gitzelmann R (1981) The diagnosis of hereditary fructose intolerance. Helv Paediatr Acta 36:297–316
7. Haan EA, Danks DM, Grimes A, Hoogenraad NJ (1982) Carrier detection in ornithine transcarbamylase deficiency. J Inherit Metab Dis 5:37–40
8. Burlina AB, Ferrari V, Dionisi-Vici C et al. (1992) Allopurinol challenge in children. J Inherit Metab Dis 15:707–712
9. Spada M, Guardamagna O, Rabier D et al. (1994) Recurrent episodes of bizarre behaviour in a boy with ornithine transcarbamylase deficiency: diagnostic failure of protein loading and allopurinol challenge tests. J Pediatr 125:249–251
10. Costa CCG, Tavares de Almeida I, Jakobs C et al. (1999) Dynamic changes of plasma acylcarnitine levels induced by fasting and sunflower oil challenge test in normal children. Pediatr Res 46:440–444
11. Ponzone A, Guardamagna O, Spada M et al. (1993) Differential diagnosis of hyperphenylalaninemia by a combined phenylalanine-tetrahydrobiopterin loading test. Eur J Pediatr 152:665–661
12. Muntau AC, Röschinger MD, Habich M et al. (2002) Tetrahydrobiopterin as an alternative treatment for mild phenylketonuria. N Engl J Med 347:2122–2132
13. Weglage J, Grenzebach A, Teefelen-Heithoff V et al. (2002) Tetrahydrobiopterin responsiveness in a large series of phenylketonuria patients. J Inherit Metab Dis 25:321–322
14. Kono N, Tarui S (1990) The exercise test. In: Fernandes J, Saudubray J-M, Tada K (eds) Inborn metabolic diseases. Diagnosis and treatment. Springer, Berlin Heidelberg New York
15. Kono N, Mineo I, Shimizu T et al. (1986) Increased plasma uric acid after exercise in muscle phosphofructokinase deficiency. Neurology 36:106–108
16. Kronick JB, Scriver CR, Goodyer PR Kaplan PB (1983) A perimortem protocol for suspected genetic disease. Pediatrics 71:960–963
17. Helweg-Larsen K (1993) Postmortem protocol. Acta Paediatr [Suppl] 389:77–79
18. Poggi F, Rabier D, Vassault A et al. (1994) Protocole d'investigations métaboliques dans les maladies héréditaires du métabolisme. Arch Pediatr 1:667–673
19. Rashed MS, Ozand PT, Bennett J et al. (1995) Inborn errors of metabolism diagnosed in sudden death cases by acylcarnitine analysis of postmortem bile. Clin Chem 41:1109–1114
20. Brookes JAS, Hall-Craggs MA, Sams VR, Lees WR (1996) Non-invasive perinatal necropsy by magnetic resonance imaging. Lancet 348:1139–1141

4 Emergency Treatments

Viola Prietsch, Hélène Ogier de Baulny, Jean-Marie Saudubray

As soon as the diagnosis of a metabolic disorder is suspected a plan for the emergency management should be made. As stated in Chap. 1, both the presentation and management mainly depend on the pathophysiology involved. This chapter focuses on the main clinical presentations in neonates and children for those inborn errors of metabolism for which emergency treatment may be life saving and outlines the first steps of such treatment until the definite diagnosis is known. The subsequent management of patients is addressed in the specific chapters.

In neonates, the main clinical presentations are as follows:

1. *Neurological deterioration*: This is the commonest presentation and most often caused by maple syrup urine disease (MSUD), branched-chain organic acidurias (BCOAs), and urea-cycle defects (UCDs). Treatment must be started immediately to avoid severe cerebral sequelae.
2. *Liver failure*: Galactosemia, hereditary fructose intolerance, and tyrosinemia type I are amenable to emergency treatment.
3. *Hypoglycemia:* Blood glucose levels must be corrected immediately. The three groups of disorders usually implicated are hyperinsulinism, glycogen storage disease (GSD), and mitochondrial β-oxidation defects.
4. *Cardiac failure:* In neonates the only treatable disorders are fatty-acid-oxidation defects.
5. *Primary hyperlactatemia:* This is associated with a general lack of cellular energy and may be due to different enzymatic defects. Some patients may benefit from high dose vitamin treatment.
6. *Intractable convulsions:* Vitamin responsiveness (pyridoxine, pyridoxal phosphate, folinic acid, biotin) must be assessed systematically.

In older children all these clinical situations can also arise. Any type of coma or acute psychiatric symptoms can, in particular, be the presenting sign of a metabolic disorder. In addition, children may present with recurrent attacks of unexplained dehydration, abdominal pain, muscle pain and myolysis, or peripheral neuropathy.

Such situations require careful and urgent biochemical investigation. Emergency treatment should be started concurrently and subsequently modified accordingly. Good collaboration with the metabolic laboratory is indispensable. The results of all laboratory investigations relevant to the diagnosis of metabolic disorders for which specific emergency therapy exists should be available within 24 h.

4.1 General Principles

4.1.1 Supportive Care

Many patients, especially newborn infants, will require ventilatory and circulatory support. Most will need rehydration

and correction of electrolyte, calcium and phosphate imbalance, but such treatments, despite their importance, should not delay the start of specific therapeutic measures. Patients with a metabolic crisis frequently suffer from septicaemia, which can result in persistent catabolism and lead to therapeutic failure. Consequently, infections must be prevented, thoroughly investigated for and, if present, treated.

4.1.2 Nutrition

Whatever the disease, nutrition is extremely important; both the method of administration and the composition of feeds must be rapidly considered. Oral nutrition is preferable if the condition and clinical status allows for it. Continuous enteral tube feeding can be temporarily useful in many patients whose initial condition is poor. Total parenteral nutrition (TPN) is the method of choice in those cases where effective enteral nutrition is precluded (such as by intestinal intolerance, high-energy or high-glucose requirements or invasive techniques required for toxin removal).

Little is known about the absolute energy requirements during metabolic decompensation. Theoretically, at least the age-related recommended daily energy should be provided. Some information on energy metabolism in inborn errors of metabolism has been provided by indirect calorimetry [1, 2]. Resting energy expenditure has been shown to increase at least 30–40% during metabolic decompensation [1]. On the other hand, resting energy expenditure in patients with disorders of propionate metabolism in a stable metabolic state has been shown to be reduced to about 80% [2], and physical activity that accounts for 25–30% of total daily energy expenditure in children [1] is minimal in the immobile patient. This might explain the clinical observation that the energy required to promote anabolism is often lower than expected [3]. In particular the anabolic effect of insulin [4] may reduce energy requirements.

4.1.3 Specific Therapies

Specific therapies can be used for certain disorders. These mainly comprise substrates that enable the excretion of ammonia by alternate pathways (► Chap. 20), carnitine and vitamin supplementation and administration of additional specific drugs (► Appendix of Chap. 5). All intensive care units should ensure that these are readily available.

4.1.4 Extracorporeal Procedures for Toxin Removal

For those disorders associated with acute metabolic toxicity, such as BCOAs and UCDs, extracorporeal procedures to remove toxins are necessary when less invasive methods are insufficient. Of the available techniques hemofiltration (HF) and hemodialysis (HD) are far more efficient than exchange transfusion (ET) or peritoneal dialysis (PD).

4.2   Emergency Management of Particular Clinical Presentations

As previously stated, the management depends on the pathophysiology and the main clinical presentation.

4.2.1   Neurological Deterioration

MSUD, BCOAs and UCDs are the most common treatable disorders causing an acute toxic encephalopathy and should always be considered, particularly in a newborn infant who presents with a sepsis-like illness following an asymptomatic period. In caring for neonates with BCOAs and UCDs, there are three main risks: overhydration, cerebral edema and acute protein malnutrition [5–7]. The management of children (late-onset coma) is essentially similar to that of neonates [6, 7].

Supportive Care

Neurological Deterioration with Ketoacidosis

This is the usual presentation in patients with BCOAs including MSUD. In general, in addition to disease specific therapies, patients require supportive care, procedures for the removal of toxins, and high-energy and protein-free nutrition.

From a practical point of view, there are two situations that should be considered:

- Some patients may not initially appear seriously unwell and may have only a mild acidosis (pH > 7.20, HCO_3 > 15), mild to moderate dehydration (<10% of birth-weight), and a normal or slightly raised blood ammonia (<400 µmol/l). Blood glucose, lactate, calcium and cell count are normal. This is frequently the early presentation in *maple syrup urine disease* and in *methylmalonic, propionic and isovaleric acidurias*, when recognised early.
- In other patients, the situation appears more severe. This is especially the case for patients with *organic acidurias* where the diagnosis has been delayed for a few days. They present with severe ketoacidosis (pH <7.10, HCO_3 <10 mEq/l), are seriously dehydrated (> 10% of birth-weight), and may have hyperammonaemia (> 400 µmol/l), mild hyperlactacidaemia (<5mmol/l), hypo- or hyperglycaemia, hypocalcaemia, leukopenia and thrombocytopenia.

Neurological Deterioration with Hyperammonaemia

This is most commonly due to primary disorders of the urea cycle. Affected neonates have an acute neurological deterioration with vasomotor instability, apnoeas, and fits. Biochemically, they have a respiratory alkalosis, with plasma ammonia levels more than 400 µmol/l and often very much higher. All other routine laboratory tests are normal and, in particular, ketonuria is not usually present. As a general rule, the treatment is similar to the previous group. However, newborn infants with UCDs have a very poor outlook and even in those who receive the most aggressive treatment, the majority of survivors will be handicapped [8]. Infants treated prospectively do better, but there may still be significant complications [9, 10]. Thus, the wisdom of starting treatment should be carefully considered. Some children with organic acidurias diagnosed late, may have a similar presentation with severe hyperammonaemia without ketoacidosis [11]. The need for urgent management and unfortunately the poor prognosis, are the same as for UCDs.

Mildly Affected Patients

These neonates should be hydrated over a 24-h period, while a procedure for toxin removal is prepared. Hydration can be performed using a standard 5–10% glucose solution containing 34 mmol/l of Na^+ (2 g/l of NaCl), and 20 mmol/l of K^+ (1.5 g/l of KCl). High-calorie, protein-free nutrition should be started in parallel, using carbohydrates and lipids to provide 100 Kcal/kg/day. Initially, for the 24 to 36-h period needed to test gastric tolerance, parenteral and enteral nutrition are used together. The requirement for toxin removal is dependent on the diagnosis, the levels of metabolites, and the short term clinical course. In order to prevent acute protein malnutrition, the protein-free diet must not be used for more than two days. Once the levels of toxic metabolites have decreased, natural proteins are introduced using measured amounts of infant formula (▶ »Enteral Nutrition«).

Severely Affected Patients

Neonates with severe ketoacidosis present with intracellular dehydration that is often underestimated. In this situation, aggressive rehydration with hypotonic fluids and alkalisation may cause or exacerbate pre-existing cerebral oedema. Therefore, rehydration should be planned over a 48-h period, with an infusion of less than 150 ml/kg/24 h, that contains an average concentration of 70–85 mmol/l of Na^+ (4-5 g/l of NaCl), 30–40 mmol/l of K^+ (2–3 g/l of KCl) and 5% glucose. Acidosis can be partially corrected with i.v. bicarbonate, especially if it does not improve with the first measures of toxin removal. However, aggressive therapy with repeated boluses of i.v. bicarbonate may induce hypernatremia, cerebral oedema, and even cerebral haemorrhage [12–14]. In order to compensate for bicarbonate consumption, sodium bicarbonate may be substituted for one-quarter to one-half of the sodium requirements during the first 6–12 h of rehydration. To prevent precipitation with calcium, the bicarbonate solution should be connected to the infusion line with an Y connector. These supportive measures are applied in parallel with a procedure for toxin removal that, in addition to the dialysis of the toxic organic acids, can compensate for some of the fluid and electrolytic imbalance and allow for nutritional support.

Nutrition

Parenteral Feeding

Total parenteral nutrition (TPN) is the method of choice in infants with severe illness who are at a high risk of gastric intolerance. The amino acid free TPN solution is suitable for the first 48-h; protein must then be added using a commercially available amino acid-solution. NaCl and KCl should be progressively decreased to 2 g/l and 1.5 g/l, respectively. Initially, amino acids are introduced in an amount sufficient to meet the minimal daily requirements, and then titrated according to biochemical monitoring. The method is safe if the amino acid solution is evenly distributed over the whole day [15, 16]. The minimal isoleucine requirement in neonates is at least equal to that of valine however, many i.v. amino acid solutions provide less of the former than the latter. Consequently, when the TPN solution only provides the minimal requirement for L-valine, an additional oral supplementation of L-isoleucine (25–100 mg/day) is often necessary.

Enteral Feeding

As soon as the enteral feed is available the switch from parenteral nutrition is scheduled over a 4-day to 5-day period, using continuous nasogastric tube feeding [3, 17]. In order to ensure gastric tolerance, small volumes are given initially, for example 60 ml/day and then increased every 24 h until the full fluid requirement is met. While enteral feeds are increased, the parenteral infusion rate is decreased reciprocally. Ondansetron (0.15 mg/kg in 15 min i.v., up to 3 times daily) may be tried if there is persistent vomiting. In terms of the formulation of feeds, the first step is to progressively increase the amount of protein given to reach the desired daily requirements using human milk or infant formula. Next, calories are slowly added using either glucose polymer and lipids or a commercially available protein-free powder. Minerals, vitamins and micronutrients are also given. Addition of an amino acid mixture, if necessary, is the final step, since it increases the osmolarity of the solution and can induce diarrhea. However, in MSUD, a branched chain free amino acid mixture is always required. During this process, the volume of water is increased to cover the requirement for age and weight.

In mild decompensation enteral nutrition may be sufficient to result in a rapid clinical and biochemical recovery [17]. In this situation the composition of the enteral formula is initially based on a glucose-lipid mixture. However, to prevent acute protein malnutrition, a protein-free diet should not be used for more than 2 days. The diet should provide 130–150 kcal/kg/day. Micronutrients, osmolarity,

and renal solute load must be assessed in order to provide the recommended dietary allowance (RDA) and prevent diarrhea and dehydration. Depending on the disorder, an appropriate amino acid mixture can be added to cover the protein requirement. The latter is an absolute requirement in maple syrup urine disease [3, 15, 18]. Once the toxic metabolites have normalized, natural proteins are introduced using measured amounts of infant formula. Attention must be paid to both the total protein and essential amino acid requirements. For patients with an inborn error blocking an amino acid catabolic pathway, intake of natural protein and essential amino acids must provide the minimal safe requirements (protein accretion + non-urinary losses), which are 50–60% below the normal requirements (protein accretion + non-urinary losses + urinary losses), and consequently less than the RDA [19]. These minimal requirements represent the basis for initiation of a protein-controlled diet. Next, the natural protein and amino acid intakes are adjusted for growth and according to the specific biochemical control. The final step is transition to appropriate long-term dietary treatment.

Specific Therapies

Enhancing Anabolism: Insulin

Due to its anabolic effect insulin is used to suppress severe catabolism; however, this will only be achieved if dehydration and acidosis are also corrected. Infusion of insulin in high doses (0.2–0.3 IU/kg/h) used in association with large amounts of glucose provided by TPN may be useful [4, 20, 21]. The dose of insulin must be adjusted frequently in order to control glycemia. Sustained normalization of blood glucose levels, which is an indirect marker of effective anabolism, allows for insulin withdrawal. Human growth hormone has been useful in promoting anabolism in a variety of organic acidopathies but is unlikely be effective in the acute situation.

Alternate Pathways

Neurological damage is primarily related to the duration and the severity of hyperammonaemia, consequently ammonia must be removed as rapidly as possible. L-arginine is an essential amino acid in all disorders of the urea cycle (except arginase deficiency) and is administered together with sodium benzoate and/or sodium phenylacetate, the latter providing alternative pathways for nitrogen excretion by conjugation with glycine and glutamine, respectively [22]. Enteral sodium phenylbutyrate is employed to provide a source of phenylacetate. There has been some debate as to whether sodium benzoate or sodium phenylbutyrate/-acetate should be used for detoxification of ammonia before the diagnosis is known in organic acidopathies as there is the theoretical risk of additional intramitochondrial coenzyme A depletion [23, 24]. However, in many metabolic centres these drugs are regularly used (especially propionic aciduria), without apparent adverse effects [11, 25]. In

N-acetylglutamate synthetase deficiency, N-carbamylglutamate has become available as the treatment of choice. It may also be efficacious in hyperammonemia due to N-acetylglutamate synthetase inhibition by acyl-coenzyme A in organic acidurias [26].

L-carnitine is given to compensate for secondary carnitine deficiency caused by urinary excretion of carnitine-bound organic acids [27, 28]. As a rule, L-carnitine supplementation is never contraindicated in these disorders. Only if a long chain fatty acid oxidation defect is suspected should the administration of carnitine be avoided, at least as a bolus, because of the acute accumulation of toxic long-chain acylcarnitines and the risk of a fatal cardiac arrhythmia.

Vitamin Therapy

Megadoses of specific vitamins should be systematically tested in each case of a potentially vitamin-dependent disorder (► Chap. 5). Vitamin responsiveness is more likely in late-onset forms than in those presenting in the newborn period. As the response may be masked by the simultaneous use of other therapies, the trial should be repeated later in a stable metabolic period and compared with in vitro studies.

Biotin is the treatment of choice in both holocarboxylase synthetase and biotinidase deficiency. Hydroxocobalamin should be tried in all cases of methylmalonic aciduria, riboflavin in glutaric aciduria type I and II and thiamine in MSUD [29]. In any severe metabolic decompensation accompanied by insufficient food intake and severe lactic acidemia a trial with thiamine should also be performed [30, 31].

Additional Drugs

In methylmalonic aciduria forced diuresis and alkalinisation of urine with sodium bicarbonate helps to eliminate methylmalonic acid. In propionic and methylmalonic aciduria, metronidazole suppresses intestinal bacterial propionate production. In isovaleric aciduria and methylcrotonyl CoA carboxylase deficiency, glycine can be used in combination with carnitine to promote the excretion of glycine conjugates and is particularly useful for long-term treatment. In the emergency treatment, carnitine alone is adequate and essential to compensate for secondary carnitine deficiency [27, 32].

Extracorporeal Toxin-Removal Procedures

In some cases, the situation deteriorates so rapidly that extracorporeal toxin-removal procedures become necessary. Such treatment should be considered if the ammonia concentration exceeds 400 µmol/l and/or if ammonia levels do not decrease adequately with conservative treatment. This is often the case in multiorgan failure as alternative pathway therapy requires intact hepatic and renal function for the formation and excretion of conjugates. In all case of neonatal hyperammonemic coma, the dialysis team should

be informed immediately. MSUD may require extracorporeal detoxification if leucine levels exceed 20 mg/dl (1500 µmol/l).

The choice of the technique is highly influenced by local facilities and experience. Continuous venovenous hemodialysis (HD) [8, 33–35], hemofiltration (HF) [36, 38] and hemodiafiltration (HDF) [34] have been shown to be more effective than exchange transfusion and peritoneal dialysis. Extracorporeal membrane oxygenation has been used in driving HD and HF [39]. If such management is unavailable locally, the patient should be transferred to a specialist centre. The advantages and disadvantages of the respective techniques in the emergency treatment of various acute metabolic disorders are as follows:

Exchange Transfusion

ET has only a transient effect and it should only be used in combination with other methods or when it can be undertaken repeatedly or continuously [40–42].

Peritoneal Dialysis

Manual PD requires minimal technical expertise, can be rapidly initiated in any pediatric intensive care unit and can be effective in newborns [40, 43]. The main cause of failure is poor splanchnic blood flow secondary to shock and septicemia. PD appears to be far less effective in older children due to a lower peritoneal area relative to body weight.

Continuous Hemofiltration

HF consists of a low-resistance extracorporeal circuit connected to a small-fiber hemofilter that is permeable to water and non-protein-bound small solutes [44]. The ultrafiltrate of plasma is concurrently replaced by an electrolyte and TPN solution. HDF increases solute removal by the addition of diffusive transport from a dialysis solution flowing upstream through the ultrafiltrate compartment of the hemofilter [45]. The advantages of HF and NDF are logistical simplicity, good tolerance in neonates or infants who present with hemodynamic instability, multiorgan failure, and a hypercatabolic state, and the ability to use a large volume of TPN without the risk of overhydration. Nevertheless, these procedures require a pediatric intensive care unit with staff trained in the techniques of extracorporeal circulation [34, 36–38, 46, 47].

Hemodialysis

HD is a very effective and rapid method of removing small solutes [8, 35, 41, 48]. However, multiple dialysis sessions are most often necessary due to a rebound in the circulation of toxic metabolites. In addition, clearance is hampered by vascular instability [33, 41, 45].

Assessment of Biochemical Progress

In order to evaluate the efficiency of toxin removal it is necessary to undertake regular biochemical monitoring in blood, urine and dialysate or ultrafiltrate within set timed periods. Blood glucose, plasma electrolytes, and calcium should be corrected when necessary. Regular blood cell counts are also important since, in the organic acidurias, neutropenia and thrombocytopenia may be present or may develop after the initiation of therapy and may require specific treatment. Repeated assessments for septicemia must be undertaken and treatment initiated as soon as there is any suspicion of infection.

4.2.2 Liver Failure

Liver failure is a predominant finding in galactosemia, hereditary fructose intolerance (HFI) and tyrosinemia type I and requires urgent and specific treatment. Neonatal and late-onset forms of these disorders may present with acute deterioration, vomiting, seizures, dehydration, hypoglycemia, liver failure and tubulopathy. A number of abnormalities are associated with advanced liver disease including mellituria, hyperammonemia, hyperlactatemia, hypoglycemia, hypertyrosinemia and hypermethionemia. Tyrosinemia type I rarely presents before the third week of life. Galactosemia usually presents in the newborn period but HFI should not become manifest until after weaning since fructose is not normally part of infant formulas. As soon as these disorders are considered, galactose, fructose and protein must be excluded from the diet (with normal intake of all other nutrients) while waiting for confirmation of the diagnosis. When galactosemia or HFI is confirmed, protein can be reintroduced (▶ also Chap. 7 and 9). When tyrosinemia type I is confirmed, treatment with NTBC and a low-phenylalanine and tyrosine diet must be started to prevent production of toxic metabolites (▶ also Chap. 18).

4.2.3 Neonatal Hypoglycemia

Whatever the cause of hypoglycemia, blood glucose levels must be corrected immediately with a glucose bolus (0.5–1 g/kg) followed by a continuous infusion. However, because pathological metabolites may become normal quickly with therapy, adequate samples for metabolic studies (acylcarnitines, glucose, insulin, free fatty acids and ketone bodies) should be obtained first. Glucose should then be started via a peripheral i.v. line, initially at 150 ml/kg/day of a 10% solution (~10 mg/kg/min). Observation of the patient's glucose requirement to maintain normoglycemia is useful for both diagnosis and management. A glucose supply at a rate equivalent to hepatic glucose production (7–8 mg/kg/min in the newborn) is usually sufficient for disorders such as GSD I and disorders of gluconeogenesis. Patients with congenital hyperinsulinism will require much higher rates (10–20 mg/kg/min).

GSD Type I and Fructose-1,6-Bisphosphatase Deficiency

In these disorders, fasting hypoglycemia is associated with hyperlactatemia and metabolic acidosis. In GSD type III, a moderate hyperlactatemia is observed after glucose administration. As soon as the blood values have returned to normal, continuous enteral feeding is substituted for the glucose infusion. At first a milk-based formula containing maltodextrin as source of carbohydrate is used. Giving a normal energy intake for age in which 50 to 60% of the energy is supplied by carbohydrates, this allows for a glucose infusion of 10–12 mg/kg/min. This diet can subsequently be changed to suit the diagnosis (▶ Chap. 6 and 9).

Neonatal Hyperinsulinism

This disorder presents with recurrent hypoglycemia without ketoacidosis. The newborn requires a continuous supply of glucose that exceeds the capacities of the peripheral i.v. route and continuous enteral feeding. Consequently, central venous catheterization is unavoidable. In cases of persistent hypoglycemia, treatment with glucagon and/or diazoxide can be started. The emergency treatment of neonatal hyperinsulinism is discussed in Chap. 10.

Fatty-Acid-Oxidation Defects

FAO defects cause severe energy deprivation, and can be suspected in both newborns and children who present with fasting hypoglycemia and/or an acute deterioration associated with lethargy, hepatomegaly and liver failure, cardiac dysrhythmia, and high blood creatine-kinase, lactate and uric acid levels. These are serious disorders that may require resuscitation. In order to suppress lipolysis it is at first necessary to give an i.v. solution providing 10–12 mg/kg/min of glucose (120–150 ml/kg/day of a 12–15% glucose solution), preferably in combination with insulin. The initial diet should be fat-free. Medium chain triglycerides (2–3 g/kg/day) can be of advantage in long chain FAO defects as a fuel for the compromised energy metabolism especially in the heart. However, supplementation should be postponed until the exact site of the defect is known. Hypocarnitinemia is usually present. The efficacy and safety of carnitine supplementation is still controversial, except for carnitine transporter defect, where it is life saving. In long chain FAO defects there is a risk of forming toxic acylcarnitines, although severe secondary carnitine deficiency may require cautious oral treatment. In other disorders of FAO there may be benefit from early supplementation of carnitine to compensate for primary or secondary carnitine deficiency and to promote the excretion of fatty acylcarnitine esters (▶ also Chap. 13). However the outcome for patients with MCAD deficiency appears to be excellent without such treatment.

4.2.4 Cardiac Failure

The only treatable disorders presenting with cardiac failure in the neonatal period are the mitochondrial FAO defects associated with cardiomyopathy or conduction abnormalities. In addition to the usual cardiac drugs and symptomatic treatment of cardiac failure, specific emergency treatment are as discussed above (▶ also Chap. 13).

4.2.5 Primary Hyperlactatemia

Whatever the enzyme defect, most newborns with primary hyperlactatemia present with acute ketoacidosis and dehydration requiring supportive care similar to that described for the BCOA. Usually, this treatment is sufficient to reduce the lactate to a level that does not cause a severe metabolic acidosis. In some cases, sustained hyperlactatemia is due to a high glucose infusion and can be corrected by using a 5% or even a 2.5% glucose i.v. solution. Thus, none of these patients require any procedures for toxin removal.

Few strategies are of proven efficacy in congenital lactic acidaemia. A trial should be performed with thiamine (cofactor for the pyruvate dehydrogenase [PDH] complex), riboflavin (cofactor for complex I) and biotin (cofactor for pyruvate carboxylase). Secondary carnitine deficiency is treated with L-carnitine. It is essential to correct metabolic acidosis with sodium bicarbonate, or, if the sodium level exceeds 160 mmol/l, with trometamol. Dichloroacetate (50 mg/kg/day in one or two divided doses), an inhibitor of PDH kinase, can be an effective means to lower lactate accumulation in both PDH and respiratory-chain disorders [49]. It has, however, little effect on the clinical status. In one patient with the French phenotype of pyruvate carboxylase deficiency, the early administration of triheptanoin was life saving [50] (▶ Chaps. 12, 15 and 27).

4.2.6 Intractable Convulsions

When seizures are the preponderant or presenting sign, pyridoxine, pyridoxal phosphate [51], biotin and folinic acid [52] must be systematically tested. Familial hypomagnesaemia with secondary hypocalcaemia should considered and if present treated with enteral magnesium supplementation. Disorders of methyl group transfer (including methylenetetrahydrofolate reductase deficiency and disorders of cobalamin metabolism) may require treatment with hydroxycobalamin, folic acid, pyridoxine, betaine or methionine, depending on the underlying enzymatic defect. GLUT1 deficiency can be treated with a ketogenic diet (▶ Chap. 11).

In suspected metabolic disorders those drugs that may inhibit mitochondrial function should be used only in acute emergencies where no other effective treatment is available. These include the antiepileptic drugs sodium valproate and chloralhydrate.

4.3 Final Considerations

Once the patient is discharged from hospital precautions must be taken to avoid further episodes of decompensation. Parents must be aware of possible causes and be taught to recognise the early signs and when to initiate the first steps of the emergency treatment at home [53]. Every patient should be supplied with an emergency card detailing their particular management scheme to be followed both at home and in the primary care hospital. If there are recurrent episodes of decompensation, the insertion of a gastrostomy and/or a port-a-cath system should be considered.

References

1. Bodamer OA, Hoffmann GF, Visser GH et al. (1997) Assessment of energy expenditure in metabolic disorders. Eur J Pediatr 156 [Suppl 1]:S24–28 (Review)
2. Feillet F, Bodamer OA, Dixon MA, Sequeira S, Leonard JV (2000) Resting energy expenditure in disorders of propionate metabolism. J Pediatr 136:659–663
3. Nyhan WL, Rice-Kelts M, Klein J, Barshop BA (1998) Treatment of the acute crisis in maple syrup urine disease. Arch Pediatr Adolesc Med 152:593–598
4. Wendel U, Langenbeck U, Lombeck I, Bremer HJ (1982) Maple syrup urine disease–therapeutic use of insulin in catabolic states. Eur J Pediatr 139:172–175
5. Ogier de Baulny H (2002) Management and emergency treatments of neonates with a suspicion of an inborn error of metabolism. Semin Neonatol 7:17–26
6. Prietsch V, Lindner M, Zschocke J, Nyhan WL, Hoffmann GF (2002) Emergency management of inherited metabolic diseases. J Inherit Metab Dis 25:531–546
7. Ogier de Baulny H, Benoist JF, Rigal O et al. (2005) Methylmalonic and propionic acidemias: management and outcome. J Inherit Metab Dis 28:415–423
8. Picca S, Dionisi-Vici C, Abeni D et al. (2001) Extracorporeal dialysis in neonatal hyperammonemia: modalities and prognostic indicators. Pediatr Nephrol 16:862–867
9. Maestri NE, Hauser ER, Bartholomew R, Brusilow SW (1991) Prospective treatment of urea cycle disorders. J Pediatr 119:923–928
10. Brusilow SW, Maestri NE (1996) Urea cycle disorders: diagnosis, physiopathology, and therapy. Adv Pediatr 43:127–170
11. Walter JH, Wraith JE, Cleary MA (1995) Absence of acidosis in the initial presentation of propionic acidaemia. Arch Dis Child Fetal Neonatal Ed 72(3):F197–199
12. Dave P, Curless R, Steinman L (1984) Cerebellar hemorrhage complicating methylmalonic and propionic acidemia. Arch Neurol 41:1293–1296
13. Surtees R, Leonard JV (1989) Acute metabolic encephalopathy: A review of causes, mechanisms and treatment. J Inherit Metab Dis 12[Suppl 1]:42–54
14. Orban T, Mpofu C, Blackensee D (1994) Severe CNS bleeding followed by a good clinical outcome in the acute neonatal form of isovaleric aciduria. J Inherit Metab Dis 17:755–756

15. Berry GT, Heidenreich R, Kaplan P et al. (1991) Branched-chain amino acid-free parenteral nutrition in the treatment of acute metabolic decompensation in patients with maple syrup urine disease. N Engl J Med 324:175–179

16. Khaler SG, Millington DS, Cederbaum SD et al. (1989) Parenteral nutrition in propionic and methylmalonic acidemia. J Pediatr 115:235–241

17. Parini R, Sereni LP, Bagozzi DC et al. (1993) Nasogastric drip feeding as the only treatment of neonatal maple syrup urine disease. Pediatrics 92:280–283

18. Thompson GN, Francis DE, Halliday D (1991) Acute illness in maple syrup urine disease: dynamics of protein metabolism and implications for management. J Pediatr 119:35–41

19. Ruch T, Kerr D (1982) Decreased essential aminoacid requirements without catabolism in phenylketonuria and maple syrup urine disease. Am J Clin Nutr 35:217–228

20. Biggemann B, Zass R, Wendel U (1993) Postoperative metabolic decompensation in maple syrup urine disease is completely prevented by insulin. J Inherit Metab Dis 16:912–913

21. Leonard JV, Umpleby AM, Naughten EM, Boroujerdy MA, Sonksen PH (1983) Leucine turnover in maple syrup urine disease. J Inherit Metab Dis 6 [Suppl 2]:117–118

22. Feillet F, Leonard JV (1998) Alternative pathway therapy for urea cycle disorders. J Inherit Metab Dis 21[Suppl 1]:101–111 (Review)

23. Griffith AD, Cyr DM, Egan SG, Tremblay GC (1989) Inhibition of pyruvate carboxylase by sequestration of coenzyme A with sodium benzoate. Arch Biochem Biophys 15/269:201–207

24. Kalbag SS, Palekar AG (1988) Sodium benzoate inhibits fatty acid oxidation in rat liver: effect on ammonia levels. Biochem Med Metab Biol 40:133–142

25. Petrowski S, Nyhan WL, Reznik V et al. (1987) Pharmacologic amino acid acylation in the acute hyperammonemia of propionic acidemia. J Neurogenet 4:87–96

26. Gebhardt B, Vlaho S, Fischer D et al. (2003) N-carbamylglutamate enhances ammonia detoxification in a patient with decompensated methylmalonic aciduria. Mol Genet Metab 79:303–304

27. Chalmers RA, Roe CR, Stacey TE, Hoppel CL (1984) Urinary excretion of l-carnitine and acylcarnitines by patients with disorders of organic acid metabolism: evidence for secondary insufficiency of l-carnitine. Pediatr Res 18:1325–1328

28. Roe CR, Millington DS, Maltby DA, Bohan TP, Hoppel CL (1984) L-carnitine enhances excretion of propionyl coenzyme A as propionylcarnitine in propionic acidemia. J Clin Invest 73:1785–1788

29. Fernhoff PM, Lubitz D, Danner DJ et al. (1985) Thiamine response in maple syrup urine disease. Pediatr Res 19:1011–1006

30. Matern D, Seydewitz HH, Lehnert W, Niederhoff H, Leititis JU, Brandis M (1996) Primary treatment of propionic acidemia complicated by acute thiamine deficiency. J Pediatr 129:758–760

31. Mayatepek E, Schulze A (1999) Metabolic decompensation and lactic acidosis in propionic acidaemia complicated by thiamine deficiency. J Inherit Metab Dis 22:189–190

32. Roe CR, Millington DS, Maltby DA, Kahler SG, Bohan TP (1984) L-carnitine therapy in isovaleric acidemia. J Clin Invest 74:2290–2295

33. Schaefer F, Straube E, Oh J, Mehls O, Mayatepek E (1999) Dialysis in neonates with inborn errors of metabolism. Nephrol Dial Transplant 14:910–918

34. Jouvet P, Poggi F, Rabier D et al. (1997) Continuous venovenous haemodiafiltration in the acute phase of neonatal maple syrup urine disease. J Inherit Metab Dis 20:463–472

35. Rutledge SL, Havens PL, Haymond MW et al. (1990) Neonatal hemodialysis: effective therapy for the encephalopathy of inborn errors of metabolism. J Pediatr 116:125–128

36. Ring E, Zobel G, Stoeckler S (1990) Clearance of toxic metabolites during therapy for inborn errors of metabolism. J Pediatr 117:349–350

37. Falk MC, Knight JF, Roy LP et al. (1994) Continuous venovenous haemofiltration in the acute treatment of inborn errors of metabolism. Pediatr Nephrol 8:330–333

38. Thompson GN, Butt WW, Shann FA et al. (1991) Continuous venovenous hemofiltration in the management of acute decompensation in inborn errors of metabolism. J Pediatr 118:879–884

39. Summar M, Pietsch J, Deshpande J, Schulman G (1996) Effective hemodialysis and hemofiltration driven by an extracorporeal membrane oxygenation pump in infants with hyperammonemia. J Pediatr 128:379–382

40. Saudubray JM, Ogier H, Charpentier C et al. (1984) Neonatal management of organic acidurias clinical update. J Inherit Metab Dis 7:2–9

41. Donn SM, Swartz RD, Thoene JG (1979) Comparison of exchange transfusion, peritoneal dialysis, and hemodialysis for the treatment of hyperammonemia in an anuric newborn infant. J Pediatr 95:67–70

42. Wendel U, Langenbeck U, Lombeck I, Bremer HJ (1982) Exchange transfusion in acute episodes of maple syrup urine disease: studies on branched-chain amino and keto acids. Eur J Pediatr 138:293–296

43. Goertner L, Leupold D, Pohlandt F, Bartmann P (1989) Peritoneal dialysis in the treatment of metabolic crises caused by inherited disorders of organic and amino acid metabolism. Acta Pediatr Scand 78:70–6711

44. Alexander SR (1990) Continuous arteriovenous hemofiltration. In: Levin DL, Morris FC (eds) Essential of pediatric intensive care. Quality Medical, St Louis, Missouri, pp 1022–1048

45. Gouyon JB, Desgres J, Mousson C (1994) Removal of branched-chain amino acids by peritoneal dialysis, continuous arteriovenous hemofiltration, and continuous arteriovenous hemodialysis in rabbits: implications for maple syrup urine disease treatment. Pediatr Res 35:357–361

46. Casadevall I, Ogier H, Germain JF et al. (1992) Hemofiltration arterioveineuse continue: prise en charge d'un cas de leucinose nonatale. Arch Fr Pediatr 49:803–805

47. Sperl W, Geiger R, Maurer H et al. (1992) Continuous arteriovenous haemofiltration in a neonate with hyperammonaemic coma due to citrullinemia. J Inherit Metab Dis 15:158–159

48. Wiegand C, Thompson T, Bock GH, Mathis RK (1980) The management of life -threatening hyperammonemia: a comparison of several therapeutic modalities. J Pediatr 96:142–144

49. Stacpoole PW, Barnes CL, Hurbanis MD, Cannon SL, Kerr DS (1997) Treatment of congenital lactic acidosis with dichloroacetate. Curr Top Arch Dis Child 77:535–541

50. Mochel F, de Lonlay P, Touati G et al. (2005) Pyruvate carboxylase deficiency: clinical and biochemical response to anaplerotic diet therapy. Mol Genet Metab 84: 305–312

51. Mills PB, Surtees RA, Champion MP et al. (2005) Neonatal epileptic encephalopathy caused by mutations in the PNPO gene encoding pyridox(am)ine 5'-phosphate oxidase. Hum Mol Genet 14:1077–1086

52. Torres OA, Miller VS, Buist NM, Hyland K (1999) Folinic acid-responsive neonatal seizures. J Child Neurol 14:529–532

53. Dixon MA, Leonard JV (1992) Intercurrent illness in inborn errors of intermediary metabolism. Arch Dis Child 67:1387–1391

5 Treatment: Present Status and New Trends

John H. Walter, J. Ed Wraith

5.1   Introduction

Improvements in understanding of the biochemical and molecular basis of inborn errors have led to significant developments in our ability to treat many of these disorders. Such improvements, coupled with an ability to make more rapid diagnoses and advances in general medical care, particularly intensive care, are resulting in better long term prognosis for many patients. However the rarity of individual disorders has often made it difficult, or impossible, to obtain sufficient data for assessment of treatments that would be considered evidence based. This should be kept in mind when considering the efficacy of particular therapies. Anecdotal reports of improvements should be reviewed critically but equally it is important to remain open to new advances. This chapter discusses recent progress in the development of treatments. We have also included a list of medications, with recommended dosages, that are currently used in the treatment of inborn errors. Readers should refer to the relevant chapter for detailed information as to the management of specific disorders.

The clinical phenotype of most inborn errors is caused by the accumulation of substrate or other related metabolites or deficiency of products of the affected pathway. Although there is some overlap, treatments that are aimed at ameliorating these derangements can be broadly classified as follows:
1. Reducing the load on the affected pathway (substrate restriction)
2. Correcting product deficiency
3. Decreasing metabolite toxicity
4. Stimulating residual enzyme
5. Enzyme replacement

5.2   Reducing the Load on the Affected Pathway

5.2.1   Substrate Reduction by Dietary Restriction

Restrictive diets are the treatment of choice for a number of inborn errors (☐ Table 5.1). Such diets are highly efficacious in phenylketonuria (PKU), maple syrup urine disease (MSUD) and homocystinuria, disorders in which the defective enzyme's substrate can be effectively limited in the diet and in which substrate levels in the body can be monitored. Dietary therapy is less successful in disorders in which the defect is further downstream in the affected pathway – for example propionic and methylmalonic acidaemia and disorders of the urea cycle. Improvements in the understanding of basic human nutritional requirements, food technology and of the biochemical abnormalities in specific disorders have led to continued development. This is exemplified by PKU. The relative frequency of this disorder

in the developed world and the recommendation that diet should be continued into adulthood have made it commercially viable for specialist food manufacturers to invest in the necessary research. There have been some improvements in the palatability of aminoacid supplements, and in the range of available products. The need for dietary flexibility, particularly important for adolescents and adults, has led to the development of new products. Products have also been reformulated to increase the content of various minerals and trace elements, such as selenium, that have been recognised to be low in individuals on semi-synthetic diets [1]. Clearly, though, there is some way to go before these diets can be considered attractive.

Limiting the availability of ingested substrate for absorption by the gut is a further method for reducing substrate. Examples include the treatment of Wilson disease with zinc and the use of ezetimibe in familial hypercholesterolaemia. A more novel approach is under investigation for treatment for PKU using microencapsulated phenylalanine ammonia-lyase [2].

As the biochemical basis of various disorders has been determined a theoretical reason for dietary therapy has become evident in some. The efficacy of such treatments varies, for example the use of medium chain- triglycerides as a fat source in patients with very long chain acyl-CoA dehydrogenase deficiency and long chain hydroxy acyl-CoA dehydrogenase deficiency is generally accepted whereas the use of a low fat, high carbohydrate diet in medium chain acyl-CoA dehydrogenase deficiency is unnecessary [3–5]. Other disorders in which dietary therapy has been attempted but which has been found to be of limited benefit include a high cholesterol diet in Smith-Lemli-Opitz (SLO) syndrome [6], and Lorenzo's oil in X-linked adrenoleucodystrophy (ALD) [7, 8].

5.2.2   Substrate Reduction by Inhibition of Enzymes Within the Pathway

The use of NTBC in tyrosinaemia type 1 demonstrates a novel approach to inherited metabolic disease. Inhibition of 4-hydroxyphenylpyruvate dioxygenase, an enzyme proximal to fumarylacetoacetase, by NTBC, prevents the production of maleylacetoacetate, fumarylacetoacetate and succinylacetone, compounds that are the major toxic agents in this disease.

In lysosomal storage disorders (LSD), reducing the rate of production of macromolecules that normally have to be degraded inside these organelles, allows the residual activities of the patient's defective lysosome system to dispose of the toxic molecules that have already accumulated. To be effective, some residual enzyme activity has to be present and so one would presume that this approach would be more beneficial in later onset forms of LSD. Small molecule inhibitors of ceramide glucosyltransferase which catalyses

◻ **Table 5.1.** Dietary therapy

Disorder	Basis of diet	Efficacy of dietary therapy alone
Substrate restriction therapy		
Fat oxidation disorders (long chain)	Long chain fat restriction	+++
Familial hypercholesterolemia	Low saturated fat	+++
Galactosaemia	Galactose free	++++ (on liver, kidney, eyes) + (on brain, ovarian functions)
Glutaric aciduria type 1	Lysine restricted	+++
Hereditary fructose intolerance	Fructose free	+++++
Homocystinuria	Methionine restricted	++++
Lipoprotein lipase deficiency	Low saturated fat	+++
Maple syrup urine disease	Leucine, isoleucine and valine restricted	++++
Ornithine aminotransferase deficiency	Arginine restricted	+++
Organic acidaemia	Protein restricted	+
Pyruvate dehydrogenase deficiency	Low CHO	+
Phenylketonuria	Phe restricted	+++++
Refsum's disease	Phytanic acid restriction	++
Tyrosinaemia 1	Phe and tyr restriction	++
Urea cycle disorders	Protein restricted	+
Replenishing depleted products		
Glycogen storage disease	CHO enriched	+++
Providing alternative substrates		
GLUT1 deficiency	Ketogenic diet	+++

+ minimal benefit to +++++ complete or almost complete resolution of disease related problems.

the first step in glycosphingolipid biosynthesis have been developed, undergone trials in various animal models and one product, the imino sugar N-butyldeoxynojirimycin (NB-DNJ), has now entered clinical practice as Miglustat (Zavesca, Actelion). This drug which is active orally has approval for the treatment of Gaucher disease in patients unsuitable for enzyme replacement therapy [9]. As the drug is able to penetrate the blood brain barrier (unlike enzyme replacement therapy) a number of other possible therapeutic uses are being studied both in animal as well as human clinical trials.

5.3   Correcting Product Deficiency

5.3.1 Replenishing Depleted Products

Where the deficiency of an enzyme's product is important in the aetiology of clinical illness its replacement may form

the basis of treatment, for example the administration of carbohydrate in glycogen storage disease (GSD), arginine or citrulline in urea cycle disorders and tyrosine in PKU. More recent developments are the use of serine and glycine in 3-phosphoglycerate dehydrogenase deficiency [10] creatine in guanidinoacetate methyltransferase (GAMT) deficiency [11] and neurotransmitters in defects of biopterin synthesis and primary disorders of neurotransmitter metabolism (▶ Chap. 17 and 29). Cholesterol in SLO syndrome is active in increasing cholesterol plasma levels and decreasing 7,8-dehydrocholesterol accumulation but has little clinical effect.

5.3.2 Increasing Substrate Supply

Giving pharmacological amounts of substrate may be effective particularly in inborn errors of membrane transport proteins, for example L-carnitine in carnitine transporter deficiency and ornithine in triple H syndrome. Such therapy is, of course, dependent upon the substrate itself having low toxicity. Similar treatment has been attempted in Menkes disease with copper histidine (with variable effect) [12, 13] and in disorders with single enzyme deficiencies such as mannose in carbohydrate deficient glycoprotein syndrome type 1b (phosphomannose isomerase deficiency) [14].

5.3.3 Providing Alternative Substrates

In GLUT1 deficiency there is a failure of glucose to be transported across the blood brain barrier. The ketogenic diet provides an alternative fuel for the brain [15]. A ketogenic diet has also been used in pyruvate dehydrogenase (PDH) deficiency. Other examples of this form of therapy include medium chain triglycerides (MCT) in disorders of long chain fatty acid oxidation and carnitine transport and folinic acid in cerebral folate transporter disorder [16].

5.4 Decreasing Metabolite Toxicity

5.4.1 Removing Toxic Metabolites

A number of medications are used to expedite removal of normal metabolites that accumulate to toxic levels in particular inborn errors. These include well established treatments such as sodium benzoate and sodium phenylbutyrate in disorders associated with hyperammonaemia, glycine in isovaleric acidaemia and L-carnitine in organic acidaemias. Although there have been no rigorous clinical trials to demonstrate its efficacy, L-carnitine has become an established treatment for organic acidaemias where there is a reduction in plasma carnitine and an increase in the acyl: free L-carnitine ratio [17–19]. The role of L-carnitine in the treatment of medium chain acyl-CoA dehydrogenase deficiency is not established and it is our experience that children do well following diagnosis without this treatment [20]. However it remains possible that L-carnitine may improve exercise tolerance and might have a protective effect on metabolic decompensation[21]. Further studies are still necessary to clarify its role in this disorder. In carnitine transport defect, however, the effect of supplementation is dramatic with a resolution of cardiomyopathy and prevention of further episodes of hypoketotic hypoglycaemia [22]. Animal studies suggest that carnitine may also have a protective role in hyperammonaemia [23].

5.4.2 Blocking the Effects of Toxic Metabolites

Disorders in which the phenotype is related to metabolites binding to receptors may be amenable to treatments that block this effect. For example the use of N-methyl-D-aspartate (NMDA) channel agonists, such as dextromethorphan and ketamine in nonketotic hyperglycinemia (NKH), limit the neuroexcitatory effect on glycine on the NMDA receptor [24–26]. Although there have been reports of successful treatments the variable phenotype has made the efficacy difficult to access. Our experience, with a number of infants presenting with the severe neonatal form of this disorder, has not been encouraging.

5.5 Stimulating Residual Enzyme

5.5.1 Co-Enzyme Treatment

Some metabolic disorders are caused by mutations that affect the metabolism or binding of a co-enzyme or cofactor, necessary for normal enzyme activity. Treatment with the co-enzyme may lead to a complete reversal of the clinical phenotype, for example biotin in biotinidase deficiency. Most disorders with co-enzyme responsive variants show a more limited improvement, for example a majority of patients with vitamin B_{12} responsive methylmalonic acidaemia continue to produce abnormal, albeit smaller, quantities of methylmalonic acid. Other disorders may not be fully correctable because of difficulties in getting the co-enzyme to the appropriate location. This is the case in disorders of biopterin synthesis (GTP cyclohydrolase deficiency, 6 pyruvoyltetrabioterin deficiency) where oral tetrahydrobiopterin (BH_4) rapidly corrects the hyperphenylalaninaemia (HPA) by its effect on liver phenylalanine hydroxylase (PAH). However BH_4 does not easily cross the blood brain barrier and is consequently not available to tyrosine hydroxylase or tryptophan hydroxylase. The profound neurotransmitter deficiency associated with these conditions is therefore not improved by BH_4 therapy.

Many single enzyme disorders are due to mutations that prevent any enzyme production and cannot therefore improve with co-enzyme therapy. For example this is the case with probably all patients with propionic acidaemia and nearly all those with MSUD even though both propionyl CoA carboxylase and α-ketoacid dehydrogenase require a co-enzyme (biotin and thiamine, respectively). Mutations that affect protein folding and subsequently limit co-enzyme binding may be amenable to treatment with pharmaceutical doses of the co-enzyme. This is exemplified by BH_4 therapy which in some of the milder forms of HPA due to PAH deficiency may be an alternative to dietary phe restriction. In the future it may be possible to use pharmacological agents, other than co-enzymes, to rescue of conformationally-defective proteins [27]. Disorders with known co-enzyme responsive variants are listed in ◻ Table 5.2.

The appropriate form of the co-enzyme must be used for the particular disorder. For example, pyridoxal phosphate and not pyridoxine, is effective for pyridox(am)ine

⬛ **Table 5.2.** Disorders for which co-factor responsive variants have been described

Disorder	Co-factor	Therapeutic dose	Frequency of responsive variants
Methylmalonic acidaemia (CblA, CblB)	Hydroxycobalamin	1 mg IM weekly	Some
Biotinidase deficiency	Biotin	5–10 mg/day	All cases
Holocarboxylase synthetase (Multiple carboxylase deficiency)	Biotin	10–40 mg/day	Most
Glutaric aciduria type I	Riboflavin	20–40 mg/day	Rare
Homocystinuria : Classical CBS deficiency CblC MTHF deficiency	Pyridoxine Hydroxycobalamin Folic acid	50–500 mg/day 1 mg IM daily 20 mg/day	Approx. 50% Frequent Rare
Maple syrup urine disease	Thiamine	10–50 mg/day	Rare
Respiratory chain disorders	Ubiquinone	100–300 mg/day	Anecdotal evidence
Propionic acidaemia	Biotin	5–10 mg/day	Possibly never
Hyperphenylalaninaemia due to disorders of biopterin metabolism	Tetrahydrobiopterin	5–20 mg/day	All – but no improvement in CNS neurotransmitter levels
Hyperphenylalaninaemia due to PAH deficiency	Tetrahydrobiopterin	7–20 mg/day	Rare for classical PKU, more common for milder variants
Ornithine aminotransferase deficiency	Pyridoxine	300–600 mg/day	Rare
Pyridox(am)ine 5′-phosphate oxidase deficiency	Pyridoxal phosphate	10 mg/kg of pyridoxal-P 6-hourly	All
B6-responsive seizures	Pyridoxine	5–10 mg/kg/day	All
Cerebral folate deficiency syndrome	Folinic acid	0.5–1mg/kg/day	All
Thiamine responsive megaloblastic anemia	Thiamine	200 mg/day (?)	All

Cbl, cobalamin; *CBS*, cystathionine-β synthase, *PAH*, phenylalanine hydroxylase; *MTHF*, methylene tetrahydrofolate reductase.

5′-phosphate oxidase deficiency [28]; folinic acid, which is accessible to the central nervous system (CNS), rather than folic acid is required for folate responsive disorders affecting the brain; and although the active coenzymes for methylmalonyl CoA mutase and S-adenosylmethyltransferase are adenosylcobalamin and methylcobalamin, respectively, only hydroxycobalamin is effectively transported into cells and consequently used in the treatment of cbl responsive disorders affecting these enzymes.

On occasions it may be difficult to determine whether a particular patient is responsive to co-enzyme therapy. This may arise due to confounding factors that cause a concomitant decrease in a particular biochemical marker (such as the concentration of a particular metabolite) unrelated to the administration of the co-enzyme. For example the patient may have entered a recovery phase following a period of metabolic decompensation or may have recently started a dietary therapy. Standard protocols may be helpful in such situations but further assessment may be required when the patient is more stable.

Most co-enzymes are safe even in large doses and it is appropriate to treat patients, who have disorders that are known to have responsive variants, with the relevant coenzyme. The cases for giving a cocktail of various vitamins in patients before a diagnosis has been made is less strong. Disorders presenting in the newborn period, or soon after, are likely to be due to severe enzyme deficiencies that are not co-enzyme responsive. Furthermore, the majority of clinicians will have access to rapid metabolic investigations by specialist laboratories. However, if there is likely to be a delay in diagnostic investigations the intelligent use of a number of vitamins or co-factors may be indicated (⬛ Table 5.3).

5.5.2 Enzyme Enhancement Therapy

In disorders where the primary genetic defect leads to either protein misfolding or a protein trafficking defect, attempts have been made to rescue the phenotype by the use of

◨ Table 5.3. The vitamin cocktail

Biotin	10 mg/day
Thiamine	200 mg/day
Lipoic acid	100 mg/day
L-carnitine	25 mg/kg 6 hourly
Coenzyme-Q$_{10}$	5 mg/kg/day
Vitamin C	100 mg/kg/day
Riboflavin	100–300 mg/day
Pyridoxine	50–500 mg/day
Pyridoxal phospate	20 mg/kg/day
Folinic acid	20 mg/day

chemical and pharmacological chaperones. This approach to treatment has been labeled enzyme enhancement therapy (EET) [29]. Pharmacological chaperones are likely to be small molecules that are active orally and they may have a different tissue distribution from an enzyme whose delivery will be dependent on receptor mediated uptake. This advantage may also include the ability to cross the blood brain barrier a current weakness with intravenous enzyme replacement therapy (ERT).

Proof of this concept has been demonstrated in the cardiac variant of Fabry disease. Affected patients have residual α-galactosidase activity and a later, milder clinical phenotype compared to classically affected male patients. An affected patient with severe heart disease was rescued from cardiac transplantation by the use of intravenous galactose (1 g/kg) which acted as a competitive inhibitor that could bind to the active site and rescue the mutant enzyme, promoting proper folding and processing and preventing the proteosomal degradation of the mutant enzyme glycopeptides [30].

Certain enzymes may also be stimulated by specific medication. For example dichloroacetate (DCA) increases PDH activity by its inhibitory effect on PDH kinase. An open-label trial of its use in 37 patients with a variety of mitochondrial disorders suggested that it might have a beneficial effect in some patients [31].

5.6 Transplantation

5.6.1 Hematopoietic Stem Cell Transfer

Hematopoetic stem cell transfer (HSCT) readily corrects the enzyme deficiencies associated with lysosomal and peroxisomal disorders at least in cells of hematopoetic origin. An attempt has been made to treat almost all of the lysosomal disorders by this method. Unfortunately, almost all of the reported cases are either anecdotal or consist of very few cases from the same centre. The first disorder treated by this method was mucopolysaccharidosis (MPS) type I (Hurler syndrome) [32] and the greatest clinical experience exists with this disorder. There can be no doubt that a successful bone marrow transplantation (BMT) in MPS I alters the natural history of the disease, but there are considerable residual problems especially with spinal deformity and joint disease [33]. In addition BMT must be performed early (ideally < 18 months) if neurological progression is to be prevented. Whilst the risks of graft versus host disease have lessened with improvements in tissue-typing and drug therapy, primary graft rejection remains a problem in this group of patients. Other cell sources including umbilical cord blood cells are now routinely used and are giving promising results in MPS patients [34].

With other disorders the role of HSCT is less clear-cut and is probably contra-indicated in disorders with very aggressive neurodegeneration e.g. MPS III (Sanfilippo syndrome). There is some evidence that the use of umbilical-cord blood transfer from unrelated donors before the development of symptoms (neonatal period) may favorably alter the natural history of infantile Krabbe's disease [35]. The role of HSCT in metabolic disorders has been reviewed recently [36]. In the future one might expect an increased use of alternative stem cells such as bone marrow derived mesenchymal stem cells (MSCs) and the adjunctive use of ERT in disorders where recombinant enzyme is available, e.g., MPS I to try to improve outcome in these difficult patients.

5.6.2 Other Organ Transplantation

Liver transplantation has been used as a successful therapy for a number of inborn errors of metabolism including urea cycle disorders, organic acidaemias, homozygous familial hypercholesterolaemia and severe forms of GSD [37]. Whilst the indications for transplant in disorders such as Crigler-Najjar syndrome, GSD IV, or fulminant hepatic failure secondary to Wilson disease may be clear-cut, the indications in other inborn errors are not and the decision-making process is often very difficult. The mortality associated with liver transplantation is often high in patients with severe disorders of intermediary metabolism. Furthermore in organic acidaemias transplantation does not necessarily remove the risk of disease related complications occurring [38, 39]. Nevertheless for those conditions that are associated with a poor prognosis with conventional medical therapy liver transplant may be an effective treatment. Hepatocyte transplantation has also been attempted in GSD 1a and in ornithine transcarbamoylase (OTC) deficiency [40, 41].

Renal transplantation has been used for the treatment of end stage renal failure in inborn errors of metabolism such as cystinosis, however, this treatment is of limited value for those disorders where the primary defect resides primarily outside of the kidney. For example in patients with primary hyperoxaluria type I where the disorder is primarily within the liver, the transplanted kidney is exposed to oxalate resulting in a shortened graft survival (▶ Chap. 43). Combined liver-kidney transplantation has been performed in both methylmalonic acidaemia [42] and primary hyperoxaluria type I [43], although, in the latter, a pre-emptive liver transplant from a live, related donor may be most appropriate [44].

5.7 Pharmacologic Enzyme Replacement

Early attempts at pharmacologic enzyme replacement therapy (ERT) were disappointing since the need for targeting the enzymes to the tissue of interest was not appreciated. It was only after placental extracted glucocerebrosidase was modified to uncover the signalling mannose residues that efficient uptake into macrophages and subsequent disease correction was observed. Endocytosis via the mannose and mannose-6-phosphate receptor is pivotal to the success of ERT and commercially produced enzymes are modified in the manufacturing process in an attempt to maximise uptake.

Although not considered in detail in this chapter (▶ Chap. 35), the first successful ERT was with polyethylene glycol-modified adenosine deaminase in adenosine deaminase deficiency [45].

5.7.1 Gaucher Disease

ERT is proving to be the most useful of a number of new therapeutic approaches to treatment specifically for lysosomal storage disease [46]. Gaucher disease (β-glucocerebrosidase deficiency) was the first disorder for which ERT (using macrophage targeted enzyme, Cerezyme, Genzyme) has become the standard therapy to treat non-neurological manifestations of the disease. There are now many years of experience with this product and when initiated in a timely manner and in adequate dosage prevents progressive manifestations of the disease including bone marrow failure, organomegaly and bone crises [47]. Recommendations on diagnosis, evaluation and monitoring have been published for both adults [48] and children [49, 50] with type I disease. ERT is also systemically effective in type III (chronic neuronopathic) Gaucher disease but the dosage required (60–120 U/kg/2 weeks) is often higher than the dose required to obtain a satisfactory outcome in type I patients (15–60 units/kg/2 weeks). A European consensus on management of patients with type III disease has also been published [51]. In patients with acute neuronopathic (type II) Gaucher disease ERT is ineffective and this is one of the weaknesses

of all the currently available ERTs. There is no evidence that any can penetrate the blood brain barrier and treatment is therefore limited to the systemic or non-neurological manifestations of the disorders.

5.7.2 Fabry Disease

For patients with Fabry disease two recombinant enzyme products became available (Replagal, Shire and Fabrazyme, Genzyme), after both demonstrated efficacy in clinical trials [52, 53]. In Fabry disease it is important to recognize and treat patients as early as possible if they are to benefit from ERT as there will be a point in the process of tissue damage beyond which ERT will be unable to rescue the affected organ. Future objectives will also include understanding how the two available products compare and determining the optimal dose necessary to achieve the best clinical outcome.

5.7.3 Mucopolysaccharidosis Type I

Aldurazyme (Genzyme) has been licensed to treat the non-neurological manifestations of MPS type I (α-L-iduronidase deficiency) following a successful clinical trial [54]. In this study affected patients showed an improvement in pulmonary function and endurance following Aldurazyme therapy. The indications for use in patients with severe MPS I (Hurler syndrome) are limited by the product's inability to cross the blood brain barrier but there may be a role as an adjunct to hematopoetic stem cell transfer.

5.7.4 Mucopolysaccharidosis Type VI

Naglazyme (Biomarin) has been approved by both the FDA and the EMEA to treat MPS VI (Maroteaux-Lamy syndrome) following a successful clinical trial [55]. Enzyme treated patients demonstrated improvements in endurance as well as a range of other positive benefits.

5.7.5 Pompe Disease

Myozyme has been approved by the EMEA to treat both infantile and late-onset Pompe disease (GSD II) [56, 57]. In classical infantile patients ERT leads to rapid improvement in cardiomyopathy, but response in skeletal muscle is more variable.

5.7.6 Other Disorders

ERT is at an advanced stage of development for MPS II [58] where phase 3 clinical studies are in progress. Other disorders are at any earlier stage of development or are being explored in animal models. This whole area has been reviewed recently [29].

5.8 Gene Therapy

5.8.1 Gene Transfer

Initial enthusiasm for gene therapy as a potential treatment for single gene disorders has been dampened by a number of important challenges which include the targeting and efficiency of gene transfer as well as the magnitude and duration of subsequent gene expression. A great deal of work however has been achieved and there has been steady progress within the field even though there have been very few successful treatment protocols. Successful therapy for metabolic disorders must combine appropriate disease targeting with an efficient delivery system which ensures long term expression with no toxicity. This has proved to be an elusive goal, but promising results have been obtained in animal studies [59]. More success has been achieved in hematological disorders and gene therapy for adenosine deaminase deficiency has moved from early trials of safety and feasibility to more recent studies reporting on efficacy and clinical benefit [60].

5.8.2 Pharmacological Gene Therapy

Stimulating the expression of endogenous redundant genes by pharmacological agents may provide treatment for some inborn errors. In haemoglobinopathies, the production of fetal haemoglobin may be stimulated by various agents such as hydroxyurea, 5-azacytidine and sodium phenylbutyrate [61–63]. Similarly in cystic fibrosis sodium phenylbutyrate increases CTFR gene expression [64]. Kemp et al. have demonstrated that sodium phenylbutyrate, in cells from patients with X-linked ALD and from X-ALD knockout mice, increased β-oxidation and decreased VLCFAs by enhancing the expression of a peroxisomal membrane ABC transporter protein gene coding for ALD related protein (ALDRP) [65]. Furthermore, dietary treatment of the X-ALD mice with sodium phenylbutyrate led to a substantial decrease in VLCFAs in both the brain and adrenal glands. ALDRP is closely related to ALD protein (ALDP), the product of the X-ALD gene. The ALDRP gene appears to be redundant in normal individuals. Other agents have also been studied [66]. It remains to be seen whether this form of therapy may be useful in the management of human disease.

5.9 Conclusions

Although the outcome for many inborn errors remains poor there have been very encouraging developments in recent years particularly the use of ERT in lysosomal disorders. However formidable obstacles remain particularly for those disorders where there is significant in utero damage or where the CNS is primarily affected.

Appendix. Medications used in the treatment of inborn errors

Medication	Mode of action	Disorders	Recommended dose	Route	Remarks	Chapters
Agalsidase alfa (Replagal)	Recombinant analogue of human α-galactosidase A manufactured by gene activation in human fibroblast cell line	Fabry disease	0.2 mg/kg per 2 weeks	IV		39
Agalsidase beta (Fabrazyme)	Recombinant analogue of human α-galactosidase A manufactured in Chinese Hamster Ovary (CHO) cell line	Fabry disease	1.0 mg/kg per 2 weeks	IV		39
Alglucosidase alfa (Myozyme)	Recombinant analogue of human α-glucosidase manufactured in CHO cell line	Pompe disease (GSD II)	20 mg/kg per 2 weeks	IV		6
Allopurinol	Xanthine-oxidase inhibitor	Disorders leading to hyperuricaemia (PRPP synthetase superactivity; HGPRT deficiency) and APRT deficiency	Initial dosage 10–20 mg/kg per day in children and 2–10 mg/kg per day in adults	Oral		35
Ammonium tetra-thiomolybdate	Chelating agent	Wilson disease	160 mg/day in 6 divided doses	Oral		37
Betaine	Remethylates hct to meth	Classical homocystinuria Remethylation defects	100–150 mg/kg/day in 2–3 divided doses, max dose 6–9 g/day	Oral		21
Biotin	Co-factor for carboxylases Treatment of presumed trans-porter defect [67]	Biotinidase deficiency Multiple carboxylase deficiency Biotin-responsive basal ganglia disease	5–20 mg/day	Oral or IV		27
Chenodeoxycholic acid	Inhibits cholesterol 7α-hydro-xylase (rate limiting rate-limiting enzyme in bile acid biosynthesis)	3β-dehydrogenase def (3βD); Δ⁴-3-Oxosteroid 5β-Reductase Deficiency (3-ORD); cerebrotendinous xanthomatosis (CTX)	3β-D: 12–18 mg/kg/day for 1st 2 months then 9–12 mg/kg/day; 3-ORD: 8 mg/kg/day; CTX: 750 mg/day (adults)	Oral		34
Cholesterol	Replenishes cholesterol	Smith-Lemli-Opitz (SLO) syndrome	20–40 mg/kg/day in 3–4 divided doses	Oral		33
Cholestyramine	Bile acid sequestrant	Familial hypercholesterolaemia	Adults: 12–24 g /day Children: (wt in kg/70 × adult dose) in 4 divided doses	Oral	Possible vitamin A, D, and K deficiency with pro-longed treatment. Other bile acid resins include colestipol and colesevalam	32
Cholic acid		Δ⁴-3-oxosteroid 5β-reductase deficiency (3-ORD)	8 mg/kg/day	Oral		33

Medication	Mode of action	Disorders	Recommended dose	Route	Remarks	Chapters
Copper histidine	Increases intracellular copper	Menkes disease	100–200 µg Cu/day (newborn) 1 mg Cu/day in older children	SC		37
Creatine monohydrate	Replenishes creatine	Guanidinoacetate methyltransferase (GAMT) deficiency; arginine:glycine amidinotransferase (AGAT) deficiency	300–400 mg/kg/day in 3 to 6 divided doses	Oral		16
Cysteamine/phosphocysteamine	Depletes lysosomal cystine	Cystinosis	1.3 g/m²/day of free-base, given every 6 h	Oral and eye drops	Phosphocysteamine more palatable	42
Dextromethorphan	NMDA channel antagonist	NKH	5–7 mg/kg/day in 4 divided doses	Oral	Doses up to 35/mg/day have been used	24
Diazoxide	Inhibits insulin secretion	Persistent hyperinsulinism	15 mg/kg/day (newborn); 10 mg/kg/day (infants), in 3 divided doses	Oral		10
Dichloroacetate	Stimulates PDH activity by inhibiting PDH kinase	Primary lactic acidosis	50mg/kg/day in 3–4 divided doses	Oral	May cause polyneuropathy with prolonged use	12
Entacapone	Prevents the peripheral breakdown of l-dopa	Disorders of BH₄ synthesis	15 mg/kg/day in 2 to 3 divided doses	Oral		17
Ezetimibe	Inhibits cholesterol absorption	Familial hypercholesterolaemia	10 mg/day	Oral		32
Folinic acid	Provides accessible source of folate for CNS	DHPR deficiency, UMP synthase deficiency (hereditary orotic aciduria), methylene synthase deficiency, methionine synthase deficiency, hereditary folate malabsorption, some disorders of cobalamin metabolism, cerebral folate transporter defect	5–15 mg/day	Oral, IV		17, 28
Galsulfase (Naglazyme)	Recombinant analogue of human N-acetylgalactosamine 4 sulfatase manufactured in Chinese Hamster Ovary (CHO) cell line	Mucopolysaccharidosis type VI	1.0 mg/kg per week	IV		39
Gemfibrozil	Fibrates decrease TG levels; other fibrates include bezafibrate, and fenofibrate	Mixed or combined hyperlipidaemia	Adult dose: 1.2 g daily, usually in 2 divided doses; range 0.9–1.5 g daily	Oral	Can cause a myositis-like syndrome, especially with impaired renal function; combination with a statin increases risk rhabdomyolysis	32

Medication	Mode of action	Disorders	Recommended dose	Route	Remarks	Chapters
G-CSF	Stimulates granulocyte production	Neutropenia in GSD Ib, Ic	5 μg/kg once daily	SC		6
Glycine	Forms isovalerylglycine with high renal clearance	Isovaleric acidaemia	150 mg/kg/day in 3 divided doses	Oral	Up to 600 mg/kg/day during decompensation	19
Heme arginate	Inhibits 5-aminolevulinic acid synthase	Acute porphyrias	3–4 mg/kg once daily for 4 days	IV		36
Hydroxycobalamin (vitamin B₁₂)	Co-factor for methylmalonyl mutase	Disorders of cobalamin metabolism	1 mg IM weekly; oral dose 10 mg once or twice daily	IM or oral		19, 28
5-Hydroxytryptophan	Neurotransmitter replacement	Disorders of neurotransmitter synthesis	1–2mg/kg/day increasing gradually to 8–10mg/kg/day in 4 divided doses	Oral	Monitor CSF 5HIAA levels	17, 29
Imiglucerase (Cerezyme)	Recombinant analogue of human β-glucocerebrosidase manufactured in Chinese Hamster Ovary (CHO) line	Gaucher disease	Various regimens: 2.5 U/kg 3 X per week to 60 U/kg per 2 weeks For type III Gaucher disease some clinicians recommend higher dosages: 120 U/kg per 2 weeks	IV		39
Ketamine	N-methyl-D-aspartate (NDMA) channel antagonist	NKH	1–30mg/kg/day in 4 divided doses	Oral or IV		24
L-Arginine	Replenishes arginine; precursor of nitrous oxide	Urea cycle disorders; MELAS [68]	50–170 mg/kg/day (OCT and CPS def) Up to 700 mg/kg/day in AL and AS deficientiy)	Oral or IV	IV loading dose: (200mg/kg) over 90 min	20
Laronidase (Aldurazyme)	Recombinant analogue of human α-L-iduronidase manu-factured in Chinese Hamster Ovary (CHO) line	Mucopolysaccharidosis type I	100 U/kg per week	IV		39
L-carnitine	Replenishes body stores; removes toxic acyl-CoA inter-mediates from within the mitochondria	Primary and secondary carnitine deficiencies	100–200 mg/kg/day	Oral or IV	Do not use racemic mixture	13, 19, 20, 23
L-citrulline	Replenishes citrulline and arginine	Used as an alternative to arginine in CPS def and OCT def; LPI	CPS and OCT deficiency: 170 mg/kg/day or 3.8 gm/m²/day in divided doses, LPI: 100 mg/kg/day in 3–5 doses	Oral		20, 26
L-dopa	Replacement of neurotrans-mitters	Disorders of L-dopa synthesis	1–2 mg/kg/day increasing slowly to 10–12 mg/kg/day in 4 divided doses	Oral	Give as l-dopa/carbidopa (1:10 or 1:5) Monitor CSF HVA levels	17, 29

Medication	Mode of action	Disorders	Recommended dose	Route	Remarks	Chapters
L-lysine-HCl	Allows lysine absorption	Lysinuric protein intolerance	20–30 mg/kg/day in 3 divided doses	Oral		26
L-serine	Replenishes serine	3-phosphoglycerate dehydrogenase deficiency	Up to 600 mg/day in 6 divided doses	Oral		25
L-tryptophan	Increases kynurenic acid which is an endogenous antagonist of the NMDA receptor	NKH	100 mg/kg/day in 3 divided doses	Oral		24
Magnesium	Replenishes Mg	Primary hypomagnesemia with secondary hypocalcemia	0.5–1.5 ml/kg/day MgSO4 10 % solution IV; oral maintenance 0.7–3.5 mmol/kg/day elemental Mg in 3–5 divided doses	IV/oral		37
Mannose	Improves glycosylation	CDG Ib (PMI deficiency)	1 g /kg/day in 5 divided doses	Oral	Not of proven benefit in CDG Ia [69, 70]	41
Mercaptopropionyl-glycine (tiopronin)	Chelating agent	Cystinuria	15–20 mg/kg/day, up to max. of 1000 mg/day in 3 divided doses	Oral		26
Metronidazole	Reduces propionate production by gut bacteria	Propionic and methylmalonic acidaemia	7.5 mg/kg three times a day	Oral		19
Miglustat (Zavesca)	Inhibitor of glucosylceramide synthase, the first enzyme responsible for glycosphingolipid (GSL) synthesis	Gaucher disease	100 mg three times a day	Oral	Only recommended for patients with mild to moderate Gaucher disease who are unsuitable for enzyme replacement therapy	39
N-Carbamoyl-glutamate	Stimulates N-acetylglutamate synthase	N-acetylglutamate synthase deficiency Carbamoylphosphate synthetase deficiency	100–300 mg/kg/day in 4 divided doses	Oral		20
Nicotinamide	Replenishes deficiency state	Hartnup disorder	50–300 mg/day	Oral		26
Nicotinic acid (niacin)	Inhibits the release of free fatty acids from adipose tissue; increases HDL-cholesterol	Hyperlipidaemia (see Chap. 32 for indications)	Adult dose: 100–200 mg 3 times daily, gradually increased over 2–4 weeks to 1–2 g 3 times daily	Oral		32
NTBC (2-[2-nitro-4-tri-fluoro-methylben-zoyl]-1,3-cyclhexane-dione)	Inhibits 4-hydroxyphenylpyru-vate dioxygenase	Tyrosinaemia type I	1 mg/kg/day in 1–2 divided doses	Oral	Combine with low tyr, low phe diet to maintain plasma tyr < 600 µmol/l	18
Octreotide	Somatostatin analogue	Persistent hyperinsulinism	10 µg/day to 60 µg/day, given in 3 or 4 divided doses or by continuous pump	SC		10

Medication	Mode of action	Disorders	Recommended dose	Route	Remarks	Chapters
Panthenic acid	Source of co-enzyme A	Type II 3- methylglutaconic aciduria	15–150 mg/day in three divided doses		See reference [71]	19
Penicillamine	Chelating agent	Wilson disease; cystinuria	Wilson disease: up to 20 mg/kg/day in divided doses (min 500 mg/day); Cystinuria: 2g/1.73m^2	Oral		26, 37
Pyridoxine	Co-factor	Pyridoxine responsive γ cystathionase deficiency; pyridoxine responsive cystathionine β-synthase (CBS) deficiency; pyridoxine dependency with seizures; pyridoxine responsive OAT deficiency; X-linked sideroblastic anaemia; primary hyperoxaluria type 1	50–500 mg/day Pyridoxine dependency with seizures: 100 mg iv with EEG monitoring or 30 mg/kg/day for 7 days (maintenance 5–10 mg/day)	Oral	Peripheral neuropathy can occur with doses >1000 mg daily	21, 22, 29, 36, 43
Pyridoxal-Phosphate	Active co-factor	Pyridox(am)ine 5'-phosphate oxidase deficiency	40mg/kg/day in 4 divided doses	Oral		29
Riboflavin	Co-enzyme	Glutaric aciduria I, mild variants of ETF/ ETF-DH and SCAD; congenital lactic acidosis (complex 1 deficiency)	100 mg/day in 2–3 divided doses	Oral		12, 13, 15
Selegiline (l-deprenyl)	Monoamine-oxidase-B inhibitor	As adjunct to therapy with 5HT and L-dopa in BH$_4$ defects	0.1–0.25mg/day in 3 to 4 divided doses	Oral		17
Statins	HMG-CoA reductase inhibitors	Hyperlipidaemias Simvastatin has been used experimentally in SLO	See Chap. 32 for discussion regarding use of individual statins	Oral		32, 33
Sodium benzoate	Combines with glycine to form hippuric acid which has high renal clearance. Removes nitropen and reduces blood ammonia	Hyperammonaemia	250 mg/day in divided doses or by continuous IV infusion Dose may be doubled if severe hyper-ammonaemia	Oral or IV	IV loading dose: 250 mg/kg over 90 min	20
Sodium phenyl-butyrate	Converted to phenylacetate which combines with glutamine to form phenylglutamine which has high renal clearance	Hyperammonaemia	250–650 mg/kg/day; max. oral dose 20 g/day	Oral or IV		20
Tetrahydrobiopterin (BH$_4$)	Replacement of BH$_4$	Disorders of BH$_4$ synthesis or recycling; BH$_4$ responsive forms of PAH deficiency	1–3 kg/day in BH$_4$ defects; 7–20 mg/kg/day in PAH deficiency	Oral	May be contraindicated in DHPR deficiency	17
Thiamine	Co-factor	Thiamine responsive variants of MSUD, PDH deficiency and complex 1 deficiency	10–15 mg/day	Oral	Doses of up to 300 mg have been used in CLA; 500–2000 mg/day in thiamine responsive PDH?	12, 19

Medication	Mode of action	Disorders	Recommended dose	Route	Remarks	Chapters
Triethylene tetramine (trientine)	Chelating agent	Wilson disease	600 mg/day in divided doses increasing to a maximum of 2.4g/day if necessary	Oral	May reduce serum iron – iron supplements may be necessary	37
Triheptanoin	Anaplerotic substrate	VLCAD deficiency; PC deficiency	To provide 30% of total calories	Oral		12, 13
Ubiquinone (Co-enzyme Q10)		Inborn errors of CoQ₁₀ synthesis [72]	100–300 mg/day	Oral	Has been used in other mitochondrial cytopathies but unproven benefit	15
Uridine	Replenishes UMP	UMP synthase deficiency (hereditary orotic aciduria)	100–150 mg/kg/day in divided doses	Oral		35
Vigabatrin	Irreversible inhibitor of GABA transaminase	Succinic semialdehyde dehydrogenase deficiency	50–100 mg/kg/day in 2 divided doses	Oral	Monitor carefully: increases CSF GABA levels and irreversible visual field deficits possible	29
Vitamin A	Free radical scavenger	Glutathione synthetase deficiency	100 mg/kg/day	Oral		30
Vitamin C	Co-factor; antioxidant	Hawkinsinuria Tyrosinaemia III (4 hydroxyphenylpyruvate dioxygenase deficiency) Transient tyrosinaemia of the newborn Glutathione synthase deficiency	200–1000 mg/day	Oral		18, 31
Vitamin E (alpha tocopherol)	Replenishes vitamin E stores; free radical scavenger	Glutathione synthetase deficiency Abetalipoproteinaemia	10 mg/kg/day 100 mg/kg/day	Oral		30, 32
Zinc sulphate	Increases Zn; impairs Cu absorption	Acrodermatitis enteropathica (AE); Wilson disease	AE: 30–100 mg Zn/day; Wilson disease: 600 mg/day (initial adult dose), 300 mg/day (maintenance adult dose). Give in 3–4 divided doses	Oral		37

5HIAA, 5-hydroxyindoleacetic acid; *AL*, agininosuccinate lyase; *AS*, argininosuccinate synthase; *BH₄* tetrahydrobiopterin; *CBS*, cystathionine-β synthase; *CDG*, congenital defects of glycosylation; *CLA*, congenital lactic acidosis; *CNS*, central nervous system; *CPS*, carbamoyl phosphate synthetase, *G-CSF*, granulocyte colony stimulating factor; *DHA*, docosahexanoeic acid; *DHPR*, dihydropteridine reductase; *GAMT*, guanidinoacetate methyl transferase; *GSD*, glycogen storage disease; *HMG*, 3-hydroxy-3-methylglutaryl; *HVA*, homovanillic; *MMA*, methylmalonic acidaemia; *NKH*, non-ketotic hyperglycinaemia, *NMDA*, N-methyl-D-aspartate; *NTBC*, 2-(2-nitro-4-trifluoromethylbenzoyl)-1,3-cyclohexanedione; *OTC*, ornithine transcarbamoylase; *PDH*, pyruvate dehydrogenase; *PMI*, phosphomannose isomerase; *SLO*, Smith-Lelmli-Opitz; *UMP*, uridine monophosphate

References

1. Acosta PB, Stepnick Gropper S, Clarke Sheehan N et al (1987) Trace element status of PKU children ingesting an elemental diet. J Parenter Enteral Nutr 11:287-292
2. Safos S, Chang TM (1995) Enzyme replacement therapy in ENU2 phenylketonuric mice using oral microencapsulated phenylalanine ammonia-lyase: a preliminary report. Artif Cells Blood Substit Immobil Biotechnol 23:681-692
3. Brown-Harrison MC, Nada MA, Sprecher H et al (1996) Very long chain acyl-CoA dehydrogenase deficiency: successful treatment of acute cardiomyopathy. Biochem Mol Med 58:59-65
4. Pollitt RJ (1995) Disorders of mitochondrial long-chain fatty acid oxidation. J Inherit Metab Dis 18:473-490
5. Morris AA, Clayton PT, Surtees RA et al(1997) Clinical outcomes in long-chain 3-hydroxyacyl-coenzyme A dehydrogenase deficiency. J Pediatr 131:938
6. Irons M, Elias ER, Abuelo D et al (1997) Treatment of Smith-Lemli-Opitz syndrome: results of a multicenter trial. Am J Med Genet 68:311-314
7. Moser HW, Fatemi A, Zackowski K et al (2004) Evaluation of therapy of X-linked adrenoleukodystrophy. Neurochem Res 29:1003-1016
8. Moser HW, Raymond GV, Lu SE et al (2005) Follow-up of 89 asymptomatic patients with adrenoleukodystrophy treated with Lorenzo‹s oil. Arch Neurol 62:1073-1080
9. Cox TM, Aerts JM, Andria G et al (2003) The role of the iminosugar N-butyldeoxynojirimycin (miglustat) in the management of type I (non-neuronopathic) Gaucher disease: a position statement. J Inherit Metab Dis 26:513-526
10. Jaeken J, Detheux M, Van Maldergem L et al (1996) 3-Phosphoglycerate dehydrogenase deficiency: an inborn error of serine biosynthesis. Arch Dis Child 74:542-545
11. Stockler S, Hanefeld F, Frahm J (1996) Creatine replacement therapy in guanidinoacetate methyltransferase deficiency, a novel inborn error of metabolism. Lancet 348:789-790
12. Kreuder J, Otten A, Fuder H et al (1993) Clinical and biochemical consequences of copper-histidine therapy in Menkes disease. Eur J Pediatr 152:828-832
13. Kaler SG, Buist NR, H olmes CS et al (1995) Early copper therapy in classic Menkes disease patients with a novel splicing mutation [► comments]. Ann Neurol 38:921-928
14. Niehues R, Hasilik M, Alton G et al (1998) Carbohydrate-deficient glycoprotein syndrome type Ib. Phosphomannose isomerase deficiency and mannose therapy [see comments]. J Clin Invest 101:1414-1420
15. Klepper J, Diefenbach S, Kohlschutter A et al (2004) Effects of the ketogenic diet in the glucose transporter 1 deficiency syndrome. Prostaglandins Leukot Essent Fatty Acids 70:321-327
16. Ramaekers VT, Rothenberg SP, Sequeira JM et al (2005) Autoantibodies to folate receptors in the cerebral folate deficiency syndrome. N Engl J Med 352:1985-1991
17. Davies SE, Iles RA, Stacey TE et al (1991) Carnitine therapy and metabolism in the disorders of propionyl- CoA metabolism studied using 1H-NMR spectroscopy. Clin Chim Acta 204:263-277
18. De Sousa C, Chalmers RA, Stacey TE et al (1986) The response to L-carnitine and glycine therapy in isovaleric acidaemia. Eur J Pediatr 144:451-456
19. Rutledge SL, Berry GT, Stanley CA et al (1995) Glycine and L-carnitine therapy in 3-methylcrotonyl-CoA carboxylase deficiency. J Inherit Metab Dis 18:299-305
20. Walter JH (1996) L-Carnitine. Arch Dis Child 74:475-478
21. Lee PJ, Harrison EL, Jones MG et al (2005) L-carnitine and exercise tolerance in medium-chain acyl-coenzyme A dehydrogenase (MCAD) deficiency: a pilot study. J Inherit Metab Dis 28:141-152
22. Waber LJ, Valle D, Neill C et al (1982) Carnitine deficiency presenting as familial cardiomyopathy: a treatable defect in carnitine transport. J Pediatr 101:700-705
23. Igisu H, Matsuoka M, Iryo Y (1995) Protection of the brain by carnitine. Sangyo Eiseigaku Zasshi 37:75-82
24. Hamosh A, McDonald JW, Valle D et al (1992) Dextromethorphan and high-dose benzoate therapy for nonketotic hyperglycinemia in an infant [see comments]. J Pediatr 121:131-135
25. Alemzadeh R, Gammeltoft K, Matteson K (1996) Efficacy of low-dose dextromethorphan in the treatment of nonketotic hyperglycinemia. Pediatrics 97(6 Pt 1):924-926
26. Matsuo S, Inoue F, Takeuchi Y et al (1995) Efficacy of tryptophan for the treatment of nonketotic hyperglycinemia: a new therapeutic approach for modulating the N- methyl-D-aspartate receptor. Pediatrics 95:142-146
27. Ulloa-Aguirre A, Janovick JA, Brothers SP et al (2004) Pharmacologic rescue of conformationally-defective proteins: implications for the treatment of human disease. Traffic 5:821-837
28. Mills PB, Surtees RA, Champion MP et al (2005) Neonatal epileptic encephalopathy caused by mutations in the PNPO gene encoding pyridox(am)ine 5'-phosphate oxidase. Hum Mol Genet 14:1077-1086
29. Desnick RJ (2004) Enzyme replacement and enhancement therapies for lysosomal diseases. J Inherit Metab Dis 27:385-410
30. Frustaci A, Chimenti C, Ricci R et al (2001) Improvement in cardiac function in the cardiac variant of Fabry's disease with galactose-infusion therapy. N Engl J Med 345:25-32
31. Barshop BA, Naviaux RK, McGowan KA et al (2004) Chronic treatment of mitochondrial disease patients with dichloroacetate. Mol Genet Metab 83:138-149
32. Hobbs JR, Hugh-Jones K, Barrett AJ et al (1981) Reversal of clinical features of Hurler's disease and biochemical improvement after treatment by bone-marrow transplantation. Lancet 2:709-712
33. Weisstein JS, Delgado E, Steinbach LS et al (2004) Musculoskeletal manifestations of Hurler syndrome: long-term follow-up after bone marrow transplantation. J Pediatr Orthop 24:97-101
34. Staba SL, Escolar ML, Poe M et al (2004) Cord-blood transplants from unrelated donors in patients with Hurler's syndrome. N Engl J Med 350:1960-1969
35. Escolar ML, Poe MD, Provenzale JM et al (2005) Transplantation of umbilical-cord blood in babies with infantile Krabbe's disease. N Engl J Med 352:2069-2081
36. Sauer M, Grewal S, Peters C (2004) Hematopoietic stem cell transplantation for mucopolysaccharidoses and leukodystrophies. Klin Padiatr 216:163-168
37. Kayler LK, Merion RM, Lee S et al (2002) Long-term survival after liver transplantation in children with metabolic disorders. Pediatr Transplant 6:295-300
38. Chakrapani A, Sivakumar P, McKiernan PJ et al (2002) Metabolic stroke in methylmalonic acidemia five years after liver transplantation. J Pediatr 140:261-263
39. Leonard JV, Walter JH, McKiernan PJ (2001) The management of organic acidaemias: the role of transplantation. J Inherit Metab Dis 24:309-311
40. Muraca M, Gerunda G, Neri D et al (2002) Hepatocyte transplantation as a treatment for glycogen storage disease type 1a. Lancet 359:317-318
41. Horslen SP, McCowan TC, Goertzen TC et al (2003) Isolated hepatocyte transplantation in an infant with a severe urea cycle disorder. Pediatrics 111(6 Pt 1):1262-1267
42. Van't Hoff WG, Dixon M, Taylor J et al (1998) Combined liver-kidney transplantation in methylmalonic acidemia. J Pediatr 132:1043-1044

43. Jamieson NV, Watts RW, Evans DB et al (1991) Liver and kidney transplantation in the treatment of primary hyperoxaluria. Transplant Proc 23:1557-1558

44. Gruessner RW (1998) Preemptive liver transplantation from a living related donor for primary hyperoxaluria type I [letter]. N Engl J Med 338:1924

45. Hershfield MS, Buckley RH, Greenberg ML et al (1987) Treatment of adenosine deaminase deficiency with polyethylene glycol-modified adenosine deaminase. N Engl J Med 316:589-596

46. Schiffmann R, Brady RO (2002) New prospects for the treatment of lysosomal storage diseases. Drugs 62:733-742

47. Weinreb NJ, Charrow J, Andersson HC et al (2002) Effectiveness of enzyme replacement therapy in 1028 patients with type 1 Gaucher disease after 2 to 5 years of treatment: a report from the Gaucher Registry. Am J Med 113:112-119

48. Charrow J, Esplin JA, Gribble TJ et al (1998) Gaucher disease: recommendations on diagnosis, evaluation, and monitoring. Arch Intern Med 158:1754-1760

49. Baldellou A, Andria G, Campbell PE et al (2004) Paediatric non-neuronopathic Gaucher disease: recommendations for treatment and monitoring. Eur J Pediatr 163:67-75

50. Grabowski GA, Andria G, Baldellou A et al (2004) Pediatric non-neuronopathic Gaucher disease: presentation, diagnosis and assessment. Consensus statements. Eur J Pediatr 163:58-66

51. Vellodi A, Bembi B, de Villemeur TB et al (2001) Management of neuronopathic Gaucher disease: a European consensus. J Inherit Metab Dis 24:319-327

52. Schiffmann R, Kopp JB, Austin HA, III et al (2001) Enzyme replacement therapy in Fabry disease: a randomized controlled trial. JAMA 285:2743-2749

53. Eng CM, Guffon N, Wilcox WR et al (2001) Safety and efficacy of recombinant human alpha-galactosidase A – replacement therapy in Fabry's disease. N Engl J Med 345:9-16

54. Wraith JE, Clarke LA, Beck M et al (2004) Enzyme replacement therapy for mucopolysaccharidosis I: a randomized, double-blinded, placebo-controlled, multinational study of recombinant human alpha-L-iduronidase (laronidase). J Pediatr 144:581-588

55. Harmatz P, Whitley CB, Waber L et al (2004) Enzyme replacement therapy in mucopolysaccharidosis VI (Maroteaux-Lamy syndrome). J Pediatr 144:574-580

56. Van den Hout JM, Kamphoven JH, Winkel LP et al (2004) Long-term intravenous treatment of Pompe disease with recombinant human alpha-glucosidase from milk. Pediatrics 113:e448-e457

57. Winkel LP, Van den Hout JM, Kamphoven JH et al (2004) Enzyme replacement therapy in late-onset Pompe's disease: a three-year follow-up. Ann Neurol 55:495-502

58. Muenzer J, Calikoglu M, Towle D, McCandless S, Kimura A (2003) The one year experience of enzyme replacement therapy for mucopolysaccharidosis type II (Hunter syndrome). Am J Hum Genet 73:623

59. Mango RL, Xu L, Sands MS et al (2004) Neonatal retroviral vector-mediated hepatic gene therapy reduces bone, joint, and cartilage disease in mucopolysaccharidosis VII mice and dogs. Mol Genet Metab 82:4-19

60. Aiuti A, Ficara F, Cattaneo F et al (2003) Gene therapy for adenosine deaminase deficiency. Curr Opin Allergy Clin Immunol 3:461-466

61. Charache S (1993) Pharmacological modification of hemoglobin F expression in sickle cell anemia: an update on hydroxyurea studies. Experientia 49:126-132

62. Dover GJ, Humphries RK, Moore JG et al (1986) Hydroxyurea induction of hemoglobin F production in sickle cell disease: relationship between cytotoxicity and F cell production. Blood 67:735-738

63. Charache S, Dover G, Smith K et al (1983) Treatment of sickle cell anemia with 5-azacytidine results in increased fetal hemoglobin production and is associated with nonrandom hypomethylation of DNA around the gamma-delta-beta-globin gene complex. Proc Natl Acad Sci U S A 80:4842-4846

64. Rubenstein RC, Egan ME, Zeitlin PL (1997) In vitro pharmacologic restoration of CFTR-mediated chloride transport with sodium 4-phenylbutyrate in cystic fibrosis epithelial cells containing delta F508-CFTR. J Clin Invest 100:2457-2465

65. Kemp S, Wei HM, Lu JF et al (1998) Gene redundancy and pharmacological gene therapy: implications for X-linked adrenoleukodystrophy. Nat Med 4:1261-1268

66. McGuinness MC, Zhang HP, Smith KD (2001) Evaluation of pharmacological induction of fatty acid beta-oxidation in X-linked adrenoleukodystrophy. Mol Genet Metab 74:256-263

67. Zeng WQ, Al Yamani E, Acierno JS, Jr et al (2005) Biotin-responsive basal ganglia disease maps to 2q36.3 and is due to mutations in SLC19A3. Am J Hum Genet 77:16-26

68. Koga Y, Akita Y, Nishioka J et al (2005) L-arginine improves the symptoms of strokelike episodes in MELAS. Neurology 64:710-712

69. Kjaergaard S, Kristiansson B, Stibler H et al (1998) Failure of short-term mannose therapy of patients with carbohydrate-deficient glycoprotein syndrome type 1A. Acta Paediatr 87:884-888

70. Mayatepek E, Schroder M, Kohlmuller D et al (1997) Continuous mannose infusion in carbohydrate-deficient glycoprotein syndrome type I. Acta Paediatr 86:1138-1140

71. Ostman-Smith I, Brown G, Johnson A et al (1994) Dilated cardiomyopathy due to type II X-linked 3-methylglutaconic aciduria: successful treatment with pantothenic acid. Br Heart J 72:349-353

72. Quinzii C, Naini A, Salviatil et al (2006) A mutation in para-hydroxybenzoate-polyprenyl transferase (COQ2) causes primary coenzyme Q10 deficiency. Am I Humbenet 78:345-349

II Disorders of Carbohydrate Metabolism

6 The Glycogen Storage Diseases and Related Disorders

G. Peter A. Smit, Jan Peter Rake, Hasan O. Akman, Salvatore DiMauro

Glycogen Metabolism

Glycogen is a macromolecule composed of glucose units. It is found in all tissues but is most abundant in liver and muscle where it serves as an energy store, providing glucose and glycolytic intermediates (◘ Fig. 6.1). Numerous enzymes intervene in the synthesis and degradation of glycogen which is regulated by hormones.

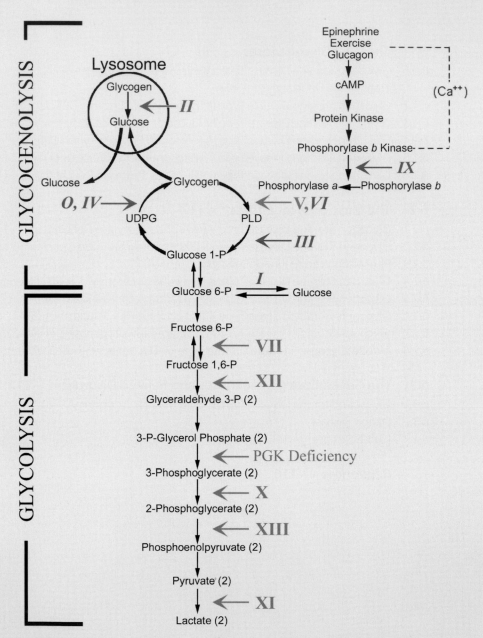

◘ **Fig. 6.1.** Scheme of glycogen metabolism and glycolysis. *PGK,* phosphoglycerate kinase; *P,* phosphate; *PLD,* phosphorylase limit dextrin; *UDPG,* uridine diphosphate glucose. Roman numerals indicate enzymes whose deficiencies cause liver (*italics*) and/or muscle glycogenoses: *0* glycogen synthase, *I,* glucose-6-phosphatase; *II,* acid maltase (α-glucosidase); *III,* debranching enzyme; *IV,* branching enzyme; *V,* myophosphorylase; *VI,* liver phosphorylase; *VII,* phosphofructokinase; *IX,* phosphorylase-b-kinase; *X,* phosphoglycerate mutase; *XI,* lactate dehydrogenase; *XII,* fructose-1,6-bisphosphate aldolase A; *XIII,* β-enolase

The liver glycogen storage disorders (GSDs) comprise GSD I, the hepatic presentations of GSD III, GSD IV, GSD VI, the liver forms of GSD IX, and GSD 0. GSD I, III, VI, and IX present similarly with hypoglycemia, marked hepatomegaly, and growth retardation. GSD I is the most severe affecting both glycogen breakdown and gluconeogenesis. In GSD Ib there is additionally a disorder of neutrophil function. Most patients with GSD III have a syndrome that includes hepatopathy, myopathy, and often cardiomyopathy. GSD VI and GSD IX are the least severe: there is only a mild tendency to fasting hypoglycemia, liver size normalises with age, and patients reach normal adult height. GSD IV manifests in most patients in infancy or childhood as hepatic failure with cirrhosis leading to end-stage liver disease. GSD 0 presents in infancy or early childhood with fasting hypoglycemia and ketosis and, in contrast, with postprandial hyperglycemia and hyperlactatemia. Treatment is primarily dietary and aims to prevent hypoglycemia and suppress secondary metabolic decompensation. This usually requires frequent feeds by day, and in GSD I and in some patients with GSD III, continuous nocturnal gastric feeding.

The muscle glycogenoses fall into two clinical groups. The first comprises GSD V, GSD VII, the muscle forms of GSD IX (VIII according to McKusick), phosphoglycerate kinase deficiency (IX according to McKusick), GSD X, GSD XI, GSD XII and GSD XIII, and is characterised by exercise intolerance with exercise-induced myalgia and cramps, which are often followed by rhabdomyolysis and myoglobinuria; all symptoms are reversible with rest. Disorders in the second group, consisting of the myopathic form of GSD III, and rare neuromuscular forms of GSD IV, manifest as sub-acute or chronic myopathies, with weakness of trunk, limb, and respiratory muscles. Involvement of other organs (erythrocytes, central or peripheral nervous system, heart, liver) is possible, as most of these enzymes defects are not confined to skeletal muscle.

Generalized glycogenoses comprise GSD II, caused by the deficiency of a lysosomal enzyme, and Danon disease due to the deficiency of a lysosomal membrane protein. Recent work on myoclonus epilepsy with Lafora bodies (Lafora disease) suggests that this is a glycogenosis, probably due to abnormal glycogen synthesis. GSD II can be treated by enzyme replacement therapy, but there is no specific treatment for Danon and Lafora disease.

The glycogen storage diseases (GSDs) and related disorders are caused by defects of glycogen degradation, glycolysis and, paradoxically, glycogen synthesis. They are all called glycogenoses, although not all affect glycogen breakdown.

Glycogen, an important energy source, is found in most tissues, but is especially abundant in liver and muscle. In the liver, glycogen serves as a glucose reserve for the maintenance of normoglycemia. In muscle, glycogen provides energy for muscle contraction.

Despite some overlap, the GSDs can be divided in three main groups: those affecting liver, those affecting muscle, and those which are generalized (◘ Table 6.1). GSDs are denoted by a Roman numeral that reflects the historical sequence of their discovery, by the deficient enzyme, or by the name of the author of the first description. The Fanconi-Bickel syndrome is discussed in Chap. 11.

6.1 The Liver Glycogenoses

The liver GSDs comprise GSD I, the hepatic presentations of GSD III, GSD IV, GSD VI, the liver forms of GSD IX, and GSD 0. GSD I, III, VI, and IX present with similar symptoms during infancy, with hypoglycemia, marked hepatomegaly, and retarded growth. GSD I is the most severe of these four conditions because not only is glycogen breakdown impaired, but also gluconeogenesis. Most patients with GSD III have a syndrome that includes hepatopathy, myopathy, and often cardiomyopathy. GSD IV manifests in most patients in infancy or childhood as hepatic failure with cirrhosis leading to end-stage liver disease. GSD VI and the hepatic forms of GSD IX are the mildest forms: there is only a mild tendency to fasting hypoglycemia, liver size normalises with age, and patients reach normal adult height. GSD 0 presents in infancy or early childhood with fasting hypoglycemia and ketosis contrasting with postprandial hyperglycemia and hyperlactatemia. The muscle forms of GSD III and IX are also discussed in this section.

6.1.1 Glycogen Storage Disease Type I (Glucose-6-Phosphatase of Translocase Deficiency)

GSD I, first described by von Gierke, comprises GSD Ia caused by deficiency of the catalytic subunit of glucose-6-phosphatase (G6Pase), and GSD Ib, due to deficiency of the endoplasmic reticulum (ER) glucose-6-phosphate (G6P) translocase. There is controversy about the existence of ER phosphate translocase deficiency (GSD Ic) and ER glucose transporter deficiency (GSD Id) as distinct entities. In this chapter, the term GSD Ib includes all GSD I non-a forms.

Clinical Presentation

A protruded abdomen, truncal obesity, rounded doll face, hypotrophic muscles, and growth delay are conspicuous clinical findings. Profound hypoglycemia and lactic acidosis occur readily and can be elicited by trivial events, such as delayed meals or reduced food intake associated with inter-

◨ Table 6.1. Main features of glycogen storage diseases and related disorders

Type or synonym	Defective enzyme or transporter	Main tissue involved	Main clinical symptoms
Liver			
Ia Von Gierke	Glucose-6-phosphatase	Liver, kidney	Hepatomegaly, short stature, hypoglycemia, lactatemia, hyperlipidemia
Ib (non-a)	Glucose-6-phosphate translocase	Liver, kidney, leucocytes	Same as Ia, neutropenia, infections
III Cori, Forbes	Debranching enzyme and sub-types	Liver, muscle	Hepatomegaly, (cardio)myopathy, short stature, hypo-glycemia
IV Andersen	Branching enzyme	Liver	Hepato(spleno)megaly, liver cirrhosis, rare neuromuscu-lar forms
VI Hers	Liver phosphorylase	Liver	Hepatomegaly, short stature, hypoglycemia
IX	Phosphorylase kinase and subtypes	Liver and/or muscle	Hepatomegaly, short stature (myopathy), hypoglycemia
0	Glycogen synthase	Liver	Hypoglycemia
Muscle			
V Mc Ardle	Myophosphorylase	Muscle	Myalgia, exercise intolerance, weakness
VII Tarui	Phosphofructokinase and variants	Muscle, erythrocytes	Myopathy, hemolytic anemia, multisystem involvement (seizures, cardiopathy)
–	Phosphoglycerate kinase	Muscle, erythrocytes, central nervous system	Exercise intolerance, hemolytic anemia convulsions
X	Phosphoglycerate mutase	Muscle	Exercise intolerance, cramps
XI	Lactate dehydrogenase	Muscle	Exercise intolerance, cramps, skin lesions
XII	Aldolase A	Muscle	Exercise intolerance, cramps
XIII	β-Enolase	Muscle	Exercise intolerance, cramps
Generalized			
II Pompe	Lysosomal α-glucosidase	Generalized in lysosomes	Hypotonia, cardio-myopathy Infantile, juvenile, adult forms
IIb Pseudo Pompe Danon	Lysosomal-associated membrane protein 2	Heart, muscle	Cardio-myopathy
Lafora	Enzyme defect not known	Polyglucosan bodies in all organs	Myoclonic epilepsy, dementia, convulsions

current illnesses. The liver functions are normal and cirrhosis does not develop. In the second or third decade, the liver's surface may become uneven and its consistency much firmer because of the development of adenomas. The kidneys are moderately enlarged. The spleen remains normal in GSD Ia but is enlarged in most patients with GSD Ib. Patients bruise easily due to impaired platelet function, and

nosebleeds may be especially troublesome. Skin xanthomas are seen in patients with severe hypertriglyceridemia, and gouty arthritis in patients with hyperuricemia. Patients may also suffer from episodes of diarrhoea or loose stools.

About one in five GSD I patients has type Ib [1]. Most patients with GSD Ib develop neutropenia before the age of 1 year, a few at an older age, and even fewer are totally spared. Patients with neutropenia show neutrophil dysfunction, including impaired motility and migration and impaired metabolic burst [2], and suffer with frequent and severe infections, which can affect the upper and lower respiratory tract, the skin, the urinary tract, or result in deep abscesses [3]. More than 75% of the GSD Ib patients show symptoms of inflammatory bowel disease (IBD), including peri-oral and peri-anal infections and protracted diarrhoea.

Metabolic Derangement

Among the enzymes involved in hepatic glycogen metabolism, G6Pase is unique since its catalytic site is situated inside the lumen of the ER. This means that its substrate, G6P, must cross the ER membrane and requires a transporter. There is still debate over different proposed models of G6Pase, over the existence of additional transporters for its products, phosphate and glucose [4, 5], and over the existence of GSD Ic (putative ER phosphate/pyrophosphate transporter deficiency), and GSD Id (putative ER glucose transporter deficiency). In particular, patients diagnosed by enzyme studies as GSD Ic have been found to have the same mutations in the G6P translocase gene as in GSD Ib (see Genetics) [6]. The description of a GSD Id patient has been withdrawn [7], and no other patient with GSD Ic has been reported [8].

Hypoglycemia occurs during fasting as soon as exogenous sources of glucose are exhausted, since the final steps in both glycogenolysis and gluconeogenesis are blocked. However, there is evidence that GSD I patients are capable of some endogenous hepatic glucose production [9], although the mechanism is still unclear. Residual G6Pase activity or the activity of non specific phosphatases may result in hydrolysis of G6P to glucose; glycogen may be degraded into glucose by amylo-1,6-glucosidase, or autophagy combined with lysosomal acid maltase activity.

Hyperlactatemia is a consequence of excess G6P that cannot be hydrolysed to glucose and is further metabolised in the glycolytic pathway, ultimately generating pyruvate and lactate. This process is intensified under hormonal stimulation as soon as the exogenous provision of glucose fails. Substrates such as galactose, fructose and glycerol need liver G6Pase to be metabolised to glucose. Consequently ingestion of sucrose and lactose results in hyperlactatemia, with only a small rise in blood glucose [10].

The serum of untreated patients has a milky appearance due to hyperlipidemia, primarily from increased triglycerides with cholesterol and phospholipids less elevated. The hyperlipidemia only partially responds to intensive dietary treatment [11, 12]. The increased concentrations of triglycerides and cholesterol are reflected in increased numbers of VLDL and LDL particles, whereas the HDL particles are decreased [13]. VLDL particles are also increased in size due to the accumulation of triglycerides. Hyperlipidemia is a result of both increased synthesis from excess of acetyl-

coenzyme A (CoA) via malonyl-CoA, and decreased serum lipid clearance [14]. Elevated hepatic G6P levels may also play a role via activation of transcription of lipogenic genes. Decreased plasma clearance is a result of impaired uptake and impaired lipolysis of circulating lipoproteins. Reduced ketone production during fasting is a consequence of the increased malonyl-CoA levels, which inhibit mitochondrial β-oxidation [15].

Hyperuricemia is a result of both increased production and decreased renal clearance. Increased production is caused by increased degradation of adenine nucleotides to uric acid, associated with decreased intra-hepatic phosphate concentration and ATP depletion [16]. Decreased renal clearance is caused by competitive inhibition of uric acid excretion by lactate [17].

Genetics

Both GSD Ia and Ib are autosomal recessive disorders. In 1993, the gene encoding G6Pase (G6PC) was identified on chromosome 17q21. Today more than 75 different mutations have been reported [18, 19]. Subsequently, the gene encoding the G6P transporter (G6PT) was identified on chromosome 11q23. More than 65 different mutations have been reported [20]. Patients formerly diagnosed by enzyme studies as GSD Ib, Ic and the putative Id shared the same mutations in G6PT [6]. Recently however, a GSD Ic patient without mutations in G6PT was described suggesting the existence of a distinct GSD Ic locus [21].

Diagnosis

GSD Ia is characterized by deficient G6Pase activity in intact and disrupted liver microsomes, whereas deficient G6Pase activity in intact microsomes, and (sub)normal G6Pase activity in disrupted microsomes, indicates a defect in the G6P transporter [22]. However, enzyme studies in liver tissue obtained by biopsy are now usually un-necessary since the diagnosis can be based on clinical and biochemical findings combined with DNA investigations in leukocytes. If patients suffer from neutropenia, recurrent infections and/or IBD, mutation analyses of G6PT should be performed first [18, 19], although in younger GSD Ib patients these findings are not always present [3]. If no mutations in G6PC or in G6PT are identified, a glucose tolerance test should be performed. A marked decrease in blood lactate concentration from an elevated level at zero time indicates a gluconeogenesis disorder, including GSD I, whereas an increase in blood lactate concentration suggests one of the other hepatic GSDs. If the suspicion of GSD I remains, enzyme assays in fresh liver tissue should be performed.

Identification of mutations in either G6PC or G6PT alleles of a GSD I index case allows reliable prenatal DNA-based diagnosis in chorionic villus samples. Carrier detection in the partners of individuals carrying a known mutation is a reliable option, since a high detection rate is observed for both G6PC and G6PT.

Treatment

Dietary Treatment

The goal of treatment is, as far as possible, to prevent hypoglycemia, thus limiting secondary metabolic derangements. Initially, treatment consisted of frequent carbohydrate-enriched meals during day and night. In 1974, continuous nocturnal gastric drip feeding (CNGDF) via a nasogastric tube was introduced, allowing both patients and parents to sleep during the night [23].

CNGDF can be used in very young infants. Both a glucose/glucose polymer solution and a formula (sucrose and lactose-free/low in GSD I) enriched with maltodextrin are suitable. There are no studies comparing these two methods. CNGDF should be started within 1 h after the last meal. Otherwise, a small oral or bolus feed should be given. Within 15 min after the discontinuation of the CNGDF, a feed should be given. CNGDF can be given using a nasogastric tube or by gastrostomy. Gastrostomy is contraindicated in GSD Ib patients because of the risk of IBD and local infections. It is advisable to use a feeding pump that accurately controls flow rate and has alarms alerting of flaws in the system. Parents need thorough teaching with meticulous explanation of technical and medical details and should feel completely confident with the feeding pump system.

In 1984 uncooked cornstarch (UCCS), from which glucose is more slowly released than from cooked starch, was introduced [24]. During the day, this prolongs the period between meals, thus improving metabolic control. Overnight, it may be used as an alternative for CNGDF. Theoretically, pancreatic amylase activity is insufficiently mature in children less than 1 year of age and therefore UCCS should not be started in these patients. Nevertheless, it may be effective and useful even in these younger children [25]. The starting dose of 0.25 g/kg bodyweight should be increased slowly to prevent side-effects, such as bowel distension, flatulence, and loose stools, although these side-effects are usually transient. Precaution is needed in GSD Ib patients since UCCS may exaggerate IBD. UCCS can be mixed in water in a starch/water ratio of 1:2. If UCCS is used overnight, no glucose should be added to avoid insulin release and an UCCS tolerance test should be performed to investigate the permissible duration of the fasting period. It has been documented that both CNGDF and UCCS can maintain normoglycemia during the night with equally favourable (short-term) results [26, 27]. UCCS is also used in daytime to prolong the fasting period.

Glucose requirements decrease with age and are calculated from the theoretical glucose production rate, which varies between 8–9 mg/kg/min in neonates and 2–3 mg/kg/min in adults. Only the required amount of glucose should be given since larger quantities of exogenous glucose may cause undesired swings in glycemia which make patients more sensitive to rebound hypoglycemia and increases peripheral body fat storage.

During infections, a frequent supply of exogenous glucose must be maintained, even though anorexia, vomiting, and diarrhoea may make this difficult. Furthermore, glucose metabolism is increased with fever. Replacement of meals and snacks by glucose polymer drinks is often needed. Nasogastric drip feeding 24 h a day may be necessary. If this is not tolerated, a hospital admission is indicated for intravenous therapy.

There is no consensus as to the extent to which lactate production from galactose and fructose should be avoided. Lactate may serve as an alternate fuel for the brain, thereby protecting patients against cerebral symptoms from reduced glucose levels [28]. Furthermore, milk products, fruits and vegetables are important sources of vitamins and minerals. On the other hand, stringent maintenance of normolactatemia by complete avoidance of lactose and fructose ingestion may lead to a more favourable outcome [29].

The dietary plan should be carefully designed and followed to provide enough essential nutrients as recommended by the WHO. Otherwise, supplementation should be started. Special attention should be directed to *calcium* (limited milk intake) and *vitamin D*. Furthermore, increased carbohydrate metabolism needs an adequate supply of *vitamin B_1*.

Prior to *elective* surgery, bleeding time (platelet aggregation) should be normalised by continuous gastric drip feeding for several days or by intravenous glucose infusion over 24–48 hours. Close peri-operative monitoring of blood glucose and lactate concentration is essential.

Pharmacological Treatment

Until recently, *(sodium)bicarbonate* was recommended to reduce hyperlactatemia. Bicarbonate also induces alkalisation of the urine, thereby diminishing the risk of urolithiasis and nephrocalcinosis. However, it was found that a progressive worsening of hypocitraturia occurs [30] so that alkalisation with *citrate* may be even more beneficial in preventing or ameliorating urolithiasis and nephrocalcinosis.

Uric acid is a potent radical scavenger and it may be a protective factor against the development of atherosclerosis [31]. Consequently, it is recommended to accept a serum uric acid concentration within the high normal range. To prevent gout and urate nephropathy, however, a *xanthine-oxidase inhibitor (allopurinol)* should be started if it exceeds this.

If persistent *microalbuminuria* is present, a (long-acting) a*ngiotensin converting enzyme (ACE) inhibitor* should be started to reduce or prevent further deterioration of renal function. *Additional blood pressure lowering drugs* should be used if blood pressure remains above the 95th percentile for age.

To reduce the risk of cholelithiasis and pancreatitis, *triglyceride-lowering drugs* (nicotinic acid, fibrates) are indicated only if severe hypertriglyceridemia persists. *Cholesterol-lowering drugs* are not indicated in younger patients.

In adult patients however, progressive renal insufficiency may worsen the hyperlipidemia and atherogenecity, and in such cases *statins* may be indicated, although there is at present no evidence of their efficacy. *Fish-oil* is not indicated since its effect on reducing serum triglyceride and cholesterol is not long lasting and it may even lead to increased lipoprotein oxidation, thereby increasing atherogenecity [32].

There is no place for *growth hormone therapy* since, although it may enhance growth, it does not improve final height. Similarly, neither are *oestrogen and testosterone* indicated to enhance pubertal development as they do not improve final height scores. Ethinyloestradiol should be avoided both because of its association with liver adenomas and its incompatibility with hyperlipidemia. *Oral contraception* is possible with high doses of progestagen from the 5th to the 25th day of the cycle or with daily administration of low doses of progestagen [33].

The benefits of prophylaxis with oral antibiotics have not been studied in neutropenic GSD Ib patients. However, prophylaxis with *cotrimoxazol* may be of benefit in symptomatic patients or in those with a neutrophil count < 500 × 10^6/l [34].

Although patients with GSD Ib and neutropenia have been treated with *granulocyte colony-stimulating factor (GCSF)* from 1989 and it is widely thought that the severity of infections decreases and IBD regresses, no unequivocal improvement in outcome has been established [35]. It is advised to limit the use of GCSF to one or more of the following indications: *(1)* a persistent neutrophil count below 200 × 10^6/l; *(2)* a single life threatening infection requiring antibiotics intravenously; *(3)* serious IBD documented by abnormal colonoscopy and through biopsies; or *(4)* severe diarrhoea requiring hospitalisation or disrupting normal life [36]. Patients generally respond to low doses (starting dose 2.5 µg/kg every other day). Data on the safety and efficacy of long-term GCSF administration are limited. The most serious frequent complication is splenomegaly including hypersplenism. Reports of acute myelogenous leukemia [37] and renal carcinoma [38] arising during long-term use of GCSF make stringent follow-up necessary. Bone marrow aspiration with cytogenetic studies before treatment and once yearly during GCSF treatment are advised, along with twice yearly abdominal ultrasound.

Follow-up, Complications, Prognosis, Pregnancy

The biomedical targets are summarised in ◘ Table 6.2 and are based on what level of abnormality constitutes an added health risk [39]. One should attempt to approach these targets as far as possible, but without reducing the quality of life. A single blood glucose assay in the clinical setting is not useful; it is preferable to make serial glucose measurements at home preprandially and in the night over 48–72 h. Lactate excretion in urine should be estimated in samples collected at home and delivered frozen [40, 41]. Serum uric acid, cholesterol and triglyceride concentrations, and venous

◘ **Table 6.2.** Biomedical targets in GSD I

1.	Preprandial blood glucose >3.5–4.0 mmol/l (adjusted to target 2)
2.	Urine lactate/creatinine ratio <0.06 mmol/mmol (or urine lactate <0.4–0.6 mmol/l)
3.	Serum uric acid concentration in high normal range for age and laboratory
4.	Venous blood base excess >–5 mmol/l and venous blood bicarbonate >20 mmol/l
5.	Serum triglyceride concentration <6.0 (<10.0 mmol/l in adult patients)
6.	Normal faecal alpha-1-antitrypsine for GSD Ib patients
7.	Body mass index <+2.0 SDS (in growing children between 0 and +2.0 SDS)

blood gases should be estimated during each outpatient visit. A good marker for the degree of apparent asymptomatic IBD activity in GSD Ib is faecal alpha-1-antitrypsine [42].

Intensive dietary treatment with improved metabolic and endocrine control has led to reduced morbidity and mortality, and improved quality of life [29]. Long-term cerebral function is normal if hypoglycemic damage is prevented. Most patients are able to lead fairly normal lives. With ageing, however, patients may develop complications of different organ systems [1, 25, 43].

Proximal and distal renal tubular as well as *glomerular functions* are at risk [44, 45]. Proximal renal tubular dysfunction is observed in patients with poor metabolic control and improves after starting intensive dietary treatment [46]. However, distal renal tubular dysfunction can occur even in patients with optimal metabolic control and may lead to hypercalciuria and hypocitraturia [47, 48]. Regular ultrasonography of the kidneys is recommended. Progressive glomerular renal disease starts with a silent period of hyperfiltration that begins in the first years of life [49]. Microalbuminuria may develop at the end of the first or in the second decade of life and is an early manifestation of the progression of renal disease [50]. Subsequently, proteinuria and hypertension may develop, with deterioration of renal function leading to end-stage renal disease in the 3rd–5th decade of life. The similarities in the natural history of renal disease in GSD I and of nephropathy in insulin dependent diabetes mellitus is striking. The pathogenesis however, is still unclear. As in diabetic nephropathy, ACE inhibitors should be started if microalbuminuria persists over a period of 3 months with a moderate dietary restriction of protein and sodium. Hemodialysis, continuous ambulatory peritoneal dialysis and renal transplantation are therapeutic options for end-stage renal disease in GSD I.

Single or multiple liver *adenomas* may develop in the second or third decade [51, 52] but remain unchanged during many years of intensive dietary treatment; a reduction in size and/or number has been observed in some patients following optimal metabolic control. Liver adenomas may cause mechanical problems and acute haemorrhage; furthermore, they may develop into carcinomas. To screen for adenomas and to follow their size and number, ultrasonography should be performed regularly. Increase in size of nodules or loss of definition of their margins necessitate further investigations, such as CT scans or MRI. In addition, serum α-fetoprotein and carcino-embryonal antigen can be used to screen for malignant transformation. However, neither CT nor MRI are highly predictive of malignant transformation, and false negative results for both tumour markers have been reported [53]. The management of liver adenomas is either conservative or surgical. In severe cases of adenomas, enucleation or partial liver resection are therapeutic options. Where there is a recurrence of adenomas or suspected malignant transformation, liver transplantation is a therapeutic option provided there are no metastases [54]. Liver transplantation also corrects glucose homeostasis, but in GSD 1b does not correct neutropenia and neutrophil dysfunction, nor does it prevent the development of renal failure [55]. Immunosuppression may worsen renal function.

Osteopenia appears to be a result of both decreased bone matrix formation and decreased mineralisation [56, 57]. Limited peak bone mass formation increases the risk of fractures later in life. It is important for normal bone formation to suppress secondary metabolic and hormonal disturbances, especially chronic lactatemia.

Anemia is observed at all ages, but especially in adolescent and adult patients [1, 43]. The anemia may be refractory to iron because of inappropriate production, by hepatic adenomas, of hepcidin, a peptide hormone that controls the release of iron from intestinal cells and macrophages [58].

Polycystic ovaries (PCOs) have been observed in adolescent and adult female patients [59]. Their pathophysiology is unresolved and their effects on reproductive function are unclear. PCOs may cause acute abdominal pain as a result of vascular disturbances. This should be differentiated from pancreatitis and haemorrhage into liver adenoma.

Despite severe hyperlipidemia, *cardiovascular morbidity and mortality* is infrequent and, when present, may be related to secondary metabolic changes caused by the progressive renal disease. The preservation of normal endothelial function [1, 43, 60] may result from diminished platelet aggregation [61], increased levels of apolipoprotein E [62], decreased susceptibility of LDL to oxidation – possibly related to the altered lipoprotein fatty acid profile in GSD Ia [32] – and increased antioxidative defences in plasma protecting against lipid peroxidation [31].

A vascular complication that may cause more morbidity and mortality in the ageing patient is *pulmonary hypertension* followed by progressive heart failure [63]. It may develop in the second decade or later. No specific treatment is available. Monitoring by ECG and cardiac ultrasonography is recommended after the first decade.

Depressive illness needing therapy is observed rather frequently in adult patients [1, 43]. Lifelong intensive dietary treatment 24 hours a day, together with the threat of serious medical problems, is a major burden for both patients and their parents.

Successful *pregnancies* have been reported [1, 33]. Close supervision and reintroduction of intensive dietary treatment is necessary.

6.1.2 Glycogen Storage Disease Type III (Debranching Enzyme Deficiency)

The release of glucose from glycogen requires the activity of both phosphorylase and glycogen debranching enzyme (GDE). GSD III, also known as Cori or Forbes disease, is an autosomal recessive disorder due to deficiency of GDE which causes storage of glycogen with an abnormally compact structure, known as phosphorylase limit dextrin. Differences in tissue expression of the deficient GDE explain the existence of various subtypes of GDS III. Most patients with GSD III have a generalized defect in which enzyme activity is deficient in liver, muscle, heart, leukocytes and cultured fibroblasts, and have a syndrome that includes both hepatic and myopathic symptoms, and often cardiomyopathy (GSD IIIa). About 15% of patients only have symptoms of liver disease and are classified as GSD IIIb. Subgroups due to the selective deficiency of either the glucosidase activity (GSD IIIc) or of the transferase activity (GSD IIId) are very rare.

Clinical Presentation
Hepatic Presentation
Hepatomegaly, short stature, hypoglycemia, and hyperlipidemia predominate in children, and this presentation may be indistinguishable from GSD I. Splenomegaly can be present, but the kidneys are not enlarged and renal function is normal. With increasing age, these symptoms improve in most GSD III patients and may disappear around puberty.

Myopathic Presentation
Clinical myopathy may not be apparent in infants or children, although some show hypotonia and delayed motor milestones. Myopathy often appears in adult life, long after liver symptoms have subsided. Adult-onset myopathies may be distal or generalised. Patients with distal myopathy develop atrophy of leg and intrinsic hand muscles, often leading to the diagnosis of motor neurone disease or peripheral neuropathy [64]. The course is slowly progressive and the myopathy is rarely crippling. The generalised myopathy tends to be more severe, often affecting respiratory muscles. In the EMG, myopathic features are mixed with irritative features (fibrillations, positive sharp waves,

myotonic discharges), a pattern that may reinforce the diagnosis of motor neurone disease in patients with distal muscle atrophy. Nerve conduction velocities are often decreased [65]. Although GDE works hand-in-hand with myophosphorylase and one would therefore expect GDE deficiency to cause symptoms similar to those of McArdle disease, cramps and myoglobinuria are exceedingly rare.

Muscle biopsy typically shows a vacuolar myopathy. The vacuoles contain PAS-positive material and corresponds to large pools of glycogen, most of which is free in the cytoplasm. However, some of the glycogen is present within lysosomes. Biochemical analysis shows greatly increased concentration of phosphorylase-limit dextrin, as expected.

In agreement with the notion that the enzyme defect is generalised, peripheral neuropathy has been documented both electrophysiologically and by nerve biopsy and may contribute to the weakness and the neurogenic features of some patients. Similarly, while symptomatic cardiopathy is uncommon, cardiomyopathy (similar to idiopathic hypertrophic cardiomyopathy) is detectable in virtually all patients with myopathy [66].

Metabolic Derangement

GDE is a bifunctional enzyme, with two catalytic activities, oligo-1,4→1,4-glucantransferase and amylo-1,6-glucosidase. After phosphorylase has shortened the peripheral chains of glycogen to about four glycosyl units, these residual stubs are removed by GDE in two steps. A maltotriosyl unit is transferred from a donor to an acceptor chain (transferase activity), leaving behind a single glucosyl unit, which is hydrolysed.

During infancy and childhood patients suffer from fasting hypoglycemia, associated with ketosis and hyperlipidemia. Serum transaminases are also increased in childhood but decrease to (almost) normal values with increasing age. In contrast to GSD I, blood lactate concentration is normal. Elevated levels of serum creatine kinase (CK) and aldolase suggest muscle involvement, but normal values do not exclude the future development of myopathy.

Genetics

The gene for GDE (GDE) is located on chromosome 1p21. At present, at least 48 different mutations in the GDE gene have been associated with GSD III. GSD IIIb is associated with mutations in exon 3, while mutations beyond exon 3 are associated with GSD IIIa. When all known GSD III mutations are taken into consideration, there is no clear correlation between the type of mutation and the severity of the disease. This makes prognostic counselling based on mutations difficult [67].

Diagnosis

Diagnosis is based on enzyme studies in leukocytes, erythrocytes and/or fibroblasts, combined with DNA investigations in leukocytes. Prenatal diagnosis is possible by identifying mutations in the GDE gene if these are already known. If not, polymorphic markers may be helpful in informative families. Prenatal diagnosis based on GDE activity in cultured amniocytes or chorionic villi is technically difficult and does not always discriminate between the carrier state and the affected fetus.

Treatment

The main goal of dietary treatment is prevention of hypoglycemia and correction of hyperlipidemia. Dietary management is similar to GSD Ia but, since the tendency to develop hypoglycemia is less marked, only some younger patients will need continued nocturnal gastric drip feeding, and a late evening meal and/or uncooked corn starch will be sufficient to maintain normoglycemia during the night. In GSD III (as opposed to GSD I), restriction in fructose and galactose is unnecessary and dietary protein intake can be increased since no renal dysfunction exists. The latter would not only have a beneficial effect on glucose homeostasis, but also on atrophic myopathic muscles.

Complications, Prognosis, Pregnancy

With increasing age, both clinical and biochemical abnormalities gradually disappear in most patients; parameters of growth normalise, and hepatomegaly usually disappears after puberty [43]. In older patients, however, liver fibrosis may develop into cirrhosis. In about 25% of these patients, liver adenoma may occur, and transformation into hepatocellular carcinoma has been described, although this risk is apparently small. Liver transplantation has been performed in patients with end-stage cirrhosis and/or hepatocellular carcinoma [55, 66].

Generally, prognosis is favourable for the hepatic form (GSD IIIb), whereas it is less favourable for GSD IIIa, because severe progressive myopathy and cardiomyopathy may develop even after a long period of latency. Currently there is no satisfactory treatment for either the myopathy or cardiomyopathy.

Successful pregnancy has been reported; regular dietary management with respect to the increasing needs for energy (carbohydrates) and nutrients is warranted [68].

6.1.3 Glycogen Storage Disease Type IV (Branching Enzyme Deficiency)

GSD IV, or Andersen Disease, is an autosomal recessive disorder due to a deficiency of glycogen branching enzyme (GBE). Deficiency of GBE results in the formation of an amylopectin-like compact glycogen molecule with fewer branching points and longer outer chains. The pathophysiological consequences of this abnormal glycogen for the liver still need to be elucidated. Patients with the classical form of GSD IV develop progressive liver disease early in life. The non-progressive hepatic variant of GSD IV is less frequent and these patients usually survive into adulthood. Besides these liver related presentations, there are rare neuromuscular forms of GSD IV.

Clinical Presentation
Hepatic Forms

Patients are normal at birth and present generally in early childhood with hepatomegaly, failure to thrive, and liver cirrhosis. The cirrhosis is progressive and causes portal hypertension, ascites, and oesophageal varices. Some patients may also develop hepatocellular carcinoma [69]. Life expectancy is limited due to severe progressive liver failure and – without liver transplantation – death generally occurs when patients are 4 to 5 years of age [70, 71].

Patients with the non-progressive form present with hepatomegaly and sometimes elevated transaminases. Although fibrosis can be detected in liver biopsies, this is apparently non-progressive. No cardiac or skeletal muscle involvement is seen. These patients have normal parameters for growth.

Neuromuscular Forms

Neuromuscular forms can be divided into four clinical presentations according to the age of onset. A neonatal form, which is extremely rare, presents as fetal akinesia deformation sequence (FADS), consisting of arthrogryposis multiplex congenita, hydrops fetalis, and perinatal death. A congenital form presents with hypotonia, cardiomyopathy, and death in early infancy. A third form manifests in childhood with either myopathy or cardiomyopathy. Lastly, the adult form may present as a myopathy or as a multisystemic disease also called adult polyglucosan body disease (APBD) [72]. APBD is characterised by progressive upper and lower motor neurone dysfunction (sometimes simulating amyotrophic lateral sclerosis), sensory loss, sphincter problems and, inconsistently, dementia. In APBD, polyglucosan bodies have been described in processes (not perikarya) of neurones and astrocytes in both grey and white matter.

Muscle biopsy in the neuromuscular forms shows the typical foci of polyglucosan accumulation, intensely PAS-positive and diastase-resistant. Similar deposits are seen in the cardiomyocytes of children with cardiomyopathy and in motor neurones of infants with Werdnig-Hoffmann-like presentation [73].

Metabolic Derangement

Hypoglycemia is rarely seen, and only in the classical hepatic form, when liver cirrhosis is advanced, and detoxification and synthesis functions become impaired. The clinical and biochemical findings under these circumstances are identical to those typical of other causes of cirrhosis, with elevated liver transaminases and abnormal values for blood clotting factors, including prothrombin and thromboplastin generation time.

Genetics

The *GBE* gene has been mapped to chromosome 3p14. Three important point mutations, R515C, F257L and R524X were found in patients with the classical progressive liver cirrhosis form [74]. In patients with the non-progressive liver form, the Y329S mutation has been reported. This mutation results in a significant preservation of GBE activity, thereby explaining the milder course of the disease [70]. Interestingly, the mutation found in patients with APBD [72] also appears to be relatively mild [74] which may explain the late onset of this disorder.

Diagnosis

The diagnosis is usually only suspected at the histological examination of a liver or muscle biopsy which shows large deposits that are periodic-acid-Schiff-staining but partially resistant to diastase digestion. Electron microscopy shows accumulation of fibrillar aggregations that are typical for amylopectin. The enzymatic diagnosis is based on the demonstration of GBE deficiency in liver, muscle, fibroblasts, or leukocytes. Prenatal diagnosis is possible using DNA mutation analysis in informative families, but difficult by measuring the enzyme activity in cultured amniocytes or chorionic villi because of high residual enzyme activity.

Treatment

There is no specific dietary treatment for GSD IV. Dietary treatment focuses on the maintenance of normoglycemia by frequent feedings and a late evening meal. Liver transplantation is the only effective therapeutic approach at present for GSD IV patients with the classic progressive liver disease [55, 71].

Complications, Prognosis, Pregnancy

The ultimate prognosis depends on the results of liver transplantation which was favourable in 13 GSD IV patients [55]. The prognosis also depends on the occurrence of amylopectin storage in extra-hepatic tissues. This risk seems to be especially high for cardiac tissue. Of 13 patients with GSD IV who underwent liver transplantation, two died from heart failure due to amylopectin storage in the myocardium [55]. A positive result of liver transplantation may be the development of systemic microchimerism, with donor cells present in various tissues. This would lead to a transfer of enzyme activity from normal to deficient cells outside the liver [70]. No pregnancies are reported in classical GSD IV.

Patients with the non-progressive liver variant have been reported to survive into their mid forties. With increasing age, liver size tends to decrease and elevated transaminases return to (nearly) normal values.

6.1.4 Glycogen Storage Disease Type VI (Glycogen Phosphorylase Deficiency)

GSD VI or Hers disease is an autosomal recessive disorder due to a deficiency of the liver isoform of glycogen phosphorylase. Phosphorylase breaks the straight chains of glycogen down to glucose-1-phosphate in a concerted action with debranching enzyme. Glucose-1-phosphate in turn is converted into glucose-6-phosphate and then into free glucose.

Clinical Presentation

GSD VI is a rare disorder with a generally benign course. Patients are clinically indistinguishable from those with liver GSD type IX caused by phosphorylase kinase (PHK) deficiency and present with hepatomegaly and growth retardation in early childhood. Cardiac and skeletal muscles are not involved. Hepatomegaly decreases with age and usually disappears around puberty. Growth usually normalises after puberty [66].

Metabolic Derangement

The tendency towards hypoglycemia is not as severe as seen in GSD I or GSD III and usually appears only after prolonged fasting in infancy. Hyperlipidemia and hyperketosis are usually mild. Lactic acid and uric acid are within normal limits.

Genetics

Three isoforms of phosphorylase are known, encoded by three different genes. The gene encoding the liver isoform, *PYGL*, is on chromosome 14q21-q22, and mutations have been described [75].

Diagnosis

Deficient phosphorylase activity can be documented in liver tissue.

Treatment

Treatment of liver phosphorylase deficiency is symptomatic, and consists of preventing hypoglycemia using a high-carbohydrate diet and frequent feedings; a late evening meal is unnecessary in most patients.

6.1.5 Glycogen Storage Disease Type IX (Phosphorylase Kinase Deficiency)

GSD IX, or phosphorylase kinase (PHK) deficiency, is the most frequent glycogen storage disease. According to the mode of inheritance and clinical presentation six different subtypes are distinguished: *(1)* X-linked liver glycogenosis (XLG or GSD IXa), by far the most frequent subtype; *(2)* combined liver and muscle PHK deficiency (GSD IXb); *(3)* autosomal liver PHK deficiency (GSD IXc); *(4)* X-linked muscle glycogenosis (GSD IXd); *(5)* autosomal muscle PHK deficiency (GSD IXe); and *(6)* heart PHK deficiency (GSD IXf) with the mode of inheritance not clear yet [75, 76], but probably due to AMP kinase mutations [76a].

Clinical Presentation
Hepatic Presentation

The main clinical symptoms are hepatomegaly due to glycogen storage, growth retardation, elevated liver transaminases, and hypercholesterolemia and hypertriglyceridemia. Symptomatic hypoglycemia and hyperketosis are only seen after long periods of fasting in young patients. The clinical course is generally benign. Clinical and biochemical abnormalities disappear with increasing age and after puberty most patients are asymptomatic [77, 78].

Myopathic Presentation

Not surprisingly, the myopathic variants present clinically similar to a mild form of McArdle disease (▶ below), with exercise intolerance, cramps, and recurrent myoglobinuria in young adults. Less frequent presentations include infantile weakness and respiratory insufficiency or late-onset weakness. Muscle morphology shows subsarcolemmal deposits of normal-looking glycogen.

Metabolic Derangement

The degradation of glycogen is controlled both in liver and in muscle by a cascade of reactions resulting in the activation of phosphorylase. This cascade involves the enzymes adenylate cyclase and PHK. PHK is a decahexameric protein composed of four subunits, α, β, γ, and δ: the α and β subunits are regulatory, the γ subunit is catalytic, and the δ subunit is a calmodulin and confers calcium sensitivity to the enzyme. The hormonal activating signals for glycogenolysis are glucagon for the liver and adrenaline for muscle. Glucagon and adrenaline activate the membrane-bound adenylate cyclase, which transforms ATP into cyclic AMP (cAMP) and interacts with the regulatory subunit of the cAMP-dependent protein kinase, resulting in phosphorylation of PHK. Ultimately, this activated PHK transforms glycogen phosphorylase into its active conformation, a process which is defective in GSD type IX.

Genetics

Two different isoforms of the α subunit (α_L for liver and α_M for muscle) are encoded by two different genes on the X chromosome (*PHKA2* and *PHKA1* respectively), while the β subunit (encoded by *PHKB*), two different isoforms of the γ subunit (γ_T for testis/liver and γ_M for muscle, encoded by *PKHG2* and *PKHG1*, respectively), and three isoforms of calmodulin (*CALM1*, *CALM2*, *CALM3*) are encoded by autosomal genes. The *PHKA2* gene has been mapped to chromosome Xp22.2-p22.1, the *PHKB* gene to chromosome 16q12-q13, and the *PKHG2* gene to chromosome 16p12-p11 [75, 79, 80].

The most common hepatic variant, XLG or GSD IXa (resulting from *PHKA2* mutations), comprises two different entities: XLG 1, the classical type, and XLG 2, the less common variant. In XLG 1 the PHK activity is deficient in liver and decreased in blood cells. In XLG 2, PHK activity is normal in liver, erythrocytes and leukocytes. Therefore, normal PHK activity in erythrocytes or even liver tissue does not exclude XLG. This phenomenon may be explained by the fact that XLG 2 is due to minor mutations with regulatory effects on PHK activity, which is not decreased in vitro [75, 79, 80].

The predominance of affected men with the myopathic presentation suggested that the X-linked α_M isoform may be involved predominantly, a concept bolstered by reports of mutations in the *PHKA1* gene in two patients [81, 82]. However, a thorough molecular study of six myopathic patients, five men and one woman, revealed only one novel mutation in *PHKA1*, whereas no pathogenic mutations were found in any of the six genes (*PHKA1*, *PHKB*, *PHKG1*, *CALM1*, *CALM2*, *CALM3*) encoding muscle subunits of PHK in the other five patients [83]. This surprising result suggested that most myopathic patients with low PHK activity either harbor elusive mutations in PHK genes or mutations in other unidentified genes [76a].

Diagnosis

As stated above, assays of PHK in various tissues may not allow for a definitive diagnosis. Where possible, this should be based on the identification of mutations within the different PHK genes.

Treatment and Prognosis

Treatment of the hepatic form is symptomatic, and consists of preventing hypoglycemia using a high-carbohydrate diet and frequent feedings; a late evening meal is unnecessary except for young patients.

Growth improves without specific treatment with age. XLG patients have a specific growth pattern characterised by initial growth retardation, a late growth spurt, and complete catch-up in final height occurring after puberty [84, 78]. Prognosis is generally favourable for the hepatic types, and more uncertain for the myopathic variants.

6.1.6 Glycogen Storage Disease Type 0 (Glycogen Synthase Deficiency)

Although this rarely diagnosed enzyme defect leads to decreased rather than increased liver glycogen, it causes symptoms that resemble hepatic glycogenosis.

Clinical Presentation

The first symptom of GSD 0 is fasting hypoglycemia which appears in infancy or early childhood. Nevertheless, patients can remain asymptomatic. Recurrent hypoglycemia often leads to neurological symptoms. Developmental delay is seen in a number of GSD 0 patients and is probably associated with these periods of hypoglycemia typically occurring in the morning before breakfast. The size of the liver is normal, although steatosis is frequent. Some patients display stunted growth, which improves after dietary measures to protect them from hypoglycemia. The small number of patients reported in the literature may reflect underdiagnosis, since the symptomatology is usually mild and the altered metabolic profile is not always interpreted correctly [85–87].

Metabolic Derangement

GSD 0 is caused by a deficiency of glycogen synthase (GS), a key-enzyme of glycogen synthesis. Consequently, patients with GS deficiency have decreased liver glycogen concentration, resulting in fasting hypoglycemia. This is associated with ketonemia, low blood lactate concentrations, and mild hyperlipidemia. Post-prandially, there is often a characteristic reversed metabolic profile, with hyperglycemia and elevated blood lactate.

Genetics

The gene that encodes GS, *GYS2,* is located on chromosome 12p12.2, and several mutations are known [86, 87].

Diagnosis

Patients with GSD 0 may be misdiagnosed as having diabetes mellitus, especially when glucosuria and ketonuria are also present. Diagnosis of GSD 0 is based on the demonstration of decreased hepatic glycogen content and deficiency of the GS enzyme in a liver biopsy or by DNA analysis. Demonstration of pathological mutations in DNA material from extra-hepatic sources makes the diagnosis possible even without a liver biopsy.

Treatment and Prognosis

Treatment is symptomatic, and consists of preventing hypoglycemia with a high-carbohydrate diet, frequent feedings and, in young patients, late evening meals. Although most patients have normal intellect, developmental delay may follow repeated periods of hypoglycemia. Tolerance to fasting improves with age. Increased energy consumption during pregnancy with reoccurrence of hypoglycemia has been reported [86].

6.2   The Muscle Glycogenoses

At rest, muscle utilizes predominantly fatty acids. During submaximal exercise, it additionally uses energy from blood glucose, mostly derived from liver glycogen. In contrast, during very intense exercise, the main source of energy is anaerobic glycolysis following breakdown of muscle glycogen. When the latter is exhausted, fatigue ensues. Enzyme defects within the pathway affect muscle function.

6.2.1 Glycogen Storage Disease Type V (Myophosphorylase Deficiency)

Clinical Presentation

GSD V, decribed in 1951 by McArdle is characterised by exercise intolerance, with myalgia and stiffness or weakness of exercising muscles, which is relieved by rest. Two types of exertion are more likely to cause symptoms: brief intense isometric exercise, such as pushing a stalled car, or less intense but sustained dynamic exercise, such as walking in the snow. Moderate exercise, for example walking on level ground, is usually well tolerated. Strenuous exercise often results in painful cramps, which are true contractures as the shortened muscles are electrically silent. An interesting constant phenomenon is the second wind that affected individuals experience when they rest briefly at the first appearance of exercise-induced myalgia. Myoglobinuria (with the attendant risk of acute renal failure) occurs in about half of the patients. Electromyography (EMG) can be normal or show non-specific myopathic features at rest, but documents electrical silence in contracted muscles. As in most muscle glycogenoses, resting serum CK is consistently elevated in McArdle patients. After carnitine palmitoyl transferase II (CPT II) deficiency, McArdle disease is the second most common cause of recurrent myoglobinuria in adults [88].

Clinical variants of McArdle disease include the fatal infantile myopathy described in a few cases, and fixed weakness in older patients [65]. However, some degree of fixed weakness does develop in patients with typical McArdle disease as they grow older and is associated with chronically elevated serum CK levels.

Metabolic Derangement

There are three isoforms of glycogen phosphorylase: brain/heart, liver, and muscle, all encoded by different genes. GSD V is caused by deficient myophosphorylase activity.

Genetics

GSD V is an autosomal recessive disorder. The gene for the muscle isoform (PYGM) has been mapped to chromosome 11q13. The number of known pathogenic mutations has rapidly increased to over 40 [89]. By far the most common mutation in Caucasians is the R49X mutation, which accounts for 81% of the alleles in British patients [90], and 63% of alleles in U.S. patients [91]. This mutation, however, has never been described in Japan, where a single codon deletion 708/709 seems to prevail [92].

No genotype:phenotype correlations have been detected. Patients with the same genotype may have very different clinical manifestations, not entirely explained by different lifestyles. A study of 47 patients with GSD V for associated insertion/deletion polymorphism in the angiotensin-converting enzyme (ACE) revealed a good correlation between clinical severity and number of ACE genes harbouring a deletion [93].

Diagnosis

The forearm ischemic exercise (FIE) test is informative but is being abandoned as it is neither reliable, reproducible, nor specific, and is painful. Alternative diagnostic tests include a non-ischemic version of the FIE [94], and a cycle test based on the unique decrease in heart rate shown by McArdle patients between the 7th and the 15th minute of moderate exercise, a reflection of the second wind phenomenon [95]. Muscle histochemistry shows subsarcolemmal accumulation of glycogen that is normally digested by diastase. A specific histochemical stain for phosphorylase can be diagnostic except when the muscle specimen is taken too soon after an episode of myoglobinuria. Myophosphorylase analysis of muscle provides the definitive answer, but muscle biopsy may be avoided altogether in Caucasian patients by looking for the common mutation (R49X) in genomic DNA. The presence of the mutation – even only in one allele – establishes the diagnosis.

Treatment

There is no specific therapy. Probably, the most important therapy is aerobic exercise [96], although oral sucrose improved exercise tolerance, and may have a prophylactic effect when taken before planned activity. This effect is explained by the fact that sucrose is rapidly split into glucose and fructose; both bypass the metabolic block in GSD V and hence contribute to glycolysis [97].

6.2.2 Glycogen Storage Disease Type VII (Phosphofructokinase Deficiency)

Clinical Presentation

Clinically, GSD VII, first described by Tarui, is indistinguishable from McArdle disease, except for the absence of the second wind phenomenon [98]. Some laboratory results are useful in the differential diagnosis, including an increased bilirubin concentration and reticulocyte count, reflecting a compensated hemolysis. Thus, the diagnosis of PFK deficiency is based on the combination of muscle symptoms and compensated hemolytic anemia: the only other muscle glycogenosis with these features is phosphoglycerate kinase deficiency (▶ below).

There are two clinical variants, one manifesting as fixed weakness in adult life (although most patients recognise having suffered from exercise intolerance in their youth), the other affecting infants or young children, who have both generalised weakness and symptoms of multisystem involvement (seizures, cortical blindness, corneal opacifications, or cardiomyopathy) [65]. The infantile variant, in which no mutation in the PFK-M gene has been documented is probably genetically different from the typical adult myopathy.

Metabolic Derangement and Genetics

PFK is a tetrameric enzyme under the control of three autosomal genes. A gene (*PFK-M*) on chromosome 12 encodes the muscle subunit; a gene (*PFK-L*) on chromosome 21 encodes the liver subunit; and a gene (*PFK-P*) on chromosome 10 encodes the platelet subunit. Mature human muscle expresses only the M subunit and contains exclusively the M homotetramer (M4), whereas erythrocytes, which contain both the M and the L subunits, contain five isozymes: the two homotetramers (M4 and L4) plus three hybrid forms (M1L3; M2L2; M3L1). In patients with typical PFK deficiency, mutations in *PFK-M* cause total lack of activity in muscle but only partial PFK deficiency in red blood cells, where the residual activity approximates 50% and is accounted for by the L4 isozyme. At least 15 mutations have been reported in the *PFK-M* gene of patients with typical PFK deficiency [65].

Diagnosis

Muscle histochemistry shows predominantly subsarcolemmal deposits of normal glycogen, most of which stains normally with the PAS and is normally digested by diastase. Patients with PFK deficiency also accumulate increasing amounts of polyglucosan, which stains intensely with the PAS reaction but is resistant to diastase digestion and – in the electron microscope – appears composed of finely granular and filamentous material, similar to the storage material in branching enzyme deficiency and in Lafora disease (▶ below).

The lack of the histochemical reaction for PFK is suggestive, but conclusive evidence comes from the biochemical documentation of PFK deficiency (provided that the muscle specimen has been snap-frozen at the time of biopsy: PFK is notoriously labile!). Muscle biopsy can be avoided if the clinical presentation is typical and a known pathogenic mutation can be documented in blood DNA; however, there is no common mutation.

Treatment

There is no specific therapy. Contrary to McArdle disease, sucrose should be avoided, but aerobic exercise might be useful. The astute observation that patients with PFK deficiency noticed worsening of their exercise intolerance after high-carbohydrate meals was explained by the fact that glucose lowers the blood concentration of free fatty acids and ketone bodies, alternative muscle fuels.

6.2.3 Phosphoglycerate Kinase Deficiency

Phosphoglycerate kinase (PGK) is a single polypeptide encoded by a gene (*PGK1*) on Xq13 for all tissues except spermatogenic cells. Although this enzyme is virtually ubiquitous, clinical presentations depend on the isolated or associated involvement of three tissues, erythrocytes

(hemolytic anemia), central nervous system (CNS, with seizures, mental retardation, stroke), and skeletal muscle (exercise intolerance, cramps, myoglobinuria). The most common association, seen in 8 of 27 reported patients, is nonspherocytic hemolytic anemia and CNS dysfunction, followed by isolated myopathy (7 patients), isolated blood dyscrasia (6 patients), and myopathy plus CNS dysfunction (3 patients) [99]. There was only one patient with myopathy and hemolytic anemia, while two patients showed involvement of all three tissues.

The seven myopathic cases were clinically indistinguishable from McArdle disease, but muscle biopsies showed less severe glycogen accumulation [100]. Mutations in *PGK1* were identified in 4 of the 7 myopathic patients. The different involvement of single or multiple tissues remains unexplained but it may have to do with leaky mutations allowing for some residual PGK activity in some tissues.

6.2.4 Glycogen Storage Disease Type X (Phosphoglycerate Mutase Deficiency)

GSD X or phosphoglycerate mutase (PGAM) deficiency is an autosomal recessive disorder. Phosphoglycerate mutase is a dimeric enzyme: different tissues contain various proportions of a muscle (MM) isozyme, a brain (BB) isozyme, and the hybrid (MB) isoform. Normal adult human muscle has a marked predominance of the MM isozyme, whereas in most other tissues PGAM-BB is the only isozyme demonstrable by electrophoresis [65]. A gene (*PGAMM*) on chromosome 7 encodes the M subunit.

About a dozen patients with muscle PGAM deficiency have been described. The clinical picture is stereotypical: exercise intolerance and cramps after vigorous exercise, often followed by myoglobinuria. Manifesting heterozygotes have been identified in several families. The muscle biopsy shows inconsistent and mild glycogen accumulation, accompanied in one case by tubular aggregates [101]. Four different mutations in the *PGAMM* gene have been identified [65].

6.2.5 Glycogen Storage Disease Type XII (Aldolase A Deficiency)

GSD XII or aldolase A deficiency is an autosomal recessive disorder. Aldolase exists in three isoforms (A, B, and C): skeletal muscle and erythrocytes contain predominantly the A isoform, which is encoded by a gene (*ALDOA*) on chromosome 16. The only reported patient with aldolase A deficiency was a 4 1/2-year-old boy, who had episodes of exercise intolerance and weakness following febrile illnesses [102].

6.2.6 Glycogen Storage Disease Type XIII (β-Enolase Deficiency)

GSD XIII or β-enolase deficiency is an autosomal recessive disorder. β-Enolase is a dimeric enzyme and exists in different isoforms resulting from various combinations of three subunits, α, β, and γ. The β subunit is encoded by a gene (*ENO3*) on chromosome 17. GSD XIII is still represented by a single patient, a 47-year-old Italian man with adult-onset but rapidly progressive exercise intolerance and myalgia, and chronically elevated serum CK [103].

6.2.7 Glycogen Storage Disease Type XI (Lactate Dehydrogenase Deficiency)

GSD XI or lactate dehydrogenase (LDH) deficiency is an autosomal recessive disorder. Lactate dehydrogenase is a tetrameric enzyme composed of two subunits, M (or A) and H (or B) resulting in five isozymes. The gene for LDH-M (*LDHM*) is on chromosome 11.

The first case was identified on the basis of an apparently paradoxical laboratory finding: during an episode of myoglobinuria, the patient had the expected high levels of serum CK, but extremely low level of LDH. Several Japanese patients and two Caucasian patients with LDH-M deficiency have been reported. All have had exercise intolerance, cramps, with or without myoglobinuria. Skin lesions and dystocia have been described in Japanese patients [104]. Several mutations in *LDHM* have been reported.

6.2.8 Muscle Glycogen Storage Disease Type 0 (Glycogen Synthase Deficiency)

Very recently, a new muscular glycogen storage disease type 0 has been described in a child with hypertrophic cardiomyopathy and myopathy due to a homozygous stop mutation in the muscular glycogen synthase gene GYS1 [104a].

6.3 The Generalized Glycogenoses and Related Disorders

6.3.1 Glycogen Storage Disease Type II (Acid Maltase Deficiency)

In contrast with the diseases discussed hitherto in this chapter, GSD II is a lysosomal storage disorder, caused by the generalized deficiency of the lysosomal enzyme, acid maltase or α-glucosidase.

Clinical Presentation

Although the defect involves a single ubiquitous enzyme, it manifests as three different clinical phenotypes: infantile, juvenile, and adult. The *infantile form* is generalised, and usually fatal by 1 year of age. The diagnosis is suggested by the association of profound hypotonia from muscle weakness, (floppy infant syndrome), hyporeflexia and an enlarged tongue. The heart is extremely enlarged, and the electrocardiogram is characterised by huge QRS complexes and shortened PR intervals. The liver has a normal size unless enlarged by cardiac decompensation. The cerebral development is normal. The clinical course is rapidly downward, and the child dies from cardiopulmonary failure or aspiration pneumonia [105].

The *juvenile form* starts either in infancy or in childhood, presents with retarded motor milestones and causes severe proximal, truncal, and respiratory muscle weakness (sometimes with calf hypertrophy, which, in boys, can raise the suspicion of Duchenne muscular dystrophy), but shows no overt cardiac disease. Myopathy deteriorates gradually leading to death from respiratory failure in the second or third decade.

The *adult form* is also confined to muscle and mimics other myopathies with a long latency. Decreased muscle strength and weakness develop in the third or fourth decade of life. Cardiac involvement is minimal or absent. The slow, progressive weakness of the pelvic girdle, paraspinal muscles and diaphragm simulates limb-girdle muscular dystrophy or polymyositis and results in walking difficulty and respiratory insufficiency, but old age can be attained. The early and preferential involvement of truncal and respiratory muscles is an important clinical characteristic. Experience with the adult form has increased during the past few years, leading to the detection of hitherto unknown complications, such as rupture of aneurysms of cerebral arteries (due to accumulation of glycogen in vascular smooth muscle) with fatal outcome [106]. A study on the quality of life of a large cohort of adult-onset Pompe's patients confirmed that this disorder causes severe physical limitations while not impairing mental health [107].

Metabolic Derangement

The enzyme defect results in the accumulation of glycogen within the lysosomes of all tissues, but particularly in muscle and heart, resulting in muscle weakness. Serum levels of transaminases (ASAT, ALAT), CK and CK-myocardial band (in the infantile form) are elevated [105]. Intermediary metabolism is unaffected.

Genetics

Acid maltase is encoded by a gene (*GAA*) on chromosome 17q25. Over 80 pathogenic mutations in *GAA* are known. Some degree of genotype:phenotype correlation is becoming apparent, with severe mutations associated with the infantile form and leaky mutations associated with the adult variant. However, the biochemical bases for the different phenotypes remain largely unclear. Prenatal diagnosis is possible by enzyme assay or DNA analysis of chorionic villi.

Diagnosis

In the infantile form, a tentative diagnosis can be based on the typical abnormalities in the electrocardiogram. Muscle biopsy shows a severe vacuolar myopathy with accumulation of both intralysosomal and free glycogen in both the infantile and childhood variants. Another clue to the correct diagnosis in myopathic Pompe disease is the EMG, which shows, – besides myopathic features – fibrillation potentials, positive waves, and myotonic discharges, more easily seen in paraspinal muscles. Glycogen deposition may be unimpressive in adult cases, with variable involvement of different muscles. A useful histochemical stain is that for acid phosphatase, another lysosomal enzyme, which is virtually absent in normal muscle but very prominent in the lysosome-rich muscle of Pompe patients.

For confirmation, acid maltase should be determined in tissues containing lysosomes. The preferred tissues are fibroblasts or muscle, but lymphocytes may be usable. The activity of this acid maltase must be differentiated from contamination with a non-specific cytosolic neutral maltase. Residual enzyme activity is found in the adult form, whereas the enzyme is absent in the infantile form.

Treatment

Palliative therapy includes respiratory support, dietary regimens (e.g. high-protein diet), and aerobic exercise. Enzyme replacement therapy using recombinant human α-glucosidase, obtained in large quantities from rabbit milk has been used successfully. Four infants with Pompe disease were treated with spectacular results: although one patient died of an intercurrent infection at 4 years of age, all four patients showed remarkable clinical improvement in motor and cardiac function and parallel improvement in muscle morphology [108]. The same therapeutic approach was applied with success in three children with the muscular variant [109]. Before starting enzyme replacement, all three were wheelchair-bound and two were respirator-dependent. After 3 years of treatment, their pulmonary function had stabilised and their exercise tolerance had improved, and the youngest patient resumed walking independently. Alglucosidase alfa (Myozyme), a recombinant analog of human α-glucosidase manufactured in CHO cell lines, has now been approved by the EMEA for use in both the infantile and later onset forms. It appears to be important to start enzyme replacement therapy as early as possible.

6.3.2 Danon Disease

Danon Disease or GSD IIb, or pseudo-Pompe disease, is an X-linked dominant lysosomal storage disease due to deficiency of LAMP-2 (lysosomal-associated membrane protein 2). The disease starts after the first decade, is extremely rare and affects cardiac and skeletal muscle. Acid maltase activity is normal, muscle biopsy shows vacuolar myopathy with vacuoles containing glycogen and cytoplasmatic degradation products [110, 111]. Some patients are mentally retarded. As expected, hemizygous females are also affected, but generally show the first symptoms at a later age. No specific therapy is available, but cardiac transplantation should be considered [112]. The gene encoding LAMP2 was mapped to Xq28 [111].

6.3.3 Lafora Disease

Clinically, Lafora disease (myoclonus epilepsy with Lafora bodies) is characterised by seizures, myoclonus, and dementia. Onset is in adolescence and the course is rapidly progressive, with death occurring almost always before 25 years of age.

The pathologic hallmark of the disease are the Lafora bodies, round, basophilic, strongly PAS-positive intracellular inclusions seen only in neuronal perikarya, especially in the cerebral cortex, substantia nigra, thalamus, globus pallidus, and dentate nucleus. Polyglucosan bodies are also seen in muscle, liver, heart, skin, and retina, showing that Lafora disease is a generalised glycogenosis. However, the obvious biochemical suspect, branching enzyme, is normal.

Linkage analysis localised the gene responsible for Lafora disease (*EPM2A*) to chromosome 6q24 and about 30 pathogenic mutation have been identified [113]. The protein encoded by *EPM2A*, dubbed laforin, may play a role in the cascade of phosphorylation/dephosphorylation reactions controlling glycogen synthesis and degradation.

References

1. Rake JP, Visser G, Labrune P et al (2002) Glycogen storage disease type I: diagnosis, management, clinical course and outcome. Results of the European Study on Glycogen Storage Disease Type I (ESGSD I). Eur J Pediatr 161[Suppl 1]:S20-S34
2. Kuijpers TW, Maianski NA, Tool AT et al (2003) Apoptotic neutrophils in the circulation of patients with glycogen storage disease type 1b (GSD1b). Blood 101:5021-5024
3. Visser G, Rake JP, Fernandes J et al (2000) Neutropenia, neutrophil dysfunction, and inflammatory bowel disease in glycogen storage disease type Ib: results of the European Study on Glycogen Storage Disease type I. J Pediatr 137:187-191
4. Foster JD, Nordlie RC (2002) The biochemistry and molecular biology of the glucose-6-phosphatase system. Exp Biol Med (Maywood) 227:601-608
5. Waddell ID, Burchell A (1993) Identification, purification and genetic deficiencies of the glucose-6-phosphatase system transport proteins. Eur J Pediatr 152[Suppl 1]: S14-S17
6. Veiga-da-Cunha M, Gerin I, Chen YT et al (1999) The putative glucose 6-phosphate translocase gene is mutated in essentially all cases of glycogen storage disease type I non-a. Eur J Hum Genet 7: 717-723
7. Burchell A (1998) A reevaluation of GLUT 7. Biochem J 331:973
8. Melis D, Havelaar AC, Verbeek E et al (2004) NPT4, a new microsomal phosphate transporter: mutation analysis in glycogen storage disease type Ic. J Inherit Metab Dis 27: 725-733

9. Collins JE, Bartlett K, Leonard JV, Ayynsley-Green A (1990) Glucose production rates in type 1 glycogen storage disease. J Inherit Metab Dis 13:195-206

10. Fernandes J (1974) The effect of disaccharides on the hyperlactacidaemia of glucose-6-phosphatase-deficient children. Acta Paediatr Scand 63: 695-698

11. Fernandes J, Alaupovic P, Wit JM (1989) Gastric drip feeding in patients with glycogen storage disease type I: its effects on growth and plasma lipids and apolipoproteins. Pediatr Res 25: 327-331

12. Greene HL, Swift LL, Knapp HR (1991) Hyperlipidemia and fatty acid composition in patients treated for type IA glycogen storage disease. J Pediatr 119:398-403

13. Alaupovic P, Fernandes J (1985) The serum apolipoprotein profile of patients with glucose-6-phosphatase deficiency. Pediatr Res 19:380-384

14. Bandsma RH, Smit GP, Kuipers F (2002) Disturbed lipid metabolism in glycogen storage disease type 1. Eur J Pediatr 161[Suppl 1]:S65-S69

15. Fernandes J, Pikaar NA (1972) Ketosis in hepatic glycogenosis. Arch DisChild 47: 41-46

16. Greene HL, Wilson FA, Hefferan P et al (1978) ATP depletion, a possible role in the pathogenesis of hyperuricemia in glycogen storage disease type I. J Clin Invest 62:321-328

17. Cohen JL, Vinik A, Faller J, Fox IH (1985) Hyperuricemia in glycogen storage disease type I. Contributions by hypoglycemia and hyperglucagonemia to increased urate production. J Clin Invest 75: 251-257

18. Matern D, Seydewitz HH, Bali D, Lang C, Chen YT (2002) Glycogen storage disease type I: diagnosis and phenotype/genotype correlation. Eur J Pediatr 161[Suppl 1]: S10-S19

19. Rake JP, ten Berge AM, Visser G et al (2000) Glycogen storage disease type Ia: recent experience with mutation analysis, a summary of mutations reported in the literature and a newly developed diagnostic flow chart. Eur J Pediatr 159:322-330

20. Chou JY, Matern D, Mansfield BC, Chen YT (2002) Type I glycogen storage diseases: disorders of the glucose-6-phosphatase complex. Curr Mol Med 2:121-143

21. Lin B, Hiraiwa H, Pan CJ, Nordlie RC, Chou JY (1999) Type-1c glycogen storage disease is not caused by mutations in the glucose-6-phosphate transporter gene. Hum Genet 105:515-517

22. Narisawa K, Otomo H, Igarashi Y et al (1983) Glycogen storage disease type 1b: microsomal glucose-6-phosphatase system in two patients with different clinical findings. Pediatr Res 17:545-549

23. Burr IM, O'Neill JA, Karzon DT, Howard LJ, Greene HL (1974) Comparison of the effects of total parenteral nutrition, continuous intragastric feeding, and portacaval shunt on a patient with type I glycogen storage disease. J Pediatr 85:792-795

24. Chen YT, Cornblath M, Sidbury JB (1984) Cornstarch therapy in type I glycogen-storage disease. N Engl J Med 310:171-175

25. Wolfsdorf JI, Crigler JF, Jr (1999) Effect of continuous glucose therapy begun in infancy on the long-term clinical course of patients with type I glycogen storage disease. J Pediatr Gastroenterol Nutr 29:136-143

26. Smit GP, Ververs MT, Belderok B, Van Rijn M, Berger R, Fernandes J (1988) Complex carbohydrates in the dietary management of patients with glycogenosis caused by glucose-6-phosphatase deficiency. Am J Clin Nutr 48:95-97

27. Wolfsdorf JI, Keller RJ, Landy H, Crigler JF, Jr (1990) Glucose therapy for glycogenosis type 1 in infants: comparison of intermittent uncooked cornstarch and continuous overnight glucose feedings. J Pediatr 117:384-391

28. Fernandes J, Berger R, Smit GP (1984) Lactate as a cerebral metabolic fuel for glucose-6-phosphatase deficient children. Pediatr Res 18:335-339

29. Daublin G, Schwahn B, Wendel U (2002) Type I glycogen storage disease: favourable outcome on a strict management regimen avoiding increased lactate production during childhood and adolescence. Eur J Pediatr 161[Suppl 1]:S40-S45

30. Weinstein DA., Somers MJ, Wolfsdorf JI (2001) Decreased urinary citrate excretion in type 1a glycogen storage disease. J Pediatr 138:378-382

31. Wittenstein B, Klein M, Finckh B, Ullrich K, Kohlschutter A (2002) Radical trapping in glycogen storage disease 1a. Eur J Pediatr 161[Suppl 1]:S70-S74

32. Bandsma RH, Rake JP, Visser G et al (2002) Increased lipogenesis and resistance of lipoproteins to oxidative modification in two patients with glycogen storage disease type 1a. J Pediatr 140:256-260

33. Mairovitz V, Labrune P, Fernandez H, Audibert F, Frydman R (2002) Contraception and pregnancy in women affected by glycogen storage diseases. Eur J Pediatr 161[Suppl 1]:S97-101

34. Kerr KG (1999) The prophylaxis of bacterial infections in neutropenic patients. J Antimicrob Chemother 44:587-591

35. Visser G, Rake JP, Labrune P et al (2002) Granulocyte colony-stimulating factor in glycogen storage disease type 1b. Results of the European Study on Glycogen Storage Disease Type 1. Eur J Pediatr 161[Suppl 1]:S83-S87

36. Visser G, Rake JP, Labrune P et al (2002) Consensus guidelines for management of glycogen storage disease type 1b - European Study on Glycogen Storage Disease Type 1. Eur J Pediatr 161[Suppl 1]:S120-S123

37. Simmons PS, Smithson WA, Gronert GA, Haymond MW (1984) Acute myelogenous leukemia and malignant hyperthermia in a patient with type 1b glycogen storage disease. J Pediatr 105: 428-431

38. Donadieu J, Barkaoui M, Bezard F, Bertrand Y, Pondarre C, Guiband P (2000) Renal carcinoma in a patient with glycogen storage disease Ib receiving long-term granulocyte colony-stimulating factor therapy. J Pediatr Hematol Oncol 22:188-189

39. Rake JP, Visser G, Labrune P, Leonard JV, Ullrich K, Smit GP (2002) Guidelines for management of glycogen storage disease type I – European Study on Glycogen Storage Disease Type I (ESGSD I). Eur J Pediatr 161[Suppl 1]:S112-S119

40. Hagen T, Korson MS, Wolfsdorf JI (2000) Urinary lactate excretion to monitor the efficacy of treatment of type I glycogen storage disease. Mol Genet Metab 70:189-195

41. Lee PJ, Chatterton C, Leonard JV (1996) Urinary lactate excretion in type 1 glycogenosis--a marker of metabolic control or renal tubular dysfunction? J Inherit Metab Dis 19:201-204

42. Visser G, Rake JP, Kokke FT, Nikkels PG, Sauer PJ, Smit GP (2002) Intestinal function in glycogen storage disease type I. J Inherit Metab Dis 25:261-267

43. Talente GM, Coleman RA, Alter C et al (1994) Glycogen storage disease in adults. Ann Intern Med 120:218-226

44. Chen YT (1991) Type I glycogen storage disease: kidney involvement, pathogenesis and its treatment. Pediatr Nephrol 5:71-76

45. Lee PJ, Dalton RN, Shah V, Hindmarsh PC, Leonard JV (1995) Glomerular and tubular function in glycogen storage disease. Pediatr Nephrol 9:705-710

46. Chen YT, Scheinman JI, Park HK, Coleman RA, Roe CR (1990) Amelioration of proximal renal tubular dysfunction in type I glycogen storage disease with dietary therapy. N Engl J Med 323:590-593

47. Iida S, Matsuoka K, Inouse M, Tomiyasu K, Noda S (2003) Calcium nephrolithiasis and distal tubular acidosis in type 1 glycogen storage disease. Int J Urol 10:56-58

48. Restaino I, Kaplan BS, Stanley C, Baker L (1993) Nephrolithiasis, hypocitraturia, and a distal renal tubular acidification defect in type 1 glycogen storage disease. J Pediatr 122:392-396

49. Baker L, Dahlem S, Goldfarb S et al (1989) Hyperfiltration and renal disease in glycogen storage disease, type I. Kidney Int 35:1345-1350

50. Chen YT, Coleman RA, Scheinman JI, Kolbeck PC, Sidbury JB (1988) Renal disease in type I glycogen storage disease. N Engl J Med 318:7-11

51. Labrune P, Trioche P, Duvaltier I, Chevalier P, Odievre M (1997) Hepatocellular adenomas in glycogen storage disease type I and III: a series of 43 patients and review of the literature. J Pediatr Gastroenterol Nutr 24:276-279

52. Lee PJ (2002) Glycogen storage disease type I: pathophysiology of liver adenomas. Eur J Pediatr 161[Suppl 1]: S46-S49

53. Bianchi L (1993) Glycogen storage disease I and hepatocellular tumours. Eur J Pediatr 152[Suppl 1]:S63-S70

54. Labrune P (2002) Glycogen storage disease type I: indications for liver and/or kidney transplantation. Eur J Pediatr 161[Suppl 1]:S53-S55

55. Matern D, Starzl TE, Arnaout W et al (1999) Liver transplantation for glycogen storage disease types I, III, and IV. Eur J Pediatr 158[Suppl 2]:S43-S48

56. Lee PJ, Patel JS, Fewtrell M, Leonard JV, Bishop NJ (1995) Bone mineralisation in type 1 glycogen storage disease. Eur J Pediatr 154:483-487

57. Rake JP, Visser G, Huismans D et al (2003) Bone mineral density in children, adolescents and adults with glycogen storage disease type Ia: a cross-sectional and longitudinal study. J Inherit.Metab Dis 26:371-384

58. Weinstein DA, Roy CN, Fleming MD, Loda MF, Wolfsdorf JI, Andrews NC (2002) Inappropriate expression of hepcidin is associated with iron refractory anemia: implications for the anemia of chronic disease. Blood 100:3776-3781

59. Lee PJ, Patel A, Hindmarsh PC, Mowat AP, Leonard JV (1995) The prevalence of polycystic ovaries in the hepatic glycogen storage diseases: its association with hyperinsulinism. Clin Endocrinol (Oxf) 42:601-606

60. Lee PJ, Celermajer DS, Robinson J, McCarthy SN, Betteridge DJ, Leonard JV (1994) Hyperlipidaemia does not impair vascular endothelial function in glycogen storage disease type 1a. Atherosclerosis 110:95-100

61. Corby DG, Putnam CW, Greene HL (1974) Impaired platelet function in glucose-6-phosphatase deficiency. J Pediatr 85:71-76

62. Trioche P, Francoual J, Capel L, Odievre M, Lindenbaum A, Labrune P (2000) Apolipoprotein E polymorphism and serum concentrations in patients with glycogen storage disease type Ia. J Inherit Metab Dis 23:107-112

63. Humbert M, Labrune P, Simonneau G (2002) Severe pulmonary arterial hypertension in type 1 glycogen storage disease. Eur J Pediatr 161[Suppl 1]: S93-S96

64. DiMauro S, Hartwig GB, Hays A et al (1979) Debrancher deficiency: neuromuscular disorder in 5 adults. Ann Neurol 5:422-436

65. DiMauro S, Hays AP, Tsujino S (2004) Nonlysosomal glycogenosis. In: Engel AG, Franzini-Amstrong C (eds) Myology: basic and clinical. McGraw-Hill, New York, pp 1535-1558

66. Wolfsdorf JI, Weinstein DA (2003) Glycogen storage diseases. Rev Endocrinol Metab Disord 4:95-102

67. Lucchiari S, Fogh I, Prelle A et al (2002) Clinical and genetic variability in glycogen storage disease type IIIa: Seven novel AGL gene mutations in the Mediterranean area. Am J Med Genet 109:183-190

68. Lee P (1999) Successful pregnancy in a patient with type III glycogen storage disease managed with cornstarch supplements. Br J Obstet Gynaecol 106:181-182

69. de Moor RA, Schweizer JJ, van Hoek B, Wasser M, Vink R, Maaswinkel-Mooy PD (2000) Hepatocellular carcinoma in glycogen storage disease type IV. Arch Dis Child 82:479-480

70. Moses SW, Parvari R (2002) The variable presentations of glycogen storage disease type IV: a review of clinical, enzymatic and molecular studies. Curr Mol Med 2:177-188

71. Selby R, Starzl TE, Yunis E et al (1993) Liver transplantation for type I and type IV glycogen storage disease. Eur J Pediatr 152[Suppl 1]: S71-S76

72. Lossos A, Meiner Z, Barash V et al (1998) Adult polyglucosan body disease in Ashkenazi Jewish patients carrying the Tyr329Ser mutation in the glycogen-branching enzyme gene. Ann Neurol 44:867-872

73. Tay SK, Akman HO, Chung WK et al (2004) Fatal infantile neuromuscular presentation of glycogen storage disease type IV. Neuromuscul Disord 14: 253-260

74. Bao Y, Kishnani P, Wu JY, Chen HT (1996) Hepatic and neuromuscular forms of glycogen storage disease type IV caused by mutations in the same glycogen-branching enzyme gene. J Clin Invest 97:941-948

75. Hendrickx J, Willems PJ (1996) Genetic deficiencies of the glycogen phosphorylase system. Hum Genet 97:551-556

76. Huijing F, Fernandes J (1969) X-chromosomal inheritance of liver glycogenosis with phosphorylase kinase deficiency. Am J Hum Genet 21:275-284

76a. Arad M, Maron BJ, Gorham JM et al (2005) Glycogen storage disease presenting as hypertrophic cardiomyopathy. N Engl J Med 352:362-372

77. Fernandes J, Koster JF, Grose WF, Sorgedrager N (1974) Hepatic phosphorylase deficiency. Its differentiation from other hepatic glycogenoses. Arch Dis Child 49:186-191

78. Willems PJ, Gerver WJ, Berger R, Fernandes J (1990) The natural history of liver glycogenosis due to phosphorylase kinase deficiency: a longitudinal study of 41 patients. Eur J Pediatr 149:268-271

79. Hendrickx J, Dams E, Coucke P, Lee P, Fernandes J, Willems PJ (1996) X-linked liver glycogenosis type II (XLG II) is caused by mutations in PHKA2, the gene encoding the liver alpha subunit of phosphorylase kinase. Hum Mol Genet 5:649-652

80. Hendrickx J, Lee P, Keating JP et al (1999) Complete genomic structure and mutational spectrum of PHKA2 in patients with x-linked liver glycogenosis type I and II. Am J Hum Genet 64:1541-1549

81. Bruno C, Manfredi G, Andreu AL et al (1998) A splice junction mutation in the alpha(M) gene of phosphorylase kinase in a patient with myopathy. Biochem Biophys Res Commun 249:648-651

82. Wehner M, Clemens PR, Engel AG, Kilimann MW (1994) Human muscle glycogenosis due to phosphorylase kinase deficiency associated with a nonsense mutation in the muscle isoform of the alpha subunit. Hum Mol Genet 3:1983-1987

83. Burwinkel B, Hu B, Schroers A et al (2003) Muscle glycogenosis with low phosphorylase kinase activity: mutations in PHKA1, PHKG1 or six other candidate genes explain only a minority of cases. Eur J Hum Genet 11:516-526

84. Schippers HM, Smit GP, Rake JP, Visser G (2003) Characteristic growth pattern in male X-linked phosphorylase-b kinase deficiency (GSD IX). J Inherit Metab Dis 26:43-47

85. Aynsley-Green A, Williamson DH, Gitzelmann R (1977) Hepatic glycogen synthetase deficiency. Definition of syndrome from metabolic and enzyme studies on a 9-year-old girl. Arch Dis Child 52:573-579

86. Laberge AM, Mitchell GA, van de Werve G, Lambert M (2003) Long-term follow-up of a new case of liver glycogen synthase deficiency. Am J Med Genet A 120:19-22

87. Orho M, Bosshard NU, Buist NR et al (1998) Mutations in the liver glycogen synthase gene in children with hypoglycemia due to glycogen storage disease type 0. J Clin Invest 102:507-515

88. Tonin P, Lewis PJ, Servidei S, DiMauro S (1990) Metabolic causes of myoglobinuria. Ann Neurol 27:181-185

89. Martin MA, Rubio JC, Wevers RA et al (2004) Molecular analysis of myophosphorylase deficiency in Dutch patients with McArdle's disease. Ann Hum Genet 68:17-22

90. Bartram C, Edwards RH, Clague J, Beynon RJ (1993) McArdle's disease: a nonsense mutation in exon 1 of the muscle glycogen phosphorylase gene explains some but not all cases. Hum Mol Genet 2:1291-1293

91. el Schahawi M, Tsujino S, Shanske S, DiMauro S (1996) Diagnosis of McArdle's disease by molecular genetic analysis of blood. Neurology 47:579-580

92. Tsujino S, Shanske S, Carroll JE, Sabina RL, DiMauro S (1994) Two mutations, one novel and one frequently observed, in Japanese patients with McArdle's disease. Hum Mol Genet 3:1005-1006

93. Martinuzzi A, Sartori E, Fanin M et al (2003) Phenotype modulators in myophosphorylase deficiency. Ann Neurol 53:497-502

94. Kazemi-Esfarjani P, Skomorowska E, Jensen TD, Haller RG, Vissing A (2002) A nonischemic forearm exercise test for McArdle disease. Ann Neurol 52:153-159

95. Vissin J, Haller RG (2003) A diagnostic cycle test for McArdle's disease. Ann Neurol 54:539-542

96. Haller RG (2000) Treatment of McArdle disease. Arch Neurol 57:923-924

97. Vissing J, Haller RG (2003) The effect of oral sucrose on exercise tolerance in patients with McArdle's disease. N Engl J Med 349:2503-2509

98. Haller RG, Vissing J (2004) No spontaneous second wind in muscle phosphofructokinase deficiency. Neurology 62:82-86

99. Morimoto A, Ueda I, Hirashima Y et al (2003) A novel missense mutation (1060G → C) in the phosphoglycerate kinase gene in a Japanese boy with chronic haemolytic anaemia, developmental delay and rhabdomyolysis. Br J Haematol 122:1009-1013

100. Schroder JM, Dodel R, Weis J, Stefanidis I, Reichmann H (1996) Mitochondrial changes in muscle phosphoglycerate kinase deficiency. Clin Neuropathol 15:34-40

101. Vissing J, Schmalbruch H, Haller RG, Clausen T (1999) Muscle phosphoglycerate mutase deficiency with tubular aggregates: effect of dantrolene. Ann Neurol 46: 274-277

102. Kreuder J, Borkhardt A, Repp R et al (1996) Brief report: inherited metabolic myopathy and hemolysis due to a mutation in aldolase A. N Engl J Med 334:1100-1104

103. Comi GP, Fortunato F, Lucchiari S et al (2001) Beta-enolase deficiency, a new metabolic myopathy of distal glycolysis. Ann Neurol 50:202-207

104. Kanno T, Maekawa M (1995) Lactate dehydrogenase M-subunit deficiencies: clinical features, metabolic background, and genetic heterogeneities. Muscle Nerve 3:S54-S60

104a.Holme E, Kollberg G, Oldfors A et al (2005) Muscular glycogen storage disease 0 – A new disease entity in a child with hypertrophic cardiomyopathy and myopathy due to a homozygous stop mutation in the muscular glycogen synthase gene (GYS1). J Inherit Metab Dis 28[Suppl 1]:214

105. Van den Hout HM, Hop W, van Diggelen OP et al (2003) The natural course of infantile Pompe's disease: 20 original cases compared with 133 cases from the literature. Pediatrics 112:332-340

106. Makos MM, McComb RD, Hart MN, Bennett DR (1987) Alpha-glucosidase deficiency and basilar artery aneurysm: report of a sibship. Ann Neurol 22:629-633

107. Hagemans ML, Janssens AC, Winkel LP et al (2004) Late-onset Pompe disease primarily affects quality of life in physical health domains. Neurology 63:1688-1692

108. Van den Hout JM, Kamphoven JH, Winkel LP et al (2004) Long-term intravenous treatment of Pompe disease with recombinant human alpha-glucosidase from milk. Pediatrics 113:e448-e457

109. Winkel LP, Van den Hout JM, Kamphoven JH et al (2004) Enzyme replacement therapy in late-onset Pompe's disease: a three-year follow-up. Ann Neurol 55:495-502

110. Danon MJ, Oh SJ, DiMauro S et al (1981) Lysosomal glycogen storage disease with normal acid maltase. Neurology 31:51-57

111. Nishino I, Fu J, Tanji K et al (2000) Primary LAMP-2 deficiency causes X-linked vacuolar cardiomyopathy and myopathy (Danon disease). Nature 406:906-910

112. Dworzak F, Casazza F, Mora M et al (1994) Lysosomal glycogen storage with normal acid maltase: a familial study with successful heart transplant. Neuromuscul Disord 4:243-247

113. Minassian BA, Ianzano L, Meloche M et al (2000) Mutation spectrum and predicted function of laforin in Lafora's progressive myoclonus epilepsy. Neurology 55:341-346

7 Disorders of Galactose Metabolism

Gerard T. Berry, Stanton Segal, Richard Gitzelmann

»Whenever you consider a galactose disorder, stop milk feeding first and only then seek a diagnosis!«

Galactose Metabolism

Together with its 4'-epimer, glucose, galactose forms the disaccharide lactose, which is the principal carbohydrate in milk, providing 40 % of its total energy. Ingested, exogenous lactose is hydrolyzed in the small intestine to galactose (◼ Fig. 7.1a), and glucose by lactase. Galactose is mainly metabolized into galactose-1-phosphate (galactose-1-P) by galactokinase (GALK). Galactose-1-P uridyltransferase (GALT) converts uridine diphosphoglucose (UDPglucose) and galactose-1-P into uridine diphosphogalactose (UDPgalactose) and glucose-1-P. The latter is metabolized into glucose-6-P from which glucose, pyruvate and lactate are formed (not illustrated). Galactose can also be converted into galactitol by aldose reductase, and into galactonate by galactose dehydrogenase. UDPglucose (or UDP-N-acetylglucosamine) can be converted into UDPgalactose (or UDP-N-acetylgalactosomine) by UDPgalactose 4'-epimerase (GALE). The utilization of UDPgalactose in the synthesis of glycoconjugates including glycoproteins, glycolipids and glycosaminoglycans, and their subsequent degradation (◼ Fig. 7.1a) may constitute the pathways of endogenous, de novo synthesis of galactose. All four of these uridine sugar nucleotides are used for glycoconjugate synthesis. UDPglucose is also the key element in glycogen production while UDPgalactose is used for lactose synthesis. The UDPglucose pyrophosphorylase enzyme (◼ Fig. 7.1b) that is primarily responsible for interconversion of UDPglucose and glucose-1-P can catalyze, albeit in a limited way, the interconversion of UDPgalactose and galactose-1-P, and also contribute to endogenous synthesis of galactose.

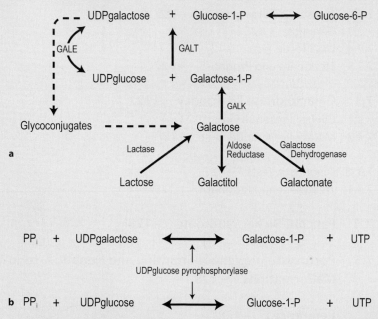

◼ **Fig. 7.1a,b.** Galactose metabolism (simplified). *GALE,* UDP galactose 4'-epimerase; *GALK,* galactokinase; *GALT,* galactose-1-P uridyltransferase; *P,* phosphate; *PP_i,* pyrophosphate; *UDP,* uridine diphosphate; *UTP,* uridine triphosphate. The pathways with multiple enzymatic steps are shown by broken lines

Three inborn errors of galactose metabolism are known. The most important is classic galactosemia due to galactose-1-phosphate uridyltransferase (GALT) deficiency. A complete or near-complete deficiency is life threatening with multiorgan involvement and long-term complications [1]. Partial deficiency is usually, but not always, benign. Uridine diphosphate galactose 4-epimerase (GALE) deficiency exists in at least two forms. The very rare profound deficiency clinically resembles classical galactosemia. The more frequent partial deficiency is usually benign. Galactokinase (GALK) deficiency is extremely rare and the most insidious, since it results in the formation of nuclear cataracts without provoking symptoms of intolerance. The Fanconi-Bickel syndrome (Chap. 11) is a congenital disorder of galactose transport due to GLUT2 deficiency leading to hypergalactosemia. Other secondary causes of impaired liver handling of galactose in the neonatal period are congenital portosystemic shunting and multiple hepatic arteriovenous malformations.

7.1 Deficiency of Galactose-1-Phosphate Uridyltransferase

7.1.2 Clinical Presentation

As over 167 mutations in the GALT gene have been identified [2–4], different forms of the deficiency exist. Infants with complete or near-complete deficiency of the enzyme (classical galactosemia) have normal weight at birth but, as they start drinking milk, lose more weight than their healthy peers and fail to regain birth weight. Symptoms appear in the second half of the first week and include refusal to feed, vomiting, jaundice and lethargy. Hepatomegaly, edema and ascites may follow. Death from sepsis, usually due to E.coli, may follow within days but it has been noted as early as 3 days of age. Symptoms are milder and the course is less precipitous when milk is temporarily withdrawn and replaced by intravenous nutrition. Nuclear cataracts appear within days or weeks and become irreversible within weeks of their appearance. Congenital cataracts and vitreous hemorrhages [5] may also be present.

In many countries, newborns with galactosemia are discovered through mass screening for blood galactose, the transferase enzyme or both; this screening is performed using dried blood spots usually collected between the second and seventh days. At the time of discovery, the first symptoms may already have appeared, and the infant may already have been admitted to a hospital, usually for jaundice. Where newborns are not screened for galactosemia or when the results of screening are not yet available, diagnosis rests on clinical awareness. It is crucial that milk feeding be stopped as soon as galactosemia is considered, and resumed only when a galactose disorder has been excluded. The presence of a reducing substance in a routine urine specimen may be the first diagnostic lead. Galactosuria is present provided the last milk feed does not date back more than a few hours and vomiting has not been excessive. However, owing to the early development of a proximal renal tubular syndrome, the acutely ill galactosemic infant may also excrete some glucose, together with an excess of amino acids. While hyperaminoaciduria may aid in the diagnosis, glucosuria often complicates it. When both reducing sugars (galactose and glucose) are present and reduction and glucose tests are done, and when the former test is strongly positive and the latter is weakly positive, the discrepancy is easily overlooked. Glucosuria is recognized, and galactosuria is missed. On withholding milk, galactosuria ceases, but amino acids in excess continue to be excreted for a few days. However, galactitol and galactonate continue to be excreted in large amounts. Albuminuria may also be an early finding that disappears with dietary lactose restriction.

Partial transferase deficiency associated with 25% residual GALT activity is usually asymptomatic. It is more frequent than classical galactosemia and is most often discovered in mass newborn screening because of moderately elevated blood galactose (free and/or total) and/or low transferase activity. In partial deficiency with only 10% residual GALT activity, there may be liver disease and mental retardation in patients left untreated during early infancy.

7.1.2 Metabolic Derangement

Individuals with a profound deficiency of GALT can phosphorylate ingested galactose but fail to metabolize galactose-1-phosphate. As a consequence, galactose-1-phosphate and galactose accumulate, and the alternate pathway metabolites, galactitol and galactonate, are formed. Cataract formation can be explained by galactitol accumulation. The pathogenesis of the hepatic, renal and cerebral disturbances is less clear but is probably related to the accumulation of galactose-1-phosphate and (perhaps) of galactitol.

7.1.3 Genetics

The mode of inheritance is autosomal recessive. The birth incidence of classical galactosemia is one in 40,000–60,000. In Ireland it is one in 10,000–20,000. The gene is situated on chromosome 9, and over 167 mutations or polymorphisms have been described [2–4; see the following website: http://www.alspac.bris.ac.uk/galtdb/genomic_seq.htm]. Some genotype-phenotype matching is available [6-13]. For instance, homozygosity for the Q188R mutation, unfortunately prevalent, has been associated with unfavorable clinical outcome [11–13]. Because transferase polymorphism abounds [2–4], partial transferase deficiency is more frequent than classical galactosemia. Many allelic variants associated with a partial enzyme defect have been reported, but the best known is the Duarte variant due to a N314D *GALT* gene mutation that exists in cis with a small deletion in the 5′ flanking region [2]. Variants such as the Q188R/N314D compound heterozygote can be distinguished by enzyme electrophoresis or DNA analysis. The N314D Duarte variant when combined with the severe Q188R mutation is almost always benign.

7.1.4 Diagnostic Tests

Diagnosis is made by assaying transferase in heparinized whole blood or erythrocyte lysates, and/or by measuring abnormally high levels of galactose-1-phosphate in red cells. Where rapid shipment of whole blood is difficult, blood dried on filter paper can also be used for a semiquantitative assay. In patients with classical galactosemia, deficiency of GALT is complete or nearly complete. It should be noted that, when an infant has received an exchange transfusion, as is often the case, assays in blood must be postponed for three to four months. In this situation, an assay of urinary galactitol will be extremely helpful. Mutation analysis of the *GALT* gene in genomic DNA isolated from leukocytes may indicate a diagnosis of GALT deficiency. In some hospitals, a blood specimen, liquid or dried on filter paper, is collected prior to every exchange transfusion. The finding of reduced transferase activity in parental blood may provide additional helpful information since, in heterozygotes, the enzyme activity in erythrocytes is approximately 50% of normal. Cultured skin fibroblasts can also be used for the enzyme assay. If taken post-mortem, liver or kidney cortex may provide diagnostic enzyme information but these specimens must be adequately collected and frozen, since in vivo cell damage and/or autolysis may result in decreased enzyme activity. Antenatal diagnosis is possible by measuring transferase activity in cultured amniotic fluid cells, biopsied chorionic villi, or amniotic fluid galactitol [14]. Restricting maternal lactose intake does not interfere with a diagnosis based on galactitol measurements in amniotic fluid.

In partial transferase deficiency, there is a spectrum of residual enzyme activities in the erythrocyte with the most common partial deficiency, the compound heterozygote Duarte/Galactosemia (D/G) defect, having approximately 25% of the normal mean activity. As a rule, erythrocyte galactose-1-phosphate is also elevated. Each newborn with partial GALT deficiency must nevertheless be observed closely, because allelic variants other than Duarte may be operative, and they may be true clinically relevant variants such as individuals of African descent with a S135L/S135L

genotype [10]. Assessment involves quantitation of plasma galactose and galactitol, of erythrocyte galactose-1-phosphate, galactitol, galactonate and GALT activity/enzyme electrophoresis (isoelectric focusing) and investigation of the parents. Galactose-tolerance tests are notoriously noxious to the child with classical galactosemia and have no place in evaluating the need for treatment of partial deficiencies.

7.1.5 Treatment and Prognosis

Treatment of the newborn with classical galactosemia consists of the exclusion of all lactose from the diet. This must be started immediately after the disorder is suspected clinically or following a positive newborn screening results even before the results of diagnostic tests are available. When a lactose-free diet is instituted early enough, symptoms disappear promptly, jaundice resolves within days, cataracts may clear, liver and kidney functions return to normal and liver cirrhosis may be prevented.

For dietary treatment, the following facts are worthy of consideration:

— From early embryonic life on, man is capable of synthesizing UDPgalactose from glucose through the epimerase reaction, which converts UDPglucose to UDPgalactose. Therefore, man does not depend on exogenous galactose. Raising a child with galactosemia on a diet completely devoid of galactose is a lofty goal of many zealous caregivers; yet, such a diet does not exist! In fact, and this is a point of contention in long-term care, an ultra-strict diet has never been shown to be safe for a patient with GALT deficiency. Utilizing an »evidence-based medicine« approach, the only therapy that is convincingly beneficial is the exclusion of lactose from the diet of a neonate and young infant with galactosemia. Single reports and anecdotal information suggest that children and/or adults may suffer cataracts [15], liver disease [16] and organic brain disease [17] with ingestion of lactose. However, there is no evidence that galactose contained in fruits and vegetables had played a role in these rare patients that may harbor non-GALT modifier genes, which render them more susceptible to complications.

— Nonetheless, milligram amounts of galactose cause an appreciable rise of galactose-1-phosphate in erythrocytes (e.g. ~500 mg of galactose in a 70 kg adult with Q188R/Q188R genotype will increase galactose-1-phosphate by 30% in 8 hours); it is possible that the same happens in sensitive tissues, such as brain, liver and kidney. However, at this time, it is impossible to define toxic tissue levels of galactose-1-phosphate and, therefore, safe amounts of dietary galactose cannot be defined. Patients with relatively increased alternate metabolic pathway activities should have greater tolerance for galactose. More and more cases are being described, albeit most are anecdotal in nature, in which a child or adult with classic galactosemia is able to ingest a normal diet without any obvious side-effects [18].

- Patients with galactosemia certainly synthesize galactose from glucose. This is also true for the fetal-placental unit. Healthy pregnant women on a lactose-restricted diet may give birth to healthy newborns whose tissues are laden with galactose-containing macromolecules. In newborns first exposed to milk, then diagnosed and treated properly, erythrocyte galactose-1-phosphate stays high for several weeks. These facts and other observations [19–25] are evidence for continuous self-intoxication [26] by the patient, a matter of concern because of some late complications such as premature ovarian failure [27–29] and central nervous system dysfunction [11, 30–32]. In adults on a strict lactose-exclusion diet, galactose intake was estimated at 20–40 mg/ day; at the same time, they produced more galactose endogenously than they consumed in their diets [21, 33]. Minimal amounts of galactose from food and hidden sources may contribute to erythrocyte galactose-1-phosphate, but only real breaks in the diet, such as with dairy products, are likely to cause a rise above 6 mg/dl. Such breaks do not cause any discomfort to the patient who, therefore, never develops aversion to galactose-containing food. The measurement of urinary galactitol for monitoring treatment has not been successful when used to identify acute effects, but may be beneficial when the ingestion is on a daily basis [33].

Treatment of the Newborn Infant

Treating newborns is comparatively easy, as adequate lactose-free soy-based formulas are available. However, there has been concern about the safety of soy-based infant formulas containing isoflavones. At present, there is no conclusive evidence of adverse effects [34]. Elimination of milk and milk products is the mainstay of treatment.

Spoon-Feeding

When spoon-feeding is started, parents must learn to know all other sources of lactose and need assistance from the pediatrician and dietitian, who must have recourse to published recommendations [35]. Parents are advised to do the following:
- Prepare meals from basic foodstuffs
- Avoid canned food, byproducts and preserves unless they are certified not to contain lactose or dairy products
- Read and reread labels and declarations of ingredients, which may change without notification
- Look out for hidden sources of galactose and lactose from milk powder, milk solids, hydrolyzed whey (a sweetener labeled as such), drugs in tablet form, toothpaste, baking additives, fillers, sausages etc.

- Support campaigns for complete food and drug labeling

Vegetables and Fruits

Parents must be trained to understand that eliminating all galactose from the diet can never be reached. The reason for this is that galactose is present in a great number of vegetables and fruits [36], as a component of galactolipids and glycoproteins, in the disaccharide melibiose and in the oligosaccharides raffinose and stachyose [37, 38]. The latter two contain galactose in alpha-galactosidic linkage not hydrolyzable by human small intestinal mucosa in vitro or in vivo [38]. They are often considered safe for consumption by patients. However, this may not be the case when the small intestine is colonized by bacteria capable of releasing galactose. Theoretically, ingestion of raffinose- and stachyose-rich vegetables (beans, peas, lentils etc.) by a patient who has diarrhea may lead to enhanced intestinal absorption of galactose. However, gastroenterologists have stated that the small intestine may be colonized even in the absence of diarrhea; obviously, the issue is not closed. In addition, the normal inhabitants of the large colon may facilitate the release of galactose from macromolecules that pass through.

Cheese

It is not generally known that Swiss cheeses of the Emmentaler, Gruyère, and Tilsiter types are galactose- and lactose-free, as these sugars are cleared by the fermenting microorganisms [39]. Other hardened cheeses may prove equally safe for patients. Calcium supplements should be prescribed before cheese is introduced to the child's diet; supplements may also be needed by older children and young adults [40]. Calcium prescriptions containing lactobionate [30] may also be a source of galactose because the beta-galactosidase of human intestinal mucosa hydrolyses lactobionate, freeing galactose [41].

Breaks of Discipline

Whether single or repeated breaks of discipline (such as occasional ice cream by a school-age child or adult with galactosemia) will cause any damage is unknown. Dietary treatment of female patients is continued during pregnancy [42].

Complications of Treated Galactosemia

Mild growth retardation, delayed speech development, verbal dyspraxia, difficulties in spatial orientation and visual perception, and mild intellectual deficit have been variably described as complications of treated galactosemia. The complete set of sequelae is not necessarily present in every patient, and the degree of handicap appears to vary widely. Ovarian dysfunction, an almost inescapable consequence of galactosemia is not prevented even by strict diet and is often signaled early in infancy or childhood by hypergonadotropism. Less than five women with the Q188R/ Q188R genotype have experienced one or more successful pregnancies and deliveries; some of them subsequently developed secondary amenorrhea. Since in female patients, the number of expected ovulatory cycles is limited, it may be wise to temporarily suppress cycles by birth-control medication, which is lifted when the young woman wishes to become pregnant. This is not an established form of therapy, in contrast with chronic estrogen and progesterone supplementation. Prescription is hampered by the fact that seemingly all drug tablets contain lactose, providing 100 mg or more of the noxious sugar per treatment day [33]. However, some female patients have received the birth-control medication containing galactose for many years without any obvious side effects [17].

Long-Term Results

Several reports have indicated the lack of effectiveness of dietary treatment on long-term complications [1, 11, 28–32, 43, 44]. It must be stressed here that said studies were retrospective, not prospective, and not multicentered using the same instruments and endpoints, and were probably marred by negative selection of patients. There has never been an adequate prospective study of patients with galactosemia to document the natural history and done in conjunction with proper dietary monitoring. More recently, the quality of life in treated patients has also been called into question [45]. Also, some patients, males in particular, manifest an introverted personality and/or depression [17].

Treatment of Partial Transferase Deficiency due to D/G genotype

Because it is impossible to decide whether partial transferase deficiency needs to be treated, some centers have adopted a pragmatic approach, prescribing a lactose-free formula to all infants discovered by newborn screening for 1-4 months after birth until erythrocyte galactose-1-phosphate levels normalize on a regular diet with lactose. Some centers will initiate this transition with a galactose challenge. For example, if at the end of a 1-week trial with a daily supplement of formula containing lactose the erythrocyte galactose-1-phosphate level is below 1 mg/dl the infant will be returned to normal nutrition. Other centers opt for 1 year of treatment and utilize a 1-month challenge with cow's milk. The utility of such treatment during early infancy is unknown, and, in fact, some centers will employ no treatment at all.

Dietary Treatment in Pregnant Woman at Risk

Based on the presumption that toxic metabolites deriving from galactose ingested by the heterozygous mother accumulate in the galactosemic fetus, mothers are often counseled to refrain from drinking milk for the duration of pregnancy. However, despite dietary restriction by the mother, galactose-1-phosphate and galactitol accumulate in the

fetus [26, 30, 46-48] and in the amniotic fluid [14]. It was hypothesized [26] that the affected fetus produces galactose-1-phosphate endogenously from glucose-1-phosphate via the pyrophosphorylase/epimerase pathway (◌ Fig. 7.1), which also provides UDPgalactose and, thus, secures the biosynthesis of galactolipids and galactoproteins indispensable for cell differentiation and growth. Since the affected fetus does not depend on (but may suffer from) the galactose he receives from his mother via the placenta, galactose restriction is the prudent stance for pregnant mothers. Affected newborns of treated mothers appear healthy at birth.

7.2 Uridine Diphosphate-Galactose 4'-Epimerase Deficiency

7.2.1 Clinical Presentation

This disorder exists in at least two forms, both of which are discovered through newborn screening using suitable tests sensitive to both galactose and galactose-1-phosphate in dried blood. In the 5 patients from 3 families with the severe form of the disorder, the enzyme defect was subtotal [49]. The newborns presented with vomiting, jaundice and hepatomegaly reminiscent of untreated classical galactosemia; one was found to have elevated blood methionine on newborn screening. All had galactosuria and hyperaminoaciduria; one had cataracts, and one had sepsis. In some, there was evidence for sensorineural deafness and/or dysmorphic features, but it is unclear whether this is related to GALE deficiency per se, as there was a high degree of consanguinity in the families of Pakistani/Asian ancestry with homozygosity for the V94M GALE gene mutation.

Infants with the mild form appear healthy [50]. The enzyme defect is incomplete; reduced stability and greater than normal requirement for the coenzyme nicotinamide adenine dinucleotide have been described [51]. Milk-fed newborns with the mild form detected in newborn screening are healthy and have neither hypergalactosemia, galactosuria nor hyperaminoaciduria.

7.2.2 Metabolic Derangement

The enzyme deficiency provokes an accumulation of UDPgalactose after milk feeding. This build-up also results in the accumulation of galactose-1-phosphate (◌ Fig. 7.1).

7.2.3 Genetics

Epimerase deficiency is inherited as an autosomal-recessive trait. The epimerase gene resides on chromosome 1 [52]. Several mutations have been identified [53–57] and characterized including the V94M mutation that was present in a homozygous form in all of the patients tested with a severe phenotype [51, 57, 58]. It is also well established that this enzyme catalyzes the conversion of UDP-N-acetylglucosamine to UDP-N-acetylgalactosamine [57]. A compound heterozygous patient (L183P/N34S) of mixed Pakistani/Caucasian ancestry with a mild form and mental retardation, that may or may not be related to the underlying GALE deficiency, has been reported [54]. As in GALT deficiency, abnormal glycosylation of proteins, that appears to be dependent, at least in part, on lactose consumption, has been reported in severe GALE deficiency [49] and is thought to be a secondary biochemical complication, not primarily related to the genetic defect.

7.2.4 Diagnostic Tests

The deficiency should be suspected when red cell galactose-1-phosphate is measurable while GALT is normal. Diagnosis is confirmed by the assay of epimerase in erythrocytes. Heterozygous parents have reduced epimerase activity, a finding that usually helps in the evaluation. Diagnosis of the severe form is based on the clinical symptoms, chemical signs and more marked deficiency of epimerase in red cells. The utility of studying the enzyme deficiency in whole white cell pellets, isolated lymphocytes and EBV-transformed lymphoblasts in potentially clinically relevant variant cases is under scrutiny [54].

7.2.5 Treatment and Prognosis

The child with the severe form of epimerase deficiency is unable to synthesize galactose from glucose and is, therefore, galactose-dependent. Dietary galactose in excess of actual biosynthetic needs will cause accumulation of UDPgalactose and galactose-1-phosphate, the latter being one presumptive toxic metabolite. When the amount of ingested galactose does not meet biosynthetic needs, synthesis of galactosylated compounds, such as galactoproteins and galactolipids, is impaired. As there is no easily available chemical parameter on which to base the daily galactose allowance (such as, e.g., blood phenylalanine in phenylketonuria) treatment is extremely difficult. Children known to suffer from the disorder have impaired psychomotor development.

Infants with the mild form of epimerase deficiency described thus far have not required treatment, but it is advisable that the family physician or pediatrician examine one or two urine specimens for reducing substances and exclude aminoaciduria within a couple of weeks after diagnosis, while the infant is still being fed milk. He should also watch the infant's psychomotor progress without, however, causing concern to the parents.

7.3   Galactokinase Deficiency

7.3.1 Clinical Presentation

Cataracts are the only consistent manifestation of the untreated disorder [58], though pseudotumor cerebri has been described [59]. Liver, kidney and brain damage, as seen in transferase deficiency, are not features of untreated galactokinase deficiency, and hypergalactosemia and galactose/galactitol/glucose diabetes are the only chemical signs.

7.3.2 Metabolic Derangement

Persons with GALK deficiency lack the ability to phosphorylate galactose (◘ Fig. 7.1). Consequently, nearly all of the ingested galactose is excreted, either as such or as its reduced metabolite, galactitol, formed by aldose reductase. As in GALT deficiency, cataracts result from the accumulation of galactitol in the lens [60], causing osmotic swelling of lens fibers and denaturation of proteins.

7.3.3 Genetics

The mode of inheritance is autosomal recessive. In most parts of Europe, in the USA and in Japan, birth incidence is in the order of one in 150,000 to one million. It is higher in the Balkan countries [61], the former Yugoslavia, Rumania and Bulgaria, where it favors Gypsies (below). In Gypsies, birth incidence was calculated as one in 2,500.

Two genes have been reported to encode galactokinase: GK1 on chromosome 17q24 [62] and GK2 on chromosome 15 [63]. Many GK1 mutations have now been described [62, 64–71]. The GK1 P28T mutation was identified as the founder mutation responsible for galactokinase deficiency in Gypsies [64, 69] and in immigrants from Bosnia in Berlin [61].

7.3.4 Diagnostic Tests

Provided they have been fed mother's milk or a lactose-containing formula prior to the test, newborns with the defect are discovered by mass screening methods for detecting elevated blood galactose. If they have been fed glucose-containing fluid, the screening test could be false-negative. Any chance finding of a reducing substance in urine, especially in children or adults with nuclear cataracts, calls for the identification of the excreted substance. In addition to galactose, galactitol and glucose may be found. Every person with nuclear cataracts ought to be examined for GALK deficiency. Final diagnosis is made by assaying GALK activity in heparinized whole blood, red cell lysates, liver or fibroblasts. Heterozygotes have intermediate activity in erythrocytes. Reports of GALK variants have appeared [58, 59].

7.3.5 Treatment and Prognosis

Treatment may be limited to the elimination of milk from the diet. Minor sources of galactose, such as milk products, green vegetables, legumes, drugs in tablet form, etc., can probably be disregarded, since it can be assumed that the small amounts of ingested galactose are either metabolized or excreted before significant amounts of galactitol can be formed. When diagnosis is made rapidly and treatment begun promptly, i.e., during the first two to three weeks of life, cataracts can clear. When treatment is late, and cataracts too dense, they will not clear completely (or at all) and must be removed surgically. In patients who have had their lenses removed, recurring cataracts may appear, originating from remnants of the posterior lens capsule. This can be avoided by continuing the diet.

As in carriers with GALT deficiency [72], the speculation [73] that heterozygosity for GALK deficiency predisposes to the formation of presenile cataracts remains unproven [74]. It has been suggested that heterozygotes restrict their milk intake [73], though scientific proof of the merits of this measure is lacking.

7.4 Fanconi-Bickel Syndrome

This is a recessively inherited disorder of glucose and galactose transport due to GLUT2 deficiency and is extremely rare. A few cases have been discovered during newborn screening for galactose in blood. For further details, ► Chap. 11.

7.5 Portosystemic Venous Shunting and Hepatic Arterio-Venous Malformations

Portosystemic bypass of splanchnic blood via ductus venosus Arantii [75] or intrahepatic shunts [76, 77] causes alimentary hypergalactosemia, which is discovered during metabolic newborn screening.

References

1. Segal S (1998) Galactosaemia today: The enigma and the challenge. J Inherit Metab Dis 21:455-471
2. Tyfield LA (2000) Galactosaemia and allelic variation at the galactose-1-phosphate uridyltransferase gene: a complex relationship between genotype and phenotype. Eur J Pediatr 159[Suppl 3]: S204-S207
3. Tyfield L, Reichardt J, Fridovich-Keil J et al (1999) Classical galactosemia and mutations at the galactose-1-phosphate uridyl transferase (GALT) gene. Hum Mutat 13:417-430
4. Elsas LJ II, Lai K (1998) The molecular biology of galactosemia. Genet Med 1:40-48
5. Levy HL, Brown AE, Williams SE et al (1996) Vitreous hemorrhage as an ophthalmic complication of galactosemia. J Pediatr 129:922-925
6. Berry GT, Nissim I, Mazur AT et al (1995) In vivo oxidation of [13C]galactose in patients with galactose-1-phosphate uridyltransferase deficiency. Biochem Mol Med 56:158-165
7. Berry GT, Singh RH, Mazur AT et al (2000) Galactose breath testing distinguishes variant and severe galactose 1-phosphate uridyltransferase genotypes. Pediatr Res 48:323-328
8. Berry GT, Leslie N, Reynolds R et al (2001) Evidence for alternate galactose oxidation in a patient with deletion of the galactose-1-phosphate uridyltransferase gene. Mol Genet Metab 72:316-321
9. Berry GT, Reynolds RA, Yager CT et al (2004) Extended [13C]galactose oxidation studies in patients with galactosemia. Mol Genet Metab 82:130-136
10. Henderson H, Leisegang F, Brown R et al (2002) The clinical and molecular spectrum of galactosemia in patients from the Cape Town region of South Africa. BMC Pediatr 2:7
11. Shield JPH, Wadsworth EJK, MacDonald A et al (2000) The relationship of genotype to cognitive outcome in galactosemia. Arch Dis Child 83:248-250
12. Guerrero NV, Singh RH, Manatunga A et al (2000) Risk factors for premature ovarian failure in females with galactosemia. J Pediatr 137:833-841
13. Webb AL, Singh RH, Kennedy MJ et al (2003) Verbal dyspraxia and galactosemia. Pediatr Res 53:396-402
14. Jakobs C, Kleijer WJ, Allen J et al (1995) Prenatal diagnosis of galactosemia. Eur J Pediatr 154[Suppl 2]:S33-S36
15. Beigi B, O'Keefe M, Bowell R et al (1993) Ophthalmic findings in classical galactosaemia – prospective study. Br J Ophthalmol 77:162-164
16. Vogt M, Gitzelmann R, Allemann J (1980) Dekompensierte Leberzirrhose infolge Galaktosämie bei einem 52jährigen Mann. Schweiz Med Wochenschr 110:1781-1783
17. Berry GT, Segal S (unpublished observations)
18. Lee PJ, Lilburn M, Wendel U et al (2003) A woman with untreated galactosemia. Lancet 362:446
19. Gitzelmann R (1969) Formation of galactose-1-phosphate from uridine diphosphate galactose in erythrocytes from patients with galactosemia. Pediatr Res 3:279-286
20. Gitzelmann R, Hansen RG (1974) Galactose biogenesis and disposal in galactosemics. Biochim Biophys Acta 372: 374-378
21. Berry GT, Nissim I, Lin Z et al (1995) Endogenous synthesis of galactose in normal men and patients with hereditary galactosaemia. Lancet 346:1073-1074
22. Berry GT, Nissim I, Gibson JB et al (1997) Quantitative assessment of whole body galactose metabolism in galactosemic patients. Eur J Pediatr 156[Suppl1]:S43-S49
23. Berry GT, Moate PJ, Reynolds RA et al (2004) The rate of de novo galactose synthesis in patients with galactose-1- phosphate uridyltransferase (GALT) deficiency. Mol Genet Metab 81:22-30
24. Schadewaldt P, Kamalanathan L, Hammen H-W et al (2004) Age dependence of endogenous galactose formation in Q188R homozygous galactosemic patients. Mol Genet Metab 81:31-44
25. Forbes GB, Barton LD, Nicholas DL et al (1998) Composition of milk produced by a mother with galactosemia. J Pediatr 113:90-91
26. Gitzelmann R, Hansen RG, Steinmann B (1975) Biogenesis of galactose, a possible mechanism of self-intoxication in galactosemia. In: Hommes FA, Van den Berg CJ (eds) Normal and pathological development of energy metabolism. Academic Press, London, p 2537

27. Kaufman FR, Kogut MD, Donnell GN et al (1981) Hypergonado-tropic hypogonadism in female patients with galactosemia. N Engl J Med 304:994-998

28. Waggoner DD, Buist NRM, Donnell GN (1990) Long-term prognosis in galactosaemia: result of a survey of a 350 cases. J Inherit Metab Dis 13:802-818

29. Gibson JB (1995) Gonadal function in galactosemics and in galac-tose-intoxicated animals. Eur J Pediatr 154[Suppl 2]:S14-S20

30. Komrower GM (1982) Galactosemia – thirty years on. The experi-ence of a generation. J Inherit Metab Dis 5[Suppl 2]:96-104

31. Schweitzer S, Shin Y, Jakobs C et al (1993) Long-term outcome in 134 patients with galactosaemia. Eur J Pediatr 152:36-43

32. Manis FR, Cohn LB, McBride Chang C et al (1997) A longitudinal study of cognitive functioning in patients with classical galactos-aemia, including a cohort treated with oral uridine. J Inherit Metab Dis 20:549-555

33. Berry GT, Palmieri M, Gross KC et al (1993) The effect of dietary fruits and vegetables on urinary galactitol excretion in galactose-1-phosphate uridyltransferase deficiency. J Inherit Metab Dis 16:91-100

34. Merritt RJ, Jenks BH (2004) Safety of soy-based infant formulas containing isoflavones: the clinical evidence. J Nutr 134:1220S-1224S

35. Gross KC, Acosta PB (1991) Fruits and vegetables are a source of galactose: implications in planning the diets of patients with galac-tosaemia. J Inherit Metab Dis 14:253-258

36. Acosta PB, Gross KC (1995) Hidden sources of galactose in the environment. Eur J Pediatr 154[Suppl 2]:S8792

37. Wiesmann U, Ros-Beutler B, Schlchter R (1995) Leguminosae in the diet: The raffinose-stachyose-question. Eur J Pediatr 154[Suppl 2]: S9396

38. Gitzelmann R, Auricchio S (1965) The handling of soya alpha-galac-tosides by a normal and a galactosemic child. Pediatrics 36:231-235

39. Steffen C (1975) Enzymatische Bestimmungsmethoden zur Erfas-sung der Grünungsvorgänge in der milchwirtschaftlichen Techno-logie. Lebensm Wiss Technol 8:16

40. Kaufman FR, Loro ML, Azen C et al (1993) Effect of hypogonadism and deficient calcium intake on bone density in patients with galactosemia. J Pediatr 123:365-370

41. Harju M (1990) Lactobionic acid as a substrate of α-galactosidases. Milchwissenschaft 45:411-415

42. Sardharwalla IB, Komrower GM, Schwarz V (1980) Pregnancy in classical galactosemia. In: Burman D, Holton JB, Pennock CA (eds) Inherited disorders of carbohydrate metabolism. MTP, Lancaster, pp 125-132

43. Gitzelmann R, Steinmann B (1984) Galactosemia: How does long-term treatment change the outcome? Enzyme 32:37-46

44. Holton JB (1996) Galactosaemia: pathogenesis and treatment. J Inherit Metab Dis 19:3-7

45. Bosch AM, Grootenhuis MA, Bakker HD et al (2004) Living with clas-sical galactosemia: health-related quality of life consequences. Pediatrics 113(5) 423-428

46. Irons M, Levy HL, Pueschel S et al (1985) Accumulation of galactose-1-phosphate in the galactosaemic fetus despite maternal milk avoidance. J Pediatr 107:261-263

47. Holton JB (1995) Effects of galactosemia in utero. Eur J Pediatr 154: S77- S81

48. Gitzelmann R (1995) Galactose-1-phosphate in the pathophysiol-ogy of galactosemia. Eur J Pediatr 154[Suppl 2]:S45-S49

49. Walter JH, Roberts RE, Besley GT et al (1999) Generalised uridine diphosphate galactose-4-epimerase deficiency. Arch Dis Child 80:374-376

50. Gitzelmann R, Steinmann B, Mitchell B et al (1976) Uridine diphos-phate galactose 4-epimerase deficiency. IV. Report of eight cases in three families. Helv Paediatr Acta 31:441-445

51. Gitzelmann R, Steinmann B (1973) Uridine diphosphate galactose 4- epimerase deficiency II. Clinical follow-up, biochemical studies and family investigation. Helv Paediatr Acta 28:497-510

52. Daude N, Gallaher TK, Zeschnigk M et al (1995) Molecular cloning, characterization, and mapping of a full-length cDNA encoding human UDP-galactose 4‹-epimerase. Biochem Mol Med 56:1-7

53. Maceratesi P, Daude N, Dallapiccola B et al (1998) Human UDP-galactose 4‹epimerase (GALE) gene and identification of five missense mutations in patients with epimerase-deficiency. Mol Genet Metab 63:26-30

54. Alano A, Almashanu S, Chinsky JM et al (1998) Molecular character-ization of a unique patient with epimerase-deficiency galactosae-mia. J Inherit Metab Dis 21:341-350

55. Wohlers TM, Christacos NC, Harreman MT et al (1999) Identification and characterization of a mutation, in the human UDP-galactose-4-epimerase gene, associated with generalized epimerase-defi-ciency galactosemia. Am J Hum Genet 64:462-470

56. Wohlers TM, Fridovich-Keil JL (2000) Studies of the V94M-sub-stituted human UDPgalactose-4-epimerase enzyme associated with generalized epimerase-deficiency galactosaemia. J Inherit Metab Dis 23:713-729

57. Schulz JM, Watson AL, Sanders R et al (2004) Determinants of func-tion and substrate specificity in human UDP-galactose epimerase. J Biol Chem 279:32796-32803

58. Gitzelmann R (1967) Hereditary galactokinase deficiency a newly recognized cause of juvenile cataracts. Pediatr Res 1:1423

59. Bosch AM, Bakker HD, VanGennip AH et al (2002) Clinical features of galactokinase deficiency: A review of the literature. J Inherit Metab Dis 25:629-634

60. Ai Y, Zheng Z, O'Brien-Jenkins A et al (2000) A mouse model of galactose-induced cataracts. Hum Mol Genet 9:1821-1827

61. Reich S, Hennermann J, Vetter B et al (2002) An unexpectedly high frequency of hypergalactosemia in an immigrant Bosnian popula-tion revealed by newborn screening. Pediatr Res 51:598-601

62. Stambolian D, Ai Y, Sidjanin D et al (1995) Cloning the galactokinase cDNA and identification of mutations in two families with cataracts. Nat Genet 10:307-312

63. Lee RT, Peterson GL, Calman AF et al (1992) Cloning of a human galactokinase gene (GK2) on chromosome 15 by complementation in yeast. Proc Natl Acad Sci USA 89:10887-10891

64. Kalaydjieva L, Perez-Lezaun A, Angelicheva D et al (1999) A founder mutation in the GK1 gene is responsible for galactokinase defi-ciency in Roma (Gypsies). Am J Hum Genet 65:1299-1307

65. Asada M, Okano Y, Imamura T et al (1999) Molecular characteriza-tion of galactokinase deficiency in Japanese patients. J Hum Genet 44:377-382

66. Kolosha V, Anoia E, deCespedes C et al (2000) Novel mutations in 13 probands with galactokinase deficiency. Hum Mutat 15:447-453

67. Hunter M, Angelicheva D, Levy HL et al (2001) Novel mutations in the GALK1 gene in patients with galactokinase deficiency. Hum Mutat 17:77-78

68. Okano Y, Asada M, Fujimoto A et al (2001) A genetic factor for age-related cataract: Identification and characterization of a novel galactokinase variant, "Osaka," in Asians. Am J Hum Genet 68:1036-1042

69. Hunter M, Heyer E, Austerlitz F et al (2002) The P28T mutation in the GALK1 gene accounts for galactokinase deficiency in Roma (Gypsy) patients across Europe. Pediatr Res 51:602-606

70. Timson DJ, Reece RJ (2003) Functional analysis of disease-causing mutations in human galactokinase. Eur J Biochem 270:1767-1774

71. Sangiuolo F, Magnani M, Stambolian D et al (2004) Biochemical characterization of two GALK1 mutations in patients with galac-tokinase deficiency. Hum Mutat 23:396

72. Karas N, Gobec L, Pfeifer V et al (2003) Mutations in galactose-1-phosphate uridyltransferase gene in patients with idiopathic presenile cataract. J Inherit Metab Dis 26:699-704

73. Stambolian D, Scarpino-Myers V, Eagle RC Jr et al (1986) Cataracts in patients heterozygous for galactokinase deficiency. Invest Ophthalmol Vis Sci 27:429-433

74. Maraini G, Fielding Hejtmancik J, Shiels A et al (2003) Galactokinase gene metations and age-related cataract. Lack of association in an Italian population. Mol Vis 9:397-400

75. Gitzelmann R, Arbenz UV, Willi UV (1992) Hypergalactosaemia and portosystemic encephalopathy due to persistence of ductus venosus Arantii. Eur J Pediatr 151:564-568

76. Matsumoto T, Okano R, Sakura N et al (1993) Hypergalactosaemia in a patient with portal-hepatic venous and hepatic arterio-venous shunts detected by neonatal screening. Eur J Pediatr 152:990-992

77. Gitzelmann R, Forster I, Willi UV (1997) Hypergalactosaemia in a newborn: Self-limiting intrahepatic portosystemic venous shunt. Eur J Pediatr 156:719-722

8 Disorders of the Pentose Phosphate Pathway

Nanda M. Verhoeven, Cornelis Jakobs

The Pentose Phosphate Pathway

The pentose phosphate pathway (◘ Fig. 8.1) is present in most cell types. Its function is twofold: the provision of ribose-5-phosphate (ribose-5-P) for ribonucleic acid synthesis and the reduction of nicotinamide adenine dinucleotide phosphate (NADP) into NADPH, a cofactor in many biosynthetic processes. To date, three inborn errors in the pentose phosphate pathway have been described.

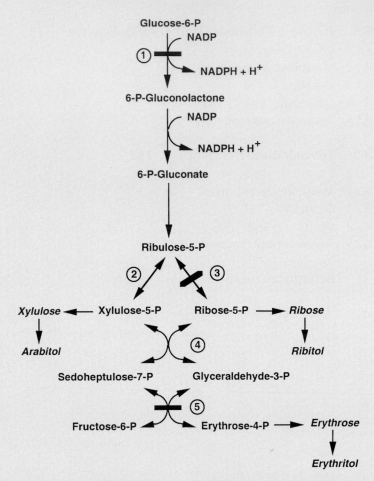

◘ **Fig. 8.1.** The pentose phosphate pathway. All sugars are D stereoisomers. The conversion of the sugar phosphates into their corresponding sugars and polyols (*italics*) has not been proven in humans. NADP, nicotinamide adenine dinucleotide phosphate; NADPH, reduced form; P, phosphate. **1**, glucose-6-phosphate dehydrogenase; **2**, ribulose-5-phosphate epimerase; **3**, ribose-5-phosphate isomerase; **4**, transketolase **5**, transaldolase. Enzyme defects are depicted by *solid bars* across the arrows.

Three inborn errors in the pentose phosphate pathway are known. In *glucose-6-phosphate dehydrogenase deficiency,* there is a defect in the first, irreversible step of the pathway. As a consequence NADPH production is decreased, making erythrocytes vulnerable to oxidative stress. Drug-and fava bean-induced haemolytic anaemia is the main presenting symptom of this defect. As this is a haematological disorder it is not discussed further.

Deficiency of ribose-5-phosphate isomerase has been described in one patient who suffered from a progressive leucoencephalopathy and developmental delay.

Transaldolase deficiency has been diagnosed in three unrelated families. All patients presented in the newborn period with liver problems. One of the patients died soon after birth from liver failure and cardiomyopathy, whereas another patient is now 15 years old and suffers from liver cirrhosis. Her neurological and intellectual development has been normal.

Essential pentosuria, due to a defect in the enzyme xylitol dehydrogenase, affects the related glucuronic acid pathway. Whereas the pentose phosphate pathway involves D stereoisomers, glucuronic acid gives rise to L-xylulose which is subsequently converted into xylitol and D-xylulose. Affected individuals excrete large amounts of L-xylulose in urine. This is a benign disorder and not discussed further.

8.1 Ribose-5-Phosphate Isomerase Deficiency

8.1.1 Clinical Presentation

The patient with ribose-5-phosphate isomerase deficiency [1] presented with developmental and speech delay. At the age of 4 years he developed epilepsy. From the age of 7 years he regressed, with deterioration of vision, speech, hand coordination, walking and seizure control. At 20 years of age he had no organomegaly or internal organ dysfunction. Neurological examination showed some spasticity, bilateral optic atrophy, nystagmus on lateral gaze, an increased masseter reflex, and mixed cerebellar/pseudobulbar dysarthria. He had prominent cerebellar ataxia in his arms and legs and mild peripheral neuropathy. He suffered from learning difficulties but his growth parameters were normal.

Magnetic resonance imaging (MRI) at 11 and 14 years of age showed extensive abnormalities of the cerebral white matter with prominent involvement of the U-fibers. The abnormal white matter had a slightly swollen appearance with some widening of the gyri. On proton magnetic resonance spectroscopy (MRS), performed at 14 years of age,

there were abnormal resonances in the 3.5–4.0 ppm region, corresponding to arabitol and ribitol.

8.1.2 Metabolic Derangement

Ribose-5-phosphate isomerase deficiency is a block in the reversible part of the pentose phosphate pathway. In theory, this defect leads to a decreased capacity to interconvert ribulose-5-phosphate and ribose-5-phosphate and results in the formation of sugars and polyols: ribose and ribitol from ribose-5-phosphate and xylulose and arabitol from ribulose-5-phosphate via xylulose-5-phosphate.

8.1.3 Genetics

The presence of two mutant alleles in the ribose-5-phosphate isomerase gene with one of these in the patient's mother (the father could not be investigated) suggest autosomal recessive inheritance.

8.1.4 Diagnostic Tests

Ribose-5-phosphate isomerase deficiency can be diagnosed by analysis of sugars and polyols in a random urine sample. Urinary ribitol and arabitol, as well as D-xylulose and ribose are highly elevated. Extremely high concentrations of these pentitols are also found in cerebrospinal fluid (CSF). Myoinositol concentration in CSF is decreased.

On in vivo MRS of the brain there are abnormal high peaks in the 3.5–4.0 ppm region. The diagnosis can be confirmed by an enzyme assay in fibroblasts or lymphoblasts, and by sequence analysis of the ribose-5-phosphate isomerase gene.

8.1.5 Treatment and Prognosis

No treatment for this defect is available hitherto. The prognosis is unclear.

8.2 Transaldolase Deficiency

8.2.1 Clinical Presentation

Transaldolase deficiency has been diagnosed in 3 families [2–4]. Clinical symptoms among these families vary, but liver disease has been present in all. All patients are from consanguineous Turkish parents. The first patient had low birth weight and presented in the neonatal period with an aortic coarctation. After surgical intervention, she had mild bleeding problems. An enlarged clitoris was noted. In the first months of life she developed hepatosplenomegaly with normal γGT and transaminases but mildly prolonged prothrombin time (PT) and activated partial prothrombin time (APTT). At the age of two years, a liver biopsy showed micronodular liver cirrhosis but without specific characteristics. At the age of ten years, her height was at the 10th percentile and her weight for height at the 2nd percentile. Her liver was 7 cm below the costal margin and her spleen size 14.5 cm. She had trombocytopenia, probably due to splenic pooling. Her γ-GT was raised on one occasion. Bile acids were elevated but bilirubin was normal. Ammonia was intermittently raised. Her neurological and intellectual development has been normal. MRI and MRS of the brain did not show abnormalities.

The second patient was born after caesarean section because of maternal HELLP syndrome (hemolysis, elevated liver enzymes and low platelet count). Birth weight was on the 50th percentile. She displayed generalized edema, moderate hypotonia and dysmorphic signs (down-slanting palpebral fissures, low-set ears and increased intermamillar distance). She had a severe coagulopathy unresponsive to intraveneous vitamin K, an elevated ammonia, unconjugated hyperbilirubinemia, hypoglycemia and low transaminases. Liver size was decreased whereas the spleen was moderately increased. The kidneys appeared normal but there was glomerular proteinuria. The heart was enlarged with marked myocardial thickening. The patient showed intractable liver failure, progressive myocardial hypertrophy, and developed respiratory failure and severe lactic acidosis. She died from bradycardic heart failure.

In the third family, three affected children with variable symptoms were diagnosed. The first child died at the age of 4 months from liver failure. The second child had hepatomegaly during the first 8 months of life with fibrosis and steatosis. At the age of 5 years his liver is still palpable but liver function has normalized. He suffers from chronic renal failure. The third child had neonatal liver problems with high transaminases and a prolonged APTT and PT, cardiomyopathy and transient renal failure. All four patients had transient dysmorphic features, cutis laxa and increased hair growth and in addition suffered from haemolytic anemia and pancytopenia. In the same family, there has been one abortion because of severe hydrops foetalis with oligohydramnios. Transaldolase in tissues of this fetus was deficient [4].

To date all children diagnosed with transaldolase deficiency have been of normal intellectual and neurological development.

8.2.2 Metabolic Derangement

Transaldolase, located in the reversible part of the pentose phosphate pathway, recycles pentose phosphates into hexose phosphates in concerted action with transketolase. Its deficiency results in the accumulation of polyols derived from the pathway intermediates: erythritol, arabitol and ribitol.

8.2.3 Genetics

The same mutation was found in the first and third families, but was different in the second family. All patients were homozygous for these specific mutations, suggesting autosomal recessive inheritance.

8.2.4 Diagnostic Tests

The diagnosis of transaldolase deficiency is suggested by elevated concentrations of erythritol, arabitol and ribitol in urine. Elevations are most striking in the neonatal period and are more subtle in older patients. In plasma and CSF, there are no or minor elevations of polyols. Transaldolase activity can be determined in fibroblasts, erythrocytes and lymphoblasts. Mutation analysis is also available.

8.2.5 Treatment and Prognosis

To date, no therapeutic options have been investigated. Liver transplantation may be performed in patients with severe liver cirrhosis.

References

1. Huck JHJ, Verhoeven NM, Struys EA et al (2004) Ribose-5-phosphate isomerase deficiency: new inborn error in the pentose phosphate pathway associated with a slowly progressive leukoencephalopathy. Am J Hum Genet 74: 745-51
2. Verhoeven NM, Huck JH, Roos B et al (2001) Transaldolase deficiency: liver cirrhosis associated with a new inborn error in the pentose phosphate pathway. Am J Hum Genet 68: 1086-1092
3. Verhoeven NM, Wallot M, Huck JHJ et al (2005) A newborn with severe liver failure, cardiomyopathy and transaldolase deficiency. J Inherit Metab Dis 28: 169-179
4. Valayannopoulos V, Verhoeven N, Salomons GS et al (2005) Transaldolase deficiency: an inborn error of the pentose phosphate pathway associated with a severe phenotype and multiorgan involvement including hydrops foetalis, cutis laxa, hepatic failure and haemolytic anaemia. J Inherit Metab Dis 28:217

9 Disorders of Fructose Metabolism

Beat Steinmann, René Santer, Georges van den Berghe

Fructose Metabolism

Fructose is one of the main sweetening agents in the human diet. It is found in its free form in honey, fruits and many vegetables, and is associated with glucose in the disaccharide sucrose in numerous foods and beverages. Sorbitol, also widely distributed in fruits and vegetables, is converted into fructose in the liver by sorbitol dehydrogenase (◘ Fig. 9.1). Fructose is mainly metabolized in the liver, renal cortex and small intestinal mucosa in a pathway composed of fructokinase, aldolase B and triokinase. Aldolase B also intervenes in the glycolytic-gluconeogenic pathway (right hand part of the scheme).

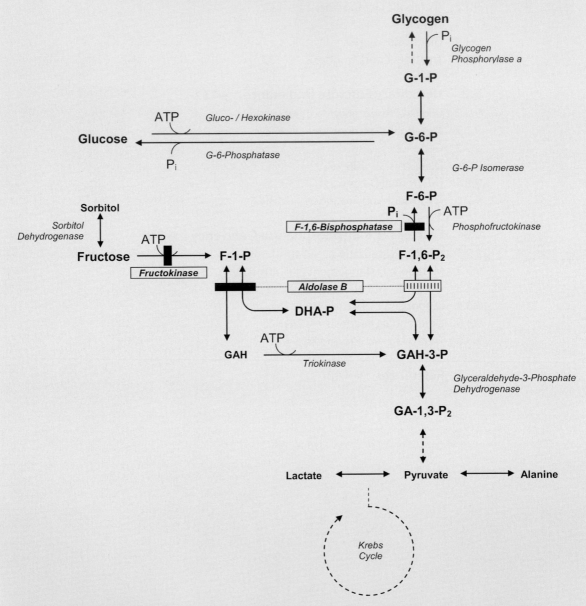

◘ Fig. 9.1. Fructose Metabolism. *DHA-P*, dihydroxyacetone phosphate; *F*, fructose; *G*, glucose; *GA*, glycerate; *GAH*, glyceraldehyde; *P*, phosphate; *Pi*, inorganic phosphate. The three enzyme defects in fructose metabolism are boxed and depicted by *solid bars* across the *arrows*; the diminished activity of aldolase B toward fructose-1,6-bisphosphate in hereditary fructose intolerance is depicted by a *broken bar*

Three inborn errors are known in the pathway of fructose metabolism depicted in ◙ Fig. 9.1. *Essential fructosuria* is a harmless anomaly characterized by the appearance of fructose in the urine after the intake of fructose-containing food. In *hereditary fructose intolerance* (HFI), fructose may provoke prompt gastrointestinal discomfort and hypoglycemia upon ingestion, symptoms that may vary from patient to patient and depend on the ingested dose. Fructose may cause liver and kidney failure when taken persistently, and its intake becomes life-threatening when given intravenously. *Fructose-1,6-bisphosphatase* (FBPase) *deficiency* is also usually considered an inborn error of fructose metabolism although, strictly speaking, it is a defect of gluconeogenesis. The disorder is manifested by the appearance of hypoglycemia and lactic acidosis (neonatally, or later during fasting or induced by fructose) and may also be life-threatening.

9.1 Essential Fructosuria

9.1.1 Clinical Presentation

Essential fructosuria is a rare non-disease; it is detected by routine screening of urine for reducing sugars [1]. It is caused by a deficiency of fructokinase, also known as ketohexokinase (KHK), the first enzyme of the main fructose pathway (◙ Fig. 9.1).

9.1.2 Metabolic Derangement

Ingested fructose is partly (10–20%) excreted as such in the urine, the rest is slowly metabolized by an alternative pathway, namely conversion into fructose-6-phosphate by hexokinase in adipose tissue and muscle.

9.1.3 Genetics

The mode of inheritance is autosomal recessive and homozygote frequency has been estimated at 1:130,000. However, since the condition is asymptomatic and harmless, it may be more prevalent than reported.

The *KHK* gene is located on chromosome 2p23.3–23.2. Tissue-specific alternative splicing results in two isoforms, ketohexokinase A, widely distributed in most fetal and adult organs but with no clear physiological role, and ketohexokinase C, expressed in adult liver, kidney and small intestine, which is affected in essential fructosuria [2]. Two mutations of the *KHK* gene, G40R and A43T, both of which alter the same conserved region of fructokinase, have been

detected and functionally characterized in a family with three compound heterozygotes [1].

9.1.4 Diagnosis

Fructose gives a positive test for reducing sugars and a negative reaction with glucose oxidase. It can be identified by various techniques, such as thin-layer chromatography, and quantified enzymatically. Fructosuria depends on the time and amount of fructose and sucrose intake and, thus, is inconstant. Fructose-tolerance tests (▶ Sect. 9.2) neither provoke an increase in blood glucose as in normal subjects, hypoglycemia or other changes as occur in HFI and FBPase deficiency, nor are metabolic changes in liver detectable by ^{31}P- magnetic resonance imaging (MRS) [3].

9.1.5 Treatment and Prognosis

Dietary treatment is not indicated, and the prognosis is excellent.

9.2 Hereditary Fructose Intolerance

9.2.1 Clinical Presentation

Infants, children and adults with hereditary fructose intolerance (HFI) are perfectly healthy and asymptomatic as long as they do not ingest food containing fructose, sucrose and/or sorbitol. Consequently, no metabolic derangement occurs during breast-feeding. The younger the child and the higher the dietary fructose load, the more severe the reaction. In the *acute* form of HFI, an affected newborn infant who is not breast-fed but receives a cow's milk formula sweetened and enriched with fructose or sucrose – formulas which should be obsolete today – is in danger of severe liver and kidney failure and death.

At weaning from breast-feeding or a fructose/saccharose-free infant formula, the first symptoms appear with the intake of fruits and vegetables [4, 5]. They are generally those of gastrointestinal discomfort, nausea, vomiting, restlessness, pallor, sweating, trembling, lethargy and, eventually, apathy, coma, jerks and convulsions. At this stage, laboratory signs may be those of acute liver failure and a generalized dysfunction of the renal proximal tubules. If the condition is unrecognized and fructose not excluded from the diet, the disease takes a *chronic*, undulating course with failure to thrive, liver disease manifested by hepatomegaly, jaundice, a bleeding tendency, edema, ascites, and signs of proximal renal tubular dysfunction. Laboratory findings are those of liver failure, proximal renal tubular dysfunction and derangements of intermediary metabolism. Note that hypoglycemia after fructose ingestion is short-lived and can be easily missed or masked by concomitant glucose intake.

HFI can be suspected in an asymptomatic infant, if the parents have excluded certain foods from the diet, having becoming aware that they are not tolerated. In older children, a distinct aversion towards foods containing fructose may develop; these feeding habits protect them but are sometimes considered as neurotic behavior. At school age, HFI is occasionally recognized when hepatomegaly or growth delay is found [6]. Some adult cases were diagnosed after developing life-threatening reactions with infusions containing fructose, sorbitol or invert sugar (a mixture of glucose and fructose obtained by hydrolysis of sucrose) when these IV solutions were still in use [7]. Because approximately half of all adults with HFI are free of caries, the diagnosis has also been made by dentists. Although several hundred patients with HFI have been identified since its recognition as an inborn error of metabolism in 1957 [4], these observations indicate that affected subjects may remain undiagnosed and still have a normal life span.

9.2.2 Metabolic Derangement

HFI is caused by deficiency of the second enzyme of the fructose pathway, aldolase B (fructose 1,6-bisphosphate aldolase) (◘ Fig. 9.1), which splits fructose-1-phosphate (F-1-P) into dihydroxyacetone phosphate and glyceraldehyde and converts the triosephosphates into glucose and lactate. Moreover, as a consequence of the high activity of fructokinase, intake of fructose results in accumulation of F-1-P. This accumulation has two major effects [8]: (i) it inhibits the production of glucose by blocking gluconeogenesis and glycogenolysis, hence inducing hypoglycemia; (ii) it provokes overutilization and hence depletion of ATP, the energy currency of the cell, and of inorganic phosphate, utilized to regenerate ATP. The latter result in an increased production of uric acid, and a series of disturbances, including inhibition of protein synthesis and ultrastructural lesions, which are responsible for the hepatic and renal dysfunction. Recently, the accumulation of F-1-P has also been shown to result in deficient glycosylation of proteins, e.g., serum transferrin, by inhibiting phosphomannose isomerase [9] (▶ Chap. 41).

Residual activity measurable with fructose-1,6-bisphosphate as substrate (see below) is mainly due to the isozyme aldolase A. Thus, glycolysis and gluconeogenesis are not impaired in the fasted state in HFI patients.

It should be noted that the IV administration of fructose to normal subjects also induces the metabolic derangements described above (including the drop in ATP and inorganic phosphate, and rise in urate) to an equivalent extent, although they are more transient than in patients with HFI, as demonstrated by ^{31}P-MRS [3]. In normal subjects, IV fructose results in increased glycemia because

of its rapid conversion into glucose. However, the equally rapid conversion of fructose into lactate may provoke metabolic acidosis. For these reasons, the use of fructose, sorbitol and invert sugar has been strongly discouraged for parenteral nutrition [10].

9.2.3 Genetics

Three different genes coding for aldolases have been identified. While isozymes A and C are mainly expressed in muscle and brain, respectively, aldolase B is the major fructaldolase of liver, renal cortex, and small intestine.

The human gene for aldolase B (*ALDOB*) has been mapped to chromosome 9q22.3. At present, according to the Human Gene Mutation Database, Cardiff (HGMD, http://uwcmml1s.uwcm.ac.uk/uwcm/mg/search/119669.html) and own data, approximately 30 causative mutations of the *ALDOB* gene have been reported. Among them, three amino acid substitutions, A150P[1], A175D, and N335K are relatively common among patients of central European descent and have been detected in 65%, 11% and 8% of mutated alleles, respectively [11, 12]. Some mutations may be found particularly in certain ethnic groups, such as the N335K and c.360-363 *del* CAAA mutations in the populations of the former Yugoslavia and Sicily, respectively. Since the three most common mutations are responsible for up to 84% of HFI cases in the European population [11, 12] and 68% of cases in North America [13] a non-invasive diagnostic approach using molecular genetic methods has been advocated and specific methods for the rapid concomitant detection of these frequent *ALDOB* mutations have been published [12, 14].

From molecular genetic neonatal screening studies in England and Germany, the prevalence of HFI has been calculated as 1:18,000 and 1:29,600, respectively [12, 15].

9.2.4 Diagnosis

Whenever HFI is suspected, fructose should be eliminated from the diet immediately. The beneficial clinical and chemical effects of withdrawal, usually seen within days, provide a first diagnostic clue. Laboratory findings will subsequently show a fall in the elevated serum transaminases and bilirubin, improved levels of blood clotting factors, and amelioration of proximal tubular dysfunction (proteinuria, mellituria, generalized hyperaminoaciduria, metabolic acidosis).

A cornerstone in the diagnosis of HFI is a careful nutritional history, with special emphasis on the time of weaning

---

[1] Note that the initiation codon ATG for methionine in *ALDOB* cDNA was ignored in previous designations and that, e.g., A150P was originally termed A149P.

when fruits and vegetables were introduced [5, 15, 16]. If the nutritional history is suggestive, or other aspects are indicative of HFI (e.g., a positive family history), the disorder should be confirmed by molecular diagnosis (▶ above) on DNA from peripheral leukocytes. This is a non-invasive approach and has the advantage over enzymatic measurement in liver tissue in that it eliminates the complication of secondarily lowered aldolase activity in a damaged liver.

If no mutation can be found despite a strong clinical and nutritional history suggestive of HFI, an enzymatic determination or a functional test should be undertaken after a few weeks of fructose exclusion. In liver biopsies from HFI patients, the capacity of aldolase to split F-1-P is reduced, usually to a few percent of normal (mean 5%, range 0–15%) [16], although residual activities as high as 30% of normal have been reported [7]. There is also a distinct (but less marked) reduction in the activity of aldolase toward fructose-1,6-bisphosphate (mean 17%, range 5-30%). As a consequence, the ratio of V_{max} towards fructose-1,6-bisphosphate versus the V_{max} towards F-1-P, which is approximately 1 in control liver, is increased to 2-∞ in HFI patients [16]. Aldolase activity is normal in blood cells, muscle, and skin fibroblasts, which contain a different isozyme, aldolase A. The enzymatic determination of aldolase B in small intestinal mucosa is discouraged. For post-mortem diagnosis, molecular studies and measurements of enzyme activity in liver and kidney cortex should be done. Of note, the level of residual activity has never been shown to correlate with the degree of tolerance to fructose.

In vivo handling of fructose is best reflected by a fructose tolerance test, in which fructose (200 mg/kg body weight) is injected as a 20% solution intravenously within 2 minutes. Blood samples are taken at 0 (2), 5, 10, 15, 30, 45, 60 and 90 minutes for determination of glucose and phosphate. In normal subjects, blood glucose increases by 0–40%, with no or minimal changes in phosphate [16]. In HFI patients, glucose and phosphate decrease within 10–20 minutes. As a rule, the decrease of phosphate precedes and occurs more rapidly than that of glucose. The test should be undertaken in a metabolic center, with careful monitoring of glucose and an indwelling catheter for the (exceptional) case of symptomatic hypoglycemia and its treatment by IV glucose administration. Oral fructose tolerance tests are not recommended, because they provoke more ill effects and are less reliable [16].

9.2.5 Differential Diagnosis

A high degree of diagnostic awareness is often needed in HFI because the spectrum of symptoms and signs is wide and nonspecific; HFI has been misdiagnosed as pyloric stenosis, gastroesophageal reflux, galactosemia, tyrosinemia, intrauterine infections, glycogen and other storage disorders, ornithine transcarbamoylase deficiency, and

later in life as Wilson disease, leukemia, and growth retardation. Fructosuria may be secondary to liver damage, e.g., in tyrosinemia. Fructose malabsorption is frequently confused with HFI. Its metabolic basis is not well understood, but it is probably due to a defective fructose transporter in the small intestine, and the ingestion of fructose, and to a considerable lesser extent of sucrose, leads to abdominal pain and diarrhea. Since this condition is diagnosed by breath hydrogen analysis after an oral load of fructose, HFI has certainly to be excluded before such a tolerance test is performed. In sucrase-isomaltase deficiency, the ingestion of sucrose results in bloating, abdominal cramps and fermentative osmotic diarrhea; free fructose, however, is well tolerated.

9.2.6 Treatment and Prognosis

In acute intoxication, intensive care may be required and supportive measures such as fresh frozen plasma may be needed. The main therapeutic step in HFI, however, is the immediate elimination of all sources of fructose from the diet. This involves the avoidance of all types of food in which fructose, sucrose and/or sorbitol occur naturally or have been added during processing. It should be borne in mind that fructose and sorbitol may be present in medications (e.g., syrups, immunoglobulin solutions, rinsing fluids, enema solutions) and infant formulas (without adequate declaration of the carbohydrate composition). In this respect, it is deplorable that European Union regulations allow infant formulae to contain up to 20% of their total carbohydrate content as sucrose [17].

Sucrose should be replaced by glucose, maltose and/or starch to prevent the fructose-free diet from containing too much fat. Despite the availability of books and online information on food composition, a dietician should be consulted and practical aspects of the diet (e.g., the considerable variability of the fructose content of different food types, and the influence of storage temperature or method of preparation and manner of cooking on bioavailability) be discussed. Substitution of vitamins, especially ascorbic acid and folates, in the form of a multivitamin preparation should be prescribed to make up for their lack of intake from fruits and vegetables.

After institution of a fructose-free diet most abnormalities disappear rapidly, except for hepatomegaly, which may persist for months or even years [18]. The reason for this is unclear. Different thresholds of fructose intake for the development of certain symptoms have appeared in the literature, ranging from 40–250 mg/kg/day as compared with an average intake of 1–2 g/kg/day in Western societies [15]. Insufficient restriction of fructose has been reported to cause isolated growth retardation, as evidenced by catch-up growth on a stricter diet [6], but this observation has been questioned [19]. It has also to be kept in mind that recommendations for maximum doses have not been validated in different genotypes and that sensitivity is known to be different in individual patients. Thus, it should be suggested to parents that they keep fructose intake as low as possible and that, at least in childhood, it should not be determined by subjective tolerance. For dietary control, the regular taking of the nutritional history is still best, as there are no good sensitive chemical parameters except, perhaps, transaminases. Quantification of carbohydrate-deficient proteins, e.g., transferrin, has been suggested for dietary monitoring; however, it is questionable whether this is a sufficiently sensitive procedure. Needless to say, patients (and their parents) should be made aware of the fact that infusions containing fructose, sorbitol or invert sugar are life-threatening, and that they should report fructose intolerance on any hospital admission. The prognosis is excellent with normal growth, intelligence and life span.

9.3 Fructose-1,6-Bisphosphatase Deficiency

9.3.1 Clinical Presentation

In about half of all cases, fructose-1,6-bisphosphatase (FBPase) deficiency presents in the first 1 to 4 days of life with severe hyperventilation caused by profound lactic acidosis and marked hypoglycemia. Later on, episodes of irritability, somnolescence or coma, apneic spells, dyspnea and tachycardia, muscular hypotonia, and moderate hepatomegaly may occur. As reported in the first patient to be described [20], such episodes are typically triggered by a febrile episode accompanied by a refusal to feed and vomiting. Attacks may also follow ingestion of large amounts of fructose (~1 g/kg body weight in one dose) especially after a period of fasting. FBPase deficiency may be life-threatening and, as in HFI, administration of IV fructose is contraindicated and may lead to death. In between attacks, patients are usually well, although mild, intermittent or chronic acidosis may exist. The frequency of the attacks decreases with age, and the majority of survivors display normal somatic and psychomotor development [21].

Most affected children experience a number of acute attacks before the diagnosis is made. Once diagnosis is established and treatment begins, the course is favorable.

In contrast to HFI, chronic ingestion of fructose does not lead to gastrointestinal symptoms – hence there is no aversion to sweet foods – or failure to thrive, and only exceptionally is there disturbed liver function.

Analysis of plasma during acute episodes reveals lactate accumulation (up to 15–25 mM) accompanied by a decreased pH and an increased lactate/pyruvate ratio (up to 30), hyperalaninemia and glucagon-resistant hypoglycemia. Hyperketonemia may be found, but in several patients ketosis has been reported to be moderate or absent (▶ below and [22]). Increased levels of free fatty acids and uric acid may also be found. Urinary analysis reveals increased lactate, alanine, glycerol, and, in most cases, ketones and glycerol-3-phosphate.

9.3.2 Metabolic Derangement

Deficiency of FBPase, a key enzyme in gluconeogenesis, impairs the formation of glucose from all gluconeogenic precursors, including dietary fructose (▣ Fig. 9.1). Consequently, maintenance of normoglycemia in patients with the defect is exclusively dependent on glucose (and galactose) intake and degradation of hepatic glycogen. Also, hypoglycemia is likely to occur when glycogen reserves are limited (as in newborns) or exhausted (as when fasting). The defect moreover provokes accumulation of the gluconeogenic substrates lactate/pyruvate, glycerol and alanine. The lactate/pyruvate ratio is usually increased, owing to secondary impairment of the conversion of 1,3-bisphosphoglycerate into glyceraldehyde-3-phosphate, resulting in accumulation of reduced nicotinamide adenine dinucleotide, the other substrate of glyceraldehyde-3-phosphate dehydrogenase (not shown in Fig. 9.1). Attention has been drawn to the fact that hyperketonemia and ketonuria, which usually accompany hypoglycemia, may be absent in some patients with FBPase deficiency [22]. This may be explained by pyruvate accumulation resulting in an increase of oxaloacetate and, hence, in the diversion of acetyl-coenzyme A (CoA) away from ketone-body formation into citrate synthesis. This, in turn, results in increased synthesis of malonyl-CoA in the cytosol. Elevated malonyl-CoA, by inhibiting carnitine-palmitoyl transferase I, prevents the entry of long-chain fatty-acyl-CoA into the mitochondria and, thereby, further reduces ketogenesis. It also promotes accumulation of fatty acids in liver and plasma, as documented in some patients.

Children with FBPase deficiency generally tolerate sweet foods, up to 2 g fructose/kg body weight per day, when given regularly distributed over the day and, in contrast to subjects with HFI, thrive on such a diet. Nevertheless, loading tests with fructose do induce hypoglycemia, as in HFI. This is caused by the inhibitory effect of the rapidly formed but slowly metabolized F-1-P on liver glycogen phosphorylase a. That higher doses of fructose are required for hypoglycemia to occur is explained by the fact that, in contrast to the aldolase-B defect, the FBPase deficiency still allows F-1-P to be converted into lactate. ^{31}P-MRS of the liver following IV administration of fructose (200 mg/kg b.w.) has documented a slower decrease in the fructose-induced accumulation of F-1-P and a delayed recovery of the ensuing depletion of Pi and ATP (both of which are signs of fructose toxicity) in patients with FBPase deficiency compared with healthy controls [3].

9.3.3 Genetics

FBPase deficiency is an autosomal-recessive disorder. Its frequency seems to be much lower than that of HFI; a first estimation of 1:350,000 has recently been reported for the Netherlands [21]. In addition to European and North American patients, many cases have been diagnosed in Japan. The high proportion of Turkish patients in our own series might simply be the result of the high rate of parental consanguinity.

There is evidence for the existence of more than one isozyme with FBPase activity in humans. The muscle isoform has different kinetic characteristics to the liver isoform and is not affected in patients with FBPase deficiency. Only the liver-type isoform gene (*FBP1*) has been cloned and characterized to date. It has been localized to chromosome 9q22.2–q22.3. *FBP1* mutations were first reported in 1995 [23], and to date 22 different mutations have been published. Among them are single nucleotide exchanges, small deletions and insertions, and one gross deletion. All regions of the gene may be affected and, with the exception of the c.961 *ins* G mutation, which has been reported to be responsible for 46% of mutated alleles in Japan [24], no single mutation is particularly frequent.

There are several FBPase-deficient patients in whom no mutation could be found affecting the coding region of *FBP1*. Therefore, it has been supposed that these patients carry mutations within the promoter region of *FBP1* or, more hypothetically, in the gene for the bifunctional enzyme which controls the concentration of fructose-2,6-bisphosphate, the main physiological regulator of FBPase [25].

9.3.5 Diagnosis

Whenever possible, the diagnosis of FBPase deficiency should be made by molecular analysis on DNA from peripheral leukocytes. If no mutation is found despite highly suggestive clinical and laboratory findings, the determination of enzymatic activity in a liver biopsy should be undertaken. In symptomatic cases, the residual activity may vary from zero to 30% of normal, indicating genetic heterogeneity of the disorder. Obligate heterozygotes have intermediate activity. Diagnosis can also be attempted in leukocytes, bearing in mind that deficient activity is diagnostic but that normal activity does not rule out FBPase deficiency in the liver [26]. Cultured skin fibroblasts, amniotic fluid cells and chorionic villi do not exhibit FBPase activity.

Loading tests with fructose (or with glycerol or alanine) or fasting tests should not form part of the initial investigations as they provide only a tentative diagnosis. However, such functional tests may be useful, and may point to a disturbance in the regulation of the fructose 6-phosphate-fructose-1,6-bisphosphate substrate cycle if mutation ana-

lysis and enzyme activity are normal despite a strong clinical and chemical suspicion of FBPase deficiency.

9.3.6 Differential Diagnosis

Other disturbances in gluconeogenesis have to be considered, including *(i)* pyruvate dehydrogenase deficiency characterized by a low lactate/pyruvate ratio and aggravation of lactic acidosis by glucose infusion; *(ii)* pyruvate carboxylase deficiency; *(iii)* phosphoenol pyruvate carboxykinase deficiency; *(iv)* respiratory chain disorders; *(v)* glycogenosis types Ia and Ib characterized by hepatonephromegaly, hyperlipidemia and hyperuricemia.

9.3.7 Treatment and Prognosis

Whenever FBPase deficiency is suspected, adequate amounts of IV or oral glucose should be given. The acute, life-threatening episodes should be treated with an IV bolus of 20% glucose followed by a continuous infusion of glucose at high rates (10-12 mg/kg/min for newborns) and bicarbonate to control hypoglycemia and acidosis.

Maintenance therapy should be aimed at avoiding fasting, particularly during febrile episodes. This involves frequent feeding, the use of slowly absorbed carbohydrates (such as uncooked starch), and a gastric drip, if necessary. In small children, restriction of fructose, sucrose and sorbitol is also recommended, as are restrictions of fat, to 20–25%, and protein to 10% of energy requirements. In the absence of any triggering effects leading to metabolic decompensation, individuals with FBPase deficiency are healthy and no carbohydrate supplements are needed.

Once FBPase deficiency has been diagnosed and adequate management introduced, its course is usually benign. Growth and psychomotor and intellectual development are unimpaired, and tolerance to fasting improves with age up to the point that the disorder does not present a problem in later life. This might be explained by an increasing capacity to store glycogen in the liver, resulting in a lesser dependence on gluconeogenesis for the maintenance of blood glucose. Many patients become obese because their concerned parents overfeed them and because later they continue these eating habits. However, under carefully observed conditions, a hypocaloric diet can lead to a considerable weight loss in obese patients without lactic acidosis and hypoglycemia [Steinmann, personal observations].

References

1. Bonthron DT, Brady N, Donaldson IA, Steinmann B (1994) Molecular basis of essential fructosuria: molecular cloning and mutational analysis of human ketohexokinase (fructokinase). Hum Mol Genet 3:1627-1631

2. Asipu A, Hayward BE, O‹Reilly J, Bonthron DT (2003) Properties of normal and mutant recombinant human ketohexokinases and implications for the pathogenesis of essential fructosuria. Diabetes 52:2426-2432

3. Boesiger P, Buchli R, Meier D, Steinmann B, Gitzelmann R (1994) Changes of liver metabolite concentrations in adults with disorders of fructose metabolism after intravenous fructose by 31P magnetic resonance spectroscopy. Pediatr Res 36:436-440

4. Froesch ER, Prader A, Labhart A, Stuber HW, Wolf HP (1957) Die hereditäre Fructoseintoleranz, eine bisher nicht bekannte kongenitale Stoffwechselstörung. Schweiz Med Wochenschr 87:1168-1171

5. Baerlocher K, Gitzelmann R, Steinmann B, Gitzelmann-Cumarasamy N (1978) Hereditary fructose intolerance in early childhood: a major diagnostic challenge. Survey of 20 symptomatic cases. Helv Paediatr Acta 33:465-487

6. Mock DM, Perman JA, Thaler MM, Morris RC Jr (1983) Chronic fructose intoxication after infancy in children with hereditary fructose intolerance. A cause of growth retardation. N Engl J Med 309:764-770

7. Lameire N, Mussche M, Baele G, Kint J, Ringoir S (1978) Hereditary fructose intolerance: a difficult diagnosis in the adult. Am J Med 65:416-423

8. Van den Berghe G (1978) Metabolic effects of fructose in the liver. Curr Top Cell Regul 13:97-135

9. Jaeken J, Pirard M, Adamowicz M, Pronicka E, Van Schaftingen E (1996) Inhibition of phosphomannose isomerase by fructose-1-phosphate: an explanation for defective N-glycosylation in hereditary fructose intolerance. Pediatr Res 40:764-766

10. Woods HF, Alberti KGMM (1972) Dangers of intravenous fructose. Lancet II:1354-1357

11. Cross NCP, DeFranchis R, Sebastio G et al (1990) Molecular analysis of aldolase B genes in hereditary fructose intolerance. Lancet I:306-309

12. Santer R, Rischewski J, von Weihe M et al (2005) The spectrum of aldolase B (ALDOB) mutations and the prevalence of hereditary fructose intolerance in Central Europe. (submitted)

13. Tolan DR, Brooks CC (1992) Molecular analysis of common aldolase B alleles for hereditary fructose intolerance in North Americans. Biochem Med Metabol Biol 48:19-25

14. Kullberg-Lindh C, Hannoun C, Lindh M (2002) Simple method for detection of mutations causing hereditary fructose intolerance. J Inherit Metab Dis 25:571-575.

15. Cox TM (2002) The genetic consequences of our sweet tooth. Nat Rev Genet 3:481-487

16. Steinmann B, Gitzelmann R (1981) The diagnosis of hereditary fructose intolerance. Helv Paediatr Acta 36:297-316

17. Commission Directive of 14 May 1991 on infant formulae and follow-on formulae (91 / 321 / EEC) (1991) No L 175 / 35-51

18. Odièvre M, Gentil C, Gautier M, Alagille D (1978) Hereditary fructose intolerance in childhood. Diagnosis, management and course in 55 patients. Am J Dis Child 132:605-608

19. Chevalier P, Trioche P, Odièvre M, Labrune P (1996) Patterns of growth in inherited fructose intolerance (Abstract). Pediatr Res 40:524

20. Baker L, Winegrad AI (1970) Fasting hypoglycemia and metabolic acidosis associated with deficiency of hepatic fructose-1,6-bisphosphatase activity. Lancet II:13-16

21. Visser G, Bakker HD, deKlerk JBC et al (2004) Natural history and treatment of fructose 1,6-diphosphatase deficiency in the Netherlands (Abstract). J Inherit Metab Dis 27[Suppl 1]:207

22. Morris AA, Deshphande S, Ward-Platt MP et al (1995) Impaired ketogenesis in fructose-1,6-bisphosphatase deficiency: a pitfall in the investigation of hypoglycemia. J Inherit Metab Dis 18:28-32

23. El-Maghrabi MR, Lange AJ, Jiang W et al (1995) Human fructose-1,6-bisphosphatase gene (FBP1): exon-intron organization, localization to chromosome bands 9q22.2-q22.3, and mutation screening in subjects with fructose-1,6-bisphosphatase deficiency. Genomics 27:520-525

24. Kikawa Y, Inuzuka M, Jin BYet al (1997) Identification of genetic mutations in Japanese patients with fructose-1,6-bisphosphatase deficiency. Am J Hum Genet 61:852-861

25. Hers HG, Van Schaftingen E (1982) Fructose 2,6-bisphosphate two years after its discovery. Biochem J 206:1-12

26. Besley GTN, Walter JH, Lewis MA, Chard CR, Addison GM (1994) Fructose-1,6-bisphosphatase deficiency: severe phenotype with normal leukocyte enzyme activity. J Inherit Metab Dis 17:333-335

10 Persistent Hyperinsulinemic Hypoglycemia

Pascale de Lonlay, Jean-Marie Saudubray

Glucose-Induced Insulin Secretion and Its Modulation

After glucose enters the pancreatic β-cell it is phosphorylated to glucose-6-phosphate by glucokinase. This enzyme, with a Km for glucose close to the concentration in blood, functions as a glucose sensor. A small change in blood glucose increases the rate of glucose metabolism, the generation of ATP by the glycolytic pathway, and the concentration of ATP relative to adenosine diphosphate. The increase in ATP results in closure of K$^+$ channels [composed of two subunits, a K$^+$-ATP channel (KIR) and the sulfonylurea receptor (SUR1)], membrane depolarization, opening of voltage-sensitive Ca^{++} channels, influx of extracellular Ca^{++}, and stimulation of insulin secretion by exocytosis from storage granules. Leucine, a potent enhancer of insulin secretion, acts by allosteric stimulation of glutamate dehydrogenase. This results in an increase in the formation of α-ketoglutarate (an intermediate of the Krebs cycle) and, hence, in elevation of ATP. Diazoxide inhibits insulin secretion by activating (opening) SUR1, whereas sulfonylureas, such as tolbutamide, stimulate insulin secretion by closing SUR1. Somatostatin and Ca^{++} antagonists inhibit insulin secretion by decreasing Ca^{++} influx.

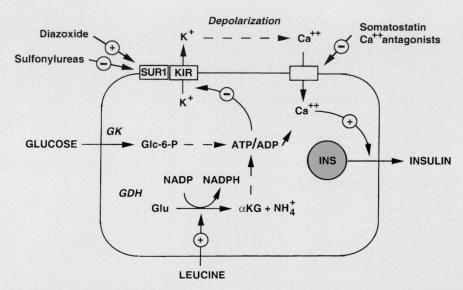

▪ **Fig.10.1.** Mechanisms of insulin secretion by the pancreatic β-cell. +, stimulation; -, inhibition; *ADP,* adenosine diphosphate; *ATP,* adenosine triphosphate; α-*KG,* α-ketoglutarate; *G-6-P,* glucose-6-phosphate; *GDH,* glutamate dehydrogenase; *GK,* glucokinase; *Glc,* glucose; *Glu,* glutamate; *Ins,* insulin; *NADP,* nicotinamide adenine dinucleotide phosphate

Hyperinsulinism can occur throughout childhood but is most common in infancy. Persistent hyperinsulinemic hypoglycemia of infancy (PHHI) is the most important cause of hypoglycemia in early infancy. The excessive secretion of insulin is responsible for profound hypoglycemia and requires aggressive treatment to prevent severe and irreversible brain damage. Onset can be in the neonatal period or later, with the severity of hypoglycemia decreasing with age. PHHI is a heterogeneous disorder with two histopathological lesions, diffuse (DiPHHI) and focal (FoPHHI), which are clinically indistinguishable. FoPHHI is sporadic and characterized by somatic islet-cell hyperplasia. DiPHHI corresponds to a functional abnormality of insulin secretion in the whole pancreas and is most often recessive although rare dominant forms can occur, usually outside the newborn period. Differentiation between focal and diffuse lesions is important because the therapeutic approach and genetic counselling differ radically. PET scanning with 18-fluoro-dopa can distinguish between focal and diffuse PHHI. A combination of glucose and glucagon is started as an emergency treatment as soon as a tentative diagnosis of PHHI is made. This is followed by diazoxide and other medication. Patients who are resistant to medical treatment require pancreatectomy; FoPHHI can be definitively cured by a limited pancreatectomy, but DiPHHI requires a subtotal pancreatectomy, following which there is a high risk of diabetes mellitus. Persistent hyperinsulinism in older children is most commonly caused by pancreatic adenoma.

10.1   Clinical Presentation

Severe hypoglycemia, with its high risk of seizures and brain damage, is the major feature of hyperinsulinism (HI) [1, 2]. The presentation varies according to the age of onset. In the neonatal period hypoglycemia is severe, occurs within 72 h after birth and is manifest in half of the patients by seizures. The majority of affected newborns are macrosomic at birth with a mean birth-weight of 3.7 kg and approximately 30% are delivered by caesarean section [1]. Other symptoms are abnormal movements, tremulousness, hypotonia, cyanosis, hypothermia and life-threatening events. In some cases, hypoglycemia is discovered from routine measurement of blood glucose. The concentration at the time of the first symptoms is often extremely low (<1 mmol/l). Hypoglycemia is persistent, occurring both in the fed and fasting state. The rate of IV glucose administration required to maintain plasma glucose above 3 mmol/l is high, with a mean of 17 mg/kg/min. The blood glucose concentration can be increased by 2 to 3 mmol/l by subcutaneous or

intramuscular administration of glucagon (0.5 mg). A mild hepatomegaly is frequently found.

Hypoglycemia from HI presenting later in infancy (1–12 months of age) has a similar clinical presentation but usually requires a lower rate of IV glucose to maintain a normal blood glucose [3]. Macrosomy at birth is common and seizures occur in approximately 50% of patients.

Hypoglycemia from HI first presenting in childhood usually occurs between 4 and 13 years of age, and is highly suggestive of a pancreatic adenoma, the histology of which is different from the focal lesion. The rate of oral or IV glucose required to maintain normal plasma glucose is lower and not all children require continuous glucose administration. Because hypoglycemia is better tolerated, the diagnosis is often delayed.

In all types of HI, facial dysmorphism with high forehead, large and bulbous nose with short columella, smooth philtrum and thin upper lip is frequently observed [4]. It should be noted that HI can be part of syndromic HIs such as Usher syndrome type Ic or congenital disorders of glycosylation [5]. A few patients with Beckwith-Widemann, Perlman or Sotos syndromes have also been described with HI. Fabricated of induced illness (FII) should be included in the differential diagnosis. Epilepsy appears to be frequent in patients with HI associated with hyperammonemia, and is not explained by hypoglycaemia [6].

Exercise-induced HI (EIHI) is a novel, autosomal dominant form of HI which has been identified in two families. The patients suffer from hypoglycaemic symptoms only when performing strenuous physical exercise [7].

Postprandial, as well as fasting hyperinsulinemic hypoglycaemia, associated with resistance to insulin has also recently been observed in a new syndrome of dominant HI in which the insulin receptor gene is implicated [8].

10.2   Metabolic Derangement

Hyperinsulinemic hypoglycemia is due to either focal or general insulin hypersecretion by the pancreas [1]. Insulin decreases plasma glucose both by inhibiting hepatic glucose release from glycogen and gluconeogenesis, and by increasing glucose uptake in muscle and fat.

PHHI is a heterogeneous disorder which can be caused by various defects in the regulation of insulin secretion by the pancreatic B-cell (◘ Fig. 10.1) [9]. These include i) channelopathies affecting either the SUR1 [10–13] or the KIR channel; ii) enzyme deficiencies including glucokinase (GK) [14], glutamate dehydrogenase (GDH) [15] and short-chain L-3-hydroxyacyl-CoA dehydrogenase (SCHAD) [16]; iii) modifications of the insulin receptor [8].

10.3   Genetics

The estimated incidence of PHHI is 1/50,000 live births but in countries with substantial consanguinity, such as Saudi Arabia, it may be as high as 1/2500 [17]. The two histological forms of HI correspond to distinct molecular entities.

Focal islet-cell hyperplasia is associated with hemi- or homozygosity of a paternally inherited mutations of the sulfonylurea-receptor (SUR1) or the inward rectifying potassium channel genes (Kir6.2) on chromosome 11p15, and loss of the maternal allele in the hyperplastic islets [18, 19]. The focal lesion probably is a sporadic event, as indicated by the somatic molecular abnormality in the pancreas and by the observation of discordant identical twins [18].

Recessive SUR1 mutations, and more rarely recessive Kir6.2 mutations, are responsible for the majority of *diffuse* and severe neonatal HI resistant to medical treatment [10–13]. Identification of these mutations should eventually allow prenatal diagnosis. Dominant SUR1 mutations are responsible for less severe HI occurring in the first year of age and are sensitive to diazoxide [20]. Dominantly expressed missense mutations of the mitochondrial matrix enzyme, GDH, cause hyperinsulism/hyperammonemia HI/HA syndrome, the second most common form of HI [15]. Dominantly expressed GK mutations are a rare cause of HI. They result in a gain of function by increased affinity of GK for glucose leading to inappropriate insulin secretion [14]. These mutations are remote from the glucose binding site and suggest an allosteric regulation defect.

Aetiology of adenoma is unknown except adenoma related to the MEN1 syndrome by dominant mutation on the MEN1 gene following menine protein deficiency [21–25]. A loss of the 11p13 region has been described in some adenomas [26] and one adenoma has been described in Bourneville's tuberous sclerosis [27].

10.4   Diagnostic Tests

The presence and severity of HI can be evaluated from the rate of glucose administration required to maintain normal glycemia and the response to medical treatment and depends on the age at presentation [3].

10.4.1   Diagnostic Criteria

The diagnostic criteria for PHHI include *(1)* fasting and post-prandial hypoglycaemia (<3 mmol/l) persisting through the first month of life and associated with hyper-insulinemia (plasma insulin >3mU/l), requiring high rates of intravenous glucose administration (>10 mg/kg/min) to maintain blood glucose >3 mmol/l; *(2)* a positive response

to SC or IM glucagon (0.5 mg followed by an increase in the blood glucose level of 2 to 3 mmol/l). Nevertheless, in infancy and childhood, normal plasma insulin and C-peptide concentrations during hypoglycemia do not exclude the diagnosis of HI and measurements must be repeated. In the absence of clearly abnormal insulin levels during a hypoglycemic episode, an 8 to 12 h fasting test, aimed at identifying inappropriately low increase of plasma levels of ketone bodies, free fatty acid and branched chain amino, acids can be helpful [28].

Hyperammonemia needs to be excluded in new patients with PHHI before deciding to pursue a more aggressive treatment since the HI/HA syndrome is usually amenable to medical or dietetic treatment. Similarly, analysis of urine organic acids and plasma acylcarnitines must be undertaken to investigate for fatty acid oxidation defects. Finally, the secondary causes of HI should be excluded, namely FII, auto-immunity, and congenital disorders of glycosylation.

10.4.2   Differentiation of Focal from Diffuse Forms

Patients who are treated surgically have to be classified according to histological criteria. The focal form, which accounts for 40% of the patients treated surgically, is defined as a focal adenomatous hyperplasia [29–34]. The lesion measures 2.5 to 7.5 mm in diameter differing from true adult-type pancreatic adenoma which is clearly limited with different topographic distribution. Diffuse PHHI shows abnormal β-cell nuclei in all sections of the whole pancreas [35]. In the absence of any distinctive clinical feature, and because preoperative classical radiology of the pancreas including echotomography, CT SCAN and NMR, is not efficient to screen for the focal form, pancreatic venous catheterization (PVS) and pancreatic arteriography were until recently the only preoperative procedures available for locating the site of insulin secretion [36–38]. These were not performed before the age of one month in order to exclude patients with transient forms, or those with hyperammonemia or with familial or consanguineous forms, which are likely to have diffuse HI. Percutaneous transhepatic catheterization is undertaken under general anesthesia, without halothane, while blood glucose is maintained between 2 and 3 mmol/l. Venous blood samples are collected in the head, isthmus, body and tail of the pancreas for measurements of plasma glucose, insulin and C-peptide levels. Patients with a focal lesion have high plasma insulin and C-peptide levels in one or several contiguous samples, with low concentrations in the remaining pancreatic samples. The patients with diffuse HI have high plasma insulin and C-peptide concentrations in all pancreatic samples.

However, the use of [18F]-labeled fluoro-L-DOPA whole-body positron emission tomography (PET) has now been evaluated for the detection of hyperfunctional islet pancreatic tissue: an abnormal focal accumulation of [^{18}F]-labeled fluoro-L-DOPA is observed in the pancreas of the patients with a focal lesion, while a diffuse uptake of the radiotracer is observed over the whole pancreas in the patients with diffuse insulin secretion [39]. We hope that this new test, an accurate non invasive technique, will replace PVS for the correct localization of focal lesion in children with congenital and persistent HI of infancy.

It has recently been suggested, although not confirmed, that a tolbutamide test could separate focal from diffuse forms of HI [40, 41].

10.5   Treatment and Prognosis

10.5.1   Medical Treatment

Treatment must be rapid and aggressive in order to prevent irreversible brain damage; this often necessitates central venous access and continuous oral alimentation using a nasogastric tube. IV glucagon given continuously (1 to 2 mg per day) can be added if blood glucose levels remain unstable despite a high glucose infusion rate.

At the same time, specific treatments must also be given. Oral diazoxide should be used to treat PHHI at a dose of 15 mg/kg/day in neonates and 10 mg/kg/day in infants, divided into 3 doses [3]. Diazoxide is usually effective in the infantile form (60% of cases in our experience), but most of those with the neonatal form are resistant to this treatment (90% of our cases).

Diazoxide efficacy is defined as the normalization of blood glucose levels (>3 mmol/l) measured before and after each meal in patients fed normally with a physiological feed and after stopping IV glucose and any other medications for at least five consecutive days. Two confirmed hypoglycemias (<3 mmol/l) in such a 24-hour glucose cycle indicates a lack of response to diazoxide and requires continuous nasogastric drip feeding and/or other measures to be restarted. Tolerance to diazoxide is usually excellent. The most frequent adverse effect is hirsutism, which can sometimes be marked and distressing in young children. Hematologic side effects and troublesome fluid retention are very rare with usual doses.

Octreotide can be tried before surgery in case of diazoxide-unresponsiveness. Doses used have varied between 10 mcg/day and 50 mcg/day, given either in 3 to 4 sub-cutaneous injections or by sub-cutaneous pump [42]. High doses may lead to worsening of the hypoglycemia by suppressing both glucagon and growth hormone. After starting treatment with octreotide treatment, many patients have vomiting and/or diarrhea and abdominal distension; however, these resolve spontaneously within 7-10 days. Steatorrhea is also common. This partially responds to oral pancreatic enzymes and remits after several weeks to months.

Gallbladder sludge can occur and necessitates routine abdominal ultrasound. Other drugs such as calcium-channel blockers (for example nifedipine) have been proposed. All these treatments, if effective in controlling blood sugar, do not need to be increased according to the weight of the patient, so that their dose usually remains unchanged.

A restricted protein diet, limiting the leucine intake to 200 mg per meal, is mandatory in HI/HA syndrome and is often effective.

10.5.2   Surgical Treatment

Surgical treatment is required when medical or dietary therapies are ineffective or when a focal form is suspected. Previously, most paediatric surgeons recommended a 95% subtotal pancreatectomy in all such cases, a procedure which is associated with a high risk of subsequent development of diabetes mellitus [43]. There is now strong evidence that DiPHHI and FoPHHI require different surgical treatment [18, 19, 44], even if the long term mental prognosis is mostly related to the duration of the initial hypoglycemia [45].

Intraoperative histology is performed to substantiate with certainty the findings of pancreatic catheterization and to guide the limits of resection in those confirmed to have FoPHHI. For these purposes, pancreatic samples must be collected from the head, the isthmus, the body and the tail of the pancreas and immediately examined microscopically. DiPHHI lesions are characterized by ß-cells with large nuclei and abundant cytoplasm in all samples. The histological analysis of focal lesions shows no abnormal ß-cell nuclei and shrunken cytoplasm giving a pattern of crowded ß-cells. In that case, additional samples are taken to localize the lesion, guided by the pancreatic catheterisation. The localization of focal forms is crucial in view of the fact that these can be located in the head of the pancreas, whereas surgeons usually resect pancreatic tissue by first removing its tail and body. After a partial pancreatectomy is performed, a further series of samples is examined to ensure that the limits of resection are within normal pancreatic tissue. A subtotal pancreatectomy is performed for diffuse lesions.

10.5.3   Prognosis

Although most of the patients treated medically remain dependent on medication, some who respond well to medical management (diazoxide and/or octreotide) may undergo a complete clinical remission, relatively rapidly (<16 months) in the case of a focal lesion and later in the diffuse form. This justifies stopping medical treatment once a year under medical supervision in order to see whether there has been a spontaneous recovery. A conservative attitude is preferable for patients with HI/HA who usually respond to diazoxide or a low leucine diet, and whose disorder is likely to recover spontaneously. Patients with FoPHHI treated by limited pancreatectomy are completely cured [44]. In contrast, in those with DiPHHI, post-operative hypoglycemia and/or diabetes mellitus or serious alteration of glucose tolerance often follows subtotal pancreatectomy despite extensive surgery [43]. Pancreatic exocrine insufficiency may be treated by pancreatic enzyme replacement. An annual investigation of residual insulin secretion, based on pre- and post-prandial plasma glucose and insulin levels at various intervals as well as measurement of glycated hemoglobin (HbAIc) and an oral glucose tolerance test (OGTT), is mandatory, as diabetes or glucose intolerance can develop later.

In conclusion, the recommended strategy for investigation is as follows:

1) Exclude a transient form (< 1 month) and secondary hyperinsulinism (FII, auto-immunity, CDG, overgrowth syndromes).
2) Assess for fasting and post-prandial hyperammonemia, an indication of HI/HA syndrome.
3) Maintain blood glucose between 3 and 6 mmol/l with appropriate methods, including continuous drip feeding, IV glucose infusion, central line catheter, continuous IV glucagon.
4) Assess diazoxide and octreotide responsiveness.
5) Having excluded transient, secondary and familial forms, and patients with hyperammonaemia, locate focal insulin secretion by transhepatic selective pancreatic venous catheterization or [18F]-labeled fluoro-L-DOPA PET.
6) Treat surgically those patients who are resistant to medical treatment and those where a focal form is strongly suspected.
7) Where surgery is required verify the histological type of lesion by intra-operative histology, and perform a sub-total pancreatectomy for a diffuse lesion, or a limited pancreatectomy for a focal lesion.
8) Undertake molecular analysis (leukocyte and pancreatic DNA) in order to provide accurate genetic counselling and prenatal diagnosis.

References

1. Stanley CA (1997) Hyperinsulinism in infants and children. Pediatr Clin North Am 44:363-374
2. Thomas CG Jr, Underwood LE, Carney CN et al (1977) Neonatal and infantile hypoglycemia due to insulin excess: new aspects of diagnosis and surgical management. Ann Surg 185:505-517
3. Touati G, Poggi-Travert F, Ogier de Baulny H et al (1998) Long-term treatment of persistent hyperinsulinaemic hypoglycaemia of infancy with diazoxide: a retrospective review of 77 cases and analysis of efficacy-predicting criteria. Eur J Pediatr 157:628-633
4. de Lonlay P, Cormier-Daire V, Fournet JC et al (2002) Facial dysmorphism in persistent hyperinsulinemic hypoglycemia. Am J Med Genet 111:130-133
5. de Lonlay P, Cuer M, Barrot S et al (1999) Hyperinsulinemic hypoglycemia as presenting symptom of carbohydrate-deficiency glycoproteins. J Pediatr 135:379-383
6. Raizen DM, Brooks-Kayal A, Steinkrauss L et al (2005) Central nervous system hyperexcitability associated with glutamate dehydrogenase gain of function mutations. J Pediatr 146:388-394
7. Otonkoski T, Kaminen N, Ustinov J et al (2003) Physical exercise-induced hyperinsulinemic hypoglycemia is an autosomal-dominant trait characterized by abnormal pyruvate-induced insulin release. Diabetes 52:199-204
8. Hojlund K, Hansen T, Lajer M et al (2004) A novel syndrome of autosomal-dominant hyperinsulinemic hypoglycemia linked to a mutation in the human insulin receptor gene. Diabetes 53:1592-1598
9. Dunne MJ, Cosgrove KE, Shepherd RM et al (2004) Hyperinsulinism in infancy: from basic science to clinical disease. Physiol Rev 84:239-275
10. Thomas PM, Cote GJ, Wohllk N et al (1995) Mutations in the sulfonylurea receptor gene in familial persistent hyperinsulinemic hypoglycemia of infancy. Science 268:426-429
11. Nestorowicz A, Wilson BA, Schoor KP et al (1996) Mutations in the sulfonylurea receptor gene are associated with familial hyperinsulinism in Ashkenazi Jews. Hum Mol Genet 5:1813-822
12. Thomas P, Ye Y, Lightner E (1996) Mutation of the pancreatic islet inward rectifier Kir6.2 also leads to familial persistent hyperinsulinemic hypoglycemia of infancy. Hum Mol Genet 5:1809-1812
13. Nestorowicz A, Inagaki N, Gonoi T et al (1997) A nonsense mutation in the inward rectifier potassium chaannel gene, Kir6.2, is associated with familial hyperinsulinism. Diabetes 46:1743-1748
14. Glaser B, Kesavan P, Heyman M et al (1998) Familial hyperinsulinism caused by an activating glucokinase mutation. N Engl J Med 338:226-230
15. Stanley CA, Lieu Y, Hsu B et al (1998) Hyperinsulinemia and hyperammonemia in infants with regulatory mutations of the glutamate dehydrogenase gene. N Engl J Med 338:1352-1357
16. Clayton PT, Eaton S, Aynsley-Green A et al (2001) Hyperinsulinism in short-chain L-3-hydroxyacyl-CoA dehydrogenase deficiency reveals the importance of beta-oxidation in insulin secretion. J Clin Invest 108:457-465
17. Thornton PS, Sumner AE, Ruchelli ED et al (1991) Familial and sporadic hyperinsulinism: histopathological findings and segregation analysis support a single autosomal recessive disorder. J Pediatr 119:721-724
18. de Lonlay P, Fournet JC, Rahier J et al (1997) Somatic deletion of the imprinted 11p15 region in sporadic persistent hyperinsulinemic hypoglycemia of infancy is specific of focal adenomatous hyperplasia and endorses partial pancreatectomy. J Clin Invest 100:802-807
19. Verkarre V, Fournet JC, de Lonlay P et al (1998) Maternal allele loss with somatic reduction to homozygosity of the paternally-inherited mutation of the SUR1 gene leads to congenital hyperinsulinism in focal islet cell adenomatous hyperplasia of the pancreas. J Clin Invest 102:1286-1291
20. Huopio H, Otonkosko T, Vauhkomen I et al (2003) A new subtype of autosomal dominant diabetes attributable to a mutation in the gene for sulfonylurea receptor 1. Lancet 361:301-307
21. Larsson C, Skogseid B, Oberg K et al (1988) Multiple endocrine neoplasia type 1 gene maps to chromosome 11 and is lost in insulinoma. Nature 332:85-87
22. Demeure MJ, Klonoff DC, Karam JH et al (1991) Insulinomas associated with multiple endocrine neoplasia type 1: the need for a different surgical approach. Surgery 110:998-1004
23. Bassett JH, Forbes SA, Pannett AA et al (1998) Characterization of mutations in patients with multiple endocrine neoplasia type 1. Am J Hum Genet 62:232-244

24. Agarwal SK, Kester MB, Debelenko LV et al (1997) Germline mutations of the MEN1 gene in familial multiple endocrine neoplasia type 1 and related states. Hum Mol Genet 6:1169-1175

25. Guru SC, Goldsmith PK, Burns AL et al (1998) Menin, the product of the MEN1 gene, is a nuclear protein. Proc Natl Acad Sci USA 95:1630-1634

26. Patel P, O'Rahilly S, Buckle V et al (1990) Chromosome 11 allele loss in sporadic insulinoma. J Clin Pathol 43:377-378

27. Kim H, Kerr A, Morehouse H (1995) The association between tuberous sclerosis and insulinoma. Am J Neuroradiol 16:1543-1544

28. Stanley CA, Baker L (1976) Hyperinsulinism in infancy: diagnosis by demonstration of abnormal response to fasting hypoglycemia. Pediatrics 57:702-711

29. Sempoux C, Guiot Y, Lefevre A et al (1998) Neonatal hyperinsulinemic hypoglycemia: heterogeneity of the syndrome and keys for differential diagnosis. J Clin Endocrinol Metab 83:1455-1461

30. Klöppel G. (1997) Nesidioblastosis. In: Soleia E, Capella C, Klöppel G (eds) Tumors of the pancreas. AFIP, Washington, pp 238-243

31. Goossens A, Gepts W, Saudubray JM et al (1989) Diffuse and focal nesidioblastosis. A clinicopathological study of 24 patients with persistent neonatal hyperinsulinemic hypoglycemia. Am J Surg Pathol 3:766-775

32. Goudswaard WB, Houthoff HJ, Koudstaal J et al (1986) Nesidioblastosis and endocrine hyperplasia of the pancreas: a secondary phenomenon. Hum Pathol 17:46-53

33. Rahier J, Fält K, Müntefering H et al (1984) The basic structural lesion of persistent neonatal hypoglycaemia with hyperinsulinism: deficiency of pancreatic D cells or hyperactivity of B cells? Diabetologia 26:282-289

34. Jaffé R, Hashida Y, Yunis EJ (1980) Pancreatic pathology in hyperinsulinemic hypoglycemia of infancy. Lab Invest 42:356-365

35. Rahier J, Sempoux C, Fournet JC et al 1998) Partial or near-total pancreatectomy for persistent neonatal hyperinsulinaemic hypoglycaemia: the pathologist's role. Histopathology 32:15-19

36. Lyonnet S, Bonnefont JP, Saudubray JM et al (1989) Localisation of focal lesion permitting partial pancreatectomy in infants. Lancet 2:671

37. Brunelle F, Negre V, Barth MO et al (1989) Pancreatic venous samplings in infants and children with primary hyperinsulinism. Pediatr Radiol 19:100-103

38. Dubois J, Brunelle F, Touati G et al (1995) Hyperinsulinism in children: diagnostic value of pancreatic venous sampling correlated with clinical, pathological and surgical outcome in 25 cases. Pediatr Radiol 25:512-516

39. Santiago-Ribeiro MJ, de Lonlay P, Delzescaux T et al (2005) Non-invasive differential diagnosis of hyperinsulinism of infancy using positron emission tomography and [18F]-fluoro-L-DOPA. J Nucl Med 46:560-566

40. Stanley CA, Thornton PS, Ganguly A et al (2004) Preoperative evaluation of infants with focal or diffuse congenital hyperinsulinism by intravenous acute insulin response tests and selective pancreatic arterial calcium stimulation. J Clin Endocrinol Metab 89:288-296

41. Giurgea I, Laborde K, Touati G et al (2004) Acute insulin responses to calcium and tolbutamide do not differentiate focal from diffuse congenital hyperinsulinism. J Clin Endocrinol Metab 89:925-929

42. Thornthon PS, Alter CA, Levitt Katz LE et al (1993) Short-and long term use of octreotide in the tretment of congenital hyperinsulinism. J Pediatr 123:637-643

43. Shilyanski J, Fisher S, Cutz E et al (1997) Is 95% pancreatectomy the procedure of choice for treatment of persistent hyperinslininemic hypoglycemia of the neonate? J Pediatr Surg 32:342-346

44. de Lonlay-Debeney P, Poggi-Travert F, Fournet JC et al (1999) Clinical aspects and course of neonatal hyperinsulinism. N Engl J Med 340:1169-1175

45. Menni P, de Lonlay P, Sevin C et al (2001) Neurologic outcomes of 90 neonates and infants with persistent hyperinsulinemic hypoglycemia. Pediatrics 107:476-479

11 Disorders of Glucose Transport

René Santer, Jörg Klepper

Glucose Transporters

D-Glucose and other monosaccharides are hydrophilic substances that cannot easily cross the lipophilic bilayer of the cell membrane. Since carbohydrates are most important for energy supply of essentially all cell types, specific transport mechanisms have evolved. Proteins embedded in the cell membrane function as hydrophilic pores that allow cellular uptake and release, and also transcellular transport of these sugars.

Glucose transporter proteins can be devided into two groups. Sodium-dependent glucose transporters (SGLTs, symporter systems, active transporters) couple sugar transport to the electrochemical gradient of sodium and hence can transport glucose against its own concentration gradient. Facilitative glucose transporters (GLUTs, uniporter systems, passive transporters) can transport monosaccharides only along an existing gradient.

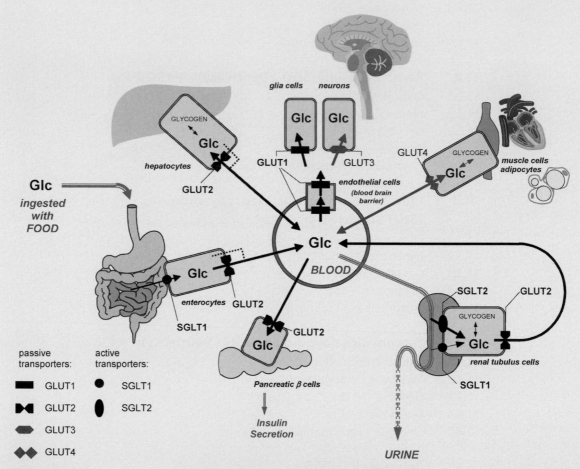

☐ **Fig. 11.1.** Overview of glucose transporters. Transport across cell membranes is depicted by arrows, and specific transporters by symbols: round for sodium-dependent, active transporters (SGLTs), and angular for facilitative, passive transporters (GLUTs). Known defects are depicted by black instead of grey transporter symbols

To date, four congenital defects of monosaccharide transport are known (◨ Fig. 11.1). Their clinical picture depends on tissue-specific expression and substrate specificity of the affected transporter. SGLT1 deficiency causes intestinal *glucose-galactose malabsorption*, a condition that presents with severe osmotic diarrhea and dehydration soon after birth. SGLT2 mutations result in isolated *renal glucosuria,* a harmless renal transport defect with normal blood glucose concentrations. In GLUT1 deficiency, also termed *glucose transporter deficiency syndrome*, clinical symptoms, usually microcephaly and an epileptic encephalopathy, are caused by impaired glucose transport at the blood brain barrier and thus into neurons and glia cells. The key finding is a low CSF glucose. *Fanconi-Bickel syndrome* is the result of a deficiency of GLUT2, an important glucose and galactose carrier within liver, kidney and pancreatic β-cells. Patients typically present with a combination of hepatic glycogen storage and a generalized renal tubular dysfunction which includes severe glucosuria.

11.1 Congenital Glucose/Galactose Malabsorption (SGLT1 Deficiency)

11.1.1 Clinical Presentation

Typically, children with congenital glucose-galactose malabsorption (GGM), caused by SGLT1 deficiency present within days after a normal pregnancy (with no polyhydramnion) and a normal birth, with bloating and profuse osmotic watery diarrhea. Stools are so loose that they may be mistaken for urine. Both breast and bottle fed infants are affected and symptoms may already begin when newborns are only given tea sweetened with glucose or polymers of glucose. As a result, patients develop severe hypertonic dehydration with fever which may be misinterpreted as a sign of an intestinal infection. If the correct diagnosis is missed and glucose and galactose are not eliminated from the diet, and if parenteral fluid administration is not available, patients die from hypovolemic shock. In typical cases, the diagnosis is considered after repeated frustrating attempts to switch from parenteral fluids to oral feeds [1]. The finding of an acidic stool pH and the detection of reducing substances in the stool is a clue to the diagnosis and most patients have mild intermittent glucosuria [2]. Chronic dehydration might be responsible for nephrolithiasis and nephrocalcinosis that develop in a number of cases [3].

11.1.2 Metabolic Derangement

A congenital defect of the sodium-dependent monosaccharide transporter SGLT1 at the apical membrane of enterocytes is the basic defect of this disorder [4]. SGLT1 contributes to the transcellular transport of these two monosaccharides which is completed by the transport out of the cell at the basolateral membrane either by facilitative diffusion or a membrane vesicle-associated transport [5]. Fructose is not a substrate for SGLT1; it is absorbed by facilitative diffusion mediated by GLUT5 both on the apical and basolateral side.

Both truncating and missense mutations of the *SGLT1* gene have been shown to result in the absence of a functioning transporter protein within the apical plasma membrane [6]. The SGLT1 protein has not yet been detected in normal human kidney; however, the fact that patients with glucose-galactose malabsorption show mild glucosuria, points to a physiological role of this transporter in renal glucose reabsorption.

11.1.3 Genetics

GGM is a relatively rare autosomal recessive disorder although the exact prevalence is unknown. *SGLT1,* located on chromosome 22q13, codes for a protein of 664 amino acids that forms 14 transmembraneous loops [7]. To date, approximately 60 different mutations have been found [4, 5] scattered all over the gene; the existence of a mutational hot spot is controversial.

11.1.4 Diagnosis

Due to its life-threatening character, GGM has to be suspected clinically and treatment has to be initiated before an ultimate diagnosis is established. The clinical stabilisation on parenteral nutrition with no foods given per os or on a fructose-based formula are in favour of the diagnosis. Oral monosaccharide tolerance tests (measuring stool pH and reducing substances, and blood glucose) combined with a hydrogen breath test can be performed. In these tests, glucose and galactose but not fructose may evoke severe clinical symptoms in affected infants but some of the test parameters may be unreliable owing to antibiotics which are frequently given to sick neonates. Glucose and galactose uptake studies on intestinal biopsies are possible but they are invasive and time-consuming. Although costly, molecular genetic studies on genomic DNA are recommended, particularly if prenatal diagnosis is likely to be requested for a future pregnancy [8].

11.1.5   Treatment and Prognosis

Whenever GGM is considered, glucose and galactose should be omitted from the diet. A formula containing fructose as the only carbohydrate is well tolerated by infants with GGM. Such a formula is easily prepared by addition of this monosaccharide to commercially available carbohydrate-free dietary products. The preparation of the diet becomes more complicated when additional foods are introduced but glucose tolerance improves with age by an as yet unknown mechanism [9]. To date, there are no long-term studies on the outcome of GGM patients and it is not clear how strict patients should adhere to the glucose- and galactose restricted diet in order not to have an increased risk of nephrolithiasis. Likewise, there is no information on long-term sequelae of a high fructose diet on liver function (▶ Chap. 9).

11.2   Renal Glucosuria (SGLT2 Deficiency)

11.2.1   Clinical Presentation

Most individuals with renal glucosuria, a congenital defect of SGLT2, are detected during a routine urine examination. Only a small number presents with polyuria and/or enuresis. Thus, renal glucosuria is an important differential diagnosis to diabetes mellitus but which is easily excluded by the presence of normal blood glucose concentrations. Renal glucosuria is an isolated defect of tubular glucose reabsorption at the proximal tubules and does not affect other glomerular or tubular kidney functions [10].

Mild renal glucosuria [0.4–5 (–10) g/1.73m^2/day] is relatively common. Individuals with a higher glucose excretion or a virtual absence of tubular glucose reabsorption (termed renal glucosuria type 0) are extremely rare.

11.2.2   Metabolic Derangement

Renal glucosuria is a non-disease; only individuals with massive glucose excretion may have a propensity to hypovolemia and hypoglycemia and can present with a delay of somatic maturation [11].

11.2.3   Genetics

Most individuals with renal glucosuria have been found to carry mutations within the *SGLT2* gene [12], located on chromosome 16p11 [13]. Its product is a low-affinity carrier that transports glucose but not galactose. Homozygosity or compound heterozygosity for *SGLT2* mutations results in the severe types of renal glucosuria whereas heterozygosity is associated with mild glucose excretion albeit not in all carriers [14]. Therefore, inheritance of renal glucosuria is best characterized as a codominant trait with variable penetrance.

To date, approximately 20 private *SGLT2* mutations have been described which are scattered all over the gene. Only a splice mutation (IVS 7 +5 g>a) was found in several kindreds of different ethnic background [14].

11.2.4   Diagnosis

Diagnosis is straight forward in patients with glucosuria and normoglycemia who do not show any other evidence of renal tubular dysfunction.

11.2.5   Treatment and Prognosis

For most cases dietary treatment is not indicated, and the prognosis, even in individuals with type 0 glucosuria, is excellent [11].

11.3   Glucose Transporter Deficiency Syndrome (GLUT1 Deficiency)

11.3.1   Clinical Presentation

GLUT1 deficiency syndrome (GLUT1-DS) typically presents as an early-onset epileptic encephalopathy. Fetal development is undisturbed; pregnancy, delivery and the neonatal period are uneventful. The majority of patients develop epilepsy within the first year of life as cerebral glucose demand increases. Seizures are of various types and frequency, often refractory to anticonvulsants, and sometimes aggravated by fasting. In infants, peculiar eye movements, staring spells, drop attacks, and cyanotic spells are the most frequent symptoms. Older children present predominantly with absences, myoclonic, and grand mal seizures [15].

Interictal EEGs in GLUT1-DS may be normal regardless of age. Occasionally, EEG recordings may show an improvement after glucose intake. In a recent study of ictal EEG features in 20 patients, focal slowing or epileptiform discharges were most prevalent in infants and generalized 2.5- to 4-Hz spike-wave patterns in older children [16]. No structural brain abnormalities are detected at neuroimaging, but PET studies may show a diminished cortical uptake with a more severe reduction in metabolism in the mesial temporal regions and thalami, accentuating a relative signal increase in the basal ganglia [17].

No dysmorphic features are observed. In early childhood, global developmental delay and a complex motor disorder become apparent. Motor milestones are delayed, speech is often significantly slurred and slowed, but all patients acquire speech and mobility. Hypotonia and ataxia result in a broad-based, unsteady gait. A substantial number of patients display additional dystonic features and elements of spasticity. Patients are of a friendly disposition and continue to make developmental progress. GLUT1-DS is not a degenerative disorder, though severe cases develop secondary microcephaly [18, 19]. Recently, variants such as transient GLUT1 deficiency and patients without epilepsy have been identified [20, 21].

11.3.2   Metabolic Derangement

GLUT1 is a membrane-spanning, glycosylated protein that provides basal glucose entry across most blood-tissue barriers. GLUT1 exclusively facilitates glucose transport across the luminal and abluminal membranes of brain capillaries representing the blood-brain barrier. Consequently, GLUT1 deficiency results in low glucose concentrations in the cerebrospinal fluid, termed hypoglycorrhachia. In addition, GLUT1 supplies glucose to neurons and glial cells [22]. Since glucose is the principal fuel for cerebral energy metabolism, GLUT1 deficiency results in impaired energy supply to the brain.

11.3.3   Genetics

The majority of patients carry heterozygous de novo mutations in the GLUT1 gene located on the short arm of chromosome 1. Mutations are distributed at random and are of various types (missense, nonsense, and splice-site mutations) including cases of haploinsufficiency [18], a case of compound heterozygosity, and one case of paternal mosaicism [19]. Phenotype-genotype correlation is yet unclear, but missense mutations may cause a milder phenotype. Autosomal dominant GLUT1-DS has been identified in three unrelated families [18]. The fact that heterozygosity for a GLUT1 mutation is sufficient to cause GLUT1-DS highlights the importance of glucose for the developing brain; presumably, homozygosity for a GLUT1 mutation is lethal in utero.

11.3.4   Diagnosis

GLUT1-DS should be suspected in any patient with hypoglycorrhachia if hypoglycemia or a central nervous system infection are absent. In patients, absolute values for cerebrospinal fluid (CSF) glucose concentrations are generally <45 mg/dl but for diagnosis CSF to blood glucose ratio is superior to the absolute concentration. This should be obtained in a non-ictal, metabolic steady state following a 4–6 h fast with blood glucose determined before the lumbar puncture to avoid stress-related hyperglycemia. Hypo-

glycorrhachia with a ratio <0.46 determined in such a way is diagnostic but the numeric value of this ratio does not correlate with clinical severity. CSF lactate is low to normal.

GLUT1-DS should be confirmed by molecular genetic methods and/or by glucose uptake studies into erythrocytes (which abundantly express GLUT1). In such studies glucose uptake is reduced to about half of the control values [23]. Again, transport kinetics do not correlate with disease severity.

11.3.5   Treatment and Prognosis

Under physiologic conditions, as a response to fasting, ketones are generated in the liver and provide an alternative fuel to the brain. This situation can be reproduced by a high-fat, low-carbohydrate ketogenic diet that was originally established for the treatment of intractable childhood epilepsy [24]. In GLUT1-DS, ketones derived from dietary fat restore brain energy metabolism. They have been shown to effectively control seizures. Most patients do not require anticonvulsant medication and their neurological function and movement disorder improve. In contrast to intractable epilepsy, it is recommended that patients with GLUT1-DS are maintained on a ketogenic diet throughout childhood and into adolescence, by which time the cerebral glucose demands decreases to adult levels.

Inhibitors of GLUT1 such as anticonvulsants (phenobarbital, chloralhydrate, diazepam), methylxanthines (theophyllin, coffeine), alcohol, and green tea should be avoided [25]. If treatment with anticonvulsants is necessary (e.g. due to non-compliance to the diet, incomplete seizure control on diet), carbamazepine or phenytoin, which have been shown not to interfere with GLUT1 function, should be considered as an add-on medication. Dietary antioxidants such as α-lipoic acid (thioctic acid, 600 to 1800 mg/day in three divided doses) have been recommended in GLUT1-DS and are still under investigation.

11.4   Fanconi-Bickel Syndrome (GLUT2 Deficiency)

11.4.1   Clinical Presentation

Patients with Fanconi-Bickel syndrome (FBS), caused by GLUT2 deficiency, typically present at an age of 3–10 months with a combination of hepatomegaly, a Fanconi-type nephropathy with severe glucosuria, a propensity to hypoglycemia in the fasted state, and glucose and galactose intolerance in the fed state [26, 27]. A few cases have presented during neonatal screening owing to hypergalactosemia [28], and cataracts have occasionally been observed in infants with FBS. At an early stage, hepatomegaly, which is caused by massive accumulation of glycogen, may not yet be present and non-specific symptoms such as fever, vomiting, chronic diarrhea, and failure to thrive may predominate. With increasing age, the clinical presentation with protuberant abdomen, moon-shaped face, and short stature, becomes more and more similar to hepatic glycogen storage diseases. Also the kidneys accumulate glycogen and their enlargement can be detected by ultrasound. Hypophosphatemic rickets are the major manifestations of tubular dysfunction resulting in joint swelling, bowing of legs and pathological fractures. FBS patients have an entirely normal mental development but growth and puberty are severely retarded [26, 27].

11.4.2   Metabolic Derangement

Fanconi-Bickel syndrome is caused by congenital deficiency or impaired function of GLUT2, a high K_m monosaccharide carrier that transports both glucose and galactose [29]. This facilitative glucose carrier is expressed in hepatocytes and at the basolateral membrane of reabsorbing cells of the proximal tubules. GLUT2 is further found at the basolateral membrane of enterocytes and within the cell membrane of pancreatic β-cells.

Intestinal uptake of glucose and galactose appear unimpaired in FBS; this has been explained by an additional transport system for glucose, a membrane vesicle-associated pathway at the basolateral membrane [5]. Postprandial hyperglycemia and hypergalactosemia are caused by impaired hepatic uptake of the two sugars. It is unclear if hyperglycemia is further exaggerated by a diminished insulin response due to an impairment of glucose sensing of β-cells. In hepatocytes GLUT2 seems to function as a glucose sensor. In the fasted state, when extracellular glucose concentration declines, the concentrations of glucose and glucose-6-phosphate within hepatocytes are inappropriately high in FBS patients. This stimulates glycogen synthesis, inhibits gluconeogenesis and glycogenolysis, and ultimately predisposes to hypoglycemia and hepatic glycogen accumulation [26].

Impaired transport of glucose out of renal tubular cells results in the accumulation of glycogen and free glucose within these cells. This impairs other transport functions resulting in a generalized tubulopathy with disproportionately severe glucosuria. The extreme amounts of glucose lost with the urine (even at times when blood glucose is low) may contribute to the propensity to develop hypoglycemia.

11.4.3   Genetics

FBS is a very rare autosomal recessive condition caused by mutations of the GLUT2 gene. More than 70% of cases come from consanguineous families [30]. The human GLUT2 gene, mapped to chromosome 3q26, codes for a 524 amino acid protein with 55% amino acid identity to GLUT1. In contrast to SGLTs, all GLUT proteins form 12 transmembranous loops within the cell membrane. The genomic structure of GLUT2 encompasses 11 exons [31] and, to date, approximately 35 different mutations scattered throughout the gene have been detected [26, 30].

11.4.4   Diagnosis

Diagnosis of FBS is suggested by the characteristic combination of an altered glucose homeostasis, hepatic glycogen accumulation, and the typical features of a Fanconi-type tubulopathy. Fasting hypoglycemia and impaired glucose and galactose tolerance may be documented during oral loading tests. Laboratory findings include mildly elevated transaminases without signs of an impaired hepatic protein synthesis or a diminished secretory function. Plasma lipids, uric acid, and lactate are elevated. If a liver biopsy is performed, both histologic and biochemical methods show an increased glycogen content; enzymatic studies of all glycogenolytic enzymes, however, are normal. Hyperaminoaciduria, hyperphosphaturia, hypercalciuria, renal tubular acidosis, mild tubular proteinuria, and polyuria are indicative of a generalized proximal tubular dysfunction. A hallmark to the diagnosis of FBS is the relatively severe glucosuria. Calculated tubular glucose reabsorption is dramatically reduced or even zero in most patients [26].

The diagnosis of FBS is ultimately confirmed by the detection of homozygosity or compound heterozygosity for GLUT2 mutations [30].

11.4.5   Treatment and Prognosis

Only symptomatic treatment is available. Measures are directed towards an improvement of glucose homeostasis and an amelioration of the consequences of renal tubulopathy. FBS patients should receive a diet with adequate caloric intake compensating for the renal glucose losses. Frequent feeds using slowly absorbed carbohydrates are recommended. Continuous carbohydrate supply by tube feeding of oligosaccharide solutions during the nights may be indicated. The administration of uncooked corn starch has a beneficial effect on metabolic control, particularly on growth [32].

Regarding tubulopathy, water and electrolytes have to be replaced in appropriate amounts. Administration of alkali may be necessary to compensate for renal tubular acidosis. Hypophosphatemic rickets requires supplementation with phosphate and vitamin D preparations. With these measures, prognosis is fairly good and some of the originally described pediatric patients have reached adulthood. The main subjective problems for these adult patients are short stature and orthopedic problems from hypophosphatemic rickets and osteomalacia. Hepatic adenomas or tumours, as described for other glycogen storage diseases, have never been observed in FBS. Metabolic decompensation with severe acidosis or renal insufficiency similar to diabetic glomerulosclerosis have been exceptional complications causing death in childhood [26].

References

1. Wright EM (1998) Genetic disorders of membrane transport. I. Glucose galactose malabsorption. Am J Physiol Gastrointest Liver Physiol 275:G879-G882
2. Meeuwisse GW (1970) Glucose-galactose malabsorption: studies on renal glucosuria. Helv Paediatr Acta 25:13-24
3. Tasic V, Slaveska N, Blau N, Santer R (2004) Nephrolithiasis in a child with glucose-galactose malabsorption. Pediatr Nephrol 19:244-246
4. Turk E, Zabel B, Mundlos S et al (1991) Glucose/galactose malabsorption caused by a defect in the Na(+)/glucose cotransporter. Nature 350:354-356
5. Santer R, Hillebrand G, Steinmann B, Schaub J (2003) Intestinal glucose transport: evidence for a membrane traffic-based pathway in humans. Gastroenterology 124:34-39
6. Martin MG, Turk E, Lostao MP et al (1996) Defects in Na(+)/glucose cotransporter (SGLT1) trafficking and function cause glucose-galactose malabsorption. Nature Genet 12:216-220
7. Hediger MA, Turk E, Wright EM (1989) Homology of the human intestinal Na+/glucose and Escherichia coli Na+/proline cotransporters. Proc Natl Acad Sci 86:5748-5752
8. Martin MG, Turk E, Kerner C et al (1996) Prenatal identification of a heterozygous status in two fetuses at risk for glucose-galactose malabsorption. Prenatal Diag 16:458-462
9. Elsas LJ, Lambe DW (1973) Familial glucose-galactose malabsorption: remission of glucose intolerance. J Pediatr 83:226-232
10. Brodehl J, Oemar BS, Hoyer PF (1987) Renal glucosuria. Pediatr Nephrol 1:502-508
11. Scholl S, Santer R, Ehrich JHH (2004) Long-term outcome of renal glucosuria type 0 – the original patient and his natural history. Nephrol Dial Transpl 19:2394-2396
12. Santer R, Kinner M, Schneppenheim R et al (2000) The molecular basis of renal glucosuria: Mutations in the gene for a renal glucose transporter (SGLT2). J Inherit Metab Dis 23[Suppl 1]:178
13. Wells RG, Pajor AM, Kanai Y et al (1992) Cloning of a human kidney cDNA with similarity to the sodium-glucose cotransporter. Am J Physiol 263:F459–465
14. Santer R, Kinner M, Lassen C et al (2003) Molecular analysis of the SGLT2 gene in patients with renal glucosuria. J Am Soc Nephrol 14:2873-2882
15. Klepper J, Voit T (2002) Facilitated glucose transporter protein type 1 (GLUT1) deficiency syndrome: impaired glucose transport into brain – a review. Eur J Pediatr 161:295-304
16. Leary LD, Wang D, Nordli DR et al (2003) Seizure characterization and electroencephalographic features in glut-1 deficiency syndrome. Epilepsia 44:701-707
17. Pascual JM, Van Heertum RL, Wang D et al (2002) Imaging the metabolic footprint of Glut1 deficiency on the brain. Ann Neurol 52:458-464
18. De Vivo DC, Wang D, Pascual JM, Ho YY (2002) Glucose transporter protein syndromes. Int Rev Neurobiol 51:259-288
19. Klepper J, Diefenbach S, Kohlschütter A, Voit T (2004) Effects of the ketogenic diet in the glucose transporter 1 deficiency syndrome. Prostaglandins Leukot Essent Fatty Acids 70:321-327
20. Klepper J, De Vivo DC, Webb DW et al (2003) Reversible infantile hypoglycorrhachia: possible transient disturbance in glucose transport? Pediatr Neurol 29:321-325
21. Overweg-Plandsoen WC, Groener JE, Wang D et al (2003) GLUT-1 deficiency without epilepsy – an exceptional case. J Inherit Metab Dis 26:559-563
22. Maher F, Vannucci SJ, Simpson IA (1994) Glucose transporter proteins in brain. FASEB J 8:1003-1011
23. Klepper J, Garcia-Alvarez M, O'Driscoll KR et al (1999) Erythrocyte 3-O-methyl-D-glucose uptake assay for diagnosis of glucose-transporter-protein syndrome. J Clin Lab Anal 13:116-121
24. Lefevre F, Aronson N (2000) Ketogenic diet for the treatment of refractory epilepsy in children: systematic review of efficacy. Pediatrics 105:E46-52
25. Klepper J (2004) Impaired glucose transport into the brain: the expanding spectrum of glucose transporter type 1 deficiency syndrome. Curr Opin Neurol 17:193-196
26. Santer R, Steinmann B, Schaub J (2002) Fanconi-Bickel syndrome – a congenital defect of facilitative glucose transport. Curr Mol Med 2:213-227
27. Santer R, Schneppenheim R, Suter D et al (1998) Fanconi-Bickel syndrome - the original patient and his natural history, historical steps leading to the primary defect, and a review of the literature. Eur J Pediatr 157:783-797
28. Müller D, Santer R, Krawinkel M et al (1997) Fanconi-Bickel syndrome presenting in neonatal screening for galactosaemia. J Inherit Metab Dis 20:607-608
29. Santer R, Schneppenheim R, Dombrowski A et al (1997) Mutations in GLUT2, the gene for the liver-type glucose transporter, in patients with Fanconi-Bickel syndrome. Nat Genet 17:324-326
30. Santer R, Groth S, Kinner M et al (2002) The mutation spectrum of the facilitative glucose transporter gene SLC2A2 (GLUT2) in patients with Fanconi-Bickel syndrome. Hum Genet 110: 21-29
31. Takeda J, Kayano T, Fukomoto H, Bell GI (1993) Organization of the human GLUT2 (pancreatic β-cell and hepatocyte) glucose transporter gene. Diabetes 42:773-777
32. Lee PJ, van't Hoff, Leonard JV (1995) Catch-up growth in Fanconi-Bickel syndrome with uncooked cornstarch. J Inherit Metab Dis 18:153-156

III Disorders of Mitochondrial Energy Metabolism

12 Disorders of Pyruvate Metabolism and the Tricarboxylic Acid Cycle

Linda J. De Meirleir, Rudy Van Coster, Willy Lissens

Pyruvate Metabolism and the Tricarboxylic Acid Cycle

Pyruvate is formed from glucose and other monosaccharides, from lactate, and from the gluconeogenic amino acid alanine (◻ Fig 12.1). After entering the mitochondrion, pyruvate can be converted into acetylcoenzyme A (acetyl-CoA) by the pyruvate dehydrogenase complex, followed by further oxidation in the TCA cycle. Pyruvate can also enter the gluconeogenic pathway by sequential conversion into oxaloacetate by pyruvate carboxylase, followed by conversion into phospho-

enolpyruvate by phosphoenolpyruvate carboxykinase. Acetyl-CoA can also be formed by fatty acid oxidation or used for lipogenesis. Other amino acids enter the TCA cycle at several points. One of the primary functions of the TCA cycle is to generate reducing equivalents in the form of reduced nicotinamide adenine dinucleotide and reduced flavin adenine dinucleotide, which are utilized to produce energy under the form of ATP in the electron transport chain.

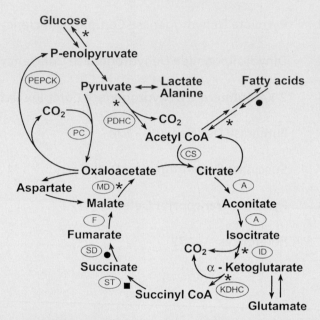

◻ **Fig. 12.1.** Overview of glucose, pyruvate/lactate, fatty acid and amino acid oxidation by the tricarboxylic acid cycle. *A,* aconitase; *CS,* citrate synthase; *F,* fumarase; *ID,* isocitrate dehydrogenase; *KDHC,* α-or 2-ketoglutarate dehydrogenase complex; *MD,* malate dehydrogenase; *PC,* pyruvate carboxylase; *PDHC,* pyruvate dehydrogenase complex; *PEPCK,* phosphoenolpyruvate carboxykinase; *SD,* succinate dehydrogenase; *ST,* succinyl coenzyme A transferase. Sites where reducing equivalents and intermediates for energy production intervene are in dicated by following symbols: *, reduced nicotinamide adenine dinucleotide; ●, reduced flavin adenine dinucleotide; ■, guanosine triphosphate

Owing to the role of pyruvate and the tricarboxylic acid (TCA) cycle in energy metabolism, as well as in gluconeogenesis, lipogenesis and amino acid synthesis, defects in pyruvate metabolism and in the TCA cycle almost invariably affect the central nervous system. The severity and the different clinical phenotypes vary widely among patients and are not always specific, with the range of manifestations extending from overwhelming neonatal lactic acidosis and early death to relatively normal adult life and variable effects on systemic functions. The same clinical manifestations may be caused by other defects of energy metabolism, especially defects of the respiratory chain (Chap. 15). Diagnosis depends primarily on biochemical analyses of metabolites in body fluids, followed by definitive enzymatic assays in cells or tissues, and DNA analysis. The deficiencies of *pyruvate carboxylase* (PC) and *phosphoenolpyruvate carboxykinase* (PEPCK) constitute defects in gluconeogenesis, and therefore fasting results in hypoglycemia with worsening lactic acidosis. Deficiency of the *pyruvate dehydrogenase complex* (PDHC) impedes glucose oxidation and aerobic energy production, and ingestion of carbohydrate aggravates lactic acidosis. Treatment of disorders of pyruvate metabolism comprises avoidance of fasting (PC and PEPCK) or minimizing dietary carbohydrate intake (PDHC) and enhancing anaplerosis. In some cases, vitamin or drug therapy may be helpful. *Dihydrolipoamide dehydrogenase* (E3) deficiency affects PDHC as well as KDHC and the branched-chain 2-ketoacid dehydrogenase (BCKD) complex (Chap. 19), with biochemical manifestations of all three disorders. The deficiencies of the TCA cycle enzymes, the *2-ketoglutarate dehydrogenase complex* (KDHC) and *fumarase*, interrupt the cycle, resulting in accumulation of the corresponding substrates. *Succinate dehydrogenase* deficiency represents a unique disorder affecting both the TCA cycle and the respiratory chain. Recently, defects of *mitochondrial transport of pyruvate* and *glutamate* (► Chap. 29) have been identified. Treatment strategies for the TCA cycle defects are limited.

12.1 Pyruvate Carboxylase Deficiency

12.1.1 Clinical Presentation

Three phenotypes are associated with pyruvate carboxylase deficiency. The patients with French phenotype (type B) become acutely ill three to forty eight hours after birth with hypothermia, hypotonia, lethargy and vomiting [1–5, 5a]. Most die in the neonatal period. Some survive but remain unresponsive and severely hypotonic, and finally succumb from respiratory infection before the age of 5 months.

The patients with North American phenotype (type A) become severely ill between two and five months of age [2, 6–8]. They develop progressive hypotonia and are unable to smile. Numerous episodes of acute vomiting, dehydration, tachypnea, facial pallor, cold cyanotic extremities and metabolic acidosis, characteristically precipitated by metabolic or infectious stress are a constant finding. Clinical examination reveals pyramidal tract signs, ataxia and nystagmus. All patients are severely mentally retarded and most have convulsions. Neuroradiological findings include subdural effusions, severe antenatal ischemia-like brain lesions and periventricular hemorrhagic cysts, followed by progressive cerebral atrophy and delay in myelination [4]. The course of the disease is generally downhill, with death in infancy.

A third form, more benign, is rare and has only been reported in a few patients [9]. The clinical course is dominated by the occurrence of acute episodes of lactic acidosis and ketoacidosis, responding rapidly to glucose 10 %, hydration and bicarbonate therapy. Despite the important enzymatic deficiency, the patients have a nearly normal cognitive and neuromotor development.

12.1.2 Metabolic Derangement

PC is a biotinylated mitochondrial matrix enzyme that converts pyruvate and CO_2 to oxaloacetate ($\blacksquare$ Fig. 12.1). It plays an important role in gluconeogenesis, anaplerosis, and lipogenesis. For gluconeogenesis, pyruvate must first be carboxylated into oxaloacetate because the last step of glycolysis, conversion of phosphoenolpyruvate to pyruvate, is irreversible. Oxaloacetate, which cannot diffuse freely out of the mitochondrion, is translocated into the cytoplasm via the malate/aspartate shuttle. Once in the cytoplasm, oxaloacetate is converted into phosphoenol-pyruvate by phosphoenol-pyruvate carboxykinase (PEPCK), which catalyzes the first committed step of gluconeogenesis.

The anaplerotic role of PC, i.e. the generation of Krebs cycle intermediates from oxaloacetate, is even more important. In severe PC deficiency, the lack of Krebs cycle intermediates lowers reducing equivalents in the mitochondrial matrix. This drives the redox equilibrium between 3-OH-butyrate and acetoacetate into the direction of acetoacetate, thereby lowering the 3-OH-butyrate/acetoacetate ratio [6]. Aspartate, formed in the mitochondrial matrix from oxaloacetate by transamination, also decreases. As a consequence, the translocation of reducing equivalents between cytoplasm and mitochondrial matrix by the malate/aspartate shuttle is impaired. This drives the cytoplasmic redox equilibrium between lactate and pyruvate into the direction of lactate, and the lactate/pyruvate ratio increases. Reduced Krebs cycle activity also plays a role in the increase of lactate and pyruvate. Since aspartate is required for the urea cycle, plasma ammonia can also go up. The energy deprivation induced by PC deficiency has been postulated to impair astrocytic buffering capacity against excitotoxic insults and to compromise microvascular morphogenesis and auto-regulation, leading to degeneration of white matter [4].

The importance of PC for lipogenesis derives from the condensation of oxaloacetate with intramitochondrially produced acetyl-CoA into citrate, which can be translocated into the cytoplasm where it is cleaved to oxaloacetate and acetyl-CoA, used for the synthesis of fatty acids. Deficient lipogenesis explains the widespread demyelination of the cerebral and cerebellar white matter and symmetrical par-aventricular cavities around the frontal and temporal horns of the lateral ventricles, the most striking abnormalities re-ported in the few detailed neuropathological descriptions of PC deficiency [1, 4].

PC requires biotin as a cofactor. Metabolic derange-ments of PC deficiency are thus also observed in biotin-responsive multiple carboxylase deficiency (▶ Chap. 27).

12.1.3   Genetics

PC deficiency is an autosomal recessive disorder. More than half of the patients with French phenotype have absence of PC protein, a tetramer formed by 4 identical subunits with MW of 130 kD, and of the corresponding mRNA. The patients with North American phenotype generally have cross-reacting material (CRM-positive) [2], as does the patient with the benign variant of PC deficiency [9]. Muta-tions have been detected in patients of both types A and B. In Canadian Indian populations with type A disease, 11 Ojibwa and 2 Cree patients were homozygous for a mis-sense mutation A610T; two brothers of Micmac origin were homozygous for a transversion M743I [8]. In other families, various mutations were found.

12.1.4   Diagnostic Tests

The possibility of PC deficiency should be considered in any child presenting with lactic acidosis and neurological abnor-malities, especially if associated with hypoglycemia, hyper-ammonemia, or ketosis. In neonates, a high lactate/pyruvate ratio associated with a low 3-OH-butyrate/acetoacetate ratio and hypercitrullinemia is nearly pathognomonic [5a]. Dis-covery of cystic periventricular leucomalacia at birth associ-ated with lactic acidosis is also highly suggestive. Typically, blood lactate increases in the fasting state and decreases af-ter ingestion of carbohydrate.

In patients with the French phenotype, blood lactate concentrations reach 10–20 mM (normal <2.2 mM) with lactate/pyruvate ratios between 50 and 100 (normal <28). In patients with the North American phenotype, blood lactate is 2–10 mM with normal or only moderately increased lac-tate/pyruvate ratios (<50). In the patients with the benign type, lactate can be normal, and only increase (usually above

10 mM) during acute episodes. Overnight blood glucose concentrations are usually normal but decrease after a 24 h fast. Hypoglycemia can occur during acute episodes of meta-bolic acidosis. Blood 3-OH-butyrate is increased (0.5–2.7 mM, normal <0.1) and 3-OH-butyrate/acetoacetate ratio is decreased (<2, normal 2.5–3).

Hyperammonemia (100–600 μM, normal <60) and an increase of blood citrulline (100–400 μM, normal <40), lysine and proline, contrasting with low glutamine, are constant findings in patients with the French phenotype [5a]. Plasma alanine is usually normal in the French phe-notype, but increased (0.5–1.4 mM, normal <0.455) in all reported patients with the North-American phenotype. During acute episodes, aspartate can be undetectably low [9].

In cerebrospinal fluid (CSF), lactate, the lactate/pyruvate ratio and alanine are increased and glutamine is decreased. Urine organic acid profile shows, besides large amounts of lactate, pyruvate and 3-OH-butyrate, an increase of α-keto-glutarate.

Measurement of the activity of PC is preferentially per-formed on cultured skin fibroblasts [6]. Assays can also be performed in postmortem liver, in which the activity of PC is 10-fold higher than in fibroblasts, but must be interpreted with caution because of rapid postmortem degradation of the enzyme. PC has low activity in skeletal muscle, which makes this tissue not useful for assay. PC activity in fibro-blasts is severely decreased, to less than 5% of normal, in all patients with the French phenotype, varies from 5 to 23% of controls in patients with the North American phenotype, and is less than 10% of controls in patients with the benign variant.

Prenatal diagnosis of PC deficiency is possible by mea-surement of PC activity in cultured amniotic fluid cells [10], direct measurement in chorionic villi biopsy specimens [3], or DNA analysis when the familial mutations are known.

12.1.5   Treatment and Prognosis

Since acute metabolic crises can be detrimental both phy-sically and mentally, patients should be promptly treated with intravenous 10% glucose. Thereafter, they should be instructed to avoid fasting. Some patients with persistent lactic acidosis may require bicarbonate to correct acidosis. One patient with French phenotype was treated with high doses of citrate and aspartate [5]. Lactate and ketones di-minished and plasma aminoacids normalized, except for arginine. In the CSF, glutamine remained low and lysine elevated, precluding normalization of brain chemistry. An orthotopic hepatic transplantation completely reversed ketoacidosis and the renal tubular abnormalities, and de-creased lactic acidemia in a patient with a severe phenotype, although concentrations of glutamine in CSF remained low [11]. Recently, one patient with French phenotype treated early by triheptanoin in order to restore anaplerosis, improved dramatically [12]. Biotin [1,6], thiamine, dichloroacetate, and a high fat or high carbohydrate diet provide no clinical benefits.

The prognosis of patients with PC deficiency depends on the severity of the defect. Patients with minimal residual PC activity usually do not live beyond the neonatal period, but some children with very low PC activity have survived beyond the age of 5 years. Those with milder defects might survive and have neurological deficits of varying degrees.

12.2   Phosphoenolpyruvate Carboxykinase Deficiency

12.2.1   Clinical Presentation

Phosphoenolpyruvate carboxykinase (PEPCK) deficiency was first described by Fiser et al. [13]. Since then, only 5 additional patients have been reported in the literature [14]. This may be explained, as discussed below, by observations that have led to the conclusion that PEPCK deficiency might be a secondary finding, which should be interpreted with utmost caution.

Patients reported to be PEPCK deficient presented, as those with PC deficiency, with acute episodes of severe lactic acidosis associated with hypoglycemia. Onset of symptoms is neonatal or after a few months. Patients display mostly progressive multisystem damage with failure to thrive, muscular weakness and hypotonia, developmental delay with seizures, spasticity, lethargy, microcephaly, hepatomegaly with hepatocellular dysfunction, renal tubular acidosis and cardiomyopathy. The clinical picture may also mimic Reye syndrome [15, 16].

Routine laboratory investigations during acute episodes show lactic acidosis and hypoglycemia, acompanied by hyperalaninemia and, as documented in some patients, by absence of elevation of ketone bodies. Liver function and blood coagulation tests are disturbed, and combined hypertriglyceridemia and hypercholesterolemia have been reported. Analysis of urine shows increased lactate, alanine and generalized aminoaciduria.

12.2.2   Metabolic Derangement

PEPCK is located at a crucial metabolic crossroad of carbohydrate, amino acid, and lipid metabolism (◘ Fig. 12.1). This may explain the multiple organ damage which seems to be caused by its deficiency. Since, by converting oxaloacetate into phosphoenolpyruvate, PEPCK plays a major role in gluconeogenesis, its deficiency should impair conversion of pyruvate, lactate, alanine, and TCA intermediates into glucose, and hence provoke lactic acidosis, hyperalaninemia and hypoglycemia. PEPCK exists as two separate isoforms, mitochondrial and cytosolic, which are encoded by two distinct genes. The deficiency of mitochondrial PEPCK, which intervenes in gluconeogenesis from lactate, should have more severe consequences than that of cytosolic PEPCK, which is supposed to play a role in gluconeogenesis from alanine.

12.2.3   Genetics

The cDNA encoding the cytosolic isoform of PEPCK in humans has been sequenced and localized to human chromosome 20. However, in accordance with the findings discussed below, no mutations have been identified.

12.2.4   Diagnostic Tests

The diagnosis of PEPCK deficiency is complicated by the existence of separate mitochondrial and cytosolic isoforms of the enzyme. Optimally, both isoforms should be assayed in a fresh liver sample after fractionation of mitochondria and cytosol. In cultured fibroblasts, most of the PEPCK activity is located in the mitochondrial compartment, and low PEPCK activity in whole-cell homogenates indicates deficiency of the mitochondrial isoform.

Deficiency of cystosolic PEPCK has been questioned because synthesis of this isoform is repressed by hyperinsulinism, a condition which was also present in a patient with reported deficiency of cytosolic PEPCK [15]. Deficiency of mitochondrial PEPCK has been disputed because in a sibling of a PEPCK-deficient patient who developed a similar clinical picture, the activity of PEPCK was found normal [16]. Further studies showed a depletion of mitochondrial DNA in this patient [17] caused by defective DNA replication [18]. The existence of PEPCK deficiency thus remains to be firmly established.

12.2.5   Treatment and Prognosis

Patients with suspected PEPCK deficiency should be treated with intravenous glucose and sodium bicarbonate during acute episodes of hypoglycemia and lactic acidosis. Fasting should be avoided, and cornstarch or other forms of slow-release carbohydrates need to be provided before bedtime. The long-term prognosis of patients with reported PEPCK deficiency is usually poor, with most subjects dying of intractable hypoglycemia or neurodegenerative disease.

Structure and Activation/Deactivation System of the Pyruvate Dehydrogenase Complex

PDHC, and the two other mitochondrial α- or 2-keto-acid dehydrogenases, KDHC and the BCKD complex, are similar in structure and analogous or identical in their specific mechanisms. They are composed of three components: E1, α- or 2-ketoacid dehydrogenase; E2, dihydrolipoamide acyltransferase; and E3, dihydrolipoamide dehydrogenase. E1 is specific for each complex, utilizes thiamine pyrophosphate, and is composed of two different subunits, E1α and E1β. The E1 reaction results in decarboxylation of the specific α- or 2-keto-acid. For the PDHC, the E1 component is the rate-limiting step, and is regulated by phosphorylation/dephosphorylation catalyzed by two enzymes, E1 kinase (inactivation) and E1 phosphatase (activation). E2 is a transacetylase that utilizes covalently bound lipoic acid. E3 is a flavoprotein common to all three 2-keto-acid dehydrogenases. Another important structural component of the PDHC is E3BP, E3 binding protein, formerly protein X. This component has its role in attaching E3 subunits to the core of E2.

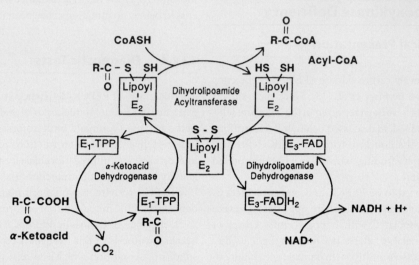

Fig. 12.2. Structure of the α- or 2-ketoacid dehydrogenase complexes, pyruvate dehydrogenase complex (PDHC), 2-ketoglutarate dehydrogenase complex (KDHC) and the branched-chain α-ketoacid dehydrogenase complex (BCKD). *CoA,* coenzyme A; *FAD,* flavin adenine dinucleotide; *NAD,* nicotinamide adenine dinucleotide; *R,* methyl group (for pyruvate, PDHC) and the corresponding moiety for KDHC and BCKD; *TPP,* thiamine pyrophosphate

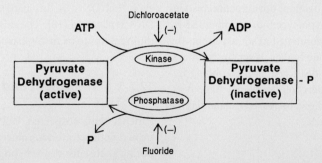

Fig. 12.3. Activation/deactivation of PDHE1 by dephosphorylation/phosphorylation. Dichloroacetate is an inhibitor of E1 kinase and fluoride inhibits E1 phosphatase. *ADP,* adenosine diphosphate; *P,* inorganic phosphate

12.3 Pyruvate Dehydrogenase Complex Deficiency

12.3.1 Clinical Presentation

More than 200 cases of pyruvate dehydrogenase complex (PDHC) deficiency have been reported [19–21], the majority of which involves the α subunit of the first, dehydrogenase component (E1) of the complex (◘ Fig. 12.2) which is X encoded. The most common features of PDHE1α deficiency are delayed development and hypotonia, seizures and ataxia. Female patients with PDHE1α deficiency tend to have a more homogeneous and more severe clinical phenotype than boys [22].

In hemizygous males, three presentations are encountered: neonatal lactic acidosis, Leigh's encephalopathy, and intermittent ataxia. These correlate with the severity of the biochemical deficiency and the location of the gene mutation. Severe neonatal lactic acidosis, associated with brain dysgenesis, such as corpus callosum agenesis, can evoke the diagnosis. In Leigh's encephalopathy, quantitatively the most important group, initial presentation, usually within the first five years of life, includes respiratory disturbances/apnoea or episodic weakness and ataxia with absence of tendon reflexes. Respiratory disturbances may lead to apnea, dependence on assisted ventilation, or sudden unexpected death. Intermittent dystonic posturing of the lower limbs occurs frequently. A moderate to severe developmental delay becomes evident within the next years. A very small subset of male patients is initially much less severely affected, with intermittent episodic ataxia after carbohydrate-rich meals, progressing slowly over years into mild Leigh's encephalopathy.

Females with PDHE1α deficiency tend to have a more uniform clinical presentation, although with variable severity, depending on variable lyonisation. This includes dysmorphic features, microcephaly, moderate to severe mental retardation, and spastic di- or quadriplegia, resembling non progressive encephalopathy. Dysmorphism comprises a narrow head with frontal bossing, wide nasal bridge, upturned nose, long philtrum and flared nostrils and may suggest fetal alcohol syndrome. Other features are low set ears, short fingers and short proximal limbs, simian creases, hypospadias and an anteriorly placed anus. Seizures are encountered in almost all female patients. These appear within the first six months of life and are diagnosed as infantile spasms (flexor and extensor) or severe myoclonic seizures. Brain MRI frequently reveals severe cortical/subcortical atrophy, dilated ventricles and partial to complete corpus callosum agenesis [23]. Severe neonatal lactic acidosis can be present. The difference in the presentation of PDHE1α deficiency in boys and girls is exemplified by observations in a brother and sister pair with the same mutation but completely different clinical features [22].

Neuroradiological abnormalities such as corpus callosum agenesis and dilated ventricles or in boys basal ganglia and midbrain abnormalities are often found. Neuropathology can reveal various degrees of dysgenesis of the corpus callosum. This is usually associated with other migration defects such as the absence of the medullary pyramids, ectopic olivary nuclei, abnormal Purkinje cells in the cerebellum, dysplasia of the dentate nuclei, subcortical heterotopias and pachygyria [24].

Only a few cases with PDHE1β deficiency have been reported [25]. These patients present with early onset lactic acidosis and severe developmental delay. Seven cases of E1-phosphatase deficiency (◘ Fig.12.3) have been identified [26], among which two brothers with hypotonia, feeding difficulties and delayed psychomotor development [27]. A few cases of PDHE2 (dihydrolipoamide transacetylase) deficiency have been reported recently [28]. The main clinical manifestations of E3BP (formerly protein X) deficiency are hypotonia, delayed psychomotor development and prolonged survival [29]. Often more slowly progressive, it also comprises early onset neonatal lactic acidosis associated with subependymal cysts and thin corpus callosum.

12.3.2 Metabolic Derangement

Defects of PDHC provoke conversion of pyruvate into lactate rather than in acetyl-CoA, the gateway for complete oxidation of carbohydrate via the TCA cycle (◘ Fig.12.1). The conversion of glucose to lactate yields less than one tenth of the ATP that would be derived from complete oxidation of glucose via the TCA cycle and the respiratory chain. Deficiency of PDHC thus specifically interferes with production of energy from carbohydrate oxidation, and lactic acidemia is aggravated by consumption of carbohydrate.

PDHC deficiency impairs production of reduced nicotinamide adenine dinucleotide (NADH) but, unlike respiratory chain defects, does not hamper oxidation of NADH. PDHC deficiency thus does not modify the NADH/NAD$^+$ ratio in the cell cytosol, which is reflected by a normal L/P ratio. In contrast, deficiencies of respiratory chain complexes I, III, and IV are generally characterized by a high L/P ratio because of impaired NADH oxidation.

12.3.3 Genetics

All components of PDHC are encoded by nuclear genes, and synthesized in the cytoplasm as precursor proteins that are imported into the mitochondria, where the mature proteins are assembled into the enzyme complex. Most of the genes that encode the various subunits are autosomal, except the E1α-subunit gene which is located on chromosome Xp22.3. Therefore, most cases of PDHC deficiency are

X-linked. To date, over 80 different mutations of the E1α subunit of PDHC have been characterized in some 130 unrelated families [30]. About half of these are small deletions, insertions, or frame-shift mutations, and the other half are missense mutations. While the consequences of most of the mutations on enzyme structure and function are not known, some affect highly conserved amino acids that are critical for mitochondrial import, subunit interaction, binding of thiamine pyrophosphate, dephosphorylation, or catalysis at the active site. No null E1α mutations have been identified in males, suggesting that such mutations are likely to be lethal. In males with recurrent E1α mutations disease there is still a variable phenotypic expression.

Only two defects of the E1β subunit have been identified [25]. The molecular basis of E3-binding protein (E3BP) deficiency has been characterized in 13 cases. Half of the patients have splicing errors, others have frameshift or nonsense mutations [31]. Recently mutations in E2 [28] and in the pyruvate dehydrogenase phosphatase gene (PDP1) [27] have been identified.

In about 25 % of cases the mother of a child with PDHE1α deficiency was a carrier of the mutation [30]. Therefore, since most cases of PDHC deficiency appear to be the consequence of new E1α mutations, the overall rate of recurrence in the same family is low. Based on measurement of PDHC activity in chorionic villus samples and/or cultured amniocytes obtained from some 30 pregnancies in families with a previously affected child, three cases of reduced activity were found. However, PDHC activities in affected females might overlap with normal controls. Therefore, prenatal testing of specific mutations determined in the proband is the most reliable method. Molecular analysis is also the preferred method for prenatal diagnosis in families at risk for E1β and E3BP deficiency.

12.3.4 Diagnostic Tests

The most important laboratory test for initial recognition of PDHC deficiency is measurement of blood and CSF lactate and pyruvate. Quantitative analysis of plasma amino acids and urinary organic acids may also be useful. Blood lactate, pyruvate and alanine can be intermittently normal, but, characteristically, an increase is observed after an oral carbohydrate load. While L/P ratio is as a rule normal, a high ratio can be found if the patient is acutely ill, if blood is very difficult to obtain, or if the measurement of pyruvate (which is unstable) is not done reliably. The practical solution to avoid these artifacts is to obtain several samples of blood, including samples collected under different dietary conditions (during an acute illness, after overnight fasting, and postprandially after a high-carbohydrate meal). Glucose-tolerance or carbohydrate-loading tests are not necessary for a definite diagnosis. In contrast to deficiencies of PC or PEPCK, fasting hypoglycaemia is not an expected

feature of PDHC deficiency, and blood lactate and pyruvate usually decrease after fasting. CSF for measurement of lactate and pyruvate (and possibly organic acids) is certainly indicated, since there may be a normal blood lactate and pyruvate, and only elevation in CSF [32].

The most commonly used material for assay of PDHC is cultured skin fibroblasts. PDHC can also be assayed in fresh lymphocytes, but low normal values might make the diagnosis difficult. Molecular analysis of the PDHE1α gene in girls is often more efficient than measuring the enzyme activity. If available, skeletal muscle and/or other tissues are useful. When a patient with suspected but unproven PDHC deficiency dies, it is valuable to freeze samples of different origin such as skeletal muscle, heart muscle, liver, and/or brain, ideally within 4 h post-mortem [33]. A skin biopsy to be kept at 4°C in a physiological solution can be useful. PDHC is assayed by measuring the release of $^{14}CO2$ from [1-^{14}C]-pyruvate in cell homogenates and tissues [34]. PDHC activity should be measured at low and high TPP concentrations to detect thiamine-responsive PDHC deficiency [35]. PDHC must also be activated (dephosphorylated; ◘ Fig. 12.3) in part of the cells, which can be done by pre-incubation of whole cells or mitochondria with dichloroacetate (DCA, an inhibitor of the kinase; ◘ Fig.12.3). In E1-phosphatase deficiency there is a deficiency in native PDH activity, but on activation of the PDH complex with DCA, activity becomes normal [27]. The three catalytic components of PDHC can be assayed separately. Immunoblotting of the components of PDHC can help distinguish if a particular protein is missing. In females with PDHE1α deficiency, X inactivation can interfere with the biochemical analysis [32]. E3BP, which anchors E3 to the E2 core of the complex, can only be evaluated using immunoblotting, since it has no catalytic activity [29].

12.3.5 Treatment and Prognosis

The general prognosis for individuals with PDHC deficiency is poor, and treatment is not very effective. Experience with early prospective treatment to prevent irreversible brain injury is lacking. Perhaps the most rational strategy for treating PDHC deficiency is the use of a ketogenic diet [36]. Oxidation of fatty acids, 3-hydroxybutyrate, and acetoacetate are providers of alternative sources of acetyl-CoA. Wexler et al. compared the outcome of males with PDHC deficiency caused by identical E1 mutations and found that the earlier the ketogenic diet was started and the more severe the restriction of carbohydrates, the better the outcome of mental development and survival [37]. Sporadic cases of improvement under ketogenic diet have been published. Thiamine has been given in variable doses (500–2000 mg/day), with lowering of blood lactate and apparent clinical improvement in some patients [38].

DCA offers another potential treatment for PDHC deficiency. DCA, a structural analogue of pyruvate, inhibits E1 kinase, thereby keeping any residual E1 activity in its active (dephosphorylated) form (❏ Fig. 12.3). DCA can be administered without apparent toxicity (about 50 mg/kg/day). Over 40 cases of congenital lactic acidosis due to various defects (including PDHC deficiency) were treated with DCA in uncontrolled studies, and most of these cases appeared to have some limited short-term benefit [39]. Chronic DCA treatment was shown to be beneficial in some patients, improving the function of PDHC, and this has been related to specific DCA-sensitive mutations [40]. Sporadic reports have also shown beneficial effect of concomitant DCA and high dose thiamine (500 mg). A ketogenic diet and thiamine should thus be tried in each patient. DCA can be added if lactic acidosis is important, especially in acute situations.

12.4 Dihydrolipoamide Dehydrogenase Deficiency

12.4.1 Clinical Presentation

Approximately 20 cases of E3 deficiency have been reported [41–43]. Since this enzyme is common to all the 2-ketoacid dehydrogenases (❏ Fig. 12.2), E3 deficiency results in multiple 2-ketoacid-dehydrogenase deficiency and should be thought of as a combined PDHC and TCA cycle defect. E3 deficiency presents with severe and progressive hypotonia and failure to thrive, starting in the first months of life. Metabolic decompensations are triggered by infections. Progressively hypotonia, psychomotor retardation, microcephaly and spasticity occur. Some patients develop a typical picture of Leigh's encephalopathy. A Reye-like picture with liver involvement and myopathy with myoglobinuria without mental retardation is seen in the Ashkenazi Jewish population [44].

12.4.2 Metabolic Derangement

Dihydrolipoyl dehydrogenase (E3) is a flavoprotein common to all three mitochondrial α-ketoacid dehydrogenase complexes (PDHC, KDHC, and BCKD; ❏ Fig. 12.2). The predicted metabolic manifestations are the result of the deficiency state for each enzyme: increased blood lactate and pyruvate, elevated plasma alanine, glutamate, glutamine, and branched-chain amino acids (leucine, isoleucine, and valine), and increased urinary lactic, pyruvic, 2-ketoglutaric, and branched-chain 2-hydroxy- and 2-keto acids.

12.4.3 Genetics

The gene for E3 is located on chromosome 7q31-q32 [45] and the deficiency is inherited as an autosomal recessive trait. Mutation analysis in 13 unrelated patients has revealed eleven different mutations [46–50]. A G194C mutation is the major cause of E3 deficiency in Ashkenazi Jewish patients [51]. The most reliable method for prenatal diagnosis is through mutation analysis in DNA from chorionic villous samples (CVS) in previously identified families.

12.4.4 Diagnostic Tests

The initial diagnostic screening should include analyses of blood lactate and pyruvate, plasma amino acids, and urinary organic acids. However, the pattern of metabolic abnormalities is not seen in all patients or at all times in the same patient, making the diagnosis more difficult. In cultured skin fibroblasts, blood lymphocytes, or other tissues, the E3 component can be assayed using a spectrophotometric method.

12.4.5 Treatment and Prognosis

There is no dietary treatment for E3 deficiency, since the affected enzymes effect carbohydrate, fat, and protein metabolism. Restriction of dietary branched-chain amino acids was reportedly helpful in one case [52]. DL-lipoic acid has been tried but its effect remains controversial [51].

12.5 2-Ketoglutarate Dehydrogenase Complex Deficiency

12.5.1 Clinical Presentation

Isolated deficiency of the 2-ketoglutarate dehydrogenase complex (KDHC) has been reported in ten children in several unrelated families [53–55]. As in PDHC deficiency, the primary clinical manifestations included developmental delay, hypotonia, ataxia, opisthotonos and, less commonly, seizures and extrapyramidal dysfunction. On magnetic resonance imaging (MRI) bilateral striatal necrosis can be found [56]. All patients presented in the neonatal period and early childhood.

In one patient the clinical picture was milder [55]. This patient had suffered from mild perinatal asphyxia. During the first months of life, he developed opisthotonus and axial hypertonia, which improved with age. 2-Ketoglutaric acid (2-KGA) was intermittently increased in urine, but not in plasma and CSF. Diagnosis was confirmed in cultured skin fibroblasts. Surendam et al. [57] presented three families with the clinical features of DOOR syndrome (onychoosteodystrophy, dystrophic thumbs, sensorineural deafness), increased urinary levels of 2-KGA, and decreased activity of the E1 component of KDHC.

12.5.2   Metabolic Derangement

KDHC is a 2-ketoacid dehydrogenase that is analogous to PDHC and BCKD (◨ Fig. 12.2). It catalyzes the oxidation of 2-KGA to yield CoA and NADH. The E1 component, 2-ketoglutarate dehydrogenase, is a substrate-specific dehydrogenase that utilizes thiamine and is composed of two different subunits. In contrast to PDHC, the E1 component is not regulated by phosphorylation/dephosphorylation. The E2 component, dihydrolipoyl succinyl-transferase, is also specific to KDHC and includes covalently bound lipoic acid. The E3 component is the same as for PDHC. An E3-binding protein has not been identified for KDHC. Since KDHC is integral to the TCA cycle, its deficiency has consequences similar to that of other TCA enzyme deficiencies.

12.5.3   Genetics

KDHC deficiency is inherited as an autosomal recessive trait. The E1 gene has been mapped to chromosome 7p13-14 and the E2 gene to chromosome 14q24.3. The molecular basis of KDHC deficiencies has not yet been resolved. While prenatal diagnosis of KDHC should be possible by measurement of the enzyme activity in CVS or cultured amniocytes, this has not been reported.

12.5.4   Diagnostic Tests

The most useful test for recognizing KDHC deficiency is urine organic acid analysis, which can show increased excretion of 2-KGA with or without concomitantly increased excretion of other TCA cycle intermediates. However, mildly to moderately increased urinary 2-KGA is a common finding and not a specific marker of KDHC deficiency. Some patients with KDHC deficiency also have increased blood lactate with normal or increased L/P ratio. Plasma glutamate and glutamine may be increased. KDHC activity can be assayed through the release of $^{14}CO2$ from [1-^{14}C]-2-ketoglutarate in crude homogenates of cultured skin fibroblasts, muscle homogenates and other cells and tissues [53].

12.5.5   Treatment and Prognosis

There is no known selective dietary treatment that bypasses KDHC, since this enzyme is involved in the terminal steps of virtually all oxidative energy metabolism. Thiamine-responsive KDHC deficiency has not been described.

12.6   Fumarase Deficiency

12.6.1   Clinical Presentation

Approximately 26 patients with fumarase deficiency have been reported. The first case was described in 1986 [58]. Onset started at three weeks of age with vomiting and hypotonia, followed by development of microcephaly (associated with dilated lateral ventricles), severe axial hypertonia and absence of psychomotor progression.

Until the publication of Kerrigan [59] only 13 patients were described, all presenting in infancy with a severe encephalopathy and seizures, with poor neurological outcome. Kerrigan reported on 8 patients from a large consanguineous family. All patients had a profound mental retardation and presented as a static encephalopathy. Six out of 8 developed seizures. The seizures were of various types and of variable severity, but several patients experienced episodes of status epilepticus. All had a relative macrocephaly (in contrast to previous cases) and large ventricles. Dysmorphic features such as frontal bossing, hypertelorism and depressed nasal bridge were noted.

Neuropathological changes include agenesis of the corpus callosum with communicating hydrocephalus as well as cerebral and cerebellar heterotopias. Polymicrogyria, open operculum, colpocephaly, angulations of frontal horns, choroid plexus cysts, decreased white matter, and a small brainstem are considered characteristic [59].

12.6.2   Metabolic Derangement

Fumarase catalyzes the reversible interconversion of fumarate and malate (◨ Fig. 12.1). Its deficiency, like other TCA cycle defects, causes: (i) impaired energy production caused by interrupting the flow of the TCA cycle and (ii) potential secondary enzyme inhibition associated with accumulation in various amounts of metabolites proximal to the enzyme deficiency such as fumarate, succinate, 2-KGA and citrate (◨ Fig. 12.1).

12.6.3   Genetics

Fumarase deficiency is inherited as an autosomal recessive trait. A single gene, mapped to chromosome 1q42.1, and the same mRNA, encode alternately translated transcripts to generate a mitochondrial and a cytosolic isoform [60]. A variety of mutations have been identified in several unrelated families [60–63]. Prenatal diagnosis is possible by fumarase assay and/or mutational analysis in CVS or cultured amniocytes [62]. Heterozygous mutations in the fumarase gene are associated with a predisposition to cutaneous and uterine leiomyomas and to kidney cancers [64].

12.6.4 Diagnostic Tests

The key finding is increased urinary fumaric acid, sometimes associated with increased excretion of succinic acid and 2-KGA. Mild lactic acidosis and mild hyperammonemia can be seen in infants with fumarase deficiency, but generally not in older children. Other diagnostic indicators are an increased lactate in CSF, a variable leucopenia and neutropenia.

Fumarase can be assayed in mononuclear blood leukocytes, cultured skin fibroblasts, skeletal muscle or liver, by monitoring the formation of fumarate from malate or, more sensitively, by coupling the reaction with malate dehydrogenase and monitoring the production of NADH [58].

12.6.5 Treatment and Prognosis

There is no specific treatment. While removal of certain amino acids that are precursors of fumarate could be beneficial, removal of exogenous aspartate might deplete a potential source of oxaloacetate. Conversely, supplementation with aspartate or citrate might lead to overproduction of toxic TCA cycle intermediates.

12.7 Succinate Dehydrogenase Deficiency

12.7.1 Clinical Presentation

Succinate dehydrogenase (SD) is part of both the TCA cycle and the respiratory chain. This explains why SD deficiency resembles more the phenotypes associated with defects of the respiratory chain. The clinical picture of this very rare disorder [65–69] can include: Kearns-Sayre syndrome, isolated hypertrophic cardiomyopathy, combined cardiac and skeletal myopathy, generalized muscle weakness with easy fatigability, and early onset Leigh encephalopathy. It can also present with cerebellar ataxia and optic atrophy and tumor formation in adulthood. Profound hypoglycemia was seen in one infant [70].

SD deficiency may also present as a compound deficiency state that involves aconitase and complexes I and III of the respiratory chain. This disorder, found only in Swedish patients, presents with life-long exercise intolerance, myoglobinuria, and lactic acidosis, with a normal or increased L/P ratio at rest and a paradoxically decreased L/P ratio during exercise [68].

12.7.2 Metabolic Derangement

SD is part of a larger enzyme unit, complex II (succinate-ubiquinone oxidoreductase) of the respiratory chain. Complex II is composed of four subunits. SD contains two of these subunits, a flavoprotein (Fp, SDA) and an iron-sulfur protein (Ip, SDB). SD is anchored to the membrane by two additional subunits, C and D. SD catalyzes the oxidation of succinate to fumarate (◘ Fig. 12.1) and transfers electrons to the ubiquinone pool of the respiratory chain.

Theoretically, TCA-cycle defects should lead to a decreased L/P ratio, because of impaired production of NADH. However, too few cases of SD deficiency (or other TCA-cycle defects) have been evaluated to determine whether this is a consistent finding. Profound hypoglycemia, as reported once, might have resulted from the depletion of the gluconeogenesis substrate, oxaloacetate [70]. The combined SD/aconitase deficiency found only in Swedish patients, appears to be caused by a defect in the metabolism of the iron-sulfur clusters common to these enzymes [69].

12.7.3 Genetics

Complex II is unique among the respiratory chain complexes in that all four of its subunits are nuclear encoded. The flavoprotein and iron-sulfur-containing subunits of SD (A and B) have been mapped to chromosomes 5p15 and 1p35-p36, respectively, while the two integral membrane proteins (C and D) have been mapped to chromosomes 1q21 and 11q23. Homozygous and compound heterozygous mutations of SDA have been identified in several patients [67, 70–72]. In two sisters with partial SDA deficiency and late onset neurodegenerative disease with progressive optic atrophy, ataxia and myopathy, only one mutation was found, suggesting a dominant pattern of transmission [72].

Mutations in SDB, SDC or SDD cause susceptibility to familial pheochromocytoma and familial paraganglioma [73]. This suggests that SD genes may act as tumor suppression genes.

12.7.4 Diagnostic Tests

In contrast to the other TCA cycle disorders, SD deficiency does not always lead to a characteristic organic aciduria. Many patients, especially those whose clinical phenotypes resemble the patients with respiratory chain defects, do not exhibit the expected succinic aciduria and can excrete variable amounts of lactate, pyruvate, and the TCA cycle intermediates, fumarate and malate [70].

Diagnostic confirmation of a suspected SD deficiency requires analysis of SD activity itself, as well as complex-II (succinate-ubiquinone oxidoreductase) activity, which reflects the integrity of SD and the remaining two subunits of this complex. These enzyme assays can be accomplished using standard spectrophotometric procedures. Magnetic resonance spectroscopy provides a characteristic pattern with accumulation of succinate [74].

12.7.5   Treatment and Prognosis

No effective treatment has been reported. Although SD is a flavoprotein, riboflavin-responsive defects have not been described.

12.8   Pyruvate Transporter Defect

12.8.1   Clinical Presentation

Only one patient has been completely documented [75]. Neonatal lactic acidosis in a female baby from consanguineous parents was associated with generalized hypotonia and facial dysmorphism. MRI of the brain revealed cortical atrophy, periventricular leukomalacia and calcifications. Progressive microcephaly, failure to thrive and neurological deterioration led to death at the age of 19 months. Selak et al. [76] described four patients with hypotonia, developmental delay, seizures and ophthalmological abnormalities and found decreased respiration rates in mitochondria with pyruvate, but not with other substrates, suggesting a decreased entry of pyruvate into the mitochondria.

12.8.2   Metabolic Derangement

The pyruvate carrier mediates the proton symport of pyruvate across the inner mitochondrial membrane. Consequently, the metabolic derangement should be the same as in pyruvate dehydrogenase deficiency.

12.8.3   Diagnostic Tests

As in PDHC deficiency, high lactate and pyruvate are found with normal lactate/pyruvate ratio. To evidence the transport defect, $[2\text{-}^{14}C]$ pyruvate oxidation is measured in both intact and digitonin-permeabilized fibroblasts. Oxidation of $[2\text{-}^{14}C]$ pyruvate is severely impaired in intact cells but not when digitonin allows pyruvate to bypass the transport step by disrupting the inner mitochondrial membrane.

12.8.4   Genetics

Chromosome localization and cDNA sequence of the pyruvate carrier is still unknown. Prenatal diagnosis on CVS can be done by the biochemical method [75].

12.8.5   Treatment and Prognosis

No treatment is known at this moment.

Acknowledgement. We would like to acknowledge the authors of previous editions, D. Kerr, I. Wexler and A. Zinn, for the basis of this chapter.

References

1. Saudubray JM, Marsac C, Charpentier C et al (1976) Neonatal congenital lactic acidosis with pyruvate carboxylase deficiency in two siblings. Acta Paediatr Scand 65:717-724
2. Robinson BH, Oei J, Sherwood WG et al (1984) The molecular basis for the two different clinical presentations of classical pyruvate carboxylase deficiency. Am J Hum Genet 36:283-294
3. Van Coster RN, Janssens S, Misson JP et al (1998) Prenatal diagnosis of pyruvate carboxylase deficiency by direct measurement of catalytic activity on chorionic villi samples. Prenat Diagn 18: 1041-1044
4. Brun N, Robitaille Y, Grignon A et al (1999) Pyruvate carboxylase deficiency: prenatal onset of ischemia-like brain lesions in two sibs with the acute neonatal form. Am J Med Genet 84:94-101
5. Ahmad A, Kahler SG, Kishnani PS et al (1999) Treatment of pyruvate carboxylase deficiency with high doses of citrate and aspartate. Am J Med Genet 87: 331-338
5a. Garcia-Gazorla A, Rabier D, Touati G et al (2006) Pyruvate carboxylase deficiency: metabolic characteristics and new neurological aspects. Ann Neurol 59:121-127
6. De Vivo DC, Haymond MW, Leckie MP et al (1977) The clinical and biochemical implications of pyruvate carboxylase deficiency. J Clin Endocrinol Metab 45:1281-1296
7. Robinson BH, Taylor J, Sherwood WG (1980) The genetic heterogeneity of lactic acidosis: occurrence of recognizable inborn errors of metabolism in a pediatric population with lactic acidosis. Pediatr Res 14:956-962
8. Carbone MA, MacKay N, Ling M et al (1998) Amerindian pyruvate carboxylase deficiency is associated with two distinct missense mutations. Am J Hum Genet 62:1312-1319
9. Van Coster RN, Fernhoff PM, De Vivo DC (1991) Pyruvate carboxylase deficiency: A benign variant with normal development. Pediatr Res 30:1-4
10. Marsac C, Augereau Ch, Feldman G et al (1982) Prenatal diagnosis of pyruvate carboxylase deficiency. Clin Chim Acta 119:121-127
11. Nyhan WL, Khanna A, Barshop BA et al (2002) Pyruvate carboxylase deficiency -- insights from liver transplantation. Mol Genet Metab 77:143-149
12. Mochel F, Delonlay P, Touati G et al (2005) Pyruvate carboxylase deficiency: clinical and biochemical response to anaplerotic diet therapy. Mol Genet Metab 84:305-312
13. Fiser RH, Melsher HL, Fiser DA (1974) Hepatic phosphoenolpyruvate carboxylase (PEPCK) deficiency. A new cause of hypoglycemia in childhood. Pediatr Res 10:60
14. Clayton PT, Hyland K, Brand M, Leonard JV (1986) Mitochondrial phosphoenolpyruvate carboxykinase deficiency. Eur J Pediatr 145:46-50
15. Vidnes J, Sovik O (1976) Gluconeogenesis in infancy and childhood. III. Deficiency of the extramitochondrial form of hepatic phosphoenolpyruvate carboxykinase in a case of persistent neonatal hypoglycaemia. Acta Paediatr Scand 65:301-312
16. Leonard JV, Hyland K, Furukawa N, Clayton PT (1991) Mitochondrial phosphoenolpyruvate carboxykinase deficiency. Eur J Pediatr 150:198-199

17. Bodnar AG, Cooper JM, Holt LJ et al (1993) Nuclear complementation restores mtDNA levels in cultured cells from a patient with mtDNA depletion. Am J Hum Genet 53:663-669

18. Bodnar AG, Cooper JM, Leonard JV, S chapira AH (1995) Respiratory-deficient human fibroblasts exhibiting defective mitochondrial DNA replication. Biochem J 305:817-822

19. Robinson BH, MacKay N, Chun K, Ling M (1996) Disorders of pyruvate carboxylase and the pyruvate dehydrogenase complex. J Inherit Metab Dis 19:452-462

20. Kerr DS, Wexler ID, Tripatara A, Patel MS (1996) Defects of the human pyruvate dehydrogenase complex. In: Patel MS, Roche T, Harris RA (eds) Alpha keto acid dehydrogenase complexes. Birkhauser, Basel, pp 249-270

21. Otero LJ, Brown RM, Brown GK (1998) Arginine 302 mutations in the pyruvate dehydrogenase E1alpha subunit gene: identification of further patients and in vitro demonstration of pathogenicity. Hum Mutat 12:114-121

22. De Meirleir L, Specola N, Seneca S, Lissens W (1998) Pyruvate dehydrogenase E1alpha deficiency in a family: different clinical presentation in two siblings. J Inherit Metab Dis 21:224-226

23. De Meirleir L (2002) Defects of pyruvate metabolism and the Krebs cycle. J Child Neurol 17[Suppl 3]:3S26-33

24. Michotte A, De Meirleir L, Lissens W et al (1993) Neuropathological findings of a patient with pyruvate dehydrogenase E1 alpha deficiency presenting as a cerebral lactic acidosis. Acta Neuropathol (Berl) 85:674-678

25. Brown RM, Head RA, Boubriak II et al (2004) Mutations in the gene for the E1β subunit: a novel cause of pyruvate dehydrogenase deficiency. Hum Genet 115:123-127

26. Ito M, Kobashi H, Naito E et al (1992) Decrease of pyruvate dehydrogenase phosphatase activity in patients with congenital lactic acidemia. Clin Chim Acta 209:1-7

27. Cameron J, Mai M, Levandovsky N et al (2004) Identification of a novel mutation in the catalytic subunit 1 of the pyruvate dehydrogenase phosphatase (PDP1) gene in two brothers. BBA 1657:38

28. Brown RM, Head RA, Clayton PT, Brown GK (2004) Dihydrolipoamide acetyltransferase deficiency. J Inherit Metab Dis 27:S1, 125

29. Brown RM, Head RA, Brown GK (2002) Pyruvate dehydrogenase E3 binding protein deficiency. Hum Genet 110:187-191

30. Lissens W, De Meirleir L, Seneca et al (2000) Mutations in the X-linked pyruvate dehydrogenase (E1) α subunit gene (PDHA1) in patients with pyruvate dehydrogenase complex deficency. Hum Mutat 15:209-219

31. Aral B, Benelli C, Ait-Ghezala G et al (1997) Mutations in PDX1, the human lipoyl-containing component X of the pyruvate dehydrogenase-complex gene on chromosome 11p1, in congenital lactic acidosis. Am J Hum Genet 61:1318-1326

32. De Meirleir L, Lissens W, Denis R et al (1993) Pyruvate dehydrogenase deficiency: clinical and biochemical diagnosis. Pediatr Neurol 9:216-220

33. Kerr DS, Berry SA, Lusk MM et al (1988) A deficiency of both subunits of pyruvate dehydrogenase which is not expressed in fibroblasts. Pediatr Res 24:95-100

34. Sheu KFR, Hu CWC, Utter MF (1981) Pyruvate dehydrogenase complex activity in normal and deficient fibroblasts. J Clin Invest 67:1463-1471

35. Naito E, Ito M, Yokota I et al (2002) Diagnosis and molecular analysis of three male patients with thiamine-responsive pyruvate dehydrogenase complex deficiency. J Neurol Sci 201:33-37

36. Falk RE, Cederbaum SD, Blass JP et al (1976) Ketogenic diet in the management of pyruvate dehydrogenase deficiency. Pediatrics 58:713-721

37. Wexler ID, Hemalatha SG, McConnell J et al (1997) Outcome of pyruvate dehydrogenase deficiency treated with ketogenic diets. Studies in patients with identical mutations. Neurology 49:1655-1661

38. Naito E, Ito M, Yokota I et al (2002) Thiamine-responsive pyruvate dehydrogenase deficiency in two patients caused by a point mutation (F2005L and L216F) within the thiamine pyrophosphate binding site. Biochim Biophys Acta 1588:79-84

39. Stacpoole PW, Barnes CL, Hurbanis MD et al (1997) Treatment of congenital lactic acidosis with dichloroacetate: a review. Arch Pediatr Adolesc Med 77:535-541

40. Fouque F, Brivet M, Boutron A et al (2003) Differential effect of DCA treatment on the pyruvate dehydrogenase complex in patients with severe PDHC deficiency. Pediatr Res 53:793-799

41. Elpeleg ON, Ruitenbeek W, Jakobs C et al (1995) Congenital lactic-acidemia caused by lipoamide dehydrogenase deficiency with favorable outcome. J Pediatr 126:72-74

42. Elpeleg ON, Shaag A, Glustein JZ et al (1997) Lipoamide dehydrogenase deficiency in Ashkenazi Jews: an insertion mutation in the mitochondrial leader sequence. Hum Mutat 10:256-257

43. Grafakou O, Oexle K, van den Heuvel L et al (2003) Leigh syndrome due to compound heterozygosity of dihydrolipoamide dehydrogenase gene mutations. Description of the first E3 splice site mutation. Eur J Pediatr 162:714-718

44. Shaag A, Saada A, Berger I et al (1999) Molecular basis of lipoamide dehydrogenase deficiency in Ashkenazi Jews. Am J Med Genet 82:177-182

45. Scherer SW, Otulakowski G, Robinson BH, Tsui LC (1991) Localization of the human dihydrolipoamide dehydrogenase gene (DLD) to 7q31 > q32. Cytogenet Cell Genet 56:176-177

46. Elpeleg ON, Shaag A, Glustein JZ et al (1997) Lipoamide dehydrogenase deficiency in Ashkenazi Jews: an insertion mutation in the mitochondrial leader sequence. Hum Mutat 10:256-257

47. Hong YS, Kerr DS, Liu TC et al (1997) Deficiency of dihydrolipoamide dehydrogenase due to two mutant alleles (E340K and G101del). Analysis of a family and prenatal testing. Biochim Biophys Acta 1362:160-168

48. Hong YS, Kerr DS, Craigen WJ et al (1996) Identification of two mutations in a compound heterozygous child with dihydrolipoamide dehydrogenase deficiency. Hum Mol Genet 5:1925-1930

49. Shany E, Saada A, Landau D et al (1999) Lipoamide dehydrogenase deficiency due to a novel mutation in the interface domain. Biochim Biophys Res Comm 262:163-166

50. Cerna L, Wenchich L, Hansikova H et al (2001) Novel mutations in a boy with dihydrolipoamide dehydrogenase deficiency. Med Sci Monit 7:1319-1325

51. Hong YS, Korman SH, Lee J et al (2003) Identification of a common mutation (Gly194Cys) in both Arab Moslem and Ashkenazi Jewish patients with dihydrolipoamide dehydrogenase (E3) deficiency: possible beneficial effect of vitamin therapy. J Inherit Metab Dis 26:816-818

52. Sakaguchi Y, Yoshino M, Aramaki S et al (1986) Dihydrolipoyl dehydrogenase deficiency: a therapeutic trial with branched-chain amino acid restriction. Eur J Pediatr 145:271-274

53. Bonnefont JP, Chretien D, Rustin P et al (1992) Alpha-ketoglutarate dehydrogenase deficiency presenting as congenital lactic acidosis. J Pediatr 121:255-258

54. Rustin P, Bourgeron T, Parfait B et al (1997) Inborn errors of the Krebs cycle: a group of unusual mitochondrial diseases in human. Biochim Biophys Acta 1361:185-197

55. Dunckelman RJ, Ebinger F, Schulze A et al (2000) 2-ketoglutarate dehydrogenase deficiency with intermittent 2-ketoglutaric aciduria. Neuropediatrics 31:35-38

56. Al Aqeel A, Rashed M, Ozand PT et al (1994) A new patient with alpha-ketoglutaric aciduria and progressive extrapyramidal tract disease. Brain Dev 16[Suppl]:33-37

57. Surendran S, Michals-Matalon K, Krywawych S et al (2002) DOOR syndrome: deficiency of E1 component of the 2-oxoglutarate dehydrogenase complex. Am J Med Genet 113:371-374

58. Zinn AB, Kerr DS, Hoppel CL (1986) Fumarase deficiency: a new cause of mitochondrial encephalomyopathy. N Engl J Med 315:469-475

59. Kerrigan JF, Aleck KA, Tarby TJ et al (2000) Fumaric aciduria: clinical and imaging features. Ann Neurol 47:583-588

60. Coughlin EM, Chalmers RA, Slaugenhaupt SA et al (1993) Identification of a molecular defect in a fumarase deficient patient and mapping of the fumarase gene. Am J Hum Genet 53:86-89

61. Bourgeron T, Chretien D, Poggi-Bach J et al (1994) Mutation of the fumarase gene in two siblings with progressive encephalopathy and fumarase deficiency. J Clin Invest 93:2514-2518

62. Coughlin EM, Christensen E, Kunz PL et al (1998) Molecular analysis and prenatal diagnosis of human fumarase deficiency. Mol Genet Metab 63:254-262

63. Remes AM, Filppula SA, Rantala H et al (2004) A novel mutation of the fumarase gene in a family with autosomal recessive fumarase deficiency J Mol Med 82:550-554

64. Gross KL, Panhuysen CI, Kleinman MS et al (2004) Involvement of fumarate hydratase in nonsyndromic uterine leiomyomas: genetic linkage analysis and FISH studies. Genes Chromosomes Cancer 41:183-190

65. Rivner MH, Shamsnia M, Swift TR et al (1989) Kearns-Sayre syndrome and complex II deficiency. Neurology 39:693-696

66. Bourgeron T, Rustin P, Chretien D et al (1995) Mutation of a nuclear succinate dehydrogenase gene results in mitochondrial respiratory chain deficiency. Nat Genet 11:144-149

67. Taylor RW, Birch-Machin MA, Schaefer J et al (1996) Deficiency of complex II of the mitochondrial respiratory chain in late-onset optic atrophy and ataxia. Ann Neurol 39:224-232

68. Haller RG, Henriksson KG, Jorfeldt L et al (1991) Deficiency of skeletal muscle succinate dehydrogenase and aconitase. Pathophysiology of exercise in a novel human muscle oxidative defect. J Clin Invest 88:1197-1206

69. Hall RE, Henriksson KG, Lewis SF et al (1993) Mitochondrial myopathy with succinate dehydrogenase and aconitase deficiency. Abnormalities of several iron-sulfur proteins. J Clin Invest 92:2660-2666

70. Van Coster R, Seneca S, Smet J et al (2003) Homozygous Gly555Glu mutation in the nuclear-encoded 70kDa Flavoprotein Gene Causes instability of the respiratory chain complex II. Am J Med Genet 120A:13-18

71. Parfait B, Chretien D, Rotig A et al (2000) Compound heterozygous mutations in the flavoprotein gene of the respiratory chain complex II in a patient with Leigh syndrome. Hum Genet 106:236-243

72. Birch-Machin MA, Taylor RW, Cochran B et al (2000) Late-onset optic atrophy, ataxia, and myopathy associated with a mutation of a complex II gene. Ann Neurol 48:330-335

73. Rustin P, Munnich A, Rotig A (2002) Succinate dehydrogenase and human diseases: new insights into a well-known enzyme Eur J Hum Genet 10:289-291

74. Brockmann K, Bjornstad A, Dechent P et al (2002) Succinate in dystrophic white matter: a proton magnetic resonance spectroscopy finding characteristic for complex II deficiency. Ann Neurol 52:38-46

75. Brivet M, Garcia-Cazorla A, Lyonnet S et al (2003) Impaired mitochondrial pyruvate importation in a patient and a fetus at risk. Mol Gen Metab 78:186-192

76. Selak MA, Grover WM, Foley CM et al (1997) Possible defect in pyruvate transport in skeletal muscle mitochondria from four children with encephalomyopathies and myopathies. International Conference on Mitochondrial Diseases, Philadelphia, Abstract 59

13 Disorders of Mitochondrial Fatty Acid Oxidation and Related Metabolic Pathways

Charles A. Stanley, Michael J. Bennett, Ertan Mayatepek

Fatty Acid Oxidation

Fatty acid oxidation (▣ Fig. 13.1) comprises four components: the carnitine cycle, the ß-oxidation cycle, the electron-transfer path, and the synthesis of ketone bodies. Long-chain free fatty acids of exogenous and endogenous origin are activated toward their coenzyme A (CoA) esters in the cytosol. These fatty acyl-CoAs enter the mitochondria as fatty acylcarnitines via the carnitine cycle. Medium- and short-chain fatty acids enter the mitochondria directly and are activated toward their CoA derivatives in the mitochondrial matrix. Each step of the four-step ß-oxidation cycle

shortens the fatty acyl-CoA by two carbons until it is completely converted to acetyl-CoA. The electron-transfer path transfers some of the energy released in the ß-oxidation to the respiratory chain, resulting in the synthesis of ATP. In the liver, most of the acetyl-CoA from fatty acid ß-oxidation cycle is used to synthesize the ketone bodies 3-hydroxybutyrate and acetoacetate. The ketones are then exported for terminal oxidation (chiefly in the brain). In other tissues, such as muscle, the acetyl-CoA enters the Krebs‹ cycle of oxidation and ATP production.

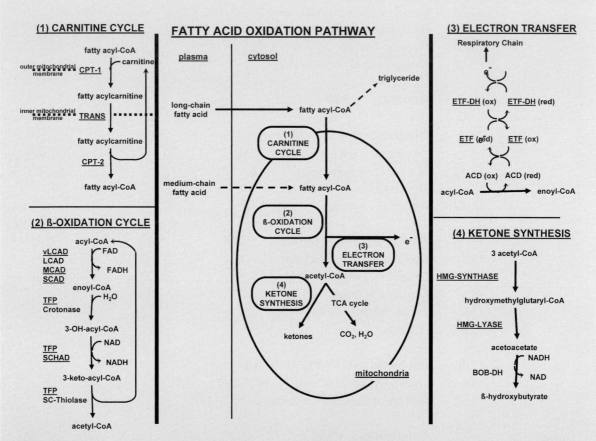

▣ **Fig. 13.1.** Mitochondrial fatty acid-oxidation pathway. In the center panel, the pathway is subdivided into its four major components, which are shown in detail in the side panels. Sites of identified defects are *underscored*. *BOB-DH*, ß-hydroxybutyrate dehydrogenase; *CoA*, coenzyme A; *CPT*, carnitine palmitoyl transferase; *ETF*, electron-transfer flavoprotein; *ETF-DH*, ETF dehydrogenase; *FAD*, flavin adenine dinucleotide; *FADH*, reduced FAD; *HMG*, 3-hydroxy-3-methylglutaryl; *LCAD*, long-chain acyl-CoA dehydrogenase; *MCAD*, medium-chain acyl-CoA dehydrogenase; *NAD*, nicotinamide adenine dinucleotide; *NADH*, reduced NAD; *SCAD*, short-chain acyl-CoA dehydrogenase; *SCHAD*, short-chain 3-hydroxyacyl-CoA dehydrogenase; *TCA*, tricarboxylic acid; *TFP*, trifunctional protein; *TRANS*, carnitine/acylcarnitine translocase; *vLCAD*, very-long-chain acyl-CoA dehydrogenase

More than a dozen genetic defects in the fatty acid oxidation pathway are currently known. Nearly all of these defects present in early infancy as acute life-threatening episodes of hypoketotic, hypoglycemic coma induced by fasting or febrile illness (for recent reviews, see [1-4]). In some of the disorders there also may be chronic skeletal muscle weakness or acute exercise-induced rhabdomyolysis and acute or chronic cardiomyopathy. Recognition of the fatty acid oxidation disorders is often difficult because patients can appear well until exposed to prolonged fasting, and screening tests of metabolites may not always be diagnostic. Rare related disorders include a transport defect of fatty acids, and secondary (as in the Sjögren-Larsson syndrome), or primary defects in the metabolism of leukotrienes.

13.1   Introduction

The oxidation of fatty acids in mitochondria plays an important role in energy production. During late stages of fasting, fatty acids provide 80% of total body energy needs through hepatic ketone body synthesis and by direct oxidation in other tissues. Long-chain fatty acids are the preferred fuel for the heart and also serve as an essential source of energy for skeletal muscle during sustained exercise. Free fatty acids are released from adipose tissue triglyceride stores and circulate bound to albumin. Their oxidation to CO_2 and H_2O by peripheral tissues spares glucose consumption and the need to convert body protein to glucose. The use of fatty acids by the liver provides energy for gluconeogenesis and ureagenesis. Equally important, the liver uses fatty acids to synthesize ketones, which serve as a fat-derived fuel for the brain, and thus further reduce the need for glucose utilization.

13.2   Clinical Presentation

The clinical phenotypes of most of the disorders of fatty acid oxidation are very similar [1–4]. ◘ Table 13.1 presents the three major types of presentation with signs mainly of hepatic, cardiac, and skeletal muscle involvement. The individual defects are discussed below under the four components of the fatty acid oxidation pathway outlined in ◘ Fig. 13.1 and Table 13.1.

13.2.1   Carnitine Cycle Defects

Carnitine Transporter Defect (CTD). Although most of the fatty acid oxidation disorders affect the heart, skeletal muscle and liver, cardiac failure is seen as the major presenting manifestation only in CTD [5]. Over half of the known cases of CTD first presented with progressive heart failure and generalized muscle weakness. The age of onset of the cardiomyopathy or skeletal muscle weakness ranged from 12 months to 7 years. The cardiomyopathy in CTD patients is most evident on echocardiography, which shows poor contractility and thickened ventricular walls similar to that seen in endocardial fibroelastosis. Electrocardiograms may be normal or show increased T-waves. Without carnitine treatment, the cardiac failure can progress rapidly to death. The outcome is usually very good with carnitine therapy [6].

During the first years of life, extended fasting stress may provoke an attack of hypoketotic, hypoglycemic coma with or without evidence of cardiomyopathy. This may lead to sudden unexpected infant death. The hepatic presentation occurs less frequently than the myopathic presentation, because the liver has a separate transporter for carnitine and can usually maintain sufficient levels of carnitine to support ketogenesis.

Numerous mutations have been described in the organic cation transporter OCTN2, encoded by the SLC22A5 gene on 5q, which result in carnitine transporter deficiency [7, 8].

Carnitine Palmitoyltransferase-1 (CPT-1) Deficiency. CPT-1 is the rate-limiting, regulatory step for transport of fatty acids into the mitochondria. Three distinct genetic isoforms for CPT1 have been identified for liver/kidney (CPT-1A), cardiac and skeletal muscle (CPT-1B), and brain (CPT-1C). To date, only CPT-1A deficiency has been described. Patients with this defect have usually presented during the first 2 years of life with attacks of fasting hypoglycemic, hypoketotic coma [9, 10]. They do not have cardiac or skeletal muscle involvement. CPT-1A deficiency is the only fatty acid oxidation disorder with elevated plasma total carnitine levels, which is predominantly non-esterified (see below) [4]. The defect is also noteworthy for unusually severe abnormalities in liver function tests during and for several weeks after acute episodes of illness, including massive increases in serum transaminases and hyperbilirubinemia. Transient renal tubular acidosis has also been described in a patient with CPT-1 deficiency, probably reflecting the importance of fatty acids as fuel for the kidney [11]. To date, only approximately 30 families have been described in the literature or are known to us. There appear to be common mutations in individuals of Hutterite and Canadian Inuit ancestry but most mutations in the CPT1A gene are private [12–14].

Carnitine/Acylcarnitine Translocase (TRANS) Deficiency. Less than a dozen cases of this defect have been reported [15–17]. Most were severely affected with onset in the neonatal period and death occurring before three months

III

▣ Table 13.1. Inherited disorders of mitochondrial fatty acid oxidation

Defect	Clinical manifestations of defect			
	Hepatic	Cardiac	Skeletal muscle	
			Acute	Chronic
Carnitine cycle				
CTD	+	+		(+)
CPT-1	+			
Trans	+	+		+
CPT-2	+	+	(+)	+
β-Oxidation cycle				
Acyl-CoA dehydrogenases				
VLCAD	+	+	+	+
MCAD	+			
SCAD				+
3-Hydroxyacyl-CoA dehydrogenases				
LCHAD	+	+	+	
SCHAD			+	+
MCKT			+	+
DER				+
Electron transfer				
ETF	+	+	(+)	+
ETF-DH	+	+	(+)	+
Ketone synthesis				
HMG-CoA synthase	+			
HMG-CoA lyase	+			

CPT, carnitine-palmitoyl transferase; *CTD*, carnitine-transporter defect; *DER*, 2,4-dienoyl-coenzyme-A reductase; *ETF*, electron-transfer flavoprotein; *ETF-DH*, ETF dehydrogenase; *LCHAD*, long-chain 3-hydroxyacly-coenzyme-A dehydrogenase, *MCAD*, medium-chain acyl-coenzyme-A dehydrogenase; *MCKT*, medium-chain ketoacyl-CoA thiolase; *SCAD*, short-chain acyl-coenzyme-A dehydrogenase; *SCHAD*, short-chain 3-hydroxyacyl-coenzyme-A dehydrogenase; *TRANS*, carnitine/acylcarnitine translocase; *VLCAD*, very-long-chain acyl-coenzyme-A dehydrogenase

of age. Presentations included fasting hypoketotic hypoglycemia, coma, cardiopulmonary arrest, and ventricular arrhythmias. One of the children with neonatal onset survived until three years of age; he succumbed with progressive skeletal muscle weakness and liver failure that were unresponsive to intensive feeding. Two milder cases with attacks of fasting hypoketotic coma similar to MCAD deficiency have been reported.

Carnitine Palmitoyltransferase-2 (CPT-2) Deficiency. Three forms of this defect are known, a mild adult onset form characterized by exercise-induced attacks of rhabdomyolysis, which was the first of the fatty acid oxidation defects to be described in 1973 [18]. Patients with the milder adult form of CPT-2 deficiency begin to have attacks of rhabdomyolysis in the second and third decades of life. These attacks are triggered by catabolic stresses such as prolonged exercise, fasting, or cold exposure. Episodes are associated

with aching muscle pain, elevated plasma creatine kinase (CK) levels, and myoglobinuria, which may lead to renal shutdown [18].

There is also a severe neonatal onset form, which presents with life-threatening coma, cardiomyopathy, and weakness [19, 20]. Neonatal-onset CPT-2 deficiency and the severe forms of ETF/ETF-DH deficiency have been associated with congenital brain and renal malformations. An intermediate form of CPT-2 deficiency has been described which presents in infancy with fasting hypoketotic hypoglycemia but without the congenital abnormalities seen in the neonatal form.

A genotype: phenotype correlation has been established for CPT-2 deficiency. Most individuals with the late onset myopathic presentation carry high residual activity missense mutations in particular the 338C>T (S113L) mutation which is present on 60% of alleles. The severe neonatal disease is most often associated with zero activity nonsense mutations or deletions. The intermediate infantile disorder is usually associated with one copy of a severe mutation and one milder one [21–23].

13.2.2   ß-Oxidation Defects

These can be divided into acyl-CoA dehydrogenase and 3-hydroxy-acyl-CoA dehydrogenase deficiencies.

Very-long-chain Acyl-CoA Dehydrogenase (VLCAD) Deficiency. This defect was originally reported as a defect of the long-chain acyl-CoA dehydrogenase (LCAD) enzyme, before the existence of two separate enzymes capable of acting on long-chain substrates was recognized [24]. VLCAD is bound to the inner mitochondrial membrane whereas LCAD is a matrix enzyme. All of the known patients have mutations in the VLCAD gene [25]. A separate disorder of the LCAD enzyme has yet to be identified, perhaps because this enzyme acts primarily on branched chain rather than straight chain fatty acids [26]. Many of the patients with VLCAD deficiency have had severe clinical manifestations, including chronic cardiomyopathy and weakness in addition to episodes of fasting coma. Several have presented in the newborn period with life-threatening coma similar to patients with TRANS or severe CPT-2 deficiencies. However, milder cases of VLCAD deficiency have also been identified with a phenotype very similar to MCAD deficiency. As with CPT-2 deficiency, a genotype: phenotype correlation defines the severity of disease with milder disease associated with high residual activity missense mutations and severe disease associated with nonsense mutations and deletions [27]. Unlike CPT-2 deficiency, the more severe presentations do not have congenital malformations.

Medium-chain Acyl-CoA Dehydrogenase (MCAD) Deficiency. This is the single most common fatty acid oxidation disorder [1, 28]. It is also one of the least severe, with no evidence of chronic muscle or cardiac involvement. In addition, it is unusually homogeneous, because 60–80% of symptomatic patients are homozygous for a single A985G (K329E) missense mutation originating in Northern Europe [29]. The estimated incidence in Britain and the USA is 1 in 10,000 births [30].

As shown in Table 13.1, patients with MCAD deficiency have an exclusively hepatic type of presentation similar to that of CPT-1A deficiency. Affected individuals appear to be entirely normal until an episode of illness is provoked by an excessive period of fasting. This may occur with an infection that interferes with normal feeding or simply because breakfast is delayed. The first episode typically occurs between 3-24 months of age, after nocturnal feedings have ceased. A few neonatal cases have been reported in which attempted breast-feeding was sufficient

fasting stress to cause illness. Attacks become less frequent after childhood, because fasting tolerance improves with increasing body mass.

The response to fasting in MCAD deficiency illustrates many of the pathophysiologic features of the hepatic presentation of the fatty acid oxidation disorders (Fig. 13.2). No abnormalities occur during the first 12–14 h, because lipolysis and fatty acid oxidation have not yet been activated. By 16 h, plasma levels of free fatty acids have risen dramatically, but ketones remain inappropriately low, reflecting the defect in hepatic fatty acid oxidation. Hypoglycemia develops shortly thereafter, probably because of excessive glucose utilization due to the inability to switch to fat as a fuel. Severe symptoms of lethargy and nausea develop in association with the marked increase in plasma fatty acids. It should be stressed that patients with fatty acid oxidation defects can become dangerously ill before plasma glucose falls to hypoglycemic values. An acute attack in MCAD deficiency usually features lethargy, nausea, and vomiting which rapidly progresses to coma within 1–2 h. Seizures may occur and patients may die suddenly from acute cardiorespiratory arrest. They may also die or suffer permanent brain damage from cerebral edema. Up to 25% of undiagnosed MCAD deficient patients die during their first attack. Because there is no forewarning, the first episode may be misdiagnosed as Reye syndrome or sudden infant death syndrome (SIDS).

At the time of an acute attack in MCAD deficiency, the liver may be slightly enlarged or it may become enlarged during the first 24 h of treatment. Chronic cardiac and skeletal muscle abnormalities are not seen in MCAD deficiency, perhaps because the block in fatty acid oxidation is incomplete. However, the enzyme defect is probably expressed in cardiac and skeletal muscle and these organs are probably responsible for the sudden death, which may occur during attacks of illness in MCAD deficient infants and children.

Short-chain Acyl-CoA Dehydrogenase (SCAD) Deficiency. Clinical manifestations of this disorder have been primarily chronic failure to thrive, developmental regression, and acidemia rather than the acute life-threatening episodes of coma and hypoglycemia associated with most of the fatty acid oxidation disorders [31, 32]. Similar evidence of chronic toxicity occurs in other short-chain fatty acid oxidation disorders, MCKT and SCHAD deficiencies. Although a significant number of SCAD cases have been identified, the molecular basis of the disease remains unclear, since the most commonly found genetic changes appear to be two polymorphisms in the SCAD gene (625G<A and 511C<T). These are currently assumed to be susceptibility genes, which require a second, as yet unknown genetic hit before symptoms are elicited [33].

Long-chain 3-Hydroxyacyl-CoA Dehydrogenase (LCHAD)/ Mitochondrial Trifunctional Protein (TFP) Deficiencies. The mitochondrial trifunctional protein is an octomeric protein consisting of four α and four β subunits. The α-subunit contains long-chain enoyl-CoA hydratase and LCHAD activities whilst the β-subunit contains the long-chain 3-ketoacyl-CoA thiolase (LKAT) activity. Some patients have isolated long-chain 3-hydroxy acyl-CoA dehydrogenase (LCHAD) deficiency, while others are also deficient in long-chain enoyl-CoA hydratase and LKAT activities, which are generally described as being TFP deficient [34, 35]. The clinical phenotype of these defects ranges from a mild disorder that resembles MCAD deficiency to a more severe disorder that resembles VLCAD deficiency. Some patients have had retinal degeneration or peripheral neuropathy, suggesting a toxicity effect. A strong association has been demonstrated with heterozygote mothers developing acute fatty liver of pregnancy (AFLP) or hemolysis, elevated liver enzymes and low platelet count (HELLP) syndrome when carrying affected fetuses [36–38]. This severe obstetric complication may be due to toxic effects related to placental metabolism of fatty acids [39].

Short-chain 3-Hydroxyacyl-CoA Dehydrogenase (SCHAD) Deficiency. Early reports of patients with potential defects of SCHAD have appeared, but with inconsistent clinical phenotypes which might be indicative of tissue specificity for this penultimate stage of fatty acid oxidation. The first was a child with recurrent myoglobinuria and hypoglycemic coma who appeared to have SCHAD deficiency in muscle, but not fibroblasts [40]. The second report was of two children with recurrent episodes of fasting ketotic hypoglycemia who had reduced SCHAD enzyme activity in fibroblast mitochondria [41]. A third report identified three infants who died suddenly who on autopsy had evidence of hepatic lipid accumulation in whom only liver SCHAD activity was impaired [42]. None of these patients had disease-causing mutations in the SCHAD gene. It is important to note that there are now three reports of patients with hypoglycemia due to hyperinsulinism in whom mutations in the SCHAD gene *HAD* 1 have been demonstrated [43] (▶ Chap. 10).

Medium-chain 3-Ketoacyl-CoA Thiolase (MCKT) Deficiency. One case of a defect in MCKT has been reported: a baby boy who died in the newborn period after presenting on day two of life with vomiting and acidosis [44]. Terminally at two weeks of age he had rhabdomyolysis and myoglobinuria. Urine showed elevated ketones, suggesting fairly good acetyl-CoA generation from partial oxidation of long-chain fatty acids, similar to what has been noted in other defects that are specific for short-chain fatty acids.

2,4-Dienoyl-CoA Reductase (DER) Deficiency. Only a single case of DER deficiency has been reported in the pathway required for oxidation of unsaturated fatty acids [45]. The patient was hypotonic from birth and died at 4 months of age. The disorder was suspected based on low plasma total carnitine levels and urinary excretion of an unusual unsaturated fatty acylcarnitine in urine.

13.2.3   Electron Transfer Defects

ETF/ETF-DH Deficiencies. Defects in the pathway for transferring electrons from the first step in ß-oxidation to the electron transport system are grouped together [46]. They are also known as *glutaric aciduria type 2* or *multiple acyl-CoA dehydrogenase deficiencies*. These defects block not only fatty acid oxidation, but also the oxidation of branched-chain amino acids, sarcosine and lysine. Patients with severe or complete deficiencies of the enzymes present with hypoglycemia, acidosis, hypotonia, cardiomyopathy, and coma in the neonatal period. Some neonates with ETF/ETF-DH deficiencies have had congenital anomalies (polycystic kidney, midface hypoplasia). Partial deficiencies of ETF/ETF-DH are associated with milder disease, resembling MCAD or VLCAD deficiency. Some patients have been reported to respond to riboflavin supplementation, which is a co-factor for the enzymes. The urine organic acid profile is usually diagnostic, especially in the severe form of these deficiencies with large glutaric acid excretion and multiple acylglycine abnormalities.

13.2.4   Ketogenesis Defects

Genetic defects in ketone body synthesis, 3-hydroxy-3-methylglutaryl-CoA synthase and 3-hydroxy-3-methylglutaryl-CoA lyase deficiencies also present with episodes of fasting-induced hypoketotic hypoglycemia. These defects are described in ▶ Chap. 14.

13.3   Genetics

All of the genetic disorders of fatty acid oxidation that have been identified are inherited in autosomal recessive fashion. Heterozygote carriers are generally regarded as being clinically normal with the possible exception, noted above, of the occurrence of AFLP in LCHAD heterozygote mothers carrying an affected fetus. There is also a single case report of a heterozygote for a severe CPT-2 mutation who developed late onset muscle weakness [47].

Carriers of the fatty acid oxidation disorders generally show no biochemical abnormalities except for CTD, in which carriers have half normal levels of plasma total carnitine concentrations, and MCAD deficiency for which heterozygotes may have mild elevations of medium-chain acylcarnitines. Since some of the disorders, such as MCAD deficiency, may be present without having caused an attack of illness, siblings of patients with fatty acid oxidation disorders should be investigated to determine whether they might be affected.

Rapid progress has been made in establishing the molecular basis for several of the defects in fatty acid oxidation [1–4]. This has become especially useful clinically in MCAD deficiency. About 80% of symptomatic MCAD deficient patients are homozygous for a single missense mutation, A985G, resulting in a K329E amino acid substitution; 17% carry this mutation in combination with another mutation. This probably represents a founder effect and explains why most MCAD patients share a northwestern European ethnic background. Simple polymerase chain reaction (PCR) assays have been established to detect the A985G mutation using DNA from many different sources, including newborn blood spot cards. This method has been used to diagnose MCAD deficiency in a variety of circumstances including prenatal diagnosis, postmortem diagnosis of affected siblings, and for surveys of disease incidence. Similarly, the S113L mutation for myopathic CPT-2 deficiency and the G1528C mutation in the LCHAD gene have been used in a variety of assays to identify those defects.

13.4 Diagnostic Tests

Recently, the diagnostic investigation of disorders of fatty acid oxidation has been radically simplified by the assay of the plasma or urine acylcarnitine profile by tandem mass spectrometry [48] (► also Chap. 2). This method and other assays to detect disease-related abnormal metabolites are to be used first since they do not burden the patient and carry no risk. If no abnormalities are detected, a test which monitors the overall pathway of fatty acid oxidation is indicated. This diagnostic approach is followed in the discussion below.

13.4.1 Disease-Related Metabolites

Plasma Acylcarnitines. Since acyl-CoA intermediates proximal to blocks in the fatty acid oxidation pathway can be transesterified to carnitine, most of the fatty acid oxidation disorders can be detected by analysis of acylcarnitine profiles in plasma, blood spots on filter paper or, less preferably, urine [48, 49] (◘ Table 13.2).

The determination of blood acylcarnitine profiles by tandem mass spectrometry from filter paper blood spots allows detection of fatty acid oxidation disorders caused by deficiencies of MCAD, VLCAD, LCHAD/TFP, ETF/ETF-DH, SCHAD, SCAD and HMG-CoA lyase. A screening program of infants in the state of Pennsylvania using

◘ **Table 13.2.** Fatty acid-oxidation disorders with distinguishing metabolic markers

Disorder	Plasma acylcarnitines	Urinary acylglycines	Urinary organic acids
VLCAD	Tetradecenoyl-		
MCAD	Octanoyl-	Hexanoyl-	
	Decenoyl-	Suberyl-	
		Phenylpropionyl-	
SCAD	Butyryl-	Butyryl-	Ethylmalonic
LCHAD	3-Hydroxy-palmitoyl-		3-Hydroxydicarboxylic
	3-Hydroxy-oleoyl-		
	3-Hydroxy-linoleoyl-		
DER	Dodecadienoyl-		
ETF and ETF-DH	Butyryl-	Isovaleryl-	Ethylmalonic
	Isovaleryl-	Hexanoyl-	Glutaric
	Glutaryl-		Isovaleric
HMG-CoA lyase	Methylglutaryl-		3-Hydroxy-3-methylglutaric

DER, 2,4-dienoyl-coenzyme A reductase; ETF, electron-transfer flavoprotein; ETF-DH, ETF dehydrogenase; HMG-CoA, 3-hydroxy-3-methyl-glutaryl-coenzyme A; MCAD, medium-chain acyl-coenzyme A dehydrogenase; SCAD, short-chain acyl-coenzyme A dehydrogenase; VLCAD, very-long-chain acyl-coenzyme A dehydrogenase tandem mass spectrometry revealed a higher than expected incidence of MCAD deficiency approaching 1 in 5,000 [50]. In several countries and about half of the US states, expanded screening by tandem mass spectrometry methods is replacing or complementing more traditional, limited screening programs.

Plasma and Tissue Total Carnitine Concentrations. A peculiar feature of the fatty acid oxidation disorders is that all but one are associated with either decreased or increased concentrations of total carnitine in plasma and tissues [15]. In CTD, sodium-dependent transport of carnitine across the plasma membrane is absent in muscle and kidney. This leads to severe reduction (< 2-5% of normal) of carnitine in plasma and in heart and skeletal muscle and defines this disorder as the only true primary carnitine defect known to date. These levels of carnitine are low enough to impair fatty acid oxidation [5]. In CPT-1 deficiency, total carnitine levels are increased (150–200% of normal) [10]. In all of the other defects, except HMG-CoA synthase deficiency, total carnitine levels are reduced to 25-50% of normal (secondary carnitine deficiency). Thus, simple measurement of plasma total carnitine is often helpful to determine the presence of a fatty acid oxidation disorder. It should be emphasized that samples must be taken in the well-fed state with normal dietary carnitine intake because patients with disorders of fatty acid oxidation may show acute increases in the plasma total carnitine during prolonged fasting or during attacks of illness.

The basis of the carnitine deficiency in CTD has been shown to be a defect in the plasma membrane carnitine transporter activity. The reason for the increased carnitine levels in CPT-1 deficiency and the decreased carnitine levels in other fatty acid oxidation disorders has been unclear. Both phenomena can be explained by the competitive inhibitory effects of long-chain and medium-chain acylcarnitines on the carnitine transporter [51]. Thus, in patients with MCAD or TRANS deficiency, the blocks in acyl-CoA oxidation lead to accumulation of acylcarnitines which inhibit renal and tissue transport of free carnitine and result in lowered plasma and tissue concentrations of carnitine. Conversely, the inability to form long-chain acyl-CoA in CPT-1 deficiency results in less inhibition of carnitine transport from long-chain acylcarnitine than normal and therefore increases renal carnitine thresholds and plasma levels of carnitine to values greater than normal.

Urinary Organic Acids. The urinary organic acid profile is usually normal in patients with fatty acid oxidation disorders when they are well. During times of fasting or illness, all of the disorders are associated with an inappropriate dicarboxylic aciduria, i.e. urinary medium chain dicarboxylic acids are elevated, while urinary ketones are not. This reflects the fact that dicarboxylic acids, derived from partial oxidation of fatty acids in microsomes and peroxisomes, are produced whenever plasma free fatty acid concentrations are elevated. In MCAD deficiency, the amounts of dicarboxylic acids excreted are two- to fivefold greater than in normal fasting children. However, in other defects, only the ratio of ketones to dicarboxylic acids is abnormal. In a few of the disorders, specific abnormalities of urine organic acid profiles may be present (☐ Table 13.2), but are not likely to be found except during fasting stress.

Urinary Acylglycines. In MCAD deficiency, the urine contains increased concentrations of the glycine conjugates of hexanoate, suberate, (C-8 dicarboxylic acid), and phenylpropionate, which are derived from their coenzyme A esters [52]. When these are quantitated by isotope dilution-mass spectrometry, specific diagnosis of MCAD deficiency is possible even using random urine specimens. Abnormal glycine conjugates are present in urine from patients with some of the other disorders of fatty acid oxidation (☐ Table 13.2).

Plasma Fatty Acids. In MCAD deficiency, specific increases in plasma concentrations of the medium-chain fatty acids octanoate and cis-4-decenoate have been identified which can be useful for diagnosis. Abnormally elevated plasma concentrations of these fatty acids are most apparent during fasting. Elevated levels of free 3-hydroxy fatty acids are also found in both LCHAD and SCHAD deficiencies [53].

13.4.2   Tests of Overall Pathway

These include in vivo fasting study, in vitro fatty acid oxidation, and histology.

In Vivo Fasting Study. In diagnosing the fatty acid oxidation disorders, it is frequently useful to first demonstrate an impairment in the overall pathway before attempting to identify the specific site of defect. Blood and urine samples collected immediately prior to treatment of an acute episode of illness can be used for this purpose, e.g. by showing elevated plasma free fatty acid but inappropriately low ketone levels at the time of hypoglycemia. A carefully monitored study of fasting ketogenesis can provide this information (☐ Fig. 13.2). However, the provocative fasting test is potentially hazardous for affected patients and should only be done under controlled circumstances with careful supervision (▶ also Chap. 3). Some investigators have described fat-loading as an alternative means of testing hepatic ketogenesis, but this has been largely discarded with the development of acyl-carnitine profile testing by mass spectrometry [54] (▶ Chap. 2).

In Vitro Fatty Acid Oxidation. Cultured skin fibroblasts or lymphoblasts from patients can also be used to demonstrate a general defect in fatty acid oxidation using ^{14}C or

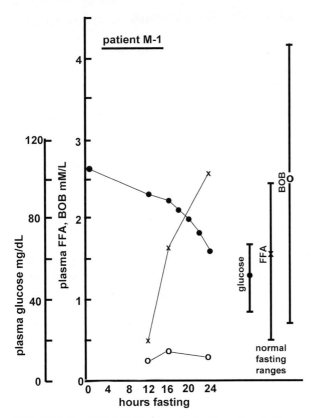

Fig. 13.2. Response to fasting in a patient with medium-chain acyl-CoA-dehydrogenase deficiency. Shown are plasma levels of glucose, free fatty acids (FFA), and β-hydroxybutyrate (BOB) in the patient, and the mean and range of values in normal children who fasted for 24 h. At 14–16 h of fasting, the patient became ill, with pallor, lethargy, nausea, and vomiting

of a defect in fatty acid oxidation. In the hepatic presentation of any of the fatty acid oxidation disorders, a liver biopsy obtained during an acute episode of illness shows an increase in neutral fat deposits which may have either a micro- or macrovesicular appearance. Between episodes, the amount of fat in liver may be normal. More severe changes, including hepatic fibrosis, have been seen in VLCAD patients who where ill for prolonged periods [55]. Patients with LCHAD deficiency may go on to develop cirrhotic changes to the liver, presumed to be a toxic effect of the accumulating 3-hydroxy fatty acids. This damage appears to reflect persistent efforts to metabolize fatty acids, since it may resolve as patients are adequately nourished. On electron microscopy, mitochondria do not show the severe swelling described in Reye syndrome, but may show minor changes such as crystalloid inclusion bodies. The fatty acid oxidation disorders which are expressed in muscle may be associated with increased fat droplet accumulation in muscle fibers and demonstrate the appearance of lipoid myopathy on biopsy.

13.4.3   Enzyme Assays

Cultured skin fibroblasts or cultured lymphoblasts have become the preferred material in which to measure the in vitro activities of specific steps in the fatty acid oxidation pathway. All of the known defects, except HMG-CoA synthase, are expressed in these cells and results of assays in cells from both control and affected patients have been reported. Because these assays are not widely available, they are most usefully applied to confirm a site of defect that is suggested by other clinical and laboratory data.

13.4.4   Prenatal Diagnosis

By assay of labeled fatty acid oxidation and/or enzyme activity in amniocytes or chorionic villi, prenatal diagnosis is theoretically possible for those disorders which are expressed in cultured skin fibroblasts, i.e. all of the currently known defects except HMG-CoA synthase deficiency. This was done in a few instances in MCAD deficiency, although molecular methods are now available for most families with severe fatty acid oxidation defects. Metabolite screening of amniotic fluid has not been useful.

Because of the greater degree of severity of LCHAD/ TFP defects prenatal diagnosis using a combination of molecular and enzymatic analysis has been used successfully to predict affected pregnancies [56]. There is also good indication for prenatal diagnosis for the severe forms of CPT-2 deficiency and ETF/ETF-DH deficiency.

[3]H-labeled substrates. In addition, different chain-length fatty acid substrates can be used with these cells to localize the probable site of defect. Very low rates of labeled fatty acid oxidation are found in CPT-1, TRANS, CPT-2, and ETF/ETF-DH deficiencies. However, high residual rates of oxidation (50–80% or more of normal) frequently make identification of the β-oxidation enzyme defects difficult.

In CTD, oxidation rates are normal unless special steps are taken to grow cells in carnitine-free media. The in vitro oxidation assays do not detect the defects in ketone synthesis. Tandem mass spectrometry using deuterated stable isotopes fatty acids has become an important method for *in vitro* testing in cultured cells. In this assay the site of the block may be indicated by the nature of labeled acylcarnitine species that accumulate in culture media and is not hindered by high residual metabolic flux. Carnitine transporter and CPT-1 deficiencies, which are not associated with accumulating acylcarnitine species, and HMG-CoA synthase deficiency, which is not expressed in fibroblasts, are not detected by this method.

Histology. The appearance of increased triglyceride droplets in affected tissues sometimes provides a clue to the presence

13.5   Treatment and Prognosis

The following sections focus on treatment of the hepatic presentation of fatty acid oxidation disorders, since this is the most life-threatening aspect of these diseases. Although there is a high risk of mortality or long-term disability during episodes of fasting-induced coma, with early diagnosis and treatment patients with most of the disorders have an excellent prognosis. The mainstay of therapy is to prevent recurrent attacks by adjusting the diet to minimize fasting stress.

13.5.1   Management of Acute Illnesses

When patients with fatty acid oxidation disorders become ill, treatment with intravenous glucose should be given immediately. Delay may result in sudden death or permanent brain damage. The goal is to provide sufficient glucose to stimulate insulin secretion to levels that will not only suppress fatty acid oxidation in liver and muscle, but also block adipose tissue lipolysis. Solutions of 10% dextrose, rather than the usual 5%, should be used at infusion rates of 10 mg/kg per min or greater to maintain high to normal levels of plasma glucose, above 100 mg/dl (5.5 mmol/l). Resolution of coma may not be immediate, perhaps because of the toxic effects of fatty acids for a few hours in mildly ill patients or as long as 1–2 days in severely ill patients.

13.5.2   Long-term Diet Therapy

It is essential to prevent any period of fasting which would be sufficient to require the use of fatty acids as a fuel. This can be done by simply ensuring that patients have adequate carbohydrate feeding at bedtime and do not fast for more than 12 h overnight. During intercurrent illnesses, when appetite is diminished, care should be taken to give extra feedings of carbohydrates during the night. In a few patients with severe defects in fatty acid oxidation who had developed weakness and/or cardiomyopathy, we have gone further to completely eliminate fasting by the addition of continuous nocturnal intragastric feedings. The use of uncooked cornstarch at bedtime might be considered as a slowly released form of glucose (for details ▶ Chap. 6), although this has not been formally tested in these disorders. Some authors recommend restricting fat intake. Although this seems reasonable in patients with severe defects, we have not routinely restricted dietary fat in milder defects such as MCAD deficiency.

13.5.3   Carnitine Therapy

In patients with CTD, treatment with carnitine improves cardiac and skeletal muscle function to nearly normal within a few months. It also corrects any impairment in hepatic ketogenesis, which may be present [5]. With oral carnitine at doses of 100 mg/kg per day, plasma carnitine levels can be maintained in the low to normal range and liver carnitine levels may be normal. However, muscle carnitine concentrations remain less than 5% of normal. Since these low levels are adequate to reverse myopathy in CTD, it appears that the threshold for defining carnitine deficiency is a tissue concentration less than approximately 5% of normal.

A possible role for carnitine therapy in those disorders of fatty acid oxidation, which are associated with secondary carnitine deficiency, remains controversial [48]. Since these disorders involve blocks at specific enzyme steps that do not involve carnitine, it is obvious that carnitine treatment cannot correct the defect in fatty acid oxidation. It has been proposed that carnitine might help to remove metabolites in these disorders, because the enzyme defects might be associated with accumulation of acyl-CoA intermediates. However, there has been no direct evidence that this is true and some evidence to the contrary has been presented [57]. In addition, as noted above, the mechanism of the secondary carnitine deficiency is not a direct one, via loss of acylcarnitines in urine, but appears to be indirect, via inhibition of the carnitine transporter in kidney and other tissues by medium or long-chain acylcarnitines. It should also be noted that the secondary carnitine deficiency could be a protective adaptation, since there is data showing that long-chain acylcarnitines may have toxic effects. Our current practice is not to recommend the use of carnitine except as an investigational drug in fatty acid oxidation disorders other than CTD.

13.5.4   Other Therapy

Since medium-chain fatty acids bypass the carnitine cycle (❏ Fig. 13.1) and enter the midportion of the mitochondrial ß-oxidation spiral directly, it is possible that they might be used as fuels in defects which block either the carnitine cycle or long-chain ß-oxidation. For example, dietary MCT was suggested to be helpful in a patient with LCHAD deficiency. The benefits of MCT have not been thoroughly investigated, but MCT clearly must not be used in patients with MCAD, SCAD, SCHAD, ETF/ETF-DH, HMG-CoA synthase, or HMG-CoA lyase deficiencies. Some patients with mild variants of ETF/ETF-DH and SCAD deficiencies have been reported to respond to supplementation with high doses of riboflavin (100 mg/day), the cofactor for these enzymes. Triheptanoin was suggested to be of benefit in three cases of vLCAD as an anaplerotic substrate, but has not yet been confirmed by controlled studies [58].

13.5.5   Prognosis

Although acute episodes carry a high risk of mortality or permanent brain damage, many patients with disorders of fatty acid oxidation can be easily managed by avoidance of prolonged fasts. These patients have an excellent long-term prognosis. Patients with chronic cardiomyopathy or skeletal muscle weakness have a more guarded prognosis, since they seem to have more severe defects in fatty acid oxidation. For example, TRANS or the severe variants of CPT-2 and ETF/ ETF-DH deficiencies frequently lead to death in the newborn period. On the other hand, the mild form of CPT-2 deficiency may remain silent as long as patients avoid exercise stress.

Leukotriene Metabolism

Leukotrienes (LTs) are a group of biologically highly active compounds derived from arachidonic acid in the 5-lipoxygenase pathway [60, 61]. Biosynthesis of LTs (◘ Fig. 13.3) is limited to a small number of human cells, including brain tissue [62]. Peroxisomes have been identified as the main cell organelle performing degradation of LTs [63, 64]. In addition to their well-known function in the mediation of inflammation and host defense, cysteinyl LTs have neuromodulatory and neuroendocrine functions in the brain [69]. The enzyme 5-lipoxygenase, which requires the presence of the 5-lipoxygenase-activating protein, catalyzes the first two committed steps in LT synthesis. The resulting unstable epoxide intermediate, LTA_4, can be further metabolized to either LTB_4 or LTC_4. The latter step is specifically catalyzed by the enzyme LTC_4 synthase and requires the presence of glutathione. LTC_4 and its metabolites LTD_4 and LTE_4 are termed cysteinyl LTs. The rate-limiting step in the synthesis of cysteinyl LTs is the conversion of LTA_4 to LTC_4, which is catalyzed by LTC_4 synthase.

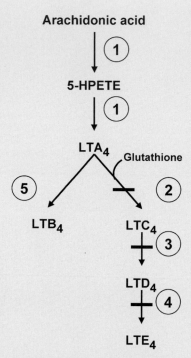

◘ **Fig. 13.3.** Pathway of leukotriene metabolism. *5-HPETE*, hydroperoxyeicosatetraenoic acid; *LT*, leukotriene; **1**, 5-lipoxygenase/FLAP (5-lipoxygenase activating protein); **2**, LTC_4 synthase; **3**, γ-glutamyl transpeptidase; **4**, dipeptidase; **5**, LTA_4 hydrolase. Enzyme defects are depicted by *solid bars*

13.6 Rare Related Disorders

13.6.1 Transport Defect of Fatty Acids

Two children with liver failure for whom a genetic defect in the transport of free fatty acids across the plasma membrane was suggested have been reported [59]. One of these children had reduced levels of long-chain free fatty acids in liver tissue; cultured fibroblasts from both showed modest reductions in both oxidation and uptake of long-chain fatty acids. Although five putative fatty acid transporters have been described, their function as carrier proteins remains speculative.

13.6.2 Defects in Leukotriene Metabolism

Leukotrienes comprise a group of biologically highly active lipid mediators derived from 20-carbon polyunsaturated fatty acids, predominantly arachidonic acid, via the 5-lipoxygenase pathway [60] (◘ Fig. 13.3). They include the cysteinyl leukotrienes (LTC_4, LTD_4, LTE_4) and the dihydroxyeicosatetraenoate, LTB_4. Synthesis of the primary cysteinyl leukotriene, LTC_4, from conjugation of the unstable LTA_4 with glutathione, is mediated by LTC_4-synthase. Stepwise cleavage of glutamate and glycine from LTC_4 by γ-glutamyl transpeptidase and membrane-bound dipeptidase yield LTD_4 and LTE_4, respectively. Biosynthesis is limited to very few human cells including mast cells, eosinophils, basophils and macrophages. Moreover, the human brain tissue also has the capacity to synthesize large amounts of leukotrienes. Beside their role as inflammatory mediators e.g. in asthma there is increasing evidence that leukotrienes may play a role as messengers or modulators of CNS activity.

A few disorders have been identified causing secondary disturbances in leukotriene elimination and degradation, e.g. defective hepatobiliary elimination of cysteinyl leukotrienes as seen in the Dubin-Johnson syndrome, altered β-oxidation in disorders of peroxisome biogenesis such as the Zellweger syndrome, and most important impaired ω-oxidation of LTB_4 in the Sjögren-Larsson syndrome.

At present, there is also evidence of three primary defects in the synthesis of cysteinyl leukotrienes representing a new group of neurometabolic diseases.

Sjögren-Larsson Syndrome (SLS). This disorder is an inborn error of fatty alcohol oxidation with an autosomal recessive mode of inheritance. The well-known clinical triad includes ichthyosis, spastic di- or tetraplegia and mental retardation [61]. The congenital ichthyosis usually brings the patient to medical attention, whereas spasticity and mental retardation become apparent later in the first or second year of life. In addition, pre-term birth, pruritus, and ocular abnormalities including a juvenile macular dystrophy occur in the majority of patients [62]. Neuroradiological findings de-

monstrate cerebral involvement including retardation of myelination and a persistent myelin deficit.

The defect in fatty alcohol oxidation in SLS is caused by deficient activity of fatty aldehyde dehydrogenase (FALDH), a transmembrane protein that is part of the microsomal enzyme complex fatty alcohol: nicotinamide-adenine (NAD^+) oxidoreductase (FAO) [63]. FALDH catalyses the oxidation of many different medium- and long-chain fatty aldehydes to the corresponding fatty acids. However, plasma fatty alcohol concentrations are not elevated in affected patients.

FALDH has also a crucial role in the ω-oxidation, i.e. inactivation of LTB_4. The biological half-life of LTB_4 is regulated by microsomal ω-oxidation to ω-hydroxy-LTB_4. Subsequent microsomal degradation of ω-hydroxy-LTB_4 yields ω-aldehyde-LTB_4 and ω-carboxy-LTB_4, respectively. Patients with SLS exhibit highly elevated urinary concentrations of LTB_4 and ω-hydroxy-LTB_4, while ω-carboxy-LTB_4 is not present [64]. In addition, fresh polymorphonuclear leukocytes are unable to convert ω-hydroxy-LTB_4 to ω-carboxy-LTB_4 [65].

Definite diagnosis of SLS requires measuring FALDH activity in cultured fibroblasts and/or mutation analysis of the FALDH gene. The FALDH gene has been mapped to chromosome 17p11.2 and many different mutations have been found in patients with SLS.

The accumulation of fatty alcohols, the modification of macromolecules by fatty aldehydes, and the presence of high concentrations of biologically active lipids, including LTB_4, have been postulated as the underlying pathophysiological mechanisms that give rise to the clinical features.

At present, there exists no curative treatment. However, the recognition of defective LTB_4 degradation in SLS gives the opportunity for a therapeutic intervention on a rational base. First results of substrate depletion therapy using the 5-lipoxygenase inhibitor zileuton showed that this compound in doses of about 600 mg three to four times daily has clear biochemical and some favourable clinical effects in SLS especially with respect to the agonizing pruritus [66].

LTC4-synthesis Deficiency. To date, LTC_4-synthesis deficiency has been identified in two infants from consanguineous parents in association with a fatal neurodevelopmental syndrome [67, 68]. Clinical symptoms included generalized muscular hypotonia, severe psychomotor retardation, microcephaly and failure to thrive. Both patients died at 6 months of age.

Biochemical findings in the patients revealed that the concentrations of LTC_4, LTD_4 and LTE_4 in CSF were below the detection limit. This profile of leukotrienes in CSF is pathognomonic for LTC_4-synthesis deficiency. In addition, LTC_4 and its metabolites were below the detection limit in both plasma and urine. Furthermore, LTC_4 could not be generated in stimulated monocytes. Moreover, [^{3}H]-LTC_4 could not be formed from [^{3}H]-LTA_4 by the patient's monocytes and platelets. The exact pathomechanisms are not clear yet. However, it is postulated that absence of LTC_4, especially in the brain, is responsible for neurological symptoms in patients with LTC_4-synthesis deficiency. A curative treatment is yet not available.

γ-Glutamyl Transpeptidase (GGT) Deficiency. At present, there have been five patients reported with GGT deficiency [69]. Clinical symptoms are inconsistent. Three of them presented with variable central nervous symptoms including mental retardation, severe behaviour problems, psychiatric symptoms or seizures whereas also asymptomatic patients have been recognized. Affected individuals have increased glutathione concentration in plasma and urine, but their cellular levels are normal. In addition to glutathionuria, they show increased levels of γ-glutamylcysteine and cysteine. Decreased activity of GGT can be demonstrated in leukocytes or fibroblasts but not in erythrocytes, which also lack this enzyme under normal conditions.

Patients with GGT deficiency are characterized by a complete deficiency of LTD_4 biosynthesis [70]. Concentrations of LTC_4 in urine and plasma are increased whereas LTD_4 and LTE_4 are below the detection limit. In addition, synthesis of LTD_4 and LTE_4 in stimulated monocytes is below the detection limit. Moreover, formation of $[^3H]$-LTD_4 from $[^3H]$-LTC_4 in monocytes is completely deficient. DNA analysis or treatment strategies are not yet available.

Membrane-bound Dipeptidase (Cysteinyl-glycinase) Deficiency. There exists only one known patient with cysteinyl-glycinuria [71] and, as it was only recently demonstrated, increased excretion of LTD_4 in urine. However, under physiological conditions LTD_4 is not detectable in human urine. Although definite enzyme measurements could not yet be performed, this 15-year old male seems to be the first patient with membrane-bound dipeptidase (cysteinyl-glycinase) deficiency. Clinical symptoms included mental retardation, motor impairment, peripheral neuropathy, foot deformity as well as partial deafness. EEG and EMG revealed abnormal results. The presumably defective enzyme is involved in the third step of the synthesis of cysteinyl leukotrienes resulting in the synthesis of LTE_4. Endogenous urinary LTE_4 represents the index metabolite for the generation of cysteinyl leukotrienes in vivo. Molecular studies or treatment options are not available yet.

References

1. Coates PM, Stanley CA (1992) Inherited disorders of mitochondrial fatty acid oxidation. Prog Liver Dis 10:123-138
2. Eaton S, Bartlett K, Pourfarzam M (1996) Mammalian mitochondrial β-oxidation. Biochem J 320:345-357
3. Wanders RJA, Vreken P, den Boer ME et al (1999) Disorders of mitochondrial fatty acyl-CoA β-oxidation. J Inherit Metab Dis 22:442-487
4. Rinaldo P, Matern D, Bennett MJ (2002)Fatty acid oxidation disorders. Annu Rev Physiol 64:477-502
5. Stanley CA, DeLeeuw S, Coates PM et al (1991) Chronic cardiomyopathy and weakness or acute coma in children with a defect in carnitine uptake. Ann Neurol 30:709-716
6. Cederbaum SD, Koo-McCoy S, Tein I et al (2002) Carnitine membrane transporter deficiency: a long-term follow up and OCTN2 mutation in the first documented case of primary carnitine deficiency. Mol Genet Metab 77:195-201
7. Nezu J-i, Tamai I, Oku A et al (1999) Primary systemic carnitine deficiency is caused by mutations in a gene encoding sodium ion-dependent carnitine transporter. Nat Genet 21:91-94
8. Wang Y, Kormann SH, Ye J et al (2001) Phenotype and genotype variation in primary carnitine deficiency. Genet Med 3:387-392
9. Demaugre F, Bonnefont J, Mitchell G et al (1988) Hepatic and muscular presentations of carnitine palmitoyl transferase deficiency: Two distinct entities. Pediatr Res 24:308-311
10. Stanley CA, Sunaryo F, Hale DE et al (1992) Elevated plasma carnitine in the hepatic form of carnitine palmitoyltransferase-1 deficiency. J Inherit Metab Dis 15:785-789
11. Falik-Borenstein ZC, Jordan SC, Saudubray JM et al (1992) Brief report: renal tubular acidosis in carnitine palmitoyltransferase type 1 deficiency. N Engl J Med 327:24-27
12. IJlst L, Mandel H, Oostheim W et al (1998) Molecular basis of hepatic carnitine palmitoyl transferase I deficiency. J Clin Invest 102:527-531
13. Gobin S, Bonnefont J-P, Prip-Buus C et al (2002) Organization of the human liver carnitine palmitoyltransferase 1 (*CPT1A*) and identification of novel mutations in hypoketotic hypoglycemia. Hum Genet 111:179-189
14. Bennett MJ, Boriack RL, Narayan S et al (2004) Novel mutations in *CPT1A* define molecular heterogeneity of hepatic carnitine palmitoylltransferase 1 deficiency. Mol Genet Metab 82:59-63
15. Stanley CA, Hale DE, Berry GT et al (1992) Brief report: a deficiency of carnitine-acylcarnitine translocase in the inner mitochondrial membrane. N Engl J Med 327:19-23
16. Pande SV, Brivet B, Slama A et al (1993) Carnitine-acylcarnitine translocase deficiency with severe hypoglycemia and auriculo ventricular block. Translocase assay in permeabilized fibroblasts. J Clin Invest 91:1247-1252
17. Chalmers RA, Stanley CA, English N, Wigglesworth JS (1997) Mitochondrial carnitine-acylcarnitine translocase deficiency presenting as sudden neonatal death. J Pediatr 131:220-225
18. DiMauro S, DiMauro PMM (1973) Muscle carnitine palmityltransferase deficiency and myoglobinuria. Science 182:929-931
19. Demaugre F, Bonnefont JP, Colonna M et al (1991) Infantile form of carnitine palmitoyltransferase II deficiency with hepatomuscular symptoms and sudden death. Physiopathological approach to carnitine palmitoyltransferase II deficiencies. J Clin Invest 87:859-864
20. Taroni F, Verderio E, Garavaglia B et al (1992) Biochemical and molecular studies of carnitine palmitoyltransferase II deficiency with hepatocardiomyopathic presentation. Prog Clin Biol Res 375:521-531
21. Taroni F, Verderio E, Dworzak F et al (1993) Identification of a common mutation in the carnitine palmitoyltransferase II gene in familial recurrent myoglobinuria patients. Nat Genet 4:314-320
22. Thuillier L, Rostane H, Droin V et al (2003) Correlation between genotype, metabolic data, and clinical presentation in carnitine palmitoyltransferase 2 (CPT 2) deficiency. Hum Mutat 21:493-501
23. Sigauke E, Rakheja D, Kitson K, Bennett MJ (2003) Carnitine palmitoyltransferase II deficiency: a clinical, biochemical, and molecular review. Lab Invest 83:1543-1554
24. Hale DE, Batshaw ML, Coates PM et al(1985) Long-chain acyl co-enzyme A dehydrogenase deficiency: an inherited cause of non-ketotic hypoglycemia. Pediatr Res 19:666-671

25. Vianey-Saban C, Divry P, Brivet M et al (1998) Mitochondrial very-long-chain acyl-coenzyme A dehydrogenase deficiency: clinical characteristics and diagnostic considerations in 30 patients. Clin Chim Acta 269:43-62

26. Wanders RJ, Denis S, Ruiter JP et al (1998) 2,6-Dimethylheptanoyl-CoA is a specific substrate for long-chain acyl-CoA dehydrogenase (LCAD): evidence for a major role of LCAD in branched-chain fatty acid oxidation. Biochim Biophys Acta 1393:35-40

27. Andresen BS, Olpin SE, Poorthuis BJ et al (1999)Clear correlation of genotype with disease phenotype in very-long-chain acyl-CoA dehydrogenase deficiency. Am J Hum Genet 64:479-494

28. Iafolla AK, Thompson RJ, Roe CR (1994) Medium-chain acyl-coenzyme A dehydrogenase deficiency: clinical course in 120 affected children. J Pediatr 124:409-415

29. Yokota I, Coates PM, Hale DE et al (1992) The molecular basis of medium chain acyl-CoA dehydrogenase deficiency: survey and evolution of 985A–G transition, and identification of five rare types of mutation within the medium chain acyl-CoA dehydrogenase gene. Prog Clin Biol Res 375:425-440

30. Yokota I, Coates PM, Hale DE et al (1991) Molecular survey of a prevalent mutation, 985A-to-G transition, and identification of five infrequent mutations in the medium-chain Acyl-CoA dehydrogenase (MCAD) gene in 55 patients with MCAD deficiency. Am J Hum Genet 49:1280-1291

31. Bhala A, Willi SM, Rinaldo P et al (1995) Clinical and biochemical characterization of short-chain acyl-coenzyme A dehydrogenase deficiency. J Pediatr 126:910-915

32. Coates PM, Hale DE, Finocchiaro G et al (1988) Genetic deficiency of short-chain acyl-coenzyme A dehydrogenase in cultured fibroblasts from a patient with muscle carnitine deficiency and severe skeletal muscle weakness. J Clin Invest 81:171-175

33. Corydon MJ, Vockley G, Rinaldo P et al (2001) Role of common variant alleles in the molecular basis of short-chain acyl-CoA dehydrogenase deficiency. Pediatr Res 49:18-23

34. Rocchiccioli F, Wanders RJ, Aubourg P et al (1990) Deficiency of long-chain 3-hydroxyacyl-CoA dehydrogenase: a cause of lethal myopathy and cardiomyopathy in early childhood. Pediatr Res 28:657-662

35. Jackson S, Kler RS, Bartlett K et al (1992) Combined defect of long-chain 3-hydroxyacyl-CoA dehydrogenase, 2-enoyl-CoA hydratase and 3-oxoacyl-CoA thiolase. Prog Clin Biol Res 375:327-337

36. Treem WR, Shoup ME, Hale DE et al(1996) Acute fatty liver of pregnancy, hemolysis, elevated liver enzymes, and low platelets syndrome, and long chain 3-hydroxyacyl-coenzyme A dehydrogenase deficiency [see comments]. Am J Gastroenterol 91:2293-2300

37. Ibdah JA, Bennett MJ, Rinaldo P et al (1999) A fetal fatty acid oxidation disorder as a cause of liver disease in pregnant women. N Engl J Med 340:1723-1731

38. Ibdah JA, Yang Z, Bennett MJ (2000) Liver disease in pregnancy and fetal fatty acid oxidation defects. Mol Genet Metab 71:182-189

39. Shekhawat P, Bennett MJ, Sadovsky Y et al (2003) Human placenta metabolizes fatty acids: implications for fetal fatty acid oxidation disorders and maternal liver diseases. Am J Physiol 284:E1098-1105

40. Tein I, Devivo DC, Hale DE et al (1991) Short-chain L-3-hydroxyacyl-CoA dehydrogenase deficiency in muscle: a new cause for recurrent myoglobinuria and encephalopathy. Ann Neurol 30:415-419

41. Bennett MJ, Weinberger MJ, Kobori JA et al (1996) Mitochondrial short-chain L-3-hydroxyacyl-coenzyme A dehydrogenase deficiency: a new defect of fatty acid oxidation. Pediatr Res 39:185-188

42. Bennett MJ, Spotswood SD, Ross KF et al (1999) Fatal hepatic short-chain L-3-hydroxyacyl-CoA dehydrogenase deficiency: clinical, biochemical and pathological studies of this recently identified disorder of mitochondrial beta-oxidation. Pediatr Dev Pathol 2:337-345

43. Clayton PT, Eaton S, Aynsley-Green A et al (2001) Hyperinsulinism in short-chain L-3-hydroxyacyl-CoA dehydrogenase deficiency reveals the importance of β-oxidation in insulin secretion. J Clin Invest 108:457-465

44. Kamigo T, Indo Y, Souri M et al (1997) Medium chain 3-ketoacyl-coenzyme A thiolase deficiency: a new disorder of mitochondrial fatty acid beta-oxidation. Pediatr Res 42:569-576

45. Roe C, Millington D, Norwood D et al (1990) 2,4-Dienoyl-coenzyme A reductase deficiency: a possible new disorder of fatty acid oxidation. J Clin Invest 85:1703-1707

46. Frerman FE, Goodman SI (1985) Deficiency of electron transfer flavoprotein or electron transfer flavoprotein:ubiquinone oxidoreductase in glutaric acidemia type II fibroblasts. Proc Natl Acad Sci USA 82:4517-4520

47. Vladutiu GD, Bennett MJ, Smail D et al (2000) A variable myopathy associated with heterozygosity for the R503C mutation in the carnitine palmitoyltransferase II gene. Mol Genet Metab 70:134-141

48. Stanley CA (1995) Carnitine disorders. Adv Pediatr 42:209-242

49. Millington DS, Terada N, Chace DH (1992) The role of tandem mass spectrometry in the diagnosis of fatty acid oxidation disorders. In: Coates PM, Tanaka K (eds) New developments in fatty acid oxidation. Wiley-Liss, New York, pp 339-354

50. Ziadeh R, Hoffman EP, Finegold DM et al (1995) Medium chain acyl-CoA dehydrogenase deficiency in Pennsylvania: neonatal screening shows high incidence and unexpected mutation frequencies. Pediatr Res 37:675-678

51. Stanley CA, Berry GT, Bennett MJ et al (1993) Renal handling of carnitine in secondary carnitine deficiency disorders. Pediatr Res 34:89-97

52. Rinaldo P, O'Shea JJ, Coates PM et al (1989) Medium-chain acyl-CoA dehydrogenase deficiency. Diagnosis by stable-isotope dilution measurement of urinary n-hexanoylglycine and 3-phenylpropionylglycine [published erratum appears in N Engl J Med 320: 1227]. N Engl J Med 1988; 319(20):1308-1313

53. Saudubray JM, Martin D, DeLonlay P (1999) Recognition and management of fatty acid oxidation defects: a series of 107 patients. J Inherit Metab Dis 22:488-502

54. Treem WR, Witzleben CA, Piccoli DA et al (1986) Medium-chain and long-chain acyl CoA dehydrogenase deficiency: clinical, pathologic and ultrastructural differentiation from Reye's syndrome. Hepatology 6:1270-1278

56. Ibdah JA, Zhao Y, Viola J et al (2001) Molecular prenatal diagnosis in families with fetal mitochondrial trifunctional protein mutations. J Pediatr 138:396-399

57. Lieu YK, Hsu BY, Price WA et al (1997) Carnitine effects on coenzyme A profiles in rat liver with hypoglycin inhibition of multiple dehydrogenases. Am J Physiol 272(3):E359-366

58. Roe CR, Sweetman L, Roe DS, David F, Brunengraber H (2002) Treatment of cardiomyopathy and rhabdomyolysis in long-chain fat oxidation disorders using an anaplerotic odd-chain triglyceride. J Clin Invest 2002:259-269

59. Odaib AA, Schneider BL, Bennett MJ et al (1998) A defect in the transporter of long-chain fatty acids associated with acute liver failure. N Engl J Med 339:1752-1757

60. Mayatepek E, Hoffmann GF (1995) Leukotrienes: biosynthesis, metabolism and pathophysiological significance. Pediatr Res 37:1-9

61. Rizzo WB (1993) Sjögren-Larsson syndrome. Semin Dermatol 12:210-218

62. van Domburg PHMF, Willemsen MAAP, Rotteveel JJ et al (1999) Sjögren-Larsson syndrome. Clinical and MRI/MRS findings in FALDH deficient patients. Neurology 52:1345-1352

63. Rizzo WB, Craft DA (1991) Sjögren-Larsson syndrome. Deficient activity of the fatty aldehyde dehydrogenase component of fatty alcohol: NAD+ oxidoreductase in cultured fibroblasts. J Clin Invest 88:1643-1648

64. Willemsen MAAP, de Jong JGN, van Domburg PHMF et al (2000) Defective inactivation of leukotriene B_4 in patients with Sjögren-Larson syndrome. J Pediatr 136:258-260
65. Willemsen MAAP, Rotteveel JJ, de Jong JGN et al (2001) Defective metabolism of leukotriene B4 in the Sjögren-Larsson syndrome. J Neurol Sci 183:61-67
66. Willemsen MAAP, Lutt M, Steijlen PM et al (2001) Clinical and biochemical effects of zileuton in patients with the Sjögren-Larsson syndrome. Eur J Pediatr 160(12):7111-7117
67. Mayatepek E, Flock B (1998) Leukotriene C_4-synthesis deficiency: a new inborn error of metabolism linked to a fatal developmental syndrome. 352:1514-1517
68. Mayatepek E, Lindner M, Zelezny R et al (1999) A severely affected infant with absence of cysteinyl leukotrirenes in cerebrospinal fluid: further evidence that leukotriene C_4-synthesis deficiency is a new neurometabolic disorder. Neuropediatrics 30:5-7
69. Ristoff E, Larsson A (1998) Patients with genetic defects in the γ-glutamyl cycle. Chem Biol Interact 111/112:113-121
70. Mayatepek E, Okun JG, Meissner T et al (2004) Synthesis and metabolism of leukotrienes in γ-glutamyl transpeptidase deficiency. J Lipid Res 45:900-904
71. Bellet H, Rejou F, Vallat C, Mion H, Dimeglio A (1999) Cystinyl-glycinuria: a new neurometabolic disorder? J Inherit Metab Dis 22:231-234

14 Disorders of Ketogenesis and Ketolysis

Andrew A.M. Morris

Ketogenesis and Ketolysis

During fasting, ketone bodies are an important fuel for many tissues, including cardiac and skeletal muscle. They are particularly important for the brain, which cannot oxidise fatty acids. The principal ketone bodies, acetoacetate and 3-hydroxybutyrate, are maintained in equilibrium by 3-hydroxybutyrate dehydrogenase; acetone is formed from acetoacetate non-enzymatically and eliminated in breath. Ketone bodies are formed in liver mitochondria, predominantly from fatty acids, but also from certain amino acids, such as leucine. For use as fuel, ketone bodies are converted to acetyl-CoA in the mitochondria of extrahepatic tissues. One of the ketolytic enzymes, mitochondrial acetoacetyl-CoA thiolase, is also involved in the breakdown of isoleucine (◘ Fig. 14.1).

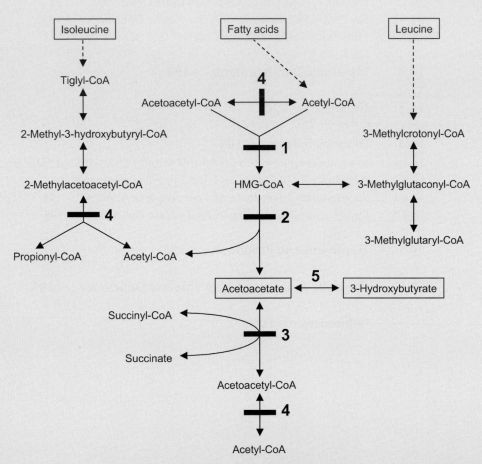

◘ **Fig. 14.1.** Biochemical pathways involving enzymes of ketogenesis and ketolysis. *HMG-CoA,* 3-hydroxy-3-methylglutaryl coenzyme A. **1,** Mitochondrial (m)HMG-CoA synthase; **2,** HMG-CoA lyase; **3,** succinyl-CoA 3-oxoacid CoA transferase (SCOT); **4,** mitochondrial acetoacetyl-CoA thiolase (T2); **5,** 3-hydroxybutyrate dehydrogenase. The enzyme defects discussed in this chapter are depicted by *solid bars* across the arrows

Disorders of ketone body metabolism present either in the first few days of life or later in childhood, during an infection or some other metabolic stress. In defects of ketogenesis, decompensation leads to encephalopathy, with vomiting and a reduced level of consciousness, often accompanied by hepatomegaly. The biochemical features – hypoketotic hypoglycaemia, with or without hyperammonaemia – resemble those seen in fatty acid oxidation disorders. In defects of ketolysis, the clinical picture is dominated by severe ketoacidosis. This is often accompanied by decreased consciousness and dehydration.

14.1   Clinical Presentation

14.1.1   Mitochondrial 3-Hydroxy-3-Methylglutaryl-CoA Synthase Deficiency

Six patients with mitochondrial 3-hydroxy-3-methylglutaryl-coenzyme A (mHMG-CoA) synthase deficiency (enzyme 1 in ◘ Fig. 14.1) have been reported since 1997. All presented with hypoglycaemia following gastroenteritis, between the ages of 9 months and 6 years [1-3]. At presentation, most patients were comatose and had hepatomegaly, which subsequently resolved. Blood lactate and ammonia concentrations were normal and urine was negative for ketone bodies. All patients recovered promptly with intravenous glucose and none suffered long-term complications.

14.1.2   3-Hydroxy-3-Methylglutaryl-CoA Lyase Deficiency

Approximately 30% of patients with HMG-CoA lyase deficiency (enzyme 2 in ◘ Fig. 14.1) present by 5 days of age, after a short initial symptom-free period. Most of the other patients present later in the first year, when they are fasted or suffer infections [4]. A few patients remain asymptomatic for a number of years [5].

Clinical features at presentation include vomiting, hypotonia and a reduced level of consciousness. Investigations show hypoglycaemia and acidosis [4]. Ketone body levels are inappropriately low but blood lactate concentrations may be markedly elevated, particularly in neonatal onset cases [6]. Many patients have hyperammonaemia, hepatomegaly and abnormal liver function tests and, in the past, cases may have been misdiagnosed as Reye's syndrome [7]. Pancreatitis is a recognised complication [8], as in other branched-chain organic acidaemias. With appropriate treatment, most patients recover from their initial episode of metabolic decompensation. Unfortunately, a number suffer neurological sequelae, including epilepsy, intellectual handicap, hemiplegia or cerebral visual loss [4, 5, 7]. Imaging may show abnormalities in the basal ganglia and focal cerebral atrophy. Even in asymptomatic patients, magnetic resonance imaging (MRI) shows diffuse mild T_2 hyperintensity of the cerebral white matter with multiple foci of more severe signal abnormality [5]. The cause of these changes is unknown; myelination may be impaired because ketone bodies are a substrate for the synthesis of myelin cholesterol [9].

14.1.3   Succinyl-CoA 3-Oxoacid CoA Transferase Deficiency

The deficiency of succinyl-CoA 3-oxoacid CoA transferase (SCOT, also known as succinyl-CoA 3-ketoacid CoA transferase; enzyme 3 in ◘ Fig. 14.1) is characterised by recurrent episodes of severe ketoacidosis. Tachypnoea is often accompanied by hypotonia and lethargy. As for HMG-CoA lyase deficiency, 30 % of patients become symptomatic within a few days of birth [10] and most of the others present later in the first year. A few patients have had cardiomegaly at the time of presentation [11]. Blood glucose, lactate and ammonia concentrations are generally normal, though there have been reports of hypoglycaemia in neonates [10] and hyperglycaemia in older children [11]. Because ketosis and acidosis are common in sick children, SCOT deficiency enters the differential diagnosis for a large number of patients.

14.1.4   Mitochondrial Acetoacetyl-CoA Thiolase Deficiency

Recurrent episodes of ketoacidosis also occur in patients with a deficiency of mitochondrial acetoacetyl-CoA thiolase (also known as 2-methyl-acetoacetyl-CoA thiolase and as β-ketothiolase, abbreviated T2; enzyme 4 in ◘ Fig. 14.1) [12]. Neonatal onset is rare. Most patients present during the first two years but one asymptomatic adult has been diagnosed [13].

Episodes of decompensation generally start with tachypnoea and vomiting, which is followed by dehydration and a falling level of consciousness [12]. A few patients have seizures. Investigations show a severe metabolic acidosis with ketonuria. Blood glucose, lactate and ammonia concentrations are normal in most cases but hyper- and hypoglycaemia have been reported [11, 12]. The high acetoacetate levels in blood and urine may cause screening tests for salicylate to give false positive results [14]. Most patients make a full recovery from episodes of decompensation but some have mental retardation, ataxia or dystonia with abnormalities in the basal ganglia on MRI [15]. A few patients have developmental delay before their first episode of ketoacidosis [15].

14.2   Metabolic Derangement

Ketone bodies are synthesised in hepatic mitochondria, primarily using acetyl-CoA derived from fatty acid oxidation (◘ Fig. 14.1). mHMG-CoA synthase catalyses the condensation of acetoacetyl-CoA and acetyl-CoA to form HMG-CoA, which is cleaved by HMG-CoA lyase to release acetyl-CoA and acetoacetate. HMG-CoA can also be derived from the amino acid, leucine. Thus, mHMG-CoA synthase and HMG-CoA lyase deficiencies both impair ketogenesis but HMG-CoA lyase deficiency also causes the accumulation of intermediates of the leucine catabolic pathway. The hypoglycaemia seen in these defects may result from impaired gluconeogenesis or from excessive glucose consumption due to the lack of ketone bodies.

Ketone body utilisation occurs in extrahepatic mitochondria, starting with the transfer of coenzyme A from succinyl-CoA to acetoacetate, catalysed by SCOT. This forms acetoacetyl-CoA, which is converted to acetyl-CoA by T2. SCOT is not expressed in liver and has no role other than ketolysis; episodic ketoacidosis is, therefore, the only consistent biochemical abnormality in SCOT deficiency. In contrast, T2 is expressed in liver and has 3 different roles. Whereas T2 promotes ketolysis in extrahepatic tissues, in liver it participates in ketogenesis by converting acetyl-CoA to acetoacetyl-CoA. Finally, T2 cleaves 2-methylacetoacetyl-CoA in the degradation pathway for isoleucine. Patients with T2 deficiency present with ketoacidosis, implying that the enzyme is more crucial in ketolysis than in ketogenesis; they also excrete intermediates of isoleucine catabolism.

14.3   Genetics

All 4 disorders are inherited as autosomal recessive traits. Their prevalence is unknown but HMG-CoA lyase deficiency is relatively common in Saudi Arabia [6]. Mutations have been identified in patients with each of the 4 disorders [3, 12, 16]. Two common mutations have been found in HMG-CoA lyase deficiency, one in the Saudi population [17] and the other in Mediterranean patients [18]. Common mutations have not been identified in the other disorders.

Prenatal diagnosis is possible using molecular techniques in families where the mutations are known or there are informative polymorphisms [19, 20]. Alternatively, prenatal diagnosis can be performed by enzyme assay in chorionic villi or cultured amniocytes [19, 20]. This is not possible for mHMG-CoA synthase and some authorities prefer not to use chorionic villi for SCOT [21]. Organic acid analysis of amniotic fluid is a third method of prenatal diagnosis for HMG-CoA lyase deficiency [19].

14.4   Diagnostic Tests

HMG-CoA lyase and T2 deficiencies are diagnosed by detecting abnormal urinary organic acids. The abnormalities are sometimes hard to detect in T2 deficiency. The organic acid profile of 2-methyl-3-hydroxybutyryl-CoA dehydrogenase deficiency, another enzyme of the isoleucine degradation pathway (discusssed in ▶ Chap. 19), resembles that of T2 deficiency and cases have been misdiagnosed as such in the past. Recognition of the other defects is even more difficult and it is likely that many cases remain undiagnosed.

14.4.1   Mitochondrial 3-Hydroxy-3-Methylglutaryl-CoA Synthase Deficiency

This diagnosis should be suspected when there is grossly impaired ketogenesis during fasting but there is normal fatty acid oxidation flux in vitro (▶ Chap. 3 and 13). Organic acid analysis typically shows massive dicarboxylic aciduria (saturated, unsaturated and hydroxy-compounds) but low levels of ketone bodies [2, 3]. Blood acylcarnitine analysis is normal. Measurement of enzyme activity requires a liver biopsy because it is only expressed at high levels in liver and testis. Moreover, enzyme assays are complicated by a cytoplasmic isoenzyme, involved in cholesterol synthesis, which normally accounts for approximately 10% of total activity. Enzymology has not been undertaken in recent patients and the diagnosis has been confirmed by mutation analysis [3].

14.4.2   3-Hydroxy-3-Methylglutaryl-CoA Lyase Deficiency

Patients with this condition excrete increased quantities of 3-hydroxy-3-methylglutaric, 3-hydroxyisovaleric, 3-methylglutaconic and 3-methylglutaric acids (◘ Fig. 14.1); 3-methylcrotonylglycine may also be present [4]. It is important to confirm the diagnosis enzymologically since a similar pattern of urinary organic acids has been found in patients with normal HMG-CoA lyase activity [22]. Assays can be undertaken on leukocytes or cultured fibroblasts as well as liver. Currently, the diagnosis is generally made following an acute presentation. There is, however, the potential to diagnose cases by neonatal screening, since 3-methylglutarylcarnitine accumulates in blood and can be detected by tandem mass spectrometry [23].

14.4.3 Succinyl-CoA 3-Oxoacid CoA Transferase Deficiency

Most patients have persistent ketonuria and a circulating concentration of ketone bodies (acetoacetate plus 3-hydroxybutyrate) above 0.2 mmol/l, even in the fed state [11]. A few patients do not have ketonuria when they are well. If a diagnostic fast is undertaken, there is an excessive rise in blood ketone body levels, sometimes to over 10 mmol/l, without hypoglycaemia [11]. Urinary organic acid analysis reveals high concentrations of 3-hydroxybutyrate, acetoacetate and sometimes 3-hydroxyisovalerate but no specific abnormalities. The diagnosis must, therefore, be confirmed by enzymology, which can be undertaken on lymphocytes or cultured fibroblasts. These assays generally show at least 20–35% apparent residual activity, due to the consumption of substrate by other enzymes [16].

14.4.4 Mitochondrial Acetoacetyl-CoA Thiolase Deficiency

Patients with T2 deficiency usually excrete increased amounts of 2-methylacetoacetate, 2-methyl-3-hydroxybutyric acid and tiglylglycine (◘ Fig. 14.1). Some patients, however, do not excrete tiglylglycine and 2-methylacetoacetate may not be detected because it is unstable. Indeed, the organic aciduria may be hard to detect even during episodes of ketoacidosis and an isoleucine load may be needed to demonstrate the abnormalities [24]. A similar organic acid picture can be seen in other disorders, such as 2-methyl-3-hydroxybutyryl-CoA dehydrogenase deficiency, but 2-methyacetoacetate is not excreted in the latter disorder (Chap. 19). Enzymological confirmation in cultured fibroblasts is, therefore, essential. Assays are complicated by the presence of 3 other thiolases that act on acetoacetyl-CoA (cytosolic and peroxisomal acetoacetyl-CoA thiolases and mitochondrial 3-oxoacyl-CoA thiolase). 2-Methylacetoacetyl-CoA is a specific substrate for T2 but it is difficult to synthesise [25]. One solution is to measure acetoacetyl-CoA thiolysis in the presence and absence of potassium, which enhances the activity of T2 but not the other enzymes [14]. Blood acylcarnitine analysis in T2 deficiency often reveals tiglylcarnitine and other abnormalities but the sensitivity and specificity are not clear [23].

14.5 Treatment and Prognosis

All these patients can decompensate rapidly in early childhood. To prevent this, fasting must be avoided and a high carbohydrate intake must be maintained during any metabolic stress, such as surgery or infection (▶ Chap. 4). Drinks containing carbohydrate should be started at the first sign of illness; hospital admission is needed if these are not tolerated or if a patient with a ketolysis disorder develops moderate or heavy ketonuria. In hospital, patients require an intravenous infusion of glucose. Acidosis is common in HMG-CoA lyase deficiency and, particularly, in the ketolysis disorders. An intravenous infusion of bicarbonate is needed if the acidosis is severe (pH<7.1) and it may be given in milder acidosis but electrolytes must be monitored frequently: there is a risk of severe and potentially fatal hypernatraemia. Extra fluids may be needed to correct dehydration, which is common in the ketolysis disorders.

A moderate protein restriction is usually recommended in HMG-CoA lyase, SCOT and T2 deficiencies, because these enzymes are directly or indirectly involved in amino acid catabolism [4, 11, 12]. A low fat diet has also been recommended [26]. Protein and fat should certainly be avoided during illness. At other times, however, dietary restriction is unnecessary in some patients [13, 16]. Carnitine supplements are often given if serum levels are low, though their value is unproven.

Patients with these disorders can die or suffer irreversible neurological damage during episodes of metabolic decompensation. Outcomes have been least good for neonatal-onset cases of HMG-CoA lyase deficiency, such as those from Saudi Arabia [6]. Once the diagnosis has been made, the outlook is much improved. Patients become more stable with age, particularly those with ketolysis defects [12]. Late complications are rare. Fatal cardiomyopathy has been reported in one patient with HMG-CoA lyase deficiency [27] and two pedigrees with T2 deficiency, though the latter cases were not enzymologically proven [28]. One patient with T2 deficiency has had a healthy child following an uncomplicated pregnancy [29].

14.6 Cytosolic Acetoacetyl-CoA Thiolase Deficiency

Cytosolic acetoacetyl-CoA thiolase (CAT) is primarily involved in the synthesis of isoprenoid compounds, such as cholesterol, rather than ketone body metabolism. Two patients with CAT deficiency have been reported [30, 31]. Both presented with mental retardation after apparently normal early development. One patient developed severe ketoacidosis on a ketogenic diet, whilst the other had persistent ketonuria that resolved on a low fat diet. No treatment had any effect on the neurological problems. One patient had low acetoacetyl-CoA thiolase activity in the cytosolic fraction from a liver biopsy. Fibroblasts from the other patient showed reduced total acetoacetyl-CoA thiolase activity with normal T2 activity, and decreased cholesterol synthesis. The human CAT cDNA has been cloned but mutations have not been reported [32].

References

1. Thompson GN, Hsu BY, Pitt JJ et al (1997) Fasting hypoketotic coma in a child with deficiency of mitochondrial 3-hydroxy-3-methylglutaryl-CoA synthase. N Engl J Med 337:1203-1207

2. Morris AA, Lascelles CV, Olpin SE et al (1998) Hepatic mitochondrial 3-hydroxy-3-methylglutaryl-coenzyme a synthase deficiency. Pediatr Res 44:392-396

3. Zschocke J, Penzien JM, Bielen R et al (2002) The diagnosis of mitochondrial HMG-CoA synthase deficiency. J Pediatr 140:778-780

4. Gibson KM, Breuer J, Nyhan WL (1988) 3-Hydroxy-3-methylglutaryl-coenzyme A lyase deficiency: review of 18 reported patients. Eur J Pediatr 148:180-186

5. van der Knaap MS, Bakker HD, Valk J (1998) MR imaging and proton spectroscopy in 3-hydroxy-3-methylglutaryl coenzyme A lyase deficiency. AJNR Am J Neuroradiol 19:378-382

6. Ozand PT, al Aqeel A, Gascon G et al (1991) 3-Hydroxy-3-methylglutaryl-coenzyme A (HMG-CoA) lyase deficiency in Saudi Arabia. J Inherit Metab Dis 14:174-188

7. Leonard JV, Seakins JW, Griffin NK (1979) beta-Hydroxy-beta-methylglutaricaciduria presenting as Reye's syndrome. Lancet 1:680

8. Wilson WG, Cass MB, Sovik O et al (1984) A child with acute pancreatitis and recurrent hypoglycemia due to 3-hydroxy-3-methylglutaryl-CoA lyase deficiency. Eur J Pediatr 142:289-291

9. Koper JW, Lopes-Cardozo M, van Golde LM (1981) Preferential utilization of ketone bodies for the synthesis of myelin cholesterol in vivo. Biochim Biophys Acta 666:411-417

10. Berry GT, Fukao T, Mitchell GA et al (2001) Neonatal hypoglycaemia in severe succinyl-CoA: 3-oxoacid CoA-transferase deficiency. J Inherit Metab Dis 24:587-595

11. Saudubray JM, Specola N, Middleton B et al (1987) Hyperketotic states due to inherited defects of ketolysis. Enzyme 38:80-90

12. Fukao T, Scriver CR, Kondo N (2001) The clinical phenotype and outcome of mitochondrial acetoacetyl-CoA thiolase deficiency (beta-ketothiolase or T2 deficiency) in 26 enzymatically proved and mutation-defined patients. Mol Genet Metab 72:109-114

13. Schutgens RB, Middleton B, vd Blij JF et al (1982) Beta-ketothiolase deficiency in a family confirmed by in vitro enzymatic assays in fibroblasts. Eur J Pediatr 139:39-42

14. Robinson BH, Sherwood WG, Taylor J et al (1979) Acetoacetyl CoA thiolase deficiency: a cause of severe ketoacidosis in infancy simulating salicylism. J Pediatr 95:228-233

15. Ozand PT, Rashed M, Gascon GG et al (1994) 3-Ketothiolase deficiency: a review and four new patients with neurologic symptoms. Brain Dev 16[Suppl]:38-45

16. Kassovska-Bratinova S, Fukao T, Song XQ et al (1996) Succinyl CoA: 3-oxoacid CoA transferase (SCOT): human cDNA cloning, human chromosomal mapping to 5p13, and mutation detection in a SCOT-deficient patient. Am J Hum Genet 59:519-528

17. Mitchell GA, Ozand PT, Robert MF et al (1998) HMG CoA lyase deficiency: identification of five causal point mutations in codons 41 and 42, including a frequent Saudi Arabian mutation, R41Q. Am J Hum Genet 62:295-300

18. Casale CH, Casals N, Pie J et al (1998) A nonsense mutation in the exon 2 of the 3-hydroxy-3-methylglutaryl coenzyme A lyase (HL) gene producing three mature mRNAs is the main cause of 3-hydroxy-3-methylglutaric aciduria in European Mediterranean patients. Arch Biochem Biophys 349:129-137

19. Mitchell GA, Jakobs C, Gibson KM et al (1995) Molecular prenatal diagnosis of 3-hydroxy-3-methylglutaryl CoA lyase deficiency. Prenat Diagn 15:725-729

20. Fukao T, Wakazono A, Song XQ et al (1995) Prenatal diagnosis in a family with mitochondrial acetoacetyl-coenzyme A thiolase deficiency with the use of the polymerase chain reaction followed by the heteroduplex detection method. Prenat Diagn 15:363-367

21. Fukao T, Song XQ, Watanabe H et al (1996) Prenatal diagnosis of succinyl-coenzyme A:3-ketoacid coenzyme A transferase deficiency. Prenat Diagn 16:471-474

22. Hammond J, Wilcken B (1984) 3-hydroxy-3-methylglutaric, 3-methylglutaconic and 3-methylglutaric acids can be non-specific indicators of metabolic disease. J Inherit Metab Dis 7[Suppl 2]:117-118

23. Rashed MS, Ozand PT, Bucknall MP et al (1995) Diagnosis of inborn errors of metabolism from blood spots by acylcarnitines and amino acids profiling using automated electrospray tandem mass spectrometry. Pediatr Res 38:324-331

24. Middleton B, Bartlett K, Romanos A et al (1986) 3-Ketothiolase deficiency. Eur J Pediatr 144:586-589

25. Middleton B, Bartlett K (1983) The synthesis and characterisation of 2-methylacetoacetyl coenzyme A and its use in the identification of the site of the defect in 2-methylacetoacetic and 2-methyl-3-hydroxybutyric aciduria. Clin Chim Acta 128:291-305

26. Thompson GN, Chalmers RA, Halliday D (1990) The contribution of protein catabolism to metabolic decompensation in 3-hydroxy-3-methylglutaric aciduria. Eur J Pediatr 149:346-350

27. Gibson KM, Cassidy SB, Seaver LH et al (1994) Fatal cardiomyopathy associated with 3-hydroxy-3-methylglutaryl-CoA lyase deficiency. J Inherit Metab Dis 17:291-294

28. Henry CG, Strauss AW, Keating JP et al (1981) Congestive cardiomyopathy associated with beta-ketothiolase deficiency. J Pediatr 99:754-757

29. Sewell AC, Herwig J, Wiegratz I et al (1998) Mitochondrial acetoacetyl-CoA thiolase (beta-ketothiolase) deficiency and pregnancy. J Inherit Metab Dis 21:441-442

30. de Groot CJ, Haan GL, Hulstaert CE et al (1977) A patient with severe neurologic symptoms and acetoacetyl-CoA thiolase deficiency. Pediatr Res 11:1112-1116

31. Bennett MJ, Hosking GP, Smith MF et al (1984) Biochemical investigations on a patient with a defect in cytosolic acetoacetyl-CoA thiolase, associated with mental retardation. J Inherit Metab Dis 7:125-128

32. Song XQ, Fukao T, Yamaguchi S et al (1994) Molecular cloning and nucleotide sequence of complementary DNA for human hepatic cytosolic acetoacetyl-coenzyme A thiolase. Biochem Biophys Res Commun 201:478-485

15 Defects of the Respiratory Chain

Arnold Munnich

The Respiratory Chain

The respiratory chain is divided into five functional units or complexes embedded in the inner mitochondrial membrane. *Complex I* [reduced nicotinamide adenine dinucleotide (NADH)-coenzyme Q (CoQ) reductase] carries reducing equivalents from NADH to CoQ and consists of 25-28 different polypeptides, seven of which are encoded by mitochondrial DNA. *Complex II* (succinate-CoQ reductase) carries reducing equivalents from reduced flavin adenine dinucleotide ($FADH_2$) to CoQ and contains five polypeptides, including the flavin adenine dinucleotide-dependent succinate dehydrogenase and a few non-heme-iron-sulfur centers. *Complex III* (reduced-CoQ-cytochrome-c reductase) carries electrons from CoQ to cytochrome c and contains 11 subunits. *Complex IV* (cytochrome-c oxidase) catalyzes the transfer of reducing equivalents from cytochrome c to molecular oxygen. It is composed of two cytochromes (a and a_3), two copper atoms, and 13 different protein subunits.

The respiratory chain catalyzes the oxidation of fuel molecules by oxygen and the concomitant energy transduction into ATP. During the oxidation process, electrons are transferred to oxygen via the energy-transducing complexes: complexes I, III, and IV for succinate and complexes III and IV for $FADH_2$ derived from the β-oxidation pathway via the electron-transfer flavoprotein (ETF) and the ETF-CoQ oxidoreductase system. CoQ (a lipoidal quinone) and cytochrome c (a low-molecular-weight hemoprotein) act as shuttles between the complexes.

The flux of electrons is coupled to the translocation of protons (H^+) into the intermembrane space at three coupling sites (complexes I, III, and IV). This creates a transmembrane gradient. *Complex V* (ATP synthase) allows protons to flow back into the mitochondrial matrix and uses the released energy to synthesize ATP. Three molecules of ATP are generated for each molecule of NADH oxidized.

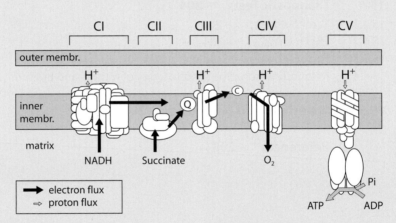

◘ **Fig. 15.1.** The mitochondrial respiratory chain. *ADP*, adenosine diphosphate; *ATP*, adenosine triphosphate; *c*, cytochrome c; *CI*, complex I (NADH-coenzyme-Q reductase); *CII*, complex II (succinate-coenzyme-Q reductase); *CIII*, complex III (reduced-coenzyme-Q-cytochrome-c reductase); *CIV*, complex IV (cytochrome-c oxidase); *CV*, complex V (ATP synthase); *NADH*, reduced nicotinamide adenine dinucleotide; *Pi*, inorganic phosphate; *Q*, coenzyme Q

Respiratory chain deficiencies have long been regarded as neuromuscular diseases only. However, *oxidative phosphorylation* (i.e. ATP synthesis by the respiratory chain) is not restricted to the neuromuscular system but proceeds in all cells that contain mitochondria (❏ Fig. 15.1). Most non-neuromuscular organs and tissues are, therefore, also dependent upon mitochondrial energy supply. Due to the twofold genetic origin of respiratory enzymes [nuclear DNA and mitochondrial (mtDNA)], a respiratory chain deficiency can theoretically give rise to any symptom in any organ or tissue at any age and with any mode of inheritance.

The diagnosis of a respiratory chain deficiency is difficult to consider initially when only one abnormal symptom is present. In contrast, this diagnosis is easier to consider when two or more seemingly unrelated symptoms are observed. The treatment, mainly dietetic, does not markedly influence the usually unfavorable course of the disease.

15.1   Clinical Presentation

Due to the ubiquitous nature of oxidative phosphorylation, a defect of the mitochondrial respiratory chain should be considered in patients presenting *(1)* with an unexplained association of neuromuscular and/or non-neuromuscular symptoms, *(2)* with a rapidly progressive course, and *(3)* with symptoms involving seemingly unrelated organs or tissues. The disease may begin at virtually any age. ❏ Table 15.1 summarizes the most frequently observed symptoms. Whatever the age of onset and the presenting symptom, the major feature is the increasing number of tissues affected in the course of the disease. This progressive organ involvement is constant, and the central nervous system is almost consistently involved in the late stage of the disease.

While the initial symptoms usually persist and gradually worsen, they may occasionally improve or even disappear as other organs become involved. This is particularly true for bone marrow and gut. Indeed, remarkable remissions of pancytopenia or watery diarrhea have been reported in infants who later developed other organ involvements. Moreover, several patients whose disease apparently started in childhood or adulthood were retrospectively shown to have experienced symptoms (transient sideroblastic anemia, neutropenia, chronic watery diarrhea, or failure to thrive) of unexplained origin in early infancy. Similarly, a benign reversible infantile myopathy with hypotonia, weakness, macroglossia, respiratory distress, and spontaneous remission within 12 years has been described.

Certain clinical features or associations are more frequent at certain ages and have occasionally been identified as distinct entities, suggesting that these associations are not fortuitous. However, considerable overlap in clinical features leads to difficulties in the classification of many patients, and the nature, clinical course, and severity of symptoms vary among (and even within) affected individuals. It is more useful to bear in mind that the diagnosis of respiratory chain deficiency should be considered regardless of the age of onset and the nature of the presenting symptom when presenting with an unexplained association of signs with a progressive course involving seemingly unrelated organs or tissues. The non-exhaustive list of clinical profiles listed below illustrates the diversity of presentations (❏ Table 15.1).

15.1.1   Fetuses

Intrauterine growth retardation (below 3rd percentile for gestational age) either isolated or associated with otherwise unexplained antenatal anomalies is retrospectively detected in 20% of respiratory enzyme chain deficient patients. Antenatal anomalies are usually multiple and include polyhydramnios, oligoamnios, arthrogryposis, decreased fetal movements, ventricular septal defects, hypertrophic cardiomyopathy, vertebral and limb defects or other visceral anomalies (VACTERL association). At variance with a number of metabolic diseases which have a symptom-free period, respiratory chain deficiency may have an antenatal expression related to the time course of the disease gene expression in the embryofetal period [1].

15.1.2   Neonates

In the neonate (age less than 1 month), the following clinical profiles are seen:
- Ketoacidotic coma with recurrent apneas, seizures, severe hypotonia, liver enlargement, and proximal tubulopathy, with or without a symptom-free period [2].
- Severe sideroblastic anemia (with or without hydrops fetalis), with neutropenia, thrombocytopenia, and exocrine pancreatic dysfunction of unexplained origin (Pearson marrow-pancreas syndrome) [3].
- Concentric hypertrophic cardiomyopathy and muscle weakness with an early onset and a rapidly progressive course (dilated cardiomyopathies are exceptional) [4].
- Concentric hypertrophic cardiomyopathy with profound central neutropenia and myopathic features in males (Barth syndrome) [5].
- Hepatic failure with lethargy, hypotonia, and proximal tubulopathy of unexplained origin [2].

□ Table 15.1. The most frequently observed symptoms in defects of the respiratory chain

Neonatal period (0–1 month)
Central nervous system
Iterative apnea, lethargy, drowsiness, near-miss
Limb and trunk hypotonia
Congenital lactic acidosis
Ketoacidotic coma
Muscle
Myopathic presentation
Muscular atrophy, hypotonia
Stiffness, hypertonia
Recurrent myoglobinuria
Poor head control, poor spontaneous movement
Liver
Hepatic failure, liver enlargement
Heart
Hypertrophic cardiomyopathy (concentric)
Kidney
Proximal tubulopathy (De Toni-Debré-Fanconi syndrome)
Infancy (1 month–2 years)
Central nervous system
Recurrent apneas, near-miss
Recurrent ketoacidotic coma
Poor head control, limb spasticity
Psychomotor regression, mental retardation
Cerebellar ataxia
»Stroke-like« episodes
Myoclonus, generalized seizures
Subacute necrotizing encephalomyopathy (Leigh syndrome)
Progressive infantile poliodystrophy (Alpers syndrome)
Muscle
Myopathic features
Muscular atrophy
Limb weakness, hypotonia
Myalgia, exercise intolerance
Recurrent myoglobinuria
Liver
Progressive liver enlargement
Hepatocellular dysfunction
Valproate-induced hepatic failure
Heart
Hypertrophic cardiomyopathy (concentric)
Kidney
Proximal tubulopathy (De Toni-Debré-Fanconi syndrome)
Tubulo-interstitial nephritis (mimicking nephronophtisis)
Nephrotic syndrome
Renal failure
Hemolytic uremic syndrome
Gut
Recurrent vomiting
Chronic diarrhea, villous atrophy
Exocrine pancreatic dysfunction
Failure to thrive
Chronic interstitial pseudo-obstruction
Endocrine
Short stature, retarded skeletal maturation
Recurrent hypoglycemia
Multiple hormone deficiency

Bone marrow
Sideroblastic anemia
Neutropenia, thrombopenia
Myelodysplastic syndrome, dyserythropoiesis
Ear
Hearing loss
Sensorineural deafness (brain stem or cochlear origin)
Eye
Optic atrophy
Diplopia
Progressive external ophthalmoplegia
Limitation of eye movements (all directions, upgaze)
»Salt-and-pepper« retinopathy, pigmentary retinal degeneration
Lid ptosis
Cataract
Skin
Mottled pigmentation of photo-exposed areas
Trichothiodystrophy
Dry, thick and brittle hair
Childhood (>2 years) and adulthood
Central nervous system
Myoclonus
Seizures (generalized, focal, drop attacks, photo-sensitivity, tonicoclonus)
Cerebellar ataxia
Spasticity
Psychomotor regression, dementia, mental retardation
»Stroke-like« episodes
Hemicranial headache, migraine
Recurrent hemiparesis, cortical blindness or hemianopsia
Leukodystrophy, cortical atrophy
Peripheral neuropathy
Muscle
Progressive myopathy
Limb weakness (proximal)
Myalgia, exercise intolerance
Recurrent myoglobinuria
Heart
Concentric hypertrophic or dilated cardiomyopathy
Different types of heart block
Endocrine
Diabetes mellitus (insulin- and non-insulin dependent)
Growth-hormone deficiency
Hypoparathyroidism
Hypothyroidism
Adrenocorticotrophin deficiency
Hyperaldosteronism
Infertility (ovarian failure or hypothalamic dysfunction)
Eye
Lid ptosis
Diplopia
Progressive external ophthalmoplegia
Limitation of eye movements (all directions, upgaze)
»Salt-and-pepper« retinopathy, pigmentary retinal degeneration
Cataract, corneal opacities
Leber hereditary optic neuroretinopathy
Ear
Sensorineural deafness
Aminoglycoside-induced ototoxicity (maternally inherited)

15.1.3 Infants

In infancy (1 month to 2 years), the clinical profiles include the following:

- Failure to thrive, with or without chronic watery diarrhea and villous atrophy; unresponsiveness to gluten-free and cow's milk protein-free diet [6].
- Recurrent episodes of acute myoglobinuria, hypertonia, muscle stiffness and elevated plasma levels of enzymes unexplained by an inborn error of glycolysis, glycogenolysis, fatty acid oxidation or muscular dystrophy [7].
- Proximal tubulopathy (de Toni-Debré-Fanconi syndrome) with recurrent episodes of watery diarrhea, rickets and mottled pigmentation of photo-exposed areas.
- A tubulo-interstitial nephritis mimicking nephronophtisis, with the subsequent development of renal failure and encephalomyopathy with leukodystrophy [8].
- Severe trunk and limb dwarfism unresponsive to growth-hormone administration, with subsequent hypertrophic cardiomyopathy, sensorineural deafness, and retinitis pigmentosa.
- Early-onset insulin-dependent diabetes mellitus with diabetes insipidus, optic atrophy, and deafness (Wolfram syndrome) [9].
- Rapidly progressive encephalomyopathy with hypotonia, poor sucking, weak crying, poor head control, cerebellar ataxia, pyramidal syndrome, psychomotor regression, developmental delay, muscle weakness, and respiratory insufficiency; occasionally associated with proximal tubulopathy and/or hypertrophic cardiomyopathy.
- Subacute necrotizing encephalomyopathy (Leigh's disease). This is a devastating encephalopathy characterized by recurrent attacks of psychomotor regression with pyramidal and extrapyramidal symptoms, leukodystrophy, and brainstem dysfunction (respiratory abnormalities). The pathological hallmark consists of focal, symmetrical, and necrotic lesions in the thalamus, brain stem, and the posterior columns of the spinal cord. Microscopically, these spongiform lesions show demyelination, vascular proliferation, and astrocytosis [10].

15.1.4 Children and Adults

In childhood (above 2 years) and adulthood, the neuro-muscular presentation is the most frequent:

- Muscle weakness with myalgia and exercise intolerance, with or without progressive external ophthalmoplegia [10].
- Ataxia, cerebellar atrophy, muscle weakness, seizures and mental retardation.
- Progressive sclerosing poliodystrophy (Alpers disease) associated with hepatic failure [11].

- Encephalomyopathy with myoclonus, ataxia, hearing loss, muscle weakness, and generalized seizures (myoclonus epilepsy, ragged red fibers, MERRF) [10].
- Progressive external ophthalmoplegia (PEO) ranging in severity from pure ocular myopathy to Kearns-Sayre syndrome (KSS). KSS is a multisystem disorder characterized by the triad (1) onset before age 20 years, (2) PEO, and (3) pigmentary retinal degeneration, plus at least one of the following: complete heart block, cerebrospinal fluid (CSF) protein levels above 100 mg/dl, or cerebellar ataxia [10].
- Mitochondrial encephalomyopathy with lactic acidosis and stroke-like episodes (MELAS). This syndrome is characterized by onset in childhood, with intermittent hemicranial headache, vomiting, proximal limb weakness, and recurrent neurological deficit resembling strokes (hemiparesis, cortical blindness, hemianopsia), lactic acidosis, and ragged red fibers (RRFs) in the muscle biopsy. Computed tomography (CT) brain scanning shows low-density areas (usually posterior), which may occur in both white and gray matter but do not always correlate with the clinical symptoms or the vascular territories. The pathogenesis of stroke-like episodes in MELAS has been ascribed to either cerebral blood-flow disruption or acute metabolic decompensation in biochemically deficient areas of the brain [10].
- Leber's hereditary optic neuroretinopathy (LHON). This disease is associated with rapid loss of bilateral central vision due to optic nerve death. Cardiac dysrythmia is frequently associated with the disease, but no evidence of skeletal muscle pathology or gross structural mitochondrial abnormality has been documented. The median age of vision loss is 20-24 years, but it can occur at any age between adolescence and late adulthood. Expression among maternally related individuals is variable, and more males are affected [10]. Kjer's autosomal dominant optic atrophy is also caused by a mitochondrial dysfunction in the dynamin-related protein OPA1 [12].
- Neurogenic muscle weakness, ataxia, retinitis pigmentosa (NARP) and variable sensory neuropathy with seizures and mental retardation or dementia [13].
- Mitochondrial myopathy and peripheral neuropathy, encephalopathy, and gastrointestinal disease manifesting as intermittent diarrhea and intestinal pseudo-obstruction (myo-neuro-gastro-intestinal encephalopathy, MNGIE) [10].
- Progressive multisystemic failure with encephalopathy, myopathy, peripheral neuropathy and renal failure.

15.2 Metabolic Derangement

As the respiratory chain transfers electrons to oxygen, a disorder of oxidative phosphorylation should result in *(1)* an increase in the concentration of reducing equivalents in both mitochondria and cytoplasm and (2) the functional impairment of the citric acid cycle, due to the excess of reduced nicotinamide adenine dinucleotide (NADH) and the lack of nicotinamide adenine dinucleotide (NAD). Therefore, an increase in the ketone body (3-hydroxybutyrate/acetoacetate) and lactate/pyruvate (L/P) molar ratios with a secondary elevation of blood lactate might be expected in the plasma of affected individuals. This is particularly true in the post-absorptive period, when more NAD is required to adequately oxidize glycolytic substrates.

Similarly, as a consequence of the functional impairment of the citric acid cycle, ketone body synthesis increases after meals, due to the channeling of acetyl-coenzyme A (CoA) towards ketogenesis. The elevation of the total level of ketone bodies in a fed individual is paradoxical, as it should normally decrease after meals, due to insulin release (paradoxical hyperketonemia).

The position of the block might differentially alter the metabolic profile of the patient. At the level of complex I it impairs the oxidation of the 30 moles of NADH formed in the citric acid cycle. In theory at least, oxidation of reduced flavin adenine dinucleotide ($FADH_2$) derived from succinate producing substrates (methionine, threonine, valine, isoleucine, and odd-numbered fatty acids) should not be altered, because it is mediated by complex II. Similarly, oxidation of $FADH_2$ derived from the first reaction of the β-oxidation pathway should occur normally, because it is mediated by the electron transfer flavoprotein coenzyme-Q-reductase system. However, complex II deficiency should not markedly alter the redox status of affected individuals

fed a carbohydrate-rich diet. A block at the level of complex III should impair the oxidation of both NAD-linked and FAD-linked substrates. Finally, given the crucial role of complex IV in the respiratory chain, it is not surprising that severe defects of cytochrome c oxidase (COX) activity cause severe lactic acidosis and markedly alter redox status in plasma.

15.3   Genetics

Any mode of inheritance may be observed in mitochondrial diseases: autosomal recessive, dominant, X-linked, maternal, or sporadic. This variability is due to the high number of genes that encode the respiratory chain proteins, 13 of which are located in the mitochondria. mtDNA encodes seven polypeptides of complex I, one of complex III (the apoprotein of cytochrome b), three of complex IV, and two of complex V. The mtDNA molecules are small (16.5kb), double-stranded, circular, and contain no introns (⬛ Fig.15.2). mtDNA has a number of unique genetic features:

- It is maternally inherited, and its mutations are, therefore, transmitted by the mother.
- It has a very high mutation rate involving both nucleotide substitutions and insertions/deletions.

During cell division, mitochondria are randomly partitioned into daughter cells. This means that, if normal and mutant mtDNA molecules are present in the mother's cells (heteroplasmy), some lineages will have only abnormal

⬛ **Fig. 15.2.** Map of the mitochondrial genome. Regions encoding cytochrome b (*cyt b*), various subunits of reduced nicotinamide adenine dinucleotide-coenzyme Q reductase (*ND*), cytochrome oxidase (*COX*), and adenosine triphosphatase (*A*), and rRNAs are indicated. Replication of the heavy strand starts in the displacement (*D*) loop at the heavy-strand origin (*OH*), and that of the light strand at OL

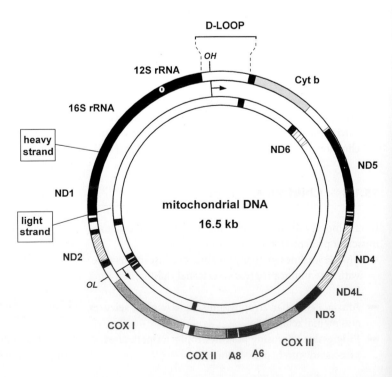

mtDNA (homoplasmy), others will have only normal mtDNA (wild type), and still others will have both normal and abnormal mtDNA. In these last cells, the phenotype will reflect the proportion of abnormal mtDNA.

The nuclear genome encodes a number of proteins involved in mtDNA replication and transcription, the protein components of the mitochondrial ribosome, multiple structural, chaperone and assembly proteins, and the remaining catalytic subunits of the respiratory enzyme complexes (other than the 13 mtDNA-encoded subunits).

15.3.1 Mutations in Mitochondrial DNA

Pathological alterations of mtDNA fall into three major classes: point mutations, rearrangements, and depletions of the number of copies.

- Point mutations result in amino acid substitutions and modifications of mRNA and tRNA. Most are heteroplasmic, maternally inherited, and associated with a striking variety of clinical phenotypes (LHON, MERRF, MELAS, NARP, Leigh syndrome, diabetes, and deafness) [14]. Interestingly, there are mutation hot spots on the mitochondrial genome as recurrent de novo mutations at specific nucleotide positions have been reported in unrelated families (ND1: G3946A; ND3: T10158C, T10191C; ND5: T12706C, G13094A, G13513A, A13514G; ND6: T14487C) [14].
- Rearrangements comprise deletions/duplications that markedly differ in size and position from patient to patient but usually encompass several coding and tRNA genes. They are usually sporadic, heteroplasmic, and unique and probably arise de novo during oogenesis or in early development (KSS, Pearson syndrome, PEO, diabetes, and deafness [14]). Occasionally, maternally transmitted mtDNA rearrangements are found [7]. In other cases, autosomal dominant transmission of multiple mtDNA deletions occurs, suggesting mutation of a nuclear gene essential for the function of the mitochondrial genome [15]. Three disease causing genes have been hitherto reported in autosomal dominant multiple mtDNA deletions, namely POLG, ANT1 and Twinkle, encoding for the mitochondrial DNA polymerase γ, the ADP/ATP translocator and a mitochondrial helicase, respectively [16-18].
- Decreased mtDNA copy number (mtDNA depletion) is a recently identified group of respiratory chain deficiencies [19]. Depletions are genetically heterogeneous autosomal recessive conditions involving either a single organ or multiple tissues. The disease causing genes identified to date in mtDNA depletion syndromes include POLG and two key enzymes in the mitochondrial nucleotide salvage pathway, deoxyguanosine kinase (DGUOK) [20] and thymidine kinase 2 (TK2) [21]. These deoxyribonucleoside kinases provide mitochondria with deoxyribonucleotides essential for mtDNA synthesis. Mutations in DGUOK and POLG have been reported in early-onset hepatic failure and encephalopathy and TK2 mutation have been described in neonatal-onset devastating skeletal myopathies (► also Chap. 35).

15.3.2 Mutations in Nuclear DNA

A few of the numerous disease-causing nuclear genes have been recently identified, including the gene of Barth syndrome (tafazzin) [22], nuclear genes for complex-I (NDUFV1, NDUFV2, NDUFS1, NDUFS3, NDUFS4, NDUFS6, NDUFS7) [43–48], complex-II (Fp subunit of succinate dehydrogenase) [29-30], and complex-IV (SURF-1, SCO1, SCO2, COX10, COX15) [31–36] deficiencies, a MNGIE gene (thymidine phosphorylase) [37], and a gene encoding coenzyme Q_{10} [37a].

15.3.3 Genetic Analysis of Respiratory Chain Deficiencies

An extensive family history, with documentation of minor signs in relatives, is of paramount importance in recognizing the mode of inheritance and in deciding on the molecular studies to be performed. Maternal inheritance indicates mtDNA mutations, autosomal dominant inheritance indicates multiple mtDNA deletions, and sporadic cases and autosomal recessive inheritance (consanguineous parents) indicate mtDNA deletions/duplications and nuclear gene mutations, respectively. Investigations require a highly specialized, experienced laboratory and should take into account the following points:

- The distribution of mutated mtDNA molecules may differ widely among tissues, accounting for the variable clinical expression and requiring investigation of the tissue that actually expresses the disease.
- mtDNA rearrangements are unstable and gradually disappear in cultured cells unless uridine is included in the culture medium, thus precluding growth under standard conditions.
- Negative results neither rule out an mtDNA mutation nor provide a clue that a nuclear mutation is involved.
- Although no clear-cut correlations between phenotypes and genotypes have been identified, certain clinical presentations hint at mutations in particular genes: Leigh syndrome with NARP, SURF1, SDH-Fp and ND1-5 mutations; Alpers syndrome with POLG [38] or DGUOK mutations; MINGIE with thymidine phosphorylase mutations, progressive external ophthalmoplegia with POLG, Twinkle or ANT1 mutations; cardioencephalomyopathy with SCO2 mutations; diabetes mellitus and deafness with MELAS mutations; Pearson syndrome and Kearns-Sayre syndrome with mtDNA deletions.

15.3.4   Genetic Counseling and Prenatal Diagnosis

The identification of certain clinical phenotypes, listed above, allows some prediction with respect to their inheritance. Moreover, it should be borne in mind that, in cases of maternal inheritance of a mtDNA mutation, risk is absent for the progeny of an affected male but is high for that of a carrier female. In this case, determination of the proportion of mutant mtDNA on chorionic villi or amniotic cells is a rational approach. Nevertheless, its predictive value remains uncertain, owing to incomplete knowledge of the tissue distribution of abnormal mtDNA, its change during development, and its quantitative relationship to disease severity.

In the absence of detectable mtDNA mutations, the measurement of the activities of respiratory enzymes in cultured amniocytes or choriocytes provides the only possibility of prenatal diagnosis, particularly since few nuclear mutations have been identified. Unfortunately, relatively few enzyme deficiencies are expressed in cultured fibroblasts of probands, even when grown with uridine. For this reason, the ongoing identification of disease-causing nuclear genes will certainly help in delivering accurate prenatal diagnoses of respiratory chain deficiencies in the future.

15.4   Diagnostic Tests

15.4.1   Screening Tests

Screening tests include the determination of lactate, pyruvate, ketone bodies, and their molar ratios in plasma as indices of oxidation/reduction status in cytoplasm and mitochondria, respectively (◨ Table 15.2). Determinations should be made before and 1 h after meals throughout the day. Blood glucose and non-esterified fatty acids should be simultaneously monitored (▸ Chap. 3). The observation of a persistent hyperlactatemia (>2.5 mM) with elevated L/P and ketone body molar ratios (particularly in the postabsorptive period) is highly suggestive of a respiratory chain deficiency. In addition, investigation of the redox status in plasma can help discriminate between the different causes of congenital lactic acidosis based on L/P and ketone body molar ratios in vivo [39]. Indeed, an impairment of oxidative phosphorylation usually results in L/P ratios above 20 and ketone body ratios above 2, whereas a defect of the pyruvate dehydrogenase (PDH) complex results in low L/P ratios (<10). Although little is known regarding tricarboxylic acid cycle disorders, it appears that these diseases also result in high L/P ratios, but ketone body molar ratios are lower in these conditions (<1) than in respiratory-chain defects (as also observed in pyruvate carboxylase deficiency; ▸ Chap.12) [40–41]. However, these diagnostic tests may fail to detect any disturbance of the redox status in plasma. Pitfalls of metabolic screening are the following:

- Hyperlactatemia may be latent in basal conditions and may only be revealed by a glucose loading test (2g/kg orally) or by determination of the redox status in the CSF. The measurement of CSF lactate and L/P ratio is useless when the redox status in plasma is altered.
- Proximal tubulopathy may lower blood lactate and increase urinary lactate. In this case, gas chromatography mass spectrometry can detect urinary lactate and citric acid cycle intermediates.
- Diabetes mellitus may hamper the entry of pyruvate into the citric acid cycle.
- Tissue-specific isoforms may be selectively impaired, barely altering the redox status in plasma (this may be particularly true for hypertrophic cardiomyopathies).
- The defect may be generalized but partial; the more those tissues with higher dependencies on oxidative metabolism suffer (such as brain and muscle), the more the oxidation/reduction status in plasma is impaired.
- The defect may be confined to complex II, barely altering (in principle) the redox status in plasma.

When diagnostic tests are negative, the diagnosis of a respiratory chain deficiency may be missed, especially when only the presenting symptom is present. By contrast, the diagnosis is easier to consider when seemingly unrelated symptoms are observed. For this reason, the investigation of patients at risk (whatever the presenting symptom) includes the systematic screening of all target organs, as multiple organ involvement is an important clue to the diagnosis (◨ Table 15.2).

15.4.2   Enzyme Assays

The observation of an abnormal redox status in plasma and/or evidence of multiple organ involvement prompts one to carry out further enzyme investigations. These investigations include two entirely distinct diagnostic procedures that provide independent clues to respiratory-chain deficiencies: polarographic studies and spectrophotometric studies.

Polarographic studies consist of the measurement of oxygen consumption by mitochondria-enriched fractions in a Clark electrode in the presence of various oxidative substrates (malate with pyruvate, malate with glutamate, succinate, palmitate, etc.). In the case of complex I deficiency, polarographic studies show impaired respiration with NADH-producing substrates, whilst respiration and phosphorylation are normal with FADH-producing substrates (succinate). The opposite is observed in the case of complex II deficiency, whereas a block at the level of complexes III or IV impairs oxidation of both NADH- and FADH-producing substrates. In complex V deficiency, res-

◻ **Table 15.2.** Screening of the respiratory chain

Standard screening tests (at least four determinations per day in fasted and 1-h-fed individuals)
  Plasma lactate
  Lactate/pyruvate molar ratio: redox status in cytoplasm
  Ketonemia (paradoxical elevation in fed individuals)
  β-hydroxy butyrate/acetoacetate molar ratio: redox status in the mitochondria
  Blood glucose and free fatty acids
  Urinary organic acids (GC-MS): lactate, ketone bodies, citric acid cycle intermediates

Provocative tests (when standard tests are inconclusive)
  Glucose test (2 g/kg orally) in fasted individuals, with determination of blood glucose, lactate, pyruvate, ketone bodies and their molar
  ratios just before glucose administration, and then every 30 min for 3–4 h (▶ Chap. 3)
  Lactate/pyruvate molar ratios in the CSF (only when no elevation of plasma lactate is observed)
  Redox status in plasma following exercise

Screening for multiple organ involvement
  Liver: hepatocellular dysfunction
  Kidney: proximal tubulopathy, distal tubulopathy, proteinuria, renal failure
  Heart: hypertrophic cardiomyopathy, heart block (ultrasound, ECG)
  Muscle: myopathic features (CK, ALAT, ASAT, histological anomalies, RRF)
  Brain: leukodystrophy, poliodystrophy, hypodensity of the cerebrum, cerebellum and the brainstem, multifocal areas of hyperintense
  signal (MELAS), bilateral symmetrical lesions of the basal ganglia and brain stem (Leigh) (EEG, NMR, CT scan)
  Peripheral nerve: distal sensory loss, hypo- or areflexia, distal muscle wasting (usually subclinical), reduced motor nerve conduction
  velocity (NCV) and denervation features (NCV, EMG, peripheral nerve biopsy showing axonal degeneration and myelinated-fiber loss)
  Pancreas: exocrine pancreatic dysfunction
  Gut: villous atrophy
  Endocrine: hypoglycemia, hypocalcemia, hypoparathyroidism, growth hormone deficiency (stimulation tests)
  Bone marrow: anemia, neutropenia, thrombopenia, pancytopenia, vacuolization of marrow precursors
  Eye: PEO, ptosis, optic atrophy, retinal degeneration (fundus, ERG, visually evoked potentials)
  Ear: sensorineural deafness (auditory evoked potentials, brain-stem-evoked response)
  Skin: trichothiodystrophy, mottled pigmentation of photo exposed areas

ALAT, alanine aminotransferase; *ASAT*, aspartate aminotransferase; *CK*, creatine kinase; *CT*, computed tomography; *ECG*, electrocardiogram; *EEG*, electroencephalogram; *EMG*, electromyogram; *ERG*, electroretinogram; *GC–MS*, gas chromatography–mass spectrometry;
MELAS, mitochondrial encephalopathy with lactic acidosis and stroke-like episodes; *NCV*, nerve conduction velocity ; *NMR*, nuclear
magnetic resonance; *PEO*, progressive external ophthalmoplegia; *RRF*, ragged red fiber

piration is impaired with various substrates, but adding the uncoupling agent 2,4-dinitrophenol or calcium ions returns the respiratory rate to normal, suggesting that the limiting step involves phosphorylation rather than the respiratory chain [42].

It is worth remembering that polarographic studies detect not only disorders of oxidative phosphorylation but also PDH deficiencies, citric acid cycle enzyme deficiencies, and genetic defects of carriers, shuttles, and substrates (including cytochrome c, cations, and adenylate), as these conditions also impair the production of reducing equivalents in the mitochondrion. In these cases, however, respiratory enzyme activities are expected to be normal.

While previous techniques required gram-sized amounts of muscle tissue, the scaled-down procedures available now allow the rapid recovery of mitochondria-enriched fractions from small skeletal muscle biopsies (100–200 mg, obtained under local anesthetic), thus making polarography feasible in infants and children [43]. Polarographic studies on intact circulating lymphocytes (isolated from 10 ml of blood on a Percoll cushion) or detergent-permeabilized cultured cells (lymphoblastoid cell lines, skin fibroblasts) are also feasible and represent a less invasive and easily reproducible diagnostic test [44]. The only limitation of these techniques is the absolute requirement of fresh material: no polarographic studies are possible on frozen material.

Spectrophotometric studies consist of the measurement of respiratory enzyme activities separately or in groups, using specific electron acceptors and donors. They do not require the isolation of mitochondrial fractions and can be carried out on tissue homogenates. For this reason, the amount of material required for enzyme assays is very small and can easily be obtained by needle biopsies of liver and kidney, and even by endomyocardial biopsies [4]. Similarly, a 25-ml flask of cultured skin fibroblasts or a lymphocyte pellet derived from a 10-ml blood sample are sufficient for extensive spectrophotometric studies. Samples should be frozen immediately and kept dry in liquid nitrogen (or at −80° C).

Particular attention should be given to apparently paradoxical cases were respiratory enzyme activities are separately normal but deficient when tested in groups (I–III,

III

II–III, III–V) as these are possible cases of coenzyme Q_{10} (CoQ_{10}) deficiency, a potentially treatable condition due to an inborn error of quinone synthesis [37a]. CoQ_{10} plays a pivotal role in the mitochondrial respiratory chain. It distributes the electrons between the various dehydrogenases and the cytochrome segments of the respiratory chain. It is in large excess compared to any other component of the respiratory chain and forms a kinetically compartmentalized pool, the redox status of which tightly regulates the activity of the dehydrogenases.

Since conclusive evidence of respiratory chain deficiency is given by enzyme assays, the question of which tissue should be investigated deserves particular attention. In principle, the relevant tissue is the one that clinically expresses the disease. When the skeletal muscle expresses the disease, the appropriate working material is a microbiopsy of the deltoid. When the hematopoietic system expresses the disease (i.e., Pearson syndrome), tests should be carried out on circulating lymphocytes, polymorphonuclear cells, or bone marrow. However, when the disease is predominantly expressed in the liver or heart, gaining access to the target organ is far less simple. Yet, a needle biopsy of the liver or an endomyocardial biopsy are usually feasible. If not, or when the disease is mainly expressed in a barely accessible organ (brain, retina, endocrine system, smooth muscle), peripheral tissues (including skeletal muscle, cultured skin fibroblasts, and circulating lymphocytes) should be extensively tested. Whichever the expressing organ, it is essential to take skin biopsies from such patients (even postmortem) for subsequent investigations on cultured fibroblasts.

It should be borne in mind, however, that the in vitro investigation of oxidative phosphorylation remains difficult regardless the tissue tested. Several pitfalls should be considered:

- A normal respiratory enzyme activity does not preclude mitochondrial dysfunction even when the tissue tested clinically expresses the disease. One might be dealing with a kinetic mutant, tissue heterogeneity, or cellular mosaicism (heteroplasmy; see below). In this case, one should carry out extensive molecular genetic analyses, test other tissues, and (possibly) repeat investigations later.
- Apart from overt misdiagnosis (i.e., confusion of Pearson syndrome with Schwachman syndrome) and false respiratory enzyme deficiencies (particularly common in non-expert centers), we are now aware of true secondary respiratory enzyme deficiency in: (1) other inborn errors of metabolism, namely propionic acidemia, TCA cycle disorders (fumarase deficiency), fatty acid oxidation disorders (long chain and 3-hydroxy long-chain acyl-CoA dehydrogenase deficiency) and mevalonate kinase deficiency); (2) primary central nervous system (CNS) disorders, particularly Friedreich ataxia, where iron load causes a free radical-mediated

iron sulfur cluster injury in the respiratory chain; (3) chromosomal microdeletions, unbalancing the stoechiometry of the respiratory chain (i.e., 1p36 deletion, 5q deletion of the NSD1 gene in the Sotos syndrome).
- The scattering of control values occasionally hampers the recognition of enzyme deficiencies, as normal values frequently overlap those found in the patients. It is helpful to express results as ratios, especially as the normal functioning of the respiratory chain requires a constant ratio of enzyme activities [45]. Under these conditions, patients whose absolute activities are in the low normal range can be unambiguously diagnosed as enzyme deficient, although this expression of results may fail to recognize generalized defects of oxidative phosphorylation.
- No reliable method for the assessment of complex I activity in circulating or cultured cells is presently available, because oxidation of NADH-generating substrates by detergent-treated or freeze-thawed control cells is variable, and the rotenone-resistant NADH cytochrome c reductase activity is very high in this tissue.
- The phenotypic expression of respiratory enzyme deficiencies in cultured cells is unstable, and activities return to normal values when cells are grown in a standard medium [46]. The addition of uridine (200 mM) to the culture medium avoids counterselection of respiratory enzyme-deficient cells and allows them to grow normally, thereby stabilizing the mutant phenotype (uridine, which is required for nucleic acid synthesis, is probably limited by the secondary deficiency of the respiratory chain-dependent dihydro-orotate dehydrogenase activity) [47].
- Discrepancies between control values may indicate faulty experimental conditions. Activities dependent on a single substrate should be consistent when tested under non-rate-limiting conditions. For example, normal succinate cytochrome c reductase activity should be twice as high as normal succinate quinone dichlorophenol-indophenol (DCPIP) reductase activity (because one electron is required to reduce cytochrome c, while two are required to reduce DCPIP).
- Incorrect freezing may result in a rapid loss of quinone-dependent activities, probably due to peroxidation of membrane lipids. Tissue samples fixed for morphological studies are inadequate for subsequent respiratory enzyme assays.

15.4.3   Histopathological Studies

The histological hallmark of mitochondrial myopathy is the RRF, which is demonstrated using the modified Gomori trichrome stain and contains peripheral and inter-myofibrillar accumulations of abnormal mitochondria. Although the diagnostic importance of pathological studies is undisputed, the absence of RRFs does not rule out the diagnosis of mitochondrial disorder [10]. Different histochemical stains for oxidative enzymes are used to analyze the distribution of mitochondria in the individual fibers and to evaluate the presence or absence of the enzymatic activities. Histochemical staining permits an estimation of the severity and heterogeneity of enzyme deficiency in the same muscle section. Myofibrillar integrity and the predominant fiber type and distribution can be evaluated with the myofibrillar adenosine triphosphatase stain. Studies using polyclonal and monoclonal antibodies directed against COX subunits are carried out in specialized centers. For analysis, the muscle specimen taken under local anesthetic must be frozen immediately in liquid-nitrogen-cooled isopentane.

15.4.4 Magnetic Resonance Spectroscopy of Muscle and Brain

Magnetic resonance spectroscopy (MRS) allows the study of muscle and brain energy metabolism in vivo. Lactate, inorganic phosphate (Pi), phosphocreatine (PCr) and intracellular pH may be measured. The Pi/PCr ratio is the most useful parameter and may be monitored at rest, during exercise, and during recovery. An increased ratio is found in most patients, and MRS is becoming a useful tool in the diagnosis of mitochondrial diseases and in the monitoring of therapeutic trials. However, the observed anomalies are not specific to respiratory enzyme deficiencies, and no correlation between MRS findings and the site of the respiratory enzyme defect can be made [10].

15.5 Treatment and Prognosis

No satisfactory therapy is presently available for respiratory chain deficiency. Treatment remains largely symptomatic and does not significantly alter the course of the disease. It includes symptomatic treatments, supplementation with cofactors, prevention of oxygen-radical damage to mitochondrial membranes, dietary recommendations, and avoidance of drugs and procedures known to have a detrimental effect.

It is advisable to avoid sodium valproate and barbiturates, which inhibit the respiratory chain and have occasionally been shown to precipitate hepatic failure in respiratory enzyme-deficient children [11]. Tetracyclines and chloramphenicol should also be avoided, as they inhibit mitochondrial protein synthesis. Due to the increasing number of tissues affected in the course of the disease, organ transplantations are exceptional (bone marrow, liver, heart).

Symptomatic treatments include: slow infusion of sodium bicarbonate during acute exacerbation of lactic acidosis, pancreatic extract administration in cases of exo-

crine pancreatic dysfunction, and repeated transfusions in cases of anemia or thrombocytopenia. Recently, administration of L-arginine, a nitric oxide precursor, has been shown to significantly decrease the frequency and severity of strokelike episodes in MELAS [48].

One recently identified condition, inborn errors of coenzyme Q_{10} (CoQ_{10}) synthesis [37a] deserves particular attention as, when recognized, this condition should be treatable by large doses of oral quinone (Ubidecarenone). In the three hitherto recognized clinical presentations (the myopathic form [49-52], the ataxic form [53] and the multisystemic form [54-56]), the respiratory enzyme activities are individually normal but they are deficient when tested in group, as CoQ_{10} acts as electron shuttle between complexes in the respiratory chain. Giving oral quinones to CoQ_{10} deficient patients restores the electron flow (5 mg/kg/day). Yet, apart from this rare situation, neither CoQ_{10} nor its analogues (Idebenone) can restore electron flow in case of respiratory chain deficiency. Oral quinones are not only useless but even possibly harmful in respiratory chain deficiency. Indeed, because quinones can divert electrons from the respiratory chain, they may become pro-oxidant and possibly deleterious if reduced quinones are not reoxidized by a normally functioning respiratory chain. The low uptake of oral quinones by CoQ_{10} sufficient cells probably limits their deleterious effect when given to respiratory enzyme deficient patients. By contrast, in Friedreich ataxia, where iron overload causes a free-radical induced iron sulfur cluster injury to an otherwise normal respiratory chain, idebenone (10 mg/kg/day) reoxidized on the respiratory chain has been shown to efficiently counteract the life threatening hypertrophic cardiomyopathy [57]. Treatment with riboflavin (100 mg/day) has been associated with improvement in a few patients with complex I deficiency myopathy. Carnitine is suggested in patients with secondary carnitine deficiency. Dichloroacetate or 2-chloropropionate administration has been proposed to stimulate pyruvate dehydrogenase (PDH) activity and has occasionally reduced the level of lactic acid [58], but detrimental effects of dichloroacetate have recently been reported.

The dietary recommendation are a high-lipid, low-carbohydrate diet in patients with complex I deficiency. Indeed, a high-glucose diet is a metabolic challenge for patients with impaired oxidative phosphorylation, especially as glucose oxidation is largely aerobic in the liver. Based on our experience, we suggest avoiding a hypercaloric diet and parenteral nutrition and recommend a low-carbohydrate diet in addition to the symptomatic treatment. Succinate (6 g/day), succinate-producing amino acids (isoleucine, methionine, threonine, and valine) or propionyl carnitine have occasionally been given to patients with complex I deficiency, as these substrates enter the respiratory chain via complex II.

References

1. von Kleist-Retzow JC, Cormier-Daire V, Viot G et al (2003) Antenatal manifestations of mitochondrial respiratory chain deficiency. J Pediatr 143:208-212
2. Cormier V, Rustin P, Bonnefont JP et al (1991) Hepatic failure in neonatal onset disorders of oxidative phosphorylation. J Pediatr 119:951-954
3. Rötig A, Cormier V, Blanche S et al (1990) Pearson's marrow-pancreas syndrome: a multisystem mitochondrial disorder in infancy. J Clin Invest 86:1601-1608
4. Rustin P, LeBidois J, Chretien D et al (1994) Endomyocardial biopsies for early detection of mitochondrial disorders in hypertrophic cardiomyopathies. J Pediatr 124:224-228
5. Bolhuis PA, Hensels GW, Hulsebos TJM et al (1991) Mapping of the locus for X-linked cardioskeletal myopathy with neutropenia and abnormal mitochondria (Barth syndrome) to Xq28. Am J Hum Genet 48:481-485
6. Cormier-Daire V, Bonnefont JP, Rustin P et al (1994) Deletion-duplication of the mitochondrial DNA presenting as chronic diarrhea with villous atrophy. J Pediatr 124:63-70
7. Saunier P, Chretien D, Wood C et al (1995) Cytochrome c oxidase deficiency presenting as recurrent neonatal myoglobinuria. Neuromuscul Disord 5:285-289
8. Rötig A, Bessis JL, Romero N et al (1991) Maternally inherited duplication of the mitochondrial DNA in proximal tubulopathy with diabetes mellitus. Am J Hum Genet 50:364-370
9. Rötig A, Cormier V, Chatelain P et al (1993) Deletion of the mitochondrial DNA in a case of early-onset diabetes mellitus, optic atrophy and deafness (DIDMOAD, Wolfram syndrome). J Clin Invest 91:1095-1098
10. Hammans SR, Morgan-Hughes JA (1994) Mitochondrial myopathies: Clinical features, investigation, treatment and genetic counselling. In: Schapira AHV, DiMauro S (eds) Mitochondrial disorders in neurology. Butterworth-Enemann, Stoneham, MA, p 4974
11. Chabrol B, Mancini J, Chretien D et al (1994) Cytochrome c oxidase defect, fatal hepatic failure and valproate: a case report. Eur J Pediatr 153:133-135
12. Delettre C, Lenaers G, Griffoin JM et al (2000) Nuclear gene OPA1, encoding a mitochondrial dynamin-related protein, is mutated in dominant optic atrophy. Nat Genet 26:207-210
13. Holt IJ, Harding AE, Petty RKH, Morgan-Hugues JA (1990) A new mitochondrial disease associated with mitochondrial DNA heteroplasmy. Am J Hum Genet 46:428-433
14. Brandon MC, Lott MT, Nguyen KC et al (2005) MITOMAP: a human mitochondrial genome database-2004 update. Nucleic Acids Res 1:33
15. Zeviani M, Servidei S, Gellera C et al (1989) An autosomal dominant disorder with multiple deletions of mitochondrial DNA starting at the D-loop region. Nature 339:309-311
16. Van Goethem G, Dermaut B, Lofgren A et al (2001) Mutation of POLG is associated with progressive external ophthalmoplegia characterized by mtDNA deletions. Nat Genet 28:211-212
17. Kaukonen J, Juselius JK, Tiranti V et al (2000) A Role of adenine nucleotide translocator 1 in mtDNA maintenance. Science 289:782-785
18. Spelbrink JN, Li FY, Tiranti V et al (2001) Human mitochondrial DNA deletions associated with mutations in the gene encoding Twinkle, a phage T7 gene 4-like protein localized in mitochondria. Nat Genet 28:223-231
19. Moraes CT, Shanske S, Trischler HJ et al (1991) mtDNA depletion with variable tissue expression: a novel genetic abnormality in mitochondrial diseases. Am J Hum Genet 48:492-501
20. Mandel H, Szargel R, Labay V et al (2001) The deoxyguanosine kinase gene is mutated in individuals with depleted hepatocerebral mitochondrial DNA. Nat Genet 29:337-341
21. Saada A, Shaag A, Mandel H et al (2001) Mutant mitochondrial thymidine kinase in mitochondrial DNA depletion myopathy. Nat Genet 29:342-344
22. Bione S, D‹Adamo P, Maestrini E et al (1996) A novel X-linked gene, G4.5 is responsible for Barth syndrome. Nat Genet 12:385-389
23. Schuelke M, Smeitink J, Mariman E et al (1999) Mutant NDUFV1 subunit of mitochondrial complex I causes leukodystrophy and myoclonic epilepsy. Nat Genet 21:260-261
24. Loeffen J, Smeitink J, Triepels R et al (1998) The first nuclear-encoded complex I mutation in a patient with Leigh syndrome. Am J Hum Genet 63:1598-1608
25. van den Heuvel L, Ruitenbeek W, Smeets R et al (1998) Demonstration of a new pathogenic mutation in human complex I deficiency: a 5-bp duplication in the nuclear gene encoding the 18-kD (AQDQ) subunit. Am J Hum Genet 62:262-268
26. Benit P, Chretien D, Kadhom N et al (2001) Large-scale deletion and point mutations of the nuclear NDUFV1 and NDUFS1 genes in mitochondrial complex I deficiency. Am J Hum Genet 68:1344-1352
27. Benit P, Beugnot R, Chretien D et al (2003) Mutant NDUFV2 subunit of mitochondrial complex I causes early onset hypertrophic cardiomyopathy and encephalopathy. Hum Mutat 21:582-586
28. Kirby DM, Salemi R, Sugiana C et al (2004) NDUFS6 mutations are a novel cause of lethal neonatal mitochondrial complex I deficiency. J Clin Invest 114:837-845
29. Bourgeron T, Rustin P, Chretien D et al (1995) Mutation of a nuclear succinate dehydrogenase gene results in mitochondrial respiratory chain deficiency. Nat Genet 11:144-149
30. Parfait B, Chretien D, Rotig A et al (2000) Compound heterozygous mutations in the flavoprotein gene of the respiratory chain complex II in a patient with Leigh syndrome. Hum Genet 106(2):236-243
31. Zhu Z, Yao J, Johns T, Fu K et al (1998) SURF1, encoding a factor involved in the biogenesis of cytochrome c oxidase, is mutated in Leigh syndrome. Nat Genet 20:337-343
32. Tiranti V, Hoertnagel K, Carrozzo R et al (1998) Mutation of SURF-1 in Leigh disease associated with cytochrome c oxidase deficiency. Am J Hum Genet 63:1609-1621
33. Valnot I, Osmond S, Gigarel N et al (2000) Mutations of the SCO1 gene in mitochondrial cytochrome c oxidase (COX) deficiency with neonatal-onset hepatic failure and encephalopathy. Am J Hum Genet 67:1104-1109
34. Papadopoulou LC, Sue CM, Davidson MM et al (1999) Fatal infantile cardioencephalomyopathy with COX deficiency and mutations in SCO2, a COX assembly gene. Nat Genet 23:333-337
35. Valnot I, von Kleist-Retzow JC, Barrientos A et al (2000) A mutation in the human heme A:farnesyltransferase gene (COX10) causes cytochrome c oxidase deficiency. Hum Mol Genet 9:1245-1249
36. Antonicka H, Mattman A, Carlson CG et al (2003) Mutations in COX15 produce a defect in the mitochondrial heme biosynthetic pathway, causing early-onset fatal hypertrophic cardiomyopathy. Am J Hum Genet 72:101-214
37. Nishino I, Spinazzola A, Hirano M (1999) Thymidine phosphorylase gene mutations in MNGIE, a human mitochondrial disorder. Science 283:689-692
37a. Quinzii C, Naini A, Salviati L et al (2006) A mutation in para-hydroxybenzoatepolyprenyl transferase (COQ2) causes primary coenzyme Q10 deficiency. Am J Hum Genet 78:345-349
38. Ferrari G, Lamantea E, Donati A et al (2005) Infantile hepatocerebral syndromes associated with mutations in the mitochondrial DNA polymerase-gamma A. Brain 128(Pt 4):723-731
39. Poggi-Travert F, Martin D, Billette de Villeneur T et al (1996) Metabolic intermediates in lactic acidosis: compounds, samples and interpretation. J Inherit Metab Dis 19:478-488
40. Bonnefont JP, Chretien D, Rustin P et al (1992) 2-ketoglutarate dehydrogenase deficiency: a rare inherited defect of the Krebs cycle. J Pediatr 121:255-258

41. Saudubray JM, Marsac C, Cathelineau C (1989) Neonatal congenital lactic acidosis with pyruvate carboxylase deficiency in two siblings. Acta Paediatr Scand 65:717-724

42. Estabrook RW (1967) Mitochondrial respiratory control and the polarographic measurement of ADP/O ratios. Methods Enzymol 10:41-47

43. Rustin P, Chretien D, Girard B et al (1994) Biochemical, molecular investigations in respiratory chain deficiencies. Clin Chim Acta 220:35-51

44. Bourgeron T, Chretien D, Rötig A et al (1992) Isolation and characterization of mitochondria from human B lymphoblastoid cell lines. Biochem Biophys Res Commun 186:16-23

45. Rustin P, Chretien D, Bourgeron T et al (1991) Assessment of the mitochondrial respiratory chain. Lancet 338:60

46. Gérard B, Bourgeron T, Chretien D et al (1992) Uridine preserves the expression of respiratory enzyme deficiencies in cultured fibroblasts. Eur J Pediatr 152:270

47. Bourgeron T, Chretien D, Rötig A et al (1993) Fate and expression of the deleted mitochondrial DNA differ between heteroplasmic skin fibroblast and Epstein-Barr virus-transformed lymphocyte cultures. J Biol Chem 268:19369-19376

48. Koga Y, Akita Y, Nishioka J et al (2005) L-arginine improves the symptoms of strokelike episodes in MELAS. Neurology 64:710-712

49. Ogasahara S, Engel AG, Frens D, Mack D (1989) Muscle coenzyme Q deficiency in familial mitochondrial encephalomyopathy. Proc Natl Acad Sci USA 86:2379-2382

50. Sobreira C, Hirano M, Shanske S et al (1997) Mitochondrial encephalomyopathy with coenzyme Q10 deficiency. Neurology 48:1238-1243

51. Boitier E, De goul F, Desguerre I et al (1998) A case of mitochondrial encephalomyopathy associated with a muscle coenzyme Q10 deficiency. J Neurol Sci 156:41-46

52. Di Giovanni S, Mirabella M, Spinazzola A et al (2001) Coenzyme Q_{10} reverses pathological phenotype and reduces apoptosis in familial CoQ10 deficiency. Neurology 57:515-518

53. Musumeci O, Naini A, Slonim AE et al (2001) Familial cerebellar ataxia with muscle coenzyme Q10 deficiency. Neurology 56:849-855

54. Rötig A, Appelkvist EL, Geromel V et al (2000) Quinone-responsive multiple respiratory-chain dysfunction due to widespread coenzyme Q10 deficiency. Lancet 356:391-395

55. Rahman S, Hargreaves I, Clayton P, Heales S (2001) Neonatal presentation of coenzyme Q10 deficiency. J Pediatr 139:456-458

56. Leshinsky-Silver E, Levine A, Nissenkorn A et al (2003) Neonatal liver failure and Leigh syndrome possibly due to CoQ-responsive OXPHOS deficiency. Mol Genet Metab 79:288-293

57. Hausse AO, Aggoun Y, Bonnet D et al (2002) Idebenone and reduced cardiac hypertrophy in Friedreich‹s ataxia. Heart 87:346-349

58. Stacpoole P, Harman EM, Curry SH et al (1983) Treatment of lactic acidosis with dichloroacetate. N Engl J Med 309:390-396

16 Creatine Deficiency Syndromes

Sylvia Stöckler-Ipsiroglu, Gajja S. Salomons

Creatine Synthesis and Transport

Creatine is synthesized by two enzymatic reactions: (1) transfer of the amidino group from arginine to glycine, yielding guanidinoacetate and catalyzed by L-arginine:glycine amidinotransferase (AGAT); (2) methylation of the amidino group in the guanidinoacetic acid molecule by S-adenosyl-L-methionine: N-guanidinoacetate methyltransferase (GAMT). Creatine synthesis primarily occurs in the kidney and pancreas which have high AGAT activity, and in liver which has high GAMT activity. From these organs of synthesis creatine is transported via the bloodstream to the organs of utilization (mainly muscle and brain), where it is taken up by a sodium and chloride dependent creatine transporter (CRTR). A human CRTR, SLC6A8, has been cloned and further characterized [1, 2].

Part of intracellular creatine is reversibly converted into the high-energy compound, creatine phosphate, by the action of creatine kinase (CK). Three cytosolic isoforms, brain type BB-CK, muscle type MM-CK and the MB-CK heterodimer, and two mitochondrial isoforms exist. Creatine and creatine phosphate are non-enzymatically converted into creatinine, with a constant daily turnover of 1.5% of body creatine. Creatinine is mainly excreted in urine and its daily excretion is directly proportional to total body creatine, and in particular to muscle mass (i.e. 20–25 mg/kg/24 h in children and adults, and lower in infants younger than 2 years).

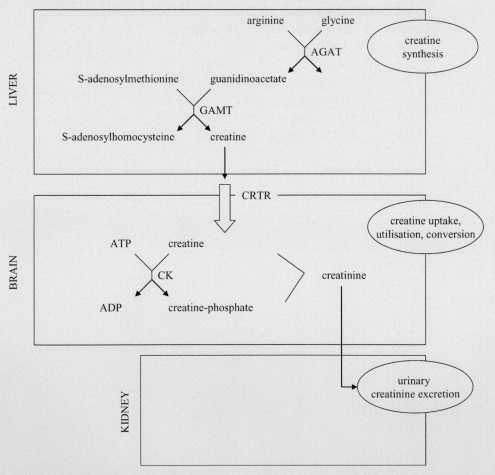

◻ Fig. 16.1. Pathway and locations of creatine/creatine phosphate metabolism. *AGAT*, arginine:glycine amidinotransferase; *CK*, creatine kinase; *CRTR*, creatine transporter (SLC6A8); *GAMT*, guanidinoacetate methyltransferase

Creatine deficiency syndromes (CDS) are a novel group of inborn errors of creatine synthesis and transport including autosomal recessive arginine:glycine amidinotransferase (AGAT) and guanidinoacetate methyltransferase (GAMT) deficiencies, and the X-linked creatine transporter (SLC6A8) deficiency. In all these disorders the common clinical hallmark is mental retardation, expressive speech delay and epilepsy; the common biochemical hallmark is cerebral creatine deficiency as detected by proton magnetic resonance spectroscopy (H-MRS). Increased levels of guanidinoacetic acid (GAA) in body fluids are pathognomonic for GAMT deficiency whereas these levels are reduced in AGAT deficiency. An increased urinary creatine/creatinine ratio is associated with SLC6A8 deficiency. Oral supplementation of creatine leads to partial restoration of the cerebral creatine pool and improvement of clinical symptoms in GAMT and AGAT deficiency. Reduction of GAA by additional dietary restriction of arginine (and supplementation of ornithine) appears to be of additional benefit for the long-term outcome of GAMT deficient patients. For SLC6A8 deficient patients no effective treatment is currently available. CDS may account for a considerable fraction of children and adults with mental retardation of unknown cause and, therefore, screening for these disorders (by urinary/plasma metabolites, brain H-MRS and/or DNA approach) should be included in the investigation of this population.

Secondary creatine deficiency can be found in OAT deficiency (▶ Chap. 22).

16.1   Clinical Presentation

Common clinical hallmarks of CDS are mental retardation and epilepsy. Mental retardation ranges from mild to severe and is characteristically associated with expressive speech delay and autistic features [3]. So far the most severe clinical phenotype has been observed in guanidinoacetate methyltransferase (GAMT) deficiency and has included intractable epilepsy and progressive extrapyramidal symptoms and signs.

16.1.1   Guanidinoacetate Methyltransferase Deficiency

The first patient with GAMT deficiency was described in 1994 [4–6]. He was considered to be normal until 4 months of age when he was noted to have developmental arrest, hypotonia, hyperkinetic (hemiballistic) extrapyramidal movements and head nodding. His electroencephalogram showed slow background activity and multifocal spike slow waves. Magnetic resonance imaging (MRI) revealed bilateral abnormalities of the globus pallidus consisting of hypointensities in T1-weighted images and of hyperintensities in T2-weighted images.

To date, 27 patients (among them 10 adults $\geq$ 18 years), are known to the authors; many of them have been published as single or groups of cases [4–13]. An overview demonstrates a broad clinical spectrum that ranges from mild to severe mental retardation, occasional drug resistant seizures and, in the most severe cases, additional extrapyramidal movement disorder and pathologic signal intensities in the basal ganglia [14].

16.1.2   Arginine:Glycine Amidino-transferase Deficiency

So far only 4 patients have been diagnosed with AGAT deficiency. Three related patients have been published and the authors are aware of one unpublished unrelated case. Two girls with mental retardation, delayed speech development, occasional seizures and brain creatine deficiency reversible with creatine supplementation [15], were finally recognized as the first cases with AGAT deficiency [16]. Subsequently, a cousin with a similar clinical presentation was identified as the third affected patient in this family [17].

16.1.3   SLC6A8 Deficiency

The first patient with SLC6A8 deficiency was reported in 2001 [18, 19]. He had mental retardation and speech delay. In the same metropolitan area three additional families were subsequently diagnosed with this disorder, all having mental retardation and severe speech delay. In three out of the 4 families, the affected males had seizures that responded well to pharmacological treatment [20]. In one of the families, two brothers (17 and 20 years old) were diagnosed with mental retardation and speech delay. Both were microcephalic and their heights were below the 5th percentile. Repeated MRI showed evidence of brain atrophy. Recently, four affected Dutch boys from two unrelated families were reported with growth retardation, mild generalized muscular hypotrophy and discrete dysmorphic facial features. Their neuropsychological tests revealed moderate mental retardation, hyperactive impulsive attention deficit and a semantic-pragmatic language disorder with oral dyspraxia [21]. More than 50 families (including ~100 affected males) with this disorder have been found [18–25]. Its prevalence is relatively high and may be responsible for ~2% of males with X-linked mental retardation [25]. However, the collection of clinical data is far from complete. Clearly, the phenotype varies widely from mild to severe mental retardation. The diagnosis has been made in both children and adults (age range 2–66 y).

Approximately 50% of the females, heterozygous for the family mutation in SLC6A8, have learning and behavioral problems [18–26]. According to skewed X-inactivation, some females may exhibit pronounced clinical manifestations similar to the male phenotype whereas others can be completely asymptomatic.

16.2 Metabolic Derangement

In CDS, the major affected organ is the brain [27]. The defects result in an almost complete lack of cerebral creatine as detected by proton magnetic resonance spectroscopy (H-MRS) (◘ Fig. 16.2). A reduction but not absence of cerebral creatine may be found in females who are heterozygous for a pathogenic mutation in SLC6A8 [26]. The prominent central nervous system (CNS) involvement in all CDS patients (also in AGAT and SLC6A8 deficiency where no toxic accumulation of substrates are found) indicates that creatine is essential for proper brain function. It could be hypothesized that apart from its role in energy storage and transmission, creatine has an additional role in brain as neurotransmitter or modulator [24]. Surprisingly, no major cardiac or skeletal muscle involvement is seen in CDS; the muscular creatine levels vary from normal [22] to reduced [11, 12], but are not absent.

In GAMT deficiency, guanidinoacetic acid accumulates in tissues and body fluids whereas creatine levels are, as expected, decreased [14, 28]. In AGAT deficiency guanidinoacetic acid is low, but, surprisingly, plasma creatine has been found to be within the normal range and urinary creatine concentration is only mildly reduced [15, 17]. The explanation for this is unknown but it could be the result of individual nutritional factors.

In SLC6A8 deficiency, guanidinoacetic acid levels are normal. In male patients the urinary creatine/creatinine ratio is raised which occurs as a result of a slight increase in urine creatine excretion (possibly caused by impaired cellular [re]uptake, including that of the renal cells) and a slight decrease in urine creatinine excretion (caused by the low intracellular creatine pool). In heterozygous females, depending on the skewing of X-inactivation, the creatine/creatinine ratio is within control range or slightly increased [18]. Creatinine in CSF may be reduced in CDS syndromes [24, 28]; however, limited information is available.

Secondary (cerebral) creatine deficiencies have been observed in OAT (ornithine aminotransferase) deficiency, triple H (hyperornithinemia, hyperammonemia, homocitrullinuria) syndrome and pyrroline-5-carboxylate (P5C) synthetase deficiency (► Chap. 22).

16.3 Genetics

The GAMT and AGAT genes are mapped at chromosome 15q15.3 and 19p13.3, respectively. Both disorders show autosomal recessive inheritance and many patients are the product of consanguineous marriages [14, 17]. The SLC6A8 gene has been mapped at Xq28. In SLC6A8 deficiency, the mother is usually a carrier of the mutation but in approximately 5% of affected families the disorder has arisen from a de novo mutation (unpublished data).

In the GAMT and SLC6A8 deficiency, many different mutations have been identified, including nonsense and missense mutations, splice errors, insertions, deletions and frameshifts [13, 24, 25, 29–31]. In the AGAT gene only 2 mutations have been found; a nonsense mutation and a splice error ([16], unpublished data). There is no evidence for a hotspot region in any of these genes; however, certain mutations appear to occur more frequently (e.g. in the GAMT gene c.327G>A and c.59G>A [14]; in the SLC6A8 gene c.319_321delCTT, and c.1221_1223delTTC).

16.4 Diagnostic Tests

16.4.1 MRS of Brain

Inborn errors of creatine metabolism (biosynthesis and transport) can be recognized by the marked reduction of the creatine signal in H-MRS of the brain (◘ Fig. 16.2). However, metabolite screening and molecular analysis remain necessary to unravel the underlying defect given the possibility of conditions causing a secondary creatine deficiency. Moreover, MRS in infants and children often requires pharmacological sedation and is not generally available as a routine investigation. Therefore, H-MRS of the brain is not appropriate for primary screening of large groups of patients even if it increasingly becomes a part of current practice for investigating mental retardation and neurological syndromes.

16.4.2 Metabolic Screening

An accurate analysis of urinary creatine/creatinine ratio and guanidinoacetic acid in body fluids is an important screening test for all CDS ([32], and references therein). Specific derivatization of guanidinoacetic acid and creatine and subsequent analysis by stable isotope gas chromatography-mass spectrometry (SID GC-MS) allows selective and sensitive measurements in a single run. Electrospray tandem-mass spectrometry is another method which allows the rapid, simultaneous determination of guanidinoacetic acid, creatine and creatinine [33, 34]. Variation of the compounds during the day is not significant indicating that a random urine sample is sufficient for the diagnosis of CDS [32].

22 Months

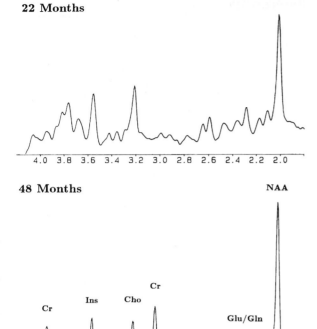

48 Months

Fig 16.2. In vivo proton magnetic resonance spectroscopy (H-MRS) of the brain of a patient with cerebral creatine deficiency due to GAMT deficiency. *Top*: complete lack of creatine resonance at 22 months of age. *Bottom*: Partial normalization of creatine spectrum at 48 months of age after 6 months of treatment with oral creatine monohydrate

A large increase in guanidinoacetic acid in urine and plasma is a valid marker for the identification of GAMT deficiency ([28, 32] and references therein), while AGAT deficiency is identified by guanidinoacetate levels below the normal range [15, 16]. Care must be taken to choose a method which is sensitive enough to differentiate reduced levels from those that are low normal. Determination of the creatine/creatinine ratio in urine has proven a valuable biochemical marker for the diagnosis of SLC6A8 deficiency mainly in males [32]. It should be noted that in approximately 10% of samples, false positive results are encountered which may be due to either a diet rich in protein (and thus creatine) or a condition associated with a reduction of urinary creatinine (e.g. from reduced muscle mass). The latter is the reason why, although creatinine in urine is expected to be reduced in all CDS, urinary creatinine excretion is unlikely to represent a reliable diagnostic marker. It is of note that, in metabolic urine screening, an overall increased concentration of amino acids and organic acids expressed as millimoles per mol of creatinine, may be a result of a decreased creatinine excretion and possibly suggesting the presence of CDS [3, 9, 28].

16.4.3 DNA Diagnostics

Detection of mutations by DNA sequence analysis, either directly or after mutation screening techniques (e.g. DGGE) are available for the 3 genes (*AGAT*, *GAMT* and *SLC6A8*) involved in CDS [16, 18, 25, 30, 31].

16.4.4 Functional Tests/Enzymatic Diagnostics

Enzyme activities in fibroblasts and/or lymphoblasts can be used to confirm GAMT and AGAT deficiency respectively [35–37]. Creatine uptake studies in cultured fibroblasts allow the confirmation of SLC6A8 deficiency at a functional level [18].

16.4.5 Prenatal Diagnosis

In GAMT deficiency it is expected that guanidinoacetate will be increased in amniotic fluid. This is the only available method for prenatal diagnosis where the pathogenic mutations have not been identified in the index patient.

A summary of diagnostic algoritms is given in reference [38].

16.5 Treatment and Prognosis

16.5.1 GAMT Deficiency

Oral creatine substitution has been effective in replenishing the cerebral creatine pool to approximately 70% of normal in all patients [7, 14]. Most have received creatine-monohydrate between 300 and 400 mg/kg/day in 3–6 divided doses. The clinical response to oral creatine supplementation alone included resolution of extrapyramidal signs and symptoms and in most patients considerable improvement of their epilepsy [14]. Additional dietary restriction of arginine has been effective in reducing accumulation of guanidinoacetate in body fluids and led to significant improvement of epilepsy in one patient who had not responded to creatine substitution alone [39]. So far there is only limited experience with additional arginine restriction [11, 12, 14, 39], but taking into account the various neurotoxic effects of guanidinoacetic acid, therapeutic reduction of this compound is likely to be important in improving the long-term outcome of patients with GAMT deficiency. Arginine is restricted to 15–25 mg/kg/day (corresponding to 0.4–0.7 g/kg/day protein intake) and supplementation of an arginine free aminoacid mixture is necessary in order to provide a sufficient nutritional aminoacid supply. Supplementation of high doses of ornithine has the potential to reduce guanidinoacetic acid synthesis by competitive inhibition of AGAT activity in vitro but, in a patient supplemented with ornithine (without arginine restriction), this approach has not been successful [28]. However, in a severely affected adult, the combined treatment including supplemention of creatine and reduction of guanidinoacetate by combined arginine restriction and ornithine supplementation led to an impressive improvement of epileptic seizures, mental capabilities and behavior [11]. In this patient, sodium benzoate was additionally given in order to prevent ammonia accumulation due to possible lack of arginine as essential aminoacid in the urea cycle.

16.5.2   AGAT Deficiency

Oral creatine supplementation (300-400 mg/kg/day) has been effective in replenishing the cerebral creatine pool (to ~60%) in the three patients reported with AGAT deficiency. The clinical response included improvement of abnormal developmental scores [15, 17]. Diagnosis and long-term observation of more patients will allow a more complete description of the natural course of the disorder and an assessment of the efficacy of early treatment.

16.5.3   SLC6A8 Deficiency

SLC6A8 deficiency appears not to be treatable by any of the approaches described above. Treatment of both males and females affected with SLC6A8 deficiency with creatine-monohydrate ([18, 19], unpublished data) has not proven to be successful. Only one heterozygous female patient with learning disability and mildly decreased creatine concentration on brain MRS showed mild improvement on neuropsychological testing after 18 weeks of treatment with creatine-monohydrate (250–750 mg/kg/day) [19]. Currently, supplementation with high doses of arginine and glycine, the primary substrates for creatine biosynthesis, combined with high doses of creatine-monohydrate, is being investigated. The rationale is that an increased cerebral uptake of both aminoacids will enhance intracerebral creatine synthesis (Mancini, van der Knaap, Salomons, unpublished). In addition, alternative strategies may be developed that facilitate creatine transport into the brain either by giving extremely high dosages of creatine (which might enhance passive cellular creatine transport or uptake via other transporters) or by modified transport via carrier peptides/molecules. However, so far, no successful treatment schemes are available.

For a detailed description of treatment protocols in CDS see reference [40].

References

1. Wyss M, Kaddurah-Daouk R (2000) Creatine and creatinine metabolism Physiol Rev 80:1107-213
2. Nash SR, Giros B, Kingsmore SF et al (1994) Cloning, pharmacological characterization, and genomic localization of the human creatine transporter. Receptors Channels 2:165-174
3. Stromberger C, Bodamer O, Stöckler-Ipsiroglu S (2003) Clinical characteristics and diagnostic clues in inborn errors of creatine metabolism. J Inherit Metab Dis 26:299-308
4. Stöckler S, Holzbach U, Hanefeld F et al (1994) Creatine deficiency in the brain: A new treatable inborn error of metabolism. Pediatr Res 36:409-413
5. Stöckler S, Isbrandt D, Hanefeld F et al (1996) Guanidinoacetate methyltransferase deficiency: the first inborn error of creatine metabolism in man. Am J Hum Genet 58:914-922
6. Stöckler S, Hanefeld F, Frahm J (1996) Creatine replacement therapy in guanidinoacetate methyltransferase deficiency, a novel inborn error of metabolism. Lancet 348:789-790
7. Schulze A, Hess T, Wevers R et al (1997) Creatine deficiency syndrome caused by guanidinoacetate methyltransferase deficiency: diagnostic tools or a new inborn error of metabolism. J Pediatr 131:626-631
8. Ganesan V, Johnson A, Connelly A et al (1997) Guanidinoacetate methyltransferase deficiency: New clinical features. Pediatr Neurol 17:155-157
9. van der Knaap MS, Verhoeven NM, Maaswinkel-Mooij P et al (2000) Mental retardation and behavioral problems as presenting signs of a creatine synthesis defect. Ann Neurol 47:540-543
10. Leuzzi V, Bianchi MC, Tosetti MC et al (2000) Brain creatine depletion: guanidinoacetate methyltransferase deficiency (improving with creatine supplementation). Neurology 55:1407-1409
11. Schulze A, Bachert P, Schlemmer H et al (2003) Lack of creatine in muscle and brain in an adult with GAMT deficiency. Ann Neurol 53:248-251
12. Ensenauer R, Thiel T, Schwab KO et al (2004) Guanidinoacetate methyltransferase deficiency: differences of creatine uptake in human brain and muscle. Mol Genet Metab 82:208-213
13. Araujo HC, Smit W, Verhoeven NM et al (2005) Guanidinoacetate methyltransferase deficiency identified in adults and a child with mental retardation. Am J Med Genet 133:122-127
14. Mercimek-Mahmutoglu S, Stöckler-Ipsiroglu S, Adami A et al (2005) Clinical, biochemical and molecular features of guanidinoacetate methyltransferase deficiency. Neurology (in press)
15. Bianchi MC, Tosetti M, Fornai F et al (2002) Reversible brain creatine deficiency in two sisters with normal blood creatine level. Ann Neurol 47:511-513
16. Item CB, Stöckler-Ipsiroglu S, Stromberger C et al (2001) Arginine: Glycine amindinotransferase (AGAT) deficiency: The third inborn error of creatine metabolism in humans. Am J Hum Genet 69:1127-1133
17. Battini R, Leuzzi V, Carducci C et al (2002) Creatine depletion in a new case with AGAT deficiency: clinical and genetic study in a large pedigree. Mol Genet Metab 77:326-331
18. Salomons GS, van Dooren SJ, Verhoeven NM et al (2001) X-linked creatine-transporter gene (SLC6A8) defect: A new creatine-deficiency syndrome. Am J Hum Genet 68:1497-1500
19. Cecil KM, Salomons GS, Ball WS Jr et al (2001) Irreversible brain creatine deficiency with elevated serum and urine creatine: a creatine transporter defect? Ann Neurol 49:401-404
20. deGrauw TJ, Salomons GS, Cecil KM et al (2002) Congenital creatine transporter deficiency. Neuropediatrics 33:232-238
21. Mancini GM, Catsman-Berrevoets CE, de Coo IF et al (2004) Two novel mutations in SLC6A8 cause creatine transporter defect and distinctive X-linked mental retardation in two unrelated Dutch families. Am J Med Genet 132:288-295

22. deGrauw TJ, Cecil KM, Byars AW et al (2003) The clinical syndrome of creatine transporter deficiency. Mol Cell Biochem 244:45-48

23. Hahn KA, Salomons GS, Tackels-Horne D et al (2002) X-linked mental retardation with seizures and carrier manifestations is caused by a mutation in the creatine-transporter gene (SLC6A8) located in Xq28. Am J Hum Genet 70:1349-1356

24. Salomons GS, van Dooren SJ, Verhoeven NM et al (2003) X-linked creatine transporter defect: an overview. J Inherit Metab Dis 26:309-318

25. Rosenberg EH, Almeida LS, Kleefstra T et al (2004) High prevalence of SLC6A8 deficiency in X-linked mental retardation. Am J Hum Genet 75:97-105

26. Cecil KM, DeGrauw TJ, Salomons GS et al (2003) Magnetic resonance spectroscopy in a 9-day-old heterozygous female child with creatine transporter deficiency. J Comput Assist Tomogr 27:44-47

27. Salomons GS, Wyss M, Jakobs C (2004) Creatine. In: Coats PM (ed) Encyclopedia of dietary supplements. Dekker, New York, pp 151-158

28. Stöckler S, Marescau B, De Deyn PP et al (1997) Guanidino compounds in guanidinoacetate methyltransferase deficiency, a new inborn error of creatine synthesis. Metabolism 46:1189-1193

29. Item CB, Stromberger C, Muhl A et al (2002) Denaturing gradient gel electrophoresis for the molecular characterization of six patients with guanidinoacetate methyltransferase deficiency. Clin Chem 48:767-769

30. Item CB, Mercimek-Mahmutoglu S, Battini R et al (2004) Characterisation of seven novel mutations in seven patients with GAMT deficiency. Hum Mutat 23:524

31. Carducci C, Leuzzi V, Carducci C et al (2000) Two new severe mutations causing guanidinoacetate methyltransferase deficiency. Mol Genet Metab 71:633-638

32. Almeida LS, Verhoeven NM, Roos B et al (2004) Creatine and guanidinoacetate: diagnostic markers for inborn errors in creatine biosynthesis and transport. Mol Genet Metab 82:214-219

33. Cognat S, Cheillan D, Piraud M et al (2004) Determination of guanidinoacetate and creatine in urine and plasma by liquid chromatography-tandem mass spectrometry. Clin Chem 50:1459-1461

34. Ilas J, Mühl A, Stöckler-Ipsiroglu S (2000) Guanidinoacetate methyltransferase (GAMT) deficiency: non-invasive enzymatic diagnosis of a newly recognized inborn error of metabolism. Clin Chim Acta 290:179-188

35. Bodamer OA, Bloesch SM, Gregg AR, Stockler-Ipsiroglu S, O`Brien WE (2001) Analysis of guanidinoacetate and creatine by isotope dilution electrospray tandem mass spectrometry. Clin Chim Acta 308:173-178

36. Verhoeven NM, Schor DS, Roos B et al (2003) Diagnostic enzyme assay that uses stable-isotope-labeled substrates to detect L-arginine:glycine amidinotransferase deficiency. Clin Chem 49:803-805

37. Verhoeven NM, Roos B, Struys EA et al (2004) Enzyme assay for diagnosis of guanidinoacetate methyltransferase deficiency. Clin Chem 50:441-443

38. Stöckler-Ipsiroglu S, Stromberger C, Item CB et al (2003) In: Blau N, Duran M, Blaskovics MF, Gibson KM (eds) Physician's guide to the laboratory diagnosis of metabolic diseases. Springer, Berlin Heidelberg New York, pp 467-480

39. Schulze A, Ebinger F, Rating D, Mayatepek E (2001) Improving treatment of guanidinoacetate methyltransferase deficiency: reduction of guanidinoacetic acid in body fluids by arginine restriction and ornithine supplementation. Mol Genet Metab 74:413-419

40. Stöckler-Ipsiroglu S, Battini R, de Grauw T, Schulze A (2006) Disorders of creatine metabolism. In: Blau N, Hoffmann GF, Leonard J, Clarke JTR (eds) Physician's guide to the treatment and follow up of metabolic diseases. Springer, Berlin Heidelberg New York (in press)

IV Disorders of Amino Acid Metabolism and Transport

17 Hyperphenylalaninaemia

John H. Walter, Philip J. Lee, Peter Burgard

Phenylalanine Metabolism

Phenylalanine (PHE), an essential aromatic aminoacid, is mainly metabolized in the liver by the PHE hydroxylase (PAH) system (◘ Fig. 17.1). The first step in the irreversible catabolism of PHE is hydroxylation to tyrosine by PAH. This enzyme requires the active pterin, tetrahydrobiopterin (BH_4), which is formed in three steps from GTP. During the hydroxylation reaction BH_4 is converted to the inactive pterin-4a-carbinolamine. Two enzymes regenerate BH_4 via q-dihydrobiopterin (qBH_2). BH_4 is also an obligate co-factor for tyrosine hydroxylase and tryptophan hydroxylase, and thus nec-essary for the production of dopamine, catecholamines, melanin, serotonin, and for nitric oxide synthase.

Defects in either PAH or the production or recycling of BH_4 may result in hyperphenylalaninaemia, as well as in deficiency of tyrosine, L-dopa, dopamine, melanin, catecholamines, and 5-hydroxytryptophan. When hydroxylation to tyrosine is impeded, PHE may be transaminated to phenylpyruvic acid (a ketone excreted in increased amounts in the urine, hence the term phenylketonuria or PKU), and further reduced and decarboxylated.

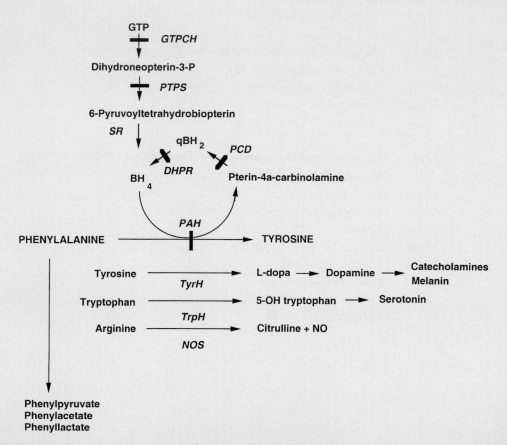

◘ **Fig. 17.1.** The phenylalanine hydroxylation system including the synthesis and regeneration of pterins and other pterin-requiring enzymes. *BH₂*, dihydrobiopterin (quinone); *BH₄*, tetrahydrobiopterin; *DHPR*, dihydropteridine reductase; *GTP*, guanosine triphosphate; *GTPCH*, guanosine triphosphate cyclohydrolase; *NO,* nitric oxide; *NOS*, nitric oxide synthase; *P,* phosphate; *PAH,* PHE hydroxylase; *PCD,* pterin-4a-carbinolamine dehydratase; *PTPS,* pyruvoyl-tetrahydrobiopterin synthase; *SR,* sepiapterin reductase; *TrpH,* tryptophan hydroxylase; *TyrH,* tyrosine hydroxylase. The enzyme defects are depicted by *solid bars* across the arrows

Mutations within the gene for the hepatic enzyme phenylalanine hydroxylase (PAH) and those involving enzymes of pterin metabolism are associated with hyperphenylalaninaemia (HPA). Phenylketonuria (PKU) is caused by a severe deficiency in PAH activity and untreated leads to permanent central nervous system damage. Dietary restriction of phenylalanine (PHE) along with aminoacid, vitamin and mineral supplements, started in the first weeks of life and continued through childhood, is an effective treatment and allows for normal cognitive development. Continued dietary treatment into adulthood with PKU is generally recommended but, as yet, there is insufficient data to know whether this is necessary. Less severe forms of PAH deficiency may or may not require treatment depending on the degree of HPA. High blood levels in mothers with PKU leads to fetal damage. This can be prevented by reducing maternal blood PHE throughout the pregnancy with dietary treatment. Disorders of pterin metabolism lead to both HPA and disturbances in central nervous system amines. Generally they require treatment with oral tetrahydrobiopterin and neurotransmitters.

17.1   Introduction

Defects in either phenylalanine hydroxylase (PAH) or the production or recycling of tetrahydrobiopterin (BH_4) may result in hyperphenylalaninaemia. Severe PAH deficiency which results in a blood phenylalanine (PHE) greater than 1200 µM when individuals are on a normal protein intake, is referred to as classical phenylketonuria (PKU) or just PKU. Milder defects associated with levels between 600 µM and 1200 µM are termed HPA and those with levels less than 600 µM but above 120 µM mild HPA. Disorders of biopterin metabolism have in the past been called malignant PKU or malignant HPA. However such disorders are now best named according to the underlying enzyme deficiency.

17.2   Phenylalanine Hydroxylase Deficiency

17.2.1   Clinical Presentation

PKU was first described by Følling in 1934 as »Imbecillitas phenylpyruvica«[1]. The natural history of the disease is for affected individuals to suffer progressive, irreversible neurological impairment during infancy and childhood [2]; untreated patients develop mental, behavioural, neurological and physical impairments. The most common outcome is severe mental retardation (IQ ≤ 50), often associated with a mousy odour (resulting from the excretion of phenylacetic

acid), eczema (20–40%), reduced hair, skin, and iris pigmentation (a consequence of reduced melanin synthesis), reduced growth and microcephaly, and neurological impairments (25% epilepsy, 30% tremor, 5% spasticity of the limbs, 80% EEG abnormalities) [3]. The brains of patients with PKU untreated in childhood have reduced arborisation of dendrites, impaired synaptogenesis and disturbed myelination. Other neurological features that can occur include pyramidal signs with increased muscle tone, hyperreflexia, Parkinsonian signs and abnormalities of gait and tics. Almost all untreated patients show behavioural problems which include hyperactivity, purposeless movements, stereotypy, aggressiveness, anxiety and social withdrawal. The clinical phenotype correlates with PHE blood levels, reflecting the degree of PAH deficiency.

17.2.2   Metabolic Derangement

Although the pathogenesis of brain damage in PKU is not fully understood it is causally related to the increased levels of blood PHE. Tyrosine becomes a semi-essential amino acid with reduced blood levels leading to impaired synthesis of other biogenic amines including melanin, dopamine, and norepinephrine. Increased blood PHE levels result in an imbalance of other large neutral amino acids (LNAA) within the brain, resulting in decreased brain concentrations of tyrosine and serotonin. The ratio of PHE levels in blood/brain is about 4:1 [4]. In addition to the effects on aminoacid transport into the brain, PHE impairs the metabolism of tyrosine hydroxylation to dopamine and tryptophan decarboxylation to serotonin. The phenylketones phenylpyruvate, phenylacetate and phenyllactate are not abnormal metabolites but appear in increased concentration and are excreted in the urine.

17.2.3   Genetics

PAH deficiency is an autosomal recessive transmitted disorder. The PAH gene is located on the long arm of chromosome 12. At the time of writing nearly 500 different mutations have been described (see http://www.pahdb.mcgill.ca). Most subjects with PAH deficiency are compound heterozygous harbouring two different mutations. Although there is no single prevalent mutation certain ones are more common in different ethnic populations. For example the R408W mutation accounts for approximately 30% of alleles within Europeans with PKU whereas in Orientals the R243Q mutation is the most prevalent accounting for 13% of alleles. Prevalence of PAH deficiency varies between different populations (for example 1 in 1 000 000 in Finland and 1 in 4 200 in Turkey). Overall global prevalence in screened populations is approximately 1 in 12 000 giving an estimated carrier frequency of 1 in 55.

Genotypes correlate well with biochemical phenotypes, pre-treatment PHE levels, and PHE tolerance [5, 6]. However due to the many other factors that effect clinical phenotype correlations between mutations and neurological, intellectual and behavioural outcome are weak. Mutation analysis is consequently of limited practical use in clinical management but may be of value in determining genotypes associated with possible BH_4 responsiveness.

17.2.4   Diagnostic Tests

Blood PHE is normal at birth in infants with PKU but rises rapidly within the first days of life. In most Western nations PKU is detected by newborn population screening. There is variation between different countries and centres in the age at which screening is undertaken (day 1 to day 10), in the methodology used (Guthrie microbiological inhibition test, enzymatic techniques, HPLC, or tandem mass spectrometry) and the level of blood PHE that is taken as a positive result requiring further investigation (120 to 240 µmol/l but with some laboratories also using a PHE/tyrosine ratio >3).

Cofactor defects must be excluded by investigation of pterins in blood or urine and DHPR in blood (▶ later). Persistent hyperphenylalaninaemia may occasionally be found in preterm and sick babies, particularly after parenteral feeding with amino acids and in those with liver disease (where blood levels of methionine, tyrosine, leucine/isoleucine and PHE are usually also raised). In some centres the diagnosis is further characterised by DNA analysis.

PAH deficiency may be classified according to the concentration of PHE in blood when patients are on a normal protein containing diet or after a standardized protein challenge [7–9]:

- classical PKU (PHE ≥1200 µmol/l; less than 1% residual PAH activity),
- hyperphenylalaninaemia (HPA) or mild PKU (PHE >600 µmol/l and <1200 µmol/l; 1–5% residual PAH activity), and
- non-PKU-HPA or mild hyperphenylalaninaemia (MHP) (PHE ≤ 600 µmol/l; >5% residual PAH activity).

Although in reality there is a continuous spectrum of severity, such a classification has some use in terms of indicating the necessity for dietary treatment.

Although rarely requested, prenatal diagnosis is possible by PAH DNA analysis on CVB or amniocentesis where the index case has had mutations identified previously.

17.2.5   Treatment and Prognosis

Principles of Treatment

The principle of treatment in PAH deficiency is to reduce the blood PHE concentration sufficiently to prevent the neuropathological effects. Blood PHE is primarily a function of residual PAH activity and PHE intake. For the majority of patients with PKU the former cannot be altered so that blood PHE must be reduced by restricting dietary PHE intake. A PHE blood level while on a normal protein containing diet defines the indication for treatment with some minor differences in cut-offs; UK (>400 µmol/l), Germany (>600 µmol/l), and USA (>360–600 µmol/l). In all published recommendations for treatment target blood PHE levels are age related. ◻ Table 17.1 shows such recommendations for UK [10], Germany [11] and the USA [12].

The degree of protein restriction required is such that in order to provide a nutritionally adequate diet a semisynthetic diet is necessary. This is composed of the following:

- Unrestricted natural foods with a very low PHE content (<30 mg/100 g; e.g. carbohydrate, fruit and some vegetables).
- Calculated amounts of restricted natural and manufactured foods with medium PHE content (>30 mg/100 g; e.g. potato, spinach, broccoli; some kinds of special bread and special pasta). In the United Kingdom

◻ **Table 17.1.** Daily phenylalanine (PHE) tolerances and target blood levels for three different recommendations

Age	PHE tolerance mg/day	Target blood PHE (µmol/l)		
		Germany	UK	USA
0–2 years	~ 130–400	40–240	120–360	120–360
3–6 years	~ 200–400	40–240	120–360	120–360
7–9 years	~ 200–400	40–240	120–480	120–360
10–12 years	~ 350–800	40–900	120–480	120–360
13–15 years	~ 350–800	40–900	120–700	120–600
Adolescents/adults	~ 450–1000	40–1200	120–700	120–900

a system of ›protein exchanges‹ is used with each 1g of natural protein representing a PHE content of approximately 50 mg.

- Calculated amounts of PHE-free amino acid mixtures supplemented with vitamins, minerals and trace elements.

Intake of these three components – including the PHE-free amino acid mixture – should be distributed as evenly as possible during the day.

Those foods with a higher concentration of PHE (e.g. meat, fish, cheese, egg, milk, yoghurt, cream, rice, corn) are not allowed. Aspartame (L-aspartyl L-phenylalanine methyl ester), a sweetener for foods (e.g. in soft-drinks) contains 50% PHE, and therefore is inappropriate in the diet of patients with PKU.

PHE free amino acid infant formulas which also contain adequate essential fatty acids, mineral and vitamins are available. Human breast milk has relatively low PHE content; in breast fed infants, PHE-free formulas are given in measured amounts followed by breast-feeding to appetite. In the absence of breast feeding a calculated quantity of a normal formula is given to provide the essential daily requirement of PHE.

With intercurrent illness, individuals may be unable to take their prescribed diet. During this period high-energy fluids may be given to counteract catabolism of body protein.

Monitoring of Treatment

The constraints of a diet that is ultimately focused at the threshold of a calculated PHE intake bears the risk of nutrient deficiency. Therefore, the treatment must be monitored by regular control of dietary intake, as well as neurological, physical, intellectual and behavioural development. ◘ Table 17.2 summarizes recommendations for monitoring treatment and outcome of PKU.

Alternative Therapies/Experimental Trials

Although dietary treatment of PKU is highly successful it is difficult and compliance is often poor, particularly as individuals reach adolescence. Hence there is a need to develop more acceptable therapies.

- *Gene therapy.* Different PAH gene transfer vehicles have been tried in the PAH^{enu2} mouse. These have included non-viral vectors, recombinant adenoviral vector, recombinant retroviral vector and recombinant adeno-associated virus vector[13]. So far none of these experiments has resulted in sustained phenotypic correction, either due to poor efficiency of gene delivery, the production of neutralizing antibodies, or the lack of cofactor in non hepatic target organs. The development of a safe and more successful gene transfer vector is still required before clinical trials in humans are likely to become possible.

◘ **Table 17.2.** Recommendations for monitoring treatment and outcome of PKU

Age	Monitoring	
	Blood PHE levels	**Clinical monitoring[1]**
0–3 years	Weekly	Every 3 months
4–6 years	Fortnightly	Every 3-6 months
7–9 years	Fortnightly	Every 6 months
10–15 years	Monthly	Every 6 months
Adolescents/ adults	2–3-monthly	Yearly

[1] Length/height, head circumference, general status of health, neurology and psychological development. When phenylalanine (PHE) levels are within the recommended range, in general no additional routine laboratory analysis is necessary. A complete fasting profile of all amino acids, minerals, vitamins and trace elements, blood count, Ca-, P-metabolism, fatty acids may be indicated in individuals with poor compliance.

- *Liver transplantation* fully corrects PAH deficiency but the risks of transplantation surgery and post transplantation immune suppressive medication are too high for it to be a realistic alternative to dietary treatment.
- *Phenylalanine ammonia lyase.* Animal experiments have been performed with a non-mammalian enzyme, PHE ammonia lyase (PAL), that converts PHE to a harmless compound, transcinnamic acid. In the PAH^{enu2} mouse enteral administration, intraperitoneal injection and recombinant *E.coli* cells expressing PAL have all led to a significant fall in blood PHE [14, 15]. However it is likely to be some time before clinical trials are attempted.
- The *large neutral aminoacids* (phenylalanine, tyrosine, tryptophan, leucine, isoleucine, and valine) compete for the same transport mechanism (the L-type amino-acid carrier) to cross the blood brain barrier. Studies in the PAH^{enu2} mouse model and in patients have reported a reduction in brain PHE levels when LNAAs (apart from PHE) have been given enterally [16, 17].
- Recently it has been shown that in certain patients *oral BH$_4$ monotherapy* (7–20 mg/kg bw) can reduce blood PHE levels into the therapeutic range [18]. Up to two-thirds of patients with mild PKU are potentially BH$_4$-responsive and might profit from cofactor treatment. PAH is a homotetrameric enzyme where each monomer has a regulatory, a catalytic, and an oligomerization domain. According to Blau and Erlandsen [19] there are four postulated mechanisms for BH$_4$-responsiveness. BH$_4$ therapy might (a) increase the binding affinity of the mutant PAH for BH$_4$, (b) protect the active tetramer

from degradation, (c) increase BH_4 biosynthesis, and (d) up-regulate PAH expression. The most likely hypothesis is that BH_4 responsiveness is multifactorial but needs further research. From experience of treatment of BH_4 deficient patients it can be expected that long-term application of BH_4 has no significant side effects. However, clinical studies are not available to demonstrate long-term therapeutic efficacy, and BH_4 is expensive and not available for all patients.

Compliance with Treatment

Compliance with treatment is most often satisfactory in infancy and childhood. However the special diet severely interferes with culturally normal eating habits, particularly in older children and adolescents and this often results in problems keeping to treatment recommendations. It has been shown that up to the age of 10 years only 40% of the sample of the German Collaborative Study of PKU has been able to keep their PHE levels in the recommended range [20] and that after the age of 10 years 50 to 80% of all blood PHE levels measured in a British & Australian sample were above recommendation [21].

Dietary treatment of PKU is highly demanding for patients and families, and is almost impossible without the support of a therapeutic team trained in special metabolic treatment. This team should consist of a dietician, a metabolic paediatrician, a biochemist running a metabolic laboratory and a psychologist skilled in the behavioural problems related to a life long diet. It is of fundamental importance that all professionals, and the families themselves, fully understand the principle and practice of the diet. The therapeutic team should be trained to work in an interdisciplinary way in a treatment centre which should care for at least 20 patients to have sufficient expertise [22].

Outcome

The outcome for PKU is dependent upon a number of variables which include the age at start of treatment, blood PHE levels in different age periods, duration of periods of blood PHE deficiency, and individual gradient for PHE transport across the blood brain barrier. Further unidentified co-modifiers of outcome are also likely. However, the most important single factor is the blood PHE level in infancy and childhood.

Longitudinal studies of development have shown that start of dietary treatment within the first 3 weeks of life with average blood PHE levels ≤ 400 µmol/l in infancy and early childhood result in near normal intellectual development, and that for each 300 µmol/l increase during the first 6 years of life IQ is reduced by 0.5 SD, and during age 5 to 10 years reduction is 0.25 SD. Furthermore, IQ at the age of 4 years is reduced by 0.25 SD for each 4 weeks delay of start of treatment and each 5 months period of insufficient PHE intake. After the age of 10 years all studies show stable IQ performance until early adulthood irrespective of PHE levels, and normal school career if compliance during the first 10 years has been according to treatment recommendations [23–26]. However, longitudinal studies covering middle and late adulthood are still lacking.

Complications in Adulthood
Neurological Abnormalities

Neuropsychological studies of reaction times demonstrate a life-long but reversible, vulnerability of the brain to increased concurrent PHE levels [27].

Nearly all patients show white matter abnormalities in brain MRI after longer periods of increased PHE levels. However, these abnormalities are not correlated to intellectual or neurological signs and are reversible after 3 to 6 months of strict dietary treatment [28].

Patients with poor dietary control during infancy show behavioural impairments such as hyperactivity, temper tantrums, increased anxiety and social withdrawal, most often associated with intellectual deficits. Well-treated subjects may show an increased risk of depressive symptoms and low self-esteem. However, without correlation to concurrent PHE levels causality of this finding remains obscure but is hypothesized to be a consequence of living with a chronic condition rather than a biological effect of increased PHE levels [29].

A very small number of adolescent and adult patients have developed frank neurological disease which has usually improved on returning to dietary treatment [30]. These individuals appear to usually have had poor control in childhood. The risk to those who have been under good control in childhood and who have subsequently relaxed their diet is probably very small. In some cases neurological deterioration has been related to severe vitamin B_{12} deficiency ($\blacktriangleright$ below) compounded by anaesthesia using nitrous oxide [31].

Dietary Deficiencies

Vitamin B_{12} deficiency can occur in adolescents and adults who have stopped their vitamin supplements but continue to restrict their natural protein intake [32]. For patients on strict diet there have been concerns regarding possible deficiencies in other vitamins and minerals including selenium, zinc, iron, retinol and polyunsaturated fatty acids. However such deficiencies are inconsistently found and it is unclear whether they are of any particular clinical significance. Low calcium, osteopenia and an increased risk of fractures have also been reported.

Diet for Life

For historical reasons clinical experience with early and strictly treated PKU does not go beyond early and middle adulthood. In view of the non-clinical life-long vulnerability of the brain to increased PHE levels, the neuropsychological findings, in particular, have been interpreted as possible markers of long-term intellectual and neurological impairments. For reasons of risk-reduction, guidelines for treatment of PKU recommend diet for life, and where this is not possible at least monitoring for life.

17.3   Maternal Phenylketonuria

17.3.1   Clinical Presentation

Although it was recognised that the offspring born to mothers with PKU are at risk of damage from the teratogenic effects of PHE over 40 years ago [33], it was not until the publication of the seminal paper by Lenke and Levy in 1980 that the maternal PKU syndrome became recognised [34]. High PHE concentrations are associated with a distinct syndrome: facial dysmorphism, microcephaly, developmental delay and learning difficulties, and congenital heart disease (❑ Table 17.3). The facial features resemble those of the fetal alcohol syndrome with small palpebral fissures, epicanthic folds, long philtrum and thin upper lip. Other malformations also can occur in higher than expected frequency e.g. cleft lip and palate, oesophageal atresia and tracheo-oesophageal fistulae, gut malrotation, bladder extrophy and eye defects.

As a result of these data, the prospective North American and German Maternal PKU Collaborative Study was initiated to assess the impact of dietary PHE restriction on the fetal outcome [35]. In the United Kingdom, data were collected within the National PKU Registry to look at the maternal PKU syndrome [36] and subsequently a Medical Research Council Working Party recommended that women with PKU should commence a diet pre-conceptually to protect against these effects [37]. The North American and German maternal PKU Collaborative Study examined the outcome of 572 pregnancies from 382 women with hyperphenylalaninaemia. It was found that optimum outcomes occur when maternal blood PHE of 120 to 360 μmol/l were achieved by 8-10 weeks gestation and subsequently maintained throughout pregnancy. The

UK data looked at 228 pregnancies and found that pre-conceptual diet improved birth head circumference, birth weight and neuropsychometric outcome at 4 and 8 years. Interestingly outcome was better in those pregnancies managed in the more experienced centres.

17.3.2   Metabolic Derangement

Fetal PHE concentrations are one and a half to twice those in the mother, due to active transport from the mother to the fetus [38]. PHE competes for placental transport with other large neutral amino acids and affects fetal development in a variety of as yet unknown ways. On the other hand, low PHE concentrations may limit fetal brain protein synthesis and be detrimental. Thus there is a need to aim to keep maternal blood PHE concentrations within a safe range. From the North American data this range is 120 to 360 μmol/l, whilst in the UK it is 100 to 250 μmol/l.

17.3.3   Management

The issue of maternal PKU needs to be addressed at an early stage with the parents of children with PKU throughout childhood. Indeed, young girls from 5 years onwards can understand a simple explanation of the problem and then as they move into the reproductive years, counselling can be directed towards them. The aim of this education is to provide them with a basic understanding of conception and PKU and the need for a strict diet ideally before conception. Genetics of PKU should be discussed, highlighting the relative low recurrence risk of 1 in 100, assuming a carrier frequency of 1 in 50. The need for close contact with the metabolic clinic into adulthood is stressed so that the young women are able to contact appropriate support in a timely fashion. Contraception must be discussed with teenage girls and reviewed frequently. If they become pregnant whilst on a normal diet, they must feel free to be able to contact the clinic immediately rather than wait until the pregnancy has proceeded for a significant length of time. Experience has shown that the most successful pregnancies are those that are planned ahead of time and in which a supportive partner is involved in the counselling process, as well as the dietary therapy.

Starting Diet for Pregnancy

Many women with PKU choosing to start a family have been on normal dietary intakes for many years because this was recommended at the time. They need, ideally, to be admitted to hospital for intensive education and institution of a PHE-restricted diet. If suitable facilities for admission are not available, they require very close supervision in their own homes or serial visits to see the dietitian. The woman, and her partner, need to be able to carefully plan menus,

❑ **Table 17.3.** Pregnancy outcome in women with classical phenylketonuria (off-diet phenylalanine >1200 μmol/l). Comparison between data of Lenke and Levy (1980) in which 0.5% pregnancies were treated [34] and Koch et al (2003)) in which 26% were treated pre-conception, 46% from the first trimester and 9% from the second trimester [35]

	Untreated [34]	Treated [35]
Mental retardation	92%	28%
Microcephaly	73%	23%
Congenital heart disease	12%	11%
Birth weight <2.5 kg	40%	21%
Spontaneous abortion	24%	17%

IV

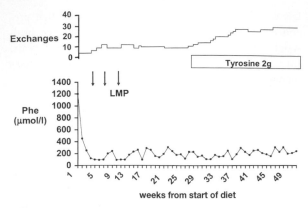

Fig. 17.2. Graph showing blood phenylalanine concentrations and protein intake during a pregnancy in a woman with PKU. An exchange represents 1 g of natural protein or 50 mg of phenylalanine. The *vertical arrows* represent the beginning of a menstrual cycle. *LMP*, last menstrual period

count and weigh protein exchanges and consume the prescribed amount of PHE and dietary substitutes. With this, blood PHE concentrations fall rapidly to the target range within 10 days providing a sense of achievement and encouragement, and hopefully determination to continue. In addition to the diet, close biochemical monitoring is required to allow adjustments to the diet which are often considerable during intercurrent illness, hyperemesis and the second and third trimesters (**□** Figure 17.2). For some women, there can be marked fluctuations in association with their menstrual cycles. The MRC guidelines suggest blood monitoring twice a week before conception and three times per week during pregnancy. For women with PKU who have always been on a PHE-restricted diet, often more education is required because they expect to be able to do the pregnancy diet easily and yet the meticulous approach needed for this may not be present. Women who have had previous pregnancies also must be warned that subsequent pregnancies may be harder to manage because they have not only to look after their own diet, but those of their other child(ren) and partner! The role of tyrosine supplementation as the pregnancy progresses is not clear, but its use is recommended by some centres.

It is a medical emergency if a woman with PKU presents pregnant whilst on a normal diet. If metabolic control can be achieved by 10–12 weeks, then fetal outcome may be satisfactory [35]. However instituting diet in this emotionally charged situation is not easy and subsequent metabolic control may not be as good for the rest of the pregnancy compared to starting diet pre-conceptually in a carefully planned fashion [39]. Termination of pregnancy needs to be discussed to take into account the timing of the pregnancy, the maternal blood PHE concentrations, the ability to lower these and the mother's wishes. If conception does not occur despite good metabolic control, relatively early referral to a reproductive medicine unit should be considered (e.g. after

6 months without contraception) for the woman and her partner to be investigated appropriately.

Ante-Natal and Obstetric Care

The woman with PKU needs to keep in close contact with the metabolic dietitian throughout the whole of the pregnancy. Review in the metabolic clinic should occur every 1–2 months to evaluate nutritional status. Routine obstetric ultrasound of the fetus is carried out at 12 and 20 weeks gestation with the latter providing a detailed anomalies scan. Serial ultrasonography is not required unless there are concerns about fetal growth. Admission into hospital may be needed if there is poor maternal weight gain, vomiting or other problems resulting in poor PHE control. Delivery should be managed in the normal manner by the local obstetric team. Breast feeding is encouraged and excess PHE in it will not harm the offspring, unless they have PKU themselves. Parents like to know the precise result from neonatal screening to exclude PKU conclusively.

17.3.4   Prognosis

Despite a recent consensus statement from the National Institutes of Health in the United States [40], many questions remain about the management of PKU into adult life. The only situation where it is quite clear that dietary intervention is of benefit in adulthood, is to protect the unborn fetus of women with PKU [41]. The data reported from the United Kingdom PKU Registry support the need for the early introduction of a PHE-restricted diet for these pregnancies and for their management in centres with experience [36] The data from North America suggest that obtaining metabolic control by 10 weeks gestation can be associated with satisfactory outcome. Overall information about untreated and treated pregnancies is consistent, providing good evidence of a graded effect of maternal PHE: birth weight and head circumference, risk of congenital anomalies, and postnatal neurodevelopment have all been shown to relate to maternal blood PHE concentrations. Even in women with milder hyperphenylalaninaemia, the risks remain proportional to the PHE levels down to the normal range [42].

Despite the evidence of beneficial effects of dietary treatment in maternal PKU, a number of questions remain unanswered. These include the effects of introducing the diet at different stages of the pregnancy; the safe and effective target concentrations of blood PHE; whether or not dietary effects as well as PHE are important; the impact of both fetal and maternal genotype; and the effects of the post-natal environment. Of interest is that some of the children from untreated pregnancies do remarkably well, whilst some from seemingly well-managed pregnancies do poorly. Close examination of these particular cases may reveal important clues regarding factors protecting against the teratogenic effects of PHE, as well as the detrimental effects of too low PHE.

Overall, the message must be that all women with PKU should be educated about the risks of maternal PKU and that PHE-restricted diet should be commenced before conception. However, improved understanding of the pathogenesis of maternal PKU is still needed to optimise care for these mothers.

17.4 Hyperphenylalaninaemia and Disorders of Biopterin Metabolism

Disorders of tetrahydrobiopterin associated with hyperphenylalaninaemia and biogenic amine deficiency include GTP cyclohydrolase I (GTPCH) deficiency, 6-pyruvoyl-tetrahydropterin synthase (PTPS) deficiency, dihydropteridine reductase (DHPR) deficiency and pterin-4a-carbinolamine dehydratase (PCD) deficiency (primapterinuria). Dopa-responsive dystonia (DRD), due to a dominant form of GTPCH deficiency, and sepiapterin reductase (SR) deficiency, also lead to CNS amine deficiency but are associated with normal blood PHE (although HPA may occur in DRD after a PHE load); these conditions are not considered further.

17.4.1 Clinical Presentation

Presentation may be in one of three ways
1. Asymptomatic. Here the infant is found to have raised PHE following newborn screening and is then investigated further for biopterin defects.
2. Symptomatic with neurological deterioration in infancy despite a low PHE diet. This will occur where no further investigations are undertaken after finding HPA in newborn screening which is wrongly assumed to be PAH deficiency.
3. Symptomatic with neurological deterioration in infancy on a normal diet. This will occur either where there has been no newborn screening for HPA or if the PHE level is sufficiently low not to have resulted in a positive screen.

Symptoms may be subtle in the newborn period and not readily apparent until several months of age. All conditions apart from PCD deficiency are associated with abnormal and variable tone, abnormal movements, irritability and lethargy, seizures, poor temperature control, progressive developmental delay, microcephaly. Cerebral atrophy and cerebral calcification can occur in DHPR deficiency. In PCD deficiency symptoms are mild and transient.

17.4.2 Metabolic Derangement

Disorders of pterin synthesis or recycling are associated with decreased activity of PAH, tyrosine hydroxylase, tryptophan hydroxylase and nitric oxide synthase (◘ Figure 17.1). The degree of hyperphenylalaninaemia, due to the PAH deficiency, is highly variable with blood PHE concentrations ranging from normal to > 2000 µmol/l. Central nervous system (CNS) amine deficiency is most often profound and responsible for the clinical symptoms. Decreased concentrations of homovanellic acid (HVA) in cerebrospinal fluid (CSF) is a measure of reduced dopamine turnover and similarly 5 hydroxyindoleacetic acid deficiency of reduced serotonin metabolism.

17.4.3 Genetics

All disorders are autosomal recessive. Descriptions of the relevant genes and a database of mutations are available on www.BH4.org.

17.4.4 Diagnostic Tests

Diagnostic protocols and interpretation of results are as follows:
1. *Urine or blood pterin analysis and blood DHPR assay*
   All infants found to have HPA on newborn screening should have blood DHPR and urine or blood pterin analysis. The interpretation of results is shown in ◘ Table 17.4.
2. *BH₄ loading test*
   An oral dose of BH_4 is given at dose of 20 mg/kg approximately 30 min before a feed. Blood samples are collected for PHE and tyrosine at 0, 4, 8 and 24 hrs. The test is positive if plasma PHE falls to normal (usually by 8 hours) with a concomitant increase in tyrosine. The rate of fall of PHE may be slower in DHPR deficiency. Blood for pterin analysis at 4 hours will confirm that the BH_4 has been taken and absorbed.

   A combined PHE (100 mg/kg) and BH_4 (20 mg/kg) loading test may be used as an alternative. This combined loading test is reported to identify BH_4 responsive PAH deficiency and discriminate between cofactor synthesis or regeneration defects and is useful if pterin analysis is not available [43].
3. *CSF neurotransmitters*
   The measurement of HVA and 5-HIAA is an essential part of the diagnostic investigation and is also subsequently required to monitor amine replacement therapy with L-dopa and 5-hydroxytrytophan (5HT). CSF must be frozen in liquid nitrogen immediately after collection and stored at –70°C prior to analysis. If blood stained, the sample should be centrifuged immediately

▢ Table 17.4. Interpretation of results of investigations in disorders of biopterin metabolism

Deficiency	Blood PHE µmol/L	Blood or urine biopterin	Blood or urine neopterin	Blood or urine primapterin	CSF 5HIAA and HVA	blood DHPR activity
PAH	>120	↑	↑	–	N	N
GTPCH	90–1200	↓↓	↓↓	–	↓	N
PTPS	240–2500	↓↓	↑↑	–	↓	N
DHPR	180–2500	↓↓	N or ↑	–	↓	↓
PCD	180–1200	↓	↑	↑↑		N

CSF, cerebrospinal fluid; *DHPR*, dihydropterin reductase; *GTPCH*, guanosine triphosphate cyclohydrolase I; *5HIAA*, 5-hydroxyindole acetic acids; *HVA*, homovanelic acid; *N*, normal; *PAH*, phenylalanine hydroxylase; *PCD*, pterin-4a-carbinolamine dehydratase; *PHE*, phenylalanine; *PTPS*, 6-pyruvoyl-tetrahydropterin synthase.

and the supernatant then frozen. The reference ranges for HVA and 5-HIAA are age related [44].

Prenatal diagnosis can be undertaken in 1st trimester following chorion villi biopsy (CVB) by mutation analysis if the mutation of the index case is already known. Analysis of amniotic fluid neopterin and biopterin in the 2nd trimester is available for all conditions. Enzyme analysis can be undertaken in fetal erythrocytes or in amniocytes in both DHPR deficiency and PTPS deficiency. GTPCH is only expressed in fetal liver tissue.

17.4.5   Treatment

For GTPCH deficiency, PTPS deficiency and DHPR deficiency the aim of treatment is to control the HPA and to correct CNS amine deficiency. In DHPR treatment with folinic acid is also required to prevent CNS folate deficiency [45]. PCD deficiency does not usually require treatment although BH$_4$ may be used initially if the child is symptomatic.

In PTPS and GPCH deficiency blood PHE responds to treatment with oral BH$_4$. In DHPR deficiency, BH$_4$ may also be effective in reducing blood PHE, however higher doses may be required than in GTPCH and PTPS deficiency and may lead to an accumulation of BH$_2$ and a possible increased risk of CNS folate deficiency [46]. It is therefore usually recommended that in DHPR deficiency HPA should be corrected by dietary means and BH$_4$ should not be given.

CNS amine replacement therapy is given as oral L-dopa with carbidopa (usually in 1:10 ratio but also available in 1:4 ratio). Carbidopa is a dopa-decarboxylase inhibitor that reduces the peripheral conversion of dopa to dopamine, thus limiting side-effects and allowing a reduced dose of L-dopa to be effective. Side-effects (nausea, vomiting, diarrhoea, irritability) may also be seen at the start of treatment. For this reason L-dopa and 5HT should initially be started

in a low dose (▢ Table 17.5) and increased gradually to the recommended maintenance dose. Further dose adjustment depends on the results of CSF HVA and 5HIAA levels. Monitoring of CSF amine levels should be 3 monthly in the first year, 6 monthly in early childhood and yearly thereafter. Where possible CSF should be collected before a dose of medication is given.

Hyperprolactinaemia occurs as a consequence of dopamine deficiency; measurement of serum prolactin can be used as a method of monitoring treatment with normal values indicating adequate replacement [47].

Selegiline (L-deprenyl) a monoamine oxidase-B inhibitor has been used as an adjunct to amine replacement therapy. It may allow a reduction in the dose of both L-dopa and 5HT and lead to an improvement in clinical symptoms [48].

More recently Entacapone, a catechol-O-methyltransferase (COMT) inhibitor (which is also licensed for use as an adjunct to co-beneldopa or co-careldopa for patients with Parkinson's disease who experience ›end-of-dose‹ deterioration) has also been reported to lead to a reduction in the requirements for L-dopa of up to 30% [49].

17.4.6   Outcome

Without treatment the natural history for GTPCH, 6PTS and DHPR deficiency is poor with progressive neurological disease and early death. The outcome with treatment depends upon the age at diagnosis and phenotypic severity. Most children with GTPCH and 6PTS deficiency have some degree of learning difficulties despite satisfactory control. Patients with DHPR deficiency if started on diet, amine replacement therapy and folinic acid within the first months of life can show normal development and growth.

◘ Table 17.5. Medication used in the treatment of disorders of biopterin metabolism

Drug	Dose (oral)	Frequency	GTPCH	PTPS	PCD	DHPR
BH$_4$	1–3 mg/kg/day	Once daily	+	+	±	–
5HT	1–2 mg/kg/day increasing by 1–2 mg/kg/day every 4–5 days up to maintenance dose of 8 to 10 mg/kg/day	Give in 4 divided doses; final maintenance dose dependent of results of CNS neurotransmitters	+	+	–	+
L-dopa (as combined preparation with carbidopa)	1–2 mg/kg/day increasing by 1–2 mg/kg/day every 4–5 days up to maintenance dose of 10 to 12 mg/kg/day	Give in 4 divided doses; final maintenance dose dependent of results of CNS neurotransmitters	+	+	–	+
Selegiline (l-deprenyl)	0.1–0.25mg/day	3 to 4 divided doses (as adjunct to 5HT & L-dopa – see text)	±	±	–	±
Entacapone	15mg/kg/day	in 2 to 3 divided doses	±	±	–	±
calcium folinate (folinic acid)	15 mg/day	once daily	–	–	–	+

BH$_4$, tetrahydrobiopterin; *CNS*, central nervous system; *DHPR*, dihydropterin reductase; *GTPCH*, guanosine triphosphate cyclohydrolase I; *5HT*, 5-hydroxytrytophan; *PCD*, pterin-4a-carbinolamine dehydratase; *PTPS*, 6-pyruvoyl-tetrahydropterin synthase.

References

1. Folling I (1994) The discovery of phenylketonuria. Acta Paediatr Suppl 407:4-10
2. The Maternal Phenylketonuria Collaborative Study: a status report (1994) Nutr Rev 52:390-393
3. Pietz J, Benninger C, Schmidt H et al (1988) Long-term development of intelligence (IQ) and EEG in 34 children with phenylketonuria treated early. Eur J Pediatr 147:361-367
4. Kreis R (2000) Comments on in vivo proton magnetic resonance spectroscopy in phenylketonuria. Eur J Pediatr 159[Suppl 2]:S126-S128
5. Guldberg P, Rey F, Zschocke J et al (1998) A European Multicenter Study of Phenylalanine Hydroxylase Deficiency: Classification of 105 mutations and a general system for genotype-based prediction of metabolic phenotype. Am J Hum Genet 63:71-79
6. Lichter Konecki U, Rupp A, Konecki DS et al (1994) Relation between phenylalanine hydroxylase genotypes and phenotypic parameters of diagnosis and treatment of hyperphenylalaninaemic disorders. German Collaborative Study of PKU. J Inherit Metab Dis 17:362-365
7. Bartholome K, Lutz P, Bickel H (1975) Determination of phenylalanine hydroxylase activity in patients with phenylketonuria and hyperphenylalaninemia. Pediatr Res 9:899-903
8. Trefz FK, Bartholome K, Bickel H et al (1981) In vivo residual activities of the phenylalanine hydroxylating system in phenylketonuria and variants. J Inherit Metab Dis 4:101-102
9. Scriver CR, Kaufman S (2001) Hyperphenylalaninemia: phenylalanine hydroxylase deficiency. In: Scriver CR, Beaudet AL, Sly WS, Valle D (eds) The metabolic and molecular bases of inherited disease. McGraw-Hill, New York, pp 1667-1724
10. Anonymous (1993) Recommendations on the dietary management of phenylketonuria. Report of Medical Research Council Working Party on Phenylketonuria. Arch Dis Child 68:426-427
11. Burgard P, Bremer HJ, Buhrdel P et al (1999) Rationale for the German recommendations for phenylalanine level control in phenylketonuria 1997. Eur J Pediatr 158:46-54
12. National Institutes of Health Consensus Development Conference Statement: phenylketonuria: screening and management, October 16-18, 2000 (2001) Pediatrics 108:972-982
13. Ding Z, Harding CO, Thony B (2004) State-of-the-art 2003 on PKU gene therapy. Mol Genet Metab 81:3-8
14. Sarkissian CN, Shao Z, Blain F et al (1999) A different approach to treatment of phenylketonuria: phenylalanine degradation with recombinant phenylalanine ammonia lyase. Proc Natl Acad Sci U S A 96:2339-2344
15. Liu J, Jia X, Zhang J et al (2002) Study on a novel strategy to treatment of phenylketonuria. Artif Cells Blood Substit Immobil Biotechnol 30:243-257
16. Koch R, Moseley KD, Yano S et al (2003) Large neutral amino acid therapy and phenylketonuria: a promising approach to treatment. Mol Genet Metab 79:110-113
17. Matalon R, Surendran S, Matalon KM et al (2003) Future role of large neutral amino acids in transport of phenylalanine into the brain. Pediatrics 112(6 Pt 2):1570-1574
18. Muntau AC, Roschinger W, Habich M et al (2002) Tetrahydrobiopterin as an alternative treatment for mild phenylketonuria. N Engl J Med 347:2122-2132
19. Blau N, Erlandsen H (2004) The metabolic and molecular bases of tetrahydrobiopterin-responsive phenylalanine hydroxylase deficiency. Mol Genet Metab 82:101-111
20. Burgard P, Schmidt E, Rupp A et al (1996) Intellectual development of the patients of the German Collaborative Study of children treated for phenylketonuria. Eur J Pediatr 155[Suppl 1]:S33-S38
21. Walter JH, White FJ, Hall SK et al (2002) How practical are recommendations for dietary control in phenylketonuria? Lancet 360:55-57
22. Camfield CS, Joseph M, Hurley T et al (2004) Optimal management of phenylketonuria: a centralized expert team is more successful than a decentralized model of care. J Pediatr 145:53-57
23. Smith I, Beasley MG, Ades AE (1990) Intelligence and quality of dietary treatment in phenylketonuria. Arch Dis Child 65:472-478

24. Smith I, Beasley MG, Ades AE (1991) Effect on intelligence of relaxing the low phenylalanine diet in phenylketonuria. Arch Dis Child 66:311-316

25. Burgard P, Link R, Schweitzer-Krantz S (2000) Phenylketonuria: evidence-based clinical practice. Summary of the roundtable discussion. Eur J Pediatr 159[Suppl 2]:S163-S168

26. Lundstedt G, Johansson A, Melin L et al (2001) Adjustment and intelligence among children with phenylketonuria in Sweden. Acta Paediatr 90:1147-1152

27. Welsh M, Pennington B (2000) Phenylketonuria. In: Yeates KO, Ris MD, Taylor HG (eds) Pediatric neuropsychology. Guildford Press, New York, pp 275-299

28. Cleary MA, Walter JH, Wraith JE et al (1995) Magnetic resonance imaging in phenylketonuria: reversal of cerebral white matter change. J Pediatr 127:251-255

29. Feldmann R, Denecke J, Pietsch M et al (2002) Phenylketonuria: no specific frontal lobe-dependent neuropsychological deficits of early-treated patients in comparison with diabetics. Pediatr Res 51:761-765

30. Thompson AJ, Smith I, Brenton D et al (1990) Neurological deterioration in young adults with phenylketonuria. Lancet 336:602-605

31. Lee P, Smith I, Piesowicz A et al (1999) Spastic paraparesis after anaesthesia. Lancet 353:554

32. Robinson M, White FJ, Cleary MA et al (2000) Increased risk of vitamin B12 deficiency in patients with phenylketonuria on an unrestricted or relaxed diet. J Pediatr 136:545-547

33. Discussion of Armstrong MD (1957) The relation of biochemical abnormality to the development of mental defect in phenylketonuria. Columbus Ohio: Ross Laboratories

34. Lenke RR, Levy HL (1980) Maternal phenylketonuria and hyperphenylalaninaemia. An international survey of the outcome of of untreated and treated pregnancies. N Engl J Med 303:1202-1208

35. Koch R, Hanley W, Levy H et al (2003) The Maternal Phenylketonuria International Study: 1984-2002. Pediatrics 112(6 Pt 2):1523-1529

36. Lee PJ, Ridout D, Walter JH et al (2005) Maternal phenylketonuria: report from the United Kingdom Registry 1978-97. Arch Dis Child 90:143-146

37. Phenylketonuria due to phenylalanine hydroxylase deficiency: an unfolding story. Medical Research Council Working Party on Phenylketonuria (1993) BMJ 306:115-119

38. Soltesz G, Harris D, Mackenzie IZ et al (1985) The metabolic and endocrine milieu of the human fetus and mother at 18-21 weeks of gestation. I. Plasma amino acid concentrations. Pediatr Res 19:91-93

39. Lee PJ, Lilburn M, Baudin J (2003) Maternal phenylketonuria: experiences from the United Kingdom. Pediatrics 112(6 Pt 2):1553-1556

40. American Academy of Pediatrics: Maternal phenylketonuria (2001) Pediatrics 107:427-428

41. Koch R, Hanley W, Levy H et al (2000) Maternal phenylketonuria: an international study. Mol Genet Metab 71:233-239

42. Levy HL, Waisbren SE, Guttler F et al (2003) Pregnancy experiences in the woman with mild hyperphenylalaninemia. Pediatrics 112(6 Pt 2):1548-1552

43. Ponzone A, Guardamagna O, Spada M et al (1993) Differential diagnosis of hyperphenylalaninaemia by a combined phenylalanine-tetrahydrobiopterin loading test. Eur J Pediatr 152:655-661

44. Hyland K, Surtees RA, Heales SJ et al (1993) Cerebrospinal fluid concentrations of pterins and metabolites of serotonin and dopamine in a pediatric reference population. Pediatr Res 34:10-14

45. Smith I, Hyland K, Kendall B (1985) Clinical role of pteridine therapy in tetrahydrobiopterin deficiency. J Inherit Metab Dis 8[Suppl 1]:39-45

46. Hyland K (1993) Abnormalities of biogenic amine metabolism. J Inherit Metab Dis 16:676-690

47. Spada M, Ferraris S, Ferrero GB et al (1996) Monitoring treatment in tetrahydrobiopterin deficiency by serum prolactin. J Inherit Metab Dis 19:231-233

48. Schuler A, Kalmanchey R, Barsi P et al (2000) Deprenyl in the treatment of patients with tetrahydrobiopterin deficiencies. J Inherit Metab Dis 23:329-332

49. Ponzone A, Spada M, Ferraris S et al (2004) Dihydropteridine reductase deficiency in man: from biology to treatment. Med Res Rev 24:127-150

18   Disorders of Tyrosine Metabolism

Anupam Chakrapani, Elisabeth Holme

Tyrosine Metabolism

Tyrosine is one of the least soluble amino acids, and forms characteristic crystals upon precipitation. It derives from two sources, diet and hydroxylation of phenylalanine (◘ Fig. 18.1). Tyrosine is both glucogenic and ketogenic, since its catabolism, which proceeds predominantly in the liver cytosol, results in the formation of fumarate and acetoacetate. The first step of tyrosine catabolism is conversion into 4-hydroxyphenylpyruvate by cytosolic tyrosine aminotransferase. Transamination of tyrosine can also be accomplished in the liver and in other tissues by mitochondrial aspartate aminotransferase, but this enzyme plays only a minor role under normal conditions. The penultimate intermediates of tyrosine catabolism, maleylacetoacetate and fumarylacetoacetate, can be reduced to succinylacetoacetate, followed by decarboxylation to succinylacetone. The latter is the most potent known inhibitor of the heme biosynthetic enzyme, 5-aminolevulinic acid dehydratase (porphobilinogen synthase, ◘ Fig. 36.1).

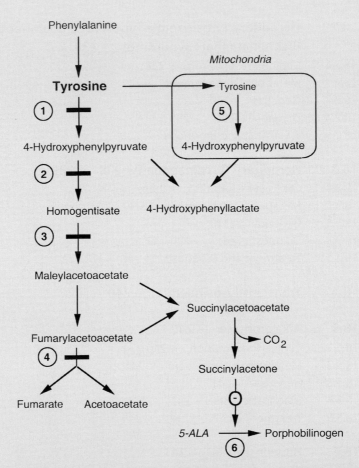

◘ **Fig. 18.1.** The tyrosine catabolic pathway. **1**, Tyrosine aminotransferase (deficient in tyrosinaemia type II); **2**, 4-hydroxyphenylpyruvate dioxygenase (deficient in tyrosinaemia type III, hawkinsinuria, site of inhibition by NTBC); **3**, homogentisate dioxygenase (deficient in alkaptonuria); **4**, fumarylacetoacetase (deficient in tyrosinaemia type I); **5**, aspartate aminotransferase; **6**, 5-aminolevulinic acid (5-ALA) dehydratase (porphobilinogen synthase). Enzyme defects are depicted by *solid bars* across the arrows

Five inherited disorders of tyrosine metabolism are known, depicted in Fig. 18.1. Hereditary tyrosinaemia type I is characterised by progressive liver disease and renal tubular dysfunction with rickets. Hereditary tyrosinaemia type II (Richner-Hanhart syndrome) presents with keratitis and blisterous lesions of the palms and soles. Tyrosinaemia type III may be asymptomatic or associated with mental retardation. Hawkinsinuria may be asymptomatic or presents with failure to thrive and metabolic acidosis in infancy. In alkaptonuria symptoms of osteoarthritis usually appear in adulthood. Other inborn errors of tyrosine metabolism include oculocutaneous albinism caused by a deficiency of melanocyte-specific tyrosinase, converting tyrosine into DOPA-quinone; the deficiency of tyrosine hydroxylase, the first enzyme in the synthesis of dopamine from tyrosine; and the deficiency of aromatic L-amino acid decarboxylase, which also affects tryptophan metabolism. The latter two disorders are covered in ▶ Chap. 29.

18.1 Hereditary Tyrosinaemia Type I (Hepatorenal Tyrosinaemia)

18.1.1 Clinical Presentation

The clinical manifestations of tyrosinaemia type 1 are very variable and an affected individual can present at any time from the neonatal period to adulthood. There is considerable variability of presentation even between members of the same family.

Clinically, tyrosinaemia type 1 may be classified based on the age at onset of symptoms, which broadly correlates with disease severity: an »acute« form that manifests before 6 months of age with acute liver failure; a »subacute« form presenting between 6 months and 1 year of age with liver disease, failure to thrive, coagulopathy, hepatosplenomegaly, rickets and hypotonia; and a more »chronic« form that presents after the first year with chronic liver disease, renal disease, rickets, cardiomyopathy and/or a porphyria-like syndrome. Treatment of tyrosinaemia type 1 with NTBC in the last decade (▶ Sect. 18.1.5) has dramatically altered its natural history.

Hepatic Disease

The liver is the major organ affected in tyrosinaemia 1, and is a major cause of morbidity and mortality. Liver disease can manifest as acute hepatic failure, cirrhosis or hepatocellular carcinoma; all three conditions may occur in the same patient. The more severe forms of tyrosinaemia type 1 present in infancy with vomiting, diarrhoea, bleeding diathesis, hepatomegaly, jaundice, hypoglycaemia, edema and ascites. Typically, liver synthetic function is most affected

and in particular, coagulation is markedly abnormal compared with other tests of liver function. Sepsis is common and early hypophosphataemic bone disease may be present secondary to renal tubular dysfunction. Acute liver failure may be the initial presenting feature or may occur subsequently, precipitated by intercurrent illnesses as »hepatic crises« which are associated with hepatomegaly and coagulopathy. Mortality is high in untreated patients [1].

Chronic liver disease leading to cirrhosis eventually occurs in most individuals with tyrosinaemia 1 – both as a late complication in survivors of early-onset disease and as a presenting feature of the later-onset forms. The cirrhosis is usually a mixed micromacronodular type with a variable degree of steatosis [2]. There is a high risk of carcinomatous transformation within these nodules [1, 3]. Unfortunately, the differences in size and fat content of the nodules make it difficult to detect malignant changes (▶ Sect. 18.1.5).

Renal Disease

A variable degree of renal dysfunction is detectable in most patients at presentation, ranging from mild tubular dysfunction to renal failure. Proximal tubular disease is very common and can become acutely exacerbated during hepatic crises. Hypophosphataemic rickets is the most common manifestation of proximal tubulopathy but generalised aminoaciduria, renal tubular acidosis and glycosuria may also be present [4]. Other less common renal manifestations include distal renal tubular disease, nephrocalcinosis and reduced glomerular filtration rates.

Neurological Manifestations

Acute neurological crises can occur at any age. Typically, the crises follow a minor infection associated with anorexia and vomiting, and occur in two phases: an active period lasting 1–7 days characterised by painful parasthesias and autonomic signs that may progress to paralysis, followed by a recovery phase over several days [5]. Complications include seizures, extreme hyperextension, self-mutilation, respiratory paralysis and death.

Other Manifestations

Cardiomyopathy is a frequent incidental finding, but may be clinically significant [6]. Asymptomatic pancreatic cell hypertrophy may be detected at presentation, but hyperinsulinism and hypoglycaemia are rare [7].

18.1.2 Metabolic Derangement

Tyrosinaemia type 1 is caused by a deficiency of the enzyme fumarylacetoacetate hydrolase (FAH), which is mainly expressed in the liver and kidney. The compounds immediately upstream from the FAH reaction, maleylacetoacetate (MAA) and fumarylacetoacetate (FAA), and their derivatives, succinylacetone (SA) and succinylacetoacetate (SAA) accumulate and have important pathogenic effects. The effects of FAA and MAA occur only in the cells of the organs in which they are produced; these compounds are not found in body fluids of patients. On the other hand, their derivatives, SA and SAA are readily detectable in plasma and urine and have widespread effects.

FAA, MAA and SA disrupt sulfhydryl metabolism by forming glutathione adducts, thereby rendering cells susceptible to free radical damage [8, 9]. Disruption of sulfhydryl metabolism is also believed to cause secondary deficiency of two other hepatic enzymes, 4-hydroxyphenylpyruvate dioxygenase and methionine adenosyltransferase, resulting in hypertyrosinemia and hypermethioninemia. Additionally, FAA and MAA are alkylating agents and can disrupt the metabolism of thiols, amines, DNA and other important intracellular molecules. As a result of these widespread effects on intracellular metabolism, hepatic and renal cells exposed to high levels of these compounds undergo either apoptotic cell death or a significant alteration of gene expression [10–12]. In patients who have developed cirrhosis, self-induced correction of the genetic defect and the enzyme abnormality occurs within some nodules [13]. The clinical expression of hepatic disease may correlate inversely with the extent of mutation reversion in regenerating nodules [14]. The mechanisms that underlie the development of hepatocellular carcinoma within nodules are poorly understood.

SA is a potent inhibitor of the enzyme 5-ALA dehydratase. 5-ALA, a neurotoxic compound, accumulates and is excreted at high levels in patients with tyrosinemia type 1 and is believed to cause the acute neurological crises seen during decompensation [5]. SA is also known to disrupt renal tubular function, heme synthesis and immune function [15–17].

18.1.3 Genetics

Hereditary tyrosinaemia type I is inherited as an autosomal recessive trait. The FAH gene has been localised to 15q 23–25 and more than 40 mutations have been reported [18]. The most common mutation, IVS12+5(g-a), is found in about 25 % of the alleles worldwide and is the predominant mutation in the French-Canadian population in which it accounts for >90 % of alleles. Another mutation, IVS6-1(g-t) is found in around 60 % of alleles in patients from the Mediterranean area. Other FAH mutations are common within certain ethnic groups: W262X in Finns, D233V in Turks, and Q64H in Pakistanis. There is no clear genotype-phenotype correlation [19]; spontaneous correction of the mutation within regenerative nodules may influence the clinical phenotype [14]. A pseudodeficiency mutation, R341W, has been reported in healthy individuals who have in vitro FAH activity indistinguishable from patients with type 1 tyrosinaemia [20]. The frequency of this mutation in various populations is unknown but it has been found in many different ethnic groups.

18.1.4 Diagnostic Tests

In symptomatic patients, biochemical tests of liver function are usually abnormal. In particular, liver synthetic function is severely affected – coagulopathy and/or hypoalbuminemia are often present even if other tests of liver function are normal. In most acutely ill patients, α-fetoprotein levels are greatly elevated. A Fanconi-type tubulopathy is often present with aminoaciduria, phosphaturia and glycosuria, and radiological evidence of rickets may be present.

Elevated levels of succinylacetone in dried blood spots, plasma or urine are pathognomonic of tyrosinaemia type 1. Other metabolite abnormalities that are suggestive of the diagnosis include elevated plasma levels of tyrosine, phenylalanine and methionine, reduced erythrocyte 5-aminolevulinate dehydratase activity and increased urinary 5-ALA excretion.

Confirmation of the diagnosis requires either enzyme assay or mutation analysis. FAH assays may be performed on liver biopsy, fibroblasts, lymphocytes or dried blood spots [21–23]. Falsely elevated enzyme results may be obtained on liver biopsy if a reverted nodule is inadvertently assayed. Enzyme assay results should therefore be interpreted in the context of the patients' clinical and biochemical findings.

Newborn Screening

Screening using tyrosine levels alone has been used in the past and has resulted in very high false positive and false negative rates [24]. More recently, methods based on the inhibitory effects of SA on porphobilinogen synthase, either alone or in combination with tyrosine levels have successfully reduced false-positive rates [25]. Molecular screening is possible in populations in which one or few mutations account for the majority of cases.

Prenatal Diagnosis

If the causative mutations in a pregnancy at risk are known, antenatal diagnosis is best performed by mutation analysis on chorionic villus sampling (CVS) or amniocytes. Alternative methods include FAH assay on CVS [26] or amniocytes [27] and determination of SA levels in amniotic fluid [28]. However, FAH is expressed at low levels in chorionic tissue and interpretation of results may be difficult. Assay for elevated SA levels in amniotic fluid is very reliable and can be performed as early as 12 weeks; however, occasional affected pregnancies have reported normal SA amniotic fluid levels [29]. When mutation analysis is not available for prenatal diagnosis, we currently use a strategy combining initial screening for the common pseudodeficiency mutation and FAH assay on CVS at 10 weeks; pregnancies that have low FAH activity on CVS subsequently undergo amniocentesis for amniotic fluid SA levels at 11–12 weeks for confirmation.

18.1.5 Treatment and Prognosis

Historically, tyrosinaemia type I was treated with a tyrosine and phenylalanine restricted diet, with or without liver transplantation. In 1992 a new drug, 2-(2-nitro-4-trifluoro-methylbenzoyl)-1,3-cyclohexanedione (NTBC), a potent inhibitor of 4-hydroxyphenylpyruvate dioxygenase was introduced (◘ Fig. 18.1, enzyme 2); it has revolutionised the treatment of type 1 tyrosinaemia and is now the mainstay of therapy [30].

NTBC

The rationale for the use of NTBC is to block tyrosine degradation at an early step so as to prevent the production of toxic down-stream metabolites such as FAA, MAA and SA; the levels of tyrosine, 4-hydroxyphenyl-pyruvate and -lactate concomitantly increase (◘ Fig. 18.1). The Gothenberg multicentre study provides the major experience of NTBC treatment in tyrosinaemia type 1 [31]. Over 300 patients have been treated; of these, over 100 have been treated for over 5 years. NTBC acts within hours of administration and has a long half-life of about 54 hours [32]. In patients presenting acutely with hepatic decompensation, rapid clinical improvement occurs in over 90% with normalisation of prothrombin time within days of starting treatment. Other biochemical parameters of liver function may take longer to normalise: α-fetoprotein concentrations may not normalise for up to several months after starting treatment. NTBC is recommended in an initial dose of 1 mg/kg body weight per day [31]. Individual dose adjustment is subsequently based on the biochemical response and the plasma NTBC concentration. Dietary restriction of phenylalanine and tyrosine is necessary to prevent the known adverse effects of hypertyrosinaemia (see tyrosinaemia type II). We currently aim to maintain tyrosine levels between 200 and 400 μmol/l using a combination of a protein-restricted diet and phenylalanine and tyrosine free amino acid mixtures.

A small proportion of acutely presenting patients (<10%) do not respond to NTBC treatment; in these patients, coagulopathy and jaundice progress and mortality is very high without urgent liver transplantation.

Adverse events of NTBC therapy have been few. Transient thrombocytopenia and neutropenia and transient eye symptoms (burning/photophobia/corneal erosion/corneal clouding) have been reported in a small proportion of patients [31]. The short- to medium-term prognosis in responders appears to be excellent. Hepatic and neurological decompensations are not known to occur on NTBC treatment, and clear deterioration of chronic liver disease is rare. Renal tubular dysfunction responds well to NTBC therapy, but long-standing renal disease may be irreversible. Neurological crises are rarely seen in patients treated with NTBC.

The risk of hepatocellular carcinoma appears to be much reduced in patients started early on NTBC treatment. In particular, the risk is very low if treatment is commenced before 6 months of age. In patients started on NTBC after 6 months of age, the risk of developing hepatocellular carcinoma increases with the age at which treatment is introduced; if NTBC is introduced after 2 years of age, the risk may not be much different from that in historical controls (◘ Table 18.1). It remains to be determined whether early NTBC treatment can prevent liver cancer in the long term. Studies on the animal mouse models suggest that late hepatocellular carcinoma may occur even if NTBC treatment is started at birth [10, 33]; careful long-term vigilance is therefore necessary in all patients.

The long-term neuropsychological outcome of NTBC-treated patients with tyrosinaemia type 1 is also unclear. Many patients appear to have significant learning difficulties; cognitive deficits affecting performance abilities more than verbal abilities have been found in many patients on psychological testing [34]. The etiology of these cognitive deficits is uncertain; whether they are related to NTBC

◘ **Table 18.1.** Risk of hepatocellular carcinoma (HCC) in tyrosinemia type 1

	Number of patients	Age (in years) at assessment	Patients developing HCC (%)
Pre-NTBC			
Weinberg et al [3]	43	>2	16 (37%)
Van Spronsen et al [1]	55	2–12	10 (18%)
Gothenberg NTBC study (2004) Treatment started at:			
< 6 months	180	2–13	1 (0.6 %)
6–12 months	61	2–12	1 (1.6%)
1–2 years	44	2–12	3 (7%)
2–7 years	65	2–19	14 (21%)
> 7 years	26	7–31	9 (35%)

treatment or high tyrosine levels or are a feature of tyrosinemia 1 per se is unknown.

Monitoring of patients on NTBC treatment should include regular blood tests for liver function, blood counts, clotting studies, alpha fetoprotein, SA, plasma PBG synthase activity, 5-ALA, NTBC levels and amino acid profile; tests of renal tubular and glomerular function; urinary SA and 5-ALA; and hepatic imaging by ultrasound and CT/MRI. Blood levels of phenylalanine and tyrosine should be frequently monitored and the diet supervised closely.

Liver Transplantation

Liver transplantation has been used for over two decades in treating type 1 tyrosinaemia, and appears to cure the hepatic and neurological manifestations [35, 36]. However, even in optimal circumstances, it is associated with approximately 5–10% mortality and necessitates lifelong immunosuppressive therapy. Therefore, at present liver transplantation in type 1 tyrosinaemia is restricted to patients with acute liver failure who fail to respond to NTBC therapy, and in patients with suspected hepatocellular carcinoma. Currently, there is no non-invasive way of reliably detecting malignancy within hepatic nodules. Regular monitoring of plasma α-fetoprotein levels and of hepatic architecture on computerized tomography (CT) or magnetic resonance imaging (MRI) are essential; liver transplantation has to be considered if these investigations suggest malignant transformation. Other situations in which liver transplantation may be considered relate to the irreversible manifestations of chronic liver disease, such as severe portal hypertension, growth failure and poor quality of life.

The long-term impact of liver transplantation on renal disease in tyrosinaemia type 1 is not fully known. Tubular dysfunction improves in most patients, but does not always normalise. Glomerular function generally remains stable but may be affected by nephrotoxic immunotherapy [34, 37]. Urinary SA excretion is much reduced after liver transplantation but does not normalise, presumably due to continued renal production [38]; whether this affects renal function long-term and predisposes to renal malignancy is unknown. In patients with severe hepatic and renal disease, combined liver and kidney transplantation should be considered.

Dietary Treatment

Before the advent of NTBC therapy, dietary protein restriction was the only available treatment for tyrosinaemia type 1 apart from liver transplantation. Dietary treatment was helpful in relieving the acute symptoms and perhaps slowing disease progression, but it did not prevent the acute and chronic complications including hepatocellular carcinoma. Currently, dietary therapy alone is not recommended, but is used in conjunction with NTBC therapy to prevent the complications related to hypertyrosinaemia. Ocular and dermatological complications are not believed to occur

below plasma tyrosine levels of 800 µmol/l; however, lower levels (200–400 µmol/l) are usually recommended due to possible effects of hypertyrosinaemia on cognitive outcome. Whether dietary treatment is used alone or in conjunction with NTBC, the principle is the same: natural protein intake is restricted to provide just enough phenylalanine plus tyrosine to keep plasma tyrosine levels <400 µmol/l; the rest of the normal daily protein requirement is given in the form of a phenylalanine- and tyrosine-free amino acid mixture. Some patients develop very low phenylalanine levels with this regimen and may require phenylalanine supplements [39].

Supportive Treatment

In the acutely ill patient supportive treatment is essential. Clotting factors, albumin, electrolytes and acid/base balance should be closely monitored and corrected as necessary. Tyrosine and phenylalanine intake should be kept to a minimum during acute decompensation. Addition of vitamin D, preferably 1,25 hydroxy vitamin D_3 or an analogue, may be required to treat rickets. Infections should be treated aggressively.

Pregnancy

To date, no published data on pregnancies in patients on NTBC treatment is available; one pregnancy in a liver-transplanted tyrosinaemia type 1 patient has had a favourable outcome [40].

18.2 Hereditary Tyrosinaemia Type II (Oculocutaneous Tyrosinaemia, Richner-Hanhart Syndrome)

18.2.1 Clinical Presentation

The disorder is characterised by ocular lesions (about 75% of the cases), skin lesions (80%), and neurological complications (60%), or any combination of these [41]. The disorder usually presents in infancy but may become manifest at any age.

Eye symptoms are often the presenting problem and may start in the first months of life with photophobia, lacrimation and intense burning pain [42]. The conjunctivae are inflamed and on slit-lamp examination herpetic-like corneal ulcerations are found. The lesions stain poorly with fluorescein. In contrast with herpetic ulcers, which are usually unilateral, the lesions in tyrosinaemia type II are bilateral. Neovascularisation may be prominent. Untreated, serious damage may occur with corneal scarring, visual impairment, nystagmus and glaucoma.

Skin lesions specifically affect pressure areas and most commonly occur on the palms and soles [43, 44]. They begin as blisters or erosions with crusts and progress to painful, nonpruritic hyperkeratotic plaques with an erythematous rim, typically ranging in diameter from 2 mm to 3 cm.

The neurological complications are highly variable: some patients are developmentally normal whilst others have variable degrees of developmental retardation. More severe neurological problems, including microcephaly, seizures, self-mutilation and behavioural difficulties have also been described [45].

It should be noted that the diagnosis of tyrosinaemia type II has only been confirmed by enzymatic and/or molecular genetic analysis in a minority of the described cases and it is possible that some of the patients actually have tyrosinaemia type III.

18.2.2   Metabolic Derangement

Tyrosinaemia type II is due to a defect of hepatic cytosolic tyrosine aminotransferase (◨ Fig. 18.1, enzyme 1). As a result of the metabolic block, tyrosine concentrations in serum and cerebrospinal fluid are markedly elevated. The accompanying increased production of the phenolic acids 4-hydroxyphenyl-pyruvate, -lactate and -acetate (not shown in ◨ Fig. 18.1) may be a consequence of direct deamination of tyrosine in the kidneys, or of tyrosine catabolism by mitochondrial aminotransferase (◨ Fig. 18.1). Corneal damage is thought to be related to crystallization of tyrosine in the corneal epithelial cells, which results in disruption of cell function and induces an inflammatory response. Tyrosine crystals have not been observed in the skin lesions. It has been suggested that excessive intracellular tyrosine enhances cross-links between aggregated tonofilaments and modulates the number and stability of microtubules [46]. As the skin lesions occur on pressure areas, it is likely that mechanical factors also play a role. The etiology of the neurological manifestations is unknown, but it is believed that hypertyrosinaemia may have a role in pathogenesis.

18.2.3   Genetics

Tyrosinaemia type II is inherited as an autosomal recessive trait. The gene is located at 16q22.1-q22.3. Twelve different mutations have so far been reported in the tyrosine aminotransferase gene [35]. Prenatal diagnosis has not been reported.

18.2.4   Diagnostic Tests

Plasma tyrosine concentrations are usually above 1200 μmol/l. When the tyrosinaemia is less pronounced a diagnosis of tyrosinaemia type III should be considered (▶ Sect. 18.3). Urinary excretion of the phenolic acids 4-hydroxyphenyl-pyruvate, -lactate, -acetate is highly elevated and N-acetyl-

tyrosine and 4-tyramine are also increased. The diagnosis can be confirmed by enzyme assay on liver biopsy or by mutation analysis.

18.2.5   Treatment and Prognosis

Treatment consists of a phenylalanine and tyrosine-restricted diet, and the skin and eye symptoms resolve within weeks of treatment [44, 47]. Generally, skin and eye symptoms do not occur at tyrosine levels < 800 μmol/l; however, as hypertyrosinaemia may be involved in the pathogenesis of the neurodevelopmental symptoms, it may be beneficial to maintain much lower levels [48]. We currently aim to maintain plasma tyrosine levels of 200–400 μmol/l using a combination of a protein-restricted diet and a phenylalanine and tyrosine free amino acid mixture. Growth and nutritional status should be regularly monitored.

Pregnancy

There have been several reports of pregnancies in patients with tyrosinaemia type II: some have suggested that untreated hypertyrosinaemia may result in fetal neurological abnormalities such as microcephaly, seizures and mental retardation [45, 49, 50]; however, other pregnancies have reported normal fetal outcome [45, 51]. In view of the uncertainty regarding possible fetal effects of maternal hypertyrosinaemia, dietary control of maternal tyrosine levels during pregnancy is recommended [50].

18.3   Hereditary Tyrosinaemia Type III

18.3.1   Clinical Presentation

Only 13 cases of tyrosinaemia type III have been described and the full clinical spectrum of this disorder is unknown [52]. Many of the patients have presented with neurological symptoms including intellectual impairment, ataxia, increased tendon reflexes, tremors, microcephaly and seizures; some have been detected by the finding of a high tyrosine concentration on neonatal screening. The most common long-term complication has been intellectual impairment, found in 75% of the reported cases. None of the described cases have developed signs of liver disease in the long-term. Eye and skin lesions have not been reported so far, but as oculocutaneous symptoms are known to occur in association with hypertyrosinaemia it is reasonable to be aware of this possibility.

18.3.2   Metabolic Derangement

Tyrosinaemia type III is due to deficiency of 4-hydroxyphenylpyruvate dioxygenase (HPD) (◨ Fig. 18.1, enzyme 2), which is expressed in liver and kidney. As a result of the enzyme block there is an increased plasma tyrosine concentration and increased excretion in urine of 4-hydroxy-phenyl-pyruvate and its derivatives 4-hydroxyphenyl-lactate and 4-hydroxyphenyl-acetate. The aetiology of the neurological symptoms is not known, but they may be related to hypertyrosinaemia as in tyrosinaemia types 1 and 2.

18.3.3   Genetics

Tyrosinaemia type III follows autosomal recessive inheritance. The HPD gene has been localised to 12q24-qter and 5 mutations associated with tyrosinaemia III have been described [35]. There is no apparent genotype-phenotype correlation; some patients with enzymatically defined HPD deficiency do not have identifiable mutations in the HPD gene [52, 53].

18.3.4   Diagnostic Tests

Elevated plasma tyrosine levels of 300–1300 μmol/l have been found in the described cases at diagnosis. Elevated urinary excretion of 4-hydroxyphenyl-pyruvate, -lactate and -acetate usually accompanies the increased plasma tyrosine concentration. Diagnosis can be confirmed by enzyme assay in liver or kidney biopsy specimens or by mutation analysis.

18.3.5   Treatment and Prognosis

At present, tyrosinemia type III appears to be associated with intellectual impairment in some cases, but not in others. It is unknown whether lowering plasma tyrosine levels will alter the natural history. Amongst the patients described, the cases detected by neonatal screening and treated early appear to have fewer neurological abnormalities than those diagnosed on the basis of neurological symptoms [52]; whether this is due to ascertainment bias or due to therapeutic intervention is unclear. Until there is a greater understanding of the etiology of the neurological complications of tyrosinaemia type III, it is reasonable to treat patients with a low-phenylalanine and tyrosine diet. We currently recommend maintaining plasma tyrosine levels between 200 and 400 μmol/l. No pregnancy data is available to date.

18.4   Transient Tyrosinaemia

Transient tyrosinaemia is one of the most common amino acid disorders, and is believed to be caused by late fetal maturation of 4-hydroxyphenylpyruvate dioxygenase (◘ Fig. 18.1, enzyme 2). It is more common in premature infants than full term newborns. The level of protein intake is an important etiological factor: the incidence of transient tyrosinaemia has fallen dramatically in the last 4 decades, concomitant with a reduction in the protein content of newborn formula milks. Transient tyrosinaemia is clinically asymptomatic. Tyrosine levels arc extremely variable, and can exceed 2000 μmol/l. Hypertyrosinaemia usually resolves spontaneously by 4–6 weeks; protein restriction to less than 2 g/kg/day with or without vitamin C supplementation results in more rapid resolution in most cases. Although the disorder is generally considered benign, some reports have suggested that it may be associated with mild intellectual deficits in the long term [54, 55]. However, large systematic studies have not been performed.

18.5   Alkaptonuria

18.5.1   Clinical Presentation

Some cases of alkaptonuria are diagnosed in infancy due to darkening of urine when exposed to air. However, clinical symptoms first appear in adulthood. The most prominent symptoms relate to joint and connective tissue involvement; significant cardiac disease and urolithiasis may be detected in the later years [56].

The pattern of joint involvement resembles osteoarthritis. In general, joint disease tends to be worse in males than in females. The presenting symptom is usually either limitation of movement of a large joint or low back pain starting in the 3rd or 4th decade. Spinal involvement is progressive and may result in kyphosis, limited spine movements and height reduction. On X-ray examination, narrowing of the disk spaces, calcification and vertebral fusion may be evident. In addition to the spine, the large weight-bearing joints such as the hips, knees and ankles are usually involved. Radiological abnormalities may range from mild narrowing of the joint space to destruction and calcification. Synovitis, ligament tears and joint effusions have also been described. The small joints of the hands and feet tend to be spared. Muscle and tendon involvement is common: thickened Achilles tendons may be palpable, and tendons and muscles may be susceptible to rupture with trivial trauma. The clinical course is characterised by episodes of acute exacerbation and progressive joint disability; joint replacement for chronic pain may be required. Physical disability increases with age and may become very severe by the 6th decade.

A greyish discoloration (ochre on microscopic examination, thus the name ochronosis) of the sclera and the ear cartilages usually appears after 30 years of age. Subsequently, dark coloration of the skin particularly over the nose, cheeks and in the axillary and pubic areas may become evident. Cardiac involvement probably occurs in most patients eventually; aortic or mitral valve calcification or regurgitation and coronary artery calcification is evident on CT scan and echocardiography in about 50% of patients by the 6th decade [56]. A high frequency of renal and prostatic stones has also been reported.

18.5.2 Metabolic Derangement

Alkaptonuria was the first disease to be interpreted as an inborn error of metabolism in 1902 by Garrod [57]. It is caused by a defect of the enzyme homogentisate dioxygenase (❏ Fig. 18.1, enzyme 3), which is expressed mainly in the liver and the kidneys. There is accumulation of homogentisate and its oxidised derivative benzoquinone acetic acid, the putative toxic metabolite and immediate precursor to the dark pigment, which gets deposited in various tissues. The relationship between the pigment deposits and the systemic manifestations is not known. It has been proposed that the pigment deposit may act as a chemical irritant [58]; alternatively, inhibition of some of the enzymes involved in connective tissue metabolism by homogentisate or benzoquinone acetic acid may have a role in pathogenesis [59].

18.5.3 Genetics

Alkaptonuria is an autosomal recessive disorder. The gene for homogentisate oxidase has been mapped to chromosome 3q2, and over 40 mutations have been identified [35]. The estimated incidence is between 1:250 000 and 1:1 000 000 live births.

18.5.4 Diagnostic Tests

Alkalinisation of the urine from alkaptonuric patients results in immediate dark brown coloration of the urine. Excessive urinary homogentisate also results in a positive test for reducing substances. Gas chromatography – mass spectrometry (GC-MS) based organic acid screening methods can specifically identify and quantify homogentisic acid. Homogentisate may also be quantified by HPLC [60] and by specific enzymatic methods [61].

18.5.5 Treatment and Prognosis

A number of different approaches have been used to attempt treatment. Dietary restriction of phenylalanine and tyrosine intake reduces homogentisate excretion, but compliance is a major problem as the diagnosis is usually made in adults [62]. Ascorbic acid prevents the binding of ^{14}C-homogentisic acid to connective tissue in rats [63] and reduces the excretion of benzoquinone acetic acid in urine [64]. Administration of the drug NTBC also reduces urinary

homogentisate excretion; the concomitant hypertyrosinaemia requires dietary adjustment to prevent ocular, cutaneous and neurological complications [56]. None of these therapies have been subjected to long-term clinical trials, and currently, no treatment can be recommended as being effective in preventing the late effects of alkaptonuria.

To date, there is no published data on pregnancies in patients with alkaptonuria.

18.6 Hawkinsinuria

18.6.1 Clinical Presentation

This rare condition, which has only been described in four families [65–67], is characterised by failure to thrive and metabolic acidosis in infancy. After the first year of life the condition appears to be asymptomatic. Early weaning from breastfeeding seems to precipitate the disease; the condition may be asymptomatic in breastfed infants.

18.6.2 Metabolic Derangement

The abnormal metabolites produced in hawkinsinuria (hawkinsin (2-cysteinyl-1,4-dihydroxycyclohexenylacetate) and 4-hydroxycycloxylacetate) are thought to derive from an incomplete conversion of 4-hydroxyphenylpyruvate to homogentisate caused by a defect 4-hydroxyphenylpyruvate dioxygenase (❏ Fig. 18.1, enzyme 2). Hawkinsin is thought to be the product of a reaction of an epoxide intermediate with glutathione, which may be depleted. The metabolic acidosis is believed to be due to 5-oxoproline accumulation secondary to glutathione depletion.

18.6.3 Genetics

Unlike most other inborn errors of metabolism, hawkinsinuria shows autosomal dominant inheritance. The molecular basis of the condition is unknown. It is believed that a specific mutation or a limited number of mutations in the 4-hydroxyphenylpyruvate dioxygenase gene can partially disrupt enzyme activity and lead to the production of hawkinsin and 4-hydroxycyclohexylacetate. Neither the enzymatic defect nor the molecular genetics have been studied in detail.

18.6.4 Diagnostic Tests

Identification of urinary hawkinsin or 4-hydroxycyclohexylacetate by GC-MS is diagnostic [67]. Hawkinsin is a ninhydrin-positive compound, which appears between urea and threonine in ion-exchange chromatography of urine amino acids [68]. Increased excretion of 4-hydro-xycyclohexylacetate is detected on urine organic acids analysis. In addition to hawkinsinuria there may be moderate tyrosinaemia, increased urinary 4-hydroxyphenylpyruvate and 4-hydroxyphenyllactate, metabolic acidosis and 5-oxo-prolinuria during infancy. 4-Hydroxycyclohexylacetate is usually detectable only after infancy.

18.6.5   Treatment and Prognosis

Symptoms in infancy respond to a return to breastfeeding or a diet restricted in tyrosine and phenylalanine along with vitamin C supplementation. The condition is asymptomatic after the first year of life and affected infants are reported to have developed normally.

References

1. van Spronsen FJ, Thomasse Y, Smit GP et al (1994) Hereditary tyrosinemia type I: a new clinical classification with difference in prognosis on dietary treatment. Hepatology 20:1187-1191
2. Russo PA, Mitchell GA, Tanguay RM (2001) Tyrosinemia: a review. Pediatr Dev Pathol 4:212-221
3. Weinberg AG, Mize CE, Worthen HG (1976) The occurrence of hepatoma in the chronic form of hereditary tyrosinemia. J Pediatr 88:434-438
4. Forget S, Patriquin HB, Dubois J et al (1999) The kidney in children with tyrosinemia: sonographic, CT and biochemical findings. Pediatr Radiol 29:104-108
5. Mitchell G, Larochelle J, Lambert M et al (1990) Neurologic crises in hereditary tyrosinemia. N Engl J Med 322:432-437
6. Edwards MA, Green A, Colli A, Rylance G (1987) Tyrosinaemia type I and hypertrophic obstructive cardiomyopathy. Lancet 1:1437-1438
7. Lindberg T, Nilsson KO, Jeppsson JO (1979) Hereditary tyrosinaemia and diabetes mellitus. Acta Paediatr Scand 68:619-620
8. Manabe S, Sassa S, Kappas A (1985) Hereditary tyrosinemia. Formation of succinylacetone-amino acid adducts. J Exp Med 162:1060-1074
9. Jorquera R, Tanguay RM (1997) The mutagenicity of the tyrosine metabolite, fumarylacetoacetate, is enhanced by glutathione depletion. Biochem Biophys Res Commun 232:42-48
10. Grompe M (2001) The pathophysiology and treatment of hereditary tyrosinemia type 1. Semin Liver Dis 21:563-571
11. Endo F, Sun MS (2002) Tyrosinaemia type I and apoptosis of hepatocytes and renal tubular cells. J Inherit Metab Dis 25:227-234
12. Tanguay RM, Jorquera R, Poudrier J, St Louis M (1996) Tyrosine and its catabolites: from disease to cancer. Acta Biochim Pol 43:209-216
13. Kvittingen EA, Rootwelt H, Berger R, Brandtzaeg P (1994) Self-induced correction of the genetic defect in tyrosinemia type I. J Clin Invest 94:1657-1661
14. Demers S, I, Russo P, Lettre F, Tanguay RM (2003) Frequent mutation reversion inversely correlates with clinical severity in a genetic liver disease, hereditary tyrosinemia. Hum Pathol 34:1313-1320
15. Roth KS, Carter BE, Higgins ES (1991) Succinylacetone effects on renal tubular phosphate metabolism: a model for experimental renal Fanconi syndrome. Proc Soc Exp Biol Med 196:428-431
16. Giger U, Meyer UA (1983) Effect of succinylacetone on heme and cytochrome P450 synthesis in hepatocyte culture. FEBS Lett 153:335-338
17. Tschudy DP, Hess A, Frykholm BC, Blease BM (1982) Immunosuppressive activity of succinylacetone. J Lab Clin Med 99:526-532
18. Stenson PD, Ball EV, Mort M et al (2003) Human Gene Mutation Database (HGMD): 2003 update. Hum Mutat 21:577-581
19. Poudrier J, Lettre F, Scriver CR et al (1998) Different clinical forms of hereditary tyrosinemia (type I) in patients with identical genotypes. Mol Genet Metab 64:119-125
20. Rootwelt H, Brodtkorb E, Kvittingen EA (1994) Identification of a frequent pseudodeficiency mutation in the fumarylacetoacetase gene, with implications for diagnosis of tyrosinemia type I. Am J Hum Genet 55:1122-1127
21. Kvittingen EA, Halvorsen S, Jellum E (1983) Deficient fumarylacetoacetate fumarylhydrolase activity in lymphocytes and fibroblasts from patients with hereditary tyrosinemia. Pediatr Res 17:541-544
22. Kvittingen EA, Jellum E, Stokke O (1981) Assay of fumarylacetoacetate fumarylhydrolase in human liver-deficient activity in a case of hereditary tyrosinemia. Clin Chim Acta 115:311-319
23. Laberge C, Grenier A, Valet JP, Morissette J (1990) Fumarylacetoacetase measurement as a mass-screening procedure for hereditary tyrosinemia type I. Am J Hum Genet 47:325-328
24. Halvorsen S (1980) Screening for disorders of tyrosine metabolism. In: Bickel H, Guthrie R, Hammersen G (eds) Neonatal screening for inborn errors of metabolism. Springer, Berlin Heidelberg New York, pp 45-57
25. Pollitt RJ, Green A, McCabe CJ et al (1997) Neonatal screening for inborn errors of metabolism: cost, yield and outcome. Health Technol Assess 1:37-47
26. McCormack MJ, Walker E, Gray RG et al (1992) Fumarylacetoacetase activity in cultured and non-cultured chorionic villus cells, and assay in two high-risk pregnancies. Prenat Diagn 12:807-813
27. Kvittingen EA, Steinmann B, Gitzelmann R et al (1985) Prenatal diagnosis of hereditary tyrosinemia by determination of fumarylacetoacetase in cultured amniotic fluid cells. Pediatr Res 19:334-337
28. Jakobs C, Stellaard F, Kvittingen EA et al (1990) First-trimester prenatal diagnosis of tyrosinemia type I by amniotic fluid succinylacetone determination. Prenat Diagn 10:133-134
29. Poudrier J, Lettre F, St Louis M, Tanguay RM (1999) Genotyping of a case of tyrosinaemia type I with normal level of succinylacetone in amniotic fluid. Prenat Diagn 19:61-63
30. Lock EA, Ellis MK, Gaskin P et al (1998) From toxicological problem to therapeutic use: the discovery of the mode of action of 2-(2-nitro-4-trifluoromethylbenzoyl)-1,3- cyclohexanedione (NTBC), its toxicology and development as a drug. J Inherit Metab Dis 21:498-506
31. Holme E, Lindstedt S (2000) Nontransplant treatment of tyrosinemia. Clin Liver Dis 4:805-814
32. Hall MG, Wilks MF, Provan WM et al (2001) Pharmacokinetics and pharmacodynamics of NTBC (2-(2-nitro-4- fluoromethylbenzoyl)-1,3-cyclohexanedione) and mesotrione, inhibitors of 4-hydroxyphenyl pyruvate dioxygenase (HPPD) following a single dose to healthy male volunteers. Br J Clin Pharmacol 52:169-177
33. Vogel A, van dB, I, Al Dhalimy M et al (2004) Chronic liver disease in murine hereditary tyrosinemia type 1 induces resistance to cell death. Hepatology 39:433-443
34. Dr. Julie Reed. (2004) Birmingham Children's Hospital. Personal Communication
35. Mohan N, McKiernan P, Preece MA et al (1999) Indications and outcome of liver transplantation in tyrosinaemia type 1. Eur J Pediatr 158[Suppl 2]:S49-S54
36. Wijburg FA, Reitsma WC, Slooff MJ et al (1995) Liver transplantation in tyrosinaemia type I: the Groningen experience. J Inherit Metab Dis 18:115-118

37. Laine J, Salo MK, Krogerus L et al (1995) The nephropathy of type I tyrosinemia after liver transplantation. Pediatr Res 37:640-645

38. Freese DK, Tuchman M, Schwarzenberg SJ et al (1991) Early liver transplantation is indicated for tyrosinemia type I. J Pediatr Gastroenterol Nutr 13:10-15

39. Wilson CJ, Van Wyk KG, Leonard JV, Clayton PT (2000)Phenylalanine supplementation improves the phenylalanine profile in tyrosinaemia. J Inherit Metab Dis 23:677-683

40. Dr. P McKiernan (2004) Birmingham Children's Hospital. Personal Communication

41. Buist NRM, Kennaway NG, Fellman JH (1985) Tyrosinaemia type II. In: Bickel H, Wachtel U (eds) Inherited diseases of amino acid metabolism. Thieme, Stuttgart, pp 203-235

42. Heidemann DG, Dunn SP, Bawle E, V, Shepherd DM (1989) Early diagnosis of tyrosinemia type II. Am J Ophthalmol 107:559-560

43. Paige DG, Clayton P, Bowron A, Harper JI (1992) I. Richner-Hanhart syndrome (oculocutaneous tyrosinaemia, tyrosinaemia type II). J R Soc Med 85:759-760

44. Rabinowitz LG, Williams LR, Anderson CE et al (1995) Painful keratoderma and photophobia: hallmarks of tyrosinemia type II. J Pediatr 126:266-269

45. Fois A, Borgogni P, Cioni M et al (1986) Presentation of the data of the Italian registry for oculocutaneous tyrosinaemia. J Inherit Metab Dis 9:262-264

46. Bohnert A, Anton-Lamprecht I (1982) Richner-Hanhart syndrome: ultrastructural abnormalities of epidermal keratinization indicating a causal relationship to high intracellular tyrosine levels. J Invest Dermatol 72:68-74

47. Macsai MS, Schwartz TL, Hinkle D et al (2001) Tyrosinemia type II: nine cases of ocular signs and symptoms. Am J Ophthalmol 132:522-527

48. Barr DG, Kirk JM, Laing SC (1991) Outcome in tyrosinaemia type II. Arch Dis Child 66:1249-1250

49. Cerone R, Fantasia AR, Castellano E et al (2002) Pregnancy and tyrosinaemia type II. J Inherit Metab Dis 25:317-318

50. Francis DE, Kirby DM, Thompson GN (1992) Maternal tyrosinaemia II: management and successful outcome. Eur J Pediatr 151:196-199

51. Chitayat D, Balbul A, Hani V et al (1992) Hereditary tyrosinaemia type II in a consanguineous Ashkenazi Jewish family: intrafamilial variation in phenotype; absence of parental phenotype effects on the fetus. J Inherit Metab Dis 15:198-203

52. Ellaway CJ, Holme E, Standing S, Preece MA et al (2001) Outcome of tyrosinaemia type III. J Inherit Metab Dis 24:824-832

53. Rüetschi U, Cerone R, Pérez CC et al (2000) Mutations in the 4-hydroxyphenylpyruvate dioxygenase gene (HPD) in patients with tyrosinemia type III. Hum Genet 106:654-662

54. Rice DN, Houston IB, Lyon IC et al (1998) Transient neonatal tyrosinaemia. J Inherit Metab Dis 12:13-22

55. Mamunes P, Prince PE, Thornton NH et al (1976) Intellectual deficits after transient tyrosinemia in the term neonate. Pediatrics 57:675-680

56. Phornphutkul C., Introne WJ, Perry MB et al (2002) Natural history of alkaptonuria. N Engl J Med 347:2111-2121

57. Garrod AE (1902) The incidence of alkaptonuria: a study in chemical individuality. Lancet 2:1616-1620

58. Crissy RE, Day AJ (1950) Ochronosis: a case report. J Bone Joint Surg Am 32:688

59. Murray JC, Lindberg KA, Pinnell SR (1977) In vitro inhibition of chick embryo lysyl oxidase by homogentisic acid. A proposed connective tissue defect in alkaptonuria. J Clin Invest 59:1071-1079

60. Bory C, Boulieu R, Chantin C, Mathieu M (1990) Diagnosis of alcaptonuria: rapid analysis of homogentisic acid by HPLC. Clin Chim Acta 189:7-11

61. Fernández-Cañón JM, Peñalva MA (1997) Spectrophotometric determination of homogentisate using aspergillus nidulans homogentisate dioxygenase. Anal Biochem 245:218-221

62. de Haas V, Carbasius Weber EC, de Clerk JB et al (1998) The success of dietary protein restriction of alkaptonuria patients is age-dependent. J Inherit Metab Dis 21:791-798

63. Lustberg TJ, Schulmanm JD, Seegmiller JE (2004) Decreased binding of 14-C homogentisic acid induced by ascorbic acid in connective tissue of rats with experimental alkaptonuria. Nature 228:770-771

64. Wolff JA, Barshop B, Nyhan W et al (1989) Effects of ascorbic acid in alkaptonuria: alterations in benzoquinone acetic acid and an ontogenic effect in infancy. Pediatr Res 26:140-144

65. Niederwieser A, Matasovic A, Tippett P, Danks D (1977) A new sulfur amino acid, named Hawkinsin, identified in a baby with transient tyrosinemia and her mother. Clin Chim Acta 76:345-356

66. Wilcken B, Hammond J, Howard N et al (1981) Hawkinsinuria: a dominantly inherited defect of tyrosine metabolism with severe effects in infancy. N Engl J Med 305:865-868

67. Borden M, Holm J, Leslie J et al (1992) Hawkinsinuria in two families. Am J Med Genet 44:52-56

68. Nyhan W (1984) Hawkinsinuria. In: Nyhan W (ed) Abnormalities in amino acid metabolism in clinical medicine. Appleton-Century-Crofts, Norwalk, CT

19 Branched-Chain Organic Acidurias/Acidemias

Udo Wendel, Hélène Ogier de Baulny

Catabolism of Branched-Chain Amino Acids

The three essential branched-chain amino acids (BCAAs), leucine, isoleucine and valine, are initially catabolized by a common pathway (◼ Fig. 19.1). The first reaction, which occurs primarily in muscle, involves reversible transamination to 2-oxo (or keto) acids and is followed by oxidative decarboxylation to coenzyme A (CoA) derivatives by branched-chain oxo- (or keto-) acid dehydrogenase (BCKD). The latter enzyme is similar in structure to pyruvate dehydrogenase (◼ Fig. 12.2). Subsequently, the degradative path-

ways of BCAA diverge. Leucine is catabolized to acetoacetate and acetyl-CoA which enters the Krebs cycle. The final step in the catabolism of isoleucine involves cleavage into acetyl-CoA and propionyl-CoA, which also enters the Krebs cycle via conversion into succinyl-CoA. Valine is also ultimately metabolized to propionyl-CoA. Methionine, threonine, fatty acids with an odd number of carbons, the side chain of cholesterol, and bacterial gut activity also contribute to the formation of propionyl-CoA.

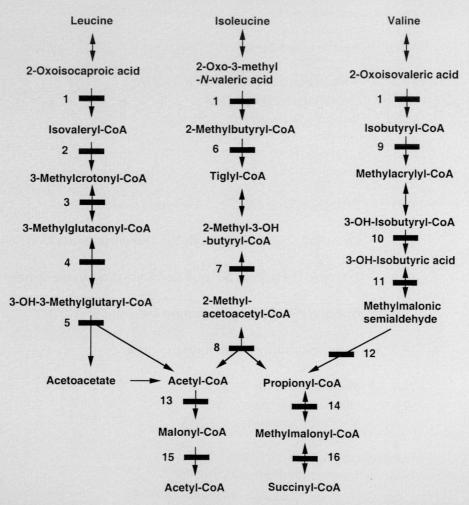

◼ **Fig. 19.1.** Pathways of branched-chain amino acid catabolism. **1**, Branched-chain 2-keto acid dehydrogenase complex; **2**, isovaleryl-coenzyme A (CoA) dehydrogenase; **3**, 3-methylcrotonyl-CoA carboxylase; **4**, 3-methylglutaconyl-CoA hydratase; **5**, 3-hydroxy-3-methylglutaryl-CoA lyase; **6**, short/branched-chain acyl-CoA dehydrogenase; **7**, 2-methyl-3-hydroxybutyryl-CoA dehydrogenase; **8**, 2-methylacetoacetyl-CoA thiolase; **9**, isobutyryl-CoA dehydrogenase; **10**, 3-hydroxyisobutyryl-CoA deacylase; **11**, 3-hydroxyisobutyric acid dehydrogenase; **12**, methylmalonic semialdehyde dehydrogenase; **13**, acetyl-CoA carboxylase (cytosolic); **14**, propionyl-CoA carboxylase; **15**, malonyl-CoA decarboxylase; **16**, methylmalonyl-CoA mutase. Enzyme defects are indicated by *solid bars*

Branched-chain organic acidurias or organic acidemias are a group of disorders that result from an abnormality of specific enzymes involving the catabolism of branched-chain amino acids (BCAAs). Collectively, the most commonly encountered are maple syrup urine disease (MSUD), isovaleric aciduria (IVA), propionic aciduria (PA) and methylmalonic aciduria (MMA). They can present clinically as a severe neonatal onset form of metabolic distress, an acute, intermittent, late-onset form, or a chronic progressive form presenting as hypotonia, failure to thrive, and developmental delay. Other rare disorders involving leucine, isoleucine, and valine catabolism are 3-methylcrotonyl glycinuria, 3-methyl-glutaconic (3-MGC) aciduria, short/branched–chain acyl-CoA dehydrogenase deficiency, 2-methyl-3-hydro-xybutyryl-CoA dehydrogenase deficiency, isobutyryl-CoA dehydrogenase deficiency, 3-hydroxyisobutyric aciduria, and malonic aciduria. All the disorders can be diagnosed by identifying acylcarnitines and other organic acid compounds in plasma and urine by gas chromatography-mass spectrometry (GC-MS) or tandem MS and all can be detected by newborn screening using tandem MS.

19.1 Maple Syrup Urine Disease, Isovaleric Aciduria, Propionic Aciduria, Methylmalonic Aciduria

19.1.1 Clinical Presentation

Children with maple syrup urine disease (MSUD), isovaleric aciduria (IVA), propionic aciduria (PA), or methylmalonic aciduria (MMA) have many clinical and biochemical symptoms in common. There are three main clinical presentations:

1. A severe neonatal-onset form with metabolic distress
2. An acute, intermittent, late-onset form
3. A chronic, progressive form presenting as hypotonia, failure to thrive, and developmental delay

In addition, prospective data gathered by newborn screening programs, mainly using tandem MS and the systematic screening of siblings of subjects with an abnormal newborn screening result, have demonstrated the relative frequency of asymptomatic forms.

Severe Neonatal-Onset Form
General Presentation
The general presentation of this form is that of a toxic encephalopathy with either ketosis or ketoacidosis (type I or II in the classification of neonatal inborn errors of metabolism

in ▶ Chap. 1). An extremely evocative clinical setting is that of a full-term baby born after a normal pregnancy and delivery who, after an initial symptom-free period, undergoes relentless deterioration with no apparent cause and unresponsive to symptomatic therapy. The interval between birth and clinical symptoms may range from hours to weeks, depending on the nature of the defect, and may be related to the timing of the sequential catabolism of carbohydrates, proteins, and fats. Typically, the first signs are poor feeding and drowsiness, followed by unexplained progressive coma. There may be cerebral edema with a bulging fontanel, arousing suspicion of a central nervous system (CNS) infection. At a more advanced stage, neuro-vegetative dysregulation with respiratory distress, hiccups, apneas, bradycardia, and hypothermia may appear. In the comatose state, most patients have characteristic changes in muscle tone and exhibit involuntary movements. Generalized hypertonic episodes with opisthotonus, boxing, or pedalling movements and slow limb elevations, spontaneously or upon stimulation, are frequently observed. Another pattern is that of axial hypotonia and limb hypertonia with large-amplitude tremors and myoclonic jerks, which are often mistaken for convulsions. In contrast, true convulsions occur late and inconsistently. The electroencephalogram may show a burst-suppression pattern. In addition to neurological signs, patients may present with dehydration and mild hepatomegaly.

Specific Signs

Maple Syrup Urine Disease. Concomitantly with the onset of the symptoms, the patient emits an intense (sweet, malty, caramel-like) maple-syrup-like odor. In general, neonatal (classic) MSUD does not display pronounced abnormalities on routine laboratory tests. Patients are not severely dehydrated, have no metabolic acidosis, no hyperammonemia or only a slight elevation (<130 µmol/l), no blood lactate accumulation, and the blood cell count is normal. The main laboratory abnormalities are greatly increased branched-chain amino acids (BCAAs) in plasma and the presence of 2-keto acids detected in urine with the 2,4-dinitrophenyl-hydrazine (DNPH) test.

Isovaleric Aciduria, Propionic Aciduria and Methylmalonic Aciduria. In contrast to MSUD, dehydration is a frequent finding in patients with IVA, PA, or MMA, and moderate hepatomegaly may be observed. They have metabolic acidosis (pH <7.30) with increased anion gap and ketonuria (Acetest 2–3 positive). However, ketoacidosis can be moderate and is often responsive to symptomatic therapy. Hyperammonemia is a constant finding. When the ammonia level is very high (>500 µmol/l), it can induce respiratory alkalosis and lead to the erroneous diagnosis of an urea-cycle disorder. Moderate hypocalcemia (<1.7 mmol/l) and hyperlactatemia (3–6 mmol/l) are frequent symptoms. The physician should be wary of attributing marked neurological dysfunction merely to these findings. Blood glucose can be normal, reduced, or elevated. When blood glucose level is very high (20 mmol/l) and is associated with glucosuria, ketoacidosis, and dehydration, it may mimic neonatal diabetes. Neutropenia, thrombocytopenia, non-regenerative anemia, and pancytopenia are findings frequently confused with sepsis. Among these disorders, IVA is easily recognized by its unpleasant sweaty feet odor.

In some cases, the combination of vomiting, abdominal distension, and constipation may suggest gastrointestinal obstruction. Cerebral hemorrhages have been described in a few neonates, a complication that may be linked to inappropriate correction of acidosis and may explain some poor neurological outcomes

Acute Intermittent Late-Onset Form

In approximately one fourth of the patients, the disease presents after a symptom-free period commonly longer than 1 year, and sometimes not even until adolescence or adulthood. Recurrent attacks may be frequent and, between them, the child may seem entirely normal. Onset of an acute attack may arise during catabolic stress, such as can occur with infections or following increased intake of protein-rich foods, but sometimes without an overt cause.

Neurological Presentation

Recurrent attacks of either coma or lethargy with ataxia are the main presentations of these acute late-onset forms. The most frequent variety of coma is that presenting with ketoacidosis, but exceptionally, this may be absent.

Hypoglycemia may occur in patients with MSUD while, in the other disorders, blood glucose levels are low, normal, or high. Mild hyperammonemia can be present in IVA, PA, and MMA patients. Although most recurrent comas are not accompanied by focal neurological signs, some patients may present with acute hemiplegia, hemianopsia, or symptoms and signs of cerebral edema mimicking encephalitis, a cerebrovascular accident, or a cerebral tumor. These acute neurological manifestations have frequently been preceded by other premonitory symptoms that had been missed or misdiagnosed. They include acute ataxia, unexplained episodes of dehydration, persistent and selective anorexia, chronic vomiting with failure to thrive, hypotonia, and progressive developmental delay.

Hepatic Forms

Some patients may present with a Reye-syndrome-like illness characterized by onset of coma, cerebral edema, hepatomegaly, liver dysfunction, hyperammonemia, and even macro- or microvesicular fatty infiltration of the liver. These observations emphasise the importance of complete metabolic investigations in such situations.

Hematologic and Immunologic Forms

Severe hematologic manifestations are frequent, mostly concomitant with ketoacidosis and coma, and sometimes are the presenting problem. Neutropenia is regularly observed in both neonatal and late-onset forms of IVA, PA, and MMA. Thrombocytopenia occurs mostly in infancy, and anemia occurs only in the neonatal period. Various cellular and humoral immunologic abnormalities have been described in patients presenting with recurrent infections, leading to erroneous diagnosis and management.

Chronic, Progressive Forms
Gastrointestinal Presentation

Persistent anorexia, chronic vomiting, failure to thrive, and osteoporosis (evidence of a long-standing GI disturbance) are frequent manifestations. In infants, this presentation is easily misdiagnosed as gastroesophageal reflux, cow's-milk-protein intolerance, celiac disease, late-onset chronic pyloric stenosis, or hereditary fructose intolerance, particularly if these symptoms start after weaning and diversifying food intake. Later in life, recurrent vomiting with ketosis may occur. These patients may remain undetected until an acute neurological crisis with coma leads to the diagnosis.

Chronic Neurological Presentation

Some patients present with severe hypotonia, muscular weakness, and poor muscle mass that can simulate congenital neurological disorders or myopathies. Nonspecific developmental delay, progressive psychomotor retardation, dementia, seizures, and movement disorders may also be observed during the course of the disease. However, these rather nonspecific findings are rarely the only presenting symptoms [1].

Complications
Neurological Complications

Maple Syrup Urine Disease. Acute cererebral oedema is a well recognised complication in newborn infants with MSUD and encephalopathy. In older patients with metabolic decompensation it may cause brain stem compression and unexpected death, particularly following intensive rehydration [2]; it may also develop slowly due to long-standing elevations of BCAAs. Additionally, demyelination can occur over time in those patients with poor biochemical control and persistently raised BCAAs. The areas most commonly affected are the periventricular white matter of the cerebral hemispheres, the deep cerebellar white matter, the dorsal part of the brain stem, the cerebral peduncles, the dorsal limb of the internal capsule, and the basal ganglia. The severity of dysmyelination does not correlate with signs of acute neurotoxicity and the changes are reversible with appropriate treatment [3, 4].

Propionic Aciduria and Methylmalonic Aciduria. An increasing number of patients with PA and MMA have presented with an acute or progressive extrapyramidal syndrome associated with increased signal within the basal ganglia (mostly the globus pallidus in MMA). The basal ganglia involvement may be due to edema that evolves to necrosis. In addition, magnetic resonance imaging (MRI) studies indicate cerebral atrophy and delayed myelination [5, 6]. These dramatic complications are arguments for adequate life-long dietary control, even if the patient is free of symptoms. Even in well treated patients with PA who are clinically and metabolically stable, brain lactate is elevated; this might indicate that aerobic oxidation is persistently impaired from elevated intracellular propionic metabolites [6].

Renal Complications

Renal tubular acidosis associated with hyperuricemia may be an early and presenting sign in some late-onset patients with MMA. This condition partially improves with metabolic control. Chronic renal failure is increasingly recognized in patients older than 10 years [7, 8]. The renal lesion is a tubulo-interstitial nephritis with type-4 tubular acidosis and adaptive changes secondary to the reduced glomerular filtration rate [9]. The course of the disease is usually indolent, but end-stage renal failure may develop, and dialysis and renal transplantation by the end of the second decade is likely to be necessary in many patients [10]. If the nephropathy is the complication of a chronic glomerular hyperfiltration secondary to excessive MMA excretion, minimizing and deceleration of renal injury may require strict metabolic control.

Skin Disorders

Large, superficial desquamation, alopecia, and corneal ulceration may develop in the course of late and severe decompensations in MSUD, PA, or MMA. These skin lesions have been described as a staphylococcal scalded-skin syndrome with epidermolysis or as acrodermatitis-enteropathica-like syndrome [11]. In many cases, these complications occur together with diarrhea and can be ascribed to acute protein malnutrition, especially to isoleucine deficiency.

Pancreatitis

Acute or chronic pancreatitis may complicate this group of organic acidemias and must be considered in the assessment of patients who have acute deterioration with appropriate symptoms. It has been the presenting illness in two patients with late-onset forms of IVA. In other cases, pancreatic involvement in the course of acute episodes has been associated with hyperglycemia and hypocalcemia. The increase of serum lipase and amylase is variable. Enlargement and even calcification or hemorrhage of the pancreas can be visualized by abdominal ultrasonography [12].

Cardiomyopathy

Acute cardiac failure due to cardiomyopathy may be responsible for rapid deterioration or death in cases of PA. This complication may develop as part of the presenting illness or may occur during an intercurrent metabolic decompensation [13].

19.1.2 Metabolic Derangement

Maple Syrup Urine Disease

MSUD is caused by a deficiency of the branched-chain 2-keto acid dehydrogenase (BCKD) complex, the second common step in the catabolism of the three BCAAs (⬛ Fig.19.1, enzyme 1). Like the other 2-keto acid dehydrogenases, BCKD is composed of three catalytic components (⬛ Fig. 12.2): a decarboxylase (E1), composed of E1α and E1β subunits and requiring thiamine pyrophosphate as a coenzyme, a dihydrolipoyl acyltransferase (E2), and a dihydrolipoamide dehydrogenase (E3). A deficiency of the E1 or E2 component can cause MSUD whereas a deficiency of the E3 component produces a specific syndrome (dihydrolipoamide dehydrogenase (E3) deficiency) with congenital lactic acidosis, branched-chain 2-keto aciduria and 2-keto glutaric aciduria (▶ Chap. 12). However E3 deficiency, particularly neonatal onset forms, may present with lactic acidemia alone, with elevation of branched-chain amino acids only becoming apparent weeks or months later.

The enzyme defect results in marked increases in the branched-chain 2-keto acids in plasma, urine and cerebrospinal fluid (CSF). Owing to the reversibility of the initial transamination step, the BCAAs also accumulate. Smaller amounts of the respective 2-hydroxy acids are formed. Allo-isoleucine is invariably found in blood in all classic MSUD patients and in variant forms, at least in those still without dietary treatment. This compound is endogenously formed and is a diastereomer of isoleucine.

Among the BCAA metabolites, leucine and 2-ketoisocaproic acid appear to be the most neurotoxic. In MSUD, they are always present in approximately equimolar concentrations in plasma, and may cause acute brain dysfunction when their plasma concentrations rise above 1 mmol/l. Isoleucine and valine are of lesser clinical significance. Their 2-keto-acid-to-amino-acid ratios favour the less toxic amino acids and cerebral symptoms do not occur even when the blood levels of both amino acids are extremely high.

Isovaleric Aciduria

Isovaleric aciduria is caused by a deficiency of isovaleryl-CoA dehydrogenase (IVD; ⬛ Fig. 19.1, enzyme 2) an intra-mitochondrial flavoenzyme which, similar to the acyl-CoA dehydrogenases (⬛ Fig. 13.1), transfers electrons to the respiratory chain via the electron-transfer flavoprotein (ETF)/ETF-ubiquinone oxidoreductase (ETF-QO) system. Deficiencies of the ETF/ETFQO system result in multiple acyl-CoA-dehydrogenase deficiency (MADD; synonym: glutaric aciduria type II) (Chap. 13).

The enzyme defect results in the accumulation of derivatives of isovaleryl-CoA, including free isovaleric acid, which is usually increased both in plasma and urine (although normal levels have been reported), 3-hydroxyisovaleric acid (3-HIVA), and N-isovalerylglycine. This glycine conjugate is the major derivative of isovaleryl-CoA, owing to the high affinity of the latter for glycine N-acylase. Conjugation with carnitine (catalyzed by carnitine N-acylase) results in the formation of isovalerylcarnitine.

Propionic Aciduria

Propionic aciduria is caused by a deficiency of the mitochondrial enzyme propionyl-CoA carboxylase (PCC; ◘ Fig. 19.1, enzyme 14), one of the four biotin-dependent enzymes. PCC is a multimeric protein composed of two different α- (which bind biotin) and β-PCC subunits. So far all patients with isolated PA have been biotin resistant.

PA is characterized by greatly increased concentrations of free propionic acid in blood and urine, and the presence of multiple organic acid byproducts, among which propionylcarnitine, 3-hydroxypropionate, and methylcitrate are the major diagnostic metabolites. The first is formed by acylation to carnitine. The second is formed by either β- or ω-oxidation of propionyl-CoA. Methylcitrate arises by condensation of propionyl-CoA with oxaloacetate, which is catalyzed by citrate synthase. During ketotic episodes, 3-HIVA is formed by condensation of propionyl-CoA with acetyl-CoA, followed by chemical reduction. Low concentrations of organic acids derived from a variety of intermediates of the isoleucine catabolic pathway, such as tiglic acid, tiglylglycine, 2-methyl-3-hydroxybutyrate, 3-hydroxybutyrate and propionylglycine can also be found. Due to an abnormal biotin metabolism, propionyl-CoA accumulation also occurs in multiple carboxylase deficiency (biotinidase deficiency, holocarboxylase synthetase (HCS) deficiency), resulting in defective activity of all four biotin-dependent carboxylases (▶ Chap. 27).

Methylmalonic Aciduria

Methylmalonic aciduria is caused by a deficiency of methylmalonyl-CoA mutase (MCM; ◘ Fig. 19.1, enzyme 16), a vitamin B_{12}-dependent enzyme. Deficient activity of the MCM-apoenzyme leads to MMA: Because the apomutase requires adenosylcobalamin (AdoCbl), disorders that affect AdoCbl formation cause variant forms of MMA (▶ Chap. 28).

The deficiency of MCM leads to the accumulation of methylmalonyl-CoA, resulting in greatly increased amounts of methylmalonic acid in plasma and urine. Owing to secondary inhibition of PCC, propionic acid also accumulates, and other propionyl-CoA metabolites, such as propionylcarnitine, 3-hydroxypropionic acid, methylcitrate, and 3-HIVA, are usually also found in urine. However, some mildly affected or asyptomatic patients, identified through urine organic acids screening in neonates but showing only slightly increased methylmalonic acid in blood and urine, have not shown a constant excretion of metabolites derived from propionyl-CoA.

Vitamin-B_{12} deficiency must be excluded when excessive urinary methylmalonic acid is found, particularly in a breast fed infant whose mother is either a strict vegetarian or suffers from subclinical pernicious anemia.

Secondary Metabolic Disturbances Common to PA and MMA

The accumulation of propionyl-CoA results in inhibitory effects on various pathways of intermediary metabolism, in increased levels of acylcarnitines (particularly propionyl carnitine) in blood and urine, leading to a relative carnitine deficiency, and in enhanced synthesis of odd-numbered long-chain fatty acids. Inhibition of various enzymes may explain some features such as hypoglycemia, hyperlactatemia, hyperammonemia, and hyperglycinemia. The abnormal ketogenesis which is a major cause of morbidity is not fully understood [14].

Propionate, essentially in form of propionyl-CoA, is produced in the body from three main sources: *(1)* catabolism of the amino acids isoleucine, valine, methionine and threonine, *(2)* anaerobic fermentation in the gut, and *(3)* mobilization and oxidation of odd-chain fatty acids during prolonged fasting states. It has been estimated that catabolism of amino acids contributes approximately 50% to the total propionate production, anaerobic gut bacteria 20%, and odd-chain fatty acids 30% [15]. These data, which are largely from stable isotope turnover studies, are based on a number of unproven assumptions and have not been reproduced in a more systematic manner; they are therefore questionable (for a critical review, see [10, 16]).

19.1.3   Genetics

Maple Syrup Urine Disease

MSUD is an autosomal-recessive disorder, with an incidence of 1 in 120 000 to 1 in 500 000. It is highly prevalent in the inbred Mennonite population in Pennsylvania, occuring in approximately 1 in 176 newborns. In countries where consanguineous marriages are common the frequency is also higher (about 1 in 50 000 in Turkey). About 75% of those affected suffer from the severe classic form, and the remainder suffer from the milder intermediate or intermittent variants. Over 150 different causal mutations scattered among the three *E1α*, *E1β*, and *E2* genes give rise to either classic or intermediate clinical phenotypes [17].

Isovaleric Aciduria

IVA is an autosomal-recessive disorder with extreme clinical variability for reasons that are unknown. Reported mutations in the *IVD* gene are highly heterogenous and generally, no phenotype/genotype correlation has been established. However, children with IVA diagnosed by newborn screening and carrying a 932C>T mutant allele can exhibit a milder, potentially asymptomatic phenotype [18].

Propionic Aciduria

PA is an autosomal recessive disorder with an incidence of less than 1 in 100 000. PA can result from mutations in the *PCCA* or *PCCB* genes encoding the α- and β- subunits, respectively, of propionyl-CoA carboxylase.

To date more than 50 different allelic variations in the *PCCB* gene and more than 30 in the *PCCA* gene, have been identified in different populations [19, 20]. Through the newborn screening programme in Japan a number of infants with an apparently mild phenotype and the Y435C mutation in the *PCCB* gene have been reported. The natural history of this phenotype is not yet clarified [21]. Particularly in PA, knowledge of the phenotype-genotype correlations may provide important information for the prediction of the metabolic outcome and for the implementation of treatments tailored to individual patients.

Methylmalonic Aciduria

Isolated MMA can be caused by mutations in the *MUT* locus, encoding the methylmalonyl CoA mutase (MCM) apoenzyme or by those in genes required for provision of its cofactor, 5′-deoxyadenosylcobalamin (AdoCbl). Isolated MMA is classified into several genotypic classes and complementation groups. These are designated either mut⁻ or mut⁰ (together termed mut), according to whether there is minimal or no apoenzyme activity, respectively, or cobalamin A or B (Cbl A/B) for cofactor defects. To date more than 50 disease-causing mutations from patients with mut^{0/−} MMA have been identified at the *MUT* locus [22]. MMA is an autosomal-recessive disorder. The incidences of benign and severe forms are each about 1 in 50 000. Approximately one half to 2/3 of patients has a mutase apoenzyme defect; the remaining patients are cobalamin variants. Recently the genes *MMAA* and *MMAB* for the Cbl A and Cbl B complementation groups, have been cloned and deleterious mutations in CblA and CblB patient cell lines identified [23–25]. It is speculated that the *MMAA* gene product is a component of a transporter or an accessory protein that is involved in the translocation of vitamin B12 into mitochondria. The gene product of *MMAB* gene is a cob(I)alamin adenosyltransferase. A few non CblA/CblB deficient MMA patients called CblH are known (► Chap. 28 for further details).

19.1.4 Diagnostic Tests

Only MSUD can be diagnosed by using plasma amino acid chromatography alone. IVA, PA and MMA are diagnosed by their specific urinary organic acid profiles using GC-MS or abnormal acylcarnitines on tandem MS, while amino acid chromatography displays nonspecific abnormalities, such as hyperglycinemia and hyperalaninemia. Whatever the clinical presentation, the diagnosis can be made by sending filter-paper blood specimens, fresh or frozen urine samples or 1–2 ml of fresh or frozen plasma samples for analyses to an experienced laboratory. Specific loading tests are not necessary. Newborn screening for this group of organic acidurias can be performed by tandem MS [26–28]. An increased leucine/ isoleucine peak in blood spots taken at 24 or 36 h of age requires immediate notification to the sender. The abnormal acylcarnitine found in PA and MMA is propionylcarnitine (C3-carnitine) and in IVA is isovaleryl-carnitine (C5-carnitine) [27].

Enzymatic studies are useful for diagnostic confirmation. Around the 14th week of gestation (second trimester), reliable and rapid prenatal diagnosis of IVA, PA, and MMA can be performed by the direct measurement of metabolites in amniotic fluid using GC-MS, stable-isotope-dilution techniques, or tandem MS. First trimester diagnosis, using direct enzyme assay or assays of the DNA in those families in which the mutations are known, can be performed in fresh or cultured chorionic villi. This can also be done in cultured amniotic cells taken in the second trimester. Prenatal diagnosis of MSUD relies exclusively on enzyme assays in chorionic villi or in cultured amniocytes as well as on mutational analysis.

19.1.5 Treatment and Prognosis

CNS dysfunction can be prevented or at least minimized by early diagnosis and emergency treatment. Neonatal-onset forms frequently require early toxin removal (► Chap. 4). Thereafter dietary restriction, which is necessary to limit the production of organic acids and their metabolites and other specific treatments, are required both for survivors of the early onset forms and those with late-onset disease. For both groups it is essential that episodes of metabolic decompensation are recognised and treated sufficiently early; parents must be taught to recognise early warning signs and manage their child appropriately.

Principles of Long-Term Dietary Treatment

Long-term dietary treatment is aimed at reducing the accumulation of toxic metabolites while, at the same time, maintaining normal physical development and nutritional status and preventing catabolism. Some patients tolerate normal foods; others need only minimal restriction or can even regulate the diet themselves. However, many need very specific food allowances, implying stringent dietary restrictions that will be necessary for life.

The cornerstone of treatment is the limitation of one or more essential amino acids which, if present in excess, are either toxic or precursors of organic acids. Precise prescriptions are established for the daily intake of amino acids, protein, and energy. The diet must provide the recommended daily allowance (RDA) and the estimated safe and adequate daily dietary intakes of minerals and vitamins and follow principles of paediatric dietetics [29].

Protein/Amino Acid Prescriptions

Requirements for BCAAs and protein vary widely from patient to patient and in the same patient, depending on the nature and severity of the disorder, other therapies prescribed (stimulation of an alternate pathway), growth rate, state of health, and feeding difficulties. Individual requirements must be estimated for each child by frequent monitoring of clinical and metabolic status. The balance between protein malnutrition and metabolic disequilibrium can be difficult to maintain in severe PA and MMA and needs to be kept under regular review especially after an acute metabolic decompensation or after a change in the diet.

Within this group of organic acidurias, only in MSUD is the diet directly related to the intake of an amino acid, that is leucine in mg amounts. Natural protein, which contains leucine, must be severely restricted in an age-dependent manner to only 1/2 to 1/10 of the normal recommended daily requirement. Consequently, in order to meet the protein RDA for the patient's age, a large supplement of BCAA-free amino acid mixture as a protein substitute is necessary. In IVA it is sufficient to restrict natural protein to, or somewhat more than, the recommended minimum daily requirements; a special amino acid mixture free of leucine is rarely needed. In neonatal PA and MMA dietary protein is generally restricted to the adequate age related safe levels. Restriction of specific amino acids has not been proved to be useful. Although controversial, a limited, relatively small amount of an amino acid mixture free of valine, isoleucine, methionine and threonine can be added to the diet to supply additional nitrogen and other essential and nonessential amino acids in order to promote a protein-sparing anabolic effect [29a].

The prescribed amounts of leucine or natural protein are provided by natural foods. Breast milk or standard infant formula is used in young infants. For toddlers and children solids are introduced, using serving lists and lists of amino acid content in foods. In all protein restricted diets, high-protein foods (eggs, meat, dairy products), apart from milk, are generally avoided, since the lower percentage of amino acids in vegetable protein (as compared with that in animal protein) makes it easier to satisfy the appetites of children.

Energy and Micronutrient Prescriptions

Energy requirements vary widely and may be greater than normal to ensure that essential amino acids are not degraded to provide energy or nitrogen for the biosynthesis of nitrogenous metabolites. Reduction of energy intake below the individual's requirements results in a decreased growth rate and a metabolic imbalance. The energy requirement is met through natural foods, special amino acid formulas, and additional fat and carbohydrates from other sources including protein-free modular feeds. Distribution of energy intake from protein, carbohydrates, and lipids should approach the recommended percentages. The diet must be asessed for minerals, vitamins, and trace elements and, if incomplete, supplemented with an appropriate commercial preparation. Hyperosmolarity of formulas should be avoided by offering sufficient fluids.

Evaluation of Clinical and Nutritional Status

This comprises regular evaluation of weight, length, and head circumference which should follow growth percentiles appropriate for the patient's age. Nutritional status is also judged by blood cell count, hemoglobin and hematocrit, plasma protein and albumin, iron and ferritin, evaluation of calcium/phosphate metabolism, and plasma amino acid profile. The metabolic and nutritional statuses are both evaluated weekly during the first month of therapy, once a month during the first year, and later every 3–6 months. In patients treated with a low-protein diet without an added amino acid mixture, measurement of urea excretion is an easy means to evaluate anabolism [29a, 30]. Regular assessment of developmental progress provides the opportunity for psychological support, as social and emotional needs are major elements of the overall therapy of the affected child and of the family's well being.

Specific Adjustments
Maple Syrup Urine Disease

Acute Phase Management in the Newborn. Exogenous toxin removal procedures such as hemodialysis and hemofiltration together with high-energy dietary treatment are usually advised for the reversal of acute metabolic decompensation in symptomatic newborns with the classic form of MSUD [31]. With these measures the plasma leucine level is reduced to 1 mmol/l or less within hours. During the recovery interval, oral intake of BCAA-free formula (tube feeding) should be started early and BCAA intake adjusted according to the plasma levels, which are monitored daily until the optimal equilibrium is attained. During this stage, plasma concentrations of valine and isoleucine may fall below normal and become rate limiting for protein synthesis, a situation which requires generous valine and isoleucine supplements in doses of 300–400 mg/day. Some patients have also been successfully treated using nasogastric tube alone [32]. Newborn screening for MSUD by tandem MS allows for early diagnosis and intervention and obviates the need for extracorporeal detoxification. In affected newborns, found positive on screening, the oral intake of BCAA-free formula (tube feeding) with adequate calorie supply (glucose polymer) and supplementation with isoleucine and valine (300–400 mg/day) can be sufficient to stimulate protein synthesis and to normalize plasma leucine levels within two to three days [2, 33].

Long-Term Management. Management of MSUD comprises a life-long strict and carefully adjusted semi-synthetic diet as well as acute phase treatment during episodes of catabolic stress. The dietary treatment of MSUD differs from that of other organic acidurias since the condition results in elevated plasma BCAA levels. In that respect MSUD can be regarded as an aminoacidopathy and the principles of dietary treatment are essentially those that apply to phenylketonuria. The diet consists of measured proportions of BCAA-containing foods (as natural protein) and a synthetic BCAA- free amino acid supplement which in most preparations also contains the recommended requirements for minerals and vitamins and other essential nutrients. Additional fat and carbohydrate are provided by protein free products and additional supplements. The aim of such treatments is to maintain the 2–3 h postprandial plasma BCAAs near to normal concentrations (leucine: 80–200 μmol/l; isoleucine 40–90 μmol/l; valine: 200–425 μmol/l). Since leucine is the most toxic precursor, the diet can be based on the leucine requirement with frequent adjustment according to plasma leucine levels.

In newborns with the classic severe form of MSUD, the leucine requirement is 300–400 mg/day (80–110 mg/kg/day), which is approximately 50–60 % of the leucine intake in healthy newborns. Minimum valine and isoleucine requirements are 200–250 mg/day. Apart from considerable interindividual variation, children, adolescents and adults with the classic form of MSUD tolerate about 500 to 700 mg of leucine per day. Individuals with variant forms tolerate greater amounts, and some do well on a low protein diet.

Serial monitoring of blood BCAA levels is essential in treatment of MSUD and intakes of BCAAs must frequently be titrated against plasma concentrations. Occasionally, small amounts of free valine and isoleucine must be added to those provided by natural protein, because the tolerance for leucine is lower than for the other two. When the plasma leucine levels are high and those of valine and isoleucine low, a rapid fall of leucine can only be achieved by combining a reduced leucine intake with a temporary supplement of valine and isoleucine.

In MSUD, unlike other organic acidurias, no abnormal acylcarnitines are formed and there is no increased carnitine loss; consequently no carnitine supplement is required. Although treatment with thiamine has often been advocated, it has not been of proven efficacy in any form of MSUD.

Emergency Regimen. During maintenance treatment minor illnesses such as fever, vomiting, or diarrhea result in an increase in catabolism and amino acid release from muscle protein. Neurotoxic levels of BCAAs and BCKAs are reached within hours and patients may present with apathy, ataxia, hallucinations, and eventually with fasting hypoglycemia and convulsions. High energy intake and temporary removal of natural protein from the diet, and continuing supplements of BCAA-free amino acids (with the early addition of valine and isoleucine supplements) help to limit accumulation of the branched-chain compounds. Owing to its anabolic effect, intravenous insulin (0.15–0.20 IU/kg body weight per h), combined with large amounts of glucose and with continued enteral BCAA-free amino acids, can be succesfully used to treat severe catabolic episodes. Such therapy may prevent metabolic decompensation following major surgery and trauma, and the necessity for extracorporeal toxin removal in critically ill children.

Maternal MSUD. In a women with MSUD maintaining the plasma leucine level between 100 and 300 μmol/l and plasma valine and isoleucine in the upper normal range resulted in the delivery of a healthy infant. Leucine tolerance increased progressively from the 22nd week of gestation from 350 to 2100 mg/day. The risk of metabolic decompensation in the mother during the catabolic postpartum period can be minimized by careful monitoring after delivery [34].

Liver Transplantation. Liver replacement results in a clear increase in whole body BCKD activity to at least the level of very mild MSUD variant; following transplant patients no longer require protein restricted diets and the risk of metabolic decompensation during catabolic events is apparently abolished [35; 35a].

Prognosis. Patients with MSUD are now expected to survive; they are generally healthy between episodes of metabolic imbalance, and some attend regular schools and have normal IQ score. However, the average intellectual performance is clearly below those of normal subjects and patients with early treated PKU [36]. The intellectual outcome is inversely related to the length of time after birth that plasma leucine levels remain above 1 mmol/l, and is dependent on the quality of long-term metabolic control [36–38]. With that respect inclusion of MSUD into neonatal screening programs by tandem-mass spectrometry may improve the prognosis in that disease. Normal development and normal intellectual outcome and performance can be achieved at least in prospectively treated patients [2] and if average long-term plasma leucine levels are not more than 1.5–2 times normal [38]. In addition, timely evaluation and intensive treatment of minor illnesses at any age is essential, as late death, attributed to recurrence of metabolic crises with infections, has occurred [2]. Assiduous care is also indicated for patients with variant forms in order to prevent further ketoacidotic crises after they have been diagnosed and to retain the relatively good prognosis.

IV

Isovaleric Aciduria

Acute Phase Management in the Newborn. Intensive treatment with non-specific measures (glucose infusion to provide calories and reduce endogenous protein catabolism, possibly bicarbonate infusion to control the acidosis) including exogenous toxin (and ammonia) removal may be needed in newborns who are often in a poor clinical condition, precluding the effective use of alternate pathways to enhance the removal of isovaleryl-CoA. For that the administration of intravenous L-carnitine (100–400 mg/kg/day) and oral L-glycine (250–600 mg/kg/day) are effective means of treatment. Glycine can be provided as a 100 mg/ml water solution delivered in four to eight separate doses.

Dietary Therapy. The aim of treatment is to reduce the isovaleric acid burden to a minimum and to keep the urine free of IVA and 3-hydroxy-IVA. Such a therapy consists of a low protein diet with supplemental glycine and carnitine and should be started as soon as possible after birth. In most patients the amount of protein tolerated meets the official protein requirements and a special amino acid mixture free of leucine is rarely needed. Excessive protein intake should be avoided.

Carnitine and Glycine Therapy. For supplemental therapy either oral L-carnitine (50–100 mg/kg/day) or oral L-glycine (150–300 mg/kg/day) can be used. Under stable conditions the need for both supplementations is still controversial, but it can be useful during metabolic stress when toxic isovaleryl-CoA accumulation increases the need for detoxifying agents (39). Supplementation with large doses of carnitine gives rise to an unpleasant odour in many IVA patients.

Prognosis. Prognosis is better than for the other organic acidurias. Even when compliant with treatment metabolic crises during catabolic stress can occur, making a short hospitalisation for intravenous fluid (glucose/electrolytes/buffer) necessary. With puberty metabolic crises no longer occur. Growth is normal; intellectual prognosis depends on early diagnosis and treatment and, subsequently, on long-term compliance [40]. With that respect inclusion of IVA into neonatal screening programs by tandem MS should improve the prognosis. So far (only one pregnancy published) there is no evidence, that uncomplicated maternal isovaleric aciduria has any adverse effect on the unborn child [41].

In asymptomatic individuals identified by newborn screening and showing a mild biochemical phenotype it is crucial to prospectively follow the course of the inherited metabolic disturbance, as far as possible without any therapeutic regimen in order to find out the natural history.

Propionic Aciduria and Methylmalonic Aciduria

Acute Phase Management in the Newborn. The urinary excretion of propionic acid is negligible, and no alternate urinary pathway is sufficient to effectively detoxify newborns with PA. However, that does not mean that exogenous toxin-removal procedures are inevitably required. Extracorporal detoxification such as hemo(dia)filtration and hemodialysis (peritoneal dialysis is far less efficient), together with measures to promote anabolism should be considered when neonatal illness is accompanied by severe hyperammonemia (>400 μmol/l). In contrast to PA, the efficient removal of toxin in MMA takes place via urinary excretion, because of the high renal clearance of methylmalonic acid (22 ± 9 ml/min per 1.73 m^2) which allows an excretion as high as 4–6 mmol/day. Thus, emergency treatment of the newborn with MMA, if not complicated by very high ammonia levels, mainly comprises rehydration and promotion of anabolism.

When conservative measures with high energy supply are sufficient, hyperammonemia (especially in PA) may be controlled by use of sodium benzoate without evident problems, despite of the theoretical risk of intramitochondrial CoA depletion. Recently carbamylglutamate, a carbamylphosphate synthetase activator, was successfully tested for its ability to antagonize propionyl-CoA induced hyperammonemia [42]. Metabolic decompensation in propionic aciduria may be complicated by severe lactic acidosis due to thiamine deficiency, requiring vitamin supplementation [43].

Long-Term Management. The goal of treatment is to reduce the production of methylmalonic or propionic acid by means of
- natural protein restriction
- maintaining an optimal calorie intake
- carnitine supplementation (100 gm/kg/day)
- reduction of intestinal production of propionate by metronidazole.

Dietary Management. The aim of dietary treatment is to reduce production of propionate by both the restriction of precursor amino acids using a low protein diet and avoidance of prolonged fasting to limit oxidation of odd-chain fatty acids which are liberated from triglyceride stores during lipolysis. The low protein diet must provide at least the minimum amount of protein, nitrogen and essential amino acids to meet requirements for normal growth. Figures for estimates of safe level of protein intake for infants, children and adolescents are available [29] which can be used as a guide for low protein diets. In early childhood this is often 1–1.5 g/kg/day. To improve the quality of this diet it may be supplemented with a relatively small amount of synthetic amino acids free from the precursor amino acids. However the long term value of these supplements remains uncertain and metabolic balance can be achieved without them [10,

29, 29a]. Some studies have shown that the addition of a special amino acid mixture to a severely restricted diet has no effect on growth or metabolic status and that these amino acids are mostly broken down and excreted as urea [29a]. If one of the precursor amino acids becomes limiting, as indicated by very low levels in plasma, just that specific amino acid can be supplemented.

The diet must be nutritionally complete with adequate energy intake and sufficient vitamins and minerals, in order to save the patient from serious complications associated with poor nutrition. Long fasts should be avoided. In order to prevent fasting at night nocturnal nasogastric tube feeding may be used in the first years of management.

In children with severe forms of PA and MMA anorexia and feeding problems are almost invariably present and in order to maintain a good nutritional state feeds have to be given via nasogastric tube or gastrostomy at some stage. This is essential to provide adequate dietary intake, to prevent metabolic decompensation, and to help the parents to cope with a child who is difficult to feed [10, 29, 29a].

Most patients with a late-onset form are easier to manage. Individual protein tolerance can be quite high. Even though their individual tolerance allows a less rigid protein restriction and leads to lower risk of malnutrition, these patients must be taught to immediately reduce their protein intake during intercurrent illness in order to prevent metabolic imbalance.

Vitamin Therapy. Every patient with MMA should be tested for responsiveness to vitamin B_{12}. Some late-onset forms (and, more rarely, neonatal-onset forms) of MMA are vitamin B_{12}-responsive; thus, parenteral vitamin therapy, starting with hydroxycobalamin 1000–2000 μg/day for about 10 days, must be carefully tried during a stable metabolic condition [7]. During this period 24-h urine samples are collected for an organic acid analysis. Vitamin-B_{12} responsiveness leads to a prompt and sustained decrease of propionyl-CoA byproducts. However, as biochemical results may be difficult to assess, they must later be confirmed by in vitro studies. Most B_{12}-responsive patients need only mild protein restriction or none at all; vitamin B_{12} is either given orally once per day or is administered once a week (1000–2000 μg i.m.). In some cases, i.m. hydroxycobalamin therapy can be kept in reserve for intercurrent infections.

Carnitine Therapy. Chronic oral administration of L-carnitine (100 mg/kg/day) appears to be effective not only in preventing carnitine depletion but also in allowing urinary propionylcarnitine excretion and by that reducing propionate toxicity [10].

Metronidazole Therapy. Microbial propionate production can be suppressed by antibiotics. Metronidazole, an antibiotic that inhibits anaerobic colonic flora has been found to be specifically effective in reducing urinary excretion of propionate metabolites by 40% in MMA and PA patients. Long-term metronidazole therapy (at a dose of 10–20 mg/kg once daily for ten consecutive days each month) may be of significant clinical benefit [10]. This intermittent administration may prevent the known side effects of the drug, such as leukopenia, peripheral neuropathy, and pseudomembranous colitis.

Growth Hormone. Growth hormone (GH) induces protein anabolism. It is contraindicated in the acutely ill patient but potentially useful in the long-term for those growing poorly [10]. There is a place for recombinant human GH treatment as an adjuvant therapy in some patients with MMA and PA, mainly in those with reduced linear growth but controlled long-term studies are needed [44].

Biochemical Monitoring. During the course of decompensation plasma ammonia, blood gases, lactate, glucose, uric acid, and ketones in urine should be monitored. Some groups prefer to also measure urea in urine [29a, 30]. Regular amino acid analysis (all essential amino acids and particular isoleucine) is important. Furthermore MMA in plasma or urine should be controlled in order to define in each individual patient the lowest possible level on treatment. The measurement of carnitine/acylcarnitines in blood may be useful. There may be little practical use for the assay of odd-chain fatty acids in terms of directing clinical management.

Prognosis. Around 15% of patients with MMA are vitamin B_{12}-responsive and have a mild disease and good long-term outcome [7, 8, 45]. Conversely, both vitamin-B_{12}-unresponsive patients with MMA and those with PA have severe disease and many encephalopathic episodes, mainly due to intercurrent infections [45a]. For all forms of MMA, mut^0 patients have the poorest prognosis and vitamin B_{12}-responsive CblA and mut^- patients the best [7]. Owing to earlier diagnosis and better treatment outcome for PA and MMA have improved in the last decade [29a, 45a]. Survival rates into early and mid childhood can now exceed 70%. However, morbidity, in terms of cognitive development, remains high with a majority of patients having DQ/IQ in the mild to moderate retarded range. With the improved management the frequency of growth retardation has decreased and now most patients with PA and MMA have growth curves within the normal range [29a]. Abnormal neurological signs (mainly movement disorders, chorea, dystonia) continue to increase with age [7, 8, 45–48]. Chronic progressive impairment of renal function is a frequent and serious complication that manifests in older patients with high MMA excretion [7, 8]. Renal transplantation is likely to be necessary for many patients with MMA who survive into adolescence [49]. Including PA and MMA into newborn screening programs by tandem MS may make it possible to identify the late onset forms of the diseases in the newborn period and contribute to a further improvement in the outcome in this group [45a]. There is only one report of a female with MMA who carried a pregnancy to term. The outcome was favourable despite high levels of methylmalonic acid in blood and urine [50].

Liver Transplantation. The hazardous long-term prognosis associated with the high risk of complications raises the question of other therapeutic means, such as liver transplantation, for those patients difficult to manage. Irrespective of the fact that this procedure carries its own, often severe complications, it has proven metabolic efficacy [51–53]. A number of patients of different age (from early infancy to adulthood) with PA and MMA have had liver transplants. After successful transplantation most patients have returned to a normal diet. However, some patients have developed acute decompensation and basal ganglia necrosis years after liver transplantation while on a normal diet. Despite sometimes only slight improvement in the levels of circulating propionyl-CoA metabolites, life threatening episodes of ketoacidosis were reduced to some degree. Today, it is recommended to maintain such patients on a relaxed diet and continue carnitine supplementation. There is ample experience that progressive renal failure and neurologicalal dysfunction, including metabolic stroke, are not always prevented.

Management of Intercurrent Decompensations. Acute intercurrent episodes are prevented or minimized by being aware of those situations that may induce protein catabolism. These include intercurrent infections, immunization, trauma, anesthesia and surgery, and dietary indiscretion. In all cases, the main response comprises a reduction in protein intake. All patients should have detailed instructions including a semi-emergency diet, in which natural protein intakes are reduced by half and an emergency diet, in which it is stopped. In both, energy supply is augmented using carbohydrates and lipids such as solutions based on protein-free formula base powder or a mixture of glucose polymer and lipids diluted in an oral rehydration solution. For children treated with specific amino acid mixtures, the usual supplements can be added, though one should be aware that they increase osmolarity and that their taste renders nasogastric tube feeding quite unavoidable. Their use is contraindicated in MMA and PA in cases of severe hyperammonemia. At home, the solution is given in small, frequent drinks at day and night or by nasogastric tube [54]. After 24–48 h, if the child is doing well, the usual diet is resumed within 2 or 3 days.

In cases of clinical deterioration with anorexia and/or gastric intolerance or if the child is obviously unwell, the patient must be hospitalized to evaluate the clinical status, to search for and treat intercurrent disease, and to halt protein catabolism. Emergency therapy depends on the presence of dehydration, acidosis, ketosis, and hyperammon-emia. Most often, intravenous rehydration for 12–24 h results in sufficient clinical improvement to allow for progressive renutrition with continuous enteral feeding. During this step natural protein should be introduced into the feeds to at least cover the minimal dietary requirements. The energy intakes are supplied with carbohydrates and lipids, applying the same rules as for the treatment of late-onset forms. During this stage of management, close metabolic evaluation is advised, as the condition is labile and may deteriorate, requiring adjustment of the therapy. Conversely, if the patient's condition improves quickly, their normal feeding should be restored without delay.

In cases of severe decompensation or worsening of the clinical and metabolic status, the use of total parenteral nutrition appears to be an effective means to improve metabolic imbalance and should prevent the need to use extracorporeal removal procedures. The decision to use such therapeutic means is an emergency judgement for which no objective criteria are available. Selection of the form of procedure depends ultimately on local resources and experience, keeping in mind that intervention started too late is likely to fail. For valuable information see also [14].

19.2   3-Methylcrotonyl Glycinuria

19.2.1   Clinical Presentation

The clinical phenotype ranges from neonatal onset with severe neurological involvement and even death to completely asymptomatic adults. On the whole, symptomatic patients present either in the neonatal period, or later in childhood. Some infants present with intractable seizures from the first days of life, others with feeding difficulties, failure to thrive, and hypotonia within the first weeks after weaning, and some have recurrent seizures resulting in microcephaly and developmental delay. Most patients, however, present with a Reye-like syndrome following intercurrent illness or a protein-enriched diet within the first 2 years of life, developing neurological manifestations along with hypoglycemia, ketoacidosis, hyperammonemia and very low plasma carnitine. Additional manifestations in late onset patients include muscular hypotonia, seizures, psychomotor retardation, hemiparesis (»metabolic stroke«), signs of »metabolic leucodystrophy«, and dilated cardiomyopathy. A few adult women, diagnosed following newborn screening of their infants, have complained of muscle weakness.

19.2.2   Metabolic Derangement

In 3-methylcrotonyl glycinuria, leucine catabolism is blocked by deficiency of 3-methyl crotonyl CoA carboxylase (3-MCC) (▫ Fig. 19.1, enzyme 3). This enzyme is one of the four biotin-containing carboxylases known in humans. Accumulation of 3-methylcrotonylglycine also occurs in multiple carboxylase deficiency but in contrast to 3-MCC is found together with lactic acid and derivatives of propionyl-CoA (▶ Chap. 27).

Due to the enzyme block, 3-methylcrotonyl-CoA and 3-methylcrotonic acid accumulate. Most of the accumulated acyl-CoA is conjugated with glycine to form 3-methylcrotonylglycine (MCG). In contrast, acylation of 3-methylcrotonyl-CoA with carnitine appears to be only a minor pathway. 3-Hydroxyisovalerate (3-HIVA), another major metabolite, is derived through the action of a crotonase on 3-methylcrotonyl-CoA and the subsequent hydrolysis of the CoA-ester. 3-Hydroxyisovaleryl-glycine has not been found in this condition. However, acylation with carnitine leads to the formation 3-hydroxyisovaleryl carnitine which is the major abnormal acylcarnitine found in plasma and dried blood by tandem MS techniques.

19.2.3 Genetics

3-MCC is a heteromeric enzyme consisting of α- (biotin-containing) and β-subunits. 3-MCC deficiency results from loss of function mutations in the *MCCA* and *MCCB* genes encoding these subunits. The mutations can be classified into two groups, denoted CGA and CGB. Around 40 mutations have been identified in both genes. They are associated with an almost total lack of enzyme activity in fibroblasts. The apparent biochemical severity of all the MCC mutations contrasts with the variety of the clinical phenotypes suggesting that there are other unknown cellular and metabolic factors that affect the resulting phenotypes [55, 56]. Although most patients are compound heterozygotes or homozygotes, some are heterozygotes with a dominant negative allele [56]. The introduction of tandem MS in newborn screening has revealed an unexpectedly high incidence of this disorder, which in certain areas appears to be the most frequent organic aciduria, found in 1:40 000 newborns in Germany and Australia [26, 27].

19.2.4 Diagnostic Tests

The diagnosis relies on a characteristic urinary profile of organic acids, with huge excretion of 3-HIVA and 3-methylcrotonylglycine and without the lactate, methylcitrate, and tiglylglycine found in multiple carboxlase deficiency (MCD). Supplementation with pharmacologic doses of biotin does not alter this pattern. Total and free carnitine concentrations in plasma are extremely low. The presence of 3-hydroxyisovaleryl-carnitine (C5OH) in plasma and in dried blood spots is diagnostic for 3-MCC deficiency, since it is not found in isovaleric aciduria. In other disorders, such as MCD, propionylcarnitine (3C) is also seen, and in

3-hydroxy-3-methylglutaryl CoA lyase deficiency, glutarylcarnitine is the major finding.

Since family studies and newborn screening identified a number of totally asymptomatic siblings and mothers with MCC deficiency it is advisable to search in any affected family for other MCC deficient subjects by analyses of the acylcarnitine profile in blood and organic acids in urine.

19.2.5 Treatment and Prognosis

Long term treatment of symptomatic infants based on a mildly protein-restricted diet (meeting the recommended requirements) results in a general improvement and a reduction in the number of exacerbations. It is effective in lowering the abnormal excretion of organic acids which, however, never disappears.

Glycine and carnitine therapies directed at increasing the excretion of glycine and carnitine conjugates are complementary rather than competitive means of detoxification. Glycine supplementation (175 mg/kg/day) increases the excretion of 3-MCG. Carnitine supplementation (100 mg/kg/day) corrects the very low plasma carnitine levels and increases the excretion of 3-HIVA. Family studies and newborn screening have identified a number of totally asymptomatic siblings and mothers with 3-MCC deficiency with very low carnitine concentrations in blood who never had any treatment, so that the need for treatment must be doubted. The poor prognosis described in early-onset forms presenting with neonatal seizures could be due to late diagnosis and treatment. In acute late-onset forms presenting with Reye-like syndrome all but one patient fully recovered.

19.3 3-Methylglutaconic Aciduria Type I

3-Methylglutaconic aciduria type I (3-MGA type I) has only been identified in very few individuals, who presented with a wide spectrum of clinical signs of a neurometabolic disease, ranging from no symptoms (at 2 years of age) to mild neurological impairment, severe encephalopathy with basal-ganglia involvement, quadriplegia, athetoid movement disorder, severe psychomotor retardation, and leukoencephalopathy in a 61-year-old female.

3-Methylglutaconyl (MGC)-CoA is metabolized to 3 hydroxy-3-methylglutaryl-CoA by 3-MGC-CoA hydratase (◘ Fig. 19.1, enzyme 4). Defective activity leads to 3-MGC aciduria type I, characterized by urinary excretion of 3-MGC and 3-methylglutaric acids. Both metabolites derive from accumulated 3-methylglutaconyl-CoA through hydrolysis and dehydrogenation, respectively. The combined urinary excretion of 3-MGC and 3-methylglutaric acids range from 500 to 1000 mmol/mol creatinine, of which 3-methylglutaric acid represents about 1%. The metabolic pattern also includes 3-HIVA and a normal amount of 3-hydroxy-3-methylglutaric acid.

3-MGC-CoA hydratase activity can be measured in fibroblasts. Human 3-MGC-CoA hydratase is identical to a previously described RNA-binding protein (designated AUH) possessing enoyl-CoA hydratase activity [57, 58]. Different mutations in the *AUH* gene have been identified. A subject detected by newborn screening with tandem MS has remained asymptomatic up to the present age of 2 years despite complete absence of enzyme activity in skin fibroblasts [58].

No clear therapeutic regimen has been described. Carnitine supplementation may have beneficial effects.

3-Methylglutaconic aciduria type I must be distinguished from many other conditions associated with 3-MGC aciduria, which include Barth syndrome (3-MGA type II), Costeff optic atrophy syndrome (3-MGA type III), and disorders of unknown origin, summarised as 3-MGA type IV.

In Barth syndrome, an X-linked disorder, dilated cardiomyopathy, skeletal myopathy, neutropenia, diminished statural growth and abnormal mitochondria with disturbed respiratory chain are the main components of the disease [59]. Due to an impaired cardiolipin metabolism, profound cardiolipin deficiency leads to an abnormal lipid structure of the inner mitochondrial membrane. This may result in mitochondrial leakage of metabolites of the leucine catabolic route during overload [60]. The underlying cause of the disease are mutations in the Tafazzin (*TAZ*) gene. Carrier detection and prenatal diagnosis are possible through mutation analysis [61].

3-MGA Type III (Costeff optic atrophy syndrome) has been reported in approximately 40 Iraqi Jewish individuals. It is a neuro-ophthalmologic syndrome that consists of early-onset bilateral optic atrophy and later-onset spasticity, extrapyramidal dysfunction and cognitive deficits. 3-Methylglutaconic and 3-methylglutaric acid excretion is increased. The cause of the disease is a founder mutation in Iraqi Jews of the *OPA 3* gene [62].

Finally, among a large heterogeneous group of patients suffering from variable, multisystemic diseases (3-MGC aciduria type IV, unspecified diseases), some respiratory-chain disorders have been described [63]. In the latter conditions, the urinary excretion of 3-MGC and 3-methylglutaric acid is low (10–200 mmol/mol creatinine), without elevation of 3-HIVA and 3-hydroxy-3-methylglutaric acids.

19.4   Short/Branched-Chain Acyl-CoA Dehydrogenase Deficiency

Isolated 2-methylbutyrylglycinuria caused by short/branched-chain acyl-CoA dehydrogenase (SBCAD) deficiency (◘ Fig. 19.1, enzyme 6) is a recently described autosomal recessive disorder of isoleucine metabolism [64]. A 4 months old male who presented with brain lesions due to early neonatal hypoglycemia was found to have an isolated increase of 2-methylbutyrylglycine (2-MBG) and 2-methylbutyrylcarnitine (2-MBC) in body fluids [65]. In the newborn screening by tandem MS more than 20 cases of SBCAD deficiency (all subjects of Hmong descent, an ethnic group originated from China) were identified because of elevated C5-acylcarnitine concentrations in blood spots [66]. They are homozygous for the same mutation (1165A>G) in the *SBCAD* gene which causes skipping of exon 10. All these subjects were asymptomatic and many follow no therapeutic regimen. Clinical relevance of this disorder remains in doubt and requires careful long-term follow-up of affected individuals [28].

19.5   2-Methyl-3-Hydroxybutyryl-CoA Dehydrogenase Deficiency

Since 2000, 2-methyl-3-hydroxybutyryl-CoA dehydrogenase (MHBD) deficiency has been described in approximately 10 males aged 2 to 8 years and in one male adult. All patients had an unusual neurodegenerative disease. Initially normal or moderately impaired development was followed from the second year of life by progressive deterioration with loss of mental and motor skills. In early childhood the severe neurodegenerative symptoms included rigidity, dystonic posturing, spastic diplegia, dysarthria, choreoathetoid movements, restlessness, cortical blindness, myoclonic seizures, brain atrophy, periventricular white matter and basal ganglia abnormalities – features resembling the sequelae of neonatal hypoxic-ischemic brain injury [67, 68]. All patients identified so far have had a severe progressive neurological phenotype rather than ketoacidotic attacks, in contrast to patients with a defect in the next step of isoleucine degradation due to 2-methyl-acetoacetyl-CoA thiolase deficiency [68].

MHBD deficiency (◘ Fig. 19.1, enzyme 7) is a defect of isoleucine degradation. Laboratory findings include marked elevations of urinary 2-methyl-3-hydroxybutyrate and tiglylglycine without elevation of 2-methylacetoacetate. The organic acid excretion is more pronounced after a 100 mg/kg oral isoleucine challenge. Enzyme studies have shown markedly decreased activity of MHBD in fibroblasts and lymphocytes. MHBD deficiency is caused by mutations in the X-chromosomal *HADH2* gene [69]. Adult females (mother and grandmother of patients) have shown mild to moderate developmental delay. A short-term stabilization of neurological symptoms and a biochemical response to an isoleucine-restricted diet has been observed in some patients.

The deficiency of 2-methyl-acetoacetyl-CoA thiolase (◘ Fig. 19.1, enzyme 8), also known as 3-ketothiolase or T2, is discussed in ▶ Chap. 14.

19.6   Isobutyryl-CoA Dehydrogenase Deficiency

The first patient with isobutyryl-CoA dehydrogenase (IBD) deficiency (◘ Fig.19.1, enzyme 9) was a 2 year old female who developed anemia and dilated cardiomyopathy in the second year of age. She had very low total plasma carnitine and while receiving carnitine supplement an increase in four-carbon species. Urine organic acids were normal. This disorder can be detected on the basis of elevated butyryl-carnitine/isobutyrylcarnitine (C4-carnitine) concentrations in newborn blood spots analysed by tandem MS [70]. The expansion of newborn screening has led to the identification of more subjects [28, 71]. The mitochondrial enzyme IBD catalyses the third step in the degradation of valine. It is encoded by the *ACAD 8* gene [72]. The possible clinical implication of this enzyme defect is not known and careful follow up is necessary [71].

19.7   3-Hydroxyisobutyric Aciduria

A few patients with increased excretion of 3-hydroxyisobutyric acid (3-HIBA), an intermediate of the catabolic pathways of valine and thymidine, have been identified. This condition may be linked to various enzymatic defects. Unfortunately, in most described cases, the enzymatic diagnosis has been speculative.

Clinical presentation is heterogeneous. Some patients present in infancy, with acute episodes of vomiting, lethargy, ketoacidosis, and sometimes associated hypoglycemia or hyperlactatemia. Muscle involvement can be a prominent sign with marked hypotonia and even myopathic features. In addition, hypertrophic cardiomyopathy has been reported in two cases. CNS involvement is highly variable, ranging from patients with normal development to neonates with brain dysgenesis and other congenital malformations and children with microcephaly, hypotonia, and seizures.

Several enzyme defects may underlie 3-hydroxyisobutyric aciduria including a combined deficiency of malonic, methylmalonic and ethylmalonic semialdehyde dehydrogenase (MMSDH) (◘ Fig. 19.1, enzyme 12). So far, evidence for a deficiency of MMSDH with two (homozygous) mutant alleles of the methylmalonic semialdehyde dehydrogenase *(MMSDH)* gene has been demonstrated in one of three suspicious patients [73]. In this patient [74] the metabolite profile (comprising β-alanine, 3-hydroxy-propionic, R- and S-3-aminoisobutyric, R- and S-3-hydroxyisobutyric and S-2-hydroxymethyl butyric acids) correlates with MMSDH deficiency, an inborn error specific to the L-valine catabolic pathway.

An unique patient with 3-hydroxyisobutyryl-CoA deacylase deficiency (◘ Fig.19.1, enzyme 10) in liver and fibroblasts has been identified [75]. He did not exhibit any abnormal organic aciduria, but cysteine and cysteamine conjugates of methylacrylic acid, S-(2-carboxypropyl)-cysteine and S-(2-carboxypropyl)-cysteamine were found.

19.8   Malonic Aciduria

Only a few patients with malonic aciduria have been described. This condition can be divided into two groups. In the first there is a clear deficiency of malonyl-CoA decarboxylase (MLYCD) expressed in fibroblasts and/or leukocytes; in the second group there is normal MLYCD activity despite a similar clinical and biological presentation.

19.8.1   Clinical Presentation

A neonatal form has been described in two patients, who displayed progressive lethargy, hypotonia, and hepatomegaly associated with metabolic acidosis and mild hyperammonemia. Hypoglycemia was present in one newborn, and hyperlactatemia was present in the other.

In the late-onset forms, most patients present with acute episodes of gastroenteritis, febrile seizures, unexplained lethargy associated with metabolic acidosis, and hypoglycemia. Some of these patients were previously known to be affected with a mild and nonspecific psychomotor retardation. Other children have been diagnosed following systematic screening for mental retardation and hypotonia. Cardiomyopathy has been present in three patients.

19.8.2   Metabolic Derangement

Malonic aciduria is due to deficiency of MLYCD (◘ Fig. 19.1, enzyme 15). The physiological role of this cytsolic enzyme is somewhat unclear. It could play a role in the regulation of cytoplasmic malonyl-CoA abundance and, thus, of mitochondrial fatty acid uptake and oxidation. Patients with MLYCD deficiency display a number of phenotypes that are reminiscent of mitochondrial fatty acid oxidation disorders [76]. However, in contrast to these, dicarboxylic aciduria together with ketonuria is found during catabolic episodes and the patients exhibit normal ketogenesis on acute fat-loading tests.

19.8.3   Genetics

MLYCD deficiency is an autosomal-recessive disorder. The pathogenic alleles identified so far are documented in the Human MLYCD Allelic Variant Database [77].

19.8.4   Diagnostic Tests

Diagnosis relies on a characteristic profile of urinary organic acids, in which malonic and methylmalonic acids are constant findings. In all but one patient, malonic acid was much higher than methylmalonic acid. Abnormal succinic aciduria has been found in about half of the cases, as have various dicarboxylic and glutaric acidurias.

Total and free carnitine concentrations in plasma are low, and accumulation of malonylcarnitine has been documented in a few patients. Malonyl-CoA decarboxylase deficiency can be detected on the basis of elevated malonylcarnitine concentration in newborn blood spots analysed by tandem MS [78]. MLYCD has been found to be reduced in cultured fibroblasts and in the leukocytes of seven out of 11 defective cell lines. Patients with normal enzyme activity may suffer from another enzyme defect or may have a tissue-specific isoenzyme alteration not expressed in fibroblasts.

19.8.5   Treatment and Prognosis

No rules for treatment and prognosis have been established. Carnitine supplementation corrects the carnitine deficiency and has improved the cardiomyopathy and muscle weakness in two patients. Long-term prognosis is unknown. Except for the two patients who have developed extrapyramidal signs following an acute crisis, most patients have residual mild developmental delay. There are subjects identified by newborn screening who remained asymptomatic at least during preschool age.

References

1. Gascon GC, Ozand PT, Brismar J (1994) Movement disorders in childhood organic acidurias clinical, neuroimaging, and biochemical correlations. Brain Dev 16:94-103
2. Morton DH, Strauss KA, Robinson DL et al (2002) Diagnosis and treatment of maple syrup urine disease: a study of 36 patients. Pediatrics 109:999-1008
3. Treacy E, Clow CL, Reade TR et al (1992) Maple syrup urine disease: interrelations between branched-chain amino-, oxo- and hydroxy-acids; implications for treatment; associations with CNS dysmyelination. J Inherit Metab Dis 15:121-135
4. Schoenberger S, Schweiger B, Schwahn B et al (2004) Dysmyelination in the brain of adolescents and young adults with maple syrup urine disease. Mol Genet Metab 82:69-75
5. Brismar J, Ozand PT (1994) CT and MR of the brain in disorders of the propionate and methylmalonate metabolism. Am J Neuroradiol 15:1459-1473
6. Chemelli AP, Schocke M, Sperl W et al (2000) Magnetic resonance spectroscopy (MRS) in five patients with treated propionic academia. J Magn Reson Imaging 11:596-600
7. Baumgartner ER, Viardot C (1995) Long-term follow-up of 77 patients with isolated methylmalonic acidaemia. J Inherit Metab Dis 18:138-142
8. Hörster F, Baumgartner MR, Suormala T et al (2004) Long-term outcome in methylmalonic acidurias depends on the underlying defect. J Inherit Metab Dis 27[Suppl 1]:66 (abstract)
9. Rutledge SL, Geraghty M, Mroczek E et al (1993) Tubulointerstitial nephritis in methylmalonic acidemia. Pediatr Nephrol 7:81-82
10. Leonard JV (1995) The management and outcome of propionic and methylmalonic acidaemia. J Inherit Metab Dis 18:430-434
11. Bodemer C, de Prost Y, Bachollet et al (1994) Cutaneous manifestations of methylmalonic and propionic acidaemia: a description based on 38 cases. Br J Dermatol 131:93-98
12. Kahler SG, Sherwood WG, Woolf D et al (1994) Pancreatitis in patients with organic acidemias. J Pediatr 124:239-243
13. Massoud AF, Leonard JV (1993) Cardiomyopathy in propionic acidaemia. Eur J Pediatr 152:441-445
14. Nyhan WL, Ozand PT (2000) Organic acidemias. In: Atlas of metabolic diseases. Chapman & Hall, London, pp 1-23
15. Sbai D, Narcy C, Thompson GN et al (1994) Contribution of odd-chain fatty acid oxidation to propionate production in disorders of propionate metabolism. Am J Nutr 59:1332-1337
16. Leonard JV (1996) Stable isotope studies in propionic and methylmalonic acidaemia. Eur J Pediatr 156[Suppl 1]:S67-S69
17. Aevarsson A, Chuang JL, Wynn RM et al (2000) Crystal structure of human branched-chain α-ketoacid dehydrogenase and the molecular basis of multienzyme complex deficiency in maple syrup urine disease. Structure 8:277-291
18. Ensenauer R, Vockley J, Willard JM et al (2004) A common mutation is associated with a mild, potentially asymptomatic phenotype in patients with isovaleric acidemia diagnosed by newborn screening. Am J Hum Genet 75:1136-1142
19. Perez-Cerda C, Clavero S, Perez B et al (2003) Functional analysis of PCCB mutations causing propionic acidemia based on expression studies in deficient human skin fibroblasts. Biochim Biophys Acta 1638:43-49
20. Perez B, Desviat LR, Rodriguez-Pombo P et al (2003) Propionic acidemia: identification of twenty–four novel mutations in Europe and North America. Mol Genet Metab 78:59-67
21. Yorifuji T, Kawai M, Muroi J et al (2002) Unexpectedly high prevalance of the mild form of propionic acidemia in Japan: presence of a common mutation and possible clinical implications. Hum Genet 111:161-165
22. Peters HL, Nefedov M, Lee LW et al (2002) Molecular studies in mutase-deficient (MUT) methylmalonic aciduria: identification of five novel mutations. Hum Mutat 20:406-410
23. Dobson CM, Wai T, Leclerc D et al (2002) Identification of the gene responsible for the cblA complementation group of vitamin B12-dependent methylmalonic academia is based on analysis of prokaryotic gene arrangements. Proc Natl Acad Sci USA 99:15554-15559
24. Dobson CM, Wai T, Leclerc D et al (2002) Identification of the gene responsible for the cblB complementation group of vitamin B12-dependent methylmalonic aciduria. Hum Mol Genet 11:3361-3369
25. Perez B, Martinez MA, Rincon A et al (2004) Mutational spectrum of isolated methylmalonic acidemia: Twenty-two novel allelic variants. J Inherit Metab Dis 27[Suppl 1]:67 (abstract)
26. Wilcken B, Wiley V, Hammond J, Carpenter K (2003) Screening newborns for inborn errors of metabolism by tandem mass spectrometry. N Engl J Med 348:2304-2312
27. Schulze A, Lindner M, Kohlmüller D et al (2003) Expanded newborn screening for inborn errors of metabolism by electrospray ionization-tandem mass spectrometry: results, outcome, and implications. Pediatrics 111:1399-1406
28. Rinaldo P, Tortorelli S, Matern D (2004) Recent developments and new applications of tandem mass spectrometry in newborn screening. Curr Opin Pediat 16:427-433
29. Dixon M (2001) Disorders of amino acid metabolism, organic acidemias and urea cylce defects. In: Shaw V, Lawson M (eds) Clinical paediatric dietetics, 2nd edn. Blackwell, London, pp 273-294

29a. Touati G, Valayannopoulos V, Mention K et al (2006) Methylmalonic and propionic acidurias: management without or with a few supplements of specific amino acid mixtures. J Inherit Metab Dis 29:288-299

30. Valayannopoulos V, Touati G, Rabier D et al (2004) The urinary urea/methylmalonic acid molar ratio is a simple and efficient biological marker to guide long term dietary treatment in methylmalonic aciduria patients. J Inherit Metab Dis 27[Suppl 1]:68 (abstract)

31. Jouvet P, Poggi F, Rabier D et al (1997) Continuous venovenous haemofiltration in the acute phase of neonatal maple syrup urine disease. J Inherit Metab Dis 20:463-472

32. Parini R, Sereni LP, Bagozzi DC et al (1993) Nasogastric feeding as the only treatment of neonatal maple syrup urine disease. Pediatrics 92:280-283

33. Heldt K, Schwahn B, Marquardt I et al (2005) Diagnosis of MSUD by newborn screening allows early intervention without extraneous detoxification. Mol Genet Metab 84:313-316

34. Grünewald S, Hinrichs F, Wendel U (1998) Pregnancy in a woman with maple syrup urine disease. J Inher Metab Dis 21:89-94

35. Wendel U, Saudubray JM, Bodner A, Schadewaldt P (1999) Liver transplantation in maple syrup urine disease. Eur J Pediatr 158 [Suppl 2]:S60-S64

35a. Barshop BA, Nyhan WL, Khanna A (2005) Hepatic and domino hepatic transplantation in maple syrup urine disease. J Inherit Metab Dis 28 [Suppl 1]:242 (abstract)

36. Hilliges C, Awiszus D, Wendel U (1993) Intellectual performance of children with maple urine disease. Eur J Pediatr 152:144-147

37. Kaplan P, Mazur A, Field M et al (1991) Intellectual outcome in children with maple syrup urine disease. J Pediatr 119:46-51

38. Hoffmann B, Helbling C, Schadewaldt P, Wendel U (2006) Impact of longitudinal plasma leucine levels in the intellectual outcome in patients with classic MSUD. Pediatr Res 59:17-20

39. Fries MH, Rinaldo P, Schmidt-Sommerfeld E et al (1996) Isovaleric acidemia: response to a leucine load after three weeks of supplementation with glycine, L-carnitine, and combined glycine-carnitine therapy. J Pediatr 129:449-452

40. Ensenauer R, Gruenert S, Willard J et al (2003) Natural history of isovaleric acidemia (IVA). J Inher Metab Dis 26[Suppl 2]:38 (abstract)

41. Shih VE, Aubry RH, De Grande G et al (1984) Maternal isovaleric acidemia. J Pediatr 105:77-78

42. Gebhardt B, Vlaho S, Fischer D et al (2003) N-carbamylglutamate enhances ammonia detoxification in a patient with decompensated methylmalonic aciduria. Mol Genet Metab 79:303-304

43. Matern D, Seydewitz, HH, Lehnert W et al (1996) Primary treatment of propionic acidemia complicated by acute thiamine deficiency. J Pediatr 129:758-760

44. Touati G, Ogier de Baulny H, Rabier D et al (2003) Beneficial effects of growth hormone treatment in children with methylmalonic and propionic acidurias. J Inherit Metab Dis 26[Suppl 2]:40 (abstract)

45. Nicolaides P, Leonard J, Surtees R (1998) Neurological outcome of methylmalonic acidaemia. Arch Dis Child 78:508-512

45a. Dionisi Vici C, Deodato F, Roschinger W et al (2006) »Classical« organic acidurias, propionic aciduria, methylmalonic aciduria and isovaleric aciduria: long-term outcome and effects of expanded newborn screening using tandem mass spectrometry. J Inherit Metab Dis 29:383-389

46. Surtees RAH, Matthews EE, Leonard JV (1992) Neurological outcome of propionic acidemia. Pediatr Neurol 5:334-337

47. van der Meer SB, Poggi F, Spada M et al (1994) Clinical outcome of long-term management of patients with vitamin-B12 unresponsive methylmalonic acidemia. J Pediatr 125:903-908

48. van der Meer SB, Poggi F, Spada M et al (1996) Clinical outcome and long term management of 17 patients with propionic acidaemia. Eur J Pediatr 155:205-210

49. Lubrano R, Scoppi P, Barsotti P et al (2001) Kidney transplantation in a girl with methylmalonic acidemia and end stage renal failure. Pediatr Nephrol 16:848-851

50. Diss E, Iams J, Reed N et al (1995) Methylmalonic aciduria in pregnancy: a case report. Am J Obstet Gynecol 172:1057-1059

51. Schlenzig JS, Poggi-Travert F, Laurent J et al (1995) Liver transplantation in two cases of propionic acidaemia. J Inherit Metab Dis 18:448-461

52. Leonard JV, Walter JH, McKiernan PJ (2001) The management of organic acidemias: the role of transplantation. J Inherit Metab Dis 24:309-311

53. Van't Hoff W, McKiernan PJ, Leonard JV (1999) Liver transplantation for methylmalonic acidaemia. Eur J Pediatr 158[Suppl 2]:S70-S74

54. Dixon MA, Leonard JV (1992) Intercurrent illness in inborn errors of intermediary metabolism. Arch Dis Child 67:1387-1391

55. Desviat LR, Perez-Cerda C, Perez B et al (2003) Functional analysis of MCCA and MCCB mutations causing methylcrotonylglycinuria. Mol Genet Metab 80:315-320

56. Baumgartner MR, Dantas MF, Suormala T et al (2004) Isolated 3-methylcrotonyl-CoA carboxylase deficiency: evidence for an allele-specific dominant negative effect and responsiveness to biotin therapy. Am J Hum Genet 75:790-800

57. Ijlst L, Loupatty FJ, Ruiter JP et al (2002) 3-methylglutaconic aciduria type I is caused by mutations in AUH. Am J Hum Genet 71:1463-1466

58. Ly TB, Peters V, Gibson KM et al (2003) Mutations in the AUH-gene cause 3-methylglutaconic aciduria type I. Hum Mutat 21:401-407

59. Barth PG, Valianpour F, Bowen VM et al (2004) X-linked cardioskeletal myopathy and neutropenia (Barth syndrome): an update. Am J Med Genet 126A:349-354

60. Schlame M, Kelley RI, Feigenbaum A et al (2003) Phospholipid abnormalities in children with Barth syndrome. J Am Coll Cardiol 42:1994-1999

61. Vaz FM, Houtkooper RH, Valianpour F (2003) Only one splice variant of the human TAZ gene encodes a functional protein with a role in cardiolipin metabolism. J Biol Chem 278:43089-43094

62. Anikster Y, Kleta R, Shaag A et al (2001) Type III 3-methylglutaconic aciduria (optic atrophy plus syndrome, or Costeff optic atrophy syndrome): identification of the OPA 3 gene and its founder mutation in Iraqi Jews. Am J Hum Genet 69:1218-1224

63. Broid E, Elpeleg O, Lahat E (1997) Type IV 3-methylglutaconic aciduria: a new case presenting with hepatic dysfunction. Pediatr Neurol 17:353-355

64. Andresen BS, Christensen E, Corydon TJ et al (2000) Isolated 2-methylbutyrylglycinuria caused by short/branched-chain acyl-CoA dehydrogenase deficiency: identification of a new enzyme defect, resolution of its molecular basis, and evidence for distinct acyl-CoA dehydrogenases in isoleucine and valine metabolism. Am J Hum Genet 67:1095-1103

65. Gibson KM, Burlingame TG, Hogema et al (2000) 2-Methylbutyryl-coenzyme A dehydrogenase deficiency: a new inborn error of L-isoleucine metabolism. Pediatr Res 47:830-833

66. Matern D, He M, Berry SA et al (2003) Prospective diagnosis of 2-methylbutyryl-CoA dehydrogenase deficiency in the Hmong population by newborn screening using tandem mass spectrometry. Pediatrics 112:74-78

67. Zschocke J, Ruiter JP, Brand J et al (2000) Progressive infantile neurodegeneration caused by 2-methyl-3-hydroxybutyryl-CoA dehydrogenase deficiency: a novel inborn error of branched-chain fatty acid and isoleucine metabolism. Pediatr Res 48:852-855

68. Poll-The BT, Wanders RJ, Ruiter JP et al (2004) Spastic diplegia and periventricular white matter abnormalities in 2-methyl-3-hydroxybutyryl-CoA dehydrogenase deficiency, a defect of isoleucine metabolism: differential diagnosis with hypoxic-ischemic brain diseases. Mol Genet Metab 81:295-299

IV

69. Ofman R, Ruiter JP, Feenstra M et al (2003) 2-Methyl-3-hydroxy-butyryl-CoA dehydrogenase deficiency is caused by mutations in the HADH2 gene. Am J Hum Genet 72:1300-1307

70. Koeberl DD, Young SP, Gregersen NS et al (2003) Rare disorders of metabolism with elevated butyryl- and isobutyryl-carnitine detected by tandem mass spectrometry newborn screening. Pediatr Res 54:219-223

71. Bischoff C, Christensen E, Simonsen H et al (2004) Mutations in IBD (ACAD 8) may explain elevated C4-carnitine detected by MS/MS screening in newborns. J Inherit Metab Dis 27[Suppl 1]:13 (abstract)

72. Nguyen TV, Andresen BS, Corydon TJ et al (2002) Identification of isobutyryl-CoA dehydrogenase and its deficiency in humans. Mol Genet Metab 77:68-79

73. Chambliss KL, Gray RG, Rylance G et al (2000) Molecular characterization of methylmalonate semialdehyde dehydrogenase deficiency. J Inherit Metab Dis 23 497-504

74. Pollitt RJ, Green A, Smith R (1985) Excessive excretion of β-alanine and of 3-hydroxypropionic, *R*- and *S*-3-Aminoisobutyric, *R*-and *S*-3-Hydroxyisobutyric and *S*-2-(hydroxymethyl)butyric acids probably due to a defect in the metabolism of the corresponding malonic semialdehydes. J Inherit Metab Dis 8 75-79

75. Brown GK, Huint SM, Scholem R et al (1982) Hydroxyisobutyryl-coenzyme A deacylase deficiency: a defect in valine metabolism associated with physical malformations. Pediatrics 70:532-538

76. Sacksteder KA, Morell JC, Wanders RJ et al (1999) MCD encodes peroxisomal and cytoplasmic forms of malonyl-CoA decarboxylase and is mutated in malonyl-CoA decarboxylase deficiency. J Biol Chem 274:24461-24468

77. Human MLYCD Allelic Variant Database (http://mlycd.hgu.mrc.ac.uk/)

78. Santer R, Fingerhut R, Lassker U et al (2003) Tandem mass spectrometric determination of malonylcarnitine: diagnosis and neonatal screening of malonyl-CoA carboxylase deficiency. Clin Chem 49:660-662

20 Disorders of the Urea Cycle and Related Enzymes

James V. Leonard

The Urea Cycle

The urea cycle (◘ Fig. 20.1) which, in its complete form, is only present in the liver, is the main pathway for the disposal of excess of nitrogen. This sequence of reactions, localised in part in the mitochondria and in part in the cytosol, converts the toxic ammonia and other nitrogenous compounds into the non-toxic product, urea, which is excreted in the urine. Genetic defects of each enzyme of the urea cycle are recognised and all are responsible for hyperammonaemia. Genetic defects of other metabolic pathways may also lead to secondary inhibition of the urea cycle. Alternative pathways for nitrogen excretion, namely conjugation of glycine with benzoate and of glutamine with phenylacetate can be exploited in the treatment of patients with defective ureagenesis.

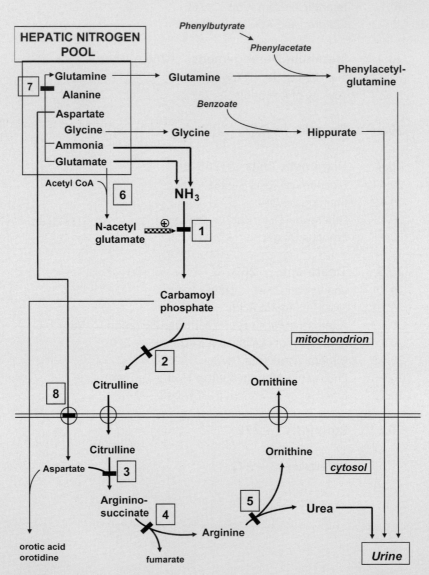

◘ **Fig 20.1.** The urea cycle and alternative pathways of nitrogen excretion. Enzymes: **1**, carbamoyl phosphate synthetase; **2**, ornithine transcarbamoylase; **3**, argininosuccinate synthetase; **4**, argininosuccinate lyase; **5**, arginase; **6**, N-acetylglutamate synthetase; 7, glutamine synthetase. **8**, Citrin (mitochondrial aspartate-glutamate carrier); + denotes stimulation. Defects are depicted by *solid bars* across the arrows

Six inherited disorders of the urea cycle are well described (◘ Fig. 20.1). These are the deficiencies of *carbamoyl phosphate synthetase (CPS), ornithine transcarbamoylase (OTC), argininosuccinate synthetase, argininosuccinate lyase, arginase*, and *N-acetylglutamate
synthetase (NAGS)*. Deficiencies of *glutamine synthetase*
and of *citrin* have also been described. All these defects
are characterised by hyperammonaemia and disordered
amino acid metabolism. The presentation is highly variable: those presenting in the newborn period usually
have an overwhelming illness that rapidly progresses
from poor feeding, vomiting, lethargy or irritability and
tachypnoea to fits, coma and respiratory failure. In infancy, the symptoms are less severe and more variable.
Poor developmental progress, behavioural problems,
hepatomegaly and gastrointestinal symptoms are common. Children and adults frequently have a chronic
neurological illness that is characterised by variable behavioural problems, confusion, irritability and episodic
vomiting. However, during any metabolic stress the patients may become acutely unwell. Arginase deficiency
has more specific symptoms, such as spastic diplegia,
dystonia, ataxia and fits. All these disorders have autosomal-recessive inheritance except ornithine transcarbamoylase deficiency, which is X-linked.

20.1 Clinical Presentation

Patients with urea-cycle disorders may present at almost
any age. However, there are certain times at which they are
more likely to develop symptoms because of metabolic
stress, such as infection precipitating protein catabolism.
These are:
- *The neonatal period.*
- *During late infancy.* Children are vulnerable during
this period because of the slowing of growth, the change
to cow's milk and weaning foods and the declining maternal antibody and consequent development of intercurrent infections.
- *Puberty.* The changing growth rate and psychosocial
factors may precipitate decompensation.

However, it must be emphasised that many patients may
present outside these periods. The patterns of the clinical
presentation of hyperammonaemia are rather characteristic
and are broadly similar for all the disorders except arginase
deficiency, which is discussed separately. The early symptoms are often non-specific and initially, therefore, the
diagnosis is easily overlooked. The most important points
in diagnosing hyperammonaemia are to think of it and to
measure the plasma ammonia concentration.

20.1.1 Neonatal Presentation

Most babies with urea cycle disorders that present in the
neonatal period are of normal birthweight and are initially
healthy but, after a short interval that can be less than
24 h, they become unwell. Common early symptoms are
poor feeding, vomiting, lethargy and/or irritability and
tachypnoea. The initial working diagnosis is almost invariably sepsis. Rather characteristically, these babies may have
a transient mild respiratory alkalosis, which can be a useful
diagnostic clue at this stage. Usually, they deteriorate rapidly,
with more obvious neurological and autonomic problems,
including changes of tone with loss of normal reflexes, vasomotor instability and hypothermia, apnoea and fits. They
may soon become unresponsive and may require full intensive care. Untreated, most babies will die, often with complications, such as cerebral or pulmonary haemorrhage, the
underlying metabolic cause for which may not be recognised. Some survive neonatal hyperammonaemia but are
invariably handicapped to a significant degree.

20.1.2 Infantile Presentation

In infancy, the symptoms are generally rather less acute
and more variable than in the neonatal period and include
anorexia, lethargy, vomiting and failure to thrive, with poor
developmental progress. Irritability and behavioural problems are also common. The liver is often enlarged but, as the
symptoms are rarely specific, the illness is initially attributed to many different causes that include gastrointestinal
disorders (gastro-oesophageal reflux, cow's milk protein
intolerance), food allergies or hepatitis. The correct diagnosis is often only established when the patient develops a
more obvious encephalopathy with changes in consciousness level and neurological signs (► below).

20.1.3 Children and Adults

At these ages, the patients commonly present with a more
obviously neurological illness.

Acute Encephalopathy
Whilst older patients often present with episodes of acute
metabolic encephalopathy, they may also have chronic
symptoms. Usually, symptoms develop following metabolic
stress precipitated by infection, anaesthesia or protein catabolism, such as that produced by the rapid involution of the
uterus in the puerperium [1]. However an obvious trigger
is not always apparent. The patients first become anorexic,
lethargic and unwell. Sometimes they are agitated and irri-
table, with behaviour problems or confusion. Vomiting and
headaches may be prominent, suggesting migraine or cyclical vomiting. Others may be ataxic as though intoxicated.

On examination, hepatomegaly may be present, particularly in those with argininosuccinic aciduria. The patients may then recover completely but, if not, they may then develop neurological problems, including a fluctuating level of consciousness, fits and (sometimes) focal neurological signs, such as hemiplegia [2] or cortical blindness. Untreated, they continue to deteriorate, becoming comatose, and they may die. The cause of death is usually cerebral oedema. Alternatively, they may recover with a significant neurological deficit.

Between episodes, the patients are usually relatively well, although some, particularly younger ones, may continue to have problems, such as vomiting or poor developmental progress. Some patients may voluntarily restrict their protein intake. In addition to those disorders already mentioned, the illness may be attributed to a wide variety of other disorders, including Reye syndrome, encephalitis, poisoning and psychosocial problems.

Chronic Neurological Illness

Learning difficulties or more obvious mental retardation are common, and some patients, particularly those with argininosuccinic aciduria, may present with relatively few symptoms apart from mental retardation and fits. About half the patients with argininosuccinic aciduria have brittle hair (trichorrhexis nodosa). Patients may present with chronic ataxia, which is worse during intercurrent infections.

Arginase Deficiency

Arginase deficiency commonly presents with spastic diplegia and, initially, a diagnosis of cerebral palsy is almost always suspected. However, the neurological abnormalities appear to be slowly progressive, although it may be difficult to distinguish this from an evolving cerebral palsy. During the course of the disease, fits, ataxia and dystonia may develop. Occasionally, patients may present with an acute encephalopathy or anticonvulsant-resistant fits [3].

Glutamine Synthetase Deficiency

Two neonates have recently been described who presented with convulsions on day 1. During the second week one patient developed a necrolytic skin disorder. Both had mild hyperammonaemia (140 µmol/l) and very low glutamine concentrations in plasma, urine and cerebrospinal fluid (CSF). A homozygous mutation was identified in the glutamine synthetase gene [4].

Citrin Deficiency

This disorder, also called citrullinaemia type II, is a deficiency of the mitochondrial aspartate-glutamate carrier. The result is an intramitochondrial deficiency of aspartate (◘ Fig. 20.1). The disorder presents at two ages: in the neonatal period with liver disease, and in adulthood with typical symptoms of hyperammonaemia [5].

20.2   Metabolic Derangement

20.2.1   The Urea Cycle

The urea cycle is the final common pathway for the excretion of waste nitrogen in mammals. The steps in the urea cycle are shown in ◘ Fig. 20.1. Ammonia is probably derived principally from glutamine and glutamate and is converted to carbamoyl phosphate by carbamoyl phosphate synthetase (CPS). This enzyme requires an allosteric activator, N-acetylglutamate, for full activity. This compound is formed by the condensation of acetyl coenzyme A (acetyl CoA) and glutamate in a reaction catalysed by N-acetyl glutamate synthetase (NAGS). Carbamoyl phosphate condenses with ornithine to form citrulline in a reaction catalysed by ornithine transcarbamoylase (OTC). The product, citrulline, condenses with aspartate to produce argininosuccinate in a reaction catalysed by argininosuccinate synthetase, and the argininosuccinate is then hydrolysed to arginine and fumarate by argininosuccinate lyase. The supply of aspartate is dependent on the mitochondrial shuttles, particularly the aspartate-glutamate carrier, *SLC25A13* [5].

The arginine is itself cleaved by arginase, releasing urea and re-forming ornithine. Within the urea cycle itself, ornithine acts as a carrier; it is neither formed nor lost. Although all the enzymes are present in the liver, there are large inter-organ fluxes. Citrulline is synthesised in the gut and metabolised to arginine in the kidney. The arginine is taken up by the liver and hydrolysed to urea and ornithine.

Each molecule of urea contains two atoms of waste nitrogen, one derived from ammonia and the other from aspartate. Regulation of the urea cycle is not fully understood, and it is likely that there are several mechanisms controlling flux through this pathway [6]. These include enzyme induction, the concentrations of substrates, intermediates and N-acetyl glutamate, and hormonal effects. Defects of each step have now been described and are listed in ◘ Table 20.1.

The plasma ammonia concentration is raised as a result of metabolic blocks in the urea-cycle. The degree to which it is elevated depends on several factors, including the enzyme involved and its residual activity, the protein intake and the rate of endogenous protein catabolism, particularly if this is increased because of infection, fever or other metabolic stresses. The values may also be falsely elevated if the specimen is not collected and handled correctly.

The concentrations of the amino acids in the metabolic pathway immediately proximal to the enzyme defect will increase, and those beyond the block will decrease (◘ Table 20.1). In addition, plasma alanine and particularly glutamine accumulate in all the disorders. The concentration of citrulline can be helpful, but it may not always be reliable, particularly during the newborn period [7].

◻ Table 20.1. Diagnostic tests in urea cycle defects

Disorder	Alternative names	Plasma amino acid concentrations	Urine orotic acid	Tissue for enzyme diagnosis	Genetics – gene (chromosome localisation)
N-acetylglutamate synthetase deficiency	NAGS deficiency	↑ glutamine ↑ alanine	N	Liver	AR – NAGS (chromosome 17q 21.31)
Carbamoyl phosphate synthetase deficiency	CPS deficiency	↑ glutamine ↑ alanine ↓ citrulline ↓ arginine	N	Liver	AR – CPS1 (chromosome 2p 35)
Ornithine transcarbamoylase deficiency	OTC deficiency	↑ glutamine ↑ alanine ↓ citrulline ↓ arginine	↑↑	Liver	X-linked – OTC (Xp21.1)
Argininosuccinic acid synthetase deficiency	Citrullinaemia	↑↑ citrulline ↓ arginine	↑	Liver/ fibroblasts	AR – ASS (chromosome 9q 34)
Argininosuccinic acid lyase deficiency	Argininosuccinic aciduria (ASA)	↑ citrulline ↑ argininosuccinic acid ↓ arginine	↑	RBC/Liver/ fibroblasts	AR – ASL (chromosome 7cen-q11.2)
Arginase deficiency	Hyperargininaemia	↑ arginine	↑	RBC/Liver	AR – ARG1 (chromosome 6q 23)

AR, autosomal recessive; *RBC*, red blood cells; *N*, normal

Orotic acid and orotidine are excreted in excess in the urine if there is a metabolic block distal to the formation of carbamoyl phosphate, as is the case in OTC deficiency, citrullinaemia, argininosuccinic aciduria and arginase deficiency (◻ Fig. 20.1). In these disorders, carbamoyl phosphate accumulates, leaves the mitochondrion and, once in the cytosol, enters the pathway for the de novo synthesis of pyrimidines. The urea cycle is also closely linked to many other pathways of intermediary metabolism, particularly the citric-acid cycle through the glutamate-aspartate shuttle (citrin).

20.2.2 Toxicity of Ammonia

Ammonia increases the transport of tryptophan across the bloodbrain barrier, which then leads to an increased production and release of serotonin. Some of the symptoms of hyperammonaemia can be explained on this basis, and the dietary tryptophan restriction reverses anorexia in some patients with urea cycle disorders [9]. Ammonia induces many other electrophysiological, vascular and biochemical changes in experimental systems, but it is not known to what extent all of these are relevant to the problems of clinical hyperammonaemia in man [10].

Using proton nuclear magnetic resonance spectroscopy, glutamine can also be shown to accumulate at high concentrations, both in experimental models and *in vivo* in man [11]. The concentrations are such that the increase in osmolality could be responsible for cellular swelling and cerebral oedema.

20.3   Genetics

The genes for all the urea-cycle enzymes including NAGS have now been mapped, isolated and fully characterised. Many mutations have been described. The most common urea cycle disorder is OTC deficiency, which is X-linked disorder and in which molecular genetic studies are particularly helpful. When the diagnosis of OTC deficiency is established, it is necessary to take a careful family history and for the mother's carrier status to be assessed, most reliably by mutation analysis. However, if the mutation is not known, the most convenient investigation is the allopurinol test, which is used to detect increased de novo synthesis of pyrimidines (▶ Chap. 35). This is easier than the protein- or alanine-loading tests and carries no risk of hyperammonaemia. However, recent studies suggest that the sensitivity and specificity are not as good as was once thought [8].

All the other conditions have autosomal-recessive inheritance.

20.4   Diagnostic Tests

20.4.1   Biochemical Tests

Routine tests are not helpful for establishing the diagnosis of hyperammonaemia. Plasma transaminases may be elevated; combined with hepatomegaly, this may lead to the erroneous diagnosis of hepatitis.

The most important diagnostic test in urea cycle disorders is measurement of the plasma ammonia concentration. Normally, this is less than 50 μmol/l but may be modestly raised as a result of a high protein intake, exercise, struggling or a haemolysed blood sample. Generally, patients who are acutely unwell with urea cycle disorders have plasma ammonia concentrations greater than 150 μmol/l, and often significantly higher. However, the concentrations may be near normal when patients are well, are early in an episode of decompensation or if they have been on a low-protein, high-carbohydrate intake for some time.

Healthy neonates have slightly higher values than older infants. If they are ill (sepsis, perinatal asphyxia etc.), plasma ammonia concentrations may increase to 180 μmol/l. Patients with inborn errors presenting in the newborn period usually have concentrations greater than 200 μmol/l, often very much greater. In that case, further investigations (particularly of the plasma amino acid and urine organic acid levels) are urgent. The following investigations should be performed:

- Repeat plasma ammonia
- Blood pH and gases
- Plasma chemistry: urea, electrolytes, glucose and creatinine
- Liver-function tests and clotting studies
- Plasma amino acids
- Urine organic acids, orotic acid and amino acids
- Plasma free and acyl carnitines

In all urea-cycle disorders, there is accumulation of glutamine and alanine and, in citrullinaemia, argininosuccinic aciduria and arginase deficiency, the changes in the amino acids are usually diagnostic (◨ Table 20.1). Orotic aciduria with raised plasma glutamine and alanine concentrations suggests OTC deficiency. The diagnosis of this and the other disorders can be confirmed by measuring enzyme activity in appropriate tissue (◨ Table 20.1) or by molecular genetic studies. The enzyme diagnosis of NAGS deficiency is not straightforward and the diagnosis is best made by molecular genetic studies [12].

20.4.2   Imaging

Patients who present with an acute encephalopathy commonly receive brain imaging at an early stage. This may show no abnormality, a localised area of altered signal or, if the patient is very seriously ill, widespread cerebral oedema [13].

Focal areas of altered signal may be identified and need to be distinguished from herpes simplex encephalitis. A careful history revealing previous episodes of encephalopathy, albeit mild, may provide vital clues. Imaging in patients who have recovered from a severe episode of hyperammonaemia usually show cerebral atrophy that may be focal, particularly in those areas in which there were altered signals during the acute illness.

20.4.3   Differential Diagnosis

The differential diagnosis of hyperammonaemia is wide, and the most common conditions are summarised in ◨ Table 20.2. In the neonatal period, the most common differential diagnoses are organic acidaemias, particularly propionic and methylmalonic acidaemia. Patients with these disorders may have had marked hyperammonaemia with minimal metabolic acidosis or ketosis. Although babies

---

◨ **Table 20.2.** Differential diagnosis of hyperammonaemia

Inherited Disorders
Urea cycle enzyme defects
  Carbamoyl phosphate synthetase deficiency
  Ornithine transcarbamoylase deficiency
  Argininosuccinate synthetase deficiency (citrullinaemia)
  Argininosuccinate lyase deficiency (argininosuccinic aciduria)
  Arginase deficiency
  N-acetylglutamate synthetase deficiency
Transport defects of urea cycle intermediates
  Lysinuric protein intolerance
  Hyperammonaemia – hyperornithinaemia – homocitrullinuria syndrome
  Citrin deficiency (citrullinaemia type II)
Organic acidaemias
  Propionic acidaemia
  Methylmalonic acidaemia and other organic acidaemias
Fatty acid oxidation disorders
  Medium chain acyl-CoA dehydrogenase deficiency
  Systemic carnitine deficiency
  Long chain fatty acid oxidation defects and other related disorders
Other inborn errors
  Pyruvate carboxylase deficiency (neonatal form)
  Ornithine aminotransferase deficiency (neonates/infants)

Acquired Disorders
Transient hyperammonaemia of the newborn
Any severe systemic illness particularly in neonates
Herpes simplex – neonates with systemic infection
Liver failure
Infection with urease positive bacteria (with urinary tract stasis)
Reye syndrome
Valproate therapy
Leukaemia therapy including therapy with asparaginase (rare)

with transient hyperammonaemia of the newborn are often born prematurely, with early onset of symptoms [14], it may be difficult to distinguish between this condition and urea-cycle disorders. All patients in whom a tentative diagnosis of Reye syndrome is made should be investigated in detail for inherited metabolic disorders, including urea-cycle disorders.

20.4.4 Prenatal Diagnosis

OTC deficiency is an X-linked disorder and prenatal diagnosis is done either by identifying the mutation or using an informative polymorphism. However, whilst the phenotype of the males can be predicted, that of the females cannot because of the random inactivation of the X chromosome. This presents a problem when counselling families, but the prognosis for females who are treated prospectively from birth is good.

All the other urea cycle disorders have autosomal-recessive inheritance and prenatal diagnosis can help most families. For CPS deficiency, prenatal diagnosis using closely linked gene markers is now possible for a substantial proportion of families. If the molecular-genetic studies are uninformative, prenatal liver biopsy is a possible alternative. Mutation analysis has to be used for NAGS deficiency. Citrullinaemia and argininosuccinic aciduria can both be diagnosed by enzyme or molecular genetic studies on chorionic villus biopsy. Argininosuccinic aciduria can also be diagnosed by measuring the argininosuccinate concentration in amniotic fluid. Arginase deficiency can be diagnosed either with molecular-genetic studies or, if they are not informative, with a fetal blood sample.

20.5 Treatment

The aim of treatment is to correct the biochemical disorder and to ensure that all the nutritional needs are met. The major strategies used are to reduce protein intake, to utilise alternative pathways of nitrogen excretion and to replace nutrients that are deficient.

20.5.1 Low-Protein Diet

Most patients require a low-protein diet. The exact quantity will depend mainly on the age of the patient and the severity of the disorder. Many published regimens suggest severe protein restriction but, in early infancy, patients may need >2 g/kg/day during phases of very rapid growth. The protein intake usually decreases to approximately 1.2–1.5 g/kg/day during pre-school years and 0.8–1 g/kg/day in late childhood. After puberty, the quantity of natural protein may be less than 0.5 g/kg/day. However, it must be empha-

sised that there is considerable variation in the needs of individual patients [15].

20.5.2 Essential Amino Acids

In the most severe variants, it may not be possible to achieve good metabolic control and satisfactory nutrition with restriction of natural protein alone. Other patients will not take their full protein allowance. In both these groups of patients, some of the natural protein may be replaced with an essential amino acid mixture, giving up to 0.7 g/kg/day. Using this, the requirements for essential amino acids can be met; in addition, nitrogen is re-utilised to synthesise non-essential amino acids, hence reducing the load of waste nitrogen.

20.5.3 Alternative Pathways for Nitrogen Excretion

In many patients, additional therapy is necessary. A major advance in this field has been the development of compounds that are conjugated to amino acids and rapidly excreted [16, 17]. The effect of the administration of these substances is that nitrogen is excreted in compounds other than urea; hence, the load on the urea cycle is reduced (◘ Fig. 20.1). The first compound introduced was sodium benzoate. Benzoate is conjugated with glycine to form hippurate, which is rapidly excreted. For each mole of benzoate given, if conjugation is complete, 1 mol of nitrogen is lost. Sodium benzoate is usually given in doses up to 250 mg/kg/day but, in acute emergencies, this can be increased to 500 mg/kg/day. The major side effects are nausea, vomiting and irritability. In neonates, conjugation may be incomplete, with increased risk of toxicity [C. Bachmann, personal communication].

The next drug used was phenylacetate, but this has now been superseded by phenylbutyrate, because the former has a peculiarly unpleasant, clinging, mousy odour. In the liver phenylbutyrate is oxidised to phenylacetate, which is then conjugated with glutamine. The resulting phenylacetylglutamine is rapidly excreted in urine; hence, if this reaction was complete 2 mol of nitrogen would be excreted for each mol of phenylbutyrate given. However recent studies indicated that phenylbutyrate is metabolized via several different pathways so that for each mol of sodium phenylbutyrate only approximately 1 mol of nitrogen is lost [18]. Phenylbutyrate is usually given as the sodium salt in doses of 250 mg/kg/day, but has been given in doses of up to 650 mg/kg/day [19]. In emergencies sodium benzoate, sodium phenylacetate (in USA Ammonul) and sodium phenylbutyrate can all be given intravenously in the same doses as oral. In a study of the side effects [20], there was a high incidence of menstrual disturbance in females. Other problems included anorexia and vomiting, but it is not easy to distinguish between the effects of the disorder and those of the medicine. Phenylbutyrate may cause a mucositis and one patient developed an oesophageal stricture (unpublished observation). Patients are often reluctant to take the medicines, and considerable ingenuity is sometimes needed to ensure that they do.

20.5.4   Replacement of Deficient Nutrients

Arginine and Citrulline

Arginine is normally a nonessential amino acid, because it is synthesised within the urea cycle. For this reason, all patients with urea-cycle disorders (except those with arginase deficiency) are likely to need a supplement of arginine to replace that which is not synthesised [21]. The aim should be to maintain plasma arginine concentrations between 50 µmol/l and 200 µmol/l. For OTC and CPS deficiencies, a dose of 50–150 mg/kg/day appears to be sufficient for most patients.

However, in severe variants of OTC and CPS, citrulline may be substituted for arginine in doses up to 170 mg/kg/day, as this will utilise an additional nitrogen molecule. Patients with citrullinaemia and argininosuccinic aciduria have a higher requirement, because ornithine is lost as a result of the metabolic block; this is replaced by administering arginine. Doses of up to 700 mg/kg/day may be needed, but this does have the disadvantage of increasing the concentrations of citrulline and argininosuccinate, respectively. The consequences of this are thought to be less important than those caused by the accumulation of ammonia and glutamine but the poor outcome of argininosuccinic aciduria may lead to a review of this therapy.

Other Medication

Citrate has long been used to provide a supply of Krebs-cycle intermediates. It is known to reduce postprandial elevation of ammonia and may be helpful in the management of argininosuccinic aciduria [22].

N-carbamyl glutamate can be used in NAGS deficiency to replace the missing compound, as it is active orally. The dose is 100–300 mg/kg/day [23]. Patients who respond may only require treatment with this compound.

Anticonvulsants may be needed for patients with urea-cycle disorders, but sodium valproate should not be used, as this drug may precipitate fatal decompensation, particularly in OTC patients [24].

20.5.5   General Aspects of Therapy

All treatments must be monitored with regular quantitative estimation of plasma ammonia and amino acids, paying particular attention to the concentrations of glutamine and essential amino acids. The aim is to keep plasma ammonia levels below 80 µmol/l and plasma glutamine levels below 800 µmol/l [25]. In practice, a glutamine concentration of 1000 µmol/l together with concentrations of essential amino acids within the normal range (▶ the algorithm, ◘ Fig. 20.2) is probably more realistic. All diets must be nutritionally complete and must meet requirements for growth and normal development.

The concept of balance of diet and medicine is important. The protein intake of the patients varies considerably, and the figures that have been given should be regarded only as a guide. The variation reflects not only the residual enzyme activity but also many other factors, including appetite and growth rate. Some patients have an aversion to protein, so it can be difficult to get them to take even their recommended intake. Consequently, they are likely to need smaller doses of sodium benzoate and phenylbutyrate. Others take more protein, and this has to be balanced by an increase in the dosages of benzoate and phenylbutyrate. Some will not take adequate quantities of sodium benzoate or sodium phenylbutyrate and, therefore, their protein intakes necessarily have to be stricter than would be needed if they took the medicines. Hence, for each patient, a balance must be found between their protein intake and the dose of their medicines to achieve good metabolic control.

Tube feeding, either by naso-gastric or gastrostomy, is often an integral part of the management to ensure a balanced diet and to give the medicines.

20.5.6   Emergency Management at Home

All patients with urea cycle disorders are at risk of acute decompensation with acute hyperammonaemia. This can be precipitated by any metabolic stress, such as fasting, a large protein load, infection, anaesthesia or surgery. For this reason, all patients should have detailed instructions of what to do when they are at risk. We routinely use a three-stage procedure [26]. If the patient is off-colour, the protein is reduced, and more carbohydrate is given. If symptoms continue, protein should be stopped and a high-energy intake given with their medication by day and night. However, if they cannot tolerate oral drinks and medicines, are vomiting or are becoming progressively encephalopathic, they should go to a hospital for assessment and intravenous therapy without delay. For further practical details, see Dixon and Leonard [26]. Patients should also have a high carbohydrate intake before any anaesthesia or surgery.

For patients who are seriously ill with hyperammonaemia, treatment is urgent. The most important and useful signs are any degree of encephalopathy and the speed of onset. It may be initially minor symptoms such as irritability but for any further alteration in conscious state treatment is urgent. Plasma ammonia concentration is not a

◻ Fig. 20.2. Guidelines for the management of patients with urea cycle disorders, primarily with OTC deficiency (not arginase deficiency). This is intended for use in patients who have been stabilised previously and should only be regarded as a guide, as many patients have individual requirements. For more detail and information about doses, please refer to the text. *EAAs*, essential aminoacids; *N*, normal; ↑/↓ increase/decrease

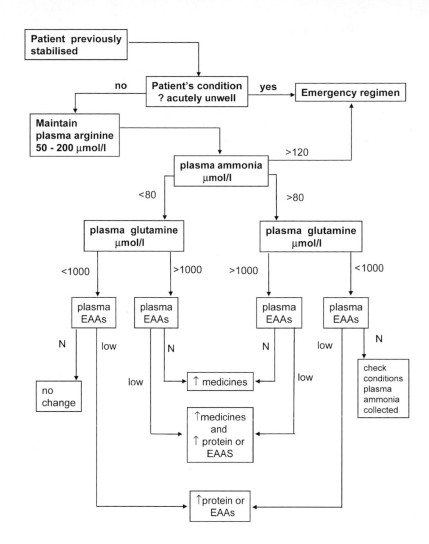

reliable guide as it may be normal in the early stages of encephalopathy and still increased when the patient is clearly improving. The steps in an emergency are listed below, and early treatment is essential (▶ Chap. 4).

20.5.7 Emergency Management in Hospital

The volumes that are given are related to age and the condition of the patient. Fluid volumes should be restricted if there is any concern about cerebral oedema.

1. Stop protein intake.
2. Give a high energy intake either orally or intravenously.
   A. Orally: (a) 10–20% soluble glucose polymer or (b) protein-free formula
   B. Intravenously: (a) 10% glucose by peripheral infusion or (b) 10–25% glucose by central venous line
3. Give sodium benzoate up to 500 mg/kg/day – orally or intravenously. If the patient has not received any medication recently, give a priming dose of 250 mg/kg in 2–4 hours then 250 mg/kg in the next 20–22 hours.

4. Give sodium phenylbutyrate up to 600 mg/kg/day – orally or intravenously. If the patient has not received any medication recently, give a priming dose of 250 mg/kg in 2–4 hours then up to 350 mg/kg in the next 20–22 hours.
5. Give L-arginine – orally or intravenously:
   A. Up to 700 mg/kg/day in citrullinaemia and argininosuccinic aciduria.
   B. Up to 150 mg/kg/day in OTC and CPS deficiencies.
6. Dialysis. If hyperammonaemia is not controlled or the medicines are not immediately available, haemofiltration (or haemodialysis/haemodiafiltration) should be started without delay. Alternatively, peritoneal dialysis may be used, but this is a less effective method for reducing hyperammonaemia.
7. Treat other conditions (sepsis, fits etc.).
8. Reduce intracranial pressure with the usual measures and maintain perfusion pressure.

For the emergency treatment of hyperammonaemia before diagnosis is known, the plan outlined may be replaced by
1. L-arginine 300 mg/kg/24 h – orally or intravenously
2. L-carnitine 200 mg/kg/24 h – orally or intravenously

20.6   Prognosis

The prognosis in these disorders is closely related to the age of the patient and their condition at the time of diagnosis. For those patients who present with symptomatic hyperammonaemia in the newborn period, the outlook is not good. Even with the most aggressive treatment, the majority of the survivors will be handicapped. Those who are treated prospectively do better, but there may still be significant complications [27]. The main factors that determine outcome are not clear but both the duration and the peak of hyperammonaemia are likely to be important [28, 29]. For these patients, there remains a serious risk of decompensation, and careful consideration should be given to early liver transplantation, which may offer the hope of a better long-term outlook [30–32]. Of those who present later, their neurological problems at the time of diagnosis are critical, as most will have already suffered neurological damage. At best, this may apparently resolve, but almost all are left with some degree of learning and neurological problems. Patients who have widespread cerebral oedema almost all die or survive with severe handicaps. By contrast, those who are treated prospectively have a better outcome.

References

1. Arn PH, Hauser ER, Thomas GH et al (1990) Hyperammonemia in women with a mutation at the ornithine carbamoyltransferase locus. A cause of postpartum coma. N Engl J Med 322:1652-1655
2. Christodoulou J, Qureshi IA, McInnes RR, Clarke JT (1993) Ornithine transcarbamylase deficiency presenting with strokelike episodes. J Pediatr 122:423-425
3. Patel JS, Van't Hoff W, Leonard JV (1994) Arginase deficiency presenting with convulsions. J Inherit Metab Dis 17:254
4. Häberle J, Gőrg B, Rutsch F et al (2005) Congenital glutamine deficiency with glutamine synthetase mutations. N Engl J Med 353:1926-1933
5. Saheki T, Kobayashi K, Iijima M et al (2004) Adult-onset type II citrullinaemia and idiopathic neonatal hepatitis caused by citrin deficiency: involvement of the aspartate glutamate carrier for urea synthesis and maintenance of the urea cycle. Mol Genet Metab 81[Suppl 1]:SS20-26
6. Newsholme EA, Leech AR (1983) Biochemistry for the medical sciences. Wiley, Chichester, pp 491-494
7. Batshaw ML, Brusilow SW (1978) Asymptomatic hyperammonaemia in low birthweight infants. Pediatr Res 12:221-224
8. Grünewald S, Fairbanks L, Genet S et al (2004) How reliable is the allopurinol load in detecting carriers for ornithine transcarbamylase deficiency? J Inherit Metab Dis 27:179-186
9. Hyman SL, Porter CA, Page TJ et al (1987) Behavior management of feeding disturbances in urea cycle and organic acid disorders. J Pediatr 111:558-562
10. Surtees RJ, Leonard JV (1989) Acute metabolic encephalopathy. J Inherit Metab Dis 12[Suppl 1]:42-54
11. Connelly A, Cross JH, Gadian DG et al (1993) Magnetic resonance spectroscopy shows increased brain glutamine in ornithine carbamoyl transferase deficiency. Pediatr Res 33:77-81
12. Caldovic L, Morizono H, Panglao MG et al (2003) Null mutations in the N-acetylglutamate synthase gene associated with acute neonatal disease and hyperammonemia. Hum Genet 112:364-368
13. Kendall B, Kingsley DPE, Leonard JV, Lingam S, Oberholzer VG (1983) Neurological features and computed tomography of the brain in children with ornithine carbamyl transferase deficiency. J Neurol Neurosurg Psychiatr 46:28-34
14. Hudak ML, Jones MD, Brusilow SW (1985) Differentiation of transient hyperammonaemia of the newborn and urea cycle enzyme defects by clinical presentation. J Pediatr 107:712-719
15. Leonard JV (2001) Nutritional treatment of urea cycle disorders. J Pediatr 138[Suppl 1]:S40-44
16. Brusilow SW, Valle DL, Batshaw ML (1979) New pathways of nitrogen excretion in inborn errors of urea synthesis. Lancet II:452-454
17. Feillet F, Leonard JV (1998) Alternative pathway therapy for urea cycle disorders. J Inherit Metab Dis 21[Suppl 1]:101-111
18. Kasumov T, Brunengraber LL, Comte B et al (2004) New secondary metabolites of phenylbutyrate in humans and rats. Drug Metab Dispos 32:10-19
19. Brusilow SW (1991) Phenylacetylglutamine may replace urea as a vehicle for waste nitrogen excretion. Pediatr Res 29:147-150
20. Wiech NL, Clissold DM, MacArthur RB (1997) Safety and efficacy of buphenyl (sodium phenylbutyrate) tablets and powder (abstract). Advances in inherited urea cycle disorders, satellite to the 7th international congress for inborn errors of metabolism, Vienna, p 25
21. Brusilow SW (1984) Arginine, an indispensible aminoacid for patients with inborn errors of urea synthesis. J Clin Invest 74:2144-2148
22. Iafolla AK, Gale DS, Roe CR (1990) Citrate therapy in arginosuccinate lyase deficiency. J Pediatr 117:102-105
23. Bachmann C, Colombo JP, Jaggi K (1982) N-acetylglutamate synthetase (NAGS) deficiency: diagnosis, clinical observations and treatment. Adv Exp Med Biol 153:39-45
24. Tripp JH, Hargreaves T, Anthony PP et al (1981) Sodium valproate and ornithine carbamyl transferase deficiency (letter). Lancet 1:1165-1166
25. Maestri NE, McGowan KD, Brusilow SW (1992) Plasma glutamine concentration: a guide to the management of urea cycle disorders. J Pediatr 121:259-261
26. Dixon MA, Leonard JV (1992) Intercurrent illness in inborn errors of intermediary metabolism. Arch Dis Child 67:1387-1391
27. Maestri NE, Hauser ER, Bartholomew D, Brusilow SW (1991) Prospective treatment of urea cycle disorders. J Pediatr 119:923-928
28. Picca S, Dionisi-Vici C, Abeni D et al (2001) Extracorporeal dialysis in neonatal hyperammonaemia: modalities and prognostic indicators. Paediatr Nephrol 16:862-867
29. Bachmann C (2003) Outcome and survival of 88 patients with urea cycle disorders: a retrospective evaluation. Eur J Pediatr 162:410-416
30. Todo S, Starzl TE, Tzakis A et al (1992) Orthotopic liver transplantation for urea cycle enzyme deficiency. Hepatology 15:419-422
31. Saudubray J-M, Touati G, DeLonlay P et al (1999) Liver transplantation in urea cycle disorders. Eur J Pediatr 158[Suppl 2]:S55-59
32. Sokal E (2006) Transplantation for inborn errors of liver metabolism. J Inherit Metab Dis 29:426-430

General Reference

Tuchman M (2001) Proceedings of a Consensus Conference For the Management of Patients with Urea Cycle disorders. J Pediatr 138[Suppl 1]:S1-S80

21   Disorders of Sulfur Amino Acid Metabolism

Generoso Andria, Brian Fowler, Gianfranco Sebastio

Metabolism of the Sulfur-Containing Amino Acids

Methionine, homocysteine and cysteine are linked by the methylation cycle (◪ Fig. 21.1, left part) and the trans-sulfuration pathway (◪ Fig. 21.1, right part). Conversion of methionine into homocysteine proceeds via methionine S-adenosyltransferase (enzyme 4). This yields S-adenosylmethionine, the methyl-group donor in a wide range of transmethylation reactions, a quantitatively important one of which is glycine N-methyltransferase (enzyme 5). These reactions also produce S-adenosylhomocysteine, which is cleaved to adenosine and homocysteine by S-adenosylhomocysteine hydrolase (enzyme 6). Depending on a number of factors, about 50% of available homocysteine is recycled into methionine. This involves methyl transfer from either 5-methyl-tetrahydrofolate (THF), catalyzed by cobalamin-requiring 5-methyl THF-homocysteine methyltransferase (enzyme 2), or betaine, catalyzed by betaine-homocysteine methyltransfcrase (enzyme 3). Homocysteine can also be condensed with serine to form cystathionine via a reaction catalyzed by pyridoxalphosphate-requiring cystathionine β-synthase (enzyme 1). Cystathionine is cleaved to cysteine and α-ketobutyrate by another pyridoxal-phosphate-dependent enzyme, γ-cystathionase (enzyme 7). The last step of the trans-sulfuration pathway converts sulfite to sulfate and is catalyzed by sulfite oxidase (enzyme 8), which requires a molybdenum cofactor.

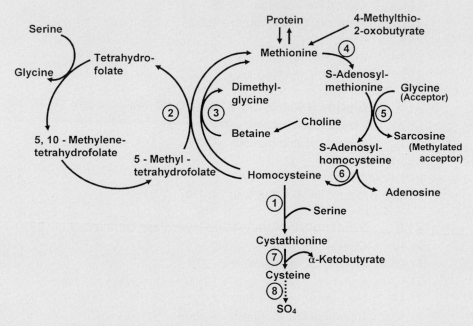

◪ **Fig. 21.1.** Metabolism of the sulfur-containing amino acids. 1, cystathionine β-synthase; 2, 5-methyltetrahydrofolate-homocysteine methyltransferase; 3, betaine-homocysteine methyltransferase; 4, methionine S-adenosyltransferase; 5, glycine N-methyltransferase; 6, S-adenosylhomocysteine hydrolase; 7, γ-cystathionase; 8, sulfite oxidase

Several inherited defects are known in the conversion of the sulfur-containing amino acid methionine to cysteine and the ultimate oxidation of cysteine to inorganic sulfate (◘ Fig. 21.1). Cystathionine β-synthase (CBS) deficiency is the most important. It is associated with severe abnormalities of four organs or organ systems: the eye (dislocation of the lens), the skeleton (dolichosteno-melia and arachnodactyly), the vascular system (thromboembolism), and the central nervous system (mental retardation, cerebro-vascular accidents). A low-methionine, high-cystine diet, pyridoxine, folate, and betaine in various combinations, and antithrombotic treatment may halt the otherwise unfavourable course of the disease. Methionine S-adenosyltransferase deficiency and γ-cystathionase deficiency usually do not require treatment. Isolated sulfite oxidase deficiency leads (in its severe form) to refractory convulsions, lens dislocation, and early death. No effective treatment exists. Combined deficiency of sulfite oxidase and xanthine oxidase is discussed in Chap. 35. Deficiencies of glycine N-methyltransferase and S-adenosylhomocysteine hydrolase have been described in a few patients.

21.1 Homocystinuria due to Cystathione β-Synthase Deficiency

21.1.1 Clinical Presentation

The eye, skeleton, central nervous system, and vascular system are all involved in the typical presentation. The patient is normal at birth and, if not treated, progressively develops the full clinical picture.

Eye

Dislocation of the ocular lens (ectopia lentis), myopia, and glaucoma are frequent, severe and characteristic complications. Retinal detachment and degeneration, optical atrophy, and cataracts may eventually appear. Myopia may precede lens dislocation, and worsens afterwards. Ectopia lentis is detected in most untreated patients from 5–10 years of age and in nearly all untreated patients by the end of the fourth decade and is often the clue to diagnosis. The dislocation is generally downwards, whereas it is usually upwards in Marfan syndrome, a phenocopy of homocystinuria caused by mutations of the fibrillin-1 gene. Once ectopia lentis has occurred, a peculiar trembling of the iris (iridodonesis) following eye or head movement may be evident.

Skeleton

Osteoporosis is almost invariably detected, at least after childhood. Frequent consequences are scoliosis and a tendency towards pathological fractures and vertebral collapse.

As in Marfan syndrome, homocystinuric patients tend to be tall, with thinning and elongation (dolichostenomelia) of long bones near puberty, enlarged metaphyses and epiphyses, especially at the knees, and arachnodactyly, present in about half the patients. Other bone deformities include genu valgum with knobbly knees, pes cavus, and pectus carinatum or excavatum. Restricted joint mobility, particularly at the extremities, contrasts with the joint laxity observed in Marfan syndrome. Abnormal X-ray findings include biconcavity and flattening of the intervertebral discs, growth arrest lines in the distal tibia, metaphyseal spicules in the hands and feet, enlarged carpal bones, retarded lunate development, and shortening of the fourth metacarpal.

Central Nervous System

Developmental delay and mental retardation affect about 60% of patients to a variable degree of severity. Seizures, electroencephalogram abnormalities, and psychiatric disturbances have also been reported in approximately half of cases. Focal neurologic signs may be a consequence of cerebro-vascular accidents.

Vascular System

Thromboembolic complications, occurring in arteries and veins of all parts of the body, constitute the major cause of morbidity and mortality. The prognosis is influenced by the site and extent of the vascular occlusion. Thrombophlebitis and pulmonary embolism are the most common vascular accidents. Thrombosis of large- and medium-sized arteries, particularly carotid and renal arteries, is a frequent cause of death. Ischemic heart disease is a less prominent feature of homocystinuria. Association with other genotypes linked to increased risk of vascular disease, such as the factor V Leiden R506Q mutation or the 677C→T mutation of the MTHFR gene, were reported to increase the risk of thrombosis in homocystinuric patients [1].

Other Features

Spontaneous pneumothorax and pancreatitis were reported to be rare findings in homocystinuric patients [2].

Clinical Variability and Natural History

The spectrum of clinical abnormalities is wide, and mild cases may only be recognized by late complications, such as thromboembolic accidents. Time-to-event curves based on detailed information on 629 patients were calculated by Mudd et al. [3] for the main clinical manifestations and mortality. Each abnormality occurred significantly earlier and at a higher rate in untreated pyridoxine-nonresponsive individuals than in untreated pyridoxine-responsive ones. The risk of thromboembolic accidents in patients undergoing surgery was relatively small, complications, six of which were lethal, being recorded in only 25 patients following 586 operations.

An Italian multicenter survey [2] revealed a strong correspondence of ectopia lentis, mental retardation, seizures, dolichostenomelia, and thrombotic accidents among affected sib-pairs, supporting a prominent role of genetic factors in determining the phenotype. Nevertheless, rare cases of intrafamilial variability have been reported. Probably, both early diagnosis and strict compliance to treatment will change the natural history of cardiovascular and mental efficiency even in pyridoxine-nonresponsive individuals.

Outcome of Pregnancies

Pyridoxine-responsive women are able to undergo pregnancies without a significant risk of malformations in the offspring. There is much less experience of outcome of pregnancies in non-responsive women. Recently, more details of 15 pregnancies in 11 women, 5 of whom were pyridoxine-nonresponsive, were reported [4]. Complications of pregnancy included preeclampsia in 2 pregnancies and superficial venous thrombosis in a third pregnancy. First-trimester spontaneous abortion was observed in 2 pregnancies. Ten pregnancies produced normal live born infants while one offspring had multiple congenital anomalies and another had Beckwith-Wiedemann syndrome. No relationship could be established between the severity of biochemical abnormalities during pregnancy and either pregnancy complications or offspring outcome. The results of this study suggest that pregnancy complications and offspring abnormalities are infrequent events. Nevertheless careful monitoring of these pregnancies is mandatory.

21.1.2   Metabolic Derangement

Cystathione β-synthase (CBS) activity can be found in many tissues, including liver, brain, pancreas, and cultured fibroblasts. In addition to the coenzyme pyridoxal phosphate, CBS also binds two other ligands, the activator S-adenosylmethionine and a heme moiety of unclear function. In vivo responsiveness to pharmacological doses of pyridoxine, present in approximately 50% of homocystinuric patients, is generally associated with the presence of a small amount of residual enzymatic activity, although exceptions to this rule are known [5].

Deficiency of CBS leads to tissue accumulation of methionine, homocysteine, and their S-adenosyl derivatives, with lack of cystathionine and low levels of cysteine. The -SH group of homocysteine readily reacts with the -SH group of a second homocysteine molecule or of other molecules, leading to the formation of a number of disulfide compounds, such as homocystine, homocysteine-cysteine mixed disulfide or protein-bound homocysteine.

The pathophysiology of CBS deficiency has not yet been completely elucidated, but accumulation of homocysteine probably plays a major role in determining some of the most relevant clinical manifestations, including generalized vascular damage and thromboembolic complications. Thromboembolism has been suggested to be the end-point of homocysteine-induced abnormalities of platelets, endothelial cells, and coagulation factors. Many underlying mechanisms have been investigated with, so far, no unifying theory proven. For example, a recent study in homocystinuric patients points to enhanced peroxidation of arachidonic acid as an important mechanism linking hyperhomocysteinemia and platelet activation in CBS patients suggesting the possible value of vitamin E treatment [6].

Among other deleterious effects homocysteine may cause abnormal cross-linking of collagen, leading to abnormalities of the skin, joints, and skeleton in patients. This mechanism seems unlikely to cause damage of the non-collagenous zonular fibers of the lens which is more likely to be due to disturbed fibrillin structure.

21.1.3   Genetics

Homocystinuria due to CBS deficiency is inherited as an autosomal recessive trait. Clinical and biochemical variations, such as pyridoxine responsiveness, are also genetically determined and related to specific mutations.

The worldwide frequency of homocystinuria has been reported to be 1 in 344 000, while that in Ireland is much higher at 1 in 65 000. However, results of screening of a small newborn population for the most common CBS mutation (I278T, ► below) suggest that the incidence in Denmark might be higher although this needs to be confirmed on a larger scale [7].

The CBS gene is located on chromosome 21 (21q22.3). Molecular studies on CBS patients have led to the characterization of more than 130 mutations most of which are private [9; and website maintained by Kraus J.P. http://uchsc.edu/sm/cbs/cbsdata/cbsmain.htm]. Only a few mutations appear to be of epidemiologic relevance. I278T which is found in some 25% of homocystinuric alleles and A114V are both associated with a pyridoxine responsive form of the disease. G307S is mostly found in CBS patients of Irish origin and is not linked to response to pyridoxine. Compound heterozygotes show a variable response to pyridoxine although the presence of the I278T mutation seems to confer pyridoxine responsiveness even in compound heterozygotes.

In at least 5% of Caucasian alleles, exon 8 displays both a 68-bp duplication of the intron-exon junction and the I278T mutation [10]. Recently, it has been shown that a peculiar nucleotidic structure generated by the 68-bp duplication allows the rescue of the wild-type sequence, so preserving the protein function [11]. The role of this polymorphic mutation as a risk factor in mild hyperhomocysteinemia and in multifactorial diseases related to this condition was also investigated with no clear conclusions.

As with many other inherited metabolic disorders the molecular approach for the diagnosis of CBS deficiency is a powerful tool but is limited due to the high proportion of 'private' mutations. In general a search for the most recurrent mutations in a given population may be rewarding but systematic screening for mutations of the entire coding region of the CBS gene is a prerequisite for reliable establishment of genotype/phenotype correlation, not least because a double mutational event has been observed on a single allele in some patients. DNA linkage analysis remains an option for prenatal diagnosis of CBS deficiency or carrier detection among family members provided that DNA from an affected subject in a genetically informative family is available.

To our knowledge, no strategy of genetic treatment has been designed so far.

21.1.4 Diagnostic Tests

Screening of urine with the cyanide-nitroprusside test often yields positive results but can also be negative and lacks specificity. Initial diagnosis is best achieved by quantitative amino acid analysis of plasma which must be immediately processed to prevent loss of disulfide amino acids by binding to protein. Increased levels of methionine, homocystine and cysteine-homocysteine disulfide, low cystine and no increase of cystathionine, is typical of CBS deficiency. Exceptional pyridoxine-responsive patients are extremely sensitive to very small supplements of pyridoxine as contained in multivitamin tablets so that false-negative results may be obtained. Determination of plasma total homocysteine, after treatment of the plasma sample with reducing agents, is useful for both preliminary diagnosis and monitoring of treatment. The determination of total plasma homocysteine including the various available methods has been comprehensively reviewed [12]. Normal plasma total homocyst(e)ine values are less than 15 µmol/l, whereas most untreated CBS patients exhibit levels above 200 µmol/l. Hyperhomocyst(e)inemia also occurs in remethylation defects, due to 5,10-methylene-tetrahydrofolate reductase deficiency and 5-methyl-THF-homocysteine-methyltransferase deficiency either isolated or due to defects in cytosolic cobalamin metabolism (▶ Chap. 28). These disorders can mostly be distinguished from CBS deficiency by the very low to normal plasma methionine level. It must be noted that, in CBS deficiency, methionine concentrations tend to decrease with age and may even be normal in some older patients. This decrease may be contributed to by folate depletion and a consequent reduced capacity for remethylation.

It needs to be borne in mind that a wide range of non-genetic causes of hyperhomocyst(e)inemia are known, including renal failure and administration of drugs such as methotrexate, trimethoprin, niacin.

Definitive diagnosis requires demonstration of greatly reduced CBS activity, usually assayed in cultured skin fibroblasts but also possible in phytohemagglutinin-stimulated lymphocytes and liver biopsies. Exceptional patients may have significant residual activity of CBS in fibroblast extracts but still show the typical abnormalities of the disease. The molecular diagnosis of CBS deficiency now provides a powerful additional approach to the diagnosis.

In many countries, newborn mass-screening programs based on detection of hypermethioninemia have been implemented. A reduced cut-off value of methionine of 67 µmol/l was proposed to decrease the high rate of false-negative results previously reported in pyridoxine-responsive patients [13].

Prenatal Diagnosis

Prenatal diagnosis of homocystinuria has been performed in at-risk pregnancies by assaying CBS in extracts of cultured amniocytes [14]. CBS activity is very low in uncultured chorionic villi from control subjects and can only be measured after culturing. In the families where mutation(s) is known, direct analysis of the CBS gene allows rapid prenatal diagnosis and, in other cases, DNA linkage analysis to the CBS locus may have diagnostic value.

Heterozygotes

On a group basis, differences between obligate heterozygotes and control subjects have been clearly demonstrated using either measurements of homocyst(e)ine in plasma after methionine loading or assay of CBS in liver biopsies, cultured skin fibroblasts, or phytohemagglutinin-stimulated lymphocytes. However, overlap of values obtained in a considerable number of obligate heterozygotes with those at the lower end of the control range limits the value of these two approaches for heterozygote testing in individual subjects.

Molecular analysis of established mutations allows heterozygote detection in individual families, and the most common CBS mutations might be considered in population screening for CBS heterozygotes. Fibroblast CBS activity compatible with a heterozygosity was found in a significant number of vascular disease patients with hyperhomocyst(e)inemia. However molecular genetic studies failed to demonstrate a causative role of CBS heterozygosity in patients affected by premature vascular disease [15].

21.1.5 Treatment and Prognosis

The aim of treatment is to reduce plasma total homocyst(e)ine levels to as close to normal as possible while maintaining normal growth rate. Plasma cystine should be kept within the normal range (67 ± 20 µmol/l) and should be supplemented if necessary (up to 200 mg/kg/day). Homocysteine levels can be lowered in a number of ways, and the best approach or combination for the individual patient will depend on the nature of the defect and social factors.

About half of patients with CBS deficiency respond, often only partially, to large oral doses of pyridoxine. In about 10% of these patients who respond fully, fasting plasma total homocysteine, methionine and cystine become normal following a period of up to a few weeks of daily administration of between a few milligrams and 1000 mg of pyridoxine. Response to the vitamin is also influenced by folate depletion, which may be due to pyridoxine administration itself. Therefore, folic acid (5–10 mg/day) should be added to the treatment. An approach to assessment of pyridoxine responsiveness is to begin with 100 mg/day and, if necessary, progressively increase this to 500–1000 mg/day with monitoring of plasma levels of methionine and total homocysteine every other day.

Since doses higher than 1000 mg/day have been associated with sensory neuropathy pyridoxine should be kept at the lowest dose able to achieve adequate metabolic control. In particular doses greater than 250 mg/day should be avoided in newborns and young infants. In patients who do not respond to pyridoxine, a low-methionine/high-cystine diet must be introduced and must be continued throughout life.

A less strict low-methionine diet may also be necessary to achieve adequate control in pyridoxine-responsive patients. Synthetic methionine-free amino acid mixtures are commercially available and are especially useful for infants. The requirement for methionine is met by small amounts of infant formula. Supplements of essential fatty acids and carbohydrates are also required if not present in the methionine-free amino acid mixture. After infancy, foods containing proteins low in methionine can be introduced, including gelatin and pulses such as lentils and soybeans. However, it should be noted that soya-modified formulas are usually enriched with methionine. In addition to pyridoxine, folate and, possibly vitamin B12, the usual vitamin and mineral supplements are recommended.

Betaine, given orally at a maximum dose of 150 mg/kg/day (6–9 g maximum in adults) is another important homocysteine-lowering agent especially useful when compliance to the diet is unsatisfactory. For older children and adults 6–9 grams of betaine are given daily divided into 3 doses. Betaine remethylates homocysteine often leading to very high methionine concentrations but with no apparent influence on the pathophysiology of the disease. However, unexplained cerebral edema has been recently described in a few cases of children receiving betaine therapy [16].

Vitamin C supplementation (1 g/day) has been shown to ameliorate endothelial dysfunction in CBS patients suggesting its possible value in reducing the long-term risk of atherothrombotic complications. The value of long-term treatment with antithrombotic agents such as dipyridamole (100 mg four times per day) either alone or combined with aspirin (100 mg/day with 100 mg aspirin/day) remains to

be proven. In the meantime, the need of such treatment should be assessed on an individual basis.

Whatever the combination of regimes employed, achievement of virtually normal total homocysteine levels is very difficult in most patients. Notwithstanding this, prevention of the severe clinical abnormalities associated with this disorder requires lifelong treatment and considerable impact on outcome has been achieved in patients for whom adequate treatment was judged as removal of free-disulfide homocystine from plasma. The results of the international survey provide a firm baseline for the evaluation of past and future therapeutic regimens [3]. When the low-methionine diet was started in the newborn period, mental retardation was prevented, the start and progression of lens dislocation were delayed, and the incidence of seizures decreased. When late-diagnosed, responsive subjects received pyridoxine treatment, the first thromboembolic episode occurred later. In fact, a normal IQ was reported in teen-aged pyridoxine-nonresponsive CBS individuals with good compliance to treatment since birth [17]. Also treatment regimens aimed at lowering plasma homocysteine significantly reduce cardiovascular risk in homocystinuric patients despite imperfect biochemical control [18].

Finally, treatment success clearly depends on early diagnosis and treatment, providing a case for mass newborn-screening.

21.2   Methionine S-Adenosyltransferase Deficiency

21.2.1   Clinical Presentation

More than 60 patients with methionine S-adenosyltransferase (MAT) deficiency have now been described, many detected by newborn screening, and the great majority have so far been symptom free suggesting a benign disorder [19]. However, neurological abnormalities and demyelination of the brain attributed to deficient formation of S-adenosylmethionine have been observed in a few patients, possibly linked to the severity of the enzyme deficiency.

21.2.2   Metabolic Derangement

This disorder is characterized by a deficiency of the hepatic form of the enzyme MAT I/III (but not the extra hepatic form, MAT II), leading to elevated methionine concentrations in tissues and physiological fluids. The product of this enzyme reaction, S-adenosylmethionine, appears not to be deficient in most cases. Alternative metabolism of methionine seems to occur above a threshold plasma methionine concentration of about 300 μM, resulting in the formation of the transamination product 4-methylthio-3-oxobutyrate and dimethyl sulfide, the latter resulting in a distinct odor of the breath.

21.2.3   Genetics

Three forms of MAT are known: MAT-I, -II, and -III. MAT-I and -III are encoded by the same gene, *MAT1A*, and correspond to tetrameric and dimeric forms of a single α1 subunit, respectively. MAT-II is encoded by a separate gene, mainly expressed in fetal liver and in kidney, brain, testis, and lymphocytes. Mutations of the *MAT1A* gene account for both autosomal recessive [20] and autosomal dominant hypermethioninemia [21]. The rarer, latter form is caused by mutation on a single allele with a dominant-negative effect.

21.2.4   Diagnostic Tests

High methionine in plasma and urine, detected by usual chromatographic methods, without increased homocyst(e)ine of the degree seen in CBS deficiency and associated with no increase of S-adenosylmethionine is suggestive of this defect, but several other causes of hypermethioninemia are possible and must be excluded. Careful interpretation of results is needed since elevated plasma total homocysteine of up to 28.6 μmol/L has been reported [20].

21.2.5   Treatment and Prognosis

Treatment is generally not indicated but, in patients with evidence of demyelination, administration of S-adenosylmethionine corrects deficiency of this compound. If the postulated association between specific mutations leading to a severe enzyme deficiency holds true [22, 23], treatment with S-adenosylmethionine may be advisable in such cases.

Four pregnancies in a woman with severe MAT I/III deficiency resulted in the birth of three normal children with fetal arrest at 10–11 weeks in the other. Four women with mild hypermethioninemia due to heterozygosity for the dominant R264H mutant allele gave birth in total to 16 normal children with just one recorded miscarriage [24].

21.3   Glycine N-Methyltransferase Deficiency

21.3.1   Clinical Presentation

Persistent isolated hypermethioninaemia associated with a history of persistent elevated plasma transaminases, and mild hepatomegaly was found in two siblings, documented from the age of 1 year and 5 years [25] and in an unrelated boy from 2 years of age [26].

21.3.2   Metabolic Derangement

The constellation of high methionine, elevated S-adenosylmethionine in plasma without elevated S-adenosylhomocysteine and sarcosine provides strong circumstantial evidence of deficiency of glycine N-methyltransferase (GMT). No direct demonstration of deficiency of this liver enzyme has been possible but mutations in the GMT gene, although yet to be expressed, strongly support GMT deficiency as a cause of these metabolic changes.

21.3.3   Genetics

The finding of compound heterozygosity for mutations in the GMT gene also occurring in either parent confirm autosomal recessive inheritance of this defect [27].

21.3.4   Diagnostic Tests

Differentiation between this defect and other forms of isolated hypermethioninemia is possible by measurement of S-adenosylmethionine and S-adenosylhomocysteine in plasma. S-adenosylmethionine levels are approximately 10- to 30-fold higher than the upper limit of controls in GMT deficiency with normal levels in methionine adenosyltransferase deficiency. Sarcosine in plasma as well as S-adenosylhomocysteine and total homocysteine are not elevated.

21.3.5   Treatment and Prognosis

It has been speculated that treatment with a low methionine diet supplemented with cystine might be beneficial but no data is available.

21.4   S-Adenosylhomocysteine Hydrolase Deficiency

21.4.1   Clinical Presentation

A single patient with this disorder was recently reported [28]. Clinical signs included severely delayed psychomotor development and severe myopathy. When diagnosed at 12.7 months development had ceased and MRI of the brain showed delayed myelination and atrophy of white matter.

21.4.2   Metabolic Derangement

Deficiency of this enzyme has been proven and leads to a block in the degradation and accumulation of S-adenosylhomocysteine as well as increased levels of S-adenosylmethionine and methionine. Elevated levels of guanidino-acetate and low levels of phosphatidylcholine and choline are compatible with inhibition of the respective transmethylases by S-adenosylhomocysteine. A number of other abnormalities such as slightly elevated total homocysteine, betaine, dimethylglycine and cystathionine remain unclear.

21.4.3   Genetics

An autosomal recessive inheritance is indicated and sequencing of the S-adenosyl homocysteine hydrolase gene revealed the presence of a maternally derived nonsense mutation and a missense mutation of paternal origin.

21.4.4   Diagnostic Tests

Differentiation between this and the two other forms of hypermethioninemia described here can be achieved by measurement of S-adenosylhomocysteine, S-adenosylmethionine and sarcosine in plasma. Each of these is elevated with approximately 100- and 30-fold elevations of S-adenosylhomocysteine and S-adenosylmethionine, respectively.

21.4.5   Treatment and Prognosis

Treatment was attempted by severely restricting methionine intake and administering phosphatidylcholine in the form of egg yolk. This resulted in lowering plasma metabolites and clinical improvement but the long term outcome remains unknown.

21.5   γ-Cystathionase Deficiency

21.5.1   Clinical Presentation

This is considered to be a benign disorder. Subjects detected without ascertainment bias are mainly asymptomatic; subjects with mental retardation have had healthy siblings with the same defect or without the defect but showing the same symptoms.

21.5.2   Metabolic Derangement

Deficiency of the pyridoxal-phosphate-requiring γ-cystathionase leads to tissue accumulation of cystathionine. Increased plasma concentrations and markedly increased excretion of cystathionine occur and N-acetylcystathionine is also excreted.

21.5.3   Genetics

Inheritance is autosomal recessive. The cystathionase gene was cloned [29] and mutant alleles in individuals with this condition have been reported [30].

21.5.4   Diagnostic Tests

High urinary excretion of cystathionine without homocystine and with normal plasma methionine points to this defect. Transient cystathioninuria in newborns is due to known secondary causes, such as vitamin-B_6 deficiency, generalized liver disease, thyrotoxicosis, and neural tumors. Milder increases of plasma and urine levels of cystathionine can also occur in the remethylation defects due to overproduction of this metabolite. While γ-cystathionase activity is certainly expressed in cultured skin fibroblasts, the level of activity is probably too small to allow reliable measurement by specific enzyme assay [31].

21.5.5   Treatment and Prognosis

Most subjects respond to administration of about 100 mg of pyridoxine daily though, as a benign disorder, it remains debatable whether treatment is needed.

21.6   Isolated Sulfite Oxidase Deficiency

21.6.1   Clinical Presentation

Characteristic findings in the severe form of this enzyme deficiency, whether isolated (approximately 20 patients reported) or due to molybdenum cofactor deficiency (more than 50 described, ► Chap. 35), are early refractory convulsions, severe psychomotor retardation, failure to thrive, microcephaly, hypotonia passing into hypertonia, lens dislocation, and early death [32]. Milder presentation has been reported.

21.6.2   Metabolic Derangement

Sulfite oxidase catalyses the last step in the oxidation of the sulfur atom of cysteine into inorganic sulfate (❏ Fig. 20.1). Its deficiency results in accumulation of the suspected toxic compound sulfite together with its detoxification products, S-sulfocysteine and thiosulfate, with reduced formation of sulfate.

21.6.3 Genetics

This autosomal recessive disease has been explained at the molecular level by cloning of the gene and characterization of mutations in several patients affected by the isolated sulfite oxidase deficiency [33, 34]. The gene for molybdenum cofactor deficiency has been localized to chromosome 6.

21.6.4 Diagnostic Tests

Increased sulfite can be detected with urine-test strips, but samples must be fresh and, in one case, no increased sulfite with normal sulfate excretion was reported. S-sulfocysteine is a stable, ninhydrin-positive diagnostic parameter that can be detected by electrophoresis or chromatography and can be quantified in urine and plasma by classical ion-exchange techniques.

Cystine levels are always very low. Thiosulfate can also be searched for by thin-layer-chromatography. The absence of xanthinuria distinguishes the isolated deficiency from the molybdenum-cofactor defect, in which xanthinuria is observed (▶ Chap. 35). Sulfite oxidase activity can be determined in cultured skin fibroblasts.

21.6.5 Treatment and Prognosis

Attempts at treatment have been mainly unsuccessful although evidence of clinical improvement in two mildly affected patients on a low-cystine and low-methionine diet has been reported [35]. Attempts to remove sulfite by binding to penicillamine were unsuccessful [36].

References

1. Kluijtmans LA, Boers GH, Verbruggen B et al (1998) Homozygous cystathionine beta-synthase deficiency, combined with factor V Leiden or thermolabile methylenetetrahydrofolate reductase in the risk of venous thrombosis. Blood 91:2015-2018
2. de Franchis R, Sperandeo MP, Sebastio G, Andria G (1998) The clinical aspects of cystathionine B-synthase deficiency: how wide is the spectrum? Italian Collaborative Study Group on Homocystinuria. Eur J Pediatr 157:867-870
3. Mudd SH, Skovby F, Levy HL et al (1985) The natural history of homocystinuria due to cystathionine R synthase deficiency. Am J Hum Genet 37:1-31
4. Levy HL, Vargas JE, Waisbren SE et al (2002) Reproductive fitness in maternal homocystinuria due to cystathionine beta-synthase deficiency. J Inherit Metab Dis 25:299-314
5. Fowler B (1985) Recent advances in the mechanism of pyridoxine-responsive disorders. J Inherit Metab Dis 8[Suppl 1]:76-83
6. Davi G, Di Minno G, Coppola A et al (2001) Oxidative stress and platelet activation in homozygous homocystinuria. Circulation 104:1124-1128
7. Gaustadnes M, Ingerslev J, Rutiger N (1999) Prevalence of congenital homocystinuria in Denmark. N Engl J Med 340:1513
8. Miles EW, Kraus JP (2004) Cystathionine beta-synthase: structure, function, regulation, and location of homocystinuria-causing mutations. J Biol Chem 279:29871-29874
9. Kraus JP, Janosik M, Kozich V et al (1999) M Cystathionine β-synthase mutations in homocystinuria. Hum Mutat 13:368-375
10. Sperandeo MP, de Franchis R, Andria G, Sebastio G (1996) A 68 bp insertion found in a homocystinuric patient is a common variant and is skipped by alternative splicing of the cystathionine β-synthase mRNA. Am J Hum Genet 59:1391-1393
11. Romano M, Marcucci R, Buratti E et al (2002) Regulation of 3' splice site selection in the 844ins68 polymorphism of the cystathionine Beta-synthase gene. J Biol Chem 277:43821-43829
12. Refsum H, Smith AD, Ueland PM et al (2004) Facts and recommendations about total homocysteine determinations: an expert opinion. Clin Chem 50:3-32
13. Peterschmitt MJ, Simmons JR, Levy HL (1999) Reduction of false negative results in screening of newborns for homocystinuria. N Engl J Med 341:1572-1576
14. Fowler B, Borresen AL, Boman N (1982) Prenatal diagnosis of homocystinuria. Lancet 2:875
15. Kozich V, Kraus E, de Franchis R et al (1995) Hyperhomocysteinemia in premature arterial disease: examination of cystathionine beta-synthase alleles at the molecular level. Hum Mol Genet 4:623-629
16. Yaghmai R, Kashani AH, Geraghty MT et al (2002) Progressive cerebral edema associated with high methionine levels and betaine therapy in a patient with cystathionine beta-synthase (CBS) deficiency. Am J Med Genet 108:57-63
17. Yap S, Rushe H, Howard PM, Naughten ER (2001) The intellectual abilities of early-treated individuals with pyridoxine-nonresponsive homocystinuria due to cystathionine beta synthase deficiency. J Inherit Metab Dis 24:437-447
18. Yap S (2003) Classical homocystinuria: vascular risk and its prevention. J Inherit Metab Dis 26:259-265
19. Mudd SH, Jenden DJ, Capdevila A et al (2000) Isolated hypermethioninemia: measurements of S-adenosylmethionine and choline. Metabolism 49:1542-1547
20. Kim SZ, Santamaria E, Jeong TE et al (2002) Methionine adenosyltransferase I/III deficiency: two Korean compound heterozygous siblings with a novel mutation. J Inherit Metab Dis 25:661-671
21. Chamberlin ME, Ubagai T, Mudd SH et al (1997) Dominant inheritance of isolated hypermethioninemia is associated with a mutation in the human methionine adenosyltransferase 1A gene. Am J Hum Genet 60:540-546
22. Chamberlin ME, Ubagai T, Mudd SH et al (1996) Demyelination of the brain is associated with methionine adenosyltransferase I/III deficiency. J Clin Invest 98:1021-1027
23. Hazelwood S, Barnardini I, Shotelersuk V et al (1998) Normal brain myelination in a patient homozygous for a mutation that encodes a severely truncated methionine adenosyltransferase I/III. Am J Med Genet 75:395-400
24. Mudd SH, Tangerman A, Stabler SP et al (2003) Maternal methionine adenosyltransferase I/III deficiency: reproductive outcomes in a woman with four pregnancies. J Inherit Metab Dis 26:443-458
25. Mudd SH, Cerone R, Schiaffino MC et al (2001) Glycine N-methyltransferase deficiency: a novel inborn error causing persistent isolated hypermethioninaemia. J Inherit Metab Dis 24:448-464
26. Augoustides-Savvopoulou P, Luka Z, Karyda S et al (2003) Glycine N-methyltransferase deficiency: a new patient with a novel mutation. J Inherit Metab Dis 26:745-759
27. Luka Z, Cerone R, Phillips JA 3rd et al (2002) Mutations in human glycine N-methyltransferase give insights into its role in methionine metabolism. Hum Genet 110:68-74
28. Baric I, Fumic K, Glenn B et al (2004) S-adenosylhomocysteine hydrolase deficiency in a human: a genetic disorder of methionine metabolism. Proc Natl Acad Sci USA 101:4234-4239

29. Lu Y, Odowd BF, Orrego H, Israel Y (1992) Cloning and nucleotide sequence of human liver cDNA encoding for cystathionine gamma-lyase. Biochem Biophys Res Commun 189:749-758

30. Wang J, Hegele RA (2003) Genomic basis of cystathioninuria (MIM 219500) revealed by multiple mutations in cystathionine γ-lyase (CTH). Hum Genet 112:404-408

31. Fowler B (1982) Transsulphuration and methylation of homo-cysteine in control and mutant human fibroblasts. Biochim Biophys Acta 721:201-207

32. Rupar CA, Gillett J, Gordon BA et al (1996) Isolated sulfite oxidase deficiency. Neuropediatrics 27:299-304

33. Garrett RM, Johnson JL, Graf TN et al (1998) Human sulfite oxidase Rl60Q: identification of the mutation in a sulfite oxidase-deficient patient and expression of the mutant enzyme. Proc Natl Acad Sci USA 95:6394-6398

34. Johnson JL, Coyne KE, Garrett RM et al (2002) Isolated sulfite oxidase deficiency: identification of 12 novel SUOX mutations in 10 patients. Hum Mutat 20:74

35. Touati G, Rusthoven E, Depondt E et al (2000) Dietary therapy in two patients with a mild form of sulphite oxidase deficiency. Evidence for clinical and biological improvement. J Inherit Metab Dis 23:45-53

36. Tardy P, Parvy P, Charpentier C et al (1989) Attempt at therapy in sulphite oxidase deficiency. J Inherit Metab Dis 12:94-95

22 Disorders of Ornithine Metabolism

Vivian E. Shih, Matthias R. Baumgartner

Ornithine Metabolism

Ornithine is an important intermediate in several metabolic pathways. The pyridoxal phosphate requiring enzyme ornithine-δ-aminotransferase (OAT) plays a pivotal role in its metabolism. During the neonatal period the flux of the OAT reaction is in the direction of ornithine synthesis. Ornithine is then converted to citrulline and arginine via the urea cycle. Later, at approximately 3–4 months of age, the OAT reaction is reversed to catabolize excess ornithine generated from arginine hydrolysis. Ornithine also plays an important role in urea synthesis (▶ also Fig. 20.1). Since both OAT and ornithine transcarbamoylase (OTC) are mitochondrial matrix enzymes, ornithine produced in the cytoplasm must be transported to the mitochondrial matrix by a specific energy-requiring transport system.

Δ^1-pyrroline-5-carboxylate synthase (P5CS), a bifunctional ATP- and NADPH-dependent mitochondrial enzyme highly active in the gut, catalyzes the reduction of glutamate to Δ^1-pyrroline-5-carboxylate (P5C), a critical step in the biosynthesis of proline, ornithine and arginine. P5C is the immediate precursor for both proline and ornithine in reactions catalyzed by P5C reductase and ornithine-δ-aminotransferase (OAT), respectively. In de novo arginine biosynthesis arginine is further metabolized by the enzymes of the urea cycle. P5CS activity is present in brain.

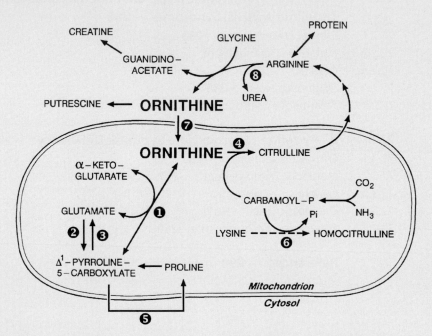

Fig. 22.1. Ornithine metabolic pathways. *Pi*, inorganic phosphate. **1**, ornithine-δ-aminotransferase (OAT); **2**, Δ^1-pyrroline-5-carboxylate synthase (P5CS); **3**, Δ^1-pyrroline-5-carboxylate dehydrogenase; **4**, ornithine transcarbamoylase (OTC); **5**, Δ^1-pyrroline-5-carboxylate reductase; **6**, lysine transcarbamoylase (the step indicated by the broken line is not well defined); **7**, mitochondrial ornithine transporter; **8**, arginase

Hyperornithinemia due to ornithine aminotransferase (OAT) deficiency is associated with gyrate atrophy of the choroid and retina. Patients usually become virtually blind by age 55. Treatment includes a low arginine diet and pharmacological doses of pyridoxine (vitamin B$_6$). Long term compliance to an arginine-restricted diet, especially when started at a young age, can slow the progression of the chorioretinal lesions and loss of vision. Rarely, neonates can present with hyperammonemic encephalopathy, hypoargininemia, and hypoornithinemia and require arginine supplementation. In the *hyperornithinemia, hyperammonemia, and homocitrullinuria (HHH) syndrome* there is a wide spectrum of clinical manifestations, most of which are related to the toxicity of hyperammonemia. Progressive spastic paraparesis is often a late complication. Patients have a marked elevation of plasma ornithine associated with hyperammonemia and increased urinary excretion of homocitrulline. HHH results from a defect in the import of ornithine into the mitochondrion and consequent urea synthesis malfunction. Treatment is similar to that for patients with urea cycle disorders.

A newly recognized disorder, *Δ^1-pyrroline-5-carboxylate synthase (P5CS) deficiency*, has been described in two siblings with progressive neurodegeneration and peripheral neuropathy, joint laxity, skin hyperelasticity and bilateral subcapsular cataracts. Their metabolic phenotype includes mild hyperammonemia, hypoornithinemia, hypocitrullinemia, hypoargininemia and hypoprolinemia. This disorder underscores the importance of low levels of amino acids as markers of metabolic disease.

22.1 Hyperornithinemia due to Ornithine Aminotransferase Deficiency (Gyrate Atrophy of the Choroid and Retina)

22.1.1 Clinical Presentation

Night blindness and myopia are usually the first symptoms in early childhood. Ocular findings include myopia, constricted visual fields, elevated dark adaptation thresholds, and very small or non-detectable electroretinographic (ERG) responses. Retinopathy can be detected before visual disturbances. Fundoscopic appearances of the chorioretinal atrophy, are illustrated in ◘ Fig. 22.2A and B.

Patients develop posterior subcapsular cataracts by the late teens and usually become virtually blind between the ages of 40 and 55 due to extensive chorioretinal atrophy. Pyridoxine responsive patients often have a milder course and maintain adequate visual acuity at older ages. Con-

siderable heterogeneity exists in the appearance of the fundus even within the same family, and siblings at the same age can show substantial differences in the severity of the ocular disease. Vitreous hemorrhage causing sudden loss of vision is a rare complication [1]. Most patients have normal intelligence.

Recent observations have identified several patients with a rare neonatal form of ornithine aminotransferase (OAT) deficiency who presented with poor feeding, failure to thrive, symptomatic neonatal hyperammonemia and orotic aciduria in the first 2 months of life and who responded to treatment for hyperammonemic encepahopathy [2–4]. These features were similar to those seen in the urea cycle disorders. The diagnosis was difficult until a subsequent rise of plasma ornithine led to further investigations.

Histopathologic study of the eye obtained post mortem from a pyridoxine responsive patient showed focal areas of photoreceptor atrophy with adjacent retinal pigment epithelial hyperplasia [5]. Electron microscopic studies revealed abnormal mitochondria in the corneal endothelium and the non-pigmented ciliary epithelium and similar, but less severe, abnormalities in the photoreceptors [5]. In addition to the ocular findings, systemic abnormalities have been reported in some patients. These include diffuse slowing on EEG, abnormal muscle histopathology, muscle weakness in some patients, abnormal ultrastructure of hepatic mitochondria [6], and peculiar fine, sparse, straight hair with microscopic abnormalities [7]. Early degenerative and atrophic brain changes that were not age-related were found by brain magnetic resonance imaging (MRI) [8] and evidence for peripheral nervous system involvement [9] was noted in half the patients studied.

22.1.2 Metabolic Derangement

Patients with gyrate atrophy of the choroid and retina (GA) have marked hyperornithinemia due to a deficiency of OAT activity (alternatively, ornithine ketoacid transaminase, OKT) [10]. The enzyme deficiency has been demonstrated in liver, muscle, hair roots, cultured skin fibroblasts and lymphoblasts. The pathophysiological mechanism of the retinal degeneration is unclear. OAT requires pyridoxal phosphate (PLP) as a cofactor. In a small number of patients, the OAT activity increased substantially when measured in the presence of high concentrations of PLP. Most of these patients showed a partial reduction of plasma ornithine when given pharmacological doses of pyridoxine (vitamin B$_6$). On the basis of these *in vitro* and *in vivo* responses, at least two variants – pyridoxine responsive and pyridoxine-nonresponsive – have been described. In rare cases, the *in vivo* response is in discord with the *in vitro* response to pyridoxine.

The four reported symptomatic neonates with OAT deficiency had increased blood ammonia and low levels

IV

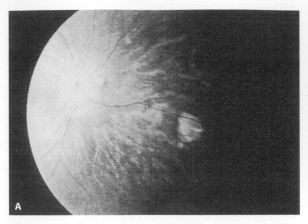
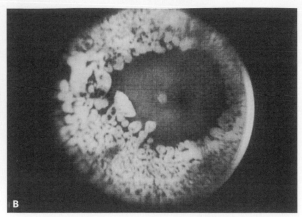

■ **Fig. 22.2.** Fundoscopic appearances of the chorioretinal atrophy showing early changes (**A**), and advanced changes (**B**)

of plasma ornithine, citrulline, and arginine and in the first several months of life with hyperornithinemia developing later in life. These observations indicate that the net flux in the OAT reaction in the newborn period is in the direction of ornithine synthesis rather than ornithine degradation. Consequently the conversion of ornithine to citrulline and arginine is reduced. The deficiency of these amino acids affects protein synthesis and urea cycle function. One infant with OAT deficiency, diagnosed prenatally, had normal plasma ornithine and arginine in cord blood but these amino acids became low at 2–4 months of age on a normal diet [11]. The blood ammonia and glutamine rose during this period. Arginine administration corrected the low plasma arginine, but hyperornithinemia developed and persisted.

Low creatine has been reported in blood, urine, and muscle [12] most likely as a result of ornithine inhibition of glycine transamidinase and the subsequent reduction of creatine biosynthesis (■ Fig. 22.1). In addition, brain MRI studies have shown reduced creatine content [13].

22.1.3   Genetics

GA is an autosomal recessive disorder and has been described in patients from various ethnic backgrounds, but its incidence is highest in the Finnish population [14]. Intermediate levels of OAT activity have been observed in skin fibroblasts from obligate heterozygotes for both pyridoxine nonresponsive and pyridoxine responsive variants. Heterozygotes for the pyridoxine responsive variant can be distinguished by a doubling of OAT activity when assayed with high concentration of PLP.

The human gene for OAT has been mapped to chromosome 10q26. Many different mutations in the OAT structural gene have been defined in GA patients of varied ethnic origins [15]. Although the incidence is highest in Finland, no common mutation has been identified.

22.1.4   Diagnostic Tests

The main biochemical finding is a plasma ornithine concentration 5- to 20-fold above normal. Patients with the pyridoxine responsive variant tend to have smaller elevations of plasma ornithine levels than patients with the pyridoxine nonresponsive variant. Urinary excretion of ornithine as well as that of lysine, arginine, and cystine is often increased when plasma ornithine is 400 μmol/l or greater. These changes are secondary to competitive inhibition by ornithine of the common renal transport shared by these amino acids. In most cases, the absence of hyperammonemia and homocitrullinuria differentiate this disorder from the HHH syndrome (■ Table 22.1). An exception to this rule is the neonatal form of OAT deficiency. Affected infants have symptomatic neonatal hyperammonemia with increased excretion of orotic acid. Amino acid analysis shows low levels of plasma ornithine, arginine, and almost no detectable citrulline in the first 2–3 months of life. Homocitrulline has been detected in the urine, but it was unclear whether it was an artifact from formula feeding. Plasma ornithine rose in these patients after 3–4 months of age, when the metabolic flux of OAT is reversed from ornithine synthesis to ornithine degradation. Since hyperornithinemia is not present in affected neonates, screening in a newborn may give false negative results and fail to detect this disorder.

For confirmation of the diagnosis, direct assay of OAT activity can be performed in extracts of skin fibroblasts or lymphoblasts. When the mutation is known, molecular analysis is appropriate for prenatal diagnosis and carrier detection.

22.1.5   Treatment and Prognosis

The goal of treatment has been to correct the amino acid abnormalities. Megavitamin B_6 and/or diet therapy have been used. Reduction of hyperornithinemia can be achieved

□ Table 22.1. Differential diagnosis of disorders involving ornithine metabolism

	OAT deficiency	P5CS deficiency	HHH syndrome	OTC deficiency
Major clinical findings	Gyrate atrophy of the choroid and retina <u>Neonatal form:</u> Failure to thrive Encephalopathy	Progressive neuro-degeneration Peripheral neuropathy Joint laxity Skin hyperelasticity Bilateral subcapsular cataracts	Mental retardation Episodic lethargy and ataxia Seizures Coagulopathy	Neonatal onset of coma Mental retardation Episodic lethargy and ataxia Aversion to protein foods
Inheritance	Autosomal recessive	Autosomal recessive	Autosomal recessive	X-linked
Major biochemical changes				
Plasma ammonia	Increased in neonatal form	Increased	Increased	Increased
Plasma ornithine	Increased Low in neonatal form	Low	Increased	Normal
Plasma citrulline and arginine	Low in neonatal form	Low	Normal	Low
Plasma proline	Normal	Low	Normal	Normal
Urine homocitrulline	*	Normal	Increased	Normal
Urine Orotic acid	Normal Increased in neonatal form	Normal	Increased	Increased

* Increased urine homocitrulline detected, but it was unclear whether it was an artifact from formula feeding.
HHH, hyperornithinemia, hyperammonemia, homocitrullinuria; *OAT*, ornithine aminotransferase; *OTC*, ornithine transcarbamoylase; *PC5S*, Δ^1-pyrroline-5-carboxylate synthase.

by dietary restriction of protein and arginine (a precursor of ornithine in foods) [12, 16, 17]. On average, food proteins contain 4–6% arginine (nuts and seeds have higher arginine content). The diet is often supplemented with a mixture of essential amino acids providing up to one-half of the nitrogen intake using products containing no arginine developed for patients with urea cycle disorders. It is cautioned that severe arginine depletion can result in hyperammonemia [17].

In the neonatal period, arginine is an essential amino acid. Infants with neonatal symptomatic hyperammonemia should be given arginine and treatment for hyperammonemia. Arginine intake in patients less than 3–4 months of age should not be restricted until plasma ornithine begins to increase.

Pharmacological dosage of pyridoxine HCl has resulted in plasma ornithine reduction in a small number of patients. Doses between 15 and 600 mg a day lowered plasma ornithine levels from 25 to 60% [18–20]. A 2-week trial of pyridoxine treatment (300–600 mg/day) is recommended for all newly diagnosed patients to determine their responsiveness.

Over 30 patients have been given a low-arginine diet long-term, some in combination with pharmacological doses of pyridoxine. Compliance to diet restriction and long term commitment and motivation are important factors influencing the outcome. A series of 17 patients on an arginine-restricted diet had plasma ornithine levels in the range of 400–500 mmol/l and showed slower loss of visual function after 13.9 years compared to 10 patients not on the diet [21] Long-term substantial reduction of plasma ornithine levels started at an early age may be beneficial in slowing the progression of chorioretinal lesions and loss of retinal function. In a study of 2 sets of siblings with gyrate atrophy who were treated with an arginine-restricted diet for 16 to 17 years, the younger sibling, who was prescribed the diet at an earlier age, demonstrated a slower progression of lesions compared with the older sibling [22]. One patient was unable to tolerate the semi-synthetic low-arginine diet and was treated with a natural food low-protein diet (0.8 g/kg/day) for 26 years with moderate reduction of plasma ornithine levels and delayed progression of chorioretinal degeneration [23].

The effects of the above therapeutic measures on vision late in life have yet to be assessed. Study of a transgenic mouse model for GA has shown that a trial of dietary arginine restriction prevented the appearance of retinopathy at the age when untreated mice developed gyrate atrophy [11]. This observation supports the importance of early diagnosis and early treatment.

Other therapeutic approaches in a small number of patients have included supplements of proline [19], creatine [12, 24], and lysine [25]. Creatine supplement corrected the muscle histopathology and phosphocreatine deficiency but did not halt progression of retinal degeneration.

A woman with GA had two uneventful pregnancies and bore two normal children prior to being diagnosed with OAT deficiency (Shih, unpublished) [21].

22.2   The Hyperornithinemia, Hyper-ammonemia, and Homocitrul-linuria (HHH) Syndrome

22.2.1   Clinical Presentation

There is wide spectrum of clinical manifestations in the HHH syndrome [26], most of which are related to hyperammonemia toxicity (◘ Table 22.1). Ocular abnormalities are notably absent. Intolerance to protein feeding, vomiting, seizures, and developmental delay from infancy are common complaints. Neonatal onset of lethargy, hypotonia, seizures with progression to coma, and death has been observed in the most severe form [27, 28]. Progressive spastic paraparesis is often a late complication [28]. Abnormal neuroimaging studies have been described. Coagulopathy, especially factor VII and X deficiencies, has been reported in several patients [29, 30].

Mildly affected adult patients may have apparently normal intelligence. Two adult siblings came to attention because of episodic neurological and psychiatric disturbances with protein intolerance [31]. One other patient was misdiagnosed with multiple sclerosis (Shih, unpublished observation).

22.2.2   Metabolic Derangement

Patients with the HHH syndrome have a marked elevation of plasma ornithine associated with hyperammonemia and increased urinary excretion of homocitrulline. The HHH syndrome is a disorder of compartmentation and its defect is in the import of ornithine into the mitochondrion (◘ Fig. 22.1), resulting in a functional deficiency of both OTC and OAT activities (◘ Table 22.1). The intramitochondrial deficiency of ornithine results in the utilization of carbamoylphosphate via alternate pathways: formation of homocitrulline from lysine (◘ Fig. 22.1) and formation of orotic acid (▶ Chap. 35, ◘ Fig. 35.3).

22.2.3   Genetics

The HHH syndrome is more frequently seen in French-Canadians than in other ethnic groups. Inheritance is autosomal recessive. The gene (ORNT1 or SLC25A15) encoding the transporter protein maps to 13q14. Λ common mutation in HHH patients of French-Canadian origin is F188Δ, a 3-bp inframe deletion [32]. The R197X mutation has been considered common in Japanese patients [33]. Obligate heterozygotes are clinically normal and cannot be identified by biochemical studies.

22.2.4   Diagnostic Tests

The HHH syndrome can be differentiated from other hyperammonemic syndromes including the urea cycle enzymopathies by laboratory findings (◘ Table 22.1). The triad of hyperornithinemia, hyperammonemia and homocitrullinuria is pathognomonic. Plasma ornithine concentration is three to ten times elevated, and tends to be somewhat lower than that seen in gyrate atrophy patients. Despite a functional deficiency of OTC activity, the plasma citrulline is often normal in the HHH syndrome.

Urine amino acid screening shows increased ornithine and homocitrulline when the plasma ornithine concentration is above 400 mmol/l. At lower plasma ornithine concentrations, homocitrullinuria may be the only urine amino acid abnormality. It is cautioned that excessive homocitrulline excretion is more commonly an artifact from canned formula. Persistent homocitrullinuria without dietary source is abnormal. Increased urinary homocitrulline has also been detected in hyperlysinemia. Orotic aciduria is common and can be induced by allopurinol challenge [31] as in patients with primary OTC deficiency (▶ Chap. 20).

The metabolic defect can best be confirmed by ^{14}C-L-ornithine incorporation assay using fibroblast monolayers [34]. The compartmentation of ornithine in HHH fibroblasts prevents the conversion of ornithine to proline and glutamate (◘ Fig. 22.1) and results in very low incorporation of radioactivity into macromolecules. Unlike the case in GA, direct measurement of OAT activity in cell extracts is normal. Prenatal diagnosis of the HHH syndrome has been achieved using the ^{14}C-ornithine incorporation assay in cultured amniocytes [27] as well as chorionic villi [35] but the ^{14}C-ornithine incorporation assay is not useful for heterozygote identification.

22.2.5   Treatment and Prognosis

Treatment is aimed at preventing ammonia toxicity and follows the principles outlined for the urea cycle disorders (▶ Chap. 20). In general, a low-protein diet and citrulline supplementation have been effective in achieving biochemical control for most patients. One patient with neonatal onset of moderate hyperammonemia responded well to treatment and had normal growth and development at 18 months of age [36]. Treatment has not prevented the late development of spastic gait [28].

Hyperammonemia during pregnancy and postpartum is a potential risk in women with the HHH syndrome. One patient developed hyperammonemia one day postpartum following two pregnancies (Shih, unpublished observation). This is likely due to endogenous nitrogen load from uterine involution. It is thus advisable to exercise caution in the postpartum dietary management of HHH patients. Another woman was treated during pregnancy with lactulose and arginine to reduce blood ammonia [37]. Offspring from both women and men with the HHH syndrome have been apparently normal.

22.3 Δ¹-Pyrroline-5-Carboxylate Synthase Deficiency

22.3.1 Clinical Presentation

The clinical information is limited to two siblings who are the product of a consanguineous union [38]. Both came to clinical attention in early infancy because of developmental delay, joint laxity, skin hyperelasticity, muscular hypotonia and failure to thrive. Bilateral subcapsular cataracts were noted at 4 years in the boy and at 20 months in his younger sister. Both siblings showed progressive deterioration in mental and motor skills after the age of 5 years resulting in severe mental retardation (IQ 50). The patients had severe hypotonia, dystonia of hands and feet, muscular wasting of limbs, pyramidal syndrome, and peripheral, predominantly axonal neuropathy with progressively decreasing motor nerve conduction velocity which left them with the inability to walk before reaching ages 15 and 21.

22.3.2 Metabolic Derangement

The metabolic phenotype described in the two siblings includes mild hyperammonemia, hypoornithinemia, hypocitrullinemia, hypoargininemia and hypoprolinemia, a pattern of metabolic abnormalities consistent with impaired proline and ornithine synthesis due to deficiency of Δ¹-pyrroline-5-carboxylate synthase (P5CS). This enzyme catalyzes an essential step in the pathways by which proline, ornithine and arginine are synthesized from glutamate. In connective tissue, there is a high proline requirement for collagen sythesis. Deficient proline synthesis may impair protein synthesis in the lens epithelium and/or fibrocytes, and it is also possible that P5C metabolism

contributes to the antioxidant defence of the lens. P5CS activity is present in the brain and proline is thought to act as an inhibitory neurotransmitter in the CNS. Thus, impaired synthesis of proline is consistent with many of the clinical abnormalities in these patients such as lax joints and hyperelastic skin, cataracts, and progressive neurodegeneration [39].

The paradoxical fasting hyperammonemia reported in one of the patients is consistent with a relative deficiency of ornithine limiting ureagenesis and ammonia detoxication in the liver. Following a meal, arginine, derived from dietary protein, temporarily corrects this deficit by producing ornithine through arginase and thus enhancing urea cycle function with the result that plasma ammonia decreases despite the nitrogen load in the meal. Notably, in this special situation, arginine becomes an essential amino acid.

22.3.3 Genetics

Deficiency of P5CS presumably is inherited as an autosomal recessive trait. The only two reported siblings are from consanguineous parents and are homozygous for a missense mutation, R84Q, which alters a conserved residue in the P5CS γ-glutamyl kinase domain. R84Q dramatically reduces P5CS activity when expressed in mammalian cells [39].

22.3.4 Diagnostic Tests

Since the abnormal metabolite profile is corrected in the fed state, the metabolic phenotype of P5CS deficiency is easily missed. The combination of low fasting levels of ornithine, citrulline, arginine and proline plus a tendency to paradoxical fasting hyperammonemia or one of the above together with a clinical phenotype of neurodegeneration with peripheral neuropathy and/or cataracts and connective tissue manifestations should suggest this disorder. Except for the low proline, the metabolite pattern is similar to that seen in neonatal OAT deficiency.

P5CS activity is undetectable in control fibroblasts [40]. Ornithine loading tests in the reported siblings resulted in transient partial correction of proline, citrulline, and arginine concentrations, and indirect enzyme studies showed reduced proline biosynthesis in fibroblasts [38], corroborating the biological significance of the metabolic block at the level of P5CS *in vivo*.

22.3.5 Treatment and Prognosis

Supplementation of the deficient amino acids seems to be a reasonable therapeutic approach. However, administration of ornithine in the two reported siblings at a late stage of the disease did not result in any clinical improvement. Early recognition would allow the opportunity for a therapeutic trial with an amino acid cocktail, such as one containing citrulline, arginine, ornithine, and proline.

References

1. Takahashi O, Hayasaka S, Kiyosawa M et al (1985) Gyrate atrophy of choroid and retina complicated by vitreous hemorrhage. Jpn J Ophthalmol 29:170-176
2. Cleary MA, Dorland L, de Koning TJ et al (2005) Ornithine aminotransferase deficiency: diagnostic difficulties in neonatal presentation. J Inherit Metab Dis 28:673-679
3. Webster M, Allen J, Rawlinson D et al (1999) Ornithine Aminotransferase Deficiency presenting with hyperammonaemia in a premature newborn. J Inherit Metab Dis 22[Suppl 1]:80
4. Champion MP, Bird S, Fensom T, Dalton RN (2002) Ornithine aminotransferase deficiency (gyrate atrophy) presenting with hyperammonaemic encephalopathy. J Inherit Metab Dis 25 [Suppl 1]:29
5. Wilson DJ, Weleber RG, Green WR (1991) Ocular clinicopathologic study of gyrate atrophy. Am J Ophthalmol 111:24-33
6. Arshinoff SA, McCulloch JC, Matuk Y et al (1979) Amino-acid metabolism and liver ultrastructure in hyperornithinemia with gyrate atrophy of the choroid and retina. Metabolism 28:979-988
7. Kaiser-Kupfer MI, Kuwabara T, Askanas V (1981) Systemic manifestations of gyrate atrophy of the choroid and retina. Ophthalmology 88:918-928
8. Valtonen M, Nanto-Salonen K, Jaaskelainen S et al (1999) Central nervous system involvement in gyrate atrophy of the choroid and retina with hyperornithinaemia. J Inherit Metab Dis 22:855-866
9. Peltola KE, Jaaskelainen S, Heinonen OJ et al (2002) Peripheral nervous system in gyrate atrophy of the choroid and retina with hyperornithinemia. Neurology 59:735-740
10. Trijbels JMF, Sengers RCA, Bakkaren JAJM et al (1977) L-Ornithine-ketoacidtransaminase deficiency in cultured fibroblasts of a patient with hyperornithinemia and gyrate atrophy of the choroid and retina. Clin Chim Acta 79:371-377
11. Wang T, Lawler AM, Steel G et al (1995) Mice lacking ornithine-delta-aminotransferase have paradoxical neonatal hypoornithinemia and retinal degeneration. Nat Genet 11:185-190
12. Vannas-Sulonen K, Simell O, Sipila I (1987) Gyrate atrophy of the choroid and retina. The ocular disease progresses in juvenile patients despite normal or near normal plasma ornithine concentration. Ophthalmology 94:1428-1433
13. Nanto-Salonen K, Komu M, Lundbom N et al (1999) Reduced brain creatine in gyrate atrophy of the choroid and retina with hyperornithinemia. Neurology 53:303-307
14. Takki K (1974) Gyrate atrophy of the choroid and retina associated with hyperornithinaemia. Br J Ophthalmol 58:3
15. Ramesh V, Gusella JF, Shih VE (1991) Molecular pathology of gyrate atrophy of the choroid and retina due to ornithine aminotransferase deficiency. Mol Biol Med 8:81-94
16. Valle D, Walser M, Brusilow SW, Kaiser-Kupfer M (1980) Gyrate atrophy of the choroid and retina: amino acid metabolism and correction of hyperornithinemia with an arginine-deficient diet. J Clin Invest 65:371-378
17. McInnes RR, Arshinoff SA, Bell L et al (1981) Hyperornithinaemia and gyrate atrophy of the retina: improvement of vision during treatment with a low-arginine diet. Lancet 1:513-516
18. Kennaway NG, Weleber RG, Buist NRM (1980) Gyrate atrophy of the choroid and retina with hyperornithinemia: Biochemical and histologic studies and repsonse to vitamin B6. Am J Hum Genet 32:529
19. Hayasaka S, Saito T, Nakajima H et al (1985) Clinical trials of vitamin B6 and proline supplementation for gyrate atrophy of the choroid and retina. Br J Ophthalmol 69:283-290
20. Shih VE, Berson EL, Gargiulo M (1981) Reduction of hyperornithinemia with a low protein, low arginine diet and pyridoxine in patients with a deficiency of ornithine- ketoacid transaminase (OKT) activity and gyrate atrophy of the choroid and retina. Clin Chim Acta 113:243-251
21. Kaiser-Kupfer MI, Caruso RC, Valle D, Reed GF (2004) Use of an arginine-restricted diet to slow progression of visual loss in patients with gyrate atrophy. Arch Ophthalmol 122:982-984
22. Kaiser-Kupfer MI, Caruso RC, Valle D (2002) Gyrate atrophy of the choroid and retina: further experience with long-term reduction of ornithine levels in children. Arch Ophthalmol 120:146-153
23. Santinelli R, Costagliola C, Tolone C et al (2004) Low-protein diet and progression of retinal degeneration in gyrate atrophy of the choroid and retina: a twenty-six-year follow-up. J Inherit Metab Dis 27:187-196
24. Heinanen K, Nanto-Salonen K, Komu M et al (1999) Creatine corrects muscle 31P spectrum in gyrate atrophy with hyperornithinemia. Eur J Clin Invest 29:1060-1065
25. Elpeleg N, Korman SH (2001) Sustained oral lysine supplementation in ornithine delta-aminotransferase deficiency. J Inherit Metab Dis 24:423-424
26. Shih VE, Efron ML, Moser HW (1969) Hyperornithinemia, hyperammonemia, and homocitrullinuria: A new disorder of amino acid metabolism associated with myoclonic seizures and mental retardation. Am J Dis Child 117:83
27. Shih VE, Laframboise R, Mandell R, Pichette J (1992) Neonatal form of the hyperornithinemia, hyperammonemia and homocitrullinuria (HHH) syndrome and prenatal diagnosis. Prenat Diagn 12:717-723
28. Salvi S, Dionisi-Vici C, Bertini E et al (2001) Seven novel mutations in the ORNT1 gene (SLC25A15) in patients with hyperornithinemia, hyperammonemia, and homocitrullinuria syndrome. Hum Mutat 18:460
29. Dionisi Vici C, Bachmann C, Gambarara M et al (1987) Hyperornithinemia-hyperammonemia-homocitrullinuria syndrome: low creatine excretion and effect of citrulline, arginine, or ornithine supplement. Pediatr Res 22:364-367
30. Smith L, Lambert MA, Brochu P et al (1992) Hyperornithinemia, hyperammonemia, homocitrullinuria (HHH) syndrome: presentation as acute liver disease with coagulopathy. J Pediatr Gastroenterol Nutr 15:431-436
31. Tuchman M, Knopman DS, Shih VE (1990) Episodic hyperammonemia in adult siblings with hyperornithinemia, hyperammonemia, and homocitrullinuria syndrome. Arch Neurol 47:1134-1137
32. Camacho JA, Obie C, Biery B et al (1999) Hyperornithinaemia-hyperammonaemia-homocitrullinuria syndrome is caused by mutations in a gene encoding a mitochondrial ornithine transporter. Nat Genet 22:151-158
33. Miyamoto T, Kanazawa N, Kato S et al (2001) Diagnosis of Japanese patients with HHH syndrome by molecular genetic analysis: a common mutation, R179X. J Hum Genet 46:260-262
34. Shih VE, Mandell R, Herzfeld A (1982) Defective ornithine metabolism in cultured skin fibroblasts from patients with the syndrome of hyperornithinemia, hyperammonemia and homocitrullinuria. Clin Chim Acta 118:149
35. Chadefaux B, Bonnefont JP, Shih VE, Saudubray JM (1989) Potential for the prenatal diagnosis of hyperornithinemia, hyperammonemia, and homocitrullinuria syndrome. Am J Med Genet 32:264

36. Zammarchi E, Ciani R, Pasquini E et al (1997) Neonatal onset of hyperornithinemia-hyperammonemia-homocitrullinuria syndrome with favourable outcome. J Pediatr 131:440-443

37. Gaye AM, Wong PWK, Kang DS et al (1983) Treatment of hyper-ornithinemia, hyperammonemia and homocitrullinuria (HHH) during pregnancy. Clin Res 31:787A

38. Baumgartner MR, Rabier D, Nassogne MC et al (2005) Delta1-pyrroline-5-carboxylate synthase deficiency: Neurodegeneration, cataracts and connective tissue manifestations combined with hyperammonemia and reduced ornithine, citrulline, arginine and proline. Eur J Pediatr 164:31-36

39. Baumgartner MR, Hu CA, Almashanu S et al (2000) Hyperammon-emia with reduced ornithine, citrulline, arginine and proline: a new inborn error caused by a mutation in the gene encoding delta(1)-pyrroline-5-carboxylate synthase. Hum Mol Genet 9:2853-2858

40. Wakabayashi Y, Yamada E, Hasegawa T, Yamada R (1991) Enzymo-logical evidence for the indispensability of small intestine in the synthesis of arginine from glutamate. I. Pyrroline-5-carboxylate synthase. Arch Biochem Biophys 291:1-8

23 Cerebral Organic Acid Disorders and Other Disorders of Lysine Catabolism

Georg F. Hoffmann

Catabolism of Lysine, Hydroxylysine, and Tryptophan

Lysine, hydroxylysine and tryptophan are degraded within the mitochodrion initially via separate pathways but which then converge in a common pathway starting with 2-aminoadipic and 2-oxoadipic acids (■ Fig. 23.1). The initial catabolism of lysine proceeds mainly via the bifunctional enzyme, 2-aminoadipic-6-semialdehyde synthase (enzyme 1). A small amount of lysine is catabolized via pipecolic acid and the peroxisomal enzyme, pipecolic acid oxidase (enzyme 2). Hydroxylysine enters this pathway after phosphorylation by hydroxylysine kinase (enzyme 3). 2-Aminoadipic-6-semialdehyde is converted into glutaryl-CoA by antiquitin (enzyme 4, deficient in B$_6$ responsive seizures) and a second bifunctional enzyme, 2-aminoadipate aminotransferase/2-oxoadipate dehydrogenase (enzyme 5). Glutaryl-CoA is converted into crotonyl-CoA by a third bifunctional enzyme, glutaryl-CoA dehydrogenase/glutaconyl-CoA decarboxylase (enzyme 6). This enzyme tranfers electrons to flavin adenine dinucle-

otide (FAD) and hence to the respiratory chain (■ Fig. 13.1) via electron transfer protein (ETF)/ETF-dehydrogenase (ETF-DH). From the five distinct enzyme deficiencies identified in the degradation of lysine, only enzymes 4 and 6 have proven relevance as neurometabolic disorders. Glutaric aciduria type I is caused by the isolated deficiency of glutaryl-CoA dehydrogenase/glutaconyl-CoA decarboxylase. Glutaric aciduria type II, caused by ETF/ETF-DH deficiencies, is discussed in ► Chap. 13. Pipecolic acid oxidase deficiency is discussed in ► Chap. 40 and antiquitin deficiency in ► Chap. 29.

The metabolic origins and fates of L- and D-2-hydroxyglutaric acids have not been completely unravelled in mammals. Yet, L- and D-2-hydroxyglutaric aciduria have recently been shown to be caused by deficiencies of specific dehydrogenases. Aspartoacylase irreversibly splits N-acetylaspartic acid into acetate and aspartate (not illustrated).

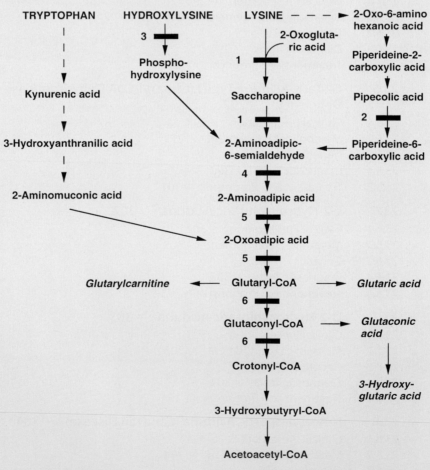

■ **Fig. 23.1.** Tryptophan, hydroxylysine and lysine catabolic pathways. *CoA*, coenzyme A. **1**, 2-aminoadipic-**6**-semialdehyde synthase; **2**, pipecolic acid oxidase; **3**, hydroxylysine kinase; **4**, antiquitin; **5**, 2-aminoadipate aminotransferase/2-oxoadipate dehydrogenase; **6**, glutaryl-CoA dehydrogenase/glutaconyl-CoA decarboxylase. Enzyme deficiencies are indicated by *solid bars* across the arrows

Seven inborn errors are described in this chapter: *Hyperlysinemia I/hyperlysinemia II or saccharopinuria, hydroxylysinuria and 2-amino-/2-oxo-adipic aciduria* may all have no clinical significance, but some patients are retarded and show variable neurological abnormalities.

Glutaric aciduria type I (GA I, synonym glutaryl-CoA dehydrogenase deficiency) causes severe neurometabolic disease. The first months may be uneventful with only subtle neurological abnormalities and/or macrocephaly, but progressive cerebral atrophy or subdural hemorrhages on neuroimaging. Between age 6 to 18 months most untreated patients suffer an acute encephalopathy resulting in irreversible destruction of susceptible brain regions, in particular the striatum, a dystonic-dyskinetic syndrome and, ultimately, often early death. Restriction of lysine, administration of L-carnitine and timely vigorous treatment during intercurrent illness is able to completely prevent or at least halt the disease.

L-2-Hydroxyglutaric aciduria shows an insidious onset with delay of unsupported walking and speech, febrile convulsions, and macrocephaly. Over the years, severe mental retardation and cerebellar ataxia develop with or without dystonia, pyramidal signs, and seizures.

D-2-Hydroxyglutaric aciduria causes severe early-onset epileptic encephalopathy with neonatal seizures, lack of psychomotor development and early death. Some patients exhibit milder neurological symptoms such as mild developmental delay, delayed speech and febrile convulsions.

N-Acetylaspartic aciduria (synonyms: aspartoacylase deficiency, spongy degeneration of the brain, Van Bogaert-Bertrand disease or Canavan disease) is an infantile degenerative disease primarily affecting the cerebral white matter. It commonly manifests with poor head control and hypotonia at 2-4 months, macrocephaly, marked developmental delay, optic nerve atrophy, progressive spasticity, opisthotonic posturing, seizures and death in childhood.

23.1 Introduction

A group of organic acid disorders presents exclusively with progressive neurological symptoms of ataxia, epilepsy, myoclonus, extrapyramidal symptoms, metabolic stroke, and macrocephaly [1]. The core »cerebral« organic acid disorders are glutaric aciduria type I, D-2-hydroxyglutaric aciduria, L-2-hydroxyglutaric aciduria, 4-hydroxybutyric aciduria and *N*-acetylaspartic aciduria. Strikingly, in all these disorders the pathological compounds that accumulate

share structural similarities with the excitatory amino acid glutamate (D-2-, L-2-, 3-hydroxyglutarate, glutarate) or have been suggested to be neurotransmitters/neuromodulators (γ-hydroxybutyrate, *N*-acetylaspartylglutamate) [2]. Evidence from in vitro and in vivo studies is growing that these acids indeed interfere with glutamatergic or gamma amino butyric acid (GABA)-ergic neurotransmission or impair energy metabolism. Delayed myelination or progressive demyelination, basal ganglia and cerebellum pathology, the main pathologies in cerebral organic acid disorders, are also characteristic of mitochondrial disorders suggesting at least partial common pathological mechanisms.

Patients with cerebral organic acid disorders often suffer a diagnostic odyssey and may even remain undiagnosed. Metabolic hallmarks such as hypoglycemia, metabolic acidosis or lactic acidemia, the usual concomitants of disorders of organic acid metabolism, are generally absent. Furthermore, elevations of diagnostic metabolites may be small and therefore missed on »routine« organic acid analysis, e.g. in glutaric aciduria type I. The correct diagnosis requires an increased awareness of these disorders by the referring physician as well as the biochemist in the metabolic laboratory. Diagnostic clues can be derived from neuroimaging findings (◘ Fig. 23.2). Progressive disturbances of myelination, cerebellar atrophy, frontotemporal atrophy, hypodensities and/or infarcts of the basal ganglia and any symmetrical (fluctuating) pathology, apparently independent of defined regions of vascular supply, are suggestive of cerebral organic acid disorders.

In contrast to the cerebral organic acid disorders, the other know defects of lysine and hydroxylysine degradation all appear to be rare biochemical variants of human metabolism without clinical significance.

23.2 Hyperlysinemia/Saccharopinuria

23.2.1 Clinical Presentation

Hyperlysinemia/saccharopinuria appears to be a rare »nondisease«. About half of the patients described were detected incidentally and are healthy [3]. Symptoms have included psychomotor retardation, epilepsy, spasticity, ataxia and short stature. Individual patients have been described with joint laxity and spherophakia, respectively.

23.2.2 Metabolic Derangement

Hyperlysinemia/saccharopinuria is caused by deficiency of the bifunctional protein 2-aminoadipic semialdehyde synthase, the first enzyme of the main route of lysine degradation [4]. The two functions, lysine-2-oxoglutarate reductase and saccharopine dehydrogenase, may be differently affected by mutations. Most often, both activities are severely reduced, resulting predominantly in hyperlysinemia and hyperlysinuria, accompanied by relatively mild sacccharopinuria (hyperlysinemia I). In hyperlysinemia II or saccharopinuria [5], there is a relatively more pronounced decrease in saccharopine dehydrogenase activity with residual activity of lysine-2-oxoglutarate reductase causing a predominant excretion of sacccharopine.

Failure to remove the ε-amino group results in an overflow of the minor lysine degradation pathway with removal of the α-amino group by oxidative deamination. The oxoacid cyclizes and is reduced to pipecolic acid. As a consequence, hyperpipecolatemia is regularly observed in hyperlysinemia.

Hyperlysinuria can also result from impaired renal tubular transport, often as part of a genetic transport defect of dibasic amino acids (▶ Chap. 26), but in this situation it occurs without hyperlysinemia.

23.2.3 Genetics

Hyperlysinemia/saccharopinuria follows an autosomal recessive inheritance. The gene has been characterized and a homozygous out-of-frame 9bp deletion identified in an affected boy [4].

23.2.4 Diagnostic Tests

The initial observation in patients with hyperlysinemia/saccharopinuria is an impressive lysinuria up to 15 000 mmol/mol creatinine (controls <70). Detailed amino acid analysis will reveal additional accumulation of saccharopine, homoarginine, 2-aminoadipic acid and pipecolic acid [6]. Elevations of the same metabolites can be documented in other body fluids such as plasma and cerebrospinal fluid (CSF) with high lysine as the predominant abnormality (up to 1700 μmol/l in plasma, controls < 200, and up to 270 μmol/l in CSF, controls < 28).

The deficiency of 2-aminoadipic semialdehyde synthase can be ascertained in fibroblasts and tissue biopsies by determining the overall degradation of [1-^{14}C] lysine to ^{14}CO$_2$. Specific assays for lysine 2-oxoglutarate reductase and saccharopine dehydrogenase have been described [7]. Molecular diagnosis has become possible [4].

23.2.5 Treatment and Prognosis

Long-term dietary restriction of lysine has no benefit. As patients do not suffer from metabolic decompensations, specific interventions during intercurrent illnesses are not necessary. As hyperlysinemia/saccharopinuria is a benign condition it is not associated with any increase in morbidity or mortality.

23.3 Hydroxylysinuria

Hydroxylysinuria and concomitant hydroxylysinemia has been identified in only six patients, all of whom showed some degree of mental retardation [8]. No further clinical and/or biochemical studies were reported. The abnormality can be assumed to be caused by a defect of hydroxylysine kinase.

23.4 2-Amino-/2-Oxo-Adipic Aciduria

23.4.1 Clinical Presentation

As with hyperlysinemia/saccharopinemia, 2-amino-/2-oxo-adipic aciduria is probably of no clinical significance. Over 20 patients are known, of whom more than half are asymptomatic ([9]; Hoffmann, unpublished observation). Symptoms include psychomotor retardation, muscular hypotonia, epilepsy, ataxia and failure to thrive.

23.4.2 Metabolic Derangement

The metabolic profile is heterogeneous with most patients showing elevations of all three metabolites, whereas some excrete only 2-aminoadipic acid [10]. Normally 2-aminoadipic acid is deaminated to 2-oxoadipic acid by a mitochondrial 2-aminoadipate aminotransferase. 2-Oxoadipic acid is also formed from the degradation of tryptophan. 2-Oxoadipic acid is further metabolized to glutaryl-CoA by oxidative decarboxylation, probably differently from that of oxoglutarate.

Isolated 2-aminoadipic aciduria is probably caused by a deficiency of 2-aminoadipate aminotransferase, and combined 2-amino/2-oxoadipic aciduria by a deficiency of the 2-oxoadipate dehydrogenase complex. However a deficiency of neither enzyme has yet been shown directly.

2-Aminoadipic acid shows a complex excitatory amino acid synaptic pharmacology, which may be related to the neurological symptoms.

23.4.3 Genetics

Autosomal recessive inheritance is implied by the pedigrees and by the finding that parents can not be biochemically differentiated from controls.

23.4.4 Diagnostic Tests

Patients are diagnosed by demonstrating variable elevations of 2-aminoadipic acid on amino acid chromatography and/or of 2-oxoadipic and 2-hydroxyadipic acids on urinary organic acid analysis. Plasma lysine may be twofold elevated and urinary glutaric acid up to 50 mmol/mol of creatinine (controls < 9) [11].

Oral loading with lysine or tryptophan (100 mg/kg body weight) increases the pathological metabolites two- to sixfold [10]. Conclusive molecular or enzymatic analyses have not yet been achieved.

23.4.5 Treatment and Prognosis

As 2-amino-/2-oxoadipic aciduria is likely a non-disease, it does not determine morbidity or mortality. Patients do not suffer from metabolic decompensations, and specific interventions during intercurrent illnesses do not appear necessary. Administration of pharmacological doses of vitamins B_1 and B_6 had no effect on the levels of pathological metabolites [10]. Dietary restriction of lysine also failed to correct the biochemical abnormalities in some patients [10] and has no proven long-term benefit.

23.5 Glutaric Aciduria Type I (Glutaryl-CoA Dehydrogenase Deficiency)

23.5.1 Clinical Presentation

Glutaric aciduria type I should be strongly considered in the differential diagnosis of any infant with macrocephaly together with progressive atrophic changes on computerized tomography (CT) or nuclear magenetic resonance (NMR) (◘ Fig. 23.2a–d) and/or acute profound dyskinesia or subacute motor delay accompanied by increasingly severe choreoathetosis and dystonia [11–17, 17a]. In many patients macrocephaly is present at or shortly after birth and precedes the severe neurological disease. An important clue to early diagnosis is the observation of a pathologically increased head growth crossing the percentiles and peaking at the age of 3-6 months. Furthermore, affected babies often present additional »soft« neurological symptoms of hypotonia with prominent head lag, irritability, jitteriness, and feeding difficulties. Neonatal posture and tone may persist until 4 to 8 months of age. During febrile illnesses or after immunizations, hypotonia is often aggravated and unusual movements and postures of hands appear. All these symptoms are still reversible and of little prognostic significance. Neuroimaging studies have been performed in a number of patients during this »presymptomatic« period revealing the characteristic findings of frontotemporal atrophy (95% of all patients) delayed myelination, and high-signal intensity in the dentate nucleus (◘ Fig. 23.2a). The clinical significance of enlarged subdural fluid spaces in infants with glutaric aciduria type I is the unprotected crossing of these spaces by bridging veins. Such infants are prone to suffer acute subdural hemorrhages including retinal hemorrhages after minor head trauma, particularly around the first birthday when starting to walk (◘ Fig. 23.2b). Parents of children with glutaric aciduria type 1 have been wrongly charged with child abuse because of chronic or acute subdurals and/or hemorrhages [18].

On average at an age of 9 months ≈ 75% of patients suffer an acute brain injury usually associated with an upper respiratory and/or gastrointestinal infection, but the encephalopathic crisis may also develop in association with fasts required for surgery, after routine immunizations, or following minor head traumas [12]. 87% of the encephalopathic crises occur by age 24 months. They have not yet been described at school age or in older children. Neurological functions are often acutely lost, including their ability to sit, pull to standing, head control, suck and swallow reflexes. The infants appear alert with profound hypotonia of the neck and trunk, stiff arms and legs, and twisting (athetoid) movements of hands and feet. There may also be generalized seizures. Mostly, there are no or only mild metabolic derangements. A severe dys-/hypotonic movement disorder develops. At this point the distinctive clinical picture of a dystonic-dyskinetic syndrome in an alert looking child with relatively well preserved intellectual functions and a prominent forehead may be recognized. If the underlying metabolic disorder remains undiagnosed, additional cerebral systems are slowly but progressively affected. A generalized cerebral atrophy emerges (◘ Figs. 23.2c and d), giving rise to pyramidal tract signs and mental retardation. Impaired chewing and swallowing, vomiting and aspiration as well as increased energy demand due to increased muscle tone frequently results in failure to thrive and malnutrition. Kyphoscoliosis and chest wall dystonia can cause restrictive lung disease. Early death (≈40% of symptomatic patients by the age of 20 years) may occur in the course of intercurrent pneumonia and respiratory failure, during hyperpyrexic crises or suddenly without warning.

Although the majority of patients present with characteristic symptoms and disease course, the natural history of glutaric aciduria type I can be variable even within families. A minority of patients (≤20%) presents with developmental delay from birth and (progressive) dystonic cerebral palsy. Some individuals, mainly diagnosed in adolescence or adulthood during family studies, have not developed neurological disease despite never having been treated. Finally, adult onset-type glutaric aciduria type I has recently been described in five previously unaffected adolescent/adult patients presenting with leukoencephalopathy [19, 20], suggesting an additional distinct clinical manifestation of this disease.

23.5.2 Metabolic Derangement

Glutaric aciduria type I is caused by a deficiency of glutaryl-CoA dehydrogenase, a mitochondrial flavin adenine

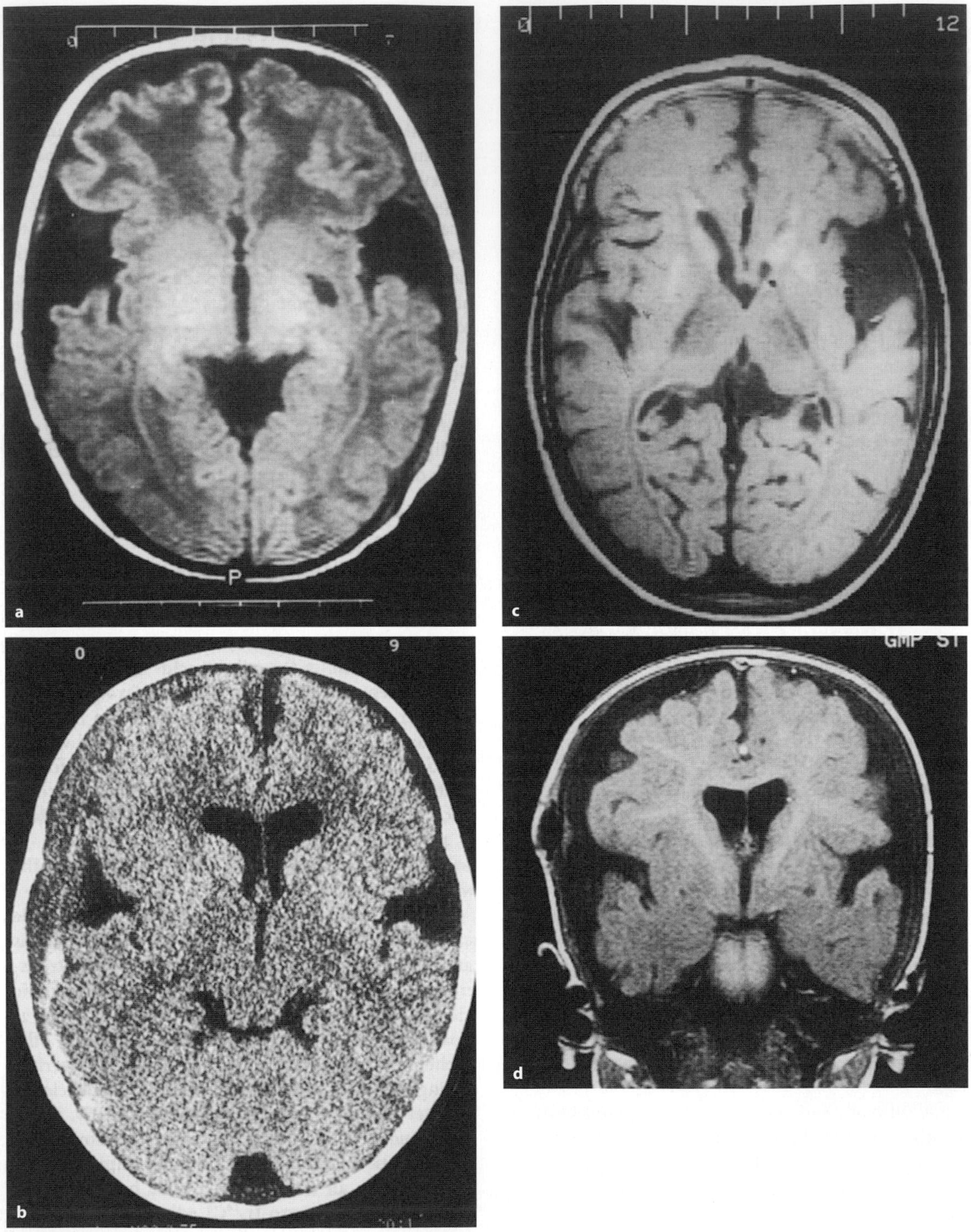

IV

□ **Fig. 23.2.** Neuroimaging findings, which are characteristic for cerebral organic acid disorders. **a** Transversal NMR image of a 2-month-old presymptomatic boy with glutaric aciduria type I deficiency showing enlargement of CSF spaces anterior to the temporal lobes with marked extension of sylvian fissures (frontotemporal atrophy). Spin echo technique (1.0 tesla): time of repetition 660 ms, time of echo 20 ms, slice thickness 5.5 mm. **b** CT scan of a 9-month-old presymptomatic boy with glutaric aciduria type I. In addition to frontotemporal atrophy subdural effusions and hematomas causing midline shift are visible. There is no pathology of the basal ganglia, and the child continued to develop normally. **c** Transversal NMR image of a 2-year-old boy with glutaric aciduria type I, who suffered an encephalopathic crisis at the age of 11 months, illustrating regression of the temporal lobes, more pronounced on the left side, with dilatation of the Sylvian fissures anterior to them as well as dilated insular cisterns (frontotemporal atrophy), delayed myelination and bilateral hypodensities of the basal ganglia. Spin echo technique (1.0 tesla): time of repetition 2660 ms, time of echo 100 ms, slice thickness 5 mm. **d** Coronal section of CNS of a 8-month-old girl with glutaric aciduria type I with dilatation of CSF spaces prominently at the Sylvian fissures by NMR. At the age of 6 months a ventricularperitoneal shunt was inserted in an attempt to drain »bilateral subdural hygromas« without knowledge of the underlying metabolic disorder. The catabolic ►

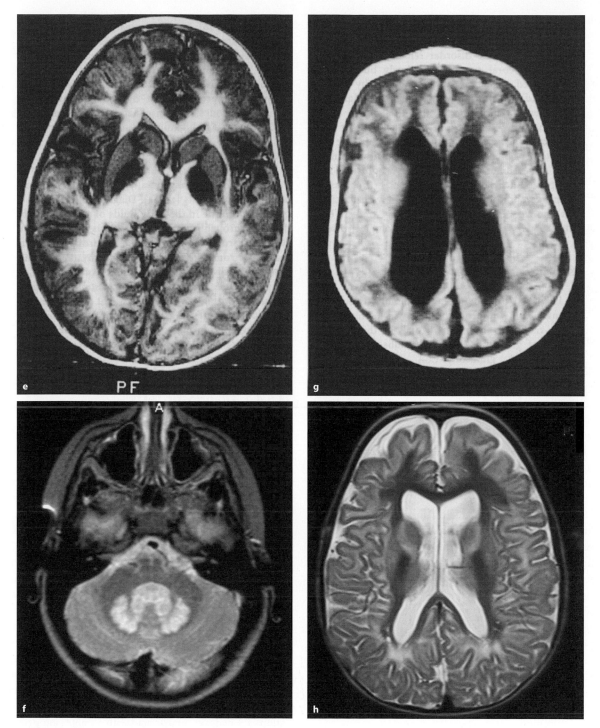

◻ Fig. 23.2 (continued)

stress following the operation resulted in an encephalopathic crisis and consecutively destruction of the basal ganglia and a severe dystonic-dyskinetic syndrome. Spin echo technique (1.0 tesla): time of repetition 380 ms, time of echo 25 ms, slice thickness 6 mm. **e** Transversal NMR image of a 3-year-old boy with L-2-hydroxyglutaric aciduria, illustrating characteristic involvement of the subcortical white matter (U fibres), the N. caudati and the putamina. In this boy the periventricular white matter was also affected early on. Spin echo technique (1.0 tesla): time of repetition 2660 ms, time of echo 100 ms, slice thickness 5 mm. **f** Axial T2-weighted MRI of an 8.5-year-old boy with L-2-hydroxyglutaric aciduria,

illustrating symmetric involvement of the dentate nuclei. Reproduced with permission from Kölker et al [2]. **g** Transversal NMR image of a 2-month-old girl with D-2-hydroxyglutaric aciduria. Please note delayed myelination and considerable occipitally pronounced enlargement of lateral ventricles. Spin echo technique (1.0 tesla): time of repetition 618 ms, time of echo 15 ms, slice thickness 6 mm. **h** Axial fast spin echo image of a 6.5-year-old girl suffering from *N*-acetylaspartic aciduria. Note the marked discrepancy between the severely affected subcortical white matter and the relatively spared central white matter, at least frontally. Reproduced with permission from Kölker et al [2] dinucleotide requiring enzyme, which catalyzes the dehydrogenation of glutaryl-CoA as well as the subsequent decarboxylation of glutaconyl-CoA to crotonyl-CoA (■ Fig. 23.1). In glutaric aciduria type I, part of the accumulating glutaryl-CoA is esterified with carnitine by carnitine acyltransferase leading to an increased ratio of acylcarnitines to free carnitine in plasma and urine. Glutarylcarnitine is excreted, contributing to secondary carnitine deficiency [11]. Patients with glutaric aciduria type I often show increased urinary excretion of dicarboxylic acids, 2-oxoglutarate and succinate, indicative of disturbed mitochondrial function [11].

Secondary carnitine deficiency is probably a major causative factor of metabolic crises. These can present with hypoglycemia and variable metabolic acidosis and may quickly progress to a Reye-like syndrome (14% in one series [14]). Metabolic crises can develop at any age and respond well to intravenous therapy with glucose, carnitine and bicarbonate [11, 14, 21].

The mechanism of age-specific destruction of specific cerebral structures in glutaric aciduria type I has been subject of intense debates and generated different hypotheses [22]: competitive inhibition of glutamate decarboxylase, the key enzyme in the biosynthesis of the inhibitory neurotransmitter GABA [23] or an increased catabolism of tryptophan via the kynurenine pathway [24], the latter supported by the observation that a key enzyme of this pathway, indolamine-2,3-dioxygenase, is up-regulated by interferons during acute viral infections, which could contribute to an encephalopathic crisis.

The most substantiated evidence points to an excitotoxic sequence [22], initiated by glutaric and 3-hydroxyglutaric acids which exhibit structural similarities to the excitotoxic amino acid glutamate. Massive activation of glutaminergic neurons from cortex to putamen via the caudate nucleus can be anticipated during encephalopathic crises. The inhibitory output of the thalamus would explain the severe muscular hypotonia and also the temporary improvement of dystonia during further catabolic episodes as a consequence of overstimulation of the remaining glutaminergic neurons. Finally, maturation-dependent changes in the expression of neuronal glutamate receptors and thereby vulnerability to 3-hydroxyglutaric and glutaric acid toxicity are likely responsible for the specific time course and localization of neurological disease.

23.5.3   Genetics

Glutaric aciduria type I is an autosomal recessive disorder. Results of newborn screening programs in various regions and cohorts worldwide enrolling approximately 2.5 million newborns world-wide give an overall mean frequency of 1:106 900 [27]. The disease is very frequent in certain communities such as the Amish people in Pennsylvania (homo-

zygous for the A421V mutation, incidence of 1:300–400) [17], the Saulteaux/Ojibway Indians in Canada (homozygous for the splice site mutation IVS-1$^{+5\ g>t}$ 1:300) [28] or the Irish travellers (homozygous for the E365K mutation) [16].

More than 150 different disease-causing mutations in the glutaryl-CoA dehydrogenase gene have been identified so far [31, 13] and unpublished data. There is a correlation between genotype and biochemical phenotype in that specific mutations with significant residual enzyme activity may be associated with low excretions of metabolites in heterozygous patients who carry a severe mutation on the other allele. However, no correlation between genotype and clinical phenotype has yet been found. Single common mutations are found in genetically homogenous communities (see above), but glutaric aciduria type I in general is genetically quite heterogeneous: the most frequent mutation in Caucasians, R402W, has been identified on 10–20% of alleles [31]. Apart from three short or single nucleotide deletions, the great majority of mutations are single base changes that are frequently found at hypermutable CpG sites.

23.5.4   Diagnostic Tests

Patients with glutaric aciduria type I are generally diagnosed by urinary organic acid analysis [11, 25]. Repeated and quantitative urinary organic acid analyses may be necessary. Additional diagnostic hints can be obtained by finding carnitine deficiency in serum and/or a pathologically increased ratio of acylcarnitines to free carnitine in serum and urine. Elevations of glutarylcarnitine in body fluids of patients can be detected through acylcarnitine analysis [27]. Application to analyses of blood spots (Guthrie cards) has enabled the recent inclusion of glutaric acidurias type I and type II into several neonatal screening programs. However, individuals with deficiency of glutaryl-CoA dehydrogenase and severe characteristic neurological disease but with only slight or inconsistent elevations of glutaric acid or glutarylcarnitine have been diagnosed in increasing numbers [25, 27, 26]. Furthermore, elevated urinary excretion of glutaric acid can also be found in a number of other disease states mostly related to mitochondrial dysfunction [11]. Determination of 3-hydroxyglutaric acid with a stable isotope dilution assay in urine has the highest diagnostic power [25]. It proved to be reliably elevated except for patients from first nations of Oji-Cree descent from North-eastern Manitoba and North-western Ontario [28].

Loading tests, e.g. with lysine or prolonged fasting tests provoking catabolism may be extremely harmful and should be avoided. Alternatively, safe loading tests can be performed in vitro using lymphoblasts or peripheral mononuclear blood cells [29]. Ultimately, demonstration of two known pathogenic mutations or enzyme analysis of glutaryl-CoA dehydrogenase is the only method that can establish the diagnosis of glutaric aciduria type I with certainty in diagnostically problematic cases. Glutaryl-CoA dehydrogenase activity can be determined in tissues, cultured fibroblasts, peripheral leukocytes, amniocytes and chorionic villi cells [30]. This procedure is justified in family studies or whenever there is strong clinical suspicion.

Carrier detection is possible by enzyme assay, though the results are not always unequivocal [30] and by molecular means in families in which the mutations are already known. Reliable *prenatal diagnosis* can be offered by enzyme assay [30], determination of glutaric acid by stable isotope dilution gas chromatography–mass spectrometry (GC-MS) assay in amniotic fluid [25], and by molecular analysis [31].

23.5.5 Treatment and Prognosis

Three decades after the first description of glutaric aciduria type I, approximately 350–400 patients have been identified world-wide and major progress has been achieved in the prevention of acute striatal necrosis and neurologic sequelae, if diagnosis is made early and treatment is started before manifestation of acute encephalopathic crises. Early diagnosis and treatment of the asymptomatic child is essential as current therapy has little effect upon the brain injured child. If an encephalopathic crisis occurred, only 5% of patients recovered completely, 75% remained handicapped of whom 40% died early. However, all concepts for diagnostic work-up, monitoring, and treatment are solely experience-based, and even with early diagnosis 10–30% of patients do not or only incompletely benefit from the current management [14, 16, 17].

Five different therapeutic measures are generally employed:

(1) Emergency Treatment
Emergency treatment must start before onset of severe neurological signs, which already indicate the manifestation of neuronal damage. During intercurrent illnesses, especially gastrointestinal infections, treatment should consist of frequent high carbohydrates feeds and increased carnitine supplementation, followed by high-dose intravenous glucose and carnitine, if necessary [11, 21]. If mixtures of free amino acids devoid of lysine are used, these are offered orally, in addition. If the temperature rises >38.5 °C (101 F), antipyretics must be administered, e.g. ibuprofen (10–15 mg/kg body wt per dose, 3-4 doses daily). All patients with glutaric aciduria type I should be supplied with an emergency card. Frequent visits and regular information and training of parents may help to prevent lapses or mistakes. This concept must be strictly followed for the first 6 years of life. After this age emergency treatment is individually adjusted.

(2) Oral Supplementations with Carnitine and Riboflavin
Carnitine should be supplemented lifelong. Published data suggests that most, but not all, patients developing an encephalopathic crisis had no carnitine supplementation. Riboflavin responsiveness has never been demonstrated in any patient with glutaric aciduria type I. If riboflavin is tried, responsiveness should be investigated by giving riboflavin in increasing doses from 50 to 300 mg and monitoring total glutaric acid in 24-hour-urine samples. In evaluating the response unrelated high daily variations of the urinary excretion of glutarate must be taken into account.

(3) Dietary Treatment
Most patients with glutaric aciduria type I are treated by restriction of natural protein in general or of lysine in particular, supplemented with a lysine free amino acid mixture (◘ Table 23.1). Application of lysine-free amino acid mixtures minimizes the risk for malnutrition, allows a reliable control of protein and lysine intake and, most importantly, has proven the best long-term results. It should therefore be followed during the vulnerable period for acute encephalopathic crises, i.e. the first 6 years of life [17a]. Tryptophan contributes only ≤20% to total body glutarate production. The intake of tryptophan should only be reduced, if consequently and reliably monitored, which is not possible by regular amino acid analysis. Concentrations of tryptophan directly modulate production of serotonin in the CNS. Using diets low in tryptophan we observed side effects such as sleeplessness, ill temper, irritability, and loss of appetite, which could be improved by isolated tryptophan supplementation [14]. There are only anecdotal data about the value of protein restriction beyond six years of age. However, protein excesses should be avoided.

Special efforts to supply adequate calories are often necessary in patients with motor dysfunction and swallowing difficulties. This may require nasogastric or gastrostomy feeding. We have also observed that an improved nutritional status is paralleled by a reduction of the dystonic/dyskinetic syndrome.

(4) Neuropharmacological Agents
Several of these have been tried to ameliorate neurological symptoms in patients with glutaric aciduria type I. Clomethiazole was found useful in severe cases of hyperpyrexia. In our experience, baclofen (Lioresal, 1-2 mg/kg daily) or benzodiazepines (diazepam, 0.1–1 mg/kg daily) reduce involuntary movements and improve motor function, probably mostly through muscle relaxation. In some patients their use and dosage are limited by worsening of truncal hypotonia. The patient's head should be kept in the midline position, as this allows maximum mobility and minimizes dystonia.

In patients with residual motor function anticholinergics, such as trihexyphenidyl, may improve choreoathetosis [17]. Valproic acid is contraindicated as it effectively competes with glutaric acid for esterification with L-carnitine and may promote disturbances in the mitochondrial acetyl-CoA/CoA ratio. Vigabatrine has been commonly used in the past. It showed little to no effect [15] but is still used by a number of patients, which in view of the severe side effects should be carefully re-evaluated on an individual basis. There are anecdotal reports of sustained improvement with experimental therapies including botulinum toxin injections and a baclofen pump [32]. Considering the severe neurological disease surprisingly little information is available on the effects of other neuropharmacological agents; medications listed in ◘ Table 23.1 could be empirically employed.

(5) Nonspecific Multiprofessional Support
This is of utmost importance since despite the severe motor handicap, intellectual functions are preserved until late into the course of the disease. Affected patients require the full recourses of a multidisciplinary specialist institution. The social integration of patients can be greatly improved using Bliss boards and, in particular, language computers. As involuntary movements of orofacial muscles may be severe, feeding difficulties can become a major problem. Increased

muscular tension and sweating, common findings in glutaric aciduria type I, require a high intake of calories and water. Percutaneous gastrostomy can lead to a dramatic improvement in nutritional status, a marked decrease in psychological tension associated with feeding, a reduction in the burden of care for families and even a reduction in the dystonia/dyskinesia. As a final remark, neurosurgical interventions of subdural hygromas and hematomas in infants and toddlers with glutaric aciduria type I should be avoided, if at all possible (◘ Fig. 23.2d).

As the risk for encephalopathic crises subsides after 4 to 5 years of age, the rationale of dietary treatment using lysine free amino acid mixtures beyond 6 years of age is uncertain. In some symptomatic patients, movement disorders were aggravated by excessive intake of protein and could be reversed after reduction of protein intake. Furthermore, five undiagnosed and untreated patients, have presented with leukoencephalopathy as late-onset disease in adolescence or adulthood [20]. However, it is not yet known whether dietary treatment can prevent chronic neurodegenerative changes. Emergency measures during intercurrent illnesses may also be partially relaxed in older children. Carnitine supplementation must be followed for life.

23.6 L-2-Hydroxyglutaric Aciduria

23.6.1 Clinical Presentation

The initial description of L-2-hydroxyglutaric aciduria was followed by a number of reports from all over the world illustrating previous mis- and under diagnosis. Most patients with L-2-hydroxyglutaric aciduria follow a characteristic disease courses [33–36]. In infancy and early childhood mental and psychomotor development appears normal or only slightly retarded. Thereafter seizures, progressive ataxia, pyramidal tract signs, slight extrapyramidal signs and progressive mental retardation become the most obvious clinical findings. Progressive macrocephaly is present in about half of the patients. The IQ in teenagers is ≈40–50. Sometimes mental deterioration is rapidly progressive, and a single patient with fatal neonatal outcome has been described [37].

In L-2-hydroxyglutaric aciduria the neuroimaging findings are very specific [33, 38]. The subcortical white matter appears mildly swollen with some effacement of gyri. The progressive loss of arcuate fibers is combined with a severe cerebellar atrophy and increased signal densities of dentate nuclei, globi pallidi and less frequently the nuclei caudati and putamina (◘ Figs. 23.2e and f) on T2-weighted images, while the thalamus shows decreased signal densities.

◘ **Table 23.1.** Maintenance therapy in patients with glutaric aciduria type I

Measures	Infants	Children <6 years	Children >6 years	Adults
Diet				
Natural Protein (g/kg b.w./day)	1.8–1.0	1.4–1.1	1.5–1.1	1.0
Amino Acid Mixture (g/kg b.w./day)	1.0–0.8	1.0–0.8	n.a.	n.a.
Lysine (mg/kg b.w./day)	100–90	80–50	n.a.	n.a.
Tryptophan (mg/kg b.w./day)	≥20–17	≥17–13	n.a.	n.a.
Energy (kcal/kg b.w./day)	120	95–80	70-60	50–40
Supplementations				
L-Carnitine (mg/kg b.w./day)	100	50–100	50–100	50

Neuropharmaceuticals (Patients with neurological disease):
Baclofen, Clonazepam, Diazepam, Triheyphenidyl, Memantine, Haloperidol, L-Dopa/Levodopa, Glutamine.
Do not Use Valproic Acid or Vigabatrin.

Multiprofessional Support of Patient and Family

n.a., not applicable.

23.6.2 Metabolic Derangement

Quantitative analysis of organic acids revealed higher elevations of L-2-hydroxyglutaric acid in CSF than in plasma [39]. In addition a number of hydroxydicarboxylic acids (glycolate, glycerate, 2,4-dihydroxybutyrate, citrate and isocitrate) were only found elevated in CSF, pointing to a specific disturbance of brain metabolism. Another consistent biochemical finding is an increase of lysine in blood and CSF.

A gene encoding a putative FAD-dependent L-2-hydroxyglutarate dehydrogenase, first tentatively identified in human liver [33], has been recently found defective in L-2-hydroxyglutaric aciduria and its gene identified and mutations demonstrated [40, 41]. It was concluded that L-2-hydroxyglutaric acid is normally converted into 2-oxoglutarate, while the origin of L-2-hydroxyglutaric acid remains uncertain.

23.6.3 Genetics

L-2-hydroxyglutaric aciduria is an autosomal recessive disorder. Heterozygotes display no detectable clinical or biochemical abnormalities, but can now be ascertained by molecular diagnosis in informative families.

23.6.4 Diagnostic Tests

L-2-hydroxyglutarate is found elevated in all body fluids [33, 42]. In addition, lysine is slightly elevated in cerebrospinal fluid as well as protein in the absence of pleocytosis. Differentiation between the two isomers of 2-hydroxyglutarate is indispensable for diagnosis. *Prenatal diagnosis* is possible utilizing accurate determination of L-2-hydroxyglutarate by stable isotope dilution GC-MS assay in amniotic fluid [42, 36] as well as molecular diagnosis.

23.6.5 Treatment and Prognosis

To date there is no rational therapy for L-2-hydroxyglutaric aciduria. Epilepsy can generally be controlled by standard medications. The oldest known patients are over 30 years of age. They are bedridden and severely retarded.

23.7 D-2-Hydroxyglutaric Aciduria

23.7.1 Clinical Presentation

Patients with D-2-hydroxyglutaric aciduria exhibit a more variable phenotype than patients with L-2-hydroxyglutaric aciduria. The clinical spectrum varies from neonatal onset,

severe seizures, lack of psychomotor development and early death [43] to mild developmental delay and no symptoms at all [44]. An international survey of 17 patients revealed a continuous spectrum between these extremes with most patients suffering from a severe early onset epileptic encephalopathy, while a substantial subgroup showed mild symptoms or were even asymptomatic [45]. Clinical and neuroradiological symptoms of the severely affected patients were quite uniform. Severe, often intractable seizures started in early infancy. The babies were severely hypotonic. Conscious levels varied from irritability to stupor. Cerebral visual failure was uniformly present. Psychomotor development appeared almost absent. A third of the severely affected patients suffered from cardiomyopathy. Less severely affected patients exhibited mostly mild neurological symptoms including slight developmental delay, delayed speech and febrile convulsions.

In the severely affected patients neuroimaging uniformly revealed disturbed and delayed gyration, myelination and opercularization, ventriculomegaly, more pronounced of the occipital horns, and cysts over the head of the caudate nucleus (❏ Fig. 23.2g). Enlarged prefrontal spaces and subdural effusions in some patients were further reminiscent of the neuroimaging findings in glutaric aciduria type I.

23.7.2 Metabolic Derangement

Patients show highly elevated levels of D-2-hydroxyglutaric acid in all body fluids with no apparent correlation to the clinical phenotype. In addition Krebs cycle intermediates are found elevated in the urine of some patients, as well as GABA in CSF [45]. The disorder has recently been shown to be due to a deficiency of D-2-hydroxyglutaric acid dehydrogenase, an enzyme that converts D-2-hydroxyglutaric acid to 2-oxoglutaric acid [46, 47, 47a]. The enzyme is homologous to FAD-dependent D-lactate dehydrogenase. The origin of D-2-hydroxyglutaric acid is still not completely resolved. It might be formed from 2-oxoglutaric acid as part of a metabolic cycle, yet to be described, or arise as an intermediate in the conversion of 5-aminolevulinic acid to 2-oxoglutaric acid [43, 44]. The neurodegeneration in D-2-hydroxyglutaric aciduria could be linked to an excitotoxic sequence. D-2-hydroxyglutaric acid directly activates N-methyl-D-aspartate (NMDA) receptors, and in addition significantly increased cellular calcium levels and inhibited ATP synthesis, but without affecting the electron-transferring complexes I–IV [21].

23.7.3 Genetics

D-2-hydroxyglutaric aciduria is an autosomal recessive disorder. Pathogenic mutations have been found in mildly as well as in severely affected patients. Heterozygotes display no detectable clinical or biochemical abnormalities, but can now be ascertained by molecular diagnosis in families known to be at risk [47].

23.7.4   Diagnostic Tests

D-2-hydroxyglutaric acid is found elevated from 120–26 000 mmol/mol of creatinine (controls <17) in urine, from 3–660 µmol/l (controls <0.9) in plasma, and from 3–320 µmol/l (controls <0.34) in cerebrospinal fluid [42, 45]. In addition, GABA was found elevated in CSF and intermediates of energy metabolism in urine (lactic, succinic, malic, and 2-oxoglutaric acids) in some patients. Differentiation between the two isomers of 2-hydroxyglutarate is essential for diagnosis. *Prenatal* diagnosis has been successfully performed by accurate determination of D-2-hydroxyglutarate by stable isotope dilution GC-MS assay in amniotic fluid ([42], C. Jacobs and M.H. Rhead, personal communication) as well as by molecular diagnosis.

A few patients with combined D- and L-2-hydroxyglutaric aciduria have been described [49]. It is unclear if they represent a third clinical and/or biochemical entity. However, in these patients *prenatal* diagnosis is not reliable using metabolite determination by stable isotope dilution GC-MS assay in amniotic fluid ([42], C. Jakobs, personal communication).

D-2-Hydroxyglutaric acid can also be elevated in multiple acyl-CoA dehydrogenase deficiency (glutaric aciduria type II), but which can be distinguished by the classical urine organic acid profile found in the latter disorder (► Chap. 13).

23.7.5   Treatment and Prognosis

To date there is no rational therapy for D-2-hydroxyglutaric aciduria. Attempts of riboflavine and L-carnitine supplementation had no benefits. Seizures can be very difficult to control, and patients have died early with profound developmental delay. In general the clinical course does not appear progressive, if affected children do not develop an early onset epileptic encephalopathy.

23.8   *N*-Acetylaspartic Aciduria (Canavan Disease)

23.8.1   Clinical Presentation

N-Acetylaspartic aciduria mostly manifests at 2–4 months of age with head lag, hypotonia and macrocephaly, progressing to marked developmental delay, seizures, optic nerve atrophy, progressive spasticity and opisthotonic posturing [50]. At birth the head circumference may not be remarkably increased; however, in the majority of cases it increases pathologically after six months of age crossing the percentiles with obvious macrocephaly by 1 year. In the second year of life seizures often develop together with irritability and sleep disturbance. Hypotonicity gives way to spasticity reminiscent of cerebral palsy. Impaired chewing and swallowing, problems with gastroesophageal reflux, vomiting and aspiration can result in recurrent infections and failure to thrive. Death usually occurs in a few years although survival in a vegetative state or near vegetative state may extend to the second decade.

The most consistent findings on magnetic resonance imaging (MRI) studies are diffuse abnormalities of white matter. Although not always present and not uniform [51], MRI usually shows symmetric diffuse low signal intensity on T1-weighed images and high signal intensity on T2-weighed images (◘ Fig. 23.2h).

Like in L-2-hydroxyglutaric aciduria, the neuropathology of Canavan disease is characterized by a progressive loss of myelinated arcuate fibers [52]. Detailed histopathological descriptions at autopsy are available. White matter is characteristically soft and gelatinous. The spongy or vacuolization changes are clearly seen in the lower layers of the gray matter and in the subcortical white matter, with the more central white matter relatively spared.

Most patient follow the described disease course, which is also termed the infantile form. Rare clinical variants with different disease courses were described as congenital, i.e. presenting at or shortly after birth, or as juvenile forms, i.e. presenting after 5 years of age [51].

23.8.2   Metabolic Derangement

The disease is caused by aspartoacylase deficiency leading to the accumulation of N-acetylaspartic acid in brain, CSF, plasma, and urine.

23.8.3   Genetics

N-Acetylaspartic aciduria is transmitted in an autosomal recessive manner. It is a pan-ethnic disease with a higher frequency among Askenazi Jews, most of whom carry two specific mutations, a missense mutation, E285A, accounting for 84%, and a nonsense mutation, Y231X, accounting for 13.4% [54]; the frequency of these two mutations makes carrier screening possible [55]. In non-Jewish patients the mutations are diverse and mostly private.

23.8.4   Diagnostic Tests

The diagnosis is best established by determining *N*-acetylaspartic acid in the urine by organic acid analysis. Hundred-fold elevations are pathognomonic for *N*-acetylaspartic aciduria, which can be confirmed by demonstrating the enzyme deficiency in fibroblasts and/or mutation analysis. Borderline elevated levels of *N*-acetylaspartic acid are sometimes found in different cases of white matter disease and can cause diagnostic confusion. Prenatal diagnosis can be problematic, as the assay of aspartoacylase in amniocytes is not reliable [53]. A combination of mutation analysis together with the exact quantitation of *N*-acetylaspartic acid in the amniotic fluid is recommended.

23.8.5   Treatment and Prognosis

No rational treatment exists for *N*-acetylaspartic aciduria. Management is symptomatic and palliative. One focus is physical therapy to prevent contractures. Special care is needed to avoid aspiration with feedings. Many patients need tube or gastrostomy feeding. Seizures can be difficult to control. The prognosis is grim with death usually occurring in the first decade of life.

References

1. Hoffmann GF, Gibson KM, Trefz FK et al (1994) Neurological manifestations of organic acid disorders. Eur J Pediatr 153[Suppl 1]:S94-100
2. Kölker S, Mayatepek E, Hoffmann GF (2002) White matter disease in cerebral organic acid disorders: clinical implications and suggested pathomechanisms. Neuropediatrics 33:225-231
3. Dancis J, Hutzler J, Ampola MG et al (1983) The prognosis of hyperlysinemia: an interim report. Am J Hum Genet 35:438-442
4. Sacksteder KA, Biery BJ, Morell JC et al (2000) identification of the α-aminoadipic semialdehyde synthase gene, which is defective in familial hyperlysinemia. Am J Hum Genet 66:1736-1743
5. Carson NAJ (1969) Sacccharopinuria: a new inborn error of lysine metabolism. In: Allen JD, Holt KS, Ireland JT, Pollitt RJ (eds) Enzymopenic anemias, lysosomes, and other papers. Proc 6th Symposium of SSIEM. Livingstone, Edinburgh, pp 163-173
6. Przyrembel H (2002) Disorders of ornithine, lysine and tryptophan. In: Blau N, Duran M, Blaskovics M, Gibson KM (eds) Physician's guide to the laboratory diagnosis of inherited metabolic disease. Chapman & Hall, London, pp 277-300
7. Dancis J, Hutzler J, Cox RP (1979) Familial hyperlysinemia: enzyme studies, diagnostic methods, comments on terminology. Am J Hum Genet 31:290-299
8. Goodman SI, Browder JA, Hiles RA, Miles BS (1972) Hydroxylysinemia. A disorder due to a defect in the metabolism of free hydroxylysin. Biochem Med 6:344-354
9. Przyrembel H, Bachmann D, Lombeck I et al (1975) Alpha-ketoadipic aciduria, a new inborn error of lysine metabolism; biochemical studies. Clin Chim Acta 58:257-269
10. Casey RE, Zaleski WA, Philp M, Mendelson IS (1978) Biochemical and clinical studies of a new case of α-aminoadipic aciduria. J Inherit Metab Dis 1:129-135
11. Baric I, Zschoke J, Christensen E et al (1998) Diagnosis and management of glutaric aciduria type I. J Inherit Metab Dis 21:326-340
12. Bjugstad KB, Goodman SI, Freed CR (2000) Age at symptom onset predicts severity of motor impairment and clinical outcome of glutaric acidemia type I. J Pediatr 137:681-686
13. Busquets C, Merinero B, Christensen E et al (2000) Glutaryl-CoA dehydrogenase deficiency in Spain: evidence of two groups of patients, genetically, and biochemically distinct. Pediatr Res 48:315-322
14. Hoffmann GF, Athanassopoulos S, Burlina A et al (1996) Clinical course, early diagnosis, treatment and prevention of disease in glutaryl-CoA dehydrogenase deficiency. Neuropediatrics 27:115-123
15. Kyllerman M, Skjeldal O, Christensen E et al (2004) Long-term follow-up, neurological outcome and survival rate in 28 Nordic patients with glutaric aciduria type 1. Eur J Paediatr Neurol 8:121-129
16. Naughten ER, Mayne PD, Monavari AA et al (2004) Glutaric aciduria type I, outcome in the Republic of Ireland. J Inherit Metab Dis 27:917-920
17. Strauss KA, Puffenberger EG, Robinson DL, Morton DH (2003) Type I glutaric aciduria, part 1: natural history of 77 patients. Am J Med Genet 121C:38-52
17a. Kölker S, Garbade SF, Greenberg CR et al (2006) Natural history, outcome and treatment efficacy in children and adults with glutaryl-CoA dehydrogenase deficiency. Pediatr Res 59:840-847
18. Morris AAM, Hoffmann GF, Naughten ER et al (1999) Glutaric aciduria and suspected child abuse. Arch Dis Childh 80:404-405
19. Bähr O, Mader I, Zschocke J et al (2002) Adult onset glutaric aciduria type I presenting with leukoencephalopathy. Neurology 59:1802-1804
20. Külkens S, Harting I, Sauer S et al (2005) Late-onset neurologic disease in glutaryl-CoA dehydrogenase deficiency. Neurology 64:2142-2144
21. Kölker S, Greenberg C, Lindner M et al (2004) Emergency treatment in glutaryl-CoA dehydrogenase deficiency. J Inherit Metab Dis 27:893-902
22. Kölker S, Koeller DM, Okun JG, Hoffmann GF (2004) Pathomechanisms of neurodegeneration in glutaryl-CoA dehydrogenase deficiency. Ann Neurol 55:7-12
23. Stokke O, Goodman SI, Moe PG (1976) Inhibition of brain glutamate decarboxylase by glutarate, glutaconate, and β-hydroxyglutarate: Explanation of the symptoms in glutaric aciduria? Clin Chim Acta 66:411-415
24. Heyes MP (1987) Hypothesis: A role for quinolinic acid in the neuropathology of glutaric aciduria type I. Can J Neurol Sci 14:441-443
25. Baric I, Wagner L, Feyh P, Liesert M, Buckel W, Hoffmann GF (1999) Sensitivity and specificity of free and total glutaric and 3-hydroxyglutaric acids measurements by stable isotope dilution assays for the diagnosis of glutaric aciduria type I. J Inherit Metab Dis 22:867-882
26. Nyhan WL, Zschocke J, Hoffmann GF et al (1999) Glutaryl-CoA dehydrogenase deficiency presenting as 3-hydroxyglutaric aciduria. Mol Genet Metab 66:199-204
27. Lindner M, Kölker S, Schulze A et al (2004) Neonatal screening for glutaryl-CoA dehydrogenase deficiency. J Inherit Metab Dis 27:851-859
28. Greenberg CR, Prasad AN, Dilling LA et al (2002) Outcome of the first three years of a DNA-based neonatal screening program for glutaric acidemia type 1 in Manitoba and Northwestern Ontario, Canada. Mol Gen Metabol 75: 70-78
29. Schulze-Bergkamen A, Okun JG, Spiekerkötter U et al (2005) Quantitative acylcarnitine profiling in peripheral blood mononuclear cells using in vitro loading with palmitic and 2-oxoadipic acids: Biochemical confirmation of fatty acid oxidation and organic acid disorders. Pediatr Res 58:873-880
30. Christensen E (1993) A fibroblast glutaryl-CoA dehydrogenase assay using detritiation of ^{3}H-labelled glutaryl-CoA: application in the genotyping of the glutaryl-CoA dehydrogenas locus. Clin Chim Acta 220:71-80

31. Goodman SI, Stein DE, Schlesinger S et al (1998) Glutaryl-CoA dehydrogenase mutations in glutaric acidemia (type I): Review and report of thirty novel mutations. Hum Mutat 12:141-144

32. Burlina AP, Zara G, Hoffmann GF et al (2004) Management of movement disorders in glutaryl-CoA dehydrogenase deficiency. Anticholinergic drugs and botulinum toxin as additional therapeutic options. J Inherit Metab Dis 27:911-915

33. Barth PG, Hoffmann GF, Jaeken J et al (1993) L-2-Hydroxyglutaric acidemia: clinical and biochemical findings in 12 patients and preliminary report on L-2-hydroxyacid dehydrogenase. J Inherit Metab Dis 16:753-761

34. de Klerk JBC, Huijmans JGM, Stroink H et al (1997) L-2-Hydroxyglutaric aciduria: clinical heterogeneity versus biochemical homogeneity in a sibship. Neuropediatrics 28:314-317

35. Barbot C, Fineza I, Diogo L et al (1997) L-2-Hydroxyglutaric aciduria: clinical, biochemical and magnetic resonance imaging in six Portuguese pediatric patients. Brain Dev 19:268-273

36. Moroni I, D‹Incerti L, Farina L et al (2000) Clinical, biochemical and neuroradiological findings in L-2-hydroxyglutaric aciduria. Neurol Sci 21:103-108

37. Chen E, Nyhan WL, Jakobs C et al (1996) L-2-Hydroxyglutaric aciduria: neuropathological correlations and first report of severe neurodegenerative disease and neonatal death. J Inherit Metab Dis 19:335-343

38. D'Incerti L, Farina L, Moroni I et al (1998) L-2-Hydroxyglutaric aciduria: MRI in seven cases. Neuroradiology 40:727-733

39. Hoffmann GF, Jakobs C, Holmes B et al (1995) Organic acids in cerebrospinal fluid and plasma of patients with L-2-hydroxyglutaric aciduria. J Inherit Metab Dis 18:189-193

40. Rzem R, Veiga-da-Cunha M, Noel G et al (2004) A gene encoding a putative FAD-dependent L-2-hydroxyglutarate dehydrogenase is mutated in L-2-hydroxyglutaric aciduria. Proc Natl Acad Sci U S A 101:16849-16854

41. Topcu M, Jobard F, Halliez S et al (2004) L-2-Hydroxyglutaric aciduria: identification of a mutant gene C14orf160, localized on chromosome 14q22.1. Hum Mol Genet 13:2803-2811

42. Gibson KM, Schor DSM, Kok RM et al (1993) Stable-isotope dilution analysis of D- and L-2-hydroxyglutaric acid: application to the detection and prenatal diagnosis of D- and L-2-hydroxyglutaric acidemias. Pediatr Res 34:277-280

43. Gibson KM, Craigen W, Herman GE, Jakobs C (1993) D-2-Hydroxyglutaric aciduria in a newborn with neurological abnormalities: a new neurometabolic disorder? J Inherit Metab Dis 16:497-500

44. Nyhan WL, Shelton GD, Jakobs C et al (1995) D-2-Hydroxyglutaric aciduria. J Child Neurol 10:137-142

45. van der Knaap MS, Jakobs C, Hoffmann GF et al (1999) D-2-Hydroxyglutaric aciduria. Biochemical marker or clinical disease entity? Ann Neurol 45:111-119

46. Achouri Y, Noel G, Vertommen D et al (2004) Identification of a dehydrogenase acting on D-2-hydroxyglutarate. Biochem J 381:35-42

47. Struys EA, Salomons GS, Achouri Y et al (2005) Mutations in the D-2-hydroxyglutarate dehydrogenase gene cause D-2-hydroxyglutaric aciduria. Am J Hum Genet 76:358-360

47a. Struys EA (2006) D-2-Hydroxyglutaric aciduria: unravelling the biochemical pathway and the genetic defect. J Inherit Metab Dis 29:21-29

48. Kölker S, Pawlak V, Ahlemeyer B et al (2002) NMDA receptor activation and respiratory chain complex V inhibition contribute to neurodegeneration in D-2-hydroxyglutaric aciduria. Eur J Neurosci 16:21-28

49. Muntau AC, Röschinger W, Merkenschlager A et al (2000) Combined D-2 and L-2-hydroxyglutaric aciduria with neonatal onset encephalopathy: A third biochemical variant of 2-hydroxyglutaric aciduria? Neuropediatrics 31:137-140

50. Matalon R, Michals K, Kaul R (1995) Canavan disease: from spongy degeneration to molecular analysis. J Pediatr 127:511-517

51. Toft PB, Geiß-Holtorff R, Rolland MO et al (1993) Magnetic resonance imaging in juvenile Canavan disease. Eur J Pediatr 152:750-753

52. Brismar J, Brismar G, Gascon G, Ozand P (1990) Canavan disease: CT and MR imaging of the brain. Am J Neuroradiol 11:805-810

53. Besley GTN, Elpeleg ON, Shaag A et al (1999) Prenatal diagnosis of Canavan disease: Problems and dilemmas. J Inherit Metab Dis 22:263-266

54. Matalon R, Michals-Matalon K (1998) Molecular basis of Canavan disease. Eur J Paediatr Neurol 2:69-76

55. Howell VM, Proos AL, LaRue D et al (2004) Carrier screening for Canavan disease in Australia. J Inherit Metab Dis 27:289-290

24 Nonketotic Hyperglycinemia (Glycine Encephalopathy)

Olivier Dulac, Marie-Odile Rolland

Glycine Metabolism

Glycine, the simplest of the amino acids, is abundant in nearly all animal proteins and enters into more biosynthetic routes than any other. Formation of glycine conjugates plays an important role in the detoxification of various compounds, including those that accumulate in certain inborn errors of metabolism. The catabolism of glycine involves several pathways, among which the glycine-cleavage system (GCS) is of major importance. This multienzyme complex degrades glycine into NH_3 and CO_2 and, thereby, also converts tetrahydrofolate into 5,10-methylene tetrahydrofolate. The latter compounds are also involved in the interconversion of serine and glycine, catalyzed by serine hydroxymethyl transferase.

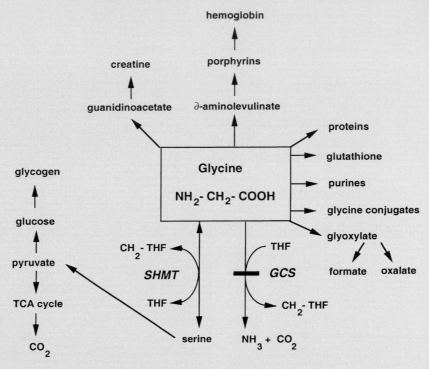

Fig. 24.1. Pathways of glycine metabolism. *CH₂-THF*, 5,10-methylene tetrahydrofolate; *GCS*, glycine cleavage system; *SHMT*, serine hydroxymethyl transferase; *TCA*, tricarboxylic acid; *THF*, tetrahydrofolate. Enzyme defect is depicted by *solid bar*

Nonketotic hyperglycinemia (NKH) or glycine encepha-lopathy is an autosomal recessive disorder character-ized by a rapidly progressive course in the neonatal period or early infancy. Symptoms include muscular hypotonia, seizures, apneic attacks, lethargy and coma. Most patients die within a few weeks, whereas survi-vors show severe psychomotor retardation. Increased glycine concentrations in plasma, urine, and cerebro-spinal fluid are biochemical features of the disorder. The primary lesion is a defect in the glycine cleavage system (GCS) (◘ Fig. 24.1). Although this was first de-monstrated in the liver, involvement within the brain is responsible for the clinical expression. No specific treat-ment is available. Prenatal diagnosis is possible by deter-mining the activity of GCS in chorionic villi.

24.1   Clinical Presentation

Nonketotic hyperglycinemia (NKH) is usually classified into two main clinical types: neonatal and late-onset. The *neonatal type* is the most common. In the series of Tada and Kure [1], 28 of 32 cases (87%) presented in the newborn period.

24.1.1   Neonatal NKH

The phenotype is characteristic so that the diagnosis can be suspected on the basis of the clinical and electroencephalo-graphic (EEG) features. A severe encephalopathy, without ketosis or acidosis, occurs after an apparent symptom free interval although some infants have brain malformations such as dysgenesis of the corpus callosum and gyral mal-formations [2] which may be detected prenataly [3] or evi-dent at birth. Most, however, appear normal at birth but within the first few hours develop a progressive encepha-lopathy characterised by lethargy, axial and limb hypotonia, and a depressed Moro response. Respiration becomes in-creasingly irregular culminating in apneic attacks, by which time the infant is in a deep coma and exhibits myoclonus, and eventually tonic or clonic seizures. Hiccups are often present.

Initially the EEG shows a burst-suppression (BS) pat-tern, consisting of periods of high amplitude activity lasting 1–3 Hz that arise periodically from a hypoactive back-ground (◘ Fig. 24.2) and without spacio-temporal differ-entiation. The bursts are mostly asynchronous over both hemispheres and comprise irregular slow waves, sharp waves and spikes. BS is present from birth and precedes clinical symptoms. It disappears by the end of the first month of life, changing into hypsarrhythmia. Magnetic resonance imaging (MRI) shows progressive cortical atrophy, callosal thinning, and delayed myelination, par-ticularly in the parietal lobes, with high signal in the pyra-midal tracts, middle cerebellar pedicles and dentate nuclei. The development of acute hydrocephalus requiring shunt-ing in early infancy has been reported [4]. On magnetic resonance spectroscopy, lactate and creatine are increased, and levels of N-acetylaspartate and myoinositol-glycine may be prognostic indicators [5]. Pulmonary hypertension has been documented in 4 patients [6].

Neonatal hypotonia gradually evolves into spasticity. Most patients die at between 6 days and 5 years of age. Neu-ropathology shows spongiosis and myelin vacuolation. The patients who survive are severely retarded [7].

Transient neonatal NKH has been described in a few newborns with symptoms indistinguishable from those of the neonatal type [8, 9]. Plasma and cerebrospinal fluid (CSF) glycine concentrations are initially elevated to those seen in neonatal NKH but return to normal by 8 weeks of age. Five of six patients had no neurological sequelae after 6 months to 13 years follow-up. However, one had severe developmental delay at 9 months of age. A favourable out-come in this group of patients may be related to a high re-sidual activity of the glycine cleavage system (GCS): 32% of wild type for the A802V mutation in the GLDC gene [10] or heterozygocity [11], and vigorous therapeutic interven-tion in the neonatal period, when the brain is particularly vulnerable to glycine. One variant with the homozygous A802V mutation was characterized by the persistence of metabolic abnormalities, although the clinical condition improved dramatically in the patients receiving sodium benzoate in the neonatal period [10].

24.1.2   Late Onset NKH

The *late-onset type* patients have no abnormal symptoms or signs in the neonatal period but thereafter develop non-specific neurological symptoms to varying degrees. The age of onset ranges from infancy to late adulthood. Three Japanese cases, all retaining 6-8% activity of the GCS, exhib-ited infantile hypotonia, mental retardation and episodes of disturbed behaviour in childhood and adolescence (aggres-siveness and intractable Attention Deficit Hyperactivity Disorder, ADHD) [12]. Several cases suffered from pulmo-nary hypertension from the first months of life, with rapidly progressive neurological deterioration following vaccina-tion, including hypotonia, pyramidal signs and loss of cog-nitive functions, but neither seizures nor EEG abnormalities [13]. Cranial MRI showed bilateral cystic leukodystrophy with thalamic involvement, extending from frontal to oc-cipital lobes. These patients died before 2 years of age; their neuropathology was somewhat similar to vanishing white matter with a cavitated leukoclastic encephalopathy, involv-ing both hemispheres with preserved U fibres and vacu-olated demyelination involving the corpus callosum, cere-

Fig. 24.2. Electroencephalogram of a 6-week-old patient with nonketotic hyperglycinemia, showing a »burst-suppression« pattern

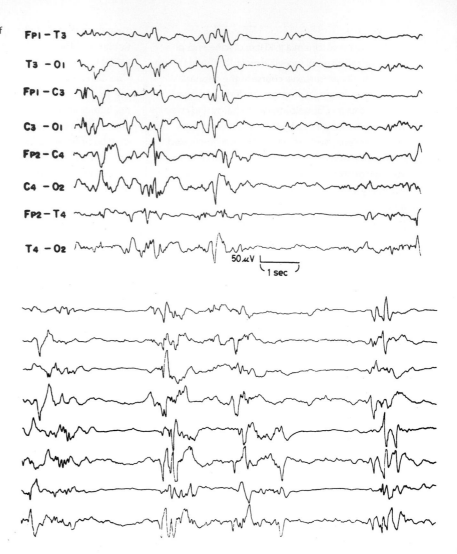

bellar peduncules, medial lemniscus and pyramidal tracts [13, 14]. One patient with late onset nonketotic hyperglycinemia underwent pregnancy with success and the neuropsychometric values of her offspring were average [15]. One 66-year-old woman developed mental deterioration in school age, and gait disturbance with dysarthria and bradykinesia in her 40s. MRI showed hypoplasia of the corpus callosum with cerebral and cerebellar atrophy. Elevated values of glycine were found in blood, CSF and urine [16].

24.2   Metabolic Derangement

The glycine cleavage system (GCS), a mitochondrial enzyme complex, is made of four individual constituents. These are a *P-protein* (pyridoxal phosphate dependent glycine decarboxylase, GLDC), a *T-protein* (tetrahydrofolate requiring aminomethyltransferase, AMT), an *H-protein* (glycine cleavage system hydrogen carrier protein, GCSH, containing lipoic acid), and an *L-protein*, lipoamide dehydrogenase. These four specific proteins allow the degradation of glycine in the liver, kidney and brain (Fig. 24.1). From the study of a large number of patients, it seems that most patients with the neonatal form of the disease have a very low GCS activity and that late onset patients have some residual activity. The overall activity measured in vitro is usually, but not always, lower in P-protein deficiency than in T-protein deficiency. The mechanism for transient NKH is unclear, because of the lack of enzymatic data. Immaturity of one or more of the components of the GCS or deficiency of any of its cofactors is postulated. Since 2000, the glycine-CO_2 exchange reaction, a new assay performed on tissue obtained from liver biopsy, has allowed identification of the deficient protein [17]. About 75% of patients with NKH have a defect in the P-protein, and the remainder have a defect in the T-protein. L- or H-protein deficiencies are apparently very rare. Initial classification into probable P- or T-protein defects allows a rational search for mutations in the appropriate gene.

The pathophysiologic mechanism(s) of glycine encephalopathy remain obscure. They may be related to the role of glycine as an adjuvant to the N-methyl-D-aspartate (NMDA) receptor, which plays a major role in ontogenesis of the brain cortex (for a review, see [18]): the GCS is abundant in rat neuronal stem cells in the neuroepithelium [19]. Glycine also plays a major role in the developing brain's neuronal excitability [20] and could account for the myoclonic epileptic encephalopathy. Indeed, the 3 other presently identified conditions that produce neonatal myoclonus with BS are also likely to activate of the NMDA receptor by causing an increase of glutamate concentration in the synaptic cleft: pyridoxine and pyridoxal phosphate deficiencies that prevent the transformation of glutamate into gamma-amino-butyric acid (GABA) and therefore increase the level of glutamate, and the glutamate transporter defect [38]. In experimental models, a slight increase in glutamate when GABA is still excitatory has devastating consequences with the development of a pattern similar to neonatal myoclonic encephalopathy [20a]. In the spinal cord, glycine is the major inhibitory neurotransmitter leading to hypotonia [21].

24.3   Genetics

NKH is transmitted as an autosomal recessive trait. The AMT, GLDC and GCSH genes have been cloned [22, 23]. In the AMT gene, three recurrent mutations, IVS 7-1G>A, R320H and 296H, are found in about 10% of the deficient alleles. Approximately 30 other AMT mutations have been identified in single cases [24]. About 50 different mutations of the GLDC gene have been collected to date by Applegarth and Toone [24], including large deletions, missense, nonsense, splicing site and frame shift mutations. Most of these are private [25]. In Finland the majority of patients carry a S564I mutation [26]. Only R515S, T269M and A389V mutations were found in a few alleles from patients tested in Europe and Canada; all others are private. Only one mutation in the GCSH gene has been reported [27]. Patients in whom no mutation or only one mutation has been found, despite sequencing of the P-, T- and H-protein genes, have been reported [28]. Available data suggest that in nonconsanguineous families the patient is likely to be a compound heterozygote. Since most reported mutations seem to be rare or private, phenotype could not be predicted from genotype.

24.4   Diagnostic Tests

When NKH is suspected clinically, plasma amino acids should be analysed, in the absence of valproate treatment. If an isolated elevation of glycine is found (control values 125–320 μmol/l), an organic acidemia with ketotic hyperglycinemia (most commonly propionic or methylmalonic acidemia) must be excluded by urinary organic acid and/or plasma acylcarnitine analysis. If no abnormal metabolites are found, glycine level should then be measured simultaneously in plasma and in CSF (provided the CSF sample is non traumatic since cells would increase glycine values; control values < 10 μmol/l). In NKH all other amino acids are unremarkable, remaining within normal values. The diagnosis of NKH is based on finding of either an increased absolute value of glycine in CSF or an increased CSF to plasma glycine ratio (control values <0.02). In classical neonatal NKH this ratio is very high (>0.08), whereas it is only slightly elevated (0.04–0.10) or even normal in late onset, milder or atypical cases (Rabier, personal communication). However, no prediction of the deficient protein or of the prognosis can be derived from these values. In all cases, the diagnosis requires consistent clinical and biochemical findings. Discordant cases require enzymatic confirmation.

In order to confirm the diagnosis by measurement of overall GCS activity and to identify the deficient protein by the glycine-CO_2 exchange reaction, a liver biopsy with about 80 mg of tissue is necessary. Neither fibroblasts nor leucocytes can be used for these assays. However, overall GCS activity is detectable in lymphoblasts obtained from B lymphocytes infected and transformed using Epstein-Barr virus [29]. When the deficient protein has been identified, sequencing of the coding region and the intron/exon boundaries can be performed on the appropriate gene.

The identification of carriers is unreliable by enzymatic analysis in lymphoblasts, and can only be done by molecular genetic testing, once the mutations have been identified in the proband.

24.5   Differential Diagnosis

Burst-suppression (BS) is not specific to NKH. When occurring with myoclonus, most often it remains undiagnosed although there is familial recurrence in over 10% of cases [30]. BS was first reported as an acute event in ischemic encephalopathy [31, 32], and may be difficult to distinguish from fragmented hypsarrhythmia when it occurs only in sleep [33]. The combination of neonatal seizures with BS is fairly frequent and its etiology is variable, ranging from inborn errors of metabolism to malformations [34]. The presence of fragmentary and erratic myoclonus is a major component of metabolic causes of BS, but it may be missing as can be the case sometimes in glycine encephalopathy. The suppression component of BS with metabolic disease is often particularly long, up to 20 seconds [34].

BS is also a frequent finding in pyridoxine dependency [35] and part of the clinical pattern of pyridox(am)ine phosphate oxidase deficiency, a condition sensitive to pyridoxal phosphate although pyridoxine has no effect [36]. Both pyridoxine dependency and pyridox(am)ine phosphate oxidase deficiency may be associated with a transient raise in CSF glycine [37] which may lead to diagnostic confusion with NKH. In one consanguinous family, three siblings developed neonatal myoclonus and for one the EEG pattern was consistent with BS at the age of 3 years [38]. On the other hand, BS has been occasionally described in rare cases of D-glyceric, methylmalonic and propionic acidemias, sulfite and xanthine oxidase deficiencies, carbamyl phosphate synthetase deficiency with hyperammonemia, and Menkes disease [34, 39, 40]. Malformations associated with BS include Aicardi malformation, schizencephaly, porencephaly, hemimegalencephaly [39, 35] and olivo-dentate dysplasia [34, 41]. However, these are most often associated with tonic seizures rather than myoclonus.

On MRI, a particular leucodystrophy called vanishing white matter may be confused with NKH since in both conditions glycine may be increased to similar values [42]. Diffusion sequence shows low apparent diffusion quotient values [43].

24.6   Prenatal Diagnosis

Since no effective treatment is available for NKH, prenatal diagnosis is frequently requested. GCS activity is present in chorionic villi, but not in amniocytes or chorionic villi in culture. More than 500 prenatal diagnoses have been performed by measuring GCS activity on crude chorionic villi tissue. False, unexplained negative results have been reported in about 1% of cases [44] with mutations in AMT or GLDC genes found subsequently in 4 cases. Moreover, there is an uninterpretable range where low normal control values and affected foetuses with residual activity overlap [45]. Molecular diagnosis can now be offered when the proband and parents have been investigated and mutations identified. This analysis is performed on DNA extracted from foetal cells obtained by either amniocentesis (14–16 weeks) or chorionic villus sampling (10–12 weeks). If only one mutation is known, the affected gene may be identified and linkage analysis using linkage markers on the chromosome can be undertaken. Measurement of amniotic fluid glycine concentration and the glycine/serine ratio is unreliable because control and affected values overlap [46].

24.7   Treatment

Treatment with sodium benzoate is usually ineffective [7], with the possible exception of rare transient cases [10]. Pantothenic acid administration has been proposed, because it is the precursor of coenzyme A that activates sodium benzoate [47]. In a late onset case, a combination of a low-protein diet, sodium benzoate and imipramine was reported to have been effective [48].

The probable contribution of NMDA receptor activation has promoted therapeutic trials with compounds that reduce NMDA transmission, such as dextrometorphan,

tryptophan and ketamine [49]. However, the latter were disappointing, probably because the epileptic encephalopathy begins long before birth.

Acknowledgement. We are grateful to Daniel Rabier for useful suggestions regarding the biochemical aspects.

References

1. Tada K, Kure S (1993) Non-ketotic hyperglycinaemia: molecular lesion, diagnosis and pathophysiology. J Inherit Metab Dis 16:691-703
2. Dobyns WB (1989) Agenesis of the corpus callosum and gyral malformations are frequent manifestations of nonketotic hyperglycinemia. Neurology 39:817-820
3. Paupe A, Bidat L, Sonigo P et al (2002) Prenatal diagnosis of hypoplasia of the corpus callosum in association with non-ketotic hyperglycinemia. Ultrasound Obstet Gynecol 20:616-619
4. Hoover-Fong JE, Shah S, Van Hove JL et al (2004) Natural history of nonketotic hyperglycinemia in 65 patients. Neurology 63:1847-1853
5. Huisman TA, Thiel T, Steinmann B et al (2002) Proton magnetic resonance spectroscopy of the brain of a neonate with nonketotic hyperglycinemia: in vivo-in vitro (ex vivo) correlation. Eur Radiol 12:858-861
6. Cataltepe S, van Marter LJ, Kozakewich H et al (2000) Pulmonary hypertension associated with nonketotic hyperglycinaemia. J Inherit Metab Dis 23:137-144
7. Chien YH, Hsu CC, Huang A et al (2004) Poor outcome for neonatal-type nonketotic hyperglycinemia treated with high-dose sodium benzoate and dextromethorphan. J Child Neurol 19:39-42
8. Luder AS, Davidson A, Goodman SI, Greene CL (1989) Transient nonketotic hyperglycinemia in neonates. J Pediatr 114:1013-1015
9. Zammarchi E, Donati MA, Ciani F (1995) Transient neonatal non-ketotic hyperglycinemia: a 13-year follow-up. Neuropediatrics 26:328-330
10. Korman SH, Boneh A, Ichinohe A et al (2004) Persistent NKH with transient or absent symptoms and a homozygous GLDC mutation. Ann Neurol 56:139-143
11. Kure S, Kojima K, Ichinohe A et al (2002) Heterozygous GLDC and GCSH gene mutations in transient neonatal hyperglycinemia. Ann Neurol 52:643-646
12. Dinopoulos A, Kure S, Chuck G et al (2004) Atypical non-ketotic hyperglycinemia: 3 cases with GLDC mutations. J Inherit Metab Dis 27[Suppl 1]:62
13. Del Toro M, Macaya A, Moreno et al (2004) Rapidly progressive infantile leukoencephalopathy associated with nonketotic hyperglycinemia and pulmonary hypertension. J Inherit Metab Dis 27 [Suppl 1]:61
14. Bekiesiniska-Figatowska M, Rokicki D, Walecki J (2001) MRI in nonketotic hyperglycinaemia: case report. Neuroradiology 43:792-793
15. Ellaway CJ, Mundy H, Lee PJ (2001) Successful pregnancy outcome in atypical hyperglycinaemia. J Inherit Metab Dis 24:599-600
16. Hasegawa T, Shiga Y, Matsumoto A et al (2002) [Late-onset nonketotic hyperglycinemia: a case report, in Japanese]. No To Shinkei 54:1068-1072
17. Toone JR, Applegarth DA, Coulter-Mackie MB, James ER (2000) Biochemical and molecular investigations of patients with nonketotic hyperglycinemia. Mol Genet Metab 70:116-121
18. Wasterlain CG, Shirasaka Y (1994) Seizures, brain damage and brain development. Brain Dev 16:279-295

19. Sato K, Yoshiada S, Fujiwara K et al (1991) Glycine cleavage system in astrocytes. Brain Res 567:64-70

20. Ben Ari Y, Khazipov R, Leinekugel X et al (1997) GABA-A, NMDA and AMPA receptors: a developmentally regulated »menage à trois«. Trends Neurosci 20:523-529

20a. Milh M, Becq H, Villeneuve N et al (2006) Inhibition of glutamate transporters results in suppression-burst pattern and partial seizures in the newborn rat. Epilepsia (in press)

21. Aprison MH, Werman R (1965) The distribution of glycine in cat spinal cord and roots. Life Sci 4(21):2075-2083

22. Kure S, Narisawa K, Tada K (1991) Structural and expression analyses of normal and mutant mRNA encoding glycine decarboxylase: three-base deletion in mRNA causes nonketotic hyperglycinemia. Biochem Biophys Res Commun 174:1176-1182

23. Hyasaka K, Nanao K, Takada G et al (1992) Isolation and sequence determination of cDNA encoding human T-protein of the glycine cleavage system. Biochem Biophys Res Commun 192:766-771

24. Applegarth DA, Toone JR (2004) Glycine encephalopathy (nonketotic hyperglycinaemia): review and update. J Inherit Metab Dis 27:417-422

25. Applegarth DA, Toone JR (2001) Nonketotic hyperglycinemia (glycine encephalopathy): Laboratory diagnosis. Mol Genet Metab 74:139-146

26. Kure S, Takayanagi M, Kurihara Y et al (1999) Nonketotic hyperglycinemia: mutation spectra of the GLDC and AMT gene in Finnish and non-Finnish populations. Am J Hum Genet 65:A2406

27. Toone JR, Applegarth DA, Kure S et al (2002) Novel mutations in the P-protein (glycine decarboxylase) gene in patients with glycine encephalopathy (non-ketotic hyperglycinemia). Mol Genet Metab 76:243-249

28. Van Hove JLK, Mahieu V, Schollen E (2004) Prognosis in nonketotic hyperglycinemia. J Inherit Metab Dis 26:71

29. Kure S, Narisawa K, Tada K (1992) Enzymatic diagnosis of nonketotic hyperglycinemia with lymphoblasts. J Pediatr 120:95-98

30. Vigevano F, Maccagnani F, Bertini E et al (1982) Encefalopatia mioclonica precoce associata ad alti livelli di acido propioico nel siero. Boll Lega It Epil 39:181-182

31. Dreyfus-Brissac C, Cukier F (1969) Le tracé paroxystique: sa valeur pronostique selon le degré de prématurité. Rev Neuropsychiatr Infant 17:795-802

32. Pampiglione G (1962) Electroencephalographic studies after cardiorespiratory resuscitation. Proc R Soc Med 55:653-657

33. Lombroso CT (1990) Early myoclonic encephalopathy, early infantile epileptic encephalopathy, and benign and severe infantile myoclonic epilepsies: a critical review and personal contributions. J Clin Neurophysiol 7:380-408

34. Schlumberger E, Dulac O, Plouin P (1992) Syndrome of neonatal epilepsy. Epilepsy syndromes in childhood and adolescence. London, Paris, Rome: Libbey J:35-42

35. Nabbout R, Soufflet C, Plouin P, Dulac O (1999) Pyridoxine dependent epilepsy: a suggestive electroclinical pattern. Arch Dis Child Fetal Neonatal Ed 81:F125-F129

36. Clayton PT, Surtees RA, DeVile C et al (2003) Neonatal epileptic encephalopathy. Lancet 361:1614

37. Maeda T, Inutsuka M, Goto K, Izumi T (2000) Transient nonketotic hyperglycinemia in an asphyxiated patient with pyridoxine-dependent seizures. Pediatr Neurol 22:225-227

38. Molinari F, Raas-Rothschild A, Rio M et al (2005) Impaired mitochondrial glutamate transport in autosomal recessive neonatal myoclonic epilepsy. Am J Hum Genet 76:334-339

39. Aukett A, Bennett MJ, Hosking GP (1988) Molybdenum co-factor deficiency: an easily missed inborn error of metabolism. Dev Med Child Neurol 30:531-535

40. Dalla Bernardina B, Dulac O, Fejerman N et al (1983) Early myoclonic epileptic encephalopathy (E.M.E.E.). Eur J Pediatr 140:248-252

41. Ohtahara S (1978) Clinico-electrical delineation of epileptic encephalopathies in childhood. Asian Med J 21:499-509

42. Van der Knaap MS, Wevers RA, Kure S et al (1999) Increased cerebrospinal fluid glycine: a biochemical marker for a leukoencephalopathy with vanishing white matter. J Child Neurol 14:728-731

43. Sener RN (2003) The glycine peak in brain diseases. Comput Med Imaging Graph 27:297-305

44. Applegarth DA, Toone JR, Rolland MO et al (2000) Non-concordance of CVS and liver glycine cleavage enzyme in three families with non-ketotic hyperglycinaemia (NKH) leading to false negative prenatal diagnoses. Prenat Diagn 20:367-370

45. Vianey-Saban C, Chevalier-Porst F, Froissart R, Rolland MO (2003) Prenatal Diagnosis of nonketotic Hyperglycinemia: a 13-year experience, from enzymatic to molecular analysis. J Inherit Metab Dis 26[Suppl 2]:82

46. Garcia-Munoz MJ, Belloque J, Merinero B et al (1989) Non-ketotic hyperglycinaemia: glycine/serine ratio in amniotic fluid – an unreliable method for prenatal diagnosis. Prenat Diagn 9:473-476

47. Palekar A (2000) Effect of panthotenic acid on hippurate formation in sodium benzoate-treated HepG2 cells. Pediatr Res 48:357-359

48. Wiltshire EJ, Poplawski NK, Harrison JR, Flechter JM (2000) Treatment of late-onset nonketotic hyperglycinaemia: effectiveness of imipramine and benzoate. J Inherit Metab Dis 23:15-21

49. Matsuo S, Inoue F, Takeuchi Y et al (1995) Efficacy of tryptophan for the treatment of nonketotic hyperglycinemia: a new therapeutic approach for modulating the N-methyl-D-aspartate receptor. Pediatrics 95:142-146

25 Disorders of Proline and Serine Metabolism

Jaak Jaeken

IV

Proline and Serine Metabolism

Proline and serine are non-essential amino acids. Unlike all other amino acids (except hydroxyproline), *proline* has no primary amino group (it is termed an imino acid) and uses, therefore, a specific system of enzymes for its metabolism (◘ Fig. 25.1). Δ1-Pyrroline 5-carboxylate (P5-C) is both the immediate precursor and the degradation product of proline. The P5-C/proline cycle transfers reducing/oxidizing potential between cellular organelles. Due to its pyrrolidine ring, proline (together with hydroxyproline) contributes to the structural stability of proteins, particularly collagen, with its high proline and hydroxyproline content.

Serine also has important functions besides its role in protein synthesis. It is a precursor of a number of compounds (partly illustrated in ◘ Fig. 25.2), including D-serine, glycine, cysteine, serine phospholipids, sphingomyelins, and cerebrosides. Moreover, it is a major source of N^5,N^{10}-methylene-tetrahydrofolate (THF) and of other one-carbon donors that are required for the synthesis of purines and thymidine. Serine is synthesized *de novo* from a glycolytic intermediate, 3-phosphoglycerate and can also be synthesized from glycine by reversal of the reaction catalyzed by serine hydroxymethyltransferase, which thereby converts N^5,N^{10}-methylene-THF into THF (▶ also Fig. 24.1).

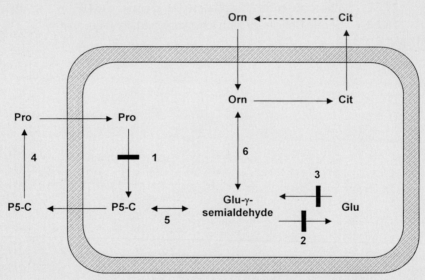

◘ **Fig. 25.1.** Proline metabolism. *Shaded area* represents mitochondrial membrane. *Cit*, citrulline; *Glu*, glutamine; *Orn*, ornithine; *Pro*, proline; *P5-C*, Δ¹-pyrroline 5-carboxylate. **1**, Proline oxidase (deficient in hyperprolinemia type 1); **2**, P5-C dehydrogenase (deficient in hyperprolinemia type 2); **3**, P5-C synthase; **4**, P5-C reductase; **5**, non-enzymatic reaction; **6**, ornithine aminotransferase (deficient in gyrate atrophy). *Bars* across arrows indicate defects of proline metabolism

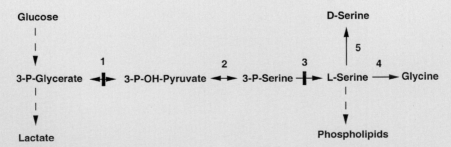

◘ **Fig. 25.2.** Pathway of de novo serine synthesis. *P*, phosphate; **1**, 3-phosphoglycerate dehydrogenase; **2**, 3-phosphohydroxy-pyruvate transaminase; **3**, 3-phosphoserine phosphatase; **4**, serine hydroxymethyltransferase (utilizes tetrahydrofolate); **5**, serine racemase. Glycine is synthesized from serine, but also from other sources. *Bars* across arrows indicate the known defects in serine synthesis

Three disorders of *proline* metabolism are known: two in its catabolism (hyperprolinemia type I due to proline oxidase deficiency and hyperprolinemia type II due to Δ1-pyrroline 5-carboxylate dehydrogenase deficiency) and one in its synthesis (Δ1-pyrroline 5-carboxylate synthase deficiency). Hyperprolinemia type I is mostly considered a non-disease, while hyperprolinemia type II seems to be associated with a disposition to recurrent seizures. The deficiency of the proline-synthesizing enzyme, $Δ^1$-pyrroline 5-carboxylate synthase, which also intervenes in ornithine synthesis, is described in ► Chap. 22.

Three disorders of *serine* metabolism are known. Two are in its biosynthesis: namely, 3-phosphoglycerate dehydrogenase deficiency and phosphoserine phosphatase deficiency. Patients with 3-phosphoglycerate dehydrogenase deficiency are affected with congenital microcephaly, psychomotor retardation and intractable seizures and partially respond to L-serine or L-serine and glycine. One patient with an association of Williams syndrome and phosphoserine phosphatase deficiency has been reported. Another, unexplained serine disorder has been reported in a patient with decreased serine in body fluids, ichthyosis and polyneuropathy but no central nervous system manifestations. There was a spectacular response to L-serine.

25.1 Inborn Errors of Proline Metabolism

25.1.1 Proline Oxidase Deficiency (Hyperprolinemia Type I)

Clinical Presentation

Hyperprolinemia type I is a very rare disorder, generally considered a benign trait but recent work suggests that it may be associated with a subset of schizophrenic patients [1–4].

Metabolic Derangement

Hyperprolinemia type I is caused by a deficiency of proline oxidase (a mitochondrial inner-membrane enzyme), which catalyses the conversion of proline into P5-C (◻ Fig. 25.1, enzyme 1). Hence, in hyperprolinemia type I, there are increased levels of proline in plasma (usually not above 2000 μM; normal range 100–450 μM), urine and cerebrospinal fluid (CSF). Hyperprolinemia (as high as 1000 μM) is also observed as a secondary phenomenon in hyperlactatemia, possibly due to inhibition of proline oxidase by lactic acid. Remarkably, and in contrast with hyperprolinemia type II, heterozygotes have hyperprolinemia.

Genetics

The mode of inheritance is autosomal recessive. *PRODH*, the gene encoding proline oxidase, maps to 22q11, in the region deleted in the velocardiofacial syndrome/DiGeorge syndrome. At least 16 missense mutations have been identified [4, 5].

Diagnostic Tests

The diagnosis is made by amino acid analysis. Direct enzyme assay is not possible, since the enzyme is not present in leukocytes or skin fibroblasts. Mutation analysis is thus necessary to confirm the diagnosis [4].

Treatment and Prognosis

Since the prognosis is generally excellent, dietary treatment is not indicated.

25.1.2 Δ¹-Pyrroline 5-Carboxylate Dehydrogenase Deficiency (Hyperprolinemia Type II)

Clinical Presentation

This is a relatively benign disorder, though a disposition to recurrent seizures is highly likely [2].

Metabolic Derangement

Hyperprolinemia type II is caused by a deficiency of P5-C dehydrogenase, a mitochondrial inner-membrane enzyme, which intervenes in the conversion of proline into glutamate (◻ Fig. 25.1, enzyme 2). Hence, in hyperprolinemia type II, there are increased levels of proline in plasma (usually exceeding 2000 μM; normal range 100–450 μM), urine and CSF, as well as of P5-C. Heterozygotes do not have hyperprolinemia. Evidence has been presented that the accumulating P5-C is a vitamin B6 antagonist (due to adduct formation) and that the seizures in this disorder may be due at least in part to vitamin B6 inactivation [6, 6a].

Genetics

This is an autosomal-recessive disease. The gene *ALDH4A1* maps to 1p36. Mutations have recently been reported in four patients (two frame-shift mutations and two missense mutations) [7].

Diagnostic Tests

The accumulation of P5-C in physiological fluids differentiates type-II and type-I hyperprolinemia. This compound can be qualitatively identified by its reactivity with ortho-aminobenzaldehyde and can be quantitatively measured by several specific assays [2]. P5-C-dehydrogenase activity can be measured in skin fibroblasts and leukocytes.

IV

Treatment and Prognosis

The benign character of the disorder does not justify dietary treatment (which, in any case, would be very difficult). Seizures are B_6 responsive.

25.2   Inborn Errors of Serine Metabolism

25.2.1   3-Phosphoglycerate Dehydrogenase Deficiency

Clinical Presentation

At least nine patients belonging to four families with this disease (first reported in 1996) are known [8, 9]. They presented at birth with microcephaly and developed pronounced psychomotor retardation, severe spastic tetraplegia, nystagmus, and intractable seizures (including hypsarrythmia).

In addition, one patient showed congenital bilateral cataract, two siblings growth retardation and hypogonadism, and two other siblings megaloblastic anemia. Magnetic resonance imaging of the brain revealed cortical and subcortical hypotrophy and evidence of disturbed myelination.

Metabolic Derangement

The deficiency of 3-phosphoglycerate dehydrogenase, the first step of serine biosynthesis (◘ Fig. 25.2, enzyme 1), causes decreased concentrations of serine and, to a lesser extent, of glycine in CSF and in fasting plasma. Serine thus becomes an essential amino acid in these patients. A significant accumulation of the substrate, 3-phosphoglycerate, is unlikely since it is an intermediate of the glycolytic pathway. Therefore, the deficiency of brain serine seems to be the main determinant of the disease. Serine plays a major role in the synthesis of important brain and myelin constituents, such as proteins, glycine, cysteine, serine phospholipids, sphingomyelins and cerebrosides.

In the two patients with megaloblastic anemia, decreased methyltetrahydrofolate was found in CSF. This can be explained by the fact that serine is converted into glycine by a reaction that forms methylenetetrahydrofolate (◘ Fig. 24.1), which is further reduced to methyltetrahydrofolate (► Chap. 28).

Genetics

This is an autosomal-recessive disease. The gene for 3-phosphoglycerate dehydrogenase has been mapped to 1q12. Two missense mutations have been identified [10]. Prenatal diagnosis is only possible by mutation analysis as there is a lack of data on enzyme activity in chorionic villi and amniocytes [11].

Diagnostic Tests

The diagnosis should be suspected in patients with encephalopathy comprising congenital microcephaly. Plasma amino acids must be measured in the fasting state (range in patients: 28–64 µM; normal range: 70–187 µM), since serine and glycine levels can be normal after feeding. In CSF, serine levels are always decreased (6–8 µM; control range 35–80 µM), as are glycine levels, but to a lesser extent. The diagnosis is confirmed by finding a deficient activity of 3-phosphoglycerate dehydrogenase in fibroblasts (reported residual activities from 6–22%).

Treatment and Prognosis

Treatment with L-serine has a beneficial effect on the convulsions, spasticity, feeding and behaviour of these patients. Oral L-serine treatment (up to 600 mg/kg/day in six divided doses) corrected the biochemical abnormalities in all reported patients and abolished the convulsions in most patients, even in those in whom many anti-epileptic treatment regimens had failed previously. During treatment with L-serine, a marked increase in the white matter volume was observed, and in some patients a progression of myelination [12]. In two patients, convulsions stopped only after adding glycine (200 mg/kg/day).

In a girl diagnosed prenatally, because of decelerating head growth, L-serine was given to the mother at 190 mg/kg/day in 3 divided doses from the 27th week of gestation. This normalized fetal head growth and with subsequent postnatal therapy the girl showed normal psychomotor development at the age of 3 years [11].

25.2.2   Phosphoserine Phosphatase Deficiency

Decreased serine levels were found in plasma (53–80 µM; normal range 70–187 µM) and CSF (18 µM; control range 27–57 µM) of one patient with Williams syndrome [13]. Phosphoserine phosphatase activity in lymphoblasts and fibroblasts amounted to about 25% of normal (◘ Fig. 25.2, enzyme 3). Oral serine normalized plasma and CSF levels of this amino acid and seemed to have some beneficial clinical effect. The gene was mapped to 7p11, and the patient was found to be a compound heterozygote for two missense mutations, excluding a link with Williams syndrome [14].

25.2.3   Serine Deficiency with Ichthyosis and Polyneuropathy

A remarkable new serine-deficiency syndrome has been discovered by De Klerk et al. [15] in a 15-year-old girl. She had ichthyosis from the first year of life, growth retardation from the age of 6 years and presented at the age of 14 years with walking difficulties and areflexia, symptoms of an axonal polyneuropathy. Psychomotor development and magnetic resonance imaging of the brain were normal. Fasting plasma and CSF serine levels were decreased but the CSF glycine level slightly increased. Oral ingestion of serine (400 mg/kg/day) cured the skin lesions and the polyneuropathy. It is hypothesized that this patient exhibits an increased conversion of serine into glycine, possibly due to hyperactivity of serine hydroxymethyltransferase (■ Fig. 25.2, enzyme 4).

References

1. Scriver CR, Schafer IA, Efron ML (1961) New renal tubular amino acid transport system and a new hereditary disorder of amino acid metabolism. Nature 192:672

2. Aral B, Kamoun P (1997) The proline biosynthesis in living organisms. Amino Acids 13:189-217

3. Jacquet H, Raux G, Thibaut F et al (2002) PRODH mutations and hyperprolinemia in a subset of schizophrenic patients. Hum Mol Genet 11:2243-2249

4. Bender HU, Almasham S, Steel G et al (2005) Functional consequences of *PRODH* missense mutations. Am J Hum Genet 76:409-420

5. Jaeken J, Goemans N, Fryns J-P et al (1996) Association of hyperprolinemia type I and heparin cofactor II deficiency with CATCH 22 syndrome: evidence for a contiguous gene syndrome locating the proline oxidase gene. J Inherit Metab Dis 19:275-277

6. Farrant RD, Walker V, Mills GA et al (2000) Pyridoxal phosphate deactivation by pyrroline-5-carboxylic acid. Increased risk of vitamin B6 deficiency and seizures in hyperprolinemia type II. J Biol Chem 276:15107-15116

6a. Clayton PT (2006) B_6-Responsive disorders: a model of vitamin dependency. J Inherit Metab Dis 29:317-326

7. Geraghty MT, Vaughn D, Nicholson AJ et al (1998) Mutations in the delta 1-pyrroline 5-carboxylate dehydrogenase gene cause type II hyperprolinemia. Hum Mol Genet 7:1411-1415

8. Jaeken J, Detheux M, Van Maldergem L et al (1996) 3-Phosphoglycerate dehydrogenase deficiency: an inborn error of serine biosynthesis. Arch Dis Child 74:542-545

9. de Koning TJ, Klomp LWJ (2004) Serine-deficiency syndromes. Curr Opin Neurol 17:197-204

10. Klomp LW, de Koning TJ, Malingre HE et al (2000) Molecular characterization of 3-phosphoglycerate dehydrogenase deficiency – a neurometabolic disorder associated with reduced L-serine biosynthesis. Am J Hum Genet 67:1389-1399

11. de Koning TJ, Klomp LW, van Oppen AC et al (2004) Prenatal and early postnatal treatment in 3-phosphoglycerate-dehydrogenase deficiency. Lancet 364:2221-2222

12. de Koning TJ, Jaeken J, Pineda M et al (2000) Hypomyelination and reversible white matter attenuation in 3-phosphoglycerate dehydrogenase deficiency. Neuropediatrics 31:287-292

13. Jaeken J, Detheux M, Fryns J-P et al (1997) Phosphoserine phosphatase deficiency in a patient with Williams syndrome. J Med Genet 34:594-596

14. Veiga-da-Cunha M, Collet JF, Prieur B et al (2004) Mutations responsible for 3-phosphoserine phosphatase deficiency. Eur J Hum Genet 12:163-166

15. De Klerk JB, Huijmans JGM, Catsman-Berrevoets CE et al (1996) Disturbed biosynthesis of serine; a second phenotype with axonal polyneuropathy and ichthyosis. Abstracts of the annual meeting of the Erfelijke Stofwisselingsziekten Nederland, Maastricht, 13–15 Oct 1996

26 Transport Defects of Amino Acids at the Cell Membrane: Cystinuria, Lysinuric Protein Intolerance and Hartnup Disorder

Kirsti Näntö-Salonen, Olli G. Simell

IV

Transepithelial Transport of Amino Acids

Epithelial cells in, e.g., renal tubules and intestinal mucosa, utilise several different amino acid transport systems (■ Fig. 26.1). Each system prefers groups of amino acids with certain physicochemical properties. Cystine and the structurally related dibasic amino acids lysine, arginine and ornithine, are transported from the intestinal or renal tubular lumen into the epithelial cells by an *apical transporter ([1]: system b$^{0,+}$)* in exchange for neutral amino acids. The dibasic amino acids are then transported from the epithelial cell into the tissues by a *basolateral dibasic amino acid transporter ([2]: system y$^+$L)* in exchange for neutral amino acids and sodium. Both these transporters are heteromers of a heavy subunit (N-glycosylated type 2 membrane glyco-

protein) and a light subunit (nonglycosylated polytopic membrane protein) linked by a disulfide bridge. A third *transporter system for neutral amino acids* [3] is expressed only at the luminal border of the epithelial cells. It transports alanine, asparagine, citrulline, glutamine, histidine, isoleucine, leucine, phenylalanine, serine, threonine, tryptophan, tyrosine and valine into the epithelial cells.

Cystinuria, lysinuric protein intolerance and Hartnup disorder are caused by defects of the *luminal cystine/dibasic amino acid transporter* [1] the *antiluminal dibasic amino acid transporter* [2], and the *neutral amino acid transporter* [3], respectively.

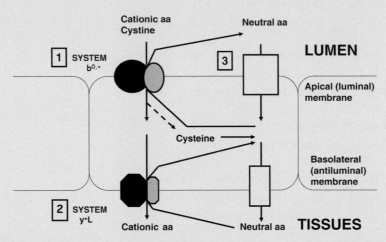

■ **Fig. 26.1.** Simplified schematic representation of cationic and neutral amino acid transport in epithelial cells. *aa*, amino acids. Modified from [1]

Inherited defects in amino acid transport at the cell membrane are usually expressed as selective renal aminoaciduria, i.e., the concentration of the affected amino acids is high in the urine while it is normal or low in plasma. Intestinal absorption of the affected amino acids is also almost always impaired. The clinical symptoms thus result from excess amounts of certain amino acids in the urine or lack of them in the tissues. Consequently, in *cystinuria* renal stones may be formed because of high urinary concentration of poorly soluble cystine. In *lysinuric protein intolerance (LPI)*, the transporter defect for the dibasic cationic amino acids leads to poor intestinal absorption and urinary loss of arginine, ornithine and, particularly, lysine. Deficiencies of arginine and ornithine, intermediates of the urea cycle, lead to hyperammonemia and protein intolerance, and insufficient supply of lysine probably plays a major role in the growth retardation and skeletal and immunological manifestations of LPI. The pellagra-like dermatitis and ataxia in *Hartnup disorder* are attributed to deficiency of tryptophan, the precursor of niacin synthesis.

26.1   Cystinuria

26.1.1   Clinical Presentation

Cystinuria is linked with a life-long risk of urolithiasis [2]. It is responsible for 1–2% of kidney stones in adults and 6–8% in children. Occasional patients never develop any problems, but others may have recurrent symptoms from early childhood. Acute episodes of abdominal or lower-back pain, hematuria, pyuria or spontaneous passing of stones may be the presenting sign. Symptomatic stones often appear in clusters between long asymptomatic periods [3]. Recurrent urinary tract infections, urinary obstruction and finally, renal insufficiency are possible complications. Cystine stones are usually radio-opaque and also visible on ultrasonography.

With increasing knowledge of the genetics of cystinuria, its classification is changing from phenotypic (based on urinary amino acid excretion in obligate heterozygotes) to mutation-based. Cystinuria associated with severe neurological findings or Prader-Willi like syndrome suggests a contiguous gene deletion on chromosome 2p16 [4, 4a] (▶ Sect. 26.1.3).

26.1.2   Metabolic Derangement

In cystinuria, the high-affinity luminal transporter (system $b^{0,+}$; ◘ Fig. 26.1) for cystine and dibasic amino acids in the epithelial cells of the proximal renal tubulus and in jejunal mucosa is defective. The defect leads to poor absorption of cystine in the intestine and its poor reabsorption in the kidney. Whereas normally 99% of the filtered cystine is reabsorbed, homozygotes with cystinuria excrete 600–1400 mg of cystine per day. The intratubular cystine concentration may exceed the threshold for cystine solubility, and crystals and stones may be formed. Sufficient amounts of cystine and dibasic amino acids are apparently absorbed from the intestine via alternative mechanisms, e.g. in oligopeptide form, since no signs of deficiency have been described.

26.1.3   Genetics

The average incidence of cystinuria is 1:7000 but varies considerably between different populations. Cystinuria can be classified into two subtypes: »type I« cystinuria, the pure autosomal recessive form of the disease that represents over 60% of the cases, and »non-type 1« cystinuria, inherited in a dominant mode with incomplete penetrance. As obligate heterozygote carriers, the parents of »type I« patients have normal urinary amino acid profile, while the cystine excretion of the parents of »non-type 1« patients ranges from normal to clearly elevated.

»Type I« cystinuria is caused by mutations in the SLC3A1 gene on chromosome 2p. The gene encodes the heavy subunit of the amino acid transporter, rBAT. More than 80 mutations in the SLC3A1 gene have been reported. »Non-type I« cystinuria results from mutations in the SLC7A9 gene on chromosome 19q. The gene product is the light subunit of the transporter, $b^{0,+AT}$. Over 50 mutations have been described. A patient may also have inherited a »type 1« allele from one parent and a »non-type I« allele from the other. Most of the mutations of these two genes have been detected only in single patients, and the distribution of the more frequent ones is associated with ethnic background [5–7]. A new genetics-based classification for cystinuria has been suggested: type A for SLC3A1 homozygotes (corresponding to »type I« cystinuria), type B for SLC7A9 homozygotes (most of which represent »non-type I« cystinuria), and type AB for the mixed type [8]. While SLC3A1 and SLC7A9 mutations including recently reported unbalanced genomic rearrangements explain most of the cases, it is possible that other genes may be involved [9]. A recessive contiguous gene deletion of chromosome 2p16 associated with cystinuria and mitochondrial disease has been described in a large kindred [4] and another named hypotonia cystinuria syndrome, mimicking Prader-Willi syndrome, is due to mutation in the PREPL gene [4a].

26.1.4   Diagnostic Tests

A positive urinary nitroprusside test and analysis of urinary amino acids lead to the diagnosis. Homozygotes with cystinuria excrete more than 0.1 mmol cystine/mmol creatinine (250 mg cystine/g creatinine) into the urine, but the excretion varies markedly. Plasma concentrations of cystine and the dibasic amino acids are normal or slightly decreased. Chemical analysis of the stones alone may be misleading, because mixed stones are not uncommon in cystinuria, and some stones may contain no cystine at all.

26.1.5 Treatment and Prognosis

Excessive hydration to dilute the urine and alkalinization to improve cystine solubility are the cornerstones of therapy. Moderate sodium restriction is recommended, as reduced sodium intake decreases cystine excretion. Adults should consume more than 3000 ml fluid/24 h (1.75–2 l/m^2/24 h), 500 ml of this before bedtime and, if possible, 500 ml during the night. Because cystine is much more soluble in alkaline urine (500 mg/l at pH 7.5 vs. about 250 mg/l at pH 7.0), permanent alkalinization of urine is beneficial. Sodium bicarbonate (1.5–2 mmol/kg/day in four doses) has been commonly used. In patients with normal renal function, potassium citrate would probably be preferable as it does not increase the sodium load. Restriction of dietary animal protein in order to reduce methionine intake and thus limit endogenous cystine synthesis may be helpful [10, 11].

If the standard therapy fails to prevent new or dissolve pre-existing stones, a thiol derivative is added to decrease urinary free cystine concentration by forming water-soluble disulfide compounds [12]. Daily D-penicillamine (2 g/1.73 m^2 body surface area divided in three doses; up to 3 g/day in adults) is well tolerated by most patients but may cause hypersensitivity reactions, renal problems or a variety of autoimmune syndromes. Mercaptopropionyl-glycine (tiopronin) has fewer adverse effects, but has occasionally caused glomerulopathy or hyperlipidemia. The dose (15–20 mg/kg/day up to 1000 mg/day in three doses in adults) is adjusted individually. The drug has also been successfully used (one dose every two days) as stone prophylaxis [13]. Captopril is well tolerated but may not be as effective as thiol compounds [14]. Percutaneous nephrolithotomy and extracorporeal shock-wave lithotripsy are seldom effective in stone removal, because cystine stones are extremely hard. New, minimally invasive urological techniques minimize the need for open surgery. Surgical procedures should always be combined with conservative preventive therapy.

Regular follow-up is mandatory to support compliance with the treatment, to monitor renal function and to detect developing stones early. New tools for monitoring the efficacy of therapy have been developed. Determination of cystine crystal volume in morning urine in addition to urinary pH and specific gravity [15], or direct assessment of urinary supersaturation [16] may prove helpful. Early detection of the disease by screening the family members of a patient is also essential. Homozygotes with »type I«

cystinuria frequently develop stones during the first decade of life and should perhaps be treated prophylactically from an early age. Other subtypes may have a milder course [17]. Renal function is frequently impaired as a result of stone-forming cystinuria.

26.2 Lysinuric Protein Intolerance

26.2.1 Clinical Presentation

Over 120 patients with LPI, 49 of them from Finland, have been reported or are known to us. The full natural history of LPI remains to be characterized, as most of the oldest patients are still in their forties. Newborns and infants are usually asymptomatic if they are only fed breast milk. Postprandial episodes of hyperammonemia usually emerge when the infants begin to receive formula with higher protein content, or supplementary high-protein foods [18, 19]. Hyperammonemia may present as refusal to eat, vomiting, stupor and unconsciousness. Forced tube feeding may be fatal. Strong aversion to high-protein foods with failure to thrive usually develops around the age of 1 year. The liver and spleen are moderately enlarged.

In toddlers and school-aged children, the presenting signs are most often growth failure and hepato- and splenomegaly. The children are usually hypotonic, muscular endurance is decreased, and they may have fractures after minor traumas. Neurological development is normal if severe or prolonged hyperammonemia has been avoided. Bone maturation is retarded, and there is often marked delay of puberty.

The clinical heterogeneity of LPI is obvious at adult age. Most patients are of moderately short stature, with abundant subcutaneous fat on a square-shaped trunk and thin extremities. They may have marked hepatomegaly with or without splenomegaly. Two thirds have skeletal changes, e.g., osteopenia [20, 21], but pathologic fractures seldom occur. Radiological signs of pulmonary fibrosis are common, but few patients suffer from symptomatic interstitial lung disease [22]. Mental capacity varies from normal to moderate retardation depending on previous history of hyperammonemia.

Some patients have mild normochromic or hypochromic anemia, leukopenia and thrombocytopenia, and their reticulocyte count is often slightly elevated. Serum ferritin and lactate dehydrogenase values are constantly elevated [23], while serum iron and transferrin concentrations are normal. High serum immunoglobulin-G concentrations and abnormalities in the distribution of lymphocyte subpopulations as well as in humoral immune responses [24, 25] have been reported. Varicella infections are usually severe and can be fatal [26]. Several cases of systemic lupus erythematosus have been reported [27–29]. Bone-marrow involvement with erythrophagocytosis and interstitial pulmonary disease with alveolar proteinosis are other rare complications [30–35].

Disturbed tubular function with mild proteinuria, glucosuria, phosphaturia, and tubular acidosis, microscopic hematuria and slow decrease of glomerular filtration rate have been reported in occasional patients. Systematic screening has revealed renal dysfunction of variable degree in 30% of the Finnish patients, with rapid deterioration in glomerular filtration in some cases. A few children and adults have died after a very uniform course of progressive multiorgan failure, often starting with interstitial lung involvement and alveolar proteinosis, progressive glomerulonephritis that leads to renal insufficiency, and severe bleeding diathesis [35]. One child with alveolar proteinosis went through a successful heart-lung transplantation, but died later after a recurrent disease [36].

Pregnancies of LPI patients have been complicated by toxemia, anemia or bleeding during the delivery, but many have been completed successfully without any major problems.

26.2.2   Metabolic Derangement

In LPI, transport of the dibasic cationic amino acids lysine, arginine and ornithine (system y^+L; ◘ Fig. 26.1) is defective at the basolateral membrane of epithelial cells in the renal tubules and small intestine [37, 38]. Massive amounts of lysine and more moderate amounts of arginine and ornithine are lost in the urine, and their intestinal absorption is limited, resulting in low plasma concentrations. The concentrations of neutral amino acids in plasma are slightly increased. Glutamine and alanine concentrations are often more clearly elevated due to the malfunction of the urea cycle. It is still unclear if the transport defect is expressed also in nonepithelial cells. In contrary to an earlier report [39], recent data indicate that fibroblasts from LPI patients have normal cationic amino acid transport, probably via other transporter isoforms [40]. Also erythrocytes possess intact cationic amino acid transport [41, 42]. A transport defect in hepatocytes has been postulated because of normal or elevated cationic amino acid concentrations in liver biopsy in LPI, and the possibility of abnormal cationic amino acid transport between various intracellular compartments has been speculated [43].

Hyperammonemia after protein ingestion and diminished protein tolerance in LPI resemble the symptoms of urea cycle enzyme deficiencies. The malfunction of urea cycle in LPI is best explained by functional deficiency of the intermediates arginine and ornithine in the hepatocytes [43]. Most patients develop a protective aversion to high-protein foods, which further impairs their amino acid intake and aggravates the cationic amino acid deficiencies and poor protein nutrition. As arginine is the rate-limiting precursor of nitric oxide synthesis, arginine deficiency may also result in persistently low nitric oxide concentrations that may influence vascular and immunological functions [44, 45]. Reduced availability of lysine, an essential amino acid, probably has a prominent role in the poor growth and skeletal and immunological manifestations of the patients. Occasional patients have severe carnitine deficiency [46, 47] that may be of dietary origin: the principal dietary source of carnitine is red meat consumed in very small amounts by most patients with LPI. Chronic lysine deficiency may also limit endogenous carnitine biosynthesis. The pathogenetic mechanisms of several clinical manifestations of LPI are still unknown. An excellent review on the subject has been published recently [47].

26.2.3   Genetics

LPI is a rare autosomal-recessive disease with only over 120 patient reported. The incidence is highest in Finland (1:60 000); clusters of families are also known in Italy and Japan, and sporadic cases have been reported on all continents. The gene SLC7A7 on chromosome 14q encodes the light subunit of the dibasic amino acid transporter y^+LAT-1. At least 30 different mutations spread along the entire gene have been reported [47–50]. All Finnish patients share the same founder mutation 1181-2A>T that causes a frame shift and a premature stop codon. The phenotypic variability is wide within the genetically homogeneous Finnish patients as well as in homozygous patients with other mutations, and no genotype/phenotype correlation has been established.

26.2.4   Diagnostic Tests

The diagnosis of LPI is based on the combination of increased urinary excretion and low plasma concentrations of the cationic amino acids, especially lysine. The concentrations of plasma lysine, arginine and ornithine are usually less than 80 µmol/l, 40 µmol/l, and 30 µmol/l, respectively. If plasma amino acid concentrations are exceptionally low due to very limited protein intake, urinary cationic amino acid excretion may on rare occasions be within the reference range.

Blood ammonia concentration increases after protein-rich meals or an intravenous L-alanine load. Orotic aciduria is practically always seen postprandially in untreated patients. Unspecific but consistent findings in LPI patients include elevated serum lactate dehydrogenase activity and increased ferritin concentration.

In the genetically homogeneous Finnish population, the diagnosis is easily confirmed by mutation analysis.

26.2.5   Treatment and Prognosis

The principal aims of treatment are to prevent hyperammonemia and to provide a sufficient supply of protein and essential amino acids for normal metabolism and growth. The protein tolerance in LPI can be improved with supplementary citrulline, a neutral amino acid that is also an intermediate in the urea cycle. Citrulline is readily absorbed and partially converted to arginine and ornithine. All the three amino acids improve the function of the urea cycle. Approximately 100 mg/kg/day of L-citrulline is given in 3–5 doses in association with protein-containing meals [51]. On citrulline supplementation, children usually tolerate 1.0–1.5 g/kg/day and adults 0.5–0.8 g/kg/day of protein without hyperammonemia or increased orotic acid excretion. There is marked interindividual variation in the protein tolerance, and infections, pregnancy and lactation may alter it extensively. Frequent home monitoring of blood ammonia and follow-up of urinary orotic acid excretion is necessary for optimal therapy. For patients with constantly highly elevated glutamine and glycine levels, we have added sodium benzoate or sodium phenylbutyrate to diminish the nitrogen load of the urea cycle.

Correction of lysine deficiency in LPI is complicated by its poor intestinal absorption and the resulting osmotic diarrhoea [52]. However, a carefully titrated dose of L-Lysine-HCl is able to elevate the plasma lysine concentrations to low-normal range without gastrointestinal or other side effects. We currently supplement all our patients with 20–30 mg/kg/day of the compound in three doses [53], but it is still too early to evaluate its effects. Carnitine supplementation is indicated for the patients with evident carnitine deficiency. Due to the restricted diet, the patients need regular supplementation with calcium, vitamins and trace elements, and the role of an experienced nutritionist is essential. We have recently started growth hormone therapy in one child with renal insufficiency and severe growth retardation, with a promising early response and no side effects.

In LPI, acute hyperammonemic crisis is fortunately a rare problem. All protein- and nitrogen-containing substances should be removed from the nutrition and sufficient energy should be supplied as intravenous glucose. Intravenous infusion of ornithine, arginine or citrulline, beginning with a priming dose of 0.5–1.0 mmol/kg in 5–10 min and continuing with 0.5–1.0 mmol/kg/h, will clear hyperammonemia rapidly. Sodium benzoate and sodium phenylbutyrate may also be used to utilize alternate pathways of ammonia elimination [54].

LPI patients should be immunized against Varicella zoster, and non-immunized patients should be treated immediately with acyclovir if they get the infection [26]. The treatment of the immunological and bone marrow complications including hemophagocytic lymphohistiocytosis, is still experimental. Good responses have been reported in individual cases with immunosuppressive drugs and with immunoglobulin infusion [55]. In alveolar proteinosis, bronchoalveolar lavage and steroid therapy have been effective in some cases [30–33]. Granulocyte-macrophage colony stimulating factor therapy has recently been suggested for idiopathic alveolar proteinosis. It has been tried also in one child with LPI, but was discontinued because of side effects [43]. The child later received a successful heart-lung transplantation, but alveolar proteinosis recurred in the transplanted lungs [36].

Although hyperammonemia and the associated mental retardation can be avoided with citrulline therapy, several other complications of LPI seem to develop and progress under current therapy. The accumulating knowledge of many new multisystem manifestations of LPI is changing the previous concept that it is, in most cases, a fairly benign and easily treated condition.

26.3   Hartnup Disorder

26.3.1   Clinical Presentation

The classical clinical symptoms of Hartnup disorder, pellagra-like dermatitis and neurological involvement, closely resemble those of nutritional niacin (nicotinic acid and nicotinamide) deficiency. Since the first description of the syndrome in several members of the Hartnup family in 1956 [56], an extensive number of subjects who fulfil the biochemical diagnostic criteria have been reported, mostly detected in newborn screening programs. However, most of them remain asymptomatic.

In the few patients that develop clinical symptoms, the skin lesions and neurological problems usually appear in early childhood [57] and tend to ameliorate with increasing age. Exposure to sunlight, fever, diarrhoea, inadequate diet or psychological stress may precipitate the symptoms. Pellagra-like skin changes are found on light-exposed areas, and the skin becomes scaly and rough with peripheral hypopigmentation. Intermittent cerebellar ataxia, attacks of headache, muscle pain and weakness may appear. Occasional patients present with mental retardation, seizures or psychiatric symptoms. Growth and developmental outcome of the patients are generally normal, although also low academic scores have been reported.

26.3.2   Metabolic Derangement

The pattern of hyperaminoaciduria in Hartnup disease is characteristic. Alanine, serine, threonine, valine, leucine, isoleucine, phenylalanine, tyrosine, tryptophan, histidine and citrulline and the monoamino-dicarboxylic amides asparagine and glutamine, are excreted in 5- to 20-fold excess in the urine, and their plasma concentrations are decreased or low normal. Renal clearance values of other amino acids are within the normal range. The long-sought molecular defect has recently been described: it involves a sodium-dependent and chloride-independent neutral amino acid transporter SLC6A19, shared by the affected neutral monoamino-monocarboxylic amino acids and expressed predominantly in the epithelial cells in the renal proximal tubuli and intestinal epithelium [58, 59]. The stools of the patients contain increased amounts of free amino acids, reflecting closely the urinary excretion pattern [60]. After an oral tryptophan load, the patients show smaller plasma tryptophan peaks than controls and excrete smaller amounts of tryptophan metabolites in the urine. The affected amino acids are readily absorbed as short oligopeptides but not as free amino acids. The unabsorbed amino acids in the colon are exposed to bacterial degradation. Degradation of tryptophan produces large amounts of indole compounds, which are then excreted in the urine.

The puzzling phenotypic heterogeneity of Hartnup disorder may be due to association with certain polygenic and environmental factors. The clinical manifestations that resemble niacin deficiency probably reflect deficient production of the tryptophan metabolite nicotinamide. Symptomatic disease may thus be prevented if dietary niacin intake is sufficient, or if necessary amount of tryptophan is absorbed in oligopeptide form [61].

26.3.3 Genetics

Hartnup disorder follows an autosomal-recessive pattern of inheritance. The reported incidence in newborns screened for aminoaciduria has varied from 1 in 25 000 to 1 in 45 000. Hartnup disorder is caused by mutations in the newly discovered gene encoding the neutral amino acid transporter SLC6A19 on chromosome 5p15.33 [58, 59].

26.3.4 Diagnostic Tests

The characteristic excess of neutral amino acids in the urine and their normal or low normal concentrations in plasma confirm the diagnosis. Urinary excretion of indole compounds may be within the normal range if the patient consumes normal or low amounts of dietary protein, but an oral load of l-tryptophan (100 mg/kg) in most cases leads to a supranormal increase in indole excretion.

26.3.5 Treatment and Prognosis

Dermatitis and neurological symptoms usually but not invariably disappear rapidly with oral nicotinamide (50–300 mg/day). An adequate supply of high-quality protein is probably important for prevention of the symptoms. Tryp-

tophan ethyl ester has been successfully used to circumvent the transport defect. Oral neomycin reduces intestinal degradation of tryptophan and decreases indole production; however, the role of the indole compounds in the disease has been poorly characterized. Early recognition of the condition in newborn screening programs permits adequate follow-up and prevention of symptomatic disease.

At least 14 pregnancies of women with Hartnup disorder have been reported, and maternal Hartnup disorder seems to be harmless to the foetus [62].

References

1. Chillaron J, Roca R, Valencia A et al (2001) Heteromeric amino acid transporters: biochemistry, genetics, and physiology. Am J Physiol Renal Physiol 281:F995-F1018
2. Lindell A, Denneberg T, Granerus G (1997) Studies on renal function in patients with cystinuria. Nephron 77:76-85
3. Purohit RS, Stoller ML (2004) Stone clustering of patients with cystine urinary stone formation. Urology 63:630-635
4. Parvari R, Brodyansky I, Elpeleg O et al (2001) A recessive contiguous gene deletion of chromosome 2p16 associated with cystinuria and a mitochondrial disease. Am J Hum Genet 69:869-875
4a. Jaeken J, Martens K, François I et al (2006) Deletion of PREPL; a gene encoding a putative serine oligopeptidase, in patients with hypotonia-cystinuria syndrome. Am J Hum Genet 78:38-51
5. Calonge MJ, Gasparini P, Chillaron J et al (1994) Cystinuria caused by mutations in RBAT, a gene involved in the transport of cystine. Nat Genet 6:420-425
6. Feliubadalo L, Font M, Purroy J et al (1999) Non-type I cystiuria caused by mutations in SLCA9, encoding a subunit (b0,+)AT of rBAT. International Cystinuria Consortium. Nat Genet 23:52-57
7. Leclerc D, Boutros M, Suh D et al (2002) SLC7A mutations in all three cystinuria subtypes. Kidney Int 62:1550-1559
8. Dello-Strogolo L, Pras E, Pontesilli C et al (2002) Comparison between SLC3A1 and SLC7A9 cystinuria patients and carriers: a need for a new classification. J Am Soc Nephrol 13:2547-2553
9. Schmidt C, Vester U, Wagner CA et al (2003) Significant contribution of genomic rearrangements in SLC3A1 and SLC7A9 to the etiology of cystinuria. Kidney Int 64:1564-1573
10. Barbey F, Joly D, Rieu P et al (2000) Medical treatment of cystinuria: critical reappraisal of long-term results. Clin Urology 163:1419-1423
11. Fjellstedt E, Denneberg T, Jeppsson JO, Tiselius HG (2001) A comparison of the effects of potassium citrate and sodium bicarbonate in the alkalinization of urine in homozygous cystinuria. Urol Res 29:295-302
12. Barbey F, Joly D, Rieu P et al (2000) Medical treatment of cystinuria: critical reappraisal of long-term results. Clin Urology 163:1419-1423
13. Berio A, Piazzi A (1998) Prophylaxis of cystine calculi by low dose of alpha mercaptopropionylglycine administered every other day. Panminerva Med 40:244-246
14. Chow GK, Streem SB (1996) Medical treatment of cystinuria: results of contemporary clinical practice. J Urol 156:1576-1578
15. Daudon M, Cohen-Solail F, Barbey F et al (2003) Cystine chrystal volume determination: a useful tool in the management of cystinuric patients. Urol Res 31:207-211
16. Coe FL, Clark C, Parks JH, Asplin JR (2001) Solid phase assay of urine cystine supersaturation in the presence of cystine binding drugs. J Urol 166:688-693
17. Goodyer P, Saadi I, Ong P et al (1998) Cystinuria subtype and the risk of nephrolithiasis. Kidney Int 54:56-61
18. Perheentupa J, Visakorpi JK (1965) Protein intolerance with deficient transport of basic amino acids: another inborn error of metabolism. Lancet 2:813-816

19. Simell O, Perheentupa J, Rapola J et al (1975) Lysinuric protein intolerance. Am J Med 59:229-240

20. Carpenter TO, Levy HL, Holtrop ME et al (1985) Lysinuric protein intolerance presenting as childhood osteoporosis. Clinical and skeletal response to citrulline therapy. N Engl J Med 312:290-294

21. Svedström E, Parto K, Marttinen M et al (1993) Skeletal manifestations of lysinuric protein intolerance. A follow-up study of 29 patients. Skeletal Radiol 22:11-16

22. Parto K, Svedström E, Majurin M-L et al (1993) Pulmonary manifestations in lysinuric protein intolerance. Chest 104:1176-1182

23. Rajantie J, Simell O, Perheentupa J, Siimes MA (1980) Changes in peripheral blood cells and serum ferritin in lysinuric protein intolerance. Acta Paediatr Scand 69:741-745

24. Yoshida Y, Machigashira K, Suehara M et al (1995) Immunological abnormality in patients with lysinuric protein intolerance. J Neurol Sci 134:178-182

25. Lukkarinen M, Parto K, Ruuskanen O et al (1999) B and T cell immunity in patients with lysinuric protein intolerance. Clin Exp Immunol 116:430-434

26. Lukkarinen M, Näntö-Salonen K, Ruuskanen O et al (1998) Varicella and varicella immunity in patients with lysinuric protein intolerance. J Inherit Metab Dis 21:103-111

27. Dionisi-Vici C, De Felice L, el Hachem M et al (1998) Intravenous immunoglobulin in lysinuric protein intolerance. J Inherit Metab Dis 21:95-102

28. Kamoda T, Nagai Y, Shigeta M et al (1998) Lysinuric protein intolerance and systemic lupus erythematosus. Eur J Pediatr 157:130-131

29. Aoki M, Fukao T, Fujita Y et al (2001) Lysinuric protein intolerance in siblings: complication of systemic lupus erythematosus in the elder sister. Eur J Pediatr 160: 522-523

30. Kerem E, Elpelg ON, Shalev RS et al (1993) Lysinuric protein intolerance with chronic interstitial lung disease and pulmonary cholesterol granulomas at onset. J Pediatr 123:275-278

31. DiRocco M, Garibotto G, Rossi GA et al (1993) Role of haematological, pulmonary and renal complications in the long-term prognosis of patients with lysinuric protein intolerance. Eur J Pediatr 152:437-440

32. Parto K, Kallajoki M, Aho H, Simell O (1994) Pulmonary alveolar proteinosis and glomerulonephritis in lysinuric protein intolerance: case reports and autopsy findings of four pediatric patients. Hum Pathol 25:400-407

33. Santamaria F, Parenti G, Guidi G et al (1996) Early detection of lung involvement in lysinuric protein intolerance: role of high-resolution computed tomography and radioisotopic methods. Am J Respir Crit Care Med 153:731-735

34. Duval M, Fenneteau O, Doireau V et al (1999) Intermittent hemophagocytic lymphohistiocytosis is a regular feature of lysinuric protein intolerance. J Pediatr 134:236-239

35. Parenti G, Sebastio G, Strisciuglio P et al (1995) Lysinuric protein intolerance characterised by bone marrow abnormalities and severe clinical course. J Pediatr 126:246-251

36. Santamaria F, Brancaccio G, Parenti G et al (2004) Recurrent fatal pulmonary alveolar proteinosis after heart-lung transplantation in a child with lysinuric protein intolerance. J Pediatr 145:268-272

37. Rajantie J, Simell O, Perheentupa J (1980) Basolateral membrane transport defect for lysine in lysinuric protein intolerance. Lancet 1:1219-1221

38. Rajantie J, Simell O, Perheentupa J (1981) Lysinuric protein intolerance. Basolateral transport defect in renal tubuli. J Clin Invest 67:1078-1082

39. Smith DW, Scriver CR, Tenenhouse HS, Simell O (1987) Lysinuric protein intolerance mutation is expressed in the plasma membrane of cultured skin fibroblasts. Proc Natl Acad Sci USA 84:7711-7715

40. Dall'Asta V, Bussolati O, Sala R et al (2000) Arginine transport through system y⁺L in cultured human fibroblasts: normal phenotype of cells from LPI subjects. Am J Physiol Cell Physiol 279:C1829-1837

41. Smith DW, Scriver CR, Simell O (1988) Lysinuric protein intolerance mutation is not expressed in the plasma membrane of erythrocytes. Hum Genet 80:395-396

42. Boyd CA, Deves R, Laynes R, Kudo Y et al (2000) Cationic amino acid transport through system y⁺L in erythrocytes of patients with lysinuric protein intolerance. Pflugers Arch 459:513-516

43. Rajantie J, Simell O, Perheentupa J (1983) «Basolateral« and mitochondrial membrane transport defect in the hepatocytes in lysinuric protein intolerance. Acta Paediatr Scand 72:65-70

44. Kayanoki Y, Kawata S, Kiso S et al (1999) Reduced nitric acid production by L-arginine deficiency in lysinuric protein intolerance exacerbates intravascular coagulation. Metabolism 48:1136-1140

45. Kamada Y, Nagaretani H, Tamura S et al (2001) Vascular endothelial dysfunction resulting from L-arginine deficiency in a patient with lysinuric protein intolerance. J Clin Invest 108:717-724

46. Korman SH, Raas-Rothschild A, Elpeleg O, Gutman A (2002) Hypocarnitinemia in lysinuric protein intolerance. Mol Genet Metab 76:81-83

47. Palacin M, Bertran J, Chillaron J et al (2004) Lysinuric protein intolerance: mechanisms of pathophysiology. Mol Genet Metab 81:27-37

48. Torrents D, Mykkänen J, Pineda M et al (1999) Identification of SLC7A7, encoding y⁺ LAT-1, as the lysinuric protein intolerance gene. Nat Genet 21:293-296

49. Borsani G, Bassi MT, Sperandeo MP et al (1999) SLC7A7, encoding a putative permease-related protein, is mutated in patients with lysinuric protein intolerance. Nat Genet 21:297-301

50. Sperandeo MP, Bassi MT, Riboni et al (2000) Structure of the SLC7A7 gene and mutational analysis of patients affected by lysinuric protein intolerance. Am J Hum Genet 66:92-99

51. Rajantie J, Simell O, Rapola J, Perheentupa J (1980) Lysinuric protein intolerance: a two-year trial of dietary supplementation therapy with citrulline and lysine. J Pediatr 97:927-932

52. Rajantie J, Simell O, Perheentupa J (1983) Oral administration of epsilon-N-acetyllysine and homocitrulline for lysinuric protein intolerance. J Pediatr 102:388-390

53. Lukkarinen M, Näntö-Salonen K, Pulkki K et al (2003) Oral supplementation corrects plasma lysine concentrations in lysinuric protein intolerance. Metabolism 52:935-938

54. Brusilow SW, Danney M, Waber LJ et al (1984) Treatment of episodic hyperammonemia in children with inborn errors of urea synthesis. N Engl J Med 310:1630-1634

55. Bader-Meunier B, Parez N, Muller S (2000) Treatment of hemophagocytic lymphohistiocytosis and alveolar proteinosis with cyclosporin A and steroids in a boy with lysinuric protein intolerance. J Pediatr 136:134

56. Baron DN, Dent CE, Harris H, Hart EW, Jepson JB (1956) Hereditary pellagra-like skin rash with temporary cerebellar ataxia, constant renal amino aciduria and other bizarre biochemical features. Lancet 2:421-428

57. Scriver CR, Mahon B, Levy HL et al (1987) The Hartnup phenotype: Mendelian transport disorder, multifactorial disease. Am J Hum Genet 40:401-412

58. Kleta R, Romeo E, Ristic Z et al (2004) Mutations in SLC6A19, encoding B0AT1, cause Hartnup disorder. Nat Genet 36:999-1002

59. Seow HF, Broer S, Broer A et al (2004) Hartnup disorder is caused by mutation in the gene encoding the neutral amino acid transporter SLC6A19. Nat Genet 36:1003-1007

60. Scriver CR (1965) Hartnup disease: a genetic modification of intestinal and renal transport of certain neutral alpha amino acids. N Engl J Med 273:530-532

61. Scriver CR, Mahon B, Levy HL et al (1987) The Hartnup phenotype: Mendelian transport disorder, multifactorial disease. Am J Hum Genet 40:401-412

62. Mahon BE, Levy HL (1986) Maternal Hartnup disorder. Am J Med Genet 24:513-518

V Vitamin-Responsive Disorders

27 Biotin-Responsive Disorders

Matthias R. Baumgartner, Terttu Suormala

The Biotin Cycle and Biotin-Dependent Enzymes

Biotin, a water-soluble vitamin widely present in small amounts in natural foodstuffs in which it is mostly protein-bound, is the coenzyme of four important carboxylases, involved in gluconeogenesis, fatty acid synthesis, and the catabolism of several amino acids (□ Fig. 27.1). Binding of biotin to the four inactive apo-carboxylases, catalysed by *holocarboxylase synthetase*, is required to generate the active holocarboxylases (□ Fig. 27.2). Recycling of biotin first involves proteo-lytic degradation of the holocarboxylases, yielding biotin bound to lysine (biocytin) or to short biotinyl peptides. *Biotinidase* then releases biotin from the latter compounds, which are derived from either endogenous or dietary sources.

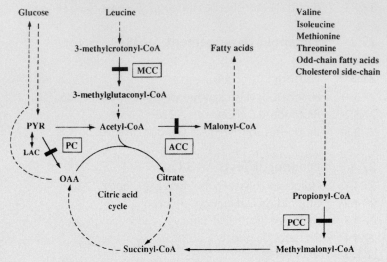

□ **Fig. 27.1.** Location of the biotin-dependent carboxylases in intermediary metabolism. *ACC*, acetyl-CoA carboxylase; *CoA*, coenzyme A; *HCS*, holocarboxylase synthetase; *LAC*, lactate; *MCC*, 3-methylcrotonyl-CoA carboxylase; *OAA*, oxaloacetate; *PC*, pyruvate carboxylase; *PCC*, propionyl-CoA carboxylase; *PYR*, pyruvate. *Full lines* indicate one enzyme, and *dotted lines* indicate that several enzymes are involved. Sites of the enzyme defects are indicated by *solid bars*

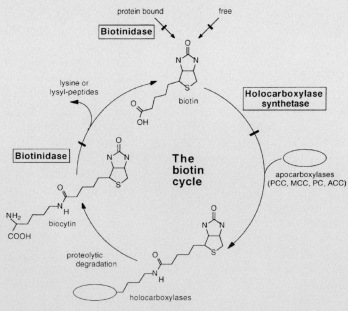

□ **Fig. 27.2.** The biotin cycle. For definitions of abbreviations, □ Fig. 27.1. Sites of the enzyme and transport defects are indicated by *solid bars*

Two inherited defects in biotin metabolism are known: *holocarboxylase synthetase* (HCS) *deficiency* and *biotinidase deficiency*. Both lead to deficiency of all biotin-dependent carboxylases, i.e. to *multiple carboxylase deficiency* (MCD). In HCS deficiency, the binding of biotin to apocarboxylases is impaired. In biotinidase deficiency, biotin depletion ensues from the inability to recycle endogenous biotin and to utilize protein-bound biotin from the diet. As the carboxylases play an essential role in the catabolism of several amino acids, in gluconeogenesis and in fatty-acid synthesis, their deficiency provokes multiple, life-threatening metabolic derangements, eliciting characteristic organic aciduria and neurological symptoms. The clinical presentation is extremely variable in both disorders. Characteristic symptoms include metabolic acidosis, hypotonia, seizures, ataxia, impaired consciousness and cutaneous symptoms, such as skin rash and alopecia. All patients with biotinidase and a majority of patients with HCS deficiency respond dramatically to oral therapy with pharmacological doses of biotin. Delayed diagnosis and treatment in biotinidase deficiency may result in irreversible neurological damage. A few patients with HCS deficiency show a partial or even no response to biotin and seem to have an impaired long-term outcome. *Acquired biotin deficiency*, which also causes MCD, is extremely rare. A defect in *biotin transport* has been reported in a single child; however the genetic defect remains unresolved to date. *Biotin-Responsive Basal Ganglia Disease* (BRBGD) is a recently described subacute encephalopathy which disappears within a few days without neurological sequelae if biotin is administered early.

27.1 Clinical Presentation

The characteristic manifestation of multiple carboxylase deficiency (MCD) is metabolic acidosis associated with neurological abnormalities and skin disease. The expression of the clinical and biochemical features is variable in both inherited disorders [1]. While patients with holocarboxylase synthetase (HCS) deficiency commonly present with the typical symptoms of MCD, those with biotinidase deficiency show a less consistent clinical picture, particularly during the early stage of the disease. The onset in biotinidase deficiency may be insidious, and the manifestation is usually very variable, neurological symptoms often being prominent without markedly abnormal organic-acid excretion or metabolic acidosis. Later-onset forms of HCS deficiency cannot be clinically distinguished from biotinidase deficiency, necessitating confirmation of the diagnosis by enzyme assay.

27.1.1 Holocarboxylase Synthetase Deficiency

Although HCS deficiency was initially termed early-onset MCD, recent experience shows that the age of onset varies widely, from a few hours after birth to 8 years of age [2, 3]. Nevertheless, about half of the patients have presented acutely in the first days of life with symptoms very similar to those observed in other severe organic acidurias, i.e., lethargy, hypotonia, vomiting, seizures and hypothermia. The most common initial clinical features consist of respiratory difficulties, such as tachypnea or Kussmaul breathing. Severe metabolic acidosis, ketosis and hyperammonaemia may lead to coma and early death. Patients with a less severe defect and later onset may also present with recurrent life-threatening attacks of metabolic acidosis and typical organic aciduria [4, 5]. Early-onset patients that recover without biotin therapy and untreated patients with a less severe defect may additionally develop psychomotor retardation, hair loss and skin lesions. The latter include an erythematous, scaly skin rash that spreads over the whole body but is particularly prominent in the diaper and intertriginous areas; alternatively, the rash may resemble seborrheic dermatitis or ichthyosis [6]. Superinfection with *Candida* may occur. Disorders of immune function have been observed with decreased T cell count and impaired in vitro and in vivo response to *Candida* antigen. Episodes of acute illness are often precipitated by catabolism during intercurrent infections or by a higher protein intake.

27.1.2 Biotinidase Deficiency

Important features are the gradual development of symptoms and episodes of remission, which may be related to increased free biotin in the diet. The full clinical picture has been reported as early as 7 weeks, but discrete neurological symptoms may occur much earlier, even in the neonatal period [7]. Neurological manifestations (lethargy, muscular hypotonia, grand mal and myoclonic seizures, ataxia) are the most frequent initial symptoms. In addition, respiratory abnormalities, such as stridor, episodes of hyperventilation and apnoea occur frequently; these may be of neurological origin [8]. Skin rash and/or alopecia are hallmarks of the disease; however, they may develop late or not at all [9, 10]. Skin lesions are usually patchy, erythematous/exudative and typically localized periorificially. Eczematoid dermatitis or an erythematous rash covering large parts of the body has also been observed, as has keratoconjunctivitis. Hair loss is usually discrete but may, in severe cases, become complete, including the eyelashes and eyebrows. Immunological dysfunction may occur in acutely ill patients. Some children with profound biotinidase deficiency may not develop symptoms until later in childhood or during adolescence [11]. Their symptoms usually are less characteristic and may include motor limb weakness, spastic paraparesis and eye problems such as loss of visual acuity and scotomata [11]. Two asymptomatic adults with profound biotinidase deficiency were ascertained after identification of their affected children by newborn screening [12]. Similarly, in two asymptomatic adolescent girls and in an asymptomatic adult male, residual plasma biotinidase activity, assessed by a sensitive assay, was between 1.2–3.1% of the mean control value, indicating that the threshold level of biotinidase activity needed for normal development is low [13, 14]. Alternatively, other factors such as modifying genes or environmental factors may protect some enzyme-deficient individuals from developing symptoms.

Because of the variability and nonspecificity of clinical manifestations, there is a great risk of a delay in diagnosis [8, 15, 16]. Late-diagnosed patients often have psychomotor retardation and neurological symptoms, such as leuko-encephalopathy, hearing loss and optic atrophy, which may be irreversible [9, 10, 15–18]. The outcome may even be fatal. One patient died at the age of 22 months, with features of Leigh syndrome proven by histopathology [8].

Metabolic acidosis and the characteristic organic aciduria of MCD are frequently lacking in the early stages of the disease. Plasma lactate and 3-hydroxyisovalerate may be only slightly elevated, whereas cerebrospinal fluid levels may be significantly higher [19]. This fact and the finding of severely decreased carboxylase activities in brain but moderately deficient activity in liver and kidney in a patient with lethal outcome [8] are in accordance with the predominance of neurological symptoms and show that, in biotinidase deficiency, the brain is affected earlier and more severely than other organs. The threat of irreversible brain damage demands that biotinidase deficiency should be considered in all children with neurological problems, even if obvious organic aciduria and/or cutaneous findings are not present. Sadly, there seems to have been little improvement in the diagnostic delay over the last 10 years [15, 17]. Therefore, neonatal screening provides the best chance of improving outcome in biotinidase deficiency. Importantly, treatment should be instituted without delay, since patients may become biotin depleted within a few days after birth [7].

27.1.3 Biotin-Responsive Basal Ganglia Disease

Biotin-responsive basal ganglia disease (BRBGD) is an autosomal recessive disorder with childhood onset that presents as a subacute encephalopathy with confusion, dysarthria and dysphagia, that progresses to severe cogwheel rigidity, dystonia, quadriparesis and, if left untreated, to death [19a]. On brain magnetic resonance imaging (MRI) examination patients display central bilateral necrosis in the head of the caudate nucleus with complete or partial involvement of the putamen. All patients diagnosed to date are of Saudi, Syrian, or Yemeni ancestry.

27.2 Metabolic Derangement

In HCS deficiency, a decreased affinity of the enzyme for biotin and/or a decreased maximal velocity lead to reduced formation of the four holocarboxylases from their corresponding inactive apocarboxylases at physiological biotin concentrations (◘ Fig. 27.2) [20–22]. In biotinidase deficiency, biotin cannot be released from biocytin and short biotinyl peptides. Thus, patients with biotinidase deficiency are unable to either recycle endogenous biotin or to use protein-bound dietary biotin (◘ Fig. 27.2) [1]. Consequently, biotin is lost in the urine, mainly in the form of biocytin [7, 23], and progressive biotin depletion occurs. Depending on the amount of free biotin in the diet and the severity of the enzyme defect, the disease becomes clinically manifest during the first months of life or later in infancy or childhood.

Deficient activity of carboxylases in both HCS and biotinidase deficiencies (◘ Fig. 27.1) results in accumulation of lactic acid and derivatives of 3-methylcrotonyl-coenzyme A (CoA) and propionyl-CoA (▶ Sect. 27.4).

Isolated inherited deficiencies of each of the three mitochondrial carboxylases, propionyl-CoA carboxylase (PCC), 3-methylcrotonyl-CoA carboxylase (MCC); (for both, ▶ Chap.19), and pyruvate carboxylase (PC; ▶ Chap.12), are also known. A single patient with an isolated defect of acetyl-CoA carboxylase (ACC, cytosolic) has been reported [24]. These isolated deficiencies are due to absence or abnormal structure of the apoenzyme and usually do not respond to biotin therapy. A patient with isolated partial MCC-deficiency and partial responsiveness to biotin therapy has recently been reported [25].

In BRBGD there is a defective cerebral transport of biotin [25a].

Acquired biotin deficiency is rare but may result from excessive consumption of raw egg white, malabsorption, long-term parenteral nutrition, hemodialysis, and long-term anticonvulsant therapy. Biotin dependency due to a defect in biotin transport has been suggested in a 3-year-old boy with normal biotinidase and nutritional biotin intake [26], but the genetic defect remains unresolved to date.

27.3 Genetics

Both HCS and biotinidase deficiency are inherited as autosomal recessive traits. HCS deficiency seems to be rarer than biotinidase deficiency. The incidences of profound (<10% residual activity) and partial (10–30% residual activity) biotinidase deficiencies are, on average, 1:112 000 and 1:129 000, respectively [27]. The incidence of combined profound and partial deficiency is about 1 in 60 000. The cDNAs for human HCS [28, 29] and biotinidase [30] have been cloned, and the corresponding genes have been mapped to human chromosomes 21q22.1 [29] and 3p25 [31], respectively. In both genes, multiple disease causing mutations have been identified.

27.3.1 Holocarboxylase Synthetase Deficiency

More than 20 different disease causing mutations have been reported [32–35]. About 2/3 of them are within the putative biotin-binding region of HCS and result in decreased affinity of the enzyme for biotin [20, 22, 32, 34, 36]; this probably accounts for the *in vivo* responsiveness to biotin therapy of these patients. The degree of abnormality of the K_m values of HCS for biotin correlates well with the time of onset and severity of illness, i.e. highest K_m with early onset and severe disease [21]. Other mutations, located outside the biotin-binding site in the N-terminal region, are associated with normal K_m but decreased V_{max} [22]. Most patients with this type of mutation also respond to biotin, although higher doses may be required and residual biochemical and clinical abnormalities may persist. Biotin responsiveness in such patients may derive from a positive effect of biotin on *HCS* mRNA transcription and thus on HCS protein, which has recently been suggested [37]. However, since this mechanism involves HCS protein itself, it requires the presence of residual HCS activity in order to work. Only one mutant allele, L216R, when present in the homozygous state, has been associated with a biotin-unresponsive, severe clinical phenotype [32]. This mutation seems to be highly prevalent in Polynesian patients of Samoan origin (David Thorburn and Callum Wilson, personal communication).

27.3.2 Biotinidase Deficiency

At least 79 different mutations have been identified in patients with profound or partial biotinidase deficiency [35, 38, 39]. The two most common mutations detected in symptomatic patients with profound deficiency in the U.S.A., accounting for about one third of the alleles, are 98-104del7ins3 and R538C [38, 40]. In contrast, in patients with profound biotinidase deficiency detected by newborn screening, three mutations – Q456H, the double-mutant allele A171T + D444H, and D252G – accounted for about half of the mutant alleles detected [38]. Strikingly, these mutations were not detected in any of the symptomatic patients [38, 40]. Furthermore, none of the symptomatic children had detectable serum biotinidase biotinyl-transferase activity while two thirds of the children identified by screening had detectable activity [41]. A comparison of

mutations in children detected by newborn screening with mutations in symptomatic children revealed four mutations comprising 59% of the mutant alleles studied [42]. Only two of these mutations occurred in both populations [42]. Thus it is possible that individuals with certain mutations in the newborn screening group may have a decreased risk of developing symptoms. Almost all individuals with partial biotinidase deficiency have the D444H mutation in combination with a mutation causing profound biotinidase deficiency on the second allele [39].

27.3.3 Biotin-Responsive Basal Ganglia Disease

BRBGD is due to mutations in SLC19A3, a gene coding for a cerebral biotin transporter related to the reduced folate and thiamine transporters [25a]. Different missense mutations have been identified.

27.4 Diagnostic Tests

A characteristic organic aciduria due to systemic deficiency of the carboxylases is the key feature of MCD. In severe cases, an unpleasant urine odour (cat's urine) may even be suggestive of the defect. MCD is reflected in elevated urinary and plasma concentrations of organic acids as follows:

- Deficiency of MCC: 3-hydroxyisovaleric acid in high concentrations, 3-methylcrotonylglycine in smaller amounts;
- Deficiency of PCC: methylcitrate, 3-hydroxypropionate, propionylglycine, tiglylglycine, propionic acid in small to moderate amounts;
- Deficiency of PC: lactate in high concentrations, pyruvate in smaller amounts.

There is no metabolic marker in BRBGD.

The majority of HCS-deficient patients excrete all of the typical organic acids in elevated concentrations, provided that the urine sample has been taken during an episode of acute illness. In contrast, in biotinidase deficiency elevated excretion of only 3-hydroxyisovalerate may be found, especially in early stages of the disease. 20 % of untreated biotinidase-deficient children had normal urinary organic acid excretion when symptomatic [10].

The measurement of carboxylase activities in lymphocytes provides direct evidence of MCD. These activities are low in HCS deficiency but may be normal in biotinidase deficiency, depending on the degree of biotin deficiency [3, 14]. The two inherited disorders can easily be distinguished by assay of biotinidase activity in serum. Today, this assay is included in the neonatal screening programs in many countries worldwide.

27.4.1 Holocarboxylase Synthetase Deficiency

- Biotin concentrations in plasma and urine are normal;
- Carboxylase activities in lymphocytes are deficient and cannot be activated by *in vitro* preincubation with biotin [1];
- Direct measurement of HCS activity requires a protein, e.g. an apocarboxylase or an apocarboxyl carrier protein of ACC as one of the substrates [21, 43]; therefore, it is not routinely performed;
- HCS deficiency can be diagnosed indirectly by demonstrating severely decreased carboxylase activities in fibroblasts cultured in a medium with low biotin concentration (10^{-10} mol/l) and by normalization (or, at least an increase) of the activities in cells cultured in media supplemented with high biotin concentrations (10^{-6}–10^{-5} mol/l) [3, 21]. It must be noted that fibroblasts of some late-onset patients may exhibit normal levels of carboxylase activities when cultured in standard media supplemented with 10% fetal calf serum, which results in a final biotin concentration of about 10^{-8} mol/l [3, 5].

27.4.2 Biotinidase Deficiency

- Biotinidase activity in plasma is absent or decreased [14, 27]. Many patients have measurable residual activity and should be evaluated for the presence of a K_m defect (▶ below);
- Symptomatic patients usually have decreased biotin concentrations in plasma and urine [7, 14], provided that an assay method that does not detect biocytin is used [44]. In addition, carboxylase activities in lymphocytes are usually decreased but are normalized within hours after either a single dose of oral biotin [7] or *in vitro* preincubation with biotin [1, 14];
- Patients excrete biocytin in urine [23], the concentration being dependent on the level of residual biotinidase activity [14];
- Carboxylase activities in fibroblasts cultured in low-biotin medium are similar to those in control fibroblasts, and are always normal in fibroblasts cultured in standard medium.

27.4.3 Acquired Biotin Deficiency

- Biotinidase activity is normal in plasma;
- Biotin concentrations are low in plasma and urine;
- Carboxylase activities in lymphocytes are decreased and are promptly normalized after a single dose of oral biotin or after preincubation with biotin in vitro [1].

27.4.4 Prenatal Diagnosis

Prenatal diagnosis of HCS deficiency is possible by enzymatic studies in cultured chorionic villi or amniotic fluid cells or by demonstration of elevated concentrations of metabolites by stable isotope dilution techniques in amniotic fluid. Organic acid analysis in milder forms of HCS deficiency may fail to show an affected fetus, necessitating enzymatic investigation in these cases [5]. Prenatal diagnosis allows rational prenatal therapy, preventing severe metabolic derangement in the early neonatal period [5, 45]. Biotinidase can be measured in chorionic villi or cultured amniotic fluid cells but, in our opinion, this is not warranted, because prenatal treatment is not necessary.

27.5 Treatment and Prognosis

With the exception of some cases of HCS deficiency, both inherited disorders can be treated effectively with oral biotin in pharmacologic doses. No adverse effects have been observed from such therapy over a more than 20-year experience of treating biotinidase deficiency [39] and, importantly, there is no accumulation of biocytin in body fluids [23], which was previously suspected to be a possible risk.

Restriction of protein intake is not necessary except in very severe cases of HCS deficiency. Acutely ill patients with metabolic decompensation require general emergency treatment in addition to biotin therapy (▶ Chap. 4).

27.5.1 Holocarboxylase Synthetase Deficiency

The required dose of biotin is dependent on the severity of the enzyme defect and has to be assessed individually [1]. Most patients have shown a good clinical response to 10–20 mg/day, although some may require higher doses, i.e. 40-200 mg/day [1, 3, 45–47]. In spite of apparently complete clinical recovery, some patients continue to excrete abnormal metabolites (particularly 3-hydroxyisovalerate), a finding that correlates inversely with the actual level of carboxylase activity in lymphocytes. Exceptionally, persistent clinical and biochemical abnormalities have been observed despite treatment with very high doses of biotin [1, 32, 45–47]. All patients with HCS deficiency have at least partially responded to pharmacological doses of biotin with the exception of those homozygous for the missense mutation L216R [32].

To date, the prognosis for most surviving, well-treated patients with HCS deficiency seems to be good, with the exception of those who show only a partial or no response to biotin [1, 32, 45–47]. Careful follow-up studies are needed to judge the long-term outcome. In one patient, followed for

9 years and treated prenatally and from the age of 3.5 months with 6 mg biotin/day, some difficulties in fine motor tasks were obvious at the age of 9 years [48]. In five Japanese patients (four families), the intelligence quotient (IQ) at the age of 5–10 years varied between 64 and 80 [45]. Four of these patients had a severe neonatal onset form, and one of them (IQ=64) was treated prenatally. Three of these patients showed recurrent respiratory infections, metabolic acidosis and organic aciduria despite high-dose (20–60 mg/day) biotin therapy. However, irreversible neurological auditory-visual deficits, as described for biotinidase deficiency, have not been reported. Prenatal biotin treatment (10 mg/day) has been reported in a few pregnancies [5, 45]. It is unclear whether prenatal treatment is essential; treatment of at-risk children immediately after birth may be sufficient.

27.5.2 Biotinidase Deficiency

Introduction of neonatal screening programs has resulted in the detection of asymptomatic patients with residual biotinidase activity [27]. Based on measurement of plasma biotinidase activity, the patients are classified into three main groups.
1. Patients with profound biotinidase deficiency, with less than 10% of mean normal serum biotinidase activity. Using a sensitive method with the natural substrate biocytin, we classify these patients further into those with complete deficiency (undetectable activity, limit of detection ~0.05% of the mean normal value) and those with residual biotinidase activity up to 10% [14].
2. Patients with partial biotinidase deficiency, with 10–30% residual activity.
3. Patients with decreased affinity of biotinidase for biocytin, i.e. Km variants [49].

Group 1

In early-diagnosed children with complete biotinidase deficiency, 5–10 mg of oral biotin per day promptly reverse or prevent all clinical and biochemical abnormalities. For chronic treatment, the same dose is recommended. Under careful clinical and biochemical control, it may be possible to reduce the daily dose of biotin to 2.5 mg. However, biotin has to be given throughout life and regularly each day, since biotin depletion develops rapidly [7].

Neonatal screening for biotinidase deficiency [27] allows early diagnosis and effective treatment. In such patients, the diagnosis must be confirmed by quantitative measurement of biotinidase activity. Treatment should be instituted without delay, since patients may become biotin deficient within a few days after birth [7].

In patients who are diagnosed late, irreversible brain damage may have occurred before the commencement of treatment. In particular, auditory and visual deficits often persist in spite of biotin therapy [9, 10, 17–19], and intel-

lectual impairment and ataxia have been observed as long-term complications [9, 15, 17, 18].

Patients with residual activity up to 10%, usually detected by neonatal screening or family studies, may remain asymptomatic for several years or even until adulthood [12–14]. According to our experience with 61 such patients (52 families), however, they show a great risk of becoming biotin deficient and should be treated with, e.g., 2.5 mg of biotin per day [14, 27, 39].

Group 2

Patients with partial biotinidase deficiency (10–30% residual activity) are mostly detected by neonatal screening and in family studies and usually remain asymptomatic. One infant with about 30% enzyme activity developed hypotonia, skin rash and hair loss during an episode of gastroenteritis at 6 months of age. This was reversed by biotin therapy [50]. We showed that among 24 patients with 14–25% serum biotinidase activity studied at the age of 8 months to 8 years, 16 patients had a subnormal biotin concentration in at least one plasma sample, with a tendency toward lower values with increasing age [51]. Therefore, it seems necessary to regularly control patients with 10-30% of residual activity and to supplement patients with borderline abnormalities with small doses of biotin, e.g., 2.5–5 mg/week.

Group 3

Among 201 patients (176 families), we found ten patients (eight families) with a K_m defect. In the routine colorimetric biotinidase assay with 0.15 mmol/l biotinyl-p-aminobenzoate as substrate, six of these patients (five families) showed profound deficiency (0.94–3% residual activity), whereas four patients (three families) showed partial deficiency (18–20% residual activity). The index patient in all five families with profound deficiency presented with a severe clinical illness [16, 49], and one of the patients with partial deficiency, although apparently asymptomatic, had marginal biotin deficiency at the age of 2 years [49]. These results show the importance of testing all patients with residual biotinidase activity for a K_m defect. They all seem to have a high risk of becoming biotin deficient and, therefore, must be treated with biotin.

27.5.3 Biotin-Responsive Basal Ganglia Disease

All clinical symptoms of BRBGD disappear within a few days with the administration of high doses of biotin (5–10 mg/kg/day) if the patient is treated early. They reappear within 1 month if biotin is discontinued. Patients diagnosed late, or who have had repeated episodes, suffer from residual symptoms such as paraparesis, mild mental retardation or dystonia [19a].

References

1. Baumgartner ER, Suormala T (1997) Multiple carboxylase deficiency: inherited and acquired disorders of biotin metabolism. Int J Vit Nutr Res 67:377-384

2. Sakamoto O, Suzuki Y, Li X et al (2000) Diagnosis and molecular analysis of an atypical case of holocarboxylase synthetase deficiency. Eur J Pediatr 159:18-22

3. Suormala T, Fowler B, Duran M et al (1997) Five patients with a biotin-responsive defect in holocarboxylase formation: evaluation of responsiveness to biotin therapy in vivo and comparative studies in vitro. Pediatr Res 41:666-673

4. Sherwood WG, Saunders M, Robinson BH et al (1982) Lactic acidosis in biotin-responsive multiple carboxylase deficiency caused by holocarboxylase synthetase deficiency of early and late onset. J Pediatr 101:546-550

5. Suormala T, Fowler B, Jakobs C et al (1998) Late-onset holocarboxylase synthetase-deficiency: pre- and post-natal diagnosis and evaluation of effectiveness of antenatal biotin therapy. Eur J Pediatr 157:570-575

6. Seymons K, De Moor A, De Raeve H, Lambert J (2004) Dermatologic signs of biotin deficiency leading to the diagnosis of multiple carboxylase deficiency. Pediatr Dermatol 21:231-235

7. Baumgartner ER, Suormala TM, Wick H, Bausch J, Bonjour JP (1985) Biotinidase deficiency associated with renal loss of biocytin and biotin. Ann NY Acad Sci 447:272-286

8. Baumgartner ER, Suormala TM, Wick H et al (1989) Biotinidase deficiency: a cause of subacute necrotizing encephalomyelopathy (Leigh syndrome). Report of a case with lethal outcome. Pediatr Res 26:260-266

9. Wastell HJ, Bartlett K, Dale G, Shein A (1988) Biotinidase deficiency: a survey of 10 cases. Arch Dis Child 63:1244-1249

10. Wolf B, Heard GS, Weissbecker KA et al (1985) Biotinidase deficiency: initial clinical features and rapid diagnosis. Ann Neurol 18:614-617

11. Wolf B, Pompionio RJ, Norrgard KJ et al (1998) Delayed onset profound biotinidase deficiency. J Pediatr 132:362-365

12. Wolf B, Norrgard KJ, Pomponio RJ et al (1997) Profound biotinidase deficiency in two asymptomatic adults. Am J Med Genet 73:5-9

13. Moeslinger D, Stockler-Ipsiroglu S, Scheibenreiter S et al (2001) Clinical and neuropsychological outcome in 33 patients with biotinidase deficiency ascertained by nationwide newborn screening and family studies in Austria. Eur J Pediatr 160:277-282

14. Suormala TM, Baumgartner ER, Wick H et al (1990) Comparison of patients with complete and partial biotinidase deficiency: biochemical studies. J Inherit Metab Dis 13:76-92

15. Grunewald S, Champion MP, Leonard JV, Schaper J, Morris AA (2004) Biotinidase deficiency: a treatable leukoencephalopathy. Neuropediatrics 35:211-216

16. Ramaekers VTH, Suormala TM, Brab M et al (1992) A biotinidase Km variant causing late onset bilateral optic neuropathy. Arch Dis Child 67:115-119

17. Weber P, Scholl S, Baumgartner ER (2004) Outcome in patients with profound biotinidase deficiency: relevance of newborn screening. Dev Med Child Neurol 46:481-484

18. Wolf B, Spencer R, Gleason T (2002) Hearing loss is a common feature of symptomatic children with profound biotinidase deficiency. J Pediatr 140:242-246

19. Duran M, Baumgartner ER, Suormala TM et al (1993) Cerebrospinal fluid organic acids in biotinidase deficiency. J Inherit Metab Dis 16:513-516

19a. Ozand PT, Gascon GG, Al Essa M et al (1998) Biotin-responsive basal ganglia disease: a novel entity. Brain 121:1267-1279

20. Aoki Y, Suzuki Y, Li X et al (1997) Characterization of mutant holocarboxylase synthetase (HCS): a Km for biotin was not elevated in a patient with HCS deficiency. Pediatr Res 42:849-854

21. Burri BJ, Sweetman L, Nyhan WL (1985) Heterogeneity in holocarboxylase synthetase in patients with biotin-responsive multiple carboxylase deficiency. Am J Hum Genet 37: 326-337

22. Sakamoto O, Suzuki Y, Li X et al (1999) Relationship between kinetic properties of mutant enzyme and biochemical and clinical responsiveness to biotin in holocarboxylase synthetase deficiency. Pediatr Res 46:671-676

23. Suormala TM, Baumgartner ER, Bausch J et al (1988) Quantitative determination of biocytin in urine of patients with biotinidase deficiency using high-performance liquid chromatography (HPLC). Clin Chim Acta 177:253-270

24. Blom W, de Muinck Keizer SM, Scholte HR (1981) Acetyl-CoA carboxylase deficiency: An inborn error of de novo fatty acid synthesis. N Engl J Med 305:465-466

25. Baumgartner MR, Dantas MF, Suormala T et al (2004) Isolated 3-methylcrotonyl-CoA carboxylase deficiency: Evidence for an allele-specific dominant negative effect and responsivness to biotin therapy. Am J Hum Genet 75:790-800

25a. Zeng WQ, Al-Yamani E, Acierno JS Jr et al (2005) Biotin-responsive basal ganglia disease maps to 2q36.3 and is due to mutations in SLC19A3. Am J Hum Genet 77:16-26

26. Mardach R, Zempleni J, Wolf B et al (2002) Biotin dependency due to a defect in biotin transport. J Clin Invest 109:1617-1623

27. Wolf B (1991) Worldwide survey of neonatal screening for biotinidase deficiency. J Inherit Metab Dis 14:923-92725

28. Leon-Del-Rio A, Leclerc D, Akerman B, Wakamatsu N, Gravel RA (1995) Isolation of cDNA encoding human holocarboxylase synthetase by functional complementation of a biotin auxotroph of Escherichia coli. Proc Natl Acad Sci USA 92:4626-4630

29. Suzuki Y, Aoki Y, Ishida Y et al (1994) Isolation and characterization of mutations in the human holocarboxylase synthetase cDNA. Nat Genet 8:122-128

30. Cole H, Reynolds TR, Lockyer JM et al (1994) Human serum biotinidase. cDNA cloning, sequence, and characterization. J Biol Chem 269:6566-6570

31. Cole H, Weremovicz S, Morton CC, Wolf B (1994) Localization of serum biotinidase (BTD) to human chromosome 3 in Band p25. Genomics 22:662-663

32. Morrone A, Malvaglia S, Donati MA et al (2002) Clinical findings and biochemical and molecular analysis of four patients with holocarboxylase synthetase deficiency. Am J Med Genet 111:10-18

33. Tang NLS, Hui J, Yong CKK et al (2003) A genomic approach to mutation analysis of holocarboxylase synthetase gene in three Chinese patients with late-onset holocarboxylase synthetase deficiency. Clin Biochem 36:145-149

34. Yang X, Aoki Y, Li X et al (2001) Structure of human holocarboxylase synthetase gene and mutation spectrum of holocarboxylase synthetase deficiency. Hum Genet 109:526-534

35. The human gene mutation database. http://archive.uwcm.ac.uk/uwcm/mg/hgmd0.html

36. Dupuis L, Campeau E, Leclerc D, Gravel RA (1999) Mecanisms of biotin responsiveness in biotin-responsive multiple carboxylase deficiency. Mol Genet Metab 66:80-90

37. Soloranza-Vargas RS, Pacheco-Alvarez D, Leon-del-Rio A (2002) Holocarboxylase synthetase is an obligate participant in biotin-mediated regulation of its own expression and of biotin-dependent carboxylases mRNA levels in human cells. PNAS 99:5325-5330

38. Hymes J, Stanley CM, Wolf B (2001) Mutations in BTD causing biotinidase deficiency. Hum Mutat 18:375-381

39. Wolf B (2003) Biotinidase deficiency: new directions and practical concerns. Curr Treat Options Neurol 5:321-328

40. Pomponio RJ, Hymes J, Reynolds TR et al (1997) Mutation in the human biotinidase gene that causes profound biotinidase deficiency in symptomatic children: molecular, biochemical, and clinical analysis. Pediatr Res 42:840-848

41. Hymes J, Fleischhauer K, Wolf B (1995) Biotinylation of histones by human serum biotinidase: assessment of biotinyl-transferase activity in sera from normal individuals and children with biotinidase deficiency. Biochem Mol Med 56:76-83

42. Norrgard KJ, Pomponio RJ, Hymes J, Wolf B (1999) Mutations causing profound biotinidase deficiency in children ascertained by newborn screening in the United States occur at different frequencies than in symptomatic children. Pediatr Res 46:20-27

43. Suzuki Y, Aoki Y, Sakamoto O et al (1996) Enzymatic diagnosis of holocarboxylase synthetase deficiency using apo-carboxyl carrier protein as a substrate. Clin Chim Acta 251:41-52

44. Baur B, Suormala T, Bernoulli C, Baumgartner ER (1998) Biotin determination by three different methods: specificity and application to urine and plasma ultrafiltrates of patients with and without disorders in biotin metabolism. Int J Vit Nutr Res 68:300-308

45. Aoki Y, Suzuki Y, Sakamoto O et al (1995) Molecular analysis of holocarboxylase synthetase deficiency: a missense mutation and a single base deletion are predominant in Japanese patients. Biochim Biophys Acta 1272:168-174

46. Santer R, Muhle H, Suormala T et al (2003) Partial response to biotin therapy in a patient with holocarboxylase synthetase deficiency: clinical, biochemical, and molecular genetic aspects. Mol Genet Metab 79:160-166

47. Wolf B, Hsia YE, Sweetman L et al (1981) Multiple carboxylase deficiency: clinical and biochemical improvement following neonatal biotin treatment. Pediatrics 68:113-118

48. Michalski AJ, Berry GT, Segal S (1989) Holocarboxylase synthetase deficiency: 9-year follow-up of a patient on chronic biotin therapy and a review of the literature. J Inherit Metab Dis 12:312-316

49. Suormala T, Ramaekers VTH, Schweitzer S et al (1995) Biotinidase Km-variants: detection and detailed biochemical investigations. J Inherit Metab Dis 18:689-700

50. Secor McVoy JR, Levy HL, Lawler M et al (1990) Partial biotinidase deficiency: clinical and biochemical features. J Pediatr 116:78-83

51. Bernoulli C, Suormala T, Baur B, Baumgartner ER (1998) A sensitive method for the determination of biotin in plasma and CSF, and application to partial biotinidase deficiency. J Inherit Metab Dis 21[Suppl 2]46:92

28   Disorders of Cobalamin and Folate Transport and Metabolism

David S. Rosenblatt, Brian Fowler

Cobalamin Transport and Metabolism

Cobalamin (cbl or vitamin B_{12}) is a cobalt-containing water-soluble vitamin that is synthesized by lower organisms but not by higher plants and animals. In the human diet, its only source is animal products in which it has accumulated by microbial synthesis. Cbl is needed for only two reactions in man, but its metabolism involves complex absorption and transport systems and multiple intracellular conversions. As methylcobalamin, it is a cofactor of the cytoplasmic enzyme methionine synthase. As adenosylcobalamin, it is a cofactor of the mitochondrial enzyme methylmalonyl-coenzyme A mutase, which is involved in the catabolism of valine, threonine and odd-chain fatty acids into succinyl-CoA, an intermediate of the Krebs cycle.

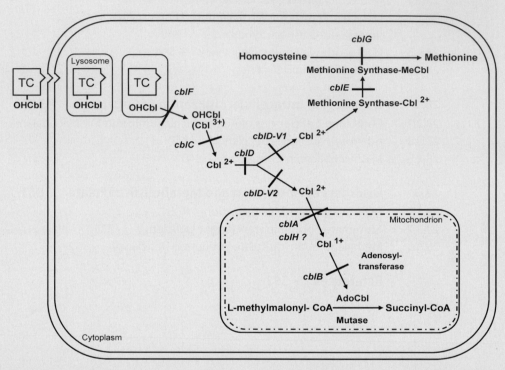

◻ **Fig. 28.1.** Cobalamin (*Cbl*) endocytosis and intracellular metabolism. The cytoplasmic, lysosomal, and mitochondrial compartments are indicated. *AdoCbl*, adenosylcobalamin; *CoA*, coenzyme A; *MeCbl*, methylcobalamin; *OHCbl*, hydroxycobalamin; *TC*, transcobalamin (previously TCII); *V1*, variant 1; *V2*, variant 2; $1^+, 2^+, 3^+$ refer to the oxidation state of the central cobalt of Cbl. Letters *A-H* refer to the sites of blocks. Enzyme defects are indicated by *solid bars*

For patients with inherited disorders affecting cobalamin (Cbl) absorption, the main clinical finding is megaloblastic anemia. Except for transcobalamin (TC) deficiency, the serum Cbl level will usually be low. Patients with disorders of intracellular Cbl metabolism show elevations of homocysteine or methylmalonic acid, either alone or in combination. The serum Cbl level is not usually low. For those disorders that affect methylcobalamin (MeCbl) formation, the major manifestations include megaloblastic anemia secondary to folate deficiency and neurological abnormalities presumably secondary to methionine deficiency or homocysteine elevation. The main findings in those disorders that affect adenosylcobalamin (AdoCbl) formation, are secondary to elevated methylmalonic acid and resultant acidosis.

Inherited disorders of cobalamin (Cbl) metabolism are divided into those involving absorption and transport and those involving intracellular utilization [1–5].

28.1 Disorders of Absorption and Transport of Cobalamin

Absorption of dietary Cbl involves first binding to a glycoprotein (R binder, haptocorrin) in the saliva. In the intestine, haptocorrin is digested by proteases, allowing the Cbl to bind to intrinsic factor (IF), which is produced in the stomach by parietal cells. Using a specific receptor, the IFCbl complex enters the enterocyte. Following release from this complex Cbl binds to transcobalamin (TC), the physiologically important circulating Cbl-binding protein, forming TC-Cbl, which is then slowly released into the portal vein. Inherited defects of several of these steps are known.

28.1.1 Hereditary Intrinsic Factor Deficiency

Clinical Presentation

Presentation is usually from one to 5 years of age but in cases of partial deficiency, can be delayed until adolescence or adulthood. Patients present with megaloblastic anemia as the main finding, together with failure to thrive, often with vomiting, alternating diarrhea and constipation, anorexia and irritability [6–8]. Hepatosplenomegaly, stomatitis or atrophic glossitis, developmental delay, and myelopathy or peripheral neuropathy may also be found.

Metabolic Derangement

IF is either absent or immunologically detectable but nonfunctional. There have been reports of IF with reduced affinity for Cbl, receptor or increased susceptibility to proteolysis [7–9].

Genetics

At least 45 patients of both sexes have been reported, and inheritance is autosomal recessive. A cDNA has been characterized, and the gene is localized on chromosome 11q13 [10]. A recently described variant of the gastric IF (GIF) gene, 68A→G, is probably not a disease causing mutation but could serve as a marker for inheritance of the disorder [11]. A 4-bp deletion (c183_186delGAAT) in the coding region of the GIF gene was identified as the cause of intrinsic factor deficiency in an 11 year-old girl with severe anemia and Cbl deficiency [12].

Diagnostic Tests

The hematological abnormalities in the defects of Cbl absorption and transport should be detected by measurement of red blood cell indices, complete blood count and bone marrow examination. Low serum Cbl levels are present. A deoxyuridine suppression test on marrow cells is useful but is not easily available in most clinical laboratories. In hereditary IF deficiency, in contrast to acquired forms of pernicious anemia, there is normal gastric acidity and normal gastric cytology. Cbl absorption, as measured by the Schilling test, is abnormal but is normalized when the labeled Cbl is mixed with a source of normal IF, such as gastric juice from an unaffected individual.

Treatment and Prognosis

IF deficiency can be treated initially with hydroxycobalamin (OHCbl, 1 mg/day intramuscularly) to replete body stores until biochemical and hematological values normalize. The subsequent dose of OHCbl required to maintain normal values may be as low as 0.25 mg every 3 months. If treatment is delayed, some neurological abnormalities may persist in spite of complete reversal of the hematological and biochemical findings.

28.1.2 Defective Transport of Cobalamin by Enterocytes (Imerslund-Gräsbeck Syndrome)

Clinical Presentation

Defective transport of Cbl by enterocytes, also known as Imerslund-Gräsbeck syndrome or megaloblastic anemia 1 (MGA1), is characterized by prominent megaloblastic anemia manifesting once fetal hepatic Cbl stores have been depleted. The disease usually appears between the ages of 1 year and 5 years, but onset may be even later [13–19]. Most patients have proteinuria and, in a few cases, this is of the tubular type, with all species of proteins represented rather than albumin alone. The literature on the renal pathology has been reviewed [20]. Although patients who excreted protein during childhood continued to excrete protein in adulthood, the renal lesions were not progressive [14]. Neurological abnormalities, such as spasticity, truncal ataxia and cerebral atrophy, may be present as a consequence of the Cbl deficiency.

Metabolic Derangement

This disorder is caused by defects of the IF-Cbl receptor, which has been recently shown to comprise two components. Cubilin was first purified as the IF-Cbl receptor from the proximal renal tubule [21–23]. Fyfe et al. demonstrated that a second component, amnionless, co-localizes with cubilin in the endocytic apparatus of polarized epithelial cells, forming a tightly bound complex that is essential for endocytic function [24]. Thus defective function of either protein may cause this disorder.

Genetics

About 250 cases have been reported and inheritance is autosomal recessive [19], with environmental factors affecting expression [22, 25]. Most patients are found in Finland, Norway, Saudi Arabia, Turkey, and among Sephardic Jews. The cubilin gene *(CUBN)* has been mapped to 10p12.1. A P1297L mutation was found in 31 of 34 disease chromosomes from 16 of 17 Finnish families segregating megaloblastic anemia [26]. Linkage studies in families from Norway, without mutations of the *CUBN* gene, led to the discovery of the amnionless gene *(AMN)*. A study of 42 MGA1 sibships confirmed *CUBN* mutations in Finnish and *AMN* mutations in Norwegian patients, and either among patients from other countries. Evidence was also provided for a possible additional MGA1 causing gene locus [27, 28].

Diagnostic Tests

In contrast to patients with IF deficiency, the Schilling test is not corrected by providing a source of human IF with the labeled Cbl [1]. The diagnosis is aided by finding low serum Cbl levels, megaloblastic anemia and proteinuria. Most of the reports in the literature do not comment on the levels of homocysteine and methylmalonic acid. Gastric morphology and pancreatic function are normal, there are no IF autoantibodies and IF levels are normal.

Treatment and Prognosis

Treatment with systemic OHCbl corrects the anemia and the neurologic findings, but not the proteinuria. As with hereditary IF deficiency, once Cbl stores are replete, low doses of systemic OHCbl may be sufficient to maintain normal hematological and biochemical values.

28.1.3   Haptocorrin (R Binder) Deficiency

Clinical Presentation

Very few cases have been described and it is not clear whether this entity has a distinct phenotype. Hematological findings are absent and neurological findings such as subacute combined degeneration of the spinal cord in one man in the fifth decade of life [29] and optic atrophy, ataxia, long-tract signs and dementia in another, may be coincidental.

Metabolic Derangement

The role of haptocorrin is uncertain but it could be involved in the scavenging of toxic Cbl analogs or in protecting methylcobalamin from photolysis [30]. Deficiency of haptocorrin has been described in isolation and in association with deficiency of other specific granule proteins such as lactoferrin [31].

Genetics

The haptocorrin gene has been cloned and mapped to chromosome 11q11-q12 [32, 33]. No mutations have been described in any patient with haptocorrin deficiency. Heterozygosity for haptocorrin deficiency appears to be associated with low serum cobalamin [34].

Diagnostic Tests

Serum Cbl levels are low, because most circulating Cbl is bound to haptocorrin. TC-Cbl levels are normal, and there are no hematologic findings of Cbl deficiency. A deficiency or absence of haptocorrin is found in plasma, saliva and leukocytes.

Treatment and Prognosis

It is uncertain whether treatment is warranted due to the lack of a clearly defined phenotype.

28.1.4   Transcobalamin Deficiency

Clinical Presentation

In TC deficiency, symptoms usually develop much earlier than in other disorders of Cbl absorption, mainly within the first few months of life. Even though the only TC in cord blood is of fetal origin, patients are not sick at birth. Presenting findings include pallor, failure to thrive, weakness and diarrhea. Although the anemia is usually megaloblastic, patients with pancytopenia or isolated erythroid hypoplasia have been described. Leukemia may be mistakenly diagnosed because of the presence of immature white cell precursors in an otherwise hypocellular marrow. Neurologic disease is not an initial finding but may develop with delayed treatment, with administration of folate in the absence of Cbl, or with inadequate Cbl treatment [35]. Neurological features include developmental delay, neuropathy, myelopathy and encephalopathy and rarely retinal degeneration [36]. Defective granulocyte function with both defective humoral and cellular immunity may occur.

Metabolic Derangement

The majority of patients have no immunologically detectable TC, although others have some detectable TC that is able to bind Cbl but lacks normal function [1, 37, 38].

Genetics

Inheritance is autosomal recessive; there have been at least 36 cases, including both twins and siblings [1, 35]. The *TC* gene has been mapped to chromosome 22q11.2-qter. Disease causing deletions, nonsense mutations, activation of an intra exonic cryptic splice site, as well as a number of polymorphic variants have been described [39–41].

Diagnostic Tests

Serum Cbl levels are not usually low, because the majority of serum Cbl is bound to haptocorrin and not to TC. However Cbl bound to TC, as reflected by the unsaturated vitamin-B_{12}-binding capacity, is low but this test must be performed before Cbl treatment is started. Since TC is involved in the transcytosis of Cbl through the enterocyte, the Schilling test may be abnormal in TC-deficient patients. In those patients in whom the Schilling test is normal, immunoreactive TC is found. Reports of levels of Cbl related metabolites are scarce and inconsistent. For example, normal plasma total homocysteine and moderately increased urine methylmalonic acid was reported in three patients and methylmalonic aciduria and homocystinuria, without specified levels in one patient [36, 42].

Study of TC synthesis in cultured fibroblasts or amniocytes allows both pre- and post-natal diagnosis in patients who do not synthesize TC [43]. DNA testing is possible for both diagnosis and heterozygote detection, in families in which the molecular defect has been identified. Recently developed assays, utilizing antibodies generated against recombinant human TC, allow reliable measurement of serum TC even in patients who have been treated with Cbl [44].

Treatment and Prognosis

Adequate treatment requires administration of oral or systemic OHCbl or cyanocobalamin (CNCbl) of 0.5–1 mg, initially daily then twice weekly, to maintain serum Cbl levels in the range of 1000–10,000 pg/ml. Intravenous Cbl is not recommended, because of the rapid loss of vitamin in the urine. Folic acid or folinic acid can reverse the megaloblastic anemia and has been used in doses up to 15 mg orally four times daily. However folates must never be given as the only therapy in TC deficiency, because of the danger of neurological deterioration.

28.2 Disorders of Intracellular Utilization of Cobalamin

A number of disorders of intracellular metabolism of Cbl have been classified as *cbl* mutants (*A-H*), based on the biochemical phenotype and on genetic complementation analysis.

28.2.1 Combined Deficiencies of Adenosylcobalamin and Methylcobalamin

Three distinct disorders are associated with functional defects in both methylmalonyl-coenzyme A (CoA) mutase and methionine synthase. They are characterized by both methylmalonic aciduria and homocystinuria.

Cobalamin F

Clinical Presentation. Of the eight known patients with *cblF* disease, seven presented in the first year of life. In this disease, a complete blood count and bone marrow examination may reveal megaloblastic anemia, neutropenia and thrombocytopenia. Other clinical findings can include failure to thrive, recurrent infections, developmental delay, lethargy, hypotonia, aspiration pneumonia, hepatomegaly and encephalopathy, pancytopenia, and heart anomalies (personal communication). The original infant girl had glossitis and stomatitis in the first week of life [45, 46]. She had severe feeding difficulties requiring tube feeding. Tooth abnormalities and dextrocardia were present. One infant died suddenly at home in the first year of life. One boy developed juvenile rheumatoid arthritis at the age of 4 years and a pigmented skin abnormality at 10 years.

Metabolic Derangement. The defect in *cblF* appears to be a failure of Cbl transport across the lysosomal membrane following degradation of TC in the lysosome. As a result, Cbl cannot be converted to either adenosylcobalamin (AdoCbl) or methylcobalamin (MeCbl). The inability of cblF patients to absorb oral Cbl suggests that IFCbl also has to pass through a lysosomal stage in the enterocyte before Cbl is released into the portal circulation.

Genetics. As both male and female patients of unaffected parents have been reported, inheritance is presumed to be autosomal recessive. The gene responsible for cblF has not been identified.

Diagnostic Tests. The serum Cbl level may be low, and the Schilling test has been abnormal in all patients tested. Usually, increased plasma total homocysteine, low to normal plasma methionine, homocystinuria and methylmalonic aciduria are found, although urine and plasma elevations of homocysteine were not reported in the original patient. Precise diagnosis of the inborn errors of Cbl metabolism requires tests in cultured fibroblasts. The incorporation of [14C] propionate into macromolecules is a good screen for the integrity of the methylmalonyl-CoA mutase reaction, and the incorporation of [14C] methyltetrahydrofolate or the conversion of labeled formate to methionine reliably measures the function of methionine synthase. The total incorporation of [57Co] CNCbl by fibroblasts and its conversion to both MeCbl and AdoCbl, can differentiate a number of the disorders. In fibroblasts from *cblF* patients, total incorporation of labeled CNCbl is elevated, but CNCbl is not converted to either AdoCbl or MeCbl. The entire label is found as free CNCbl in lysosomes. There is decreased incorporation of both labeled propionate and labeled methyltetrahydrofolate.

Treatment and Prognosis. Treatment with parenteral OHCbl (first daily and then biweekly, or even less frequently) at a dose of 1 mg/day seems to be effective in correcting the metabolic and clinical findings. Despite the fact that two Schilling tests showed an inability to absorb Cbl with or without IF, the original patient responded to oral Cbl before being switched to parenteral Cbl.

Cobalamin C

Clinical Presentation. This is the most frequent inborn error of Cbl metabolism, and several hundred patients are known [2, 47–50, 50a] (plus personal experience). Many were acutely ill in the first month of life, and most were diagnosed within the first year. The early-onset group shows feeding difficulties and lethargy, followed by progressive neurological deterioration, including hypotonia, hypertonia or both, abnormal movements or seizures, and coma. Severe pancytopenia or a non-regenerative anemia, which is not always associated with macrocytosis and hypersegmented neutrophils, but which is megaloblastic on bone-marrow examination, may be present. Patients may develop multisystem pathology, such as renal failure, hepatic dysfunction, cardiomyopathy, interstitial pneumonia or the hemolytic uremic syndrome characterized by widespread microangiopathy. Additional features include an unusual retinopathy consisting of perimacular hypopigmentation surrounded by a hyperpigmented ring and a more peripheral salt-and-pepper retinopathy sometimes accompanied by nystagmus, microcephaly and hydrocephalus. A small number of *cblC* patients were not diagnosed until after the first year of life and some as late as the end of the fourth decade of life [51–53, 53a]. The earlier-diagnosed patients in this group had findings overlapping those found in the younger-onset group. Major clinical findings in this late-onset *cblC* group included confusion, disorientation and gait abnormalities and incontinence. Macrocytic anemia was seen in only about a third of the oldest patients. Therefore, it is important to search for the *cblC* disorder by determination of metabolite levels in the presence of neurological findings alone.

Metabolic Derangement. The exact defect in the *cblC* disorder remains undefined but clearly involves an early step in intracellular Cbl processing, such as the reduction of the oxidation state of the central cobalt of Cbl from 3^+ to 2^+ following efflux of Cbl from the lysosome. Decreased activities of microsomal Cbl3$^+$ reductase, CNCbl-ligand transferase and a mitochondrial, reduced-nicotinamide-adenine-dinucleotide-linked aquacobalamin reductase have been described in fibroblast extracts but findings were not consistent [54–56]. Regardless of the exact mechanism if the reduction of Cbl does not occur, neither AdoCbl nor MeCbl can be formed, and Cbl does not bind to the two intracellular enzymes and leaves the cell.

Genetics. The gene responsible for *cblC* has been localized to chromosome 1 and recently identified [56a]. A common mutation, 271 dup A, accounts for 40% of all disease alleles. Inheritance is autosomal recessive. Prenatal diagnosis can be performed by measuring the incorporation of labeled propionate or labeled methyltetrahydrofolate and the synthesis of MeCbl and AdoCbl in cultured chorionic villus cells (but not native chorionic villus) and amniocytes and by measuring methylmalonic acid and total homocysteine levels in amniotic fluid. These methods cannot detect heterozygotes [56b].

Diagnostic Tests. Increased plasma total homocysteine, low to normal plasma methionine, homocystinuria and methylmalonic aciduria are the biochemical hallmarks of this disease. In general, methylmalonic acid levels seen are lower than those found in patients with methylmalonyl-CoA mutase deficiency, but higher than those seen in the Cbl transport defects. A complete blood count and bone marrow examination allow detection of the hematologic abnormalities.

Fibroblast studies show decreased incorporation of label from propionate, methyltetrahydrofolate (or formate) and CNCbl, and there is decreased synthesis of both AdoCbl and MeCbl. Cells fail to complement those of other *cblC* patients.

Treatment and Prognosis. Treatment with 1 mg/day OHCbl (parenteral) decreases the elevated metabolite levels, but these are not usually completely normalized. In one comprehensive study, oral OHCbl was found to be insufficient, and both folinic acid and carnitine were ineffective. Daily oral betaine (250 mg/kg/day) with twice weekly systemic OHCbl (1 mg/day) resulted in normalization of methionine and homocysteine levels and decreased methylmalonic aciduria [57]. Even though oral administration of OHCbl appears not to be effective, this route was reported to be successful in one patient [58].

Of a group of 44 patients with onset in the first year of life, 13 died, and only one patient was neurologically intact, with other survivors described as having severe or moderate impairment. Survival with mild to moderate disability was found in the patients who had a later onset [50].

Cobalamin D

Clinical Presentation. Until recently just two males from one sibship were known to belong to the *cblD* complementation group [59–61]. The elder sibling was diagnosed with behavioral problems and mild mental retardation at the age of 14 years. He had ataxia and nystagmus. Suormala et al. recently described three patients indicating heterogeneity of the *cblD* defect [62]. One patient with isolated methylmalonic aciduria presented prematurely with respiratory distress, cranial hemorrhage, necrotizing enterocolitis and convulsions but without anemia. Two unrelated patients presented with isolated homocystinuria, megaloblastic anemia and neurological changes but without metabolic decompensation.

Metabolic Derangement. The *cblD* defect can cause deficient synthesis of both AdoCbl and MeCbl together or either in isolation. This points to a multifunctional protein, or at least three different gene products involved in Cbl metabolism between the reduction of Cbl-3$^+$ and specific cobalamin coenzyme synthesis.

Genetics. All five subjects belonging to the *cblD* complementation group are male so that sex linkage cannot be excluded.

Diagnostic Tests. Methylmalonic aciduria with or without increased plasma total homocysteine and homocystinuria, or isolated methylmalonic aciduria may be found. Although the original patient showed no megaloblastic anemia, the deoxyuridine-suppression test was abnormal. In fibroblast studies findings can be similar to those of the *cblC*, *cblA* or *cblE/G* defects although differences in the severity and responsiveness to addition of OHCbl to the culture medium may be seen. This heterogeneity emphasis the necessity of complementation analysis to make a specific diagnosis in the *cbl* defects.

28.2.2   Adenosylcobalamin Deficiency

Clinical Presentation

Adenosylcobalamin (AdoCbl) deficiency comprises *cblA* and *cblB*, two disorders characterized by methylmalonic aciduria (MMA) which is often Cbl-responsive [2]. The phenotype resembles methylmalonyl-CoA mutase deficiency (► Chap. 19). Most patients have an acidotic crisis in the first year of life, many in the neonatal period. Symptoms are related to methylmalonic-acid accumulation and include vomiting, dehydration, tachypnea, lethargy, failure to thrive, developmental retardation, hypotonia and encephalopathy. The toxic levels of methylmalonic acid may result

in bone-marrow abnormalities and produce anemia, leukopenia and thrombocytopenia. Hyperammonemia, hyperglycinemia and ketonuria may be found.

Metabolic Derangement

The defect in *cblA* had been thought to lie in the reduction of the central cobalt of Cbl from the 2$^+$ to the 1$^+$oxidation state in mitochondria. The *MMAA* gene was proven to correspond to the cblA complementation group. Based on the domain characteristics of the protein sequence deduced from this gene it was proposed that the cblA protein is a transporter or an accessory protein involved in the translocation of Cbl into mitochondria [63].

A patient with all the clinical and biochemical features of *cblA* has been described, but cells from this patient complement those from other *cblA* patients. This implies that more than one step may be involved in the intramitochondrial reduction of Cbl or that intragenic complementation may occur among *cblA* lines [64].

The defect in *cblB* is deficiency of adenosyltransferase, the final intramitochondrial catalyst in the synthesis of AdoCbl [65].

Genetics

Male and female patients with *cblA* and *cblB* have been described, and parents of *cblB* patients have decreased adenosyltransferase activity, indicating autosomal-recessive inheritance. The *MMAA* gene has been localized to chromosome 4q31.1-2 [63]. It encodes a predicted polypeptide of 418 amino acid residues and its deduced sequence represents a domain structure belonging to the AAA ATPase superfamily. The precise role for the gene product has not been determined. Many mutations in the *MMAA* gene have now been described among *cblA* patients [66, 67].

The gene for adenosyltransferase has also been cloned, is localized to chromosome 12q24, and encodes a predicted protein of 250 amino acids. Examination of *cblB* patient cell lines revealed several disease causing mutant alleles, confirming that the *MMAB* gene, corresponds to the *cblB* complementation group [68, 68a].

Diagnostic Tests

Total serum Cbl is usually normal, and there is markedly increased methylmalonic aciduria (0.8–1.7 mmol/day; normal <0.04 mmol/day) but no increase of plasma total homocysteine or homocystinuria. A decrease in the level of methylmalonic-acid excretion in response to Cbl therapy is useful in distinguishing these disorders from methylmalonyl-CoA-mutase deficiency. The exact differentiation of *cblA* and *cblB* from mutase deficiency depends on fibroblast studies. In both *cblA* and *cblB* levels of methylmalonyl-CoA mutase are normal in the presence of added AdoCbl. The incorporation of labeled propionate is decreased in both *cblA* and *cblB* and is usually responsive to the addition of OHCbl to the culture medium. Uptake of labeled CNCbl is normal but there is decreased synthesis of AdoCbl. Adenosyltransferase activity is clearly deficient in *cblB*, but normal in *cblA* fibroblast extracts. Complementation analysis allows confirmation of the mutant class.

Treatment and Prognosis

Most of these patients respond to protein restriction and to OHCbl treatment, either 10 mg orally daily or 1 mg intramuscularly, once or twice weekly. For details of the planning of a protein-restricted diet, ▶ Chap. 17. Some patients appear to become resistant to Cbl treatment. Therapy with AdoCbl has been attempted in *cblB* with and without success, and it may be that AdoCbl does not reach the target enzyme intact. There have been reports of prenatal therapy with Cbl in AdoCbl deficiency. Most (90%) *cblA* patients improve on Cbl therapy, with 70% doing well long term. It must be noted that only 40% of cblB patients respond to Cbl, and their long-term survival is poorer [69].

28.2.3 Methylcobalamin Deficiency

Clinical Presentation

Methylcobalamin (MeCbl) deficiency comprises *cblE* and *cblG*. The most common clinical findings are megaloblastic anemia and neurological disease [70, 72–74]. The latter include poor feeding, vomiting, failure to thrive, cerebral atrophy, developmental delay, nystagmus, hypotonia or hypertonia, ataxia, seizures and blindness. Cerebral atrophy may be seen on imaging studies of the central nervous system, and at least one *cblE* patient showed a spinal-cord cystic lesion on autopsy. Most patients are symptomatic in the first year of life, but one *cblG* patient was not diagnosed until age 21 years and carried a misdiagnosis of multiple sclerosis [75]. Another cblG patient, who was diagnosed during his fourth decade of life, had mainly psychiatric symptoms. Two patients with minimal findings and without clear neurological features have also been reported [76].

Metabolic Derangement

The defect in *cblE* is deficiency of the enzyme, methionine synthase reductase, which is required for the activation by reductive methylation of the methionine synthase apoenzyme. The *cblG* defect is caused by deficient activity of the methionine synthase apoenzyme itself.

Genetics

There are at least 27 *cblE* and 27 *cblG* patients known. A cDNA for methionine-synthase reductase has been cloned, and mutations have been detected in *cblE* patients [77]. The methionine-synthase-reductase gene has been localized to chromosome 5p15.2–15.3. Mutations in the methionine-synthase gene have been found in *cblG* patients following

cloning of the cDNA for the gene on chromosome 1q43 [78, 79]. Patients with the *cblG* variant of methionine-synthase deficiency have null mutations [80]. Where both mutations are known in a patient, molecular analysis can be used for carrier detection in the family and for prenatal diagnosis.

Diagnostic Tests

Homocystinuria and hyperhomocysteinemia are almost always found in the absence of methylmalonic acidemia. However, one *cblE* patient had transient unexplained methylmalonic aciduria. Hypomethioninemia and cystathioninemia may be present, and there may be increased serine in the urine. A complete blood count and bone marrow examination will detect the hematological manifestations. Fibroblast extracts from *cblE* patients have normal activity of methionine synthase in the standard assay, but deficient activity can be found when the assay is performed under limiting reducing conditions [70, 76]. Cell extracts from *cblG* patients have decreased methionine synthase activity in the presence of excess reducing agent. Incorporation of labelled methyltetrahydrofolate or formation of methionine from labeled formate is decreased in cultured fibroblasts from both *cblE* and *cblG* patients. Uptake of CNCbl is normal but synthesis of MeCbl is decreased in both disorders. In some *cblG* patients (*cblG* variants) no Cbl forms are bound to methionine synthase following incubation in labeled CNCbl. Complementation analysis distinguishes *cblE* from *cblG* patients.

Treatment and Prognosis

Both of these disorders are treated with OHCbl or MeCbl, 1 mg intramuscularly, first daily and then once or twice weekly. Although the metabolic abnormalities are nearly always corrected, it is difficult to reverse the neurologic findings once they have developed. Treatment with betaine (250 mg/kg/day) has been used, and one *cblG* patient was treated with L-methionine (40 mg/kg/day) and had neurological improvement. Despite therapy, many patients with *cblG* and *cblE* show a poor outcome. In one family with *cblE*, there was successful prenatal diagnosis using cultured amniocytes, and the mother was treated with OHCbl twice per week beginning during the second trimester, and the baby was treated with OHCbl from birth. This boy has developed normally to age 14 years, in contrast to his older brother, who was not treated until after his metabolic decompensation in infancy and who is now 18 years old and has significant developmental delay. Some patients may benefit from high dose folic or folinic acid treatment.

Folate Metabolism

Folic acid (pteroylglutamic acid) is plentiful in foods such as liver, leafy vegetables, legumes and some fruit. Its metabolism involves reduction to dihydro- (DHF) and tetrahydrofolate (THF), followed by addition of a single-carbon unit, provided by histidine or serine; this carbon unit can be in various redox states (methyl, methylene, methenyl or formyl). Transfer of this single-carbon unit is essential for the endogenous formation of methionine, thymidylate (dTMP) and formylglycine-amide ribotide (FGAR) and formylaminoimidazole-carboxamide ribotide (FAICAR), two intermediates of purine synthesis. These reactions also allow regeneration of DHF and THF.

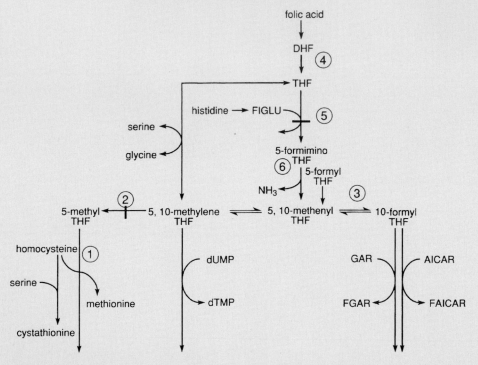

☐ **Fig. 28.2.** Folic acid metabolism. *1*, methionine synthase; *2*, methylenetetrahydrofolate reductase; *3*, methenyltetrahydro-folate cyclohydrolase; *4*, dihydrofolate reductase; *5*, glutamate formiminotransferase; *6*, formiminotetrahydrofolate cyclode-aminase; *AICAR*, aminoimidazole carboxamide ribotide; *DHF*, dihydrofolate, *dTMP*, deoxythymidine monophosphate; *dUMP*, deoxyuridine monophosphate; *FAICAR*, formylaminoimidazole carboxamide ribotide; *FGAR*, formylglycinamide ribotide; *FIGLU*, formiminoglutamate; *GAR*, glycinamide ribotide; *THF*, tetrahydro-folate. Enzyme defects are indicated by *solid bars*

Three confirmed inborn errors of folate absorption and metabolism have been described.

Hereditary folate malabsorption presents with severe megaloblastic anemia, due to the importance of dTMP and purine synthesis in hematopoiesis, and is usually associated with progressive nerurological deterioration.

Glutamate-formiminotransferase deficiency has been reported in association with various degrees of psychomotor retardation and megaloblastic anemia.

Severe *methylenetetrahydrofolate reductase (MTHFR) deficiency* presents mainly with developmental delay, often accompanied by seizures, microcephaly and findings related to cerebrovascular events. Patients typically show hyperhomocysteinemia without megaloblastic anemia.

28.3 Disorders of Absorption and Metabolism of Folate

28.3.1 Hereditary Folate Malabsorption

Clinical Presentation

This rare condition presents in the first months of life with severe megaloblastic anemia, diarrhea, stomatitis, failure to thrive and usually progressive neurological deterioration with seizures and sometimes with intracranial calcifications [81]. Peripheral neuropathy has been seen, as have partial defects in humoral and cellular immunity.

Metabolic Derangement

All patients have severely decreased absorption of oral folic acid or reduced folates, such as formyltetrahydrofolic acid (formyl-THF, folinic acid) or methyltetrahydrofolic acid. These patients provide the best evidence for the existence of a single transport system for folate at both the intestine and the choroid plexus. Transport of folates across other cell membranes is not affected in this disorder. The hematological and gastrointestinal manifestations of this disease, but not the neurological manifestations, can be reversed by pharmacologic, but relatively low levels of folate. Folate metabolism in cultured fibroblasts is normal. Recently a novel disorder was described with psychomotor retardation, spastic paraplegia, cerebellar ataxia and dyskinesia, associated with normal blood folate levels and low folate levels only in the cerebrospinal fluid (CSF) [82]. This cerebral folate deficiency syndrome has been recently found to be caused by an immune process against the cerebral folate carrier [82a].

Genetics

Several female patients are known, consanguinity has been noted in four families, and one patient's father had inter-

mediate levels of folate absorption, making autosomal-recessive inheritance likely. A cDNA for a putative intestinal folate transporter has been cloned, and it is identical to that for the reduced folate carrier [83]. To date, no report of mutations in these patients has appeared. The defect in hereditary folate malabsorption is not expressed in amniocytes or chorionic villus cells.

Diagnostic Tests

Measurement of serum, red blood cell and CSF folate levels and a complete blood count and bone marrow analysis should be performed. The most important diagnostic features are the severe megaloblastic anemia in the first few months of life, together with low serum folate levels. Measurements of related metabolite levels have been sporadically reported and inconsistently found abnormalities include increased excretion of formiminoglutamate, orotic aciduria, increased plasma sarcosine and cystathionine and low plasma methionine. Folate levels in CSF remain low even when blood levels are high enough to correct the megaloblastic anemia [84]. As mentioned, a number of patients have been reported with only neurological manifestations and low levels of CSF folate. Folate absorption may be directly investigated by measuring serum folate levels following an oral dose of between 5 and 100 mg of folic acid.

Treatment and Prognosis

High-dose oral folic acid (up to 60 mg daily) or lower parenteral doses in the physiological range correct the hematologic findings but are less effective in correcting the neurological findings and in raising the level of folate in the CSF. Both methyl-THF or folinic acid may be more effective in raising CSF levels and have been given in combination with high-dose oral folic acid. The clinical response to folates has varied among patients and, in some cases, seizures were worse after folate therapy was started. It is important to maintain blood and CSF folate in the normal range. If oral therapy does not raise CSF folate levels, parenteral therapy should be used. Intrathecal folate therapy may be considered if CSF levels of folate cannot be raised by other treatments although the required dose of folate is unknown. A recent report stresses that in some cases high oral doses of folinic acid (up to 400 mg orally daily) may eliminate the need for parenteral therapy [81]. The cerebral folate deficiency syndrome responds exclusively to folinic acid (10–20 mg/day) and not to folic acid [82a].

28.3.2 Glutamate-Formiminotransferase Deficiency

Clinical Presentation

Over a dozen patients have been described, but the clinical significance of this disorder is still unclear [4, 85, 86]. A mild and severe form has been postulated, although it is difficult to determine the importance of this distinction given the small number of patients. In the severe form of formiminotransferase deficiency there is both mental and physical retardation, abnormal electroencephalograms and dilatation of cerebral ventricles with cortical atrophy. Several of the patients had a folate-responsive megaloblastic anemia with macrocytosis and hypersegmentation of neutrophils. Patients ranged in age from 3 months to 42 years. Two had mental retardation, two had seizures and three had delayed speech as their presenting findings, and two were studied because they were the siblings of known patients. In the mild form there is no mental retardation, but there is a greater excretion of formiminoglutamate. Although mental retardation was described in most of the original patients from Japan, of the remaining eight patients, only three showed mental retardation.

Metabolic Derangement

Histidine catabolism is associated with a formimino-group transfer to THF, with the subsequent release of ammonia and the formation of 5,10-methenyl-THF. A single octameric enzyme catalyzes two different activities: glutamate formiminotransferase and formiminotetrahydrofolate cyclodeaminase. These activities are found only in the liver and kidney, and defects in either of these activities will result in formiminoglutamate excretion. It has been suggested (without any direct enzyme measurements) that the severe form of this disease is due to a block in the cyclodeaminase activity, whereas the mild form is due to a block in the formiminotransferase activity.

Genetics

Glutamate formiminotransferase deficiency has been found in both male and female children of unaffected parents. Consanguinity has not been reported; it has been presumed that the disease is inherited in an autosomal-recessive manner. Because of the lack of expression of the enzyme in cultured cells, prenatal diagnosis has not been possible, but it may be possible to measure formiminoglutamate levels in amniotic fluid. This has not been reported. The human gene has been cloned and localized to chromosome 21q22.3. Hilton et al. found mutant alleles in three patients and concluded that they represent the molecular basis for this disease, although expressed residual activity was 60% [87].

Diagnostic Tests

Elevated formiminoglutamate excretion and elevated levels of formiminoglutamate in the blood, only following a histidine load in the severe form, help to establish the diagnosis. A complete blood count and bone marrow examination may detect megaloblastic anemia. Normal to high serum folate levels are found, particularly in the mild form. Hyperhistidinemia and histidinuria have been reported. Two other metabolites that may be found in the urine are hydantoin-5-propionate, a stable oxidation product of the formiminoglutamate precursor, 4-imidazolone-5-propionate and 4-amino-5-imidazolecarboxamide, an intermediate of purine synthesis.

Treatment and Prognosis

It is not clear whether reducing formiminoglutamate excretion is of any clinical value. Although two patients in one family responded to folate therapy by reducing excretion of formiminoglutamate, six others did not. One of two patients responded to methionine supplementation. Pyridoxine and folic acid have been used to correct the megaloblastic anemia in one infant.

28.3.3 Methylenetetrahydrofolate Reductase Deficiency

This section is restricted to the severe form of this deficiency. The role of polymorphisms in methylenetetrahydrofolate reductase (MTHFR) with respect to the risk for common disease, such as neural tube defects or cardiovascular disease, is beyond the scope of this chapter (▶ [88] for a review).

Clinical Presentation

Approximately 100 patients with severe MTHFR deficiency have been described [2, 48, 85, 89–91], or are known to the authors. Most commonly, they were diagnosed in infancy, and more than half presented in the first year of life. The most common early manifestation was progressive encephalopathy with apnea, seizures and microcephaly. However, patients became symptomatic at any time from infancy to adulthood and, in the older patients, ataxic gait, psychiatric disorders (schizophrenia) and symptoms related to cerebrovascular events have been reported. An infant had extreme progressive brain atrophy, and the magnetic resonance image showed demyelination [92]. A 10-year-old boy had findings compatible with those of Angelman syndrome [93]. At least one adult with severe enzyme deficiency was completely asymptomatic. Autopsy findings have included dilated cerebral vessels, microgyria, hydrocephalus, perivascular changes, demyelination, gliosis, astrocytosis and macrophage infiltration. In some patients, thrombosis of both cerebral arteries and veins was the major cause of death. There have been reports of patients with findings similar to those seen in subacute degeneration of the spinal cord due to Cbl deficiency. Of note is the fact that MTHFR deficiency is not associated with megaloblastic anemia.

Metabolic Derangement

Methy-THF is the methyl donor for the conversion of homocysteine to methionine and, in MTHFR deficiency, the result is an elevation of total plasma homocysteine levels and decreased levels of methionine. The block in the conversion of methylene-THF to methyl-THF does not result in the trapping of folates as methyl-THF and does not interfere with the availability of reduced folates for purine and pyrimidine synthesis. This explains why patients do not have megaloblastic anemia. It is not clear whether the neuropathology in this disease results from the elevated homocysteine levels, from decreased methionine and resulting interference with methylation reactions or from some other metabolic effect. It has been reported that individuals with a severe deficiency in MTHFR may be at increased risk following exposure to nitrous oxide anesthesia [94].

Genetics

MTHFR deficiency is inherited as an autosomal-recessive disorder. There have been multiple affected children of both sexes with either unaffected parents or affected families with consanguinity. Prenatal diagnosis has been reported using amniocytes, and the enzyme is present in chorionic villi. A cDNA has been isolated, and the gene coding for MTHFR has been localized to chromosome 1p36.3. Over 50 mutations causing severe deficiency have been described, in addition to polymorphisms that result in intermediate enzyme activity and that may contribute to disease in the general population [95–101].

Diagnostic Tests

Because methyl-THF is the major circulating form of folate, serum folate levels may sometimes be low. There is a severe increase of plasma total homocysteine (often >100 μmol/l), together with plasma methionine levels ranging from zero to 18 μmol/l (mean:12 μmol/l, range of control means from different laboratories: 23–35 μmol/l). Homocystinuria is also seen, with a mean of 130 mmol/24 h and a range of 15–667 mmol/24 h. These values are much lower than are seen in cystathionine synthase deficiency. Although neurotransmitter levels have been measured in only a few patients, they are usually low. Direct measurement of MTHFR specific activity can be performed in liver, leukocytes, lymphocytes and cultured fibroblasts. In cultured fibroblasts, the specific activity is highly dependent on the stage of the culture cycle, with activity highest in confluent cells. There is a rough inverse correlation between the specific activity of the reductase in cultured fibroblasts and the clinical severity. There is a better inverse correlation between clinical severity and either the proportion of total cellular folate that is in the form of methyl-THF or the extent of labeled formate incorporation into methionine. The clinical heterogeneity in MTHFR deficiency can be seen at the biochemical level. Some of the patients have residual enzyme that is more thermolabile than the control enzyme [102]. Others have been shown to have an increased Km for NADPH [95].

Treatment and Prognosis

It is important to diagnose MTHFR deficiency early because, in the infantile forms, the only patients that have done well have been those who have been treated from birth. Early treatment with betaine following prenatal diagnosis has resulted in the best outcome [103–105]. Suggested doses have been in the range of 2–3 g/day (divided twice daily) in young infants and 6–9 g/day in children and adults. Betaine is a substrate for betaine methyltransferase, an enzyme that converts homocysteine to methionine, but is mainly active in the liver. Therefore, betaine may be expected to have the doubly beneficial effect of lowering homocysteine levels and raising methionine levels. Because betaine methyltransferase is not present in the brain, the central nervous system effects must be mediated through the effects of the circulating levels of metabolites. The dose of betaine should be modified according to plasma levels of homocysteine and methionine. Other therapeutic agents that have been used in MTHFR deficiency include folic acid or reduced folates, methionine, pyridoxine, cobalamin, and carnitine. Most of the treatment protocols omitting betaine have not been effective. Dramatic improvement was reported in a patient with severe enzyme deficiency following early introduction of methionine supplements [106].

References

1. Cooper BA, Rosenblatt DS (1987) Inherited defects of vitamin B_{12} metabolism. Ann Rev Nutr 7:291-320
2. Rosenblatt DS (2001) Inborn errors of folate and cobalamin metabolism. In: Carmel R, Jacobsen DW (eds) Homocysteine in health and disease. Cambridge University Press, New York, pp 244-258
3. Rosenblatt DS, Cooper BA (1987) Inherited disorders of vitamin B_{12} metabolism. Blood Rev 1:177-182
4. Whitehead VM, Rosenblatt DS, Cooper BA (1998) Megaloblastic anemia. In: Nathan DG, Orkin SH (eds) Hematology of infancy and childhood. Saunders, Philadelphia, pp 385-422
5. Rosenblatt DS, Cooper BA (1990) Inherited disorders of vitamin B12 utilization. Bioessays 12:331-334
6. Yang Y-M, Ducos R, Rosenberg AJ et al (1985) Cobalamin malabsorption in three siblings due to an abnormal intrinsic factor that is markedly susceptible to acid and proteolysis. J Clin Invest 76:2057-2065
7. Katz M, Mehlman CS, Allen RH (1974) Isolation and characterization of an abnormal intrinsic factor. J Clin Invest 53:1274-1283
8. Rothenberg SP, Quadros EV, Straus EW, Kapelner S (1984) An abnormal intrinsic factor (IF) molecule: A new cause of »pernicious anemia« (PA). Blood 64:41a
9. Spurling CL, Sacks MS, Jiji RM (1964) Juvenile pernicious anemia. N Engl J Med 271:995-1003
10. Hewitt JE, Gordon MM, Taggart RT et al (1991) Human gastric intrinsic factor: Characterization of cDNA and genomic clones and localization to human chromosome 11. Genomics 10:432
11. Gordon MM, Brada N, Remacha A et al (2004) A genetic polymorphism in the coding region of the gastric intrinsic factor gene (GIF) is associated with congenital intrinsic factor deficiency. Hum Mutat 23:85-91
12. Yassin F, Rothenberg SP, Rao S et al (2004) Identification of a 4-base deletion in the gene in inherited intrinsic factor deficiency. Blood 103:1515-1517

13. Grasbeck R (1972) Familial selective vitamin B12 malabsorption. N Engl J Med 287:358
14. Broch H, Imerslund O, Monn E et al (1984) Imerslund-Grasbeck anemia: A long-term follow-up study. Acta Paediatr Scand 73:248-253
15. el Mauhoub M, Sudarshan G, Aggarwal V, Banerjee G (1989) Imerslund-Grasbeck syndrome in a Libyan boy. Ann Trop Paediatr 9:180-181
16. el Bez M, Souid M, Mebazaa R, Ben Dridi MF (1992) L'anemie d'Imerslund-Grasbeck. A propos d‹un cas. Ann Pediatr (Paris) 39:305-308
17. Salameh MM, Banda RW, Mohdi AA (1991) Reversal of severe neurological abnormalities after vitamin B_{12} replacement in the Imerslund-Grasbeck syndrome. J Neurol 238:349-350
18. Kulkey O, Reusz G, Sallay P, Miltenyi M (1992) [Selective vitamin B_{12} absorption disorder (Imerslund-Grasbeck syndrome)] syndroma). Orv Hetil 133:3311-3313
19. Grasbeck R (1997) Selective cobalamin malabsorption and the cobalamin-intrinsic factor receptor. Acta Biochimica Polonica 44:725-733
20. Liang DC, Hsu HC, Huang FY, Wei KN (1991) Imerslund-Grasbeck syndrome in two brothers: renal biopsy and ultrastructure findings. Pediatr Hematol Oncol 8:361-365
21. Moestrup SK, Kozyraki R, Kristiansen M et al (1998) The intrinsic factor-vitamin B_{12} receptor and target of teratogenic antibodies is a megalin-binding peripheral membrane protein with homology to developmental proteins. J Biol Chem 273:5235-5242
22. Kozyraki R, Kristiansen M, Silahtaroglu A et al (1997) The human intrinsic factor-vitamin B_{12} receptor, *cubilin*: molecular characterization and chromosomal mapping of the gene to 10p within the autosomal recessive megaloblastic anemia (MGA1) region. Blood 91:3593-3600
23. Birn H, Verroust PJ, Nexo E et al (1997) Characterization of an epithelial ~460-kDa protein that facilitates endocytosis of intrinsic factor-vitamin B_{12} and binds receptor-associated protein. J Biol Chem 272:26497-26504
24. Fyfe JC, Madsen M, Hojrup P et al (2004) The functional cobalamin (vitamin B_{12})-intrinsic factor receptor is a novel complex of cubilin and amnionless. Blood 103:1573-1579
25. Aminoff M, Tahvanainen E, Gräsbeck R et al (1995) Selective intestinal malabsorption of vitamin B12 displays recessive mendelian inheritance: assignment of a locus to chromosome 10 by linkage. Am J Hum Genet 57:824-831
26. Aminoff M, Carter JE, Chadwick RB et al (1999) Mutations in CUBN, encoding the intrinsic factor-vitamin B12 receptor, cubilin, cause hereditary megaloblastic anaemia 1. Nat Genet 21:309-313
27. Tanner SM, Aminoff M, Wright FA et al (2003) Amnionless, essential for mouse gastrulation, is mutated in recessive hereditary megaloblastic anemia. Nat Genet 33:426-429
28. Tanner SM, Li Z, Bisson R et al (2004) Genetically heterogeneous selective intestinal malabsorption of vitamin B_{12}: founder effects, consanguinity, and high clinical awareness explain aggregations in Scandinavia and the Middle East. Hum Mutat 23:327-333
29. Carmel R (1983) R-binder deficiency. A clinically benign cause of cobalamin pseudodeficiency. J Am Med Assoc 250:1886-1890
30. Frisbie SM, Chance MR (2003) Human cobalophilin: the structure of bound methylcobalamin and a functional role in protecting methylcobalamin from photolysis. Biochemistry 32:13886-13892
31. Lin JC, Borregaard N, Liebman HA, Carmel R (2001) Deficiency of the specific granule proteins, R binder/transcobalamin I and lactoferrin, in plasma and saliva: a new disorder? Am J Med Genet 100:145-151
32. Johnston J, Bollekens J, Allen RH, Berliner N (1989) Structure of the cDNA encoding transcobalamin I, a neutrophil granule protein. J Biol Chem 264:15754
33. Johnston J, Yang-Feng T, Berliner N (1992) Genomic structure and mapping of the chromosomal gene for transcobalamin I (TCN1): comparison to human intrinsic factor [published erratum appears in Genomics Sep; 14(1):208]. Genomics 12:459-464
34. Carmel R (2003) Mild transcobalamin I (haptocorrin) deficiency and low serum cobalamin concentrations. Clin Chem 49:1367-1374
35. Hall CA (1992) The neurologic aspects of transcobalamin II deficiency. Br J Haematol 80:117
36. Souied EH, Benhamou N, Sterkers M et al (2001) Retinal degeneration associated with congenital transcobalamin II deficiency. Arch Ophthalmol 119:1076-1077
37. Haurani FI, Hall CA, Rubin R (1979) Megaloblastic anemia as a result of an abnormal transcobalamin II. J Clin Invest 64:1253-1259
38. Seligman PA, Steiner LL, Allen RH (1980) Studies of a patient with megaloblastic anemia and an abnormal transcobalamin II. N Engl J Med 303:1209-1212
39. Li N, Rosenblatt DS, Kamen BA, Seetharam S, Seetharam B (1994) Identification of two mutant alleles of transcobalamin II in an affected family. Hum Mol Genet 3:1835-1840
40. Li N, Rosenblatt DS, Seetharam B (1994) Nonsense mutations in human transcobalamin II deficiency. Biochem Biophys Res Commun 204:1111-1118
41. Namour F, Helfer A-C, Quadros EVet al (2003) Transcobalamin deficiency due to activation of an intra exonic cryptic splice site. Br J Hematol 123:915-920
42. Bibi H, Gelman-Kohan Z, Baumgartner ER, Rosenblatt DS (1999) Transcobalamin II deficiency with methylmalonic aciduria in three sisters. J Inherit Metab Dis 22:765-772
43. Rosenblatt DS, Hosack A, Matiaszuk N (1987) Expression of transcobalamin II by amniocytes. Prenat Diagn 7:35
44. Nexo E, Christensen A-L, Petersen TE, Fedosov SN (2000) Measurement of transcobalamin by ELISA. Clin Chem 46:1643-1649
45. Rosenblatt DS, Laframboise R, Pichette J, Langevin P et al (1986) New disorder of vitamin B12 metabolism (cobalamin F) presenting as methylmalonic aciduria. Pediatrics 78:51-54
46. Rosenblatt DS, Hosack A, Matiaszuk NV et al (1985) Defect in vitamin B12 release from lysosomes: newly described inborn error of vitamin B12 metabolism. Science 228:1319-1321
47. Mitchell GA, Watkins D, Melancon SB et al (1986) Clinical heterogeneity in cobalamin C variant of combined homocystinuria and methylmalonic aciduria. J Pediatr 108:410-415
48. Ogier de Baulny H, Gerard M, Saudubray JM, Zittoun J (1998) Remethylation defects: guidelines for clinical diagnosis and treatment. Eur J Pediatr 157:S77-S83
49. Traboulsi EI, Silva JC, Geraghty MT et al (1992) Ocular histopathologic characteristics of cobalamin C complementation type vitamin B12 defect with methylmalonic aciduria and homocystinuria. Am J Ophthalmol 113:269-280
50. Rosenblatt DS, Aspler AL, Shevell MI et al (1997) Clinical heterogeneity and prognosis in combined methylmalonic aciduria and homocystinuria (cblC). J Inherit Metab Dis 20:528-538
50a. Huemer M, Simma B, Fowler B et al (2005) Prenatal and postnatal treatment in cobalamin C defect. J Pediatr 147:469-472
51. Gold R, Baumgardner R, Fowler B et al (1995) Hereditary defect of cobalamin metabolism (homocystinuria and methylmalonic aciduria) of juvenile onset resembling multiple sclerosis. Neurol Neurosurg Psychiatr 60:107-108
52. Bodamer OAF, Rosenblatt DS, Appel SH, Beaudet AL (2001) Adult-onset combined methylmalonic aciduria and homocystinuria (cblC). Neurology 56:1113-1114
53. Van Hove JLK, Van Damme-Lombaerts R, Grunewald S et al (2002) Cobalamin disorder Cbl-C presenting with late-onset thrombotic microangiopathy. Am J Med Genet 111:195-201
53a. Guigonis V, Fremeaux-Bacchi V, Giraudier S (2005) Late-onset thrombocytic microangiopathy caused by cblC disease: association with a factor H mutation. Am J Kidney Dis 45:588-595

54. Watanabe F, Saido H, Yamaji R et al (1996) Mitochondrial NADH- or NADP-Linked Aquacobalamin reductase activity is low in human skin fibroblasts with defects in synthesis of cobalamin coenzymes. J Nutr 126:2947-2951

55. Pezacka EH, Rosenblatt DS (1994) Intracellular metabolism of cobalamin. Altered activities of β-axial-ligand transferase and microsomal cob(III)alamin reductase in cblC and cblD fibroblasts. In: Bath HR et al (eds) Advances in Thomas Addison's diseases. J Endocrinology, Bristol, pp 315-323

56. Pezacka EH (1993) Identification and characterization of two enzymes involved in the intracellular metabolism of cobalamin. Cyanocobalamin beta-ligand transferase and microsomal cob(III)-alamin reductase. Biochim Biophys Acta 1157:167-177

56a. Lerner-Ellis JP, Tirone JC, Pawelek PD et al (2006) Identification of the gene responsible for methylmalonic aciduria and homo-cystinuria, cblC type. Nat Genet 38:92-100

56b. Morel CF, Watkins D, Scott P et al (2005) Prenatal diagnosis for methylmalonic acidemia and inborn errors of vitamin B12 me-tabolism and transport. Mol Genet Metab 86:160-171

57. Bartholomew DW, Batshaw ML, Allen RH et al (1988) Therapeutic approaches to cobalamin-C methylmalonic acidemia and homo-cystinuria. J Pediatr 112:32-39

58. Bain MD, Jones MG, Fowler B, Besley GTN, Boxer LA, Chalmers RA (2003) Oral B12 treatment in Cbl C/D methylmalonic aciduria. J Inherit Metab Dis 26:42

59. Carmel R, Bedros AA, Mace JW, Goodman SI (1980) Congenital methylmalonic aciduria-homocystinuria with megaloblastic anemia: observations on response to hydroxocobalamin and on the effect of homocysteine and methionine on the deoxyuridine suppression test. Blood 55:570-579

60. Willard HF, Mellman IS, Rosenberg LE (1978) Genetic complemen-tation among inherited deficiencies of methylmalonyl-CoA mutase activity: Evidence for a new class of human cobalamin mutant. Am J Hum Genet 30:1-13

61. Mellman IH, Willard P, Youngdahl-Turner P, Rosenberg LE (1979) Cobalamin coenzyme synthesis in normal and mutant human fibroblasts; evidence for a processing enzyme activity deficient in cbl C cells. J Biol Chem 254:11847-11853

62. Suormala T, Baumgartner MR, Coelho D et al (2004) The cblD defect causes either isolated or combined deficiency of methyl-cobalamin and adenosylcobalamin synthesis. J Biol Chem 279:42742-42749

63. Dobson CM, Wai T, Leclerc D et al (2002) Identification of the gene responsible for the cblA complementation group of vitamin B12-responsive methylmalonic acidemia based on analysis of prokaryotic gene arrangements. Proc Natl Acad Sci USA 99:15554-15559

64. Cooper BA, Rosenblatt DS, Watkins D (1990) Methylmalonic aciduria due to a new defect in adenosylcobalamin accumulation by cells. Am J Hematol 34:115-120

65. Fenton WA, Rosenberg LE (1981) The defect in the cbl B class of human methylmalonic acidemia: deficiency of cob(I)alamin ad-enosyltransferase activity in extracts of cultured fibroblasts. Bio-chem Biophys Res Commun 98:283-289

66. Lerner-Ellis JP, Dobson CM, Wai T, Watkin D, Tirone JC, Leclerc D et al (2004) Mutations in the MMAA gene in patients with the cblA disorder of vitamin B12 metabolism. Hum Mutat 24:509-516

67. Yang X, Sakamoto O, Matsubara Y et al (2004) Mutation analy-sis of the MMAA and MMAB genes in Japanese patients with vitamin B12-responsive methylmalonic acidemia: identification of a prevalent MMAA mutation. Mol Genet Metab 82:329-333

68. Dobson CM, Wai T, Leclerc D et al (2002) Identification of the gene responsible for the cblB complementation group of vitamin B12-dependent methylmalonic aciduria. Hum Mol Genet 11:3361-3369

68a. Lerner-Ellis JP, Gradinger AB, Watkins D et al (2006) Mutation and biochemical analysis of patients belonging to the cblB com-plementation class of vitamin B12 dependent methylmalonic aciduria. Mol Genet Metab 87:219-225

69. Matsui SM, Mahoney MJ, Rosenberg LE (1983) The natural history of the inherited methylmalonic acidemias. N Engl J Med 308:857-861

70. Rosenblatt DS, Cooper BA, Pottier A et al (1984) Altered vitamin B12 metabolism in fibroblasts from a patient with megaloblastic anemia and homocystinuria due to a new defect in methionine biosynthesis. J Clin Invest 74:2149-2156

71. Gulati S, Chen Z, Brody LC, Rosenblatt DS, Banerjee R (1997) De-fects in auxillary redox proteins lead to functional methionine synthase deficiency. J Biol Chem 272:19171-19175

72. Schuh S, Rosenblatt DS, Cooper BA et al (1984) Homocystinuria and megaloblastic anemia responsive to vitamin B12 therapy. An inborn error of metabolism due to a defect in cobalamin meta-bolism. N Engl J Med 310:686-690

73. Watkins D, Rosenblatt DS (1989) Functional methionine synthase deficiency (cblE and cblG): Clinical and biochemical heterogene-ity. Am J Med Genet 34:427-434

74. Watkins D, Rosenblatt DS (1988) Genetic heterogeneity among patients with methylcobalamin deficiency: definition of two complementation groups, cblE and cblG. J Clin Invest 81:1690-1694

75. Carmel R, Watkins D, Goodman SI, Rosenblatt DS (1988) Heredi-tary defect of cobalamin metabolism (cblG mutation) presenting as a neurological disorder in adulthood. N Engl J Med 318:1738-1741

76. Vilaseca MA, Vilarinho L, Zavadakova P et al (2003) CblE type of homocysteine: mild clinical phenotype in two patients homo-zygous for a novel mutation in the MTRR gene. J Inherit Metab Dis 26:361-369

77. Leclerc D, Wilson A, Dumas R et al (1998) Cloning and mapping of a cDNA for methionine synthase reductase, a flavoprotein defec-tive in patients with homocystinuria. Proc Natl Acad Sci USA 95:3059-3064

78. Gulati S, Baker P, Li YN et al (1996) Defects in human methionine synthase in cblG patients. Hum Mol Genet 5:1859-1865

79. Leclerc D, Campeau E, Goyette P et al (1996) Human methionine synthase: cDNA cloning and identification of mutations in pa-tients of the cblG complementation group of folate/cobalamin disorders. Hum Mol Genet 5:1867-1874

80. Wilson A, Leclerc D, Saberi F et al (1997) Causal mutations in sib-lings with the cblG variant form of methionine synthase defi-ciency. Am J Hum Genet 61:A263

81. Geller J, Kronn D, Jayabose S, Sandoval C (2002) Hereditary folate malabsorption: Family report and review of the literature. Medi-cine 81:51-68

82. Ramaekers VT, Hausler M, Opladen T et al (2002) Psychomotor retardation, spastic paraplegia,cerebellar ataxia and dyskinesia associated with low 5-methyltetrahydrofolate in cerebrospinal fluide: A novel neurometabolic condition responding to folinic acid substitution. Neuropediatr 33:301-308

82a. Ramaekers VT, Rothenberg SP, Sequeira JM et al (2005) Auto-antibodies to folate receptors in the cerebral folate deficiency syndrome. N Engl J Med 352:1985-1991

83. Nguyen TT, Dyer DL, Dunning DD et al (1997) Human intestinal folate transport: cloning, expression, and distribution of comple-mentary RNA. Gastroenterology 112:783-791

84. Urbach J, Abrahamov A, Grossowicz N (1987) Congenital isolated folic acid malabsorption. Arch Dis Child 62:78-80

85. Erbe RW (1986) Inborn errors of folate metabolism. In: Blakley R, Whitehead VM (eds) Folates and pterins, vol 3:Nutritional, phar-macological and physiological aspects. Wiley, New York, pp 413-466

86. Erbe RW (1979) Genetic aspects of folate metabolism. Adv Hum Genet 9:293-354
87. Hilton JF, Christensen KE, Watkins D et al (2003) The molecular basis of glutamate formiminotransferase deficiency. Hum Mutat 22:67-73
88. Rozen R (2001) Polymorphisms of folate and cobalamin metabolism. In: Carmel R, Jacobsen DW (eds) Homocysteine in health and disease. Cambridge University Press, New York, pp 259-269
89. Visy JM, Le Coz P, Chadefaux B et al (1991) Homocystinuria due to 5,10-methylenetetrahydrofolate reductase deficiency revealed by stroke in adult siblings. Neurology 41:1313-1315
90. Haworth JC, Dilling LA, Surtees R et al (1993) Symptomatic and asymptomatic methythylenetetrahydrofolate reductase deficiency in two adult brothers. Am J Med Gen 45:572-576
91. Fowler B (1998) Genetic defects of folate and cobalamin metabolism. Eur J Pediatr 157:S60-S66
92. Sewell AC, Neirich U, Fowler B (1998) Early infantile methylenetetrahydrofolate reductase deficiency: a rare cause of progressive brain atrophy. J Inherit Metab Dis 21:22
93. Arn PH, Williams CA, Zori RT, Driscoll DJ, Rosenblatt DS (1998) Methylenetetrahydrofolate reductase deficiency in a patient with phenotypic findings of Angelman syndrome. Am J Med Genet 77:198-200
94. Selzer RR, Rosenblatt DS, Laxova R, Hogan K (2003) Adverse effect of nitrous oxide in a child with 5,10-methylenetetrahydrofolate reductase deficiency. N Engl J Med 349:45-50
95. Suormala T, Koch HG, Rummel T, Haberle J, Fowler B (2004) Methylenetetrahydrofolate reductase (MTHFR) deficiency: mutations and functional abnormalities. J Inherit Metab Dis 27:231
96. Goyette P, Christensen B, Rosenblatt DS, Rozen R (1996) Severe and mild mutations in cis for the methylenetetrahydrofolate (MTHFR) gene, and description of 5 novel mutations in MTHFR. Am J Hum Genet 59:1268-1275
97. Goyette P, Sumner JS, Milos R et al (1994) Human methylenetetrahydrofolate reductase: isolation of cDNA, mapping and mutation identification. Nat Genet 7:195-200
98. Rosenblatt DS (1994) Inborn errors of vitamin B_{12} metabolism: clinical and genetic heterogeneity. Int Pediatr 9:209-213
99. Sibani S, Leclerc D, Weisberg IS et al (2003) Characterization of mutations in severe methylenetetrahydrofolate reductase deficiency reveals an FAD-responsive mutation. Hum Mutat 21:509-520
100. Tonetti C, Saudubray J-M, Echenne B et al (2003) Relations between molecular and biological abnormalities in 11 families from siblings affected with methylenetetrahydrofolate reductase deficiency. Eur J Pediatr 162:466-475
101. Yano H, Nakaso K, Yasui K et al (2004) Mutations of the MTHFR gene (428C>T and [458G>T+459C>T]) markedly decrease MTHFR enzyme activity. Neurogenetics 5:135-140
102. Rosenblatt DS, Lue-Shing H, Arzoumanian A et al (1998) Methylenetetrahydrofolate reductase (MR) deficiency: Thermolability of residual MR activity, methionine synthase activity, and methylcobalamin levels in cultured fibroblasts. Biochem Med Met Biol 47:221-225
103. Wendel U, Bremer HJ (1983) Betaine in the treatment of homocystinuria due to 5,10-methylene THF reductase deficiency. J Pediatr 103:1007
104. Holme E, Kjellman B, Ronge E (1989) Betaine for treatment of homocystinuria caused by methylenetetrahydrofolate reductase deficiency. Arch Dis Child 64:1061-1064
105. Ronge E, Kjellman B (1996) Long term treatment with betaine in methylenetetrahydrofolate reductase deficiency. Arch Dis Child 74:239-241
106. Abeling NGGM, van Gennip AH, Blom H et al (1999) Rapid diagnosis and methionine administration: Basis for a favourable outcome in a patient with methylene-tetrahydrofolate reductase deficiency. J Inherit Metab Dis 22:240-242

VI Neurotransmitter and Small Peptide Disorders

29 Disorders of Neurotransmission

Jaak Jaeken, Cornelis Jakobs, Peter T. Clayton, Ron A. Wevers

Neurotransmitters

The neurotransmitter systems can be divided into mainly inhibitory aminoacidergic [γ-aminobutyric acid (GABA) and glycine], excitatory aminoacidergic (aspartate and glutamate), cholinergic (acetylcholine), monoaminergic (mainly adrenaline, noradrenaline, dopamine, and serotonin), and purinergic (adenosine and adenosine mono-, di-, and triphosphate). A rapidly growing list of peptides are also considered putative neurotransmitters.

GABA is formed from glutamic acid by glutamic acid decarboxylase (◘ Fig. 29.1). It is catabolized into succinic acid through the sequential action of two mitochondrial enzymes, GABA transaminase and succinic semialdehyde dehydrogenase. All these enzymes require pyridoxal phosphate as a coenzyme. Pyridoxal phosphate also intervenes in the synthesis of dopamine

and serotonin (◘ Fig. 29.2), and in many other pathways including the glycine cleavage system. A major inhibitory neurotransmitter, GABA is present in high concentration in the central nervous system, predominantly in the gray matter. GABA modulates brain activity by binding to sodium-independent, high-affinity, mostly $GABA_A$ receptors.

GLYCINE, a non-essential amino acid, is an intermediate in many metabolic processes but also one of the major inhibitory neurotransmitters in the central nervous system. The inhibitory *glycine receptors* are mostly found in the brain stem and spinal cord.

GLUTAMATE is the major excitatory neurotransmitter in the brain. Its function requires rapid uptake to replenish intracellular neuronal pools following extracellular release.

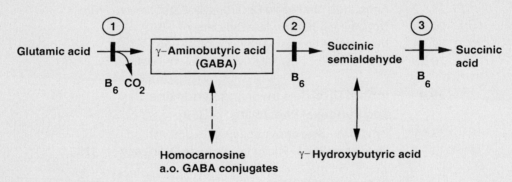

◘ **Fig. 29.1.** Brain metabolism of γ-aminobutyric acid (GABA). B_6, pyridoxal phosphate. **1,** Glutamic acid decarboxylase; **2,** GABA transaminase; **3,** succinic semialdehyde dehydrogenase.

Dotted arrow indicates reactions postulated. Enzyme defects are depicted by *solid bars*

This chapter deals mainly with inborn errors of neurotransmitter metabolism. Defects of their receptors and transporters, and disorders involving pyridoxine (vitamin B_6) and its derivative, pyridoxal phosphate, a cofactor required for the synthesis of several neurotransmitters, are also discussed.

Two defects of GABA metabolism are known: the very rare, severe, and untreatable *GABA transaminase deficiency*, and the much more frequent *succinic semialdehyde dehydrogenase (SSADH) deficiency* which, to some extent, responds to GABA transaminase inhibition. *Hyperekplexia* is a dominantly inherited defect of the α_1 subunit of the glycine receptor which causes excessive startle responses, and is treatable with clonazepam. Mutations in the γ_2-subunit of the GABA$_A$ receptor are a cause of dominantly inherited epilepsy. Disorders of the metabolism of glycine are discusssed in ► Chap. 24.

Five disorders of monoamine metabolism are discussed: *Tyrosine hydroxylase (TH) deficiency* impairs synthesis of dihydroxyphenylalanine (L-dopa), and causes an extrapyramidal disorder which responds to the latter compound. The clinical hallmark of *dopamine β-hydroxylase deficiency* is severe orthostatic hypotension with sympathetic failure. The other disorders of monoamine metabolism involve both catecholamine and serotonin metabolism. *Aromatic L-amino acid decarboxylase (AADC)* is located upstream of these intermediates. Treatment of its deficiency is more difficult and less effective. *Monoamine-oxidase A (MAO-A) deficiency*, located downstream, mainly causes behavioral disturbances; no effective treatment is known. *Guanosine triphosphate cyclohydrolase-I (GTPCH-I) deficiency* is a defect upstream of L-dopa and 5-hydroxytryptophan (5-HTP) and, therefore, can be effectively treated with these compounds.

Pyridoxine-responsive convulsions, a rare form of early or late infantile seizures, has been recently found to be caused by mutations of antiquitin, an enzyme intervening in the degradation of lysine (◘ Fig. 23.1). Recently also, defective conversion of pyridoxine to pyridoxal phosphate, due to pyridox(am)ine 5'-phosphate oxidase (PNPO) deficiency, has been identified as a cause of neonatal epilepsy.

29.1   Inborn Errors of Gamma Amino Butyric Acid Metabolism

Two genetic diseases due to a defect in brain gamma amino butyric acid (GABA) catabolism have been reported: GABA transaminase deficiency and succinic semialdehyde dehydrogenase (SSADH) deficiency (◘ Fig. 29.1).

29.1.1   Gamma Amino Butyric Acid Transaminase Deficiency

GABA transaminase deficiency was reported in 1984 in a brother and sister from a Flemish family [1]. No other patients have been identified.

Clinical Presentation

The two siblings showed feeding difficulties from birth, often necessitating gavage feeding. They had a pronounced axial hypotonia and generalized convulsions. A high-pitched cry and hyperreflexia were present during the first 6–8 months. Further evolution was characterized by lethargy and psychomotor retardation (the developmental level of 4 weeks was never attained). Corneal reflexes and the reaction of the pupils to light remained normal. A remarkable, continued acceleration of length growth was noted from birth until death. This was explained by increased fasting plasma growth hormone levels; these could be suppressed by oral glucose. In one of the patients, head circumference showed a rapid increase during the last 6 weeks of life (from the 50th to the 97th percentiles). Postmortem examination of the brain showed a spongiform leukodystrophy.

Metabolic Derangement

The cerebrospinal (CSF) and plasma concentrations of GABA, GABA conjugates, and β-alanine were increased. Liver GABA and β-alanine concentrations were normal. This metabolite pattern could be explained by a decrease in GABA transaminase activity in the liver (and lymphocytes). Intermediate levels were found in the healthy sibling, the father, and the mother. It can be assumed that the same enzymatic defect exists in the brain, since GABA transaminases of human brain and of peripheral tissues have the same kinetic and molecular properties. β-Alanine is an alternative substrate for GABA transaminase, hence its increase in this disease. In this context, it can be mentioned that the antiepileptic drug γ-vinyl-GABA (*Vigabatrin*) causes an irreversible inhibition of GABA transaminase, leading to two- to threefold increases in CSF free GABA. Interestingly, this drug also significantly decreases serum glutamic pyruvic transaminase but not glutamic oxaloacetic transaminase activity.

Genetics

The gene for GABA transaminase maps to 16p13.3 and inheritance is autosomal recessive. The patients were compound heterozygotes for two missense mutations.

Diagnostic Tests

The diagnosis requires analysis of the relevant amino acids in CSF. Due to enzymatic homocarnosine degradation, free GABA levels in the CSF show artifactual increases unless samples are deep-frozen (at –20 °C) within a few minutes when analysis is performed within a few weeks, and at –70 °C if the time until analysis is longer. Control CSF free GABA levels range from about 40 nmol/l to 150 nmol/l after the age of 1 year and are lower in younger children. Because of these low levels, sensitive techniques, such as ion-exchange chromatography and fluorescence detection [2] or a stable-isotope-dilution technique [3], must be used. Enzymatic confirmation can be obtained in lymphocytes, lymphoblasts, and liver. As for prenatal diagnosis, GABA transaminase activity is not expressed in fibroblasts, but activity is present in chorionic villus tissue [4].

Treatment and Prognosis

No clinical or biochemical response was obtained after administration of either pharmacological doses of pyridoxine (the precursor of the coenzyme of GABA transaminase) or with picrotoxin, a potent, noncompetitive GABA antagonist. The siblings died at the ages of 1 year, and 2 years and 7 months, respectively.

29.1.2   Succinic Semialdehyde Dehydrogenase Deficiency

SSADH deficiency was first reported as γ-hydroxybutyric aciduria (4-hydroxybutyric aciduria) in 1981 [5]. It has been documented in at least 350 patients [6].

Clinical Presentation

The clinical presentation is nonspecific and varies from mild to severe. It comprises psychomotor retardation, delayed speech development, hypotonia, ataxia and, less frequently, hyporeflexia, convulsions, aggressive behaviour, hyperkinesis, oculomotor apraxia, choreoathetosis, and nystagmus. Ataxia, when present, may resolve with age. MRI shows basal ganglia abnormalities, delayed myelination, and cerebellar atrophy in some patients.

Metabolic Derangement

The key feature is an accumulation of γ-hydroxybutyrate in urine, plasma, and CSF (▣ Fig. 29.1). γ-Hydroxybutyrate and GABA are neuropharmacologically active compounds. The accumulation of γ-hydroxybutyrate tends to decrease with age. Metabolites indicative of the β- and α-oxidation of γ-hydroxybutyic acid may be variably detected in the urine of SSADH-deficient patients. The identification of other metabolites in the urine of SSADH-deficient patients related to pathways of fatty acid, pyruvate, and glycine metabolism, suggests that the deficiency has metabolic consequences beyond the pathway of GABA metabolism.

Genetics

The gene for SSADH maps to chromosome 6p22, and the mode of inheritance is autosomal recessive. More than 40 disease causing mutations have been found.

Diagnostic Tests

Diagnosis is made by organic acid analysis of urine, plasma, and/or CSF. Pitfalls in this diagnosis are the instability of γ-hydroxybutyrate in urine and the variable excretion pattern of this compound which, in some patients, is only marginally increased. The enzyme deficiency can be demonstrated in lymphocytes and lymphoblasts. Residual SSADH activity measured in extracts of cultured cells has been less than 5% of control values in all patients, and parents have intermediate levels of enzyme activity [6]. SSADH activity is expressed in normal human fibroblasts, although with low activity, and in liver, kidney, and brain, and SSADH deficiency in these tissues has been demonstrated. Prenatal diagnosis can be accurately performed using both isotope-dilution mass spectrometry to measure γ-hydroxybutyric acid levels in amniotic fluid, and determination of SSADH activity in amniocytes or chorionic villus tissue.

Treatment and Prognosis

In an attempt to reduce the accumulation of γ-hydroxybutyrate, we introduced a novel treatment principle: substrate reduction by inhibition of the preceding enzymatic step. This was realized by giving vigabatrin, an irreversible inhibitor of GABA transaminase , in doses of 50–100 mg/kg/ day (divided into two daily doses) [7]. This treatment was shown to reduce CSF γ-hydroxybutyrate levels and, in the majority of patients, it was associated with variable improvement particularly of ataxia, behavior and manageability. However, the long term administration of vigabatrin should be monitored closely because this drug increases CSF (and probably also brain) GABA levels and, more importantly, because it potentially causes irreversible visual field deficits. Other agents are being actively investigated in experimental models [8].

As to prognosis, this disease can manifest a mild to severe neurological course.

29.2   Inborn Defects of Receptors and Transporters of Neurotransmitters

29.2.1   Hyperekplexia

Clinical Presentation

Hyperekplexia, or »startle disease« seems to have been reported first in 1958 [9]. Three main symptoms are required for the diagnosis [10]. The first is a generalized stiffness immediately after birth, which normalizes during the first years of life; the stiffness increases with handling and disappears during sleep. The second feature is an excessive startle reflex to unexpected stimuli (particularly auditory stimuli) from birth and which, in older children causes frequent falling. The third is a short period of generalized stiffness (during which voluntary movements are impossible) following the startle response. Associated features may occur, particularly periodic limb movements during sleep and hypnagogic myoclonus. Other symptoms are inguinal, umbilical, or epigastric herniations, congenital hip dislocation, and epilepsy.

Psychomotor development is usually normal. Brain spectroscopy has provided evidence for frontal neuronal dysfunction [11].

Metabolic Derangement

Mutations of the glycine receptor affect the wide variety of inhibitory neurotransmitter functions of glycine. Most patients show defects in the α_1 subunit of the glycine receptor [12]. In a cohort of 22 patients without a mutation in the gene coding for the latter subunit [13], a compound heterozygote was found with a missense and a splice site mutation in the gene encoding the β subunit of the glycine receptor [14].

Genetics

Hyperekplexia has, in the great majority of the patients, an autosomal-dominant inheritance with nearly complete penetrance and variable expression in most pedigrees. The gene for the glycine α_1-subunit receptor is located on chromosome 5q33. A number of mutations have been identified, most of them point mutations [15].

Diagnostic Tests

The diagnosis is based on the response to medication: the benzodiazepine clonazepam reduces the frequency and magnitude of startle responses and diminishes the frequency of falls. This drugs binds to the benzodiazepine site of the GABA$_A$ receptor [16].

Treatment and Prognosis

The stiffness decreases during the first years of life, but the excessive startle responses remain. Clonazepam significantly reduces the frequency and magnitude of the startle responses but has less effect on the stiffness. The mechanism of the beneficial effect of clonazepam is not known.

29.2.2 GABA Receptor Mutation

Clinical Presentation

Mutations in a GABA receptor have been reported first in 2001 in two large families, one with febrile seizures and generalized tonic-clonic seizures [17], the other with febrile seizures, childhood absence epilepsy, and other forms of epilepsy [18]. It seems to be a rare cause of dominant idiopathic epilepsy.

Metabolic Derangement

Mutations of the GABA receptor impede the major inhibitory neurotransmitter function of GABA, predominantly in the gray matter of the central nervous system.

Genetics

Inheritance is autosomal dominant. In both families, a different mutation was found in the γ_2-subunit of the GABA$_A$ receptor. The gene for this subunit has been mapped to chromosome 5q34.

Diagnostic Tests

The diagnosis is based on mutation analysis of the γ_2-subunit of the GABA$_A$ receptor. It should be remembered that in dominant idiopathic epilepsies mutations have also been reported in voltage-gated sodium and potassium channels and nicotinic acetylcholine receptors [19].

Treatment and Prognosis

Patients mostly respond to benzodiazepines. Prognosis depends on the type of epilepsy.

29.2.3 Mitochondrial Glutamate Transporter Defect

This recently identified disorder [20] is characterized by a severe neonatal myoclonic epilepsy syndrome with a burst-suppression EEG pattern and development of retinitis pigmentose. It is caused by a missense mutation in the gene encoding a mitochondrial glutamate transporter specifically expressed in the brain during development. The defect impairs oxidation of glutamate. Inheritance is autosomal recessive, and diagnosis is based on measurement of defective glutamate oxidation in cultured skin fibroblasts and mutation analysis. There is no specific treatment.

Monoamines

The monoamines, adrenaline, noradrenaline, dopamine, and serotonin, are metabolites of the amino acids tyrosine and tryptophan. The first step in their formation is catalyzed by amino-acid-specific hydroxylases, which require tetrahydrobiopterin (BH_4) as a cofactor. BH_4 is also a cofactor of phenylalanine hydroxylase. Its synthesis from GTP is initiated by the rate-limiting GTP cyclohydrolase-1 (GTPCH-I), which forms dihydroneopterin triphosphate (NH_2TP). L-dopa and 5-hydroxytryptophan (5-HTP) are metabolized by a common B6-dependent aromatic L-amino acid decarboxylase (AADC) into dopamine (the precursor of the catecholamines, adrenaline and noradrenaline) and serotonin (5-hydroxytryptamine), respectively. Adrenaline and noradrenaline are catabolized into vanillylmandelic acid (VMA) and 3-methoxy-4-hydroxyphenylethyleneglycol (MHPG) via monoamino oxidase A (MAO-A). This enzyme is also involved in the catabolism of both dopamine into homovanillic acid (HVA) via 3-methoxytyramine, and of serotonin into 5-hydroxyindoleacetic acid (5-HIAA). Dopaminergic modulation of ion fluxes regulates emotion, activity, behavior, nerve conduction, and the release of a number of hormones via G-protein-coupled cell-surface dopamine receptors. Serotoninergic neurotransmission modulates body temperature, blood pressure, endocrine secretion, appetite, sexual behavior, movement, emesis, and pain.

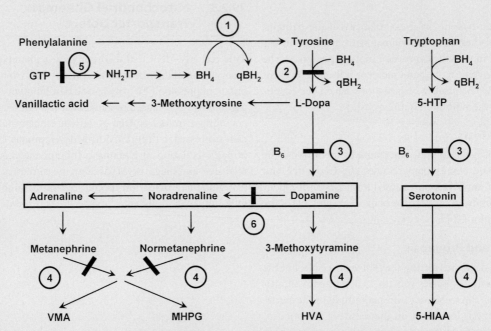

Fig. 29.2. Metabolism of adrenaline, noradrenaline, dopamine, and serotonin. *5-HIAA*, 5-hydroxyindoleacetic acid; *5-HTP*, 5-hydroxytryptophan; *B_6*, pyridoxal phosphate; *BH_4*, tetrahydrobiopterin; *L-Dopa*, L-dihydrophenylalanine; *GTP*, guanosine triphosphate; *HVA*, homovanillic acid; *MHPG*, 3-methoxy-4-hydroxyphenylethyleneglycol; *NH_2TP*, dihydroneopterin triphosphate; *qBH_2*, quininoid dihydrobiopterin; *VMA*, vanillylmandelic acid. **1**, Phenylalanine hydroxylase; **2**, tyrosine hydroxylase; **3**, aromatic L-amino acid decarboxylase; **4**, monoamine oxidase A; **5**, GTP cyclohydrolase-1; **6**, dopamine β-hydroxylase. Enzyme defects covered in this chapter are depicted by *solid bars*

29.3   Inborn Errors of Monoamine Metabolism

29.3.1   Tyrosine Hydroxylase Deficiency

Clinical Presentation

Some twenty patients with suspected or genetically confirmed tyrosine hydroxylase (TH) deficiency have been reported [21–30]. Most presented in the first year of life with truncal hypotonia or generalized hypotonia. The disease is further dominated by delayed motor development that, in combination with the hypotonia, may mimick a primary neuromuscular disorder. Hypersalivation may occur. Classical extrapyramidal signs and symptoms will generally appear at a later stage. Generalized or focal dystonia as well as severe hypokinesia with rigidity (a hypokinetic-rigid parkinsonian syndrome) can develop. The motor disorder can strongly resemble spastic paraplegia. A diurnal fluctuation of the clinical symptoms can be observed in some but not in all TH-deficient cases. The diagnosis should be considered in all patients with unexplained extrapyramidal signs or unexplained infantile floppiness and in selected cases with progressive encephalopathy.

Metabolic Derangement

TH converts tyrosine into L dopa, the direct precursor of catecholamine biosynthesis ($\square$ Fig. 29.2). This enzymatic step is rate-limiting in the biosynthesis of the catecholamines. The enzyme is expressed in the brain and in the adrenals. BH_4 is a cofactor. The biochemical hallmarks of the disease are low CSF levels of homovanillic acid (HVA) and 3-methoxy-4-hydroxy-phenylethyleneglycol (MHPG), the catabolites of dopamine and norepinephrine, respectively, with normal 5-hydroxyindoleacetic acid (5-HIAA) levels. Serotonin metabolism is unaffected.

Genetics

TH deficiency is inherited as an autosomal-recessive trait. The TH gene is located on chromosome 11p15.5. Several mutations have been described. One of these is a common mutation in the Dutch population [26].

Diagnostic Tests

The most important diagnostic test is the measurement of HVA, MHPG, and 5-HIAA in the CSF [25]. As there is a lumbosacral gradient in the concentration of HVA and 5-HIAA, care should be taken that the measurements are carried out in a standardized CSF volume fraction in a laboratory that is experienced in this field. Urinary measurements of HVA and 5-HIAA are not reliable in diagnosing the defect. Determination of amino acids in body fluids (including CSF) does not contribute diagnostically, as tyrosine and phenylalanine levels generally are normal in all body fluids of patients [25]. Direct enzyme measurement is not a diagnostic option, as there is no enzyme activity detectable in body fluids, blood cells and fibroblasts. The finding of elevated prolactin in blood as a measure of dopamine deficiency may be helpful. Normal prolactin levels do not exclude the diagnosis. Molecular genetic analysis of the TH gene is available.

Treatment and Prognosis

In most cases, TH deficiency can be treated with low-dose L-dopa in combination with a L-dopa decarboxylase inhibitor. However, the response is variable, ranging from complete remission to no improvement in relation to the progressive neurodegenerative disease. Therapy should be started with low dose L-dopa (initial dose 2–3 mg/kg per day in three divided doses) since these patients are especially prone to major side effects even on low doses (mainly irritability, dyskinesia and ballism). It may take weeks to months of careful treatment with increasing doses before a positive effect can be convincingly demonstrated. It should also be noted that it is not always possible to normalize the catecholamine levels in the CSF.

29.3.2   Aromatic L-Aminoacid Decarboxylase Deficiency

Clinical Presentation

Since its first description in 1988, aromatic L-aminoacid decarboxylase (AADC) deficiency has been reported in patients from many parts of the world [31–36]. Neonatal symptoms are reported in over half (poor sucking and feeding difficulties, lethargy, increased startle response, hypothermia, ptosis) but all patients develop neurological signs and symptoms within the first six months of life. Patients may present with a severe and progressive epileptic encephalopathy or with (extreme) truncal hypotonia and limb hypertonia. An extrapyramidal movement disorder with generalised hypotonia, limb dystonia and hypokinesia often with oculogyric crises, ptosis, autonomic dysfunction (temperature instability with hypothermia, gastro intestinal symptoms, paroxysmal sweating, impairment of sympathetic regulation of heart rate and blood pressure) and severe developmental delay will develop. Initial suspected diagnoses included epilepsy, cerebral palsy, hyperekplexia and a mitochondrial disorder.

Metabolic Derangement

AADC is implicated in two metabolic pathways: the biosynthesis of catecholamines and of serotonin ($\square$ Fig. 29.2). The activity of the homodimeric enzyme requires pyridoxal phosphate as a cofactor. A deficiency of the enzyme results in a deficiency of the catecholamines and of serotonin. The concentrations of the catabolites (HVA from dopamine, 5-HIAA from serotonin, and MHPG in the central nervous system from norepinephrine) are severely reduced in the CSF. Another biochemical hallmark of the disease is the increased concentration of metabolites upstream of the metabolic block: L-dopa, 3-methoxytyrosine, vanillyllactic acid (VLA), and 5-HTP. Often the finding of increased VLA in the urinary organic acid profile is the first important clue towards this diagnosis. In several patients a paradoxical hyperdopaminuria has been noted probably due to production of dopamine and metabolites in nonneural cells.

Genetics

AADC deficiency is inherited as an autosomal-recessive trait. The AADC gene has been assigned to chromosome 7p11. Mutations in the gene have been reported.

Diagnostic Tests

L-Dopa, 3-methoxytyrosine, VLA, and 5-HTP can be found elevated in urine, CSF and plasma. The increase of vanillactic acid may be observed in the urinary organic acid profile. Also, low concentrations of HVA, 5-HIAA, and MHPG in CSF may lead to the diagnosis. As in TH deficiency, these CSF measurements require the cooperation of a laboratory with experience in this field. AADC deficiency can easily be confirmed at the enzyme level, as all patients have also shown a deficiency of the enzyme in plasma. This pattern strongly resembles that found in pyridoxine phosphate oxidase (PNPO) deficiency but in which there are also many other metabolic disturbances including glycine and ornithine elevation (▶ Section 29.4.2).

Treatment and Prognosis

Treatment in AADC deficiency may be beneficial but the effects are limited and long-term prognosis is poor. Various strategies have been used for symptomatic treatment in individual cases. These have included cofactor supplementation in the form of vitamin B_6, MAO inhibitors (such as tranylcypromine, selegeline, phenelzine), dopamine agonists (pergolide, bromocriptine), high dose L-dopa as »substrate therapy«, serotoninergic agents (fluoxetine) or combinations of these and anticholinergic drugs (trihexylphenidyl). It has been suggested that males may respond better to therapy than females.

29.3.3   Dopamine β-Hydroxylase Deficiency

Clinical Presentation

Congenital dopamine β-hydroxylase (DBH) deficiency was first described by Man in 't Veld in 1987 [37]. Its clinical hallmark is severe orthostatic hypotension with sympathetic failure. Supine blood pressure is normal to low. This may lead to recurrent episodes of fainting and most patients complain of fatigue and impaired exercise tolerance. Symptoms become manifest in early childhood but may worsen in late adolescence. Perinatal hypoglycaemia may occur

and hypothermia and hypotension have also been found in the perinatal period. There is no obvious intellectual impairment. Additional symptoms in some patients are ptosis, weak facial musculature, hyperflexible joints, brachydactyly, high palate and sluggish deep tendon reflexes. A mild normocytic anaemia has been found.

Metabolic Derangement

DBH converts dopamine into noradrenaline. It is present in the synapses of postganglionic sympathetic neurons. A defect in the enzyme should have consequences for (nor-) adrenergic neurons and as well for the adrenals.

Genetics

The DBH gene is located on chromosome 9q34. Pathogenic mutations have been found in all known patients with symptomatic DBH deficiency.

Diagnostic Tests

The patients typically have extremely low plasma noradrenaline and adrenaline levels and increased or high-normal levels of dopamine. At the enzyme level the diagnosis can easily be confirmed by the deficiency of DBH activity in plasma. Interestingly, 4% of the population have nearly undetectable DBH activity in plasma with normal concentrations of noradrenaline and adrenaline and without clinical features of DBH deficiency. This is caused by a common allelic variant (1021 C>T) [38].

Treatment and Prognosis

Therapy with L-dihydroxyphenylserine (L-Dops) is available. This compound can be directly converted by aromatic amino acid decarboxylase into noradrenaline, thereby bypassing the defective enzyme. The prognosis on therapy is satisfactory to good.

29.3.4   Monoamine Oxidase-A Deficiency

Clinical Presentation

Monoamine oxidase-A (MAO-A) deficiency or Brunner syndrome has been identified in five generations of one Dutch family [39, 40]. Only males were affected. They showed borderline mental retardation with prominent behavioral disturbances, including aggressive and sometimes violent behavior, arson, attempted rape, and exhibitionism. The patients were non-dysmorphic and showed a tendency towards stereotyped hand movements, such as hand wringing, plucking, or fiddling. Growth was normal. Additionally, evidence has been presented that a functional polymorphism of the MAO-A gene promoter region may act as a genetic modifier of the severity of autism in males [41].

MAO exists as two isoenzymes (A and B). The genes encoding for both isoenzymes are located on the X-chromosome. Patients with a contiguous gene syndrome affecting both the *MAO-A* and *-B* genes, and also the gene responsible for Norrie disease, have been described [42]. They are severely mentally retarded and blind. Patients with only the *MAO-B* and the Norrie genes affected were also found. These patients are not mentally retarded and do not have abnormalities of catecholamine metabolites in urine. Elevated excretion of phenylethylamine as a specific MAO-B substrate is a consistent finding in patients where the *MAO-B* gene is involved in the contiguous-gene syndrome. Here, we address only the isolated deficiency of MAO-A.

Metabolic Derangement

MAO-A deficiency is a defect in the catabolism of both serotonin and the catecholamines. In patients with MAO-A deficiency, marked elevations were noted of the MAO substrates serotonin, normetanephrine, 3-methoxytyramine, and tyramine in urine (◉ Fig. 29.2). The concentrations of the metabolites downstream of the metabolic block, VMA, HVA, 5-HIAA, and MHPG, were reduced [43]. As platelet MAO-B activity was found to be normal, these results are consistent with a deficiency of MAO-A, the isoenzyme found in neural tissue.

Genetics

The locus for this X-linked inherited disease has been assigned to Xp11.21. A point mutation in the eighth exon of the *MAO-A* gene, causing a premature truncation of the protein, has been found in this family [39, 40].

Diagnostic Tests

A characteristically abnormal excretion pattern of biogenic amine metabolites is present in random urine samples of the patients. The deficiency can be diagnosed at the metabolite level by finding elevated urinary serotonin, normetanephrine, metanephrine, and 3-methoxytyramine. The ratios in urine of normetanephrine to VMA, or normetanephrine to MHPG, are abnormally high in patients with the defect [43].

The ratio HVA/VMA in urine (patients > 4) may also provide a first indication for this diagnosis. In such cases, subsequent measurement of normetanephrine remains essential. The discovery of this disorder suggests that it might be worthwhile performing systematic urinary monoamine analysis when investigating unexplained, significant, *behavior disturbances*, particularly when these occur in several male family members.

Treatment and Prognosis

No effective treatment is known at present. Both the borderline mental retardation and the behavioral abnormalities seem to be stable with time. No patient has been institutionalized because of the mental retardation.

29.3.5 Guanosine Triphosphate Cyclohydrolase-I Deficiency

Clinical Presentation

In 1994, guanosine triphosphate cyclohydrolase-I (GTPCH-I) deficiency was identified as the cause for dopa-responsive dystonia [44]. Patients with this deficiency develop symptoms during the first decade of life, sometimes as early as the first week of life. Dystonia in the lower limbs is generally considered to be the initial and most prominent symptom. Unless treated with L-dopa, the dystonia becomes generalized. Diurnal fluctuation of the symptoms with improvement after sleep is a feature in most patients. The disease may also present with focal dystonia, dystonia with a relapsing and remitting course, or hemidystonia, and may be associated with oculogyric crises, depression and migraine [44–48]. As a rule, symptomatology is asymmetric.

Metabolic Derangement

GTPCH-I is the initial and rate-limiting step in the biosynthesis of BH_4, the essential cofactor of various aromatic amino acid hydroxylases, such as phenylalanine hydroxylase, TH, and tryptophan hydroxylase (◉ Fig. 29.2) with the highest affinity for TH. Therefore, the deficiency state of the enzyme is characterized by defective biosynthesis of serotonin and catecholamines. For the other defects of biopterin metabolism, ▶ Chap. 17.

Genetics

The gene for GTPCH-I is located on chromosome 14q22.1-q22.2. GTPCH-I deficiency can be inherited as an autosomal-dominant trait with 30% penetrance. The female:male ratio is approximately 3:1. Several mutations have been found as a cause for the dominant form of the disease. These often occur as heterozygous mutations in the patient. However, there are also patients with mutations on both alleles. In such patients, the defect has been inherited as an autosomal-recessive trait.

Diagnostic Tests

Some patients with the recessively inherited form of the disease may be diagnosed through the hyperphenylalaninemia found on neonatal screening. Patients with dominantly inherited GTPCH-1 deficiency, however, escape detection during the newborn screening, as they have normal phenylalanine levels in body fluids. The following tests may be helpful in reaching the correct diagnosis:

1. Measurement of pterines especially in CSF (biopterin and neopterin)
2. Measurement of CSF HVA and 5-HIAA. The determination should be carried out by a laboratory with experience in the field, having age-related and CSF volume-fraction related reference values. A normal or slightly low CSF HVA in combination with a low or slightly low 5-HIAA may be meaningful.

3. An oral phenylalanine-loading test.
4. Mutation analysis.
5. Measurement of the enzyme activity in fibroblasts.

Treatment and Prognosis

Optimal treatment of patients would be life-long supplementation of the defective cofactor BH_4. However, in most cases, this is not feasible because of the high price of this treatment. Instead, patients have been treated successfully with a combination of low dose L-dopa, a dopa-decarboxylase inhibitor, and 5-hydroxytryptophan. There is normally a complete or near-complete response of motor problems soon after the start of the therapy. Even when the therapy is started after a diagnostic delay of several years, the results are satisfactory.

Pyridoxine and Pyridoxal-Phosphate

Pyridoxal 5′-phosphate (pyridoxal-P), the coenzyme form of vitamin B_6, is the co-factor for numerous enzymes involved in neurotransmitter metabolism. Pyridoxal-P can be formed from pyridoxine or pyridoxamine by the action of two enzymes, a kinase: pyridoxal kinase (PK), and an oxidase: pyridox(am)ine 5′-phosphate oxidase (PNPO) (Fig. 29.3). Formation of pyridoxal-P from dietary pyridoxal or dietary pyridoxal-P [which is hydrolysed prior to absorption by intestinal phosphatases (IP) or tissue non-specific alkaline phosphatase (TNS AP)] requires only pyridoxal kinase.

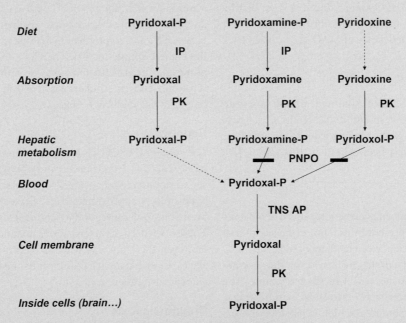

Fig. 29.3. Conversion of dietary vitamin B_6 to intracellular pyridoxal 5′-phosphate cofactor. *IP*, intestinal phosphatases; *P*, 5′-phosphate; *PK*, pyridoxal kinase; *PNPO*, pyridox(am)ine phosphate oxidase; *TNS AP*, tissue non-specific alkaline phosphatase. The enzyme block is indicated by a *solid bar*

29.4   Inborn Disorders Involving Pyridoxine and Pyridoxal Phosphate

Pyridoxine-responsive epilepsy (PRE) has been attributed to glutamate decarboxylase deficiency but this hypothesis has never been proven. Very recently, it was found to be caused by mutations of antiquitin, an enzyme intervening in the degradation of lysine (◘ Fig. 23.1). Recently also, a new disorder was identified with neonatal seizures resistant to anticonvulsants and pyridoxine, but responsive to pyridoxal phosphate. The defect was found to be in pyridox(am)ine 5′-phosphate oxidase (PNPO).

29.4.1   Pyridoxine-Responsive Epilepsy

Pyridoxine-responsive epilepsy was first reported in 1954 [49]. It is a rare cause of convulsions in early childhood, with less than 100 probands having been reported [50].

Clinical Presentation

The clinical picture of typical pyridoxine-responsive convulsions has to be differentiated from atypical presentations. The typical form satisfies the following criteria:
- Onset of convulsions before or shortly after birth
- Rapid response to pyridoxine
- Refractory to other anticonvulsants
- Dependence on a maintenance dose
- Absence of pyridoxine deficiency

The disease may start as intrauterine convulsions as early as in the fifth month of pregnancy. Some patients suffered from peripartal asphyxia, probably as a consequence of this disorder. The seizures are intermittent at onset but may proceed to status epilepticus. All types of seizures can be observed; most are long-lasting seizures and repeated status epilepticus, but brief convulsions (generalized or partial), atonic attacks, and infantile spasms can also occur. There is pronounced hyperirritability that can alternate with flaccidity. Abnormal eye movements are often reported (nystagmus, »rolling« eyes, miosis, and/or poor reaction of the pupils to light). Atypical presentations [51] differ as follows:
- Later onset of the attacks (up to the age of about 18 months)
- Prolonged seizure-free intervals without pyridoxine (for as long as 5 months)
- Need for larger pyridoxine doses in some patients

Metabolic Derangement

Pyridoxine-responsive convulsions have been attributed to brain GABA deficiency. Brain and CSF GABA were found to be low in the rare cases in which they were measured [52, 53]. It was suggested that affected patients may have a ge-netic defect at the pyridoxal phosphate binding site of the enzyme required for the synthesis of GABA, glutamic acid decarboxylase (GAD). The substrate of this enzyme, glutamate, has been reported to be increased [54] or normal [55] in CSF. However, no abnormality of the transcripts of the *GAD* gene could be identified.

In 2000, Plecko et al. detected elevated concentrations of pipecolic acid in the CSF of children with pyridoxine-dependent epilepsy [56]. This led Mills et al. [56a] to speculate that the metabolic defect in PRE is on the »brain« pathway of lysine catabolism which proceeds via pipecolic acid (◘ Fig. 23.1). Specifically, they proposed that: *(i)* the defect might lead to the accumulation of Δ^1-piperideine 6-carboxylic acid (P6C), the compound immediately after pipecolic acid in this pathway; and *(ii)* P6C might react with pyridoxal phosphate in the brain by a Knoevenagel condensation reaction and thus inactivate it. Using comparative genomics they identified a human gene, *ALDH7A1*, encoding a protein, antiquitin, which was likely to have enzyme activity as a dehydrogenase active on an equilibrium mixture of P6C and 2-amino-adipic acid 6-semialdehyde (α-aminoadipic semialdehyde, α-AASA). The *ALDH7A1* gene was located on chromosome 5q31 which had previously been shown by linkage studies to be the major locus involved in PRE [57]. Children with PRE were found to have mutations in the *ALDH7A1* gene. These mutations when expressed in CHO cells produced activity of α-AASA/P6C dehydrogenase that was < 2% of that produced by expression of wild-type DNA. It was possible to show that children with PRE had increased concentrations of α-AASA in CSF, plasma and urine, and that a chemically synthesised equilibrium mixture of P6C and α-AASA, or the urine of patients with PRE, could inactivate pyridoxal phosphate by the formation of a condensation product. Thus PRE is explained by the fact that a defect in the lysine catabolism pathway in the brain leads to accelerated loss of the active form of vitamin B_6, pyridoxal phosphate. Affected individuals need an increased intake of pyridoxine to compensate for the increased rate of loss of pyridoxal phosphate in the brain. The original hypothesis that reduced activity of GAD leads to abnormally low levels of GABA in the brain, and hence seizures, remains tenable. However the cause of reduced activity must now be recognised as pyridoxal phosphate deficiency, not mutations in *GAD*.

Genetics

In its typical form, the disease has an autosomal-recessive inheritance, and there is evidence that this also holds true for the later-onset presentation. Both the linkage studies [57] and the *ALDH7A1* mutation studies [56a] indicate that mutations in the *ALDH7A1* gene on 5q31 are the major cause of PDE. The mutations so far described include missense, nonsense and splice site mutations and a single base-pair insertion; affected individuals were either homozygous for a mutation or compound heterozygotes for two mutations.

Diagnostic Tests

It is possible that, in the future, measurement of α-AASA in the CSF, plasma or urine of patients with neonatal seizures that are difficult to control, will lead to rapid diagnosis of antiquitin deficiency. However, it is important to recognise that neonatal epileptic encephalopathy is a medical emergency and decisions usually need to be made about ongoing treatment before the results of specialised laboratory tests are available. Therefore, for the time being, the clinical and electrophysiological response to pyridoxine treatment remains an important diagnostic test. A clear-cut response to pyridoxine is most likely to be seen in a neonate in status epilepticus with clear abnormalities on EEG (e.g. burst suppression). However, a trial of pyridoxine should ideally be performed in all seizure disorders with onset before the age of 18 months and where the aetiology is unclear. A trial of pyridoxine in the neonate should always be conducted in an intensive care unit; cessation of seizures may be accompanied, for the first few hours, by neurological and respiratory depression with an isoelectric EEG. Until recently, proof of the diagnosis of PRE required temporary cessation of treatment to show that seizures recurred [50]. For the majority of children it will be possible to replace this provocation testing by measurements of α-AASA and sequencing of *ALDH7A1*.

Treatment and Prognosis

With the infant in an intensive care setting and with full EEG monitoring, 50–100 mg of pyridoxine is given intravenously. If the infant has PRE, cessation of seizures will normally occur within minutes. Apnoea, profound hypotonia and hypotension are most likely to occur in an infant who has already been loaded with antiepileptic drugs such as phenobarbitone [50]. If the pyridoxine is given enterally, fits usually stop within hours. After a single dose of 100 mg of pyridoxine, an infant will usually remain fit free for 2–5 days. Permanent control of seizures usually requires oral pyridoxine at a daily dose of 5–10 mg/kg . Untreated, PRE will usually lead to continuing anticonvulsant-resistant seizures and can be fatal (at least in the neonatal onset form). On daily pyridoxine treatment children remain seizure free (with occasional breakthrough seizures precipitated by febrile illness). They often, however, show mild to moderate learning difficulties with speech particularly affected. In one child with learning difficulties there was a significant improvement in IQ when the daily dose of pyridoxine was increased from 5 to 15 mg/kg [50]. Very high doses of pyridoxine can produce peripheral neuropathy so nerve conduction times should be monitored.

Individuals with PRE can have seizures in utero. Attempts have been made to prevent this by treating pregnant women who have a previous child affected by PRE. The daily dose of pyridoxine has varied from 9 mg to 110 mg; some authors suggest that this in utero treatment helps prevent subsequent developmental delay in the child [50].

29.4.2   Pyridox(am)ine 5′-Phosphate Oxidase Deficiency

Clinical Presentation

All 5 infants (from 3 families) with homozygous pyridox(am)ine 5′-phosphate oxidase (PNPO) deficiency were born prematurely and all but one had low Apgar scores and/or required intubation. Early acidosis was also common. Seizures commenced on the first day of life and were associated with an EEG showing a burst suppression pattern. The convulsions, which included myoclonic jerks and severe tonic clonic seizures, were resistant to conventional anticonvulsant drugs. In two patients pyridoxine treatment led to a partial improvement in clonic contractions and lip-smacking automatisms [58]. Much larger numbers of infants have been described in whom severe epilepsy has been better controlled with the use of pyridoxal-P than with the use of pyridoxine. It is not yet known whether any of these infants have mutations or polymorphisms in the PNPO gene [59].

Metabolic Derangement

CSF concentrations of the dopamine metabolite, HVA, and the serotonin metabolite, 5HIAA, were low. The CSF concentration of the L-dopa metabolite, 3-O-methyl-dopa (also known as 3-methoxytyrosine) was very high. The urinary excretion of another L-dopa metabolite, vanillactic acid was increased. These changes indicated a reduced activity of the pyridoxal-P-dependent enzyme aromatic L-aminoacid decarboxylase. Raised CSF concentrations of glycine and threonine could be explained by reduced activity of the pyridoxal-P-dependent glycine cleavage enzyme and threonine dehydratase respectively. CSF concentrations of histidine and taurine were also increased. Plasma and CSF concentrations of arginine were low. CSF concentrations of pyridoxal and pyridoxal-P were measured in 3 patients and were low in all cases.

Genetics

The *PNPO* gene is situated on chromosome 17q21.2. The newborns with epileptic encephalopathy and the biochemical changes reported above, were shown to be homozygous for missense, splice site and stop codon mutations. Expression studies showed that the splice site (IVS3-1g>a) and stop codon (X262Q) mutations were null activity mutations and that the missense mutation (R229W) markedly reduced PNPO activity.

Diagnostic Tests

The least invasive test for PNPO deficiency is measurement of urinary vanillactate but the sensitivity and specificity of this determination in neonates with seizures has yet to be determined. Apart from PNPO deficiency, no disorder has been described that is characterized by low CSF HVA and 5HIAA, with high CSF 3-methoxytyrosine, glycine, threonine, histidine and taurine, and low plasma arginine. In theory, the picture might be mimicked by another disorder affecting pyridoxal-P synthesis or transport into cells or by severe dietary B_6 deficiency. Improved methods for measurement of pyridoxine, pyridoxamine and pyridoxal and their phosphates are likely to prove important in determining of the cause of pyridoxal-P deficiency in the future. A tentative diagnosis of PNPO deficiency might be made if a neonate has seizures that respond dramatically to pyridoxal-P having failed to respond to pyridoxine (but therapeutic trials must be undertaken with great care – see below).

Treatment and Prognosis

Pyridoxal-P for intravenous use is not readily available. Fortunately, pyridoxal-P is very effective when given via a nasogastric tube (in a sick neonate) or orally following recovery from the seizures. A trial of treatment with pyridoxal-P should only be undertaken in a setting where full resuscitation and intensive care facilities are available. In a PNPO-deficient patient, nasogastric administration of 50 mg of pyridoxal-P led to cessation of seizures within an hour but this was associated with profound hypotonia and unresponsiveness and also some hypotension. His neurology only normalized after 4 days treatment. Continuing control of seizures was achieved with 10 mg/kg of pyridoxal-P 6-hourly. Without this treatment, PNPO deficiency was rapidly fatal in all affected infants. The one case who was treated with pyridoxal-P after suffering from severe seizures for 2 weeks is alive at 2y 8mo but has severe dystonia, acquired microcephaly and moderate to severe developmental delay. Treatment with L-dopa and carbidopa was helpful in controlling his dystonic spasms. The effect of earlier treatment remains to be established.

References

1. Jaeken J, Casaer P, De Cock P et al (1984) Gamma-aminobutyric acid-transaminase deficiency: a newly recognized inborn error of neurotransmitter metabolism. Neuropediatrics 15:165-169
2. Carchon HA, Jaeken J, Jansen E, Eggermont E (1991) Reference values for free gamma-aminobutyric acid determination by ion-exchange chromatography and fluorescence detection in the cerebrospinal fluid of children. Clin Chim Acta 201:83-88
3. Kok RM, Howells DW, van den Heuvel CCM et al (1993) Stable isotope dilution analysis of GABA in CSF using simple solvent extraction and electron-capture negative-ion mass fragmentography. J Inherit Metab Dis 16:508-512
4. Sweetman FR, Gibson KM, Sweetman L et al (1986) Activity of biotin-dependent and GABA metabolizing enzymes in chorionic villus samples: potential for 1st trimester prenatal diagnosis. Prenat Diagn 6:187-194
5. Jakobs C, Bojasch M, Monch E et al (1981) Urinary excretion of gamma-hydroxybutyric acid in a patient with neurological abnormalities. The probability of a new inborn error of metabolism. Clin Chim Acta 111:169-178
6. Pearl PL, Novotny EJ, Acosta MT et al (2003) Succinic semialdehyde dehydrogenase deficiency in children and adults. Ann Neurol 54[Suppl 6]:S73-S80
7. Jaeken J, Casaer P, De Cock P, François B (1989) Vigabatrin in GABA metabolism disorders. Lancet 1:1074
8. Gropman A (2003) Vigabatrin and newer interventions in succinic semialdehyde dehydrogenase deficiency. Ann Neurol 54[Suppl 6]: S66-S72
9. Kirstein I, Silfverskiold BP (1958) A family with emotionally precipitated »drop seizures«. Acta Psychiatr Neurol Scand 33:471-476
10. Tijssen MAJ, van Dijk JG, Roos RAC, Padberg GW (1995) »Startle disease«: van schrik verstijven. Ned Tijdschr Geneesk 139:1940-1944
11. Bernasconi A, Cendes F, Shoubridge EA et al (1998) Spectroscopic imaging of frontal neuronal dysfunction in hyperekplexia. Brain 121:1507-1512
12. Shiang R, Ryan SG, Zhu Y-Z et al (1993) Mutations in the α_1 subunit of the inhibitory glycine receptor cause the dominant neurologic disorder, hyperekplexia. Nat Genet 5:351-358
13. Vergouwen MN, Tijssen MA, Shiang R et al (1997) Hyperekplexia-like syndromes without mutations in the $GLRA_1$ gene. Clin Neurol Neurosurg 99:172-178
14. Rees MI, Lewis TM, Kwok JBJ et al (2002) Hyperekplexia associated with compound heterozygote mutations in the β-subunit of the human inhibitory glycine receptor (GLRB). Hum Mol Genet 11:853-860
15. Rees MI, Lewis TM, Vafa B et al (2001) Compound heterozygosity and nonsense mutations in the α_1 subunit of the inhibitory glycine receptor in hyperekplexia. Hum Genet 109:267-270
16. Tijssen MA, Schoemaker HC, Edelbroek PJ et al (1997) The effects of clonazepam and vigabatrin in hyperekplexia. J Neurol Sci 149:63-67
17. Baulac S, Huberfeld G, Gourfinkel-An I et al (2001) First genetic evidence of $GABA_A$ receptor dysfunction in epilepsy: a mutation in the γ_2 subunit gene. Nat Genet 28:46-48
18. Wallace RH, Marini C, Petrou S et al (2001) Mutant $GABA_A$ receptor γ_2 subunit in childhood absence epilepsy and febrile seizures. Nat Genet 28:49-52
19. Pandy HK, Riggs JE (2003) Channelopathies in pediatric neurology. Neurol Clin 21:765-777
20. Molinari F, Raas-Rothschild A, Rio M et al (2005) Impaired mitochondrial glutamate transport in autosomal recessive neonatal myoclonic epilepsy. Am J Hum Genet 76:334-339
21. Lüdecke B, Knappskog PM, Clayton PT et al (1996) Recessively inherited L-dopa-responsive parkinsonism in infancy caused by a point mutation (L205P) in the tyrosine hydroxylase gene. Hum Mol Genet 5:1023-1028
22. Knapskogg PM, Flatmark T, Mallet J et al (1995) Recessively inherited L-dopa-responsive dystonia caused by a point mutation (Q381K) in the tyrosine hydroxylase gene. Hum Mol Genet 4:1209-1212
23. Lüdecke B, Dworniczak B, Bartholomé K (1995) A point mutation in the tyrosine hydroxylase gene associated with Segawa's syndrome. Hum Genet 95:123-125
24. Wevers RA, de Rijk-van Andel JF, Bräutigam C et al (1999) A review on biochemical and molecular genetic aspects of tyrosine hydroxylase deficiency including a novel mutation (291delC). J Inherit Metab Dis 22:364-373
25. Bräutigam C, Wevers RA, Jansen RJT et al (1998) Biochemical hallmarks of tyrosine hydroxylase deficiency. Clin Chem 44:1897-1904
26. Van den Heuvel LPWJ, Luiten B, Smeitink JAM et al (1998) A common point mutation in the tyrosine hydroxylase gene in autosomal recessive L-dopa-responsive dystonia (DRD) in the Dutch population. Hum Genet 102:644-646

27. de Rijk-van Andel JF, Gabreëls FJM, Geurtz B et al (2000) L-dopa-responsive infantile hypokinetic rigid parkinsonism due to tyrosine hydroxylase deficiency. Neurology 55:1926-1928

28. Hoffmann GF, Assmann B, Bräutigam C et al (2003) Tyrosine hydroxylase deficiency causes progressive encephalopathy and dopa-nonresponsive dystonia. Ann Neurol 54[Suppl 6]:S56-S65

29. Furukawa Y, Kish SJ, Fahn S (2004) Dopa-responsive dystonia due to mild tyrosine hydroxylase deficiency. Ann Neurol 55:147-148

30. Schiller A, Wevers RA, Steenbergen GCH et al (2004) Long-term course of L-dopa-responsive dystonia caused by tyrosine hydroxylase deficiency. Neurology 63:1524-1526

31. Hyland K, Surtees RAH, Rodeck C, Clayton PT (1988) Aromatic L-amino acid decarboxylase deficiency: clinical features, diagnosis, and treatment of a new inborn error of neurotransmitter amine synthesis. Neurology 42:1980-1988

32. Hyland K, Clayton PT (1992) Aromatic L-amino acid decarboxylase deficiency: diagnostic methodology. Clin Chem 38:2405-2410

33. Korenke GC, Christen H-J, Hyland K et al (1997) Aromatic L-amino acid decarboxylase deficiency: an extrapyramidal movement disorder with oculogyric crises. Eur J Pediatr Neurol 2/3:67-71

34. Chang YT, Mues G, McPherson JD et al (1998) Mutations in the human aromatic L-amino acid decarboxylase gene. J Inherit Metab Dis 21[Suppl 2]:4

35. Fiumara A, Bräutigam C, Hyland K et al (2002) Aromatic L-amino acid decarboxylase deficiency with hyperdopaminuria: clinical and laboratory findings in response to different therapies. Neuropediatrics 33:203-208

36. Swoboda KJ, Saul JP, McKenna CE et al (2003) Aromatic L-amino acid decarboxylase deficiency. Overview of clinical features and outcomes. Ann Neurol 54[Suppl 6]: S49-S55

37. Man in 't Veld AJ, Boomsma F, Moleman P, Schalekamp MA (1987) Congenital dopamine-beta-hydroxylase deficiency. A novel orthostatic syndrome. Lancet 1:183-188

38. Deinum J, Steenbergen-Spanjers GC, Jansen M et al (2004) DBH gene variants that cause low plasma dopamine beta hydroxylase with or without a severe orthostatic syndrome. J Med Genet 41: e38

39. Brunner HG, Nelen MR, Breakefield XO et al (1993) Abnormal behaviour associated with a point mutation in the structural gene for monoamine oxidase A. Science 262:578-580

40. Brunner HG, Nelen MR, van Zandvoort P et al (1993) X-linked borderline mental retardation with prominent behavioural disturbance: phenotype, genetic localisation, and evidence for disturbed monoamine metabolism. Am J Hum Genet 52:1032-1039

41. Cohen IL, Liu X, Schutz C et al (2003) Association of autism severity with a monoamine oxidase A functional polymorphism. Clin Genet 64:190-197

42. Lenders JWM, Eisenhofer G, Abeling NGGM et al (1966) Specific genetic deficiencies of the A and B isoenzymes of monoamine oxidase are characterised by distinct neurochemical and clinical phenotypes. J Clin Invest 97:1010-1019

43. Abeling NGGM, van Gennip AH, van Cruchten AG et al (1998) Monoamine oxidase A deficiency: biogenic amine metabolites in random urine samples. J Neural Transm [Suppl] 52:9-15

44. Ichinose H, Ohye T, Takahashi E et al (1994) Hereditary progressive dystonia with marked diurnal fluctuation caused by mutations in the GTP cyclohydrolase I gene. Nat Genet 8:236-242

45. Bandmann O, Valante EM, Holmans P et al (1998) Dopa-responsive dystonia: a clinical and molecular genetic study. Ann Neurol 44:649-656

46. Furukawa Y, Shimadzu M, Rajput AH et al (1996) GTP-cyclohydrolase I gene mutations in hereditary progressive and dopa-responsive dystonia. Ann Neurol 39:609-617

47. Blau N, Ichinose H, Nagatsu T et al (1995) A missense mutation in a patient with guanosine triphosphate cyclohydrolase I deficiency missed in the newborn screening program. J Pediatr 126:401-405

48. Segawa M, Nomura Y, Nishiyama N (2003) Autosomal dominant guanosine triphosphate cyclohydrolase I deficiency (Segawa disease). Ann Neurol 54[Suppl 6]:S32- S45

49. Hunt AD, Stokes J, McCrory WW, Stroud HH (1954) Pyridoxine dependency: report of a case of intractable convulsions in an infant controlled by pyridoxine. Pediatrics 13:140-145

50. Baxter P (ed) (2001) Pyridoxine dependent and pyridoxine responsive conditions in paediatric neurology. Mac Keith Press for International Child Neurology Association, pp 109-165

51. Goutières F, Aicardi J (1985) Atypical presentations of pyridoxine-dependent seizures: a treatable cause of intractable epilepsy in infants. Ann Neurol 17:117-120

52. Lott IT, Coulombe T, Di Paolo RV et al (1978) Vitamin B6-dependent seizures: pathology and chemical findings in brain. Neurology 28:47-54

53. Kurlemann G, Löscher W, Dominick HC, Palm GD (1987) Disappearance of neonatal seizures and low CSF GABA levels after treatment with vitamin B6. Epilepsy Res 1:152-154

54. Baumeister FAM, Gsell W, Shin YS, Egger J (1994) Glutamate in pyridoxine-dependent epilepsy: neurotoxic glutamate concentration in the cerebrospinal fluid and its normalization by pyridoxine. Pediatrics 94:319-321

55. Kure S, Maeda T, Fukushima N et al (1998) A subtype of pyridoxine-dependent epilepsy with normal CSF glutamate concentration. J Inherit Metab Dis 21:431-432

56. Plecko B, Stöckler-Ipsiroglu S, Paschke E et al (2000) Pipecolic acid elevation in plasma and cerebrospinal fluid of two patients with pyridoxine-dependent epilepsy. Ann Neurol 48:121-125

56a. Mills PB, Struys E, Jakobs C et al (2006) Mutations in antiquitin in individuals with pyridoxine-dependent seizures. Nat Med 12:307-309

57. Cormier-Daire V, Dagonedu N, Nabbout R et al (2000) A gene for pyridoxine-dependent epilepsy maps to chromosome 5q31. Am J Hum Genet 67:991-993

58. Mills PB, Surtees RAH, Champion MP et al (2005) Neonatal epileptic encephalopathy caused by mutations in the PNPO gene encoding pyridox(am)ine 5'-phosphate oxidase. Hum Mol Genet 14:1077-1086

59. Wang HS, Kuo MF, Chou ML et al (2005) Pyridoxal phosphate is better than pyridoxine for controlling idiopathic intractable epilepsy. Arch Dis Child 90:512-515

30 Disorders in the Metabolism of Glutathione and Imidazole Dipeptides

Ellinor Ristoff, Agne Larsson, Jaak Jaeken

Glutathione and Imidazole-Dipeptide Metabolism

Glutathione is a tripeptide composed of glutamate, cysteine and glycine. It is present in almost all cells at substantial concentrations and plays an important role in several biological functions, such as free-radical scavenging, reducing reactions, amino acid transport, synthesis of proteins and DNA, and drug metabolism. Biosynthesis and catabolism of glutathione form the γ-glutamyl cycle. Glutathione synthesis involves the sequential actions of γ-glutamylcysteine synthetase and glutathione synthetase. Catabolism proceeds initially by way of γ-glutamyl transpeptidase. The γ-glutamyl residue is then released as 5-oxoproline, which is reconverted into glutamate. The biosynthesis of glutathione is feedback regulated, i.e., glutathione acts as an inhibitor of γ-glutamylcysteine synthetase.

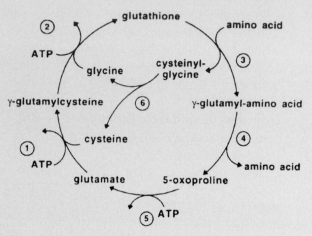

◘ Fig. 30.1. The γ-glutamyl cycle. **1**, γ-glutamyl-cysteine synthetase; **2**, glutathione synthetase; **3**, γ-glutamyl transpeptidase; **4**, γ-glutamyl cyclotransferase; **5**, 5-oxoprolinase; **6**, cysteinyl glycinase (dipeptidase)

Imidazole dipeptides derive their name from the presence of the imidazole ring of histidine. *Carnosine* (β-alanyl-histidine) is found in skeletal (but not cardiac) muscle and brain, where it may be a neurotransmitter. It is hydrolyzed by two isozymes. Cytosolic carnosinase is present in most human tissues and displays a very broad dipeptidase specificity but does not hydrolyze anserine or homocarnosine. Serum carnosinase (also found in cerebrospinal fluid) hydrolyzes carnosine and anserine but hydrolyzes homocarnosine very poorly. Its activity increases gradually with age.

Anserine (β-alanyl-1-methylhistidine) is normally absent from human tissues and body fluids but may be derived from the diet (particularly in infants, owing to their low serum carnosinase activity), or may be found in patients with serum carnosinase deficiency. Anserine is found is skeletal muscles of birds and certain mammals, such as rabbits. Its physiologic function is unclear.

Homocarnosine (γ-aminobutyryl-histidine) is a brain-specific dipeptide. It is hydrolyzed by serum but not by cytosolic carnosinase. Cerebrospinal fluid homocarnosine concentrations are higher in children (~8 μM) than in adults (~1 μM). The physiologic function of homocarnosine is unknown. It may act as a reservoir for γ-aminobutyric acid (GABA) in some parts of the brain (◘ Fig. 29.1).

Genetic defects have been described in four of the six steps of the γ-glutamyl cycle. *Glutathione synthetase deficiency* is the most frequently recognized disorder and, in its severe form, it is associated with hemolytic anemia, metabolic acidosis, 5-oxoprolinuria (pyroglutamic aciduria), central nervous system (CNS) damage and recurrent bacterial infections. *γ-Glutamylcysteine synthetase deficiency* is also associated with hemolytic anemia, and some patients with this disorder show defects of neuromuscular function and generalized aminoaciduria. *γ-Glutamyltranspeptidase deficiency* has been found in patients with CNS involvement and glutathionuria. *5-Oxoprolinase deficiency* is associated with 5-oxoprolinuria but without a clear association with other symptoms.

Serum carnosinase deficiency and *homocarnosinosis* are probably the same disorder. It is uncertain whether there is a relationship between the biochemical abnormalities and clinical symptoms. *Prolidase deficiency* causes skin lesions and recalcitrant ulceration (particularly on the lower legs) in addition to other features, such as impaired psychomotor development and recurrent infections. The severity of clinical expression is highly variable.

30.1 Disorders in the Metabolism of Glutathione

30.1.1 γ-Glutamylcysteine Synthetase Deficiency

Clinical Presentation

Eight patients with this enzyme deficiency have been identified in 6 different families to date [1–5]. A common finding among all patients is hemolytic anemia. The first two patients, German siblings, also developed spinocerebellar degeneration, peripheral neuropathy and myopathy [1]. Both patients had generalized aminoaciduria, but no other renal function defect. Treatment with sulfonamide precipitated psychosis and pronounced hemolytic anemia in one of these patients. Other symptoms found in patients with γ-glutamylcysteine synthetase deficiency are transient jaundice, reticulocytosis, hepatosplenomegaly, and learning disability [2–6].

It remains to be established if the additional symptoms with aminoaciduria in two patients and central nervous system (CNS) involvement in three patients are related to the metabolic defect.

Metabolic Derangement and Diagnostic Tests

In all patients γ-glutamylcysteine synthetase activity was low in erythrocytes, as were the levels of glutathione (GSH) in erythrocytes, leukocytes, skeletal muscle or cultured fibroblasts. Two patients have also been investigated for levels of γ-glutamylcysteine in erythrocytes that were low [5]. In one adult patient, hemolytic anemia and modest decrease in the amount of GSH in cultured lymphoblasts and fibroblasts were the only abnormalities [2]. Four patients have been found to have normal glutathione synthetase (GS) activity. One patient had markedly decreased erythrocyte survival time with a half-life that was 10–20 percent of normal [3]. The diagnosis is established by analysis of the relevant enzyme in erythrocytes or other tissues and by mutation analysis.

The γ-glutamylcysteine synthetase activity in erythrocytes in heterozygous carriers was intermediate whereas the GSH content in erythrocytes was normal [5].

Genetics

γ-Glutamylcysteine synthetase consists of two non-identical subunits encoded by two separate genes, located on chromosomes 1p21 (light or regulatory subunit) and 6p12 (heavy or catalytic subunit), respectively [7–8]. The mode of inheritance is autosomal recessive.

The two German siblings have been found to be homozygous for a 379C→T mutation, encoding for a predicted R127C amino acid change [9]. The American girl was found to be homozygous for an 1109A→T mutation predicting a H370L substitution [4]. Two affected individuals in a large Dutch kindred were found to be homozygous for a 473C→T mutation predicting a P158L substitution [5]. All mutations described are localized in the heavy subunit of γ-glutamylcysteine synthetase.

Treatment and Prognosis

The observation that sulfonamide treatment worsened the symptoms in one patient with γ-glutamylcysteine synthetase deficiency [1] and that the Dutch family with two affected individuals had chosen to refrain from eating fava beans [5], indicates that patients with this disorder should avoid food and drugs that are known to precipitate hemolytic crises in patients with glucose-6-phosphate dehydrogenase (G6PD) deficiency. The prognosis and further treatment remain to be established.

30.1.2 Glutathione Synthetase Deficiency

Previously, two forms of glutathione synthetase (GS) deficiency have been distinguished clinically: generalized and erythrocyte GS deficiency. Today we know that there is only one glutathione synthetase gene, and a new classification based on the severity of clinical signs – i.e. mild, moderate and severe GS deficiency – is more appropriate [10]. The moderate and severe forms are due to mutations which mainly affect the catalytic properties of the enzyme, and the mild form is due to mutations that mainly affect the stability of the enzyme.

VI

Clinical Presentation

GS deficiency has been recognized in more than 65 patients in about 55 families. Among these patients 7 have mild disease, 25 moderate and 33 severe disease. The clinical condition is variable and is presumably correlated to the extent of the enzyme defect. Patients with mild GS deficiency have hemolytic anemia as their only clinical symptom. Those with moderate GS deficiency show symptoms (metabolic acidosis, 5-oxoprolinuria, jaundice and hemolytic anemia) within the first few days of life. After the neonatal period, the condition is usually stabilized. During episodes of gastroenteritis and other infections, however, the patients may become critically ill due to pronounced acidosis and electrolyte imbalance. Several patients have died during such episodes. Severely affected patients have in addition progressive CNS damage, including mental retardation, ataxia, spasticity, and seizures. One patient died at the age of 28 years, and autopsy revealed atrophy of the granule cell layer of the cerebellum and focal lesions of the cortex [11]. Some severely affected patients also have increased susceptibility to bacterial infections due to defective granulocyte function.

Metabolic Derangement

Glutathione concentrations in erythrocytes and other tissues are very low; whereas γ-glutamylcysteine, the metabolite before the enzyme defect, is produced in excess due to a lack of feedback inhibition of γ-glutamylcysteine synthetase. γ-Glutamylcysteine is converted into 5-oxoproline (pyroglutamic acid) and cysteine by γ-glutamyl cyclotransferase. 5-Oxoproline is transferred to glutamate by 5-oxoprolinase, which is the rate-limiting enzyme of the γ-glutamyl cycle in many tissues. The excessive formation of 5-oxoproline exceeds the capacity of 5-oxoprolinase. Therefore, 5-oxoproline accumulates in the body fluids, causing metabolic acidosis and 5-oxoprolinuria.

The level of γ-glutamylcysteine accumulated in fibroblasts in patients with GS deficiency is similar to the level of GSH in control cells [12]. γ-Glutamylcysteine contains both reactive moieties of GSH (i.e. the γ-glutamyl and the sulfhydryl residues) and can therefore probably compensate for GSH in for example the cellular defense against oxidative stress [12].

Genetics

The defective gene is transmitted by autosomal recessive inheritance. The enzyme is a homodimer with a subunit molecular weight of 52 kDa. There is only one copy of the gene in the human genome and its location is 20.q11.2 [13]. Sixteen different missense mutations, two deletions, one insertion, one non-sense, and nine splice site mutations have been characterized in 42 patients with GS deficiency from 35 families (14–16, own unpublished results). About half of the patients are homoallelic. In another 7 patients no disease causing mutation has been identified in the coding

exons or intron-exon boundaries of the GS gene [17]. These patients had severely decreased GS activities in lysates of cultured fibroblasts, and the levels of the enzyme were undetectable using a polyclonal antibody raised against human GS. Sequence analysis revealed previously unreported splice mutations in all patients.

Diagnostic Tests

The diagnosis is usually established in a newborn infant with severe metabolic acidosis but without ketosis or hypoglycemia. In the urine, massive excretion of l-5-oxoproline (up to 1g/kg body weight/day) can be demonstrated by gas liquid chromatography. Note that 5-oxoproline is ninhydrin negative. Decreased activity of GS can be demonstrated in erythrocytes, leukocytes, or cultured skin fibroblasts. Parents usually show activity of GS intermediate between that of the patient and normal. Intracellular levels of GSH are markedly decreased. Mutation analysis can be done.

Prenatal diagnosis can be performed by analysis of 5-oxoproline in amniotic fluid, and by enzyme analysis or mutation analysis in chorion villus sample or cultured amniocytes [18,19]. The first pregnancy reported in a woman with moderate GS deficiency resulted in the birth of a normal child [20]. The pregnancy and delivery were uneventful.

Treatment and Prognosis

Treatment involves correction of the acidosis initially using parenteral sodium bicarbonate followed by oral sodium bicarbonate or citrate (up to 10 mmol/kg body weight/day). During episodes of acute infections, higher doses may be required. Vitamin E (α-tocopherol) has been shown to correct the defective granulocyte function, and should be given in doses of about 10 mg/kg body weight/day. Treatment with vitamin C (ascorbate) (100 mg/kg body weight/day) has been postulated to be of benefit. Both vitamins E and C are thought to replenish the lack of GSH as a scavenger of free radicals. Drugs that precipitate hemolytic crises in patients with G6PD deficiency should be avoided to prevent such crises. N-Acetylcysteine should not be recommended as patients with GS deficiency accumulate cysteine [12]. Cysteine is known to be neurotoxic in excessive amounts [21].

The prognosis of the patients is, at least in part, dependent on the measures taken during acute episodes. Especially during the neonatal period, it is essential to correct the metabolic acidosis and electrolyte imbalance, treat anemia and prevent excessive hyperbilirubinemia. Results from a long-term follow-up study of 28 patients from 24 families with GS deficiency indicated that early supplementation with vitamins C and E may improve the long-term clinical outcome by protecting the CNS from damage [10]. Further evidence for the benefits of treatment with vitamin E in patients with GS deficiency is the finding that vitamin E *in vitro* protects granule cells in the cerebellum of the rat brain [22]. It is, however, essential to remember that GS deficiency is a heterogeneous condition, and it is difficult to predict the outcome for individual patients. Modifying genes and environmental factors probably also affect the outcome.

30.1.3 γ-Glutamyl Transpeptidase Deficiency

Clinical Presentation

Five patients with γ-glutamyl-transpeptidase deficiency have been reported in four families [23–26]. Three of them had CNS involvement, but two siblings have apparently no signs of CNS damage. This may reflect the fact that the first three patients were identified by screening for amino acid defects in populations of mentally retarded patients. Mice with γ-glutamyl transpeptidase deficiency indicate that this enzyme deficiency is associated with CNS dysfunction and growth retardation [27, 28].

Metabolic Derangement

γ-Glutamyl transpeptidase is a membrane-bound enzyme with its active site facing the external side of the cell (ecto-enzyme), and it is the only enzyme that cleaves the γ-glutamyl bond of GSH. Patients with γ-glutamyl transpeptidase deficieny have increased GSH concentrations in plasma and urine, but the cellular levels are normal. In addition to glutathionuria, urinary levels of γ-glutamylcysteine and cysteine are increased. The patients lose up to 1 g of GSH per day via urine, which becomes a substantial source of nitrogen excretion.

Genetics

γ-Glutamyl transpeptidase deficiency is transmitted by autosomal recessive inheritance. The human genome has at least seven different loci for γ-glutamyl transpeptidase. Several of them are located on chromosome 22 [29, 30]. Mutations have yet to be identified in patients with γ-glutamyl transpeptidase deficiency.

Diagnostic Tests

Using thin-layer or paper chromatography and ninhydrin detection, the patients are identified by urinary screening for amino acid disorders. This reveals glutathionuria. Decreased activity of γ-glutamyl transpeptidase can be demonstrated in leukocytes or cultured skin fibroblasts but not in erythrocytes, which lack this enzyme under normal conditions.

Treatment and Prognosis

No specific treatment has been postulated. The prognosis must be considered serious if the patient presents with psychiatric or neurologic symptoms [25]. However, two affected siblings aged 11 years and 13 years have no signs of CNS involvement.

30.1.4 5-Oxoprolinase Deficiency

Clinical Presentation

Eight patients in six families with hereditary defects in 5-oxoprolinase have been described. The clinical symptoms that led to the discovery of the patients were not necessarily related to the metabolic defect – renal stone formation and chronic enterocolitis [31], hypoglycemia [32], and mild mental retardation [33–36].

Metabolic Derangements

The patients were identified because of 5-oxoprolinuria, excreting 4–10 g of 5-oxoproline per day. They exhibited no signs of metabolic acidosis and had normal cellular levels of GSH.

Genetics

The mode of inheritance is autosomal recessive. The mammalian enzyme has not been studied extensively but the rat enzyme is composed of two apparently identical 142 kDa subunits [37]. The location of the corresponding gene in the human genome remains to be established.

Diagnostic Tests

Urinary analysis of 5-oxoproline (pyroglutamic acid) is the initial diagnostic test, followed by determination of the corresponding enzyme activity. 5-Oxoprolinase is not present in erythrocytes and, therefore, fibroblasts or other nucleated cells must be used for the final diagnosis.

Treatment and Prognosis

No specific treatment has been proposed, and prognosis remains to be established.

30.1.5 Secondary 5-Oxoprolinuria

5-Oxoprolinuria has been described in conditions other than GS deficiency and 5-oxoprolinase deficiency. For example in severe burns or Stevens-Johnson syndrome [38], from dietary sources, e.g. certain infant formulas and tomato juice [39], in homocystinuria [40], and in urea-cycle defects, e.g. ornithine transcarbamoylase deficiency. A few patients treated with acetaminophen (paracetamol), vigabatrin or the antibiotics flucloxacillin and netimicin, have been found to excrete several grams of 5-oxoproline per day. The drugs probably interact with the γ-glutamyl cycle [41]. Very preterm infants may have transient 5-oxoprolinuria of unknown etiology [42]. Malnourished patients and pregnant women may have 5-oxoprolinuria probably because of limited availability of glycine [43].

30.2 Disorders of Imidazole Dipeptides

30.2.1 Serum Carnosinase Deficiency

Clinical Presentation

Some 30 individuals have been reported with this disorder, first described in 1967 [44, 45]. The majority of them showed mental retardation to a variable degree. Some patients had seizures and one had congenital myopathy. A few had no symptoms at all, making the relationship between the biochemical abnormalities and the clinical picture uncertain [46].

Metabolic Derangement

The deficiency of serum carnosinase activity causes a persistent carnosinuria during a meat-free diet. Several variants with abnormal kinetic properties of the enzyme have been described. In the cerebrospinal fluid (CSF) of affected persons, homocarnosine can be normal or increased.

Genetics

Inheritance is autosomal recessive. Serum carnosinase deficiency was reported in a child with 18q-syndrome, suggesting a chromosomal location of this enzyme [47].

Diagnostic Tests

The diagnosis is made by quantitative amino acid analysis of serum and/or urine after exclusion of meat from the diet. Anserine appears in the urine of these persons only after eating food containing the dipeptide. Normal persons excrete 1-methylhistidine after ingesting anserine; in serum carnosinase deficiency, there is little or no 1-methylhistidine excretion. The diagnosis is confirmed by measuring carnosinase activity in serum. It has to be noted that serum carnosinase activity may be low in other disorders, such as urea-cycle disorders and multiple sclerosis [48].

Treatment and Prognosis

No efficient treatment is available. In view of the above remarks, it is uncertain whether treatment would be necessary. There is no reason to withhold meat from the diet, because the accumulating carnosine is primarily endogenous. Prognosis is variable and does not seem to correlate with the degree of enzyme deficiency.

30.2.2 Homocarnosinosis

Clinical Presentation

This condition was described in 1976 in a Norwegian family (three of four siblings and their mother) [49], and in 2001 in a Russian boy and his father [50]. The three Norwegian siblings showed progressive spastic diplegia, mental retardation and retinitis pigmentosa, with onset between 6 and 29 years of age. Their mother, however, was symptom free. The Russian boy showed moderate psychomotor retardation and hypotonia. His father was symptom free. This makes it uncertain whether there is a relationship between the biochemical defect and the clinical symptoms.

Metabolic Derangement and Diagnostic Tests

In the CSF of the three siblings and in that of their clinically normal mother, the homocarnosine level was 30–50 times the mean of control levels. The carnosine levels were normal. Deficiency of homocarnosinase activity was found [51]. Therefore, serum carnosinase deficiency and homocarnosinosis are probably the same disorder. The diagnosis is made by quantitative amino acid analysis of the CSF.

Genetics

Inheritance in the two reported families seems to be autosomal dominant.

Treatment and Prognosis

The remarks regarding treatment and prognosis of serum carnosinase deficiency also apply here.

30.2.3 Prolidase Deficiency

Clinical Presentation

Some 40 individuals with prolidase deficiency have been reported since 1968 [52]. About a quarter of them were asymptomatic at the time of the report. The others had their first symptoms between birth and 22 years of age. All patients showed skin lesions, either mild (face, palms, soles) or severe, and had recalcitrant ulceration, particularly on the lower legs. Other features included a characteristic face, impaired motor or cognitive development and recurrent infections. Prolidase deficiency seems to be a risk factor for the development of systemic lupus erythematosus. Alternatively, patients with systemic lupus erythematosus should, where there is a family history or presentation in childhood, be specifically investigated for prolidase deficiency [53].

Metabolic Derangement

The hallmark biochemical finding is massive hyperexcretion of a large number of imidodipeptides (dipeptides with a N-terminal proline or hydroxyproline, particularly glycylproline). This is due to a deficiency of the exopeptidase prolidase (or peptidase D).

Genetics

Inheritance is autosomal recessive. The gene (PEPD) maps to chromosome 19p13.2. Several mutations have been identified [54, 55].

Diagnostic Tests

The hyperimidodipeptiduria can be detected and quantified by partition and elution chromatography and by direct chemical-ionization mass spectrometry. The finding of low or absent prolidase activity in hemolysates or in homogenates of leukocytes or fibroblasts confirms the diagnosis.

Treatment and Prognosis

Due to the rarity of the disease, experience with treatment is scarce. The skin ulcers improved with oral ascorbate, manganese (cofactor of prolidase), and an inhibitor of collagenase in one patient and with local applications of L-proline- and glycine-containing ointments in other patients. Skin grafts have been unsuccessful [56]. As to prognosis, the age of onset and the severity of clinical expression are highly variable. Studies are underway to deliver liposome-encapsulated prolidase intracellularly [57].

References

1. Konrad PN, Richards FD, Valentine WN, Paglia DE (1972) γ-Glutamyl-cysteine synthetase deficiency. A cause of hereditary hemolytic anemia. N Engl J Med 286:557-561
2. Beutler E, Moroose R, Kramer L et al (1990) Gamma-glutamylcysteine synthetase deficiency and hemolytic anemia. Blood 75:271-273
3. Hirono A, Iyori H, Sekine I et al (1996) Three cases of hereditary nonspherocytic hemolytic anemia associated with red blood cell glutathione deficiency. Blood 87:2071-2074
4. Beutler E, Gelbart T, Kondo T, Matsunaga A T (1999) The molecular basis of a case of γ-glutamylcysteine synthetase deficiency. Blood 94:2890-2894
5. Ristoff E, Augustson C, Geissler J et al (2000) A missense mutation in the heavy subunit of γ-glutamylcysteine synthetase gene causes hemolytic anemia. Blood 95:2193-2196
6. Prins H K, Oort M, Zurcher C, Beckers T (1966) Congenital non-spherocytic hemolytic anemia, associated with glutathione deficiency of the erythrocytes. Hematologic, biochemical and genetic studies. Blood 27:145-166
7. Sierra-Rivera E, Dasouki M, Summar M L et al (1996) Assignment of the human gene (GLCLR) that encodes the regulatory subunit of γ-glutamylcysteine synthetase to chromosome 1p21. Cytogenet Cell Genet 72:252-254
8. Sierra-Rivera E, Summar M L, Dasouki M et al (1995) Assignment of the gene (GLCLC) that encodes the heavy subunit of γ- glutamyl-cysteine synthetase to human chromosome 6. Cytogenet Cell Genet 70:278-279
9. Hamilton D, Wu J H, Alaoui-Jamali M, Batist G (2003) A novel missense mutation in the γ-glutamylcysteine synthetase catalytic subunit gene causes both decreased enzymatic activity and glutathione production. Blood 102:725-730
10. Ristoff E, Mayatepek E, Larsson A (2001) Long-term clinical outcome in patients with glutathione synthetase deficiency. J Pediatr 139:79-84
11. Skullerud K, Marstein S, Schrader H et al (1980) The cerebral lesions in a patient with generalized glutathione deficiency and pyroglutamic aciduria (5-oxoprolinuria). Acta Neuropathol 52:235-238
12. Ristoff E, Hebert C, Njalsson R et al (2002) Glutathione synthetase deficiency: is γ-glutamylcysteine accumulation a way to cope with oxidative stress in cells with insufficient levels of glutathione? J Inherit Metab Dis 25:577-584
13. Webb GC, Vaska VL, Gali RR et al (1995) The gene encoding human glutathione synthetase (GSS) maps to the long arm of chromosome 20 at band 11.2. Genomics 30:617-619
14. Shi ZZ, Habib GM, Rhead WJ et al. (1996) Mutations in the glutathione synthetase gene cause 5-oxoprolinuria. Nat Genet 14:361-365
15. Dahl N, Pigg M, Ristoff E et al (1997) Missense mutations in the human glutathione synthetase gene result in severe metabolic acidosis, 5-oxoprolinuria, hemolytic anemia and neurological dysfunction. Hum Mol Genet 6:1147-1152
16. Al-Jishi E, Meyer BF, Rashed MS et al (1999) Clinical, biochemical, and molecular characterization of patients with glutathione synthetase deficiency. Clin Genet 55:444-449
17. Njalsson R, Carlsson K, Winkler A et al (2003) Diagnostics in patients with glutathione synthetase deficiency but without mutations in the exons of the GSS gene. Hum Mutat 22:497
18. Erasmus E, Mienie LJ, de Vries WN et al (1993) Prenatal analysis in two suspected cases of glutathione synthetase deficiency. J Inherit Metab Dis 16:837-843
19. Manning NJ, Davies NP, Olpin SE et al (1994) Prenatal diagnosis of glutathione synthase deficiency. Prenat Diagn 14:475-478
20. Ristoff E, Augustson C, Larsson A (1999) Generalized glutathione synthetase deficiency and pregnancy. J Inherit Metab Dis 22:758-759
21. Janaky R, Varga V, Hermann A et al (2000) Mechanisms of L-cysteine neurotoxicity. Neurochem Res 25:1397-1405
22. Fonnum F, Lock EA (2000) Cerebellum as a target for toxic substances. Toxicol Lett 112-113:9-16
23. Goodman SI, Mace JW, Pollack S (1971) Serum γ-glutamyl transpeptidase deficiency. Lancet 1:234-235
24. O'Daley S (1968) An abnormal sulphydryl compound in urine. Irish J Med Sci 7:578-579
25. Wright EC, Stern J, Ersser R, Patrick AD (1980) Glutathionuria: γ-glutamyl transpeptidase deficiency. J Inherit Metab Dis 2:3-7
26. Hammond JW, Potter M, Wilcken B, Truscott R (1995) Siblings with γ-glutamyltransferase deficiency. J Inherit Metab Dis 18:82-83
27. Lieberman MW, Wiseman AL, Shi ZZ et al (1996) Growth retardation and cysteine deficiency in γ-glutamyl transpeptidase-deficient mice. Proc Natl Acad Sci U S A 93:7923-7926
28. Harding CO, Williams P, Wagner E et al (1997) Mice with genetic γ-glutamyl transpeptidase deficiency exhibit glutathionuria, severe growth failure, reduced life spans, and infertility. J Biol Chem 272:12560-12567
29. Bulle F, Mattei MG, Siegrist S et al (1987) Assignment of the human γ-glutamyl transferase gene to the long arm of chromosome 22. Hum Genet 76:283-286
30. Courtay C, Heisterkamp N, Siest G, Groffen J (1994) Expression of multiple γ-glutamyltransferase genes in man. Biochem J 297:503-508
31. Schulman JD, Goodman SI, Mace JW et al (1975) Glutathionuria: inborn error of metabolism due to tissue deficiency of γ-glutamyl transpeptidase. Biochem Biophys Res Commun 65:68-74
32. Henderson MJ, Larsson A, Carlsson B, Dear PR (1993) 5-Oxoprolin-uria associated with 5-oxoprolinase deficiency; further evidence that this is a benign disorder. J Inherit Metab Dis 16:1051-1052
33. Roesel RA, Hommes FA, Samper L (1981) Pyroglutamic aciduria (5-oxoprolinuria) without glutathione synthetase deficiency and with decreased pyroglutamate hydrolase activity. J Inherit Metab Dis 4:89-90
34. Bernier FP, Snyder FF, McLeod DR (1996) Deficiency of 5-oxoprolinase in an 8-year-old with developmental delay. J Inherit Metab Dis 19:367-368
35. Cohen LH, Vamos E, Heinrichs C et al (1997) Growth failure, encephalopathy, and endocrine dysfunctions in two siblings, one with 5-oxoprolinase deficiency. Eur J Pediatr 156:935-938

36. Mayatepek E, Hoffmann GF, Larsson A et al (1995) 5-Oxoprolinase deficiency associated with severe psychomotor developmental delay, failure to thrive, microcephaly and microcytic anaemia. J Inherit Metab Dis 18:83-84

37. Ye GJ, Breslow EB, Meister A (1996) The amino acid sequence of rat kidney 5-oxo-L-prolinase determined by cDNA cloning. J Biol Chem 271:32293-32300

38. Tham R, Nystrom L, Holmstedt B (1968) Identification by mass spectrometry of pyroglutamic acid as a peak in the gas chromatography of human urine. Biochem Pharmacol 17:1735-1738

39. Oberholzer VG, Wood CB, Palmer T, Harrison BM (1975) Increased pyroglutamic acid levels in patients on artificial diets. Clin Chim Acta 62:299-304

40. Stokke O, Marstein S, Jellum E, Lie SO (1982) Accumulation of pyroglutamic acid (5-oxoproline) in homocystinuria. Scand J Clin Lab Invest 42:361-369

41. Croal BL, Glen A C, Kelly CJ, Logan RW (1998) Transient 5-oxoprolinuria (pyroglutamic aciduria) with systemic acidosis in an adult receiving antibiotic therapy. Clin Chem 44:336-340

42. Jackson AA, Persaud C, Hall M et al (1997) Urinary excretion of 5-L-oxoproline (pyroglutamic acid) during early life in term and preterm infants. Arch Dis Child Fetal Neonatal Ed 76:F152-157

43. Jackson AA, Badaloo AV, Forrester T et al (1987) Urinary excretion of 5-oxoproline (pyroglutamic aciduria) as an index of glycine insufficiency in normal man. Br J Nutr 58:207-214

44. Perry TL, Hansen S, Tischler B et al (1967) Carnosinemia: a new metabolic disorder associated with neurological disease and mental defect. N Engl J Med 277:1219-1227

45. Gjessing LR, Lunde HA, Morkrid L, Lenner JF, Sjaastadt O (1990) Inborn errors or carnosine and homocarnosine metabolism. J Neural Transm 29[Suppl]:91-106

46. Cohen M, Hartlage PL, Krawiecki N et al (1985) Serum carnosinase deficiency: a non-disabling phenotype? J Ment Defic Res 29:383-389

47. Willi SM, Zhang Y, Hill JB et al (1997) A deletion in the long arm of chromosome 18 in a child with serum carnosinase deficiency. Pediatr Res 41:210-213

48. Wassif WS, Sherwood RA, Amir A et al (1994) Serum carnosinase activity in central nervous system disorders. Clin Chim Acta 225:57-64

49. Sjaastadt O, Berstadt J, Gjesdahl P, Gjessing L (1976) Homocarnosinosis. 2. A familial metabolic disorder associated with spastic paraplegia, progressive mental deficiency, and retinal pigmentation. Acta Neurol Scand 53:275-290

50. Kramarenko GG, Markova ED, Ivanova-Smolenskaya IA, Boldyrev AA (2001) Peculiarities of carnosine metabolism in a patient with pronounced hypercarnosinemia. Bull Exp Biol Med 132:996-999

51. Lenney JF, Peppers SC, Kucera CM, Sjaastadt O (1983) Homocarnosinosis: lack of serum carnosinase is the defect probably responsible for elevated brain and CSF homocarnosine. Clin Chim Acta 132:157-165

52. Goodman SI, Solomons CC, Muschenheim F et al (1968) A syndrome resembling lathyrism associated with iminodipeptiduria. Am J Med 45:152-159

53. Shrinath M, Walter JH, Haeney M et al (1997) Prolidase deficiency and systemic lupus erythematosus. Arch Dis Child 76:441-444

54. Ledoux P, Scriver CR, Hechtman P (1996) Expression and molecular analysis of mutations in prolidase deficiency. Am J Hum Genet 59:1035-1039

55. Lupi A, De Riso A, Della Torre S et al (2004) Characterization of a new PEPD allele causing prolidase deficiency in two unrelated patients: naturally occurring mutations as a tool to investigate structure-function relationship. J Hum Genet 49: 500-506

56. Jemec GB, Moe AT (1996) Topical treatment of skin ulcers in prolidase deficiency. Pediatr Dermatol 13:58-60

57. Perugini P, Hassan K, Genta I et al (2005) Intracellular delivery of liposome-encapsulated prolidase in cultured fibroblasts from prolidase-deficient patients. J Controlled Release 102: 181-190

31 Trimethylaminuria and Dimethylglycine Dehydrogenase Deficiency

Valerie Walker, Ron A. Wevers

Trimethylamine Metabolism

In man, trimethylamine (TMA) has a dietary origin. It is produced in the intestine (■ Fig. 31.1) by bacterial action on food, mainly choline (in lecithin), but also carnitine and TMA-N-oxide, a component of some salt water fish and shell fish. TMA is absorbed from the gut, transported to the liver and oxidised by flavin-containing monooxygenase 3 (FMO3) to TMA-N-oxide, a non-odorous product which is excreted in urine [1]. On an average Western diet, the combined urinary excretion of TMA and TMA-N-oxide of normal adults is around 50 mg per day, with less than10% as TMA [2, 3, 4].

FMO3, is one of five human microsomal FMOs which may have evolved as protection from environmental chemicals [5, 6]. It is the main FMO in liver and brain [5]. It has broad substrate specificity since, in addition to TMA, it oxidises nicotine, tyramine, drugs including tamoxifen, cimetidine, ranitidine, ketoconazole, and some xenobiotics [2,5]. Fetuses do not produce FMO3 until 15 weeks of gestation. Expression increases variably from birth, and most individuals produce significant amounts by 1 to 2 years [7].

■ **Fig. 31.1.** Metabolism of trimethylamine. *FMO3*, flavin-containing monooxygenase 3; *TMA*, trimethylamine. The enzyme defect in trimethylaminuria is indicated by a *solid bar*

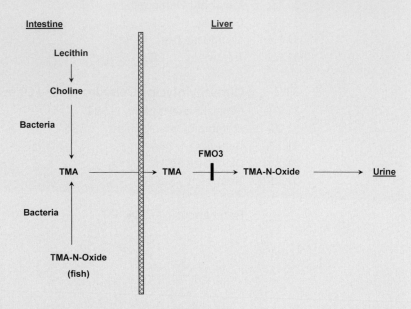

Catabolism of Choline

This process (■ Fig. 31.2) occurs within the mitochondria and involves the sequential removal of two methyl groups by dimethylglycine dehydrogenase and sarcosine dehydrogenase. These are related flavin enzymes with covalently linked FAD which use folate as cofactor. The electrons are transferred from FAD to electron transfer flavoproteins (ETF) and thence to the mitochondrial respiratory chain. The methyl groups from dimethylglycine and sarcosine are used in the formation of folate compounds (not illustrated).

■ **Fig. 31.2.** Catabolism of choline. *ETF*, electron transfer flavoprotein; *ETF-QO*, ETF-ubiquinone oxidoreductase; *FAD*, flavin adenine dinucleotide. The enzyme defect in dimethylglycine dehydrogenase deficiency is indicated by a *solid bar*

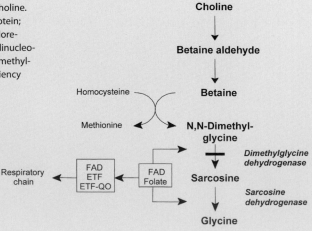

These disorders cause an unpleasant fish-like body odour. The problems are psychosocial and management centres on attempting to minimise the odour.

31.1 Trimethylaminuria (Fish Odour Syndrome)

31.1.1 Clinical Presentation

Individuals homozygous for trimethylaminuria present with an offensive body odour, usually in childhood, but sometimes as adults. Subjects may be unaware of their own odour [3, 4]. *Transient* trimethylaminuria has been reported in infancy and may be due to delayed flavin-monooxygenase 3 (FMO3) expression. The severity is variable. The odour may be episodic, becoming apparent with sweating, menstruation, and after a high dietary intake of trimethylamine (TMA) precursors [4]. In two affected breast-fed babies, the smell occurred after their mother ate eggs or fish [8]. There arc no physical abnormalities. Heterozygotes do not have the odour normally, but it may appear after a TMA load [9].

The disorder often causes psycho-social problems, with anxiety, low self-esteem, social isolation and loneliness. School performance may suffer [3, 10]. Some patients have also had adverse reactions to tyramine, sulfur-containing medications (1 patient) and labile hypertension. It is speculated that impaired FMO3 deficiency may have been contributory [3].

31.1.2 Metabolic Derangement

Trimethylaminuria is due to deficiency of FMO3. Malodorous TMA accumulates in urine and body secretions and, being volatile, is released readily into the atmosphere [3, 4]. In individuals with trimethylaminuria, TMA accounts for at least 20% of their combined TMA/TMA-N-oxide excretion and often much more. This fraction increases further with a high dietary load of TMA. Unoxidised TMA is also increased in sweat, saliva, breath and vaginal secretions. Patients with fish odour syndrome are deficient in nicotine-N-oxidation, but it is not known whether FMO3 deficiency has any other significant pharmacological or toxicological sequelae.

31.1.3 Genetics

Trimethylaminuria is an autosomal recessive disorder [3, 9] with an estimated carrier frequency in the United Kingdom of 1% [2]. The FMO3 gene is located in a cluster with the genes for FMO1 to FMO4 and FMO6 at 1q23-q25. It con-

tains one non-coding and eight coding exons [5, 11, 12]. Mutations of the gene have been identified among patients with trimethylaminuria, and found to cause severe enzyme deficiency in vitro [5, 13, 14]. In addition, there are common gene polymorphisms which are not normally associated with fish odour syndrome [5]. They might contribute to the wide inter-individual variation of TMA metabolism in the population and explain *episodic trimethylaminuria* [4, 5].

31.1.4 Diagnostic Tests

Common causes of abnormal body odour should be excluded: poor hygiene, gingivitis, urinary infections, infected vaginal discharge, as well as advanced liver and renal disease and organic acid and amino acid disorders where appropriate. Trimethylaminuria is diagnosed by determining the ratio of TMA to TMA-N-oxide in urine. Samples are collected when on a normal diet, preferably when the odour is present, and in the absence of urine infection. They must be acidified quickly (pH 2.0) and frozen. Collection of a second sample after a meal containing fresh fish has detected affected individuals (genetically proven) whose ratio was normal before loading.

For population studies to investigate the genetics of trimethylaminuria, urine was collected for 8 h after an oral TMA load (970 mg of TMA hydrochloride, equivalent to 600 mg of TMA free base) [3, 9]. TMA and TMA-N-oxide are measured by NMR spectroscopy or headspace gas chromatography-mass spectrometry. On a normal diet, normal adults and carriers display TMA/total TMA (i.e. TMA + TMA-N-oxide) ratios <10% [3, 4]. In homozygous trimethylaminuria, the ratio is >20% and usually higher (>55). There are no published data for children, but the above ratios are probably applicable.

31.1.5 Treatment

Treatment involves counselling and adjustments to diet and life-style [4, 15]. Explanation of the problem helps by providing insight. Dietary restrictions sometimes, but not invariably, reduce the odour. Affected individuals should take a low choline diet, avoiding eggs, mayonnaise, liver and other organ meats (offal), peas and soybeans and exclude sea fish and shell fish. They should choose appropriate clothing and control room ventilation to minimise sweating. Use of an acid soap (pH 5.5–6.5) may help by decreasing TMA volatility. They should avoid constipation. Long-term use of antibiotics to reduce enteric TMA production should be avoided, but a two week course of metronidazole (250 mg three times daily) or short courses of lactulose may help intermittent attacks and provide some protection for special occasions and holidays [4, 10].

31.2   Dimethylglycine Dehydrogenase Deficiency

31.2.1   Clinical Presentation

Until now, only one case has been reported. He was investigated aged 38 years for an abnormal body odour resembling fish, which was present from 5 years of age, was increased by stress and effort and caused him major social, psychological and professional problems. He also had chronic muscle fatigue with persistent elevation of creatine kinase to around 4 times normal. He had normal intelligence. His siblings and two sons were asymptomatic [16].

31.2.2   Metabolic Derangement

Dimethylglycine dehydrogenase (DMGDH) deficiency blocks choline catabolism, This causes an accumulation of around 100-fold in the patient's plasma and 20-fold in his urine of volatile N,N-dimethylglycine, and an unpleasant fish odour [16].

31.2.3   Genetics

The pedigree of the patient with DMGDH deficiency suggests autosomal recessive inheritance. The DMGDH gene is on chromosome 5q12.2–12.3 [17]. The affected patient is homozygous for a point mutation (A326G) of the DMDGH gene. From expression studies, this mutated gene codes for a stable protein lacking enzyme activity [18].

31.2.4   Diagnostic Tests

The diagnosis is made by finding raised levels of dimethylglycine in plasma and urine, preferably collected when the odour is present. Proton nuclear magnetic resonance (NMR) spectroscopy is a good method for this, and it will also detect TMA and TMA-N-oxide which are increased in trimethylaminuria, the other inherited cause of a fishy odour [16]. Dimethylglycine is not detected with gas chromatography-mass spectrometry (GC-MS) procedures using solvent extraction, used routinely in metabolic laboratories.

Normal urine excretion is age-dependent.

Reference values for dimethylglycine are as follows:
Plasma:
healthy adults	1–5 μmol/l

Urine:
infants (birth to 2 months)	<550 mmol/mol creatinine
from 2 months of age (children and adults):	<26 mmol/mol creatinine

Increased serum levels have been observed in folate deficiency (up to 10-fold), cobalamin deficiency (up to 2-fold) and renal failure (up to 2-fold) [16].

DMGDH activity is not detectable normally in blood cells and fibroblasts, and a liver biopsy would be necessary to confirm low enzyme activity, but is invasive and probably an unnecessary test for this disorder.

31.2.5   Treatment

Management is by counselling and minimising the odour by restriction of dietary choline, and avoiding excessive sweating, as outlined for trimethylaminuria. Antibiotics to modify the intestinal microflora are not indicated. The reported patient did not benefit from riboflavin supplements alone, but a trial of folate with riboflavin was suggested [16].

References

1. Tjoa S, Fennessey P (1991) The identification of trimethylamine excess in man: quantitative analysis and biological origins, Anal Biochem 197:77-82
2. Al-Waiz M, Ayesh R, Mitchell SC et al (1987) A genetic polymorphism of the N-oxidation of trimethylamine in humans. Clin Pharmacol Ther 42:588-594
3. Ayesh R, Mitchell SC, Zhang A, Smith RL (1993) The fish odour syndrome: biochemical, familial, and clinical aspects. Br Med J 307:655-657
4. Mitchell SC, Smith RL (2001) Trimethylaminuria: the fish malodor syndrome. Drug Metab Dispos 29:517-521
5. Cashman JR, Zhang J (2002) Interindividual differences of human flavin-containing monooxygenase 3: genetic polymorphisms and functional variation. Drug Metab Dispos 30:1043-1052
6. Ziegler DM (1990) Flavin-containing monooxygenases: enzymes adapted for multisubstrate specificity (Review). Trends Pharmacol Sci 11:321-324
7. Koukouritaka SB, Simpson P, Yeung CK et al (2002) Human hepatic flavin-containing monooxygenase 1 (FMO1) and 3 (FMO3) developmental expression. Pediatr Res 51:236-243
8. Lee CWG, Yu JS, Turner BB, Murray KE (1976) Trimethylaminuria: fishy odours in children. N Engl J Med 295:937-938
9. Al-Waiz M, Ayesh R, Mitchell SC et al (1989) Trimethylaminuria: the detection of carriers using a trimethylamine load test. J Inherit Metab Dis 12:80-85
10. Pike MG, King GS, Pettit BR et al (1988) Lactulose in trimethylaminuria, the fish-odour syndrome. Helv Paediatr Acta 43:345-348
11. Shephard EA, Dolphin CT, Fox MF et al (1993) Localization of genes encoding three distinct flavin-containing monooxygenases to human chromosome1q. Genomics 16:85-89
12. Dolphin CT, Riley JH, Smith RL et al (1997) Structural organization of the human flavin-containing monooxygenase 3 gene (FMO3), the favored candidate for fish odor syndrome, determined directly from genomic DNA. Genomics 46:260-267
13. Hernandez D, Addou S, Lee D et al (2003) Trimethylaminuria and a human FMO3 mutation database. Hum Mutat 22:209-213
14. Treacy EP, Akerman BR, Chow LML et al (1998) Mutations of the flavin-containing monooxygenase gene (FMO3) cause trimethylaminuria, a defect in detoxication. Hum Mol Genet 7:839-845

15. Walker V (1993) The fish odour syndrome (leader). Br Med J 307: 639-640

16. Moolenaar SH, Poggi-Bach J, Engelke UFH et al (1999) Defect in dimethylglycine dehydrogenase, a new inborn error of metabolism: NMR spectroscopy study. Clin Chem 45:459-464

17. Binzak BA, Vockley JG, Jenkins RB, Vockley J (2000) Structure and analysis of the human dimethylglycine dehydrogenase gene. Mol Genet Metab 69:181-187

18. Binzak BA, Wevers RA, Moolenaar SH et al (2001) Cloning of dimethylglycine dehydrogenase and a new human inborn error of metabolism, dimethylglycine dehydrogenase deficiency. Am J Hum Genet 68:839-847

VII Disorders of Lipid and Bile Acid Metabolism

32 Dyslipidemias

Annabelle Rodriguez-Oquendo, Peter O. Kwiterovich, Jr.

Lipoprotein Metabolism

Lipids are transported in plasma on lipoproteins, spherical particles that consist of a hydrophobic core of triglycerides and cholesteryl esters, surrounded by an amphiphilic coating of apolipoproteins, phospholipids and unesterified cholesterol. The human plasma lipoproteins are classified according to their density and electrophoretic mobility (◘ Table 32.1), and a number of species of apolipoproteins are known (◘ Table 32.2). Lipoprotein metabolism involves three major pathways, which are briefly summarized here and reviewed in more detail in the first section of the chapter.

(1) The **exogenous (intestinal) pathway** transports mainly triglycerides from the diet, but also cholesterol of both dietary and biliary origin, as chylomicrons (◘ Fig. 32.1). Lipoprotein lipase, an enzyme on the surface of capillary endothelial cells that requires apo C-II as cofactor, hydrolyzes chylomicron triglycerides into free fatty acids for uptake by muscle and fat. The resultant chylomicron remnants are taken up by the LDL receptor-related protein (LRP) in liver, where they deliver cholesterol that can be converted into bile acids, used for lipoprotein biosynthesis, or to downregulate LDL receptors and cholesterol biosynthesis.

(2) The **endogenous (hepatic) pathway** transports triglycerides and cholesterol from the liver as VLDL (◘ Fig. 32.1). In the capillaries of muscle and fat, VLDL are also hydrolyzed by lipoprotein lipase, yielding free fatty acids for uptake. Their remnants, IDL, are in part cleared from the circulation by the liver LDL receptor, and in part converted into LDL. LDL are taken up via LDL receptors by a variety of extrahepatic tissues, where they supply cholesterol mainly for membrane synthesis. Liver also takes up LDL via LDL receptors and uses their cholesterol for the synthesis of bile acids or lipoproteins.

(3) **Reverse cholesterol transport** involves release of unesterified cholesterol from cells into plasma, followed by binding to HDL, conversion by LCAT of unesterified into esterified cholesterol, and transfer of the latter via cholesteryl ester transfer protein to VLDL and ultimately IDL and LDL. HDL can also deliver cholesteryl esters directly to the liver (◘ Fig. 32.2).

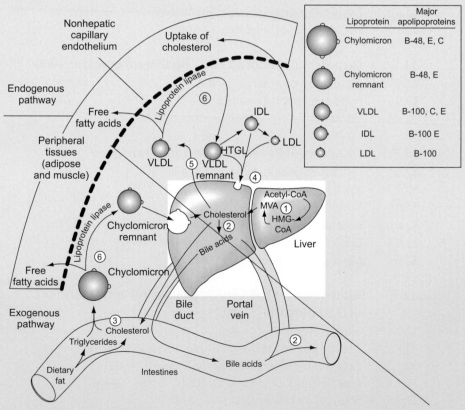

◘ **Fig. 32.1.** Pathways of exogenous and endogenous lipoprotein metabolism. The metabolism of the apoB containing lipoproteins from the intestine and the liver are depicted. The site of action of the lipid-lowering drugs are also shown. The statins (**1**), the bile acid sequestrants (**2**) and the cholesterol absorption inhibitor (**3**) all induce LDL receptors (**4**). Niacin (**5**), inhibits VLDL, IDL and LDL production. The fibric acid derivatives (**6**) enhance lipoprotein lipase activity. See text for abbreviations. Modified and reproduced with permission from Braunwald E (ed) Essential atlas of heart diseases, Appleton and Lange, Philadelphia, 1997, p 1.28

Conversion Factors

	mg/dl →	mmol/l →	mg/dl
Cholesterol	x 0.0259	x 38.6	
Triglycerides	x 0.0114	x 87.7	
Phospholipids	x 0.323	x 77.5	

32.1 Overview of Plasma Lipid and Lipoprotein Metabolism

Lipoproteins play an essential role in the delivery of free fatty acids to muscle and adipose tissue where they, respectively, serve as a fuel and are stored as triglycerides. Lipoproteins also intervene in the transfer of cholesterol from intestine to liver, from liver to other tissues, and from the latter back to the liver. The lipoprotein structure resembles a plasma membrane bilayer with hydrophilic phospholipids, apolipoproteins and some cholesterol on the outer surface, and hydrophobic triglycerides and cholesteryl esters in the core. The physical-chemical properties and composition of the major human plasma proteins are given in ◘ Table 32.1.

The plasma apolipoproteins are amphipathic proteins that interact with both the polar aqueous environment of blood, and the nonpolar core lipids. They serve various functions such as ligands for receptors, cofactors for enzymes, and structural proteins for packaging. The main characteristics of human plasma apolipoproteins are given in ◘ Table 32.2.

32.1.1 Exogenous Lipoprotein Metabolism

The exogenous pathway of lipoprotein metabolism transports dietary fats from intestine to muscle, adipose tissue and liver. After a meal is consumed, dietary lipids, mainly triglycerides (TG), cholesteryl esters, and phospholipids, are emulsified by bile acids and hydrolyzed by pancreatic lipases into their component parts, monoglyceride and free fatty acids (FFA), and unesterified cholesterol and FFA, respectively. After absorption into the intestinal cells, the monoglycerides are reconverted into TG, and incorporated together with cholesterol into chylomicrons, which contain apolipoproteins A-I, A-II, A-IV, and B-48. The assembled chylomicrons are secreted into the thoracic duct, a process that requires apo B-48. Thereafter, they enter the peripheral circulation, where they acquire apo E and apo C-I, apo C-II and apo C-III by transfer from HDL. When they enter the capillaries of skeletal muscle and adipose tissue, the chylomicrons are exposed to the enzyme lipoprotein lipase (LPL), located on the surface of the endothelial cells (◘ Table 32.3). Apo C-II is necessary for activation of LPL, provoking hydrolysis of the TG into FFA which enter muscle and adipose tissue.

The resulting chylomicron remnants, still containing cholesterol, apo B-48 and apo E, the latter acting as a ligand for the hepatic chylomicron remnant receptor, are taken up into the liver, where they deliver dietary and biliary cholesterol (◘ Fig. 32.1).

◘ **Table 32.1.** Physical-chemical properties of human plasma lipoproteins

Class	Density (g/ml)	Electrophoretic mobility	Surface components			Core lipids	
			Cholesterol	Phospholipids	Apolipoprotein	Triglycerides	Cholesteryl esters
Chylomicrons	<0.95	Remains at origin	2	7	2	86	3
VLDL	0.950–1.006	Pre-β lipoproteins	7	18	8	55	12
IDL	1.006–1.019	Slow pre-β lipoproteins	9	19	19	23	29
LDL	1.019–1.063	β-Lipoproteins	8	22	22	6	42
HDL-2	1.063–1.125	α-Lipoproteins	5	33	40	5	17
HDL-3	1.125–1.210	α-Lipoproteins	4	35	55	3	13
Lp(a)	1.040–1.090	Slow pre-β-lipoproteins					

VLDL, very low-density lipoprotein; *IDL*, intermediate-density lipoprotein; *LDL*, low density lipoprotein; *HDL*, high density lipoprotein. HDL-2 and HDL-3 are the two major subclasses of HDL. Lp(a) consists of a molecule of LDL covalently attached to a molecule of apo(a), a protein homologous to plasminogen. Its lipid composition is similar to that of LDL. Compositions are given in % (by weight).

◘ Table 32.2. Characteristics of human plasma apolipoproteins

Apolipoproteins	Major tissue sources	Functions	Molecular weight
Apo A-I Apo A-II Apo A-IV	Liver and intestine	Co-factor LCAT Not known Activates LCAT	29,016 17,414 44,465
Apo B-48	Intestine	Secretion TG from intestine	240,800
Apo B-100	Liver	Secretion TG from liver; binding ligand to LDL receptor	512,723
Apo C-I Apo C-II Apo C-III	Liver	Activates LCAT; inhibits CETP Cofactor LPL Inhibits LPL	6,630 8,900 8,800
Apo D	Many sources	Reverse cholesterol transport	19,000
Apo E	Liver	Ligand for uptake of chylomicron remnants and IDL	34,145

LCAT, lecithin cholesterol acyl transferase; *TG*, triglyceride; *LDL*, low density lipoprotein; *CETP*, cholesteryl ester transfer protein; *LPL*, lipoprotein lipase; *IDL*, intermediate density lipoprotein

◘ Table 32.3. Key enzymes and transfer proteins of plasma lipid transport

Enzyme	Major tissue source	Functions	Molecular weight
Lipoprotein lipase (LPL)	Adipose tissue Striated muscle	Hydrolyzes triglycerides and phospholipids of chylomicrons and large VLDL	50,394
Hepatic lipase (HL)	Liver	Hydrolyzes triglycerides and phospholipids of small VLDL, IDL, and HDL-2	53,222
Lecithin:cholesterol acyltransferase (LCAT)	Liver	Converts free cholesterol from cell membranes to esterified cholesterol using a free fatty acid from phosphatidylcholine on nascent (prebeta) HDL	47,090
Cholesterol ester transport protein (CETP)	Liver, spleen and adipose tissue	Transfers cholesteryl esters from HDL to apo B- containing triglyceride-rich lipoproteins Converts α HDL to pre-β HDL	74,000
Phospholipid transfer protein (PTP)	Placenta, pancreas, adipose tissue, lung	Transfers the majority of phospholipids in plasma Converts α HDL to pre-β HDL	81,000

32.1.2 Endogenous Lipoprotein Metabolism

The endogenous pathway of lipoprotein metabolism transports TG and cholesteryl esters, synthesized in the liver, to the peripheral tissues. Transport occurs under the form of VLDL with their major apolipoproteins, B-100, E, and C (I, II, III). The VLDL triglycerides are transported to tissue capillaries, where they are hydrolyzed by LPL (that also hydrolyzes chylomicrons) and its co-factor apo C-II, and thereby release FFA. The resulting VLDL remnants are further hydrolyzed, generating IDL. A portion of the IDL is cleared from the circulation via direct uptake by the liver by the binding of IDL apoE to the LDL receptor (◘ Fig. 32.1). The remaining IDL can undergo further hydrolysis by hepatic lipase (◘ Table 32.3) to yield LDL. Most of LDL are removed from the peripheral circulation by binding of apo B-100 to the liver LDL receptors.

Liver LDL receptors are clustered on the surface of the hepatocytes. They remove LDL particles by endocytosis. The cholesteryl esters are hydrolyzed to unesterified cholesterol. Overaccumulation of intrahepatic cholesterol is prevented by cholesterol-induced down-regulation of the transcription of the genes for the LDL receptor and the rate-limiting enzyme of cholesterol synthesis, hydroxymethylglutaryl coenzyme A (HMG-CoA) reductase, by inhibiting the release of transcription factors, i.e. sterol regulatory element binding proteins (SREBPs), from the cytoplasm into the nucleus [2]. Conversely, when the cholesterol pool in the liver is low, there is an increased release of SREBPs and an upregulation of LDL receptors and HMG CoA reductase.

LDL also supplies cholesterol to a variety of extrahepatic parenchymal tissues, where it is used mainly for membrane synthesis, and to adrenal cortical cells, where it serves as a precursor for steroid synthesis. Like the liver, extrahepatic tissues also have abundant LDL receptors. LDL cholesterol can also be removed via non-LDL receptor mechanisms. One class of cell surface receptors, termed scavenger receptors, takes up chemically modified LDL such as oxidized LDL (◘ Fig. 32.1), which has been generated by release of oxygen radicals from endothelial cells. Scavenger receptors are not regulated by intracellular cholesterol levels. In peripheral tissues such as macrophages and smooth muscle cells of the arterial wall, excess cholesterol accumulates within the plasma membrane, and then is transported to the endoplasmic reticulum where it is esterified to cholesteryl esters by the enzyme, acyl-CoA cholesterol acyltransferase. It is at this stage that cytoplasmic droplets are formed and that the cells are converted into foam cells (an early stage of atherogenesis). Later on, cholesteryl esters accumulate as insoluble residues in atherosclerotic plaques.

The optimal level of plasma LDL to prevent atherosclerosis and to maintain normal cholesterol homeostasis in humans is not known. At birth, the average LDL cholesterol level is 30 mg/dL. After birth, if the LDL cholesterol level is <100 mg/dl, LDL is primarily removed through the high affinity LDL receptor pathway. In Western societies, the LDL cholesterol is usually >100 mg/dl; the higher the LDL-cholesterol the greater the amount that is removed by the scavenger pathway.

While the exogenous and endogenous pathways are conceptually considered as separate pathways, an imbalance in one often produces an abnormal effect in the other. Thus, reduced LPL activity or decreased apo C-II, as well as elevated apo C-III or apo C-I, can promote hypertriglyceridemia and accumulation of remnant particles from both chylomicrons and VLDL. When the remnant particles are sufficiently small (Svedberg flotation units 20 to 60), they can enter the vascular wall and promote atherogenesis. The greater the cholesterol content of the remnants, the more atherogenic they are. This scenario can be further complicated by VLDL overproduction or by reduced LDL receptor activity.

32.1.3   Reverse Cholesterol Transport and High Density Lipoproteins

Reverse cholesterol transport (◘ Fig. 32.2) refers to the process by which unesterified or free cholesterol is removed from extrahepatic tissues, probably by extraction from cell membranes via the ATP binding cassette transporter ABCA1, and transported on HDL [3]. HDL particles are heterogeneous and differ in their percentage of apolipoproteins (A-I, A-II, and A-IV). HDL can be formed by remodeling of apolipoproteins cleaved during the hydrolysis of triglyceride-rich lipoproteins (chylomicrons, VLDL and IDL). They can also be synthesized by intestine, liver and macrophages as nascent or pre-β HDL particles that are relatively lipid-poor and disc-like in appearance. Pre-β-1

◘ **Fig. 32.2.** The pathway for HDL metabolism and reverse cholesterol transport. See text for abbreviations. Modified and reproduced with permission from Braunwald E (ed) Essential atlas of heart diseases, Appleton & Lange, Philadelphia, 1997, p 1.29

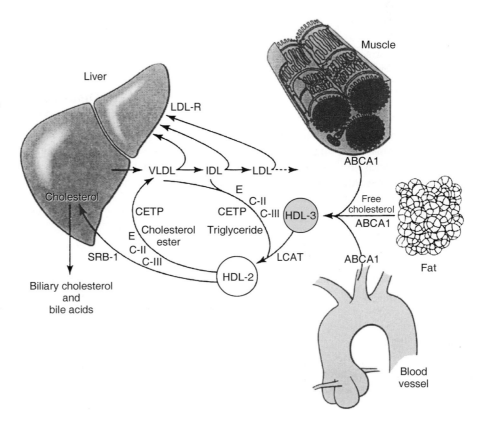

HDL is a molecular species of plasma HDL of approximately 67 kDa that contains apoA-I, phospholipids and unesterified cholesterol, and plays a major role in the retrieval of cholesterol from peripheral tissues. HDL particles possess a number of enzymes on their cell surface [4]. One enzyme, lecithin-cholesterol acyltransferase (LCAT), plays a significant role by catalyzing the conversion of unesterified to esterified cholesterol (◘ Fig. 32.2, Table 32.3). Esterified cholesterol is nonpolar and will localize in the center core of the HDL particle, allowing it to remove more unesterified cholesterol from cells. Esterified cholesterol can be transferred, via the action of cholesteryl ester transfer protein (CETP), to VLDL and IDL particles (◘ Fig. 32.2). These TG-rich lipoproteins can be hydrolyzed to LDL, which can then be cleared by hepatic LDL receptor. Another enzyme that plays a critical role in the metabolic fate of HDL is hepatic lipase (HL), which hydrolyzes the triglycerides and phospholipids on larger HDL particles (HDL-2), producing smaller HDL particles (HDL-3). Nascent HDL particles are regenerated by the action of HL and phospholipid transfer protein (PTP) (◘ Table 32.3). HDL may also deliver cholesteryl esters to the liver directly via the scavenger receptor SRB-1 (◘ Fig. 32.2) [3, 5].

A number of epidemiological studies has shown an inverse relationship between coronary artery disease (CAD) and HDL cholesterol. HDL are thought to be cardioprotective due to their participation in reverse cholesterol transport, and perhaps also by their role as an antioxidant [3]. HDL impedes LDL oxidation by metal ions, an effect that may be due to the influence of several molecules on HDL, including apoA-I, platelet-activating factor acetylhydrolase, and paraoxonase [4]. Accumulation of HDL-2, thought to be the most cardioprotective of the HDL subclasses, is favored by estrogens, which negatively regulate hepatic lipase. In contrast, progesterone and androgens, which positively regulate this enzyme, lead to increased production of HDL-3.

Clinical studies have begun to address the effect of HDL cholesterol on cardiovascular endpoints. Men in the Veterans Administration High-density Lipoprotein Intervention Trial, with known CAD and treated with gemfibrozil for approximately 5 years, had a 24% reduction in death from CAD, nonfatal myocardial infarction and stroke, compared to men treated with placebo. This risk reduction was associated with a 6% increase in HDL cholesterol, 31% decrease in triglyceride levels and no significant change in LDL cholesterol levels [6]. Further analysis using nuclear magnetic resonance spectroscopy indicated that the shift from small, dense LDL particles to larger LDL particles and an increase in HDL-3 with gemfibrozil explained a further amount of the percent reduction in CAD. In the Bezafibrate Infarction Prevention Study, bezafibrate significantly raised HDL cholesterol by 18% and reduced relative risk for nonfatal myocardial infarction and sudden death by 40% in a subpopulation of study participants with triglycerides >200 mg/dl [7].

32.1.4   Lipid Lowering Drugs

In recent years, pharmacologic manipulation of the metabolic and cellular processes of lipid and lipoprotein metabolism (◘ Figs. 32.1 and 32.2) has greatly improved the treatment of dyslipidemias. Inhibitors of the rate-limiting enzyme of cholesterol synthesis, HMG-CoA reductase, called statins, effectively decrease the intrahepatic cholesterol pool (◘ Fig. 32.1) This effect, in turn, leads to the proteolytic release of SREBPs from the cytoplasm into the nucleus where they stimulate the transcription of the LDL receptor gene, resulting in an increased uptake of plasma LDL by the liver. Resins, which sequester bile acids, prevent entero-hepatic recycling and reuptake of bile acids through the ileal bile acid transporter. More hepatic cholesterol is converted into bile acids, lowering the cholesterol pool, and thus also inducing LDL receptors (◘ Fig. 32.1). A cholesterol absorption inhibitor interferes with the uptake of cholesterol from the diet and bile by a cholesterol transporter (CT) (◘ Fig. 32.1). This decreases the amount of cholesterol delivered by the chylomicron remnants to the liver, producing a fall in the hepatic cholesterol pool and induction of LDL receptors. Niacin, or vitamin B$_3$, when given at high doses, inhibits the release of FFA from adipose tissue, decreases the hepatic production of apoB-100, leading to decreased production of VLDL, and subsequently, IDL and LDL (◘ Fig 32.1). Fibrates are agonists for peroxisome proliferator activator receptors (PPAR), which upregulate the LPL gene and repress the apo C-III gene; both of these effects enhance lipolysis of triglycerides in VLDL (◘ Fig. 32.1). Fibrates also increase apo A-I production, while niacin decreases HDL catabolism, both leading to increased HDL levels.

32.2   Disorders of Exogenous Lipoprotein Metabolism

Two disorders of exogenous lipoprotein metabolism are known. Both involve chylomicron removal.

32.2.1   Lipoprotein Lipase Deficiency

Patients with classic lipoprotein lipase (LPL) deficiency present in the first several months of life with very marked hypertriglyceridemia, often ranging between 5,000 to 10,000 mg/dl (◘ Table 32.4). The plasma cholesterol level is usually 1/10 of the triglyceride level. This disorder is often suspected because of colic, creamy plasma on the top of a hematocrit tube, hepatosplenomegaly, or eruptive xanthomas. Usually only the chylomicrons are elevated (type I phenotype) (◘ Table 32.5), but occasionally the VLDL are also elevated (type V phenotype). The disorder can present later in childhood with abdominal pain and pancreatitis, a

◻ Table 32.4. Guidelines for plasma triglyceride levels in adults

Triglyceride levels		Category
mg/dl	mmol/l	
<150	<1.71	Desirable
150–199	1.71–2.67	Borderline
200–399	2.28–4.55	Elevated
400–999	4.56–11.39	High
>1,000	>11.40	Very High

◻ Table 32.5. Lipoprotein phenotypes of hyperlipidemia

Lipoprotein phenotype	Elevated lipoprotein
Type I	Chylomicrons
Type IIa	LDL
Type IIb	LDL, VLDL
Type III	Cholesterol-enriched IDL
Type IV	VLDL
Type V	Chylomicrons, VLDL

life-threatening complication of the massive elevation in chylomicrons. Lipemia retinalis is usually present, premature atherosclerosis is uncommon.

Familial LPL deficiency is a rare, autosomal recessive condition that affects about one in one million children. Parents are often consanguineous. The large amounts of chylomicrons result from a variety of mutations in the LPL gene.

When chylomicrons are markedly increased, they can replace water (volume) in plasma, producing artifactual decreases in concentrations of plasma constituents; for example, for each 1,000 mg/dl increase of plasma triglyceride, serum sodium levels decrease between 2 and 4 meq/liter.

The diagnosis is first made by a test for post-heparin lipolytic activity (PHLA). LPL is attached to the surface of endothelial cells through a heparin-binding site. After the intravenous injection of heparin (60 units/kg), LPL is released and the activity of the enzyme is assessed in plasma drawn 45 min after the injection. The mass of LPL released can also be assessed, using an ELISA assay. Parents of LPL deficient patients often have LPL activity halfway between normal controls and the LPL deficient child. The parents may or may not be hypertriglyceridemic.

Treatment is a diet very low in fat (10–15% of calories) [8]. Lipid lowering medication is ineffective. Affected in-

fants can be given Portagen, a soybean-based formula containing medium-chain triglycerides (MCT). MCT do not require the formation of chylomicrons for absorption, since they are directly transported from the intestine to the liver by the portal vein. A subset of LPL-deficient patients with unique, possibly posttranscriptional genetic defects, respond to therapy with MCT oil or omega-3 fatty acids by normalizing fasting plasma triglycerides; a therapeutic trial with MCT oil should, therefore, be considered in all patients presenting with the familial chylomicronemia syndrome [8]. Older children may also utilize MCT oil to improve the palatability and caloric content of their diet. Care must be taken that affected infants and children get at least 1% of their calories from the essential fatty acid, linoleic acid.

32.2.2 Apo C-II Deficiency

Marked hypertriglyceridemia (TG >1,000 mg/dl) can also present in patients with a rare autosomal recessive disorder affecting apo C-II, the co-factor for LPL. Affected homozygotes have been reported to have triglycerides ranging from 500 to 10,000 mg/dl (◻ Table 32.4). Apo C-II deficiency can be expressed in childhood but is often delayed into adulthood. The disorder is suspected by milky serum or plasma or by unexplained recurrent bouts of pancreatitis. A type V lipoprotein phenotype (◻ Table 32.5) is often found, but a type I pattern may also be present. Eruptive xanthomas and lipemia retinalis may also be found. As with the LPL defect, those with apo C-II deficiency do not get premature atherosclerosis.

The diagnosis can be confirmed by a PHLA test, and measuring apo C-II levels in plasma, using an ELISA assay. Apo C-II levels are very low to undetectable. The deficiency can be corrected by the addition of normal plasma to the in vitro assay for PHLA.

Apo C-II deficiency is even rarer than LPL deficiency and caused by a variety of mutations. Obligate heterozygous carriers of apo C-II mutants usually have normal plasma lipid levels, despite a 50% reduction in apo C-II levels.

The treatment of patients with apo C-II deficiency is the same as that discussed above for LPL deficiency. Infusion of normal plasma in vivo into an affected patient will decrease plasma triglycerides levels.

32.3 Disorders of Endogenous Lipoprotein Metabolism

These diseases comprise disorders of VLDL overproduction and of LDL removal.

32.3.1   Disorders of VLDL Overproduction

Familial Hypertriglyceridemia

Patients with familial hypertriglyceridemia (FHT) most often present with elevated triglyceride levels with normal LDL cholesterol levels (type IV lipoprotein phenotype) (◘ Table 32.5). The diagnosis is confirmed by finding at least one (and preferably two or more) first degree relatives with a similar type IV lipoprotein phenotype. The VLDL levels may increase to a considerable degree, leading to hypercholesterolemia as well as marked hypertriglyceridemia (>1,000 mg/dl) and occasionally to hyperchylomicronemia (type V lipoprotein phenotype) (◘ Table 32.5). This extreme presentation of FHT is usually due to the presence of obesity and type II diabetes. Throughout this spectrum of hypertriglyceridemia and hypercholesterolemia, the LDL cholesterol levels remain normal, or low normal. The LDL particles may be small and dense, secondary to the hypertriglyceridemia, but the number of these particles is not increased (see also below).

Patients with FHT often manifest hyperuricemia, in addition to hyperglycemia. There is a greater propensity to peripheral vascular disease than CAD in FHT. A family history of premature CAD is not usually present. The unusual rarer patient with FHT who has a type V lipoprotein phenotype may develop pancreatitis.

The metabolic defect in FHT appears to be due to the increased hepatic production of triglycerides but the production of apo B-100 is not increased. This results in the enhanced secretion of very large VLDL particles that are not hydrolyzed at a normal rate by LPL and apoC-II. Thus, in FHT there is not an enhanced conversion of VLDL into IDL and subsequently, into LDL (◘ Fig. 32.1).

Diet, particularly reduction to ideal body weight, is the cornerstone of therapy in FHT. For patients with persistent hypertriglyceridemia above 400 mg/dl, treatment with fibric acid derivatives, niacin or the statins may reduce the elevated triglycerides by up to 50%. Management of type II diabetes, if present, is also an important part of the management of patients with FHT (► Sect. 32.7).

Familial Combined Hyperlipidemia and the Small Dense LDL Syndromes
Clinical Presentation

Patients with familial combined hyperlipidemia (FCHL) may present with elevated cholesterol alone (type IIa lipoprotein phenotype), elevated triglycerides alone (type IV lipoprotein phenotype), or both the cholesterol and triglycerides are elevated (type IIb lipoprotein phenotype) (◘ Table 32.5). The diagnosis of FCHL is confirmed by the finding of a first degree family member, who has a different lipoprotein phenotype from the proband. Other characteristics of FCHL include the presence of an increased number of *small, dense LDL particles,* which link FCHL to other disorders, including hyperapobetalipoproteinemia

(hyperapoB), LDL subclass pattern B, and familial dyslipidemic hypertension [9]. In addition to hypertension, patients with the small-dense LDL syndromes can also manifest hyperinsulinism, glucose intolerance, low HDL cholesterol levels, and increased visceral obesity (syndrome X).

From a clinical prospective, FCHL and other small, dense LDL syndromes clearly aggregate in families with premature CAD, and as a group, these disorders are the most commonly recognized dyslipidemias associated with premature CAD, and may account for one-third, or more, of the families with early CAD.

Metabolic Derangement

There are three metabolic defects that have been described both in FCHL patients and in those with hyperapoB: *(1)* overproduction of VLDL and apo B-100 in liver; *(2)* slower removal of chylomicrons and chylomicron remnants; and, *(3)* abnormally increased free-fatty acids (FFA) levels [9, 10].

The abnormal FFA metabolism in FCHL and hyperapo B subjects may reflect the primary defect in these patients. The elevated FFA levels indicate an impaired metabolism of intestinally derived triglyceride-rich lipoproteins in the post-prandial state and, as well, impaired insulin-mediated suppression of serum FFA levels. Fatty acids and glucose compete as oxidative fuel sources in muscle, such that increased concentrations of FFA inhibit glucose uptake in muscle and result in insulin resistance. Finally, elevated FFA may drive hepatic overproduction of triglycerides and apo B.

It has been hypothesized that a cellular defect in the adipocytes of hyperapoB patients prevents the normal stimulation of FFA incorporation into TG by a small molecular weight basic protein, called the acylation stimulatory protein (ASP) [11]. The active component in chylomicrons responsible for enhancement of ASP in human adipocytes does not appear to be an apolipoprotein, but may be transthyretin, a protein that binds retinol-binding protein and complexes thyroxin and retinol [11]. ASP also appears to be generated in vivo by human adipocytes, a process that is accentuated postprandially, supporting the hypothesis that ASP plays an important role in clearance of triglycerides from plasma and fatty acid storage in adipose tissue [11]. Recently, Cianflone and co-workers [12] reported that an orphan G protein coupled receptor (GPCR), called C5L2, bound ASP with high affinity and promoted triglyceride synthesis and glucose uptake. The functionality of C5L2 is not known, nor is it known if there might be a defect in C5L2 in some patients with hyperapoB.

A defect in the adipocytes of hyperapoB patients might explain both metabolic abnormalities of TG-rich particles in hyperapoB. Following ingestion of dietary fat, chylomicron TG is hydrolyzed by LPL, producing FFA. The defect in the normal stimulation of the incorporation of FFA into TG by ASP in adipocytes from hyperapoB patients leads to increased levels of FFA that: *(1)* flux back to the liver increasing VLDL apo B production; and, *(2)* feedback inhibit further hydrolysis of chylomicron triglyceride by LPL [9]. Alternatively, there could be a defect in stimulation of release of ASP by adipocytes, perhaps due to an abnormal transthyretin/retinol binding system [11]. In that regard, plasma retinol levels have been found to be significantly lower in FCHL patients. This may possibly also affect the peroxisome proliferator activator receptors which are retinoic acid dependent.

Kwiterovich and colleagues isolated and characterized three basic proteins (BP) from normal human serum [13]. BP I stimulates the mass of cellular triacylglycerols in cultured fibroblasts from normals about two fold, while there is a 50% deficiency in such activity in cultured fibroblasts from hyperapoB patients. In contrast, BP II abnormally stimulates the formation of unesterified and esterified cholesterol in hyperapoB cells [13]. Such an effect might further accentuate the overproduction of apolipoprotein B and VLDL in hyperapoB patients [9]. Pilot data in hyperapoB fibroblasts indicate a deficiency in the high-affinity binding of BP I, but an enhanced high-affinity binding of BP II [13]. HyperapoB fibroblasts have a baseline deficiency in protein tyrosine phosphorylation that is not reversed with BP I, but is with BP II. These observations together suggest the existence of a receptor-mediated process for BP I and BP II that involves signal transduction [13]. We postulate that a defect in a BP receptor might exist in a significant number of patients with hyperapoB and premature CAD.

Genetics

The basic genetic defect(s) in FCHL and the other small, dense LDL syndromes are not known. FCHL and these other syndromes are clearly genetically heterogeneous, and a number of genes (oligogenic effect) may influence the expression of FCHL and the small dense LDL syndromes [9, 14, 15]. In a Finnish study, Pajukanta and coworkers mapped the first major locus of FCHL to chromosome 1q21–23, and recently provided strong evidence that the gene underlying the linkage is the upstream transcription factor-1 (USF-1) gene [16]. USF-1 regulates many important genes in plasma lipid metabolism, including certain apolipoproteins and HL. Linkage of type 2 diabetes mellitus as well as FCHL to the region harboring the USF-1 gene has been observed in several different populations worldwide [17], raising the possibility that USF-1 may also contribute to the metabolic syndrome and type 2 diabetes.

Treatment and Prognosis

The treatment of FCHL and hyperapoB starts with a diet reduced in total fat, saturated fat and cholesterol. This will reduce the burden of post-prandial chylomicrons and chylomicron remnants (which may also be atherogenic). Reduction to ideal body weight may improve insulin sensitivity and decrease VLDL overproduction. Regular aerobic

exercise also appears important. Two classes of drugs, fibric acids and nicotinic acid, lower triglycerides and increase HDL and may also convert small, dense LDL to normal sized LDL. The HMG-CoA reductase inhibitors do not appear as effective as the fibrates or nicotinic acid in converting small, dense LDL into large, buoyant LDL. However, the statins are very effective in lowering LDL cholesterol and the total number of atherogenic, small, dense LDL particles. In many patients with FCHL, combination therapy of a statin with either a fibrate or nicotinic acid will be required to obtain the most optimal lipoprotein profile [9] (▶ also Sect. 32.7). Patients with the small, dense LDL syndromes appear to have a greater improvement in coronary stenosis severity on combined treatment. This appears to be associated with drug-induced improvement in LDL buoyancy.

Lysosomal Acid Lipase Deficiency: Wolman Disease and Cholesteryl Ester Storage Disease

Wolman disease is a fatal disease that occurs in infancy [18]. Clinical manifestations include hepatosplenomegaly, steatorrhea, and failure to thrive. Patients have a lifespan that is generally under one year, while those with cholesteryl ester storage disease (CESD) can survive for longer periods of time [19]. In some cases, patients with CESD have developed premature atherosclerosis.

Lysosomal acid lipase (LAL) is an important lysosomal enzyme that hydrolyzes LDL-derived cholesteryl esters into unesterified cholesterol. Intracellular levels of unesterified cholesterol are important in regulating cholesterol synthesis and LDL receptor activity. In LAL deficiency, cholesteryl esters are not hydrolyzed in lysosomes and do not generate unesterified cholesterol. In response to low levels of intracellular unesterified cholesterol, cells continue to synthesize cholesterol and apo B-containing lipoproteins. In CESD, the inability to release free cholesterol from lysosomal cholesteryl esters results in elevated synthesis of endogenous cholesterol and increased production of apo B-containing lipoproteins. Wolman disease and CESD are autosomal recessive disorders due to mutations in the LAL gene on chromosome 10.

Lovastatin reduced both the rate of cholesterol synthesis and the secretion of apo B-containing lipoproteins, leading to significant reductions in total –197 mg/dl) and LDL (–102 mg/dl) cholesterol and triglycerides (–101 mg/dl) [20].

32.3.2 Disorders of LDL Removal

These disorders, characterized by marked elevations of plasma total and LDL cholesterol, provided the initial insights into the role of LDL in human atherosclerosis. The elucidation of the molecular defects in such patients, with monogenic forms of marked hypercholesterolemia, has provided unique and paramount insights into the mechanisms underlying cholesterol and LDL metabolism and the biochemical rationale for their treatment. Here we will discuss six monogenic diseases that cause marked hypercholesterolemia: familial hypercholesterolemia (FH); familial ligand defective apo B-100 (FDB); heterozygous FH3; autosomal recessive hypercholesterolemia (ARH); sitosterolemia, and cholesterol 7-α-hydroxylase deficiency.

Familial Hypercholesterolemia (LDL Receptor Defect)

Clinical Presentation

Familial hypercholesterolemia (FH) is an autosomal dominant disorder that presents in the heterozygous state with a two- to three-fold elevation in the plasma levels of total and LDL cholesterol [1]. Since FH is completely expressed at birth and early in childhood, it is often associated with premature CAD; by age 50, about half the heterozygous FH males and 25 percent of affected females will develop CAD. Heterozygotes develop tendon xanthomas in adulthood, often in the Achilles tendons and the extensor tendons of the hands. Homozygotes usually develop CAD in the second decade; atherosclerosis often affects the aortic valve, leading to life-threatening supravalvular aortic stenosis. FH homozygotes virtually all have planar xanthomas by the age of 5 years, notably in the webbing of fingers and toes and over the buttocks.

Metabolic Derangement and Genetics

FH is one of the most common inborn errors of metabolism and affects 1 in 500 worldwide (◘ Table 32.6). FH has a higher incidence in certain populations, such as Afrikaners, Christian Lebanese, Finns and French-Canadians, due to founder effects [21]. FH is due to one of more than 900 different mutations in the LDL receptor gene [21]. About one in a million children inherit two mutant alleles for the LDL receptor, presenting with a four- to eight-fold increase in LDL cholesterol levels (FH homozygous phenotype). Based on their LDL receptor activity in cultured fibroblasts, FH homozygotes are classified into LDL receptor-negative (<2% of normal activity) or LDL receptor-defective (2–25% of normal activity) homozygotes [1]. Most FH homozygotes inherit two different mutant alleles (genetic compounds) but some have two identical LDL receptor mutations (true homozygotes). Mutant alleles may fail to produce LDL receptor proteins (null alleles), encode receptors blocked in intracellular transport between endoplasmic reticulum and Golgi (transport-defective alleles), produce proteins that cannot bind LDL normally (binding defective), those that bind LDL normally, but do not internalize LDL (internalization defects), and those that disrupt the normal recycling of the LDL receptor back to the cell surface (recycling defects) [1].

Prenatal diagnosis of FH homozygotes can be performed by assays of LDL receptor activity in cultured amniotic fluid cells, direct DNA analysis of the molecular defect(s), or by linkage analysis using tetranucleotide DNA polymorphisms.

Treatment

Treatment of FH includes a diet low in cholesterol and saturated fat that can be supplemented with plant sterols or stanols to decrease cholesterol absorption. FH heterozygotes usually respond to higher doses of HMG-CoA reductase inhibitors. However, the addition of bile acid binding sequestrants or a cholesterol absorption inhibitor (see also below) is often necessary to also achieve LDL goals. Especially in those FH heterozygotes that may be producing increased amounts of VLDL, leading to borderline hypertriglyceridemia and low HDL cholesterol levels, niacin (nicotinic acid) may be a very useful adjunct to treatement. Nicotinic acid can also be used to lower an elevated Lp (a) lipoprotein. FH homozygotes may respond somewhat to high doses of HMG-CoA reductase inhibitors and nicotinic acid, both of which decrease production of hepatic VLDL, leading to decreased production of LDL. Cholesterol absorption inhibitors also lower LDL in FH homozygotes. In the end, however, FH homozygotes will require LDL apheresis every two weeks to effect a further lowering of LDL into a range that is less atherogenic. If LDL apheresis is not sufficient, then heroic hepatic transplantation may be considered. In the future, ex vivo gene therapy for FH homozygotes may become the treatment of choice [22].

Familial Ligand-Defective Apo B

Heterozygotes with familial ligand-defective apo B (FDB) may present with normal, moderately elevated, or markedly increased LDL cholesterol levels [21] (◘ Table 32.6). Hypercholesterolemia is usually not as markedly elevated in FDB as in patients with heterozygous FH, a difference attributed to effective removal of VLDL and IDL particles through the interaction of apo E with the normal LDL receptor in FDB. About 1/20 affected patients present with tendon xanthomas and more extreme hypercholesterolemia. This disorder represents a small fraction of patients with premature CAD, i.e. no more than 1%.

In FDB patients, there is delayed removal of LDL from blood despite normal LDL receptor activity. A mutant allele produces a defective ligand binding region in apo B-100, leading to decreased binding of LDL to the LDL receptor. The most commonly recognized mutation in FDB is a missense mutation (R3500Q) in the LDL receptor-binding domain of apo B-100 [21]. The frequency of FDB heterozygotes is about 1 in 1,000 in Central Europe but appears less common in other populations (◘ Table 32.6). Since the clearance of VLDL remnants and IDL occurs through the binding of apo E, and not apo B, to the LDL (B, E) receptor, the clearance of these triglyceride enriched particles in this disorder is not affected.

Dietary and drug treatment of FDB is similar to that used for FH heterozygotes. Induction of LDL receptors will enhance the removal of the LDL particles that contain the normal apo B-100 molecules, as well as increase the removal of VLDL remnant and IDL that utilize apo E and not apo B-100 as a ligand for the LDL receptor.

Heterozygous FH3

Another form of autosomal dominant hypercholesterolemia, termed heterozygous FH3 has been described [21]. While the clinical phenotype is indistinguishable from FH heterozygotes, the disorder does not segregate with *LDLR*. The disorder results from a mutation in *PCSK9*, a gene that codes for neural apoptosis-regulated convertase 1, a member of the proteinase K family of subtilases. Further research about the function of PCSK9, and its relation to LDL metabolism, promises to provide new insights into the genetic and molecular control of marked hypercholesteromia and very high LDL levels.

Autosomal Recessive Hypercholesterolemia

Autosomal recessive hypercholesterolemia (ARH) is a rare autosomal recessive disorder characterized clinically by LDL cholesterol levels intermediate between FH heterozygotes and FH homozygotes. ARH patients often have large tuberous xanthomas but their onset of CAD is on average later than that in FH homozygotes. To date, most of the families reported have been Lebanese or Sardinian. The cholesterol levels in the parents are often normal, but can be elevated.

The ARH protein functions as an adapter linking the LDL receptor to the endocytic machinery [21]. A defect in ARH prevents internalization of the LDL receptor. Strikingly, in ARH there is normal LDL receptor activity in fibroblasts but it is defective in lymphocytes. To date at least ten mutations have been described in ARH, all involving the interruption of the reading frame, producing truncated ARH [21].

Fortunately, patients with ARH respond quite dramatically to treatment with statins, but some will also require LDL apheresis. A bile acid sequestrants or a cholesterol absorption inhibitor may be added to the statin to effect a further reduction in LDL cholesterol.

Sitosterolemia

This is a rare, autosomal, recessive trait in which patients present with normal to moderately to markedly elevated total and LDL cholesterol levels, tendon and tuberous xanthomas, and premature CAD [21]. Homozygotes manifest abnormal intestinal hyperabsorption of plant or shell fish sterols (sitosterol, campesterol, and stigmasterol) and of cholesterol. In normal individuals, plant sterols are not absorbed and plasma sitosterol levels are low (0.3 to 1.7 mg/dl) and are less than 1% of the total plasma sterol, while in homozygotes with sitosterolemia, levels of total plant sterols are elevated (13 to 37 mg/dl) and represent

7–16% of the total plasma sterols. Patients often present in childhood with striking tuberous and tendon xanthomas despite normal or FH heterozygote-like LDL cholesterol levels. The clinical diagnosis is made by documenting the elevated plant sterol levels. The parents are normocholesterolmic and have normal plant sterol levels.

Two ABC half transporters, ABCG5 and ABCG 8 [21], together normally limit the intestinal absorption of plant sterols and cholesterol and promote the elimination of these dietary sterols in the liver. Sitosterolemia is caused by two mutations in either of the two adjacent genes that encode these half-transporters (◘ Table 32.6), thereby enhancing absorption of dietary sterols, and decreasing elimination of these sterols from liver into bile. This leads to suppression of the LDL receptor gene, inhibition of LDL receptor synthesis and elevated LDL levels.

Dietary treatment is very important in sitosterolemia and primarily consists of diet very low in cholesterol and in plant sterols. Thus, in contrast to a standard low cholesterol, low saturated fat diet, plant foods with high fat, high plant sterol content such as oils and margarines, must be avoided. Bile acid binding resins, such as cholestyramine, are particularly effective in lowering plant sterol and LDL sterol concentrations. The cholesterol absorption inhibitor, ezetimibe, is also quite effective [23]. These patients respond poorly to statins.

Cholesterol 7α - Hydroxylase Deficiency

Only a few patients have been described with a deficiency in the rate limiting enzyme of bile acid synthesis, cholesterol 7α-hydroxylase that converts cholesterol into 7α-hydroxy- cholesterol (▶ Chap. 34 and ◘ Fig. 34.1). Both hypercholesterolemia and hypertriglyceridemia were reported [21]. It is postulated that this defect increases the hepatic cholesterol pool, and decreases LDL receptors. As with the sitosterolemics, these subjects were relatively resistent to statin therapy.

32.4   Disorders of Endogenous and Exogenous Lipoprotein Transport

32.4.1   Dysbetalipoproteinemia (Type III Hyperlipoproteinemia)

This disorder is often associated with premature atherosclerosis of the coronary, cerebral and peripheral arteries. Xanthomas are often present and usually are tuberoeruptive or planar, especially in the creases of the palms. Occasionally, tuberous and tendon xanthomas are found. Patients with dysbetalipoproteinemia present with elevations in both plasma cholesterol and triglycerides, usually but not always, above 300 mg/dl. The hallmark of the disorder is the presence of VLDL that migrate as beta lipoproteins (β-VLDL), rather than prebeta lipoproteins (type III lipoprotein phenotype) (◘ Table 32.5). β-VLDL reflect the accumulation of cholesterol-enriched remnants of both hepatic VLDL and intestinal chylomicrons (◘ Fig. 32.1) [24]. These remnants accumulate because of the presence of a dysfunctional apoE, the ligand for the receptor-mediated removal of both chylomicron and VLDL remnants by the liver.

There are two genetic forms of dysbetalipoproteinemia [24]. The most common form is inherited as a recessive trait. Such patients have an E_2E_2 genotype. The E_2E_2 genotype is necessary but not sufficient for dysbetalipoproteinemia. Other genetic and metabolic factors, such as overproduction of VLDL in the liver seen in FCHL, or hormonal and environmental conditions, such as hypothyroidism, low estrogen state, obesity, or diabetes are necessary for the full blown expression of dysbetalipoproteinemia. The recessive form has a delayed penetrance until adulthood and a prevalence of about 1:2000. In the rarer form of the disorder, inherited as a dominant and expressed as hyperlipidemia even in childhood, there is a single copy of another defective apo E allele [24].

The diagnosis of dysbetalipoproteinemia includes: *(1)* demonstration of E_2E_2 genotype; *(2)* performing preparative ultracentrifugation and finding the presence of β-VLDL on agarose gel electrophoresis (floating β lipoproteins); and, *(3)* a cholesterol enriched VLDL (VLDL cholesterol/triglyceride ratio > 0.30; normal ratio <0.30). LDL and HDL cholesterol levels are low or normal.

Patients with this disorder are very responsive to therapy. A low-fat diet is important to reduce the accumulation of chylomicron remnants, and reduction to ideal body weight may decrease the hepatic overproduction of VLDL particles. The drug of choice is a fibric acid derivative, but nicotinic acid and HMG-CoA reductase inhibitors may also be effective. Treatment of the combined hyperlipidemia in dysbetalipoproteinemia with a fibrate will correct both the hypercholesterolemia and hypertriglyceridemia; this effect is in contrast to treatment of FCHL with fibrates alone, which usually reduces the triglyceride level, but increases the LDL cholesterol level.

32.4.2   Hepatic Lipase Deficiency

Patients with hepatic lipase (HL) deficiency can present with features similar to dyslipoproteinemia (type III hyperlipoproteinemia) (see above), including hypercholesterolemia, hypertriglyceridemia, accumulation of triglyceride-rich remnants, planar xanthomas and premature cardiovascular disease [25]. Recurrent bouts of pancreatitis have been described. The LDL cholesterol is usually low or normal in both disorders.

HL hydrolyzes both triglycerides and phospholipids in plasma lipoproteins. As a result, HL converts IDL to LDL and HDL-2 to HDL-3, thus playing an important role in the metabolism of both remnant lipoproteins and HDL (◘ Figs. 32.1 and 32.2). HL shares a high degree of homology to LPL and pancreatic lipase.

HL deficiency is a rare genetic disorder, which is inherited as an autosomal recessive trait. The frequency of this disorder is not known, and it has been identified in only a small number of kindreds. Obligate heterozygotes are normal. The molecular defects described in HL deficiency include a single A → G substitution in intron I of the HL gene [26].

HL deficiency can be distinguished from dysbetalipoproteinemia in two ways: first, the elevated triglyceride-rich lipoproteins have a normal VLDL cholesterol/triglyceride ratio <0.3, because the triglyceride is not being hydrolyzed by HL; and second, the HDL cholesterol often exceeds the 95th percentile in HL deficiency but is low in dysbetalipoproteinemia. The diagnosis is made by a PHLA test (see above). Absent HL activity is documented by measuring total PHLA activity, and HL and LPL activity separately.

Treatment includes a low total fat diet. In one report, the dyslipidemia in HL deficiency improved on treatment with lovastatin but not gemfibrozil.

32.5   Disorders of Reduced LDL Cholesterol Levels

32.5.1   Abetalipoproteinemia

Abetalipoproteinemia is a rare, autosomal recessive disorder in patients with undetectable plasma apo B levels [27]. Patients present with symptoms of fat malabsorption and neurological problems. Fat malabsorption occurs in infancy with symptoms of failure to thrive (poor weight gain and steatorrhea). Fat malabsorption is secondary to the inability to assemble and secrete chylomicrons from enterocytes. Neurological problems begin during adolescence and include dysmetria, cerebellar ataxia, and spastic gait. Other manifestations include atypical retinitis pigmentosa, anemia (acanthocytosis) and arrhythmias.

Total cholesterol levels are exceedingly low (20 to 50 mg/dl) and no detectable levels of chylomicrons, VLDL, or LDL are present. HDL levels are measurable but low. Parents have normal lipid levels.

It was initially thought that the lack of plasma apo B levels were due to defects in the APOB gene. Subsequent studies have demonstrated no defects in the APOB gene. Immunoreactive apo B-100 is present in liver and intestinal cells. Wetterau and colleagues [28] found that the defect in synthesis and secretion of apo B is secondary to the absence of microsomal triglyceride transfer protein (MTP), a molecule that permits the transfer of lipid to apo B. MTP is a heterodimer composed of the ubiquitous multifunctional protein, protein disulfide isomerase, and a unique 97-kDa subunit. Mutations that lead to the absence of a functional 97-kDa subunit cause abetalipoproteinemia. Over a dozen mutant 97-kDa subunit alleles have been described.

Treatment of patients with abetalipoproteinemia is difficult. Steatorrhea can be controlled by reducing the intake of fat to 5 to 20 g/day. This measure alone can result in marked clinical improvement and growth acceleration. In addition, the diet should be supplemented with linoleic acid (e.g., 5 g corn oil or safflower oil/day). MCT as a caloric substitute for long-chain fatty acids may produce hepatic fibrosis, and thus MCT should be used with caution, if at all. Fat-soluble vitamins should be added to the diet. Rickets can be prevented by normal quantities of vitamin D, but 200–400 IU/kg/day of vitamin A may be required to raise the level of vitamin A in plasma to normal. Enough vitamin K (5–10 mg/day) should be given to maintain a normal prothrombin time. Neurologic and retinal complications may be prevented, or ameliorated, through oral supplementation with vitamin E (150-200 mg/kg/day). Adipose tissue rather than plasma may be used to assess the delivery of vitamin E.

32.5.2   Hypobetalipoproteinemia

Patients with hypobetalipoproteinemia often have a reduced risk for premature atherosclerosis and an increased life span. These patients do not have any physical stigmata of dyslipidemia. The concentrations of fat-soluble vitamins in plasma are low to normal. Most patients have low levels of LDL cholesterol below the 5th percentile (approximately 40 to 60 mg/dl), owing to the inheritance of one normal allele and one autosomal dominant mutant allele for a truncated apolipoprotein B. Hypobetalipoproteinemia occurs in about 1 in 2,000 people.

Over several dozen gene mutations (nonsense and frame shift mutations) have been shown to affect the full transcription of apolipoprotein B and cause familial hypobetalipoproteinemia. The various gene mutations lead to the production of truncated apolipoprotein B.

Occasionally, hypobetalipoproteinemia is secondary to anemia, dysproteinemias, hyperthyroidism, intestinal lymphangiectasia with malabsorption, myocardial infarction, severe infections, and trauma.

Plasma levels of truncated apo B are generally low and are thought to be secondary to low synthesis and secretion rates of the truncated forms of apo B from hepatocytes and enterocytes. The catabolism of LDL in hypobetalipoproteinemia also appears to be increased. The diagnosis is confirmed by demonstrating the presence of a truncated apoB in plasma.

No treatment is required. Neurologic signs and symptoms of a spinocerebellar degeneration similar to those of Friedreich ataxia and peripheral neuropathy have been found in several affected members.

32.5.3   Homozygous Hypobetalipoproteinemia

The clinical presentation of children with this disorder depends upon whether they are homozygous for null alleles in the APOB gene (i.e., make no detectable apo B) or homozygous (or compound heterozygotes) for other alleles who produce lipoproteins containing small amounts of apo B or a truncated apo B [29]. Null-allele homozygotes are similar phenotypically to those with abetalipoproteinemia (see above) and may have fat malabsorption, neurologic disease, and hematologic abnormalities as their prominent clinical presentation and will require similar treatment (▶ above). However, the parents of these children are heterozygous for hypobetalipoproteinemia. Patients with homozygous hypobetalipoproteinemia may develop less marked ocular and neuromuscular manifestations, and at a later age, than those with abetalipoproteinemia. The concentrations of fat-soluble vitamins are low.

32.6   Disorders of Reverse Cholesterol Transport

32.6.1   Familial Hypoalphalipoproteinemia

Hypoalphalipoproteinemia is defined as a low level of HDL cholesterol (<5th percentile, age and sex specific) in the presence of normal lipid levels [30]. Patients with this syndrome have a significantly increased prevalence of CAD, but do not manifest the clinical findings typical of other forms of HDL deficiency (see below). Low HDL cholesterol levels of this degree are most often secondary to disorders of triglyceride metabolism (▶ above). Consequently, primary hypoalphalipoproteinemia, although more prevalent than the rare recessive disorders including deficiencies in HDL, is relatively uncommon. In some families, hypoalphalipoproteinemia behaves as an autosomal dominant trait but the basic defect is unknown. Since it is likely that the etiology of low HDL cholesterol levels is oligogenic (significant effect of several genes), Cohen, Hobbs and colleagues [31] tested whether rare DNA sequence variants in three candidate genes, ABCA1, APOA1 and LCAT, contributed to the hypoalpha phenotype. Nonsynonymous sequence variants were significantly more common (16% versus 2%) in individuals with hypoalpha (HDL cholesterol <5th %) than in those with hyperalpha (HDL cholesterol >95th %). The variants were most prevalent in the ABCA1 gene.

32.6.2   Apolipoprotein A-I Mutations

The HDL cholesterol levels are very low (0–4 mg/dl), and the apolipoprotein A-I levels are usually <5 mg/dl. Corneal clouding is usually present in these patients. Planar xanthomas are not infrequently described; the majority, but not all, of these patients develop premature CAD [30, 32, 33].

The *APOA1* gene exists on chromosome 11 as part of a gene cluster with the *APOC3* and *APOA4* genes. A variety of molecular defects have been described in *APOA1*, including gene inversions, gene deletions, and nonsense and missense mutations. In contrast, *APOA1* structural variants, usually due to a single amino acid substitution, do not have, in most instances, any clinical consequences [33]. Despite lower HDL cholesterol levels (decreased by about one half), premature CAD is not ordinarily present. In fact, in one Italian variant, *APOA-I*$_{Milano}$, the opposite has been observed (i.e., increased longevity in affected subjects). In a recent study by Nissen et al. [34], these investigators tested proof of concept of apoA-I$_{Milano}$ by infusing recombinant apoA-I$_{Milano}$/phospholipid complexes (ETC-216) in a small group of adults between the ages of 30–75 years with acute coronary syndrome. The study participants underwent five weekly infusions of placebo, low (15 mg/kg) or high (45 mg/kg) dose of ETC-216. The primary outpoint, change of percent atheroma volume as quantified by intravascular ultrasonography, decreased 3.2% (*p*<0.02) in subjects treated with ETC-216, while the percent atheroma volume increased in the placebo group.

32.6.3   Tangier Disease

Its name is derived from the island of Tangier in the Chesapeake Bay in Virginia, USA, where Dr Donald Fredrickson described the first kindred. HDL cholesterol levels are extremely low and of an abnormal composition (HDL Tangier or T). HDL$_T$ are chylomicron-like particles on a high fat diet, which disappear when a patient consumes a low-fat diet [30].

The characteristic clinical findings in Tangier patients include the presence of enlarged orange yellow tonsils, splenomegaly and a relapsing peripheral neuropathy. The finding of orange tonsils is due to the deposition of beta carotene-rich cholesteryl esters (foam cells) in the lymphatic tissue. Other sites of foam cell deposition include the skin, peripheral nerves, bone marrow, and the rectum. Mild hepatomegaly, lymphadenopathy and corneal infiltration (in adulthood) may also occur.

The *APOA1* gene in Tangier patients is normal. The underlying defect has now been determined to be a deficiency in ABCA1, an ATP binding cassette transporter [35]. Under normal circumstances, this plasma membrane protein has been shown to mediate cholesterol efflux to nascent, apo A-I rich HDL particles (◘ Figs. 32.1 and 32.2). The presence of low HDL cholesterol in subjects with Tangier disease is due to the lack of cholesterol efflux by the deficient ABCA1 to nascent HDL and then increased catabolism of this lipid-poor HDL particle. The clinical diagnosis

of Tangier disease can be confirmed by determining the reduced efflux of cholesterol from Tangier fibroblasts onto an acceptor in the culture medium [36].

In general, patients with Tangier disease have an increased incidence of atherosclerosis in adulthood [30]. Treatment with a low fat diet diminishes the abnormal lipoprotein species that are believed to be remnants of abnormal chylomicron metabolism.

32.6.4   Lecithin-Cholesterol Acyltransferase Deficiency

Lecithin-cholesterol acyltransferase (LCAT) is an enzyme located on the surface of HDL particles and is important in transferring fatty acids from the sn-2 position of phosphatidylcholine (lecithin) to the 3-β-OH group on cholesterol (◘ Table 32.3). In this process, lysolecithin and esterified cholesterol are generated (α-LCAT). Esterification can also occur on VLDL/LDL particles (β-LCAT).

In patients with *classic LCAT deficiency*, both α- and β-LCAT activity are missing [37]. LCAT deficiency is a rare, autosomal, recessively inherited disorder. More than several dozen mutations in this gene, located on chromosome 16, have been described. The diagnosis should be suspected in patients presenting with low HDL cholesterol levels, corneal opacifications and renal disease (proteinuria, hematuria). Laboratory tests include the measurement of plasma free cholesterol to total cholesterol ratio. Levels above 0.7 are diagnostic for LCAT deficiency.

In *Fish Eye disease*, only α-LCAT activity is absent. Patients present with corneal opacifications, but do not have renal disease [37]. It has been hypothesized that the variability in clinical manifestations from patients with Fish Eye disease, compared to LCAT deficiency, may reside in the amount of total plasma LCAT activity.

To date, no therapies exist to treat the underlying defects. Patients succumb primarily from renal disease, and atherosclerosis may be accelerated by the underlying nephrosis. Thus, patients with LCAT deficiency, and other lipid metabolic disorders associated with renal disease, should be aggressively treated including a low fat diet. This includes the secondary dyslipidemia associated with the nephrotic syndrome which responds to statin therapy.

32.6.5   Cholesteryl Ester Transfer Protein Deficiency

The role of the cholesteryl ester transfer protein (CETP) in atherosclerosis has not been well defined. The CETP gene is upregulated in peripheral tissues and liver in response to dietary or endogenous hypercholesterolemia. HDL particles isolated from patients with CETP deficiency may be less effective in promoting cholesterol efflux from cultured cells.

This may be due to the increased concentration of cholesterol within the HDL particles and its inability to adsorb additional cholesterol from peripheral tissues. Some investigators have termed this type of HDL as being »dysfunctional«.

Elevated HDL cholesterol levels due to deficiency of CETP were first described in Japanese families and several mutations have been found. Increased CAD in Japanese families with CETP deficiency was primarily observed for HDL cholesterol 41–60 mg/dl; for HDL cholesterol >60 mg/dl, men with and without mutations had low CAD prevalence [38]. Thus, genetic CETP deficiency may or may not be an independent risk factor for CAD. These effects occur in spite of lower levels of apo B in CETP deficiency [39].

Due to its important role in modulating HDL levels, CETP inhibitors have been developed to raise plasma HDL cholesterol levels. De Grooth et al [40] examined the safety and efficacy of the CETP inhibitor, JTT-705, in a randomized, double-blind, placebo controlled study of 198 subjects. Study subjects entered the active treatment phase and were randomized to placebo, JTT-705 300 mg once daily, 600 mg once daily, or 900 mg once daily for 4 weeks. The activity of CETP decreased 37% in subjects taking the 900 mg dose, while HDL cholesterol levels increased in a dose-dependent manner, with a maximum rise of 34% in subjects taking the 900 mg dose. LDL cholesterol levels decreased 7% in the high dose group and triglyceride levels were unchanged. The effects of the CETP inhibitor CP-529,414 (torcetrapib) on elevating HDL cholesterol were also examined by treating adults between the ages of 18 and 55 years with placebo or torcetrapib 10, 30, 60, and 120 mg daily and 120 mg twice daily for 14 days [41]. The HDL cholesterol levels increased from 16–91% with increasing doses of this CETP inhibitor. Total cholesterol levels remained the same due to significant lowering of non-HDL cholesterol levels. In a separate study with torcetrapib, investigators found that this inhibitor effectively increased HDL cholesterol levels when given as monotherapy or in combination with atorvastatin [42].

32.6.6 Elevated Lipoprotein (a)

Lipoprotein (a) [Lp(a)] consists of one molecule of LDL whose apo B-100 is covalently linked to one molecule of apolipoprotein (a) [apo(a)] by a disulfide bond [43]. The physiological function(s) of Lp(a) are unknown. Apo(a) is highly homologous to plasminogen, and when the Lp(a) level is elevated (>30 mg/dl for total Lp(a), >10 mg/dl for Lp(a) cholesterol), apo(a) interferes with the thrombolytic action of plasmin, promoting thrombosis. Lp(a) also appears to promote atherosclerosis, particularly in some families, due to its similarity to LDL.

Apo(a) exists in a number of size isoforms, with the smaller isoforms correlating with higher plasma levels of Lp(a). Plasma levels of Lp(a) in whites tend to be lower than in blacks (median values, 1 vs 10 mg/ml, respectively). However, elevated plasma levels of Lp(a) do not correlate directly with the extent of cardiovascular disease in African-Americans. It should be emphasized that Lp(a) is often not measured accurately [43].

Niacin and estrogen can effectively lower Lp(a) levels, while the statins and fibrates do not. To date, clinical trial evidence is lacking regarding the benefit of lowering Lp(a) on the prevalence of cardiovascular disease.

32.7 Guidelines for the Clinical Evaluation and Treatment of Dyslipidemia

32.7.1 Clinical Evaluation

The patient who is being evaluated for dyslipidemia requires a thorough family history and an evaluation of current intake of dietary fat and cholesterol. The family history is reviewed for premature (before 60 years of age) cardiovascular disease (heart attacks, coronary artery bypasses, coronary angioplasties, angina) cerebrovascular (strokes, transient ischemic attacks) and peripheral vascular disease; dyslipidemia; diabetes mellitus; obesity; and, hypertension in grandparents, parents, siblings, children, and aunts and uncles. A dietary assessment is best performed by a registered dietician.

The medical history is focused on the two major complications of dyslipidemias, atherosclerotic cardiovascular disease and pancreatitis. The patient is asked about chest pain, arrhythmias, palpitations, myocardial infarction, stroke (including transient ischemic attacks), coronary artery bypass graft surgery, and balloon angioplasty. The results of past resting and stress electrocardiograms and coronary arteriography are assessed. Any history of recurrent abdominal pain, fatty food intolerance and pancreatitis is reviewed. The past and current use of lipid-lowering drugs is determined, as well as a history of an untoward reactions or side effects. The review of systems includes diseases of the liver, thyroid, and kidney, the presence of diabetes mellitus, and operations including transplantation. For women, a menstrual history, including current use of oral contraceptives and post-menopausal estrogen replacement therapy, is obtained.

The presence of other risk factors for CAD [44, 45] are systematically assessed: cigarette smoking, hypertension, low HDL cholesterol (<40 mg/dl), age (>45 years in men, >55 years in women), diabetes (CAD risk equivalent), obesity, physical inactivity and atherogenic diet. An electrocardiogram is obtained.

Height and weight are determined to assess obesity using the Quetelet (body mass) index: weight (kg)/height (m^2). An index of 30 or higher is defined as obesity and between 25 and 30 is considered overweight. Waist circumference can be measured (abnormal >40 inches in men, >35 inches in women). The physical examination includes an assessment of tendon, tuberous and planar xanthomas. The eyes are examined for the presence of xanthelasmas, corneal arcus, corneal clouding, lipemia retinalis, and atherosclerotic changes in the retinal blood vessels. The cardiovascular exam includes an examination for bruits in the carotid, abdominal, and femoral arteries, auscultation of the heart, assessment of peripheral pulses and measurement of blood pressure. The rest of the exam includes palpation of the thyroid, assessment of hepatosplenomegaly and deep tendon reflexes (which are decreased in hypothyroidism).

The clinical chemistry examination includes (at the minimum) a measurement of total cholesterol, total triglycerides, LDL cholesterol and HDL cholesterol, a chemistry panel to assess fasting blood sugar, uric acid, tests of liver and kidney function and thyroid stimulating hormone (TSH). We also assess the plasma levels of apo B and apo A-I; apo B provides an assessment of the total number of atherogenic, apolipoprotein B-containing particles, while the ratio of apo B to apo A-I when > 1.0 often indicates high risk of CAD and usually reflects an elevation in the apo B-containing particles and a depression of the apo A-I-containing particles. Other tests may be ordered when clinically indicated, such as »non-traditional« risk factors for cardiovascular disease, i.e., Lp (a) lipoprotein, homocysteine, prothrombotic factors, small-dense LDL and highly sensitive C-reactive protein (hsCRP). HbA1C is measured when a patient has known diabetes mellitus.

32.7.2   Dietary Treatment, Weight Reduction and Exercise

The cornerstone of treatment of dyslipidemia is a diet reduced in total fat, saturated fat and cholesterol [44, 45] (◨ Table 32.7). This is important to reduce the burden of post-prandial lipemia as well as to induce LDL receptors. A Step I and Step II dietary approach is often used [44] (◨ Table 32.7), but most dyslipidemic patients will require a Step II Diet. The use of a registered dietician or nutritionist is usually essential to achieving dietary goals. The addition of 400 I.U. or more of vitamin E and 500 mg or more of vitamin C is *not* currently recommended as an adjunct to diet. There is no clear evidence that such supplementations decrease risk for CAD, and in fact may impair the treatment of dyslipidemia [46].

If a patient is obese (Quetelet index >30), or overweight (Quetelet index 25–30), weight reduction will be an important part of the dietary management. This is particularly true if hypertriglyceridemia or diabetes mellitus are present.

Regular aerobic exercise is essential in most patients to help control their weight and dyslipidemia. The duration, intensity and frequency of exercise are critical. For an adult, a minimum of 1,000 calories per week of aerobic exercise is required. This usually translates into three or four sessions a week of 30 min or more, during which time the patient is in constant motion and slightly out of breath.

◨ **Table 32.6.** Major monogenic diseases that cause marked hypercholesterolemia. Modified with permission from Rader, Cohen and Hobbs [21]

Disease	Defective gene	Prevalence	LDL-C	Metabolic defect
Autosomal dominant				
FH	LDLR			Decreased LDL clearance (1^0)
Heterozygous FH		1 in 500	3X	Increased LDL production (2^0)
Homozygous FH		1 in 1 x 10^6	5X	
FDB	APOB			Decreased LDL clearance
Heterozygous FDB		1 in 1000	2X	
Homozygous FDB		1 in 4 x 10^6	3X	
FH3	PCSK9			Unknown
Heterozygous FH3		<1 in 2500	3X	
Autosomal recessive				
ARH	ARH	<1 in 5 x 10^6	4X	Decreased LDL clearance
Sitosterolemia	ABCG5 or ABCG8	<1 in 5 x 10^6	1X to 5X	Decreased cholesterol excretion(1^0) Decreased LDL clearance (2^0)

ARH, autosomal recessive hypercholesterolemia; *FDB*, familial ligand defective apoB-100; *FH*, familial hypercholesterolemia. *X* indicates the mean LDL-cholesterol (LDL-C) level in normals

32.7.3 Goals for Dietary and Hygienic Therapy

Four lipid parameters are used to define abnormal levels and determine therapeutic goals: LDL cholesterol (■ Table 32.8), triglycerides (■ Table 32.4), HDL cholesterol (low <40 mg/dl) and non-HDL cholesterol (total cholesterol minus HDL cholesterol) [44]. If the goals for LDL cholesterol are achieved with dietary management alone, drug therapy is not recommended. The recommended goal for triglycerides is a level <150 mg/dl in adults; the ideal goal is <100 mg/dl. Values of triglycerides >200 mg/dl are associated with the presence of small, dense LDL particles in 80% of patients. Low HDL cholesterol is a value <40 mg/dl. The minimum treatment goal for HDL cholesterol is >40 mg/dl.

The most recent recommendations from the National Cholesterol Education Program (NCEP) [45] offer guidelines for assessing risk and initiating treatment in patients with hypercholesterolemia. As shown in ■ Table 32.7, dietary intervention is used initially in the treatment of patients with dyslipidemia. A more aggressive reduction in the total daily allowance of saturated fat and cholesterol is used in patients with CAD or those failing to respond to the Step I diet. Patients with CAD should be placed simultaneously on the Step II diet and lipid-lowering drug therapy. Ideally, all patients should be formally counseled by a registered dictitian. Physicians should reinforce the importance of the dietary plan for their patients.

The value of pharmacologically lowering lipid levels to reduce cardiovascular event rates is well established, but the optimal level of cholesterol has not yet been determined. Several recent studies showed that intensive lowering of LDL cholesterol levels with atorvastatin 80 mg/day reduced cardiovascular event rates in patients with acute coronary syndrome [47] and slowed atherosclerotic progression [48] more than standard lipid-lowering therapy. In fact, in these studies, a target LDL cholesterol level of <70 mg/dl conferred greater benefit than a level of <100 mg/dl. Subsequent analyses from these studies showed that highly sensitive C-reactive protein (hsCRP) was an important independent predictor of events [49, 50]. Further, patients in the Heart Protection Study [51], who had CAD, diabetes, and/or hypertension, had a significant reduction in CAD events and death when treated with 40 mg of simvastatin, despite baseline LDL cholesterol levels already »at goal« <100 mg/dl.

As the result of these latest clinical trials, the NCEP has established new lipid-lowering guidelines for primary and secondary prevention of CAD [45] (■ Table 32.8). As be-

■ Table 32.7. National cholesterol education program diets: Step I and II

Step I
- Less than 30% calories as fat: less than 10% saturated, 10–15% monounsaturated, and up to 10% polyunsaturated
- 55% carbohydrates
- 15–20% protein
- Less than 300 mg cholesterol/day

Step II
- Less than 30% calories as fat: <7% saturated, 10-15% monounsaturated, and 10% polyunsaturated
- Less than 200 mg cholesterol/day

■ Table 32.8. NCEP-ATP III guidelines for LDL-lowering pharmacotherapy initiation and goals. Adapted from the National Cholesterol Education Program, Adult Treatment Panel III [44, 45]

Patient category	Initiation of drug therapy LDL cholesterol (mg/dl)	Therapeutic goal LDL cholesterol (mg/dl)
High risk CAD or CAD risk equivalents (10-year risk >20%)	≥100 (<100: consider drug options)[1]	<100 (optional goal: <70)[1]
Moderately high risk No CAD and >2 risk factors (10-year risk 10–20%)[2]	≥130 (100–129: consider drug options)[1]	<130 (optional goal: <100)[1]
Moderate risk No CAD and <2 risk factors (10-year risk ≤20%)	≥160	<130
Lower risk 0–1 risk factor	≥190 (160–189: LDL-lowering drug therapy optional)	<160

[1] Drug therapy advisable on the basis of clinical trials. The optional goal of LDL cholesterol in high risk patients is <70 mg/dl, or in those with high triglycerides (>200 mg/dl), a non-HDL cholesterol <100 mg/dl. The optional goal of LDL cholesterol in moderately risk patients is <100 mg/dl, or in those with high triglycerides, a non-HDL cholesterol <130 mg/dl.
[2] Positive risk factors for CAD are cigarette smoking, hypertension, low HDL cholesterol (<40 mg/dl), age (>45 years in men, >55 years in women), diabetes, obesity, physical inactivity and atherogenic diet).
CAD, coronary artery disease; *HDL-C*, HDL cholesterol; *LDL-C*, LDL cholesterol.

fore, the threshold of the LDL cholesterol level to initiate drug therapy and the target for treatment depends on the presence or absence of CAD, CAD risk equivalents, and associated risk factors. In this latest classification, for patients with CAD or CAD risk equivalents, the minimum target for LDL cholesterol is <100 mg/dl with an optional target of <70 mg/dl For those at moderate risk (at least two risk factors for CAD), the minimum target for LDL cholesterol is <130 mg/dl with an optional target of <100 mg/dl.

The guidelines provide recommendations for complete screening of TC, LDL cholesterol, HDL cholesterol, and TG, encouraging the use of plant sterols or stanols, and soluble fiber, and treatment using non-HDL cholesterol (total cholesterol minus HDL cholesterol) guidelines for patients with TG ≥200 mg/dl [44, 45]. For those with hypertriglyceridemia (>200 mg/dl), the optional targets for the high risk and moderate risk groups, are a non-HDL cholesterol of <100 mg/dl and <130 mg/dl, respectively.

32.7.4   Low Density Lipoprotein-Lowering Drugs

Agents which will lower LDL cholesterol include inhibitors of HMG-CoA reductase (the statins), bile acid sequestrants, cholesterol absorption inhibitors, and niacin (nicotinic acid) (◻ Table 32.9). The fibrates can also modestly reduce LDL cholesterol levels, but in hypertriglyceridemic patients with FCHL, LDL levels may stay the same or actually increase [36].

The *statins* available in Europe and the U.S.A. include atorvastatin (Lipitor), fluvastatin (Lescol), lovastatin (Mevacor), pravastatin (Pravachol), simvastatin (Zocor) and rosuvastatin (Crestor) [44, 45]. The equivalent doses are about: 5 mg rosuvastatin = 10 mg atorvastatin = 20 mg simvastatin = 40 mg lovastatin = 40 mg pravastatin = 80 mg fluvastatin. Lovastatin, simvastatin and pravastatin are derived from a biological product, while atorvastatin, fluvastatin and rosuvastatin are entirely synthetic products.

Statins undergo extensive first-pass metabolism via the hepatic portal system and typically less than 20% of these agents reaches systemic circulation [51]. In the liver the statins inhibit the rate limiting enzyme of cholesterol biosynthesis, HMG-CoA reductase, (◻ Fig. 32.1) leading to a decrease in hepatic cholesterol stores, increasing the release of SREBPs, stimulating the production of LDL receptors and lowering the LDL levels significantly. The statins also improve endothelial cell function and stabilize unstable plaques [49, 50].

Statins are generally well tolerated, and have an excellent safety profile with minimal side effects. Liver function tests (AST, ALT) should be monitored at baseline, following 6–8 weeks after initiating treatment and every 4 months for the first year. After that, patients on a stable dose of a statin can have their liver function tests monitored every six months. Consideration should be given to reducing the dosage of drug, or its discontinuation, should the liver function tests exceed 3 times the upper limits of normal. In clinical trials the discontinuation rate due to elevation of transaminases was less than 2%. Between 1/500 to 1/1,000 patients may develop myositis on a statin which can lead to life threatening rhabdomyolysis. Rhabdomyolysis is a rare event, occurring at an incidence of 1.2 per 10,000 patient-years [52]. Creatine kinase (CK) should be measured at baseline and repeated if the patient develops muscle aches and cramps. The statin is discontinued if the CK is >5x the upper limit of normal in those with symptoms of myositis, or >10x the upper limit of normal in asymptomatic patients. CK is not routinely measured in patients at follow-up since it is not predictive of who will develop rhabdomyolysis.

Three statins, lovastatin, simvastatin and atorvastatin, are metabolized by the CYP3A4 isozyme of the cytochrome P450 microsomal enzyme system, and consequently have drug interactions with other agents metabolized by CYP3A4. Inhibitors of CYP3A4 include erythromycin, fluvoxamine, grapefruit juice, itraconazole, ketoconazole, nefazodone, and sertraline. Drugs that are substrates for CYP3A4 may also increase the level of the statin in the blood and include: antiarrhythmics (lidocaine, propafenone

◻ **Table 32.9.** Effect of drug classes on plasma lipid and lipoprotein levels. Adapted and modified from Gotto AM Jr (1992) Management of lipid and lipoprotein disorders. In: Gotto AM Jr, Pownall HJ (eds) Manuel of lipid disorders. Williams & Wilkins, Baltimore, MD

Drug class	TC	LDL-C	HDL-C	TG
Statins	15–60%	20–60%	3–10%	10–30%
Bile acid resins	10–20%	15–20%	3–5%	Variable
Cholesterol absorption inhibitor	10–20%	15–20%	3–5%	5–10%
Niacin	25%	10–15%	15–35%	20–50%
Fibrates	15%	Variable	6–15%	20–50%

TC, total cholesterol; *LDL-C*, LDL cholesterol; *HDL-C*, HDL cholesterol; *TG*, triglycerides.

and quinidine), benzodiazepines, calcium channel blockers, amiodarone, carbamazepine, clozapine, cyclosporine, and nonsedating antihistamines. Statins are not safe in pregnant or nursing women, and should not be used in patients with active or chronic hepatic disease or cholestasis because of potential hepatotoxicity.

The *bile acid resins* (cholestyramine (Questran), colestipol (Colestid), and colesevalam (Welchol) do not enter the blood stream, but bind bile acids in the intestine, preventing their reabsorption (❏ Fig 32.1). More cholesterol is converted into bile acids in the liver, decreasing the cholesterol pool, increasing the proteolytic release of SREBPs, leading to upregulation of LDL receptors and lower LDL levels (❏ Table 32.9). There is a compensatory increase in hepatic cholesterol synthesis that limits the efficacy of the sequestrants. The side effects of the resins include constipation, heart burn, bloating, decreased serum folate levels, and interference of the absorption of other drugs. The second generation sequestrant, colesevalam, does not appear to interfere with the absorption of other drugs, and in general is associated with a lower prevalence of annoying side effects such as constipation, because it is given in a lower dose than the first generation sequestrants.

The *cholesterol absorption inhibitor,* ezetimibe, a 2-azetidinone, is currently the only member of this drug class. Ezetimibe inhibits the intestinal absorption of cholesterol derived from the diet and from the bile by about 50% (❏ Fig. 32.1). Ezetimibe thus reduces the overall delivery of cholesterol to the liver, decreasing hepatic cholesterol, increasing the release of SREBPs, promoting the upregulation of LDL receptor, and decreasing LDL cholesterol levels. The use of ezetimibe is associated with a compensatory increase in cholesterol biosynthesis, limiting its efficacy. The mechanism of action of ezetimibe presumably occurs through the selective inhibition of a newly discovered transporter that moves cholesterol from bile acid micelles into the cells of the jejunum [54]. The transporter is a Niemann-Pick C1-like 1 (NPC1L1) protein localized at the brush border of enterocytes [54]. Ezetimibe significantly reduces cholesterol absorption in animals homozygous for wild type NPC1L1, but has no effect in NPC1L1 knock-out mice [54]. Ezetimibe is absorbed from the intestine and in the liver is conjugated to a more active glucuronide form, which undergoes enterohepatic circulation. This process increases its elimination half-life to about 22 h. Ezetimibe is usually well-tolerated, and there are generally few drug interactions with this drug. Ezetimibe can be combined with any of the statins producing, on average, an additional 25% reduction in LDL cholesterol. Ezetimibe is also available combined with simvastatin in a single formulation (Vytorin). Ezetimibe should not be used for combination therapy with a statin in patients with active liver disease or unexplained persistent elevations in serum transaminases, or those with chronic or severe liver disease. Co-administration of ezetimibe with cholestyramine decreased the levels of ezetimibe, and co-administration with fibrates increased plasma levels of ezetimibe. Ezetimibe should not be used in patients on cyclosporine until more data are available.

Niacin (nicotinic acid) is vitamin B_3. When given in high doses, niacin becomes a lipid-altering agent. Niacin inhibits the release of free fatty acids from adipose tissue, leading to decreased delivery of FFA to liver and reduced triglyceride synthesis. As a result, the proteolysis of apo B-100 is increased, leading to decreased VLDL secretion and subsequently, to decreased IDL and LDL formation (❏ Fig. 32.1). This is associated with a decreased formation of small, dense LDL particles. Niacin also inhibits the uptake of HDL through its catabolic pathway, prolonging the half-life of HDL, and presumably increasing reverse cholesterol transport. Niacin is also the only lipid-altering drug that reduces Lp(a) lipoprotein. Niacin is commonly prescribed in those patients with the dyslipidemic triad (low HDL, elevated triglycerides and increased small, dense LDL) (❏ Table 32.9). Niacin is useful in treating FCHL and in those with isolated low HDL cholesterol. Niacin should not be used in patients with active peptic ulcer disease or liver disease. Niacin can precipitate the onset of type II diabetes mellitus or gout. In patients with borderline or elevated fasting blood sugar or uric acid levels, niacin should be used with care. Niacin is no longer contraindicated in patients with type II diabetes who are under good control. The modest increase in blood sugar with niacin can usually be compensated for by adjusting the diabetic medications. There are a number of niacin preparations available over the counter or by prescription. Immediate crystalline niacin can be purchased in most pharmacies and health food stores. The slow release niacin products and the extended release niacin (Niaspan) are available by prescription. The slow release niacin is not associated with flushing but has been reported to have a greater propensity to increase liver function tests. Niaspan also decreases flushing but the prevalence of abnormal liver function tests with Niaspan is comparable to regular niacin. Niaspan has also been combined with lovastatin (Advicor, Kos Pharmaceuticals), and can be used in those with an elevated LDL cholesterol, a reduced HDL cholesterol, and hypertriglyceridemia.

32.7.5 Triglyceride Lowering Drugs

Those drugs that can effectively lower triglycerides include nicotinic acid, fibrates, and statins (particularly when used at their highest doses). A 30 to 50% reduction in triglycerides is often achieved (❏ Table 32.9).

One theoretical advantage of niacin and fibrate therapy for hypertriglyceridemia is the improvement or shift of dense subfractions (pattern B) to lighter subfractions (pattern A) (54). The measurement of dense LDL or HDL subfractions can be made by density gradient electrophoresis or nuclear magnetic resonance spectroscopy. These different methodologies have shown the existence of a number of lipoprotein subfractions. Prospective epidemiologic studies, clinical trials, and in vitro studies have all suggested that dense LDL is more atherogenic and that a shift to lighter subfractions may reduce risk for CAD. Fibrates can also effectively lower triglyceride levels and raise HDL cholesterol [54] (◘ Table 32.9).

32.7.6   Combination Pharmacotherapy

Statin therapy is most often started initially in those with CAD or CAD risk equivalence. Depending on the LDL cholesterol response, it may be necessary to add a second drug to achieve the LDL cholesterol goal, particularly the optional goal of 70 mg/dl (◘ Table 32.8). A second drug may also be necessary because of a low HDL cholesterol, a high triglyceride, or both. Statins have been used in combination with bile acid sequestrants, fibrates, niacin and a cholesterol absorption inhibitor. Sequestrants have been paired fibrates, niacin, and ezetimibe. Niacin and fibrates have also been used together. There are ongoing studies of ezetimibe combined with either niacin or fibrates. Different combination therapies may be required either because a patient is unable to tolerate the side effects of a particular class of drug, or because a certain combination has not achieved optimal control of LDL cholesterol, HDL cholesterol, non-HDL cholesterol, or triglyceride levels. In placebo-controlled clinical trials, combination therapy has been shown to be very effective at reducing CAD. As well, combination therapy provides a complementary effect on reduction of hsCRP levels.

Abbreviations

ABC	ATP binding casette
ACAT	acyl coenzyme A:cholesterol acyltransferase
Apo	apolipoprotein
ARH	autosomal recessive hypercholesterolemia
ASP	acylation stimulatory protein
BP	basic proteins
CAD	coronary artery disease
CESD	cholesteryl ester storage disease
CETP	cholesteryl ester transfer protein
FDB	familial defective apoB-100
FCHL	familial combined hyperlipidemia
FFA	free fatty acids
FH	familial hypercholesterolemia
FH3	heterozygous FH3
FHT	familial hypertriglyceridemia
HDL	high density lipoproteins
HL	hepatic lipase
HMG-CoA	hydroxymethylglutaryl coenzyme A
HSCRP	highly sensitive C-reactive protein
IDL	intermediate density lipoproteins
LAL	lysosomal acid lipase
LCAT	lecithin:cholesterol acyltransferase
LDL	low density lipoproteins
LPL	lipoprotein lipase
LRP	LDL receptor-related protein
MCT	medium-chain triglycerides
MTP	microsomal triglyceride transfer protein
PHLA	post-heparin lipolytic activity
SREBP	sterol regulating element binding protein
TG	triglycerides
VLDL	very low density lipoproteins

References

1. Goldstein JL, Brown MS (2001) Molecular medicine. The cholesterol quartet. Science 292:1310-1312
2. Horton JD, Goldstein JL, Brown MS (2002) SREBPs: Activators of the complete program of cholesterol and fatty acid synthesis in the liver. J Clin Invest 109:1125-1131
3. Rader D (2002) High-density lipoproteins and atherosclerosis. Am J Cardiol 90(Suppl):62i-70i
4. Heinecke JW, Lusis AJ (1998) Paraoxonase-gene polymorphisms associated with coronary heart disease: Support for the oxidative damage hypothesis? Am J Hum Genet 62:36-44
5. Yesilaltay A, Kocher O, Rigotta A, Krieger M (2005) Regulation of SR-BI-mediated high-density lipoprotein metabolism by the tissue-specific adaptor protein PDZK1. Curr Opin Lipidol 16:147-152
6. Rubins HB, Robins SJ, Collins D et al (1999) Gemfibrozil for the secondary prevention of coronary heart disease in men with low levels of high-density lipoprotein cholesterol. Veterans Affairs High-Density Lipoprotein Cholesterol Intervention Trial Study Group. N Engl J Med 341:410-418
7. Bezafibrate Infarction Prevention Study Group (2000) Secondary prevention by raising HDL cholesterol and reducing triglycerides in patients with coronary artery disease: the Bezafibrate Infarction Prevention (BIP) study. Circulation 102:21-27
8. Rouis M, Dugi KA, Previato L et al (1997) Therapeutic response to medium-chain triglycerides and omega-3 fatty acids in a patient with the familial chylomicronemia syndrome. Arterioscler Thromb Vasc Biol 17:1400-1406
9. Kwiterovich Jr PO (2002) Clinical relevance of the biochemical, metabolic and genetic factors that influence low density lipoprotein heterogeneity. Am J Cardiol 90:30i-48i(Suppl 8A)
10. Millar JS, Packard CJ (1998) Heterogeneity of apolipoprotein B-100-containing lipoproteins: What we have learnt from kinetic studies. Curr Opin Lipidol 9:197-202
11. Maslowska M, Wang HW, Cianflone K (2005) Novel roles for acylation stimulatory protein/C3a desArg: a review of recent in vitro and in vivo evidence. Vitam Horm 70:309-332
12. Kalant D, Maclaren R, Cui W et al (2005) C5L2 is a functional receptor for acylation stimulatory protein. J Biol Chem 280:23936-23944
13. Motevalli M, Goldschmidt-Clermont PJ, Virgil D, Kwiterovich Jr PO (1997) Abnormal protein tyrosine phosphorylation in fibroblasts from hyperapoB subjects. J Biol Chem 272:24703-24709
14. Aouizerat BE, Allayee H, Bodnar J et al (1999) Novel genes for familial combined hyperlipidemia. Curr Opin Lipidol 10:113-122
15. Lusis AJ, Fogelman AM, Fonarow GC (2004) Genetic basis of atherosclerosis: part I: new genes and pathways. Circulation 110:1868-1873
16. Pajukanta P, Lilja HE, Sinsheimer JS et al (2004) Familial combined hyperlipidemia is associated with upstream transcription factor 1 (USF1). Nat Genet 36:371-376

17. Allayee H, Krass KL, Pajukanta P et al. (2002) Locus for elevated apolipoprotein B levels on chromosome 1p31 in families with familial combined hyperlipidemia. Circ Res 90:926-931

18. Wolman M (1995) Wolman disease and its treatment. Clin Pediatr 34:207-212

19. Beaudet AL, Ferry GD, Nichols BL, Rosenberg HS (1977) Cholesterol ester storage disease: clinical, biochemical, and pathological studies. J Pediatr 90:910-914

20. Ginsberg HN, Le NA, Short MP et al (1987) Suppression of apolipoprotein B production during treatment of cholesteryl ester storage disease with lovastatin. Implications for regulation of apolipoprotein B synthesis. J Clin Invest 80:1692-1697

21. Rader DJ, Cohen J, Hobbs HH (2003) Monogenic hypercholesterolemia: new insights in pathogenesis and treatment. J Clin Invest 111:1795-1803

22. Grossman M, Rader DJ, Muller DW et al (1995) A pilot study of ex vivo gene therapy for homozygous familial hypercholesterolemia. Nat Med 1:1148-1154

23. Salen G, von Bergmann K, Lutjohann D et al and the Multicenter Sitosterolemia Study Group (2004) Ezetimibe effectively reduces plasma plant sterols in patients with sitosterolemia. Circulation 109:766-771

24. Mahley RW, Huang Y, Rall SC Jr (1999) Pathogenesis of type III hyperlipoproteinemia (dysbetalipoproteinemia). J Lipid Res 40:1933-1949

25. Hegele RA, Little JA, Vezina C (1993) Hepatic lipase deficiency: Clinical biochemical and molecular genetic characteristics. Arterioscler Thromb 13:720-728

26. Brand K, Dugi KA, Brunzell JD (1996) A novel A→G mutation in intron I of the hepatic lipase gene leads to alternative splicing resulting in enzyme deficiency. J Lipid Res 37:1213-1223

27. Rader DJ, Brewer HB (1993) Abetalipoproteinemia. New insights into lipoprotein assembly and vitamin E metabolism from a rare genetic disease. JAMA 270:865-869

28. Wetterau JR, Aggerbeck LP, Bouma ME et al (1992) Absence of microsomal triglyceride transfer protein in individuals with abetalipoproteinemia. Science 258:999-1001

29. Gabelli C, Bilato C, Martini S et al (1996) Homozygous familial hypobetalipoproteinemia. Increased LDL catabolism in hypobetalipoproteinemia due to a truncated apolipoprotein B species, apoB-87Padova. Arterioscler Thromb Biol 16:1189-1196

30. Breslow JL (2000) Genetics of lipoprotein abnormalities associated with coronary artery disease susceptibility. Annu Rev Genet 34:233-254

31. Cohen JC, Kiss RS, Pertsemlidis A et al (2004) Multiple rare alleles contribute to low plasma levels of HDL cholesterol. Science 305:869-872

32. Bruce C, Chouinard RA Jr, Tall AR (1998) Plasma lipid transfer proteins, high-density lipoproteins, and reverse cholesterol transport. Annu Rev Nutr 18:297-330

33. von Eckardstein A, Assmann G (1998) High density lipoproteins and reverse cholesterol transport: Lessons from mutations. Atherosclerosis 137:S7-11

34. Nissen SE, Tsunoda T, Tuzcu EM et al (2003) Effect of recombinant apoA-I Milano on coronary atherosclerosis in patients with acute coronary syndromes. JAMA 290:2292-2300

35. Brewer HB, Remaley AT, Neufeld EB et al (2004) Regulation of plasma high-density lipoprotein levels by the ABCA1 transporter and the emerging role of high-density lipoprotein in the treatment of cardiovascular disease. Arterioscler Thromb Vasc Biol 24:1755-1760

36. Remaley AT, Schumacher UK, Stonik JA et al (1997) Decreased reverse cholesterol transport from Tangier disease fibroblasts. Acceptor specificity and effect of brefeldin on lipid efflux. Arterioscler Thromb Biol 17:1813-1821

37. Calabresi L, Pisciotta L, Costantin A (2005) The molecular basis of lecithin:cholesterol acyltransferase deficiency syndromes. A comprehensive study of molecular and biochemical findings in 13 unrelated Italian families. Arterioscler Thromb Vasc Biol 25:1972-1978

38. Zhong S, Sharp DS, Grove JS et al (1996) Increased coronary heart disease in Japanese-American men with mutations in the cholesteryl ester transfer protein gene despite increased HDL levels. J Clin Invest 97:2917-2923

39. Ikewaki K, Nishiwaki M, Sakamoto T et al (1995) Increased catabolic rate of low density lipoproteins in humans with cholesteryl ester transfer protein deficiency. J Clin Invest 96:1573-1581

40. de Grooth GJ, Kuivenhoven JA, Stalenhoef AF et al (2002) Efficacy and safety of a novel cholesteryl ester transfer protein inhibitor, JTT-705, in humans: a randomized phase II dose-response study. Circulation 105:2159-2165

41. Clark RW, Sutfin TA, Ruggeri RB et al (2004) Raising high-density lipoprotein in humans through inhibition of cholesteryl ester transfer protein: an initial multidose study of torcetrapib. Arterioscler Thromb Vasc Biol 24:490-497

42. Brousseau ME, Schaefer EJ, Wolfe ML et al (2004) Effects of an inhibitor of cholesteryl ester transfer protein on HDL cholesterol. N Engl J Med 350:1505-1515

43. Marcovina SM, Koschinsky ML et al (2003) Report of the National Heart, Lung and Blood Institute Workshop on Lipoprotein (a) and Cardiovascular Disease: Recent Advances and Future Directions. Clin Chem 49:1785-1786

44. NCEP: Executive Summary of The Third Report of The National Cholesterol Education Program (NCEP) Expert Panel on Detection, Evaluation, And Treatment of High Blood Cholesterol In Adults (Adult Treatment Panel III) (2001) JAMA 285:2486-2497

45. Grundy SM, Cleeman JI, Merz CN et al (2004) Implications of recent clinical trials for the National Cholesterol Education Program. Adult Treatment Panel III guidelines. Circulation 110:227-239

46. Brown BG, Zhao XO, Chait A et al (2001) Simvastatin and niacin, antioxidant vitamins, or the combination for the prevention of coronary disease. N Engl J Med 345:1583-1592

47. Cannon CP, Braunwald E, McCabe CH et al (2004) Intensive versus moderate lipid lowering with statins after acute coronary syndromes. N Engl J Med 350:1495-1504

48. Nissen SE, Tuzcu EM, Schoenhagen P et al (2004) Effect of intensive compared with moderate lipid lowering therapy on progression of coronary atherosclerosis: a randomized controlled trial. JAMA 291:1071-1080

49. Ridker PM, Cannon CP, Morrow D et al (2005) C-reactive protein levels and outcomes after therapy. N Engl J Med 352:20-28

50. Nissen SE, Tuzcu EM, Schoenhagen P et al (2005) Statin therapy, LDL cholesterol, C-reactive protein and coronary artery disease. N Engl J Med 352:29-38

51. Garcia MJ, Reinoso RF, Sanchez Navarro A, Prous JR (2003) Clinical pharmacokinetics of statins. Methods Find Exp Clin Pharmacol 25:457-481

52. Gaist D, Rodriguez LA, Huerta C et al (2001) Lipid-lowering drugs and risk of myopathy: a population-based follow-up study. Epidemiology 12:565-569

53. Altmann SW, Davis HR Jr, Zhu LJ et al (2004) Niemann-Pick C1 Like 1 protein is critical for intestinal cholesterol absorption. Science 303:1201-1204

54. Fruchart J-C, Brewer HB Jr, Leitersdorf E (1998) Consensus for the use of fibrates in the treatment of dyslipoproteinemia and coronary heart disease. Am J Cardiol 101:10S-16S

33 Disorders of Cholesterol Synthesis

Hans R. Waterham, Peter T. Clayton

Cholesterol Synthesis

Cholesterol is a major end product of the isoprenoid biosynthetic pathway, which produces numerous molecules (i.e. isoprenoids) with pivotal functions in a variety of cellular processes including cell growth and differentiation, protein glycosylation, signal transduction pathways etc. [1]. Cholesterol synthesis (◘ Fig. 33.1) starts from acetyl-coenzyme A. A series of ten enzyme reactions (not shown in detail in ◘ Fig. 33.1) leads to

the formation of squalene, which after cyclization is converted into lanosterol. Subsequent conversion of lanosterol into cholesterol is proposed to occur via two major routes involving the same enzymes which, depending on the timing of reduction of the Δ^{24} double bond, postulate either 7-dehydrocholesterol or desmosterol as the ultimate precursor of cholesterol.

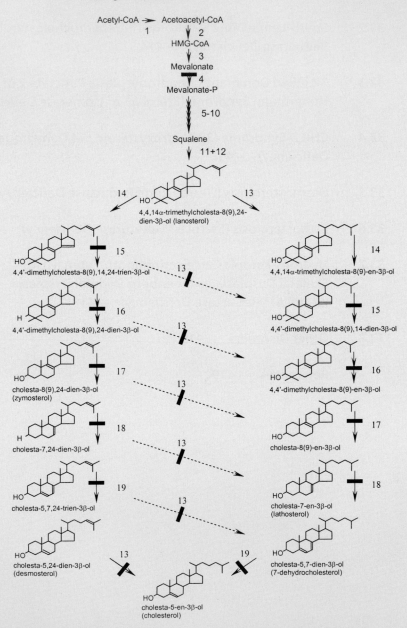

◘ **Fig. 33.1.** Pathway of isoprenoid and cholesterol synthesis. *CoA*, coenzyme A; *HMG*, 3-hydroxy-3-methylglutaryl; *P*, phosphate; *PP*, pyrophosphate. **1**, acetyl-CoA acetyltransferase; **2**, HMG-CoA synthase; **3**, HMG-CoA reductase; **4**, mevalonate kinase; **5**, mevalonate-P kinase; **6**, mevalonate-PP decarboxylase; **7**, isopentenyl-PP isomerase; **8**, geranyl-PP synthase; **9**, farnesyl-PP synthase; **10**, squalene synthase; **11**, squalene epoxidase; **12**, 2,3-oxidosqualene sterol cyclase; **13**, sterol Δ^{24}-reductase; **14**, sterol C-14 demethylase; **15**, sterol Δ^{14}-reductase; **16**, sterol C-4 demethylase complex; **17**, sterol Δ^8-Δ^7 isomerase; **18**, sterol Δ^5-desaturase; **19**, sterol Δ^7-reductase. Enzyme deficiencies are indicated by *solid bars* across the arrows

Eight distinct inherited disorders have been linked to specific enzyme defects in the isoprenoid/cholesterol biosynthetic pathway after the finding of abnormally increased levels of intermediate metabolites in tissues and/or body fluids of patients followed by the demonstration of disease-causing mutations in genes encoding the implicated enzymes. Two of these disorders are due to a defect of the enzyme mevalonate kinase and affect the synthesis of all isoprenoids. Patients with these disorders characteristically present with recurrent episodes of high fever associated with abdominal pain, vomiting and diarrhoea, (cervical) lymphadenopathy, hepatosplenomegaly, arthralgia and skin rash, and may present with additional congenital anomalies.

The remaining six enzyme defects specifically affect the synthesis of cholesterol and involve four autosomal recessive and two X-linked dominant inherited syndromes. Patients afflicted with one of these defects present with multiple congenital and morphogenic anomalies, including internal organ, skeletal and/or skin abnormalities, and/or a marked delay in psychomotor development reflecting cholesterol's pivotal role in human embryogenesis and development.

33.1 Mevalonic Aciduria and Hyper-Immunoglobulinaemia-D and Periodic Fever Syndrome (Mevalonate Kinase Deficiency)

33.1.1 Clinical Presentation

Two previously defined clinical entities are now known to be caused by a deficiency of the enzyme mevalonate kinase, i.e. classic mevalonic aciduria (MA) and the more benign hyper-IgD and periodic fever syndrome, alternatively known as Dutch-type periodic fever (HIDS). Both disorders typically present with episodes of high fever that last 3–5 days and recur in average every 4–6 weeks, and are associated with abdominal pain, vomiting and diarrhoea, (cervical) lymphadenopathy, hepatosplenomegaly, arthralgia and skin rash [2–4]. These febrile crises usually start in the first year of life and may be provoked by vaccinations, physical and emotional stress and minor trauma. In addition to these febrile crises, patients with the more severe MA may present with congenital defects such as mental retardation, ataxia, cerebellar atrophy, hypotonia, severe failure to thrive and dysmorphic features, which in the most severely affected patients may lead to death in early infancy. Current insights dictate that HIDS and MA are the mild and severe end of a clinical and biochemical continuum and that both defects should be regarded as one clinical entity, i.e. mevalonate kinase deficiency [5, 6].

33.1.2 Metabolic Derangement

Both MA and HIDS are caused by a deficiency of the enzyme mevalonate kinase (MK; enzyme 4 in ◘ Fig. 33.1) but to variable degrees: in white blood cells or cultured primary skin fibroblasts of MA patients the activity of MK is hardly measurable, while in cells of HIDS patients a residual MK activity of 2–8% of the activities in cells of healthy controls is found [5–8]. MK catalyzes the phosphorylation of mevalonate to produce 5-phosphomevalonate and is the next enzyme in the isoprenoid synthesis pathway after HMG-CoA reductase, the highly-regulated and major rate-limiting enzyme of the pathway [1]. As a consequence of the MK deficiency, high and moderately elevated levels of mevalonic acid can be detected in plasma and urine of patients with MA and HIDS, respectively. Since MK functions relatively early in the biosynthetic pathway, the synthesis of all isoprenoids will be affected to a certain extent. Yet, most of the characteristic clinical manifestations are thought to be due to a (temporary) shortage of nonsterol isoprenoid end products [6]. It may well be possible, however, that in severe MA cases a relative shortage of sterol isoprenoids during embryonic development led to some of the clinical problems.

33.1.3 Genetics

MA and HIDS are both autosomal recessively inherited and due to different mutations in the MK-encoding *MVK* gene located on chromosome 12q24 [5, 7–9]. Nearly all patients with the HIDS phenotype are compound heterozygotes for the V377I *MVK* allele, which is found exclusively in HIDS patients, and a second allele, which is found also in MA patients [9]. The V377I allele encodes an active MK enzyme, the correct assembly/maturation of which is temperature-dependent and thus responsible for the observed residual MK enzyme activity associated with the HIDS phenotype [9]. Other relatively common disease-causing mutations in the *MVK* gene are H20P, I268T and A334T. In total, more than 35 different disease-causing mutations have been identified that are widely distributed over the *MVK* gene and most of which are listed in the infevers database at http://fmf.igh.cnrs.fr/infevers. These include primarily missense, and nonsense mutations, while only a few insertions, deletions and splice site mutations have been identified.

33.1.4 Diagnostic Tests

Several diagnostic tools for laboratory analysis of the two MK deficiency disorders are available. A first test involves the analysis of mevalonic acid levels in body fluids by organic acid analysis or, preferably, by stable isotope dilution gas chromatography-mass spectrometry (GC-MS) [10]. Due to the variable degrees of MK deficiency, this test works best for MA patients, who have high levels of mevalonic acid (1–56 mol/mol creatinine in urine), but may not always be diagnostic for HIDS patients due to their rather low levels even during fever (urinary concentration 0.005-0.040 mol/mol creatinine while normally not detectable). In addition to the clinical characteristics, a diagnostic parameter of most patients with HIDS has been the continuously elevated plasma IgD (>100 IU/ml) and/or IgA levels [3]. Similar elevations have been reported also in patients with classic MA. The best diagnostic tests remain the direct measurement of MK activities in white blood cells or primary skin fibroblasts from patients [11] and molecular analysis of the *MVK* gene through sequence analysis of the coding exons plus flanking intronic sequences [9]. The latter two tests are also the first choice for prenatal diagnosis and can be performed in chorionic villi, chorionic villous cells and amniotic fluid cells. Carrier detection is best performed by molecular testing.

33.1.5   Treatment and Prognosis

There is currently no efficacious treatment for MA or HIDS available. In individual HIDS cases, clinical improvement as a result of treatment with corticosteroid, colchicine, or cyclosporin has been reported, but in the majority of patients these treatments do not have beneficial effects [12]. In a small group of HIDS patients simvastatin treatment had a positive effect on the number of days of illness [13], but treatment with similar statins in MA patients led to worsening of the clinical symptoms. Treatment of two HIDS patients with etanercept, a soluble p75 TNF alpha receptor-Fc fusion protein used for treatment of patients with tumour necrosis factor receptor associated periodic syndrome (TRAPS), led to a reduction of the frequency and severity of symptoms, but this form of treatment has not been tested in larger groups of patients [14].

The long-term outcome in HIDS is relatively benign as the clinical symptoms tend to become less frequent and less severe with age [3].

33.2   Smith-Lemli-Opitz Syndrome (7-Dehydrocholesterol Reductase Deficiency)

33.2.1   Clinical Presentation

Patients with Smith-Lemli-Opitz Syndrome (SLOS) clinically present with a large and variable spectrum of morphogenic and congenital anomalies, and constitute a clinical and biochemical continuum ranging from mild (hardly recognizable) to very severe (lethal in utero) [15–18].

Affected patients typically have a characteristic craniofacial appearance, including microcephaly, a short nose with broad nasal bridge and anteverted nares, a long filtrum, micro/retrognathia and often blepharoptosis, low-set posteriorly rotated ears, cleft or high arched palate, pale hair and broad or irregular alveolar ridges. Common limb abnormalities include cutaneous syndactyly of the 2nd and 3rd toes (>97% of cases), postaxial polydactyly and short proximally placed thumbs. Genital abnormalities may include hypospadias, cryptorchidism and ambiguous or even female external genitalia in affected boys. Also common are congenital heart defects, and renal, adrenal, lung and gastrointestinal anomalies. Additional major features are profound prenatal and postnatal growth retardation, mental retardation, feeding difficulties and behavioural problems, sleeping disorders and sunlight sensitivity. Although none of these clinical symptoms are pathognomonic for SLOS, the presence of a combination of the more common clinical features associated with SLOS should certainly prompt physicians to consider SLOS in the differential diagnosis. For more detailed reports on this topic the reader is referred to other reviews summarizing and discussing clinical aspects of SLOS [17, 18].

33.2.2   Metabolic Derangement

SLOS is caused by a deficiency of the enzyme sterol Δ^7-reductase (7-dehydrocholesterol reductase, DHCR7, enzyme 19 in ◘ Fig. 33.1), which catalyzes the reduction of the C7–C8 double bond of 7-dehydrocholesterol (cholesta-5,7-dien-3β-ol) to produce cholesterol (cholest-5-en-3β-ol), generally regarded as the predominant final step in cholesterol biosynthesis. As a consequence of the DHCR7 deficiency, low cholesterol and elevated levels of 7-dehydrocholesterol can be detected in plasma, cells and tissues of the vast majority of SLOS patients [19, 20]. In addition, elevated 8-dehydrocholesterol (cholesta-5,8(9)-dien-3β-ol) levels are detected, probably synthesized from the accumulating 7-dehydrocholesterol by the enzyme sterol Δ^8-Δ^7 isomerase functioning in the reverse direction. Several studies have shown that overall clinical severity in SLOS correlates best either with the reduction in absolute cholesterol levels or with the sum of 7-dehydrocholesterol plus 8-dehydrocholesterol expressed as the fraction of total sterol [21]. There is also evidence that the efficiency of transfer of cholesterol from mother to foetus may play a role in determining severity as inferred from the significant correlation between a patients' clinical severity score and their mother's apo E genotype [22].

33.2.3   Genetics

SLOS is the most frequently occurring defect of cholesterol biosynthesis known to date and it is inherited as an autosomal recessive trait. Dependent on the geographic region, incidences have been reported that range from 1:15,000 to 1:60,000 in Caucasians [18]. The higher incidences observed in particular in some East-European countries appear to reflect founder effects.

The *DHCR7* gene encoding 3β-hydroxysterol Δ^7-reductase is located on chromosome 11q13. Currently, over 80 different disease-causing mutations have been reported in the *DHCR7* gene of more than 200 SLOS patients analyzed at the genetic level [20, 21, 23–25]. Although mutations are distributed widely all over the gene, a few common mutations have been recognized including T93M, R404C, W151X, V326I and IVS8-1G>C. By far the most prevalent in Caucasians is the severe IVS8-1G>C splice site mutation (allele frequency of ~30%), which leads to aberrant splicing of the *DHCR7* mRNA at a cryptic splice acceptor site located 5′ of the mutated splice site resulting in the partial retention of a 134-bp intron sequence and produces no functional protein.

33.2.4 Diagnostic Tests

Laboratory diagnosis of SLOS [20] includes sterol analysis of plasma or tissues of patients by GC-MS, in which the detection of elevated levels of 7-dehydrocholesterol (and 8-dehydrocholesterol) are diagnostic. DHCR7 enzyme activities (or the lack thereof) can be measured directly in primary skin fibroblasts, lymphoblasts or tissue samples (e.g. chorionic villi) of patients using either [^{3}H]-labelled 7-dehydrocholesterol or ergosterol (converted to brassicasterol) as substrate. Alternatively, primary skin fibroblasts or lymphoblasts of patients can be cultured in lipoprotein-depleted medium to induce cholesterol biosynthesis whereupon the defect can be detected by sterol analysis using GC-MS. Finally, molecular analysis through sequence analysis of the coding exons and flanking intronic sequences of the *DHCR7* gene is performed. The latter two tests are first choice for prenatal diagnosis performed in chorionic villous cells and amniotic fluid cells, with, as a good alternative, direct molecular testing in chorionic villi. Carrier detection is most reliably performed by molecular testing.

33.2.5 Treatment and Prognosis

It is generally considered that the availability of cholesterol during development of the foetus is one of the major determinants of the phenotypic expression in SLOS [18, 22]. Since most anomalies occurring in SLOS are of early-embryonic origin, it will not be feasible to develop a postnatal therapy to entirely cure the patients. The therapy currently mostly employed aims to replenish the lowered cholesterol levels in the patients through dietary supplementation of cholesterol with or without bile acids [26]. While this treatment leads

to a substantial elevation of plasma cholesterol concentrations in patients, the plasma concentrations of 7-dehydrocholesterol and 8-dehydrocholesterol are often only marginally reduced. In general, the clinical effects of this treatment have been rather disappointing, although several reports have indicated that dietary cholesterol supplementation may improve behaviour, growth and general well-being in children with SLOS. A recent standardized study with 14 SLOS patients indicated that cholesterol supplementation had hardly any effect on developmental progress [27]. Moreover, this treatment probably does not significantly change the sterol levels in brain, which are dependent on *de novo* cholesterol synthesis due to the limited ability of cholesterol to cross the blood-brain barrier. More recently, promising results have been reported for an alternative therapeutic strategy aimed primarily at lowering of the elevated 7-dehydrocholesterol and 8-dehydrocholesterol levels through the use of simvastatin, an oral HMG-CoA reductase inhibitor [28]. Two rather mildly affected SLOS patients treated with simvastatin showed a marked decrease of 7-dehydrocholesterol and 8-dehydrocholesterol levels and a concomitant increase of cholesterol in plasma as well as cerebrospinal fluid in conjunction with promising short-term clinical improvement. The efficacy and long-term outcome of this treatment, which might be of benefit to relatively mildly affected SLOS patients, is currently being tested in a larger trial.

33.3 X-Linked Dominant Chondro-dysplasia Punctata 2 or Conradi-Hünermann Syndrome (Sterol Δ^8–Δ^7 Isomerase Deficiency)

33.3.1 Clinical Presentation

Patients with X-linked dominant chondrodysplasia punctata 2 (CDPX2), also known as Conradi-Hünermann or Happle syndrome, display skin defects ranging from ichthyosiform erythroderma in the neonate, through linear or whorled atrophic and pigmentary lesions in childhood to striated hyperkeratosis, coarse lusterless hair and alopecia in adults. These skin lesions are associated with cataracts, and skeletal abnormalities including short stature, asymmetric rhizomelic shortening of the limbs, calcific stippling of the epiphyseal regions, and craniofacial defects [29–31]. The pattern of the skin defects and probably also the variability in severity and asymmetry of the bone and eye abnormalities observed in CDPX2 patients are consistent with functional X-chromosomal mosaicism. The expression of these skin and skeletal abnormalities can be bilateral and is often asymmetric. As the defect is predominantly observed in females, CDPX2 is considered lethal in hemizygous males. However, a few affected males with aberrant karyotypes and even true hemizygotes have been identified.

33.3.2 Metabolic Derangement

CDPX2 is caused by a deficiency of the enzyme sterol Δ^8-Δ^7 isomerase (enzyme 17 in ◘ Fig. 33.1), which catalyses the conversion of cholesta-8(9)-en-3β-ol to lathosterol by shifting the double bond from the C8–C9 to the C7–C8 position [32–34]. As a consequence of the deficiency, elevated levels of cholesta-8(9)-en-3β-ol and 8-dehydrocholesterol can be detected in plasma and cells of patients, although the plasma cholesterol levels are often normal or low normal.

33.3.3 Genetics

CDPX2 is inherited as an X-linked dominant trait and due to heterozygous mutations in the *EBP* gene encoding the enzyme sterol Δ^8-Δ^7 isomerase and located on chromosome Xp11.22-23 [32, 33]. The product of the *EBP* gene, i.e. emopamil binding protein, was initially identified as a binding protein for the Ca^{2+} antagonist emopamil and high affinity acceptor for several other anti-ischemic drugs but later shown to encode for sterol Δ^8-Δ^7 isomerase. Currently, over 30 different disease-causing mutations have been identified in the *EBP* gene of primarily female patients with CDPX2. Most analyzed patients are heterozygous for a mutation that has arisen de novo (somatic mutations) in line with the sporadic nature of the disorder, but in a few cases indications for gonadal mosaicism have been obtained. Inheritance of a mutation from an affected mother usually results in a more severe expression of the disease in offspring.

33.3.4 Diagnostic Tests

Laboratory diagnosis of CDPX2 can be achieved by analysis of plasma sterols of patients (by GC-MS) to detect cholesta-8(9)-en-3β-ol [34]. Also, primary skin fibroblasts or lymphoblasts of patients can be cultured in lipoprotein-depleted medium to induce cholesterol biosynthesis whereupon the enzyme defect can be detected by sterol analysis using GC-MS. Finally, mutation analysis can be performed by sequence analysis of the coding exons and flanking intronic sequences of the *EBP* gene [32, 34]. Recently, a severe form of CDPX2 has been detected by ultrasound scan showing a small fetus, nuchal oedema, what appeared to be multiple fractures of very short long bones, and a narrow thorax. After termination of the pregnancy the diagnosis of CDPX2 was achieved using sterol analysis followed by analysis of the *EBP* gene [35]. Prenatal diagnosis by molecular analysis is possible but so far has not been reported.

33.3.5 Treatment and Prognosis

Long-term outcome of patients with CDPX2 depends on the severity of clinical symptoms. Surviving male patients usually show severe developmental delay. In contrast, the majority of affected girls show completely normal psychomotor development. Many need surgery for cataracts or scoliosis. Correction of scoliosis associated with hemidysplasia of vertebrae requires a special anterior strut graft and a posterior fusion [36].

33.4 CHILD Syndrome (3β-Hydroxysteroid C-4 Dehydrogenase Deficiency)

33.4.1 Clinical Presentation

Patients with CHILD syndrome (Congenital Hemidysplasia with Ichtyosiform erythroderma and Limb Defects) display skin and skeletal abnormalities similar to those observed in patients with CDPX2, but with a striking unilateral distribution affecting the right side of the body more often than the left in contrast to the bilateral distribution in CDPX2 patients [31, 37]. Ichthyosiform skin lesions are usually present at birth and often involve large regions of one side of the body with a sharp line of demarcation in the midline. Alopecia, nail involvement and ipsilateral limb reduction defects with calcific stippling of the epiphysis are common on the affected side. In comparison with CDPX2, patients with CHILD syndrome show no cataracts, more obvious skin lesions and more severe limb defects. Like CDPX2, CHILD is considered lethal in hemizygous males as so far hardly any males with the defect have been diagnosed.

33.4.2 Metabolic Derangement

CHILD syndrome is caused by a deficient activity of a 3β-hydroxysteroid dehydrogenase [38], which has been suggested to be part of a sterol C-4 demethylase complex [composed of a C-4 methyl oxidase, a 4α-carboxysterol-C-4 dehydrogenase (i.e. 3β-hydroxysteroid dehydrogenase) and a C-4 ketoreductase; enzyme complex 16 in ◘ Fig. 33.1] which catalyses the sequential removal of the two methyl groups at the C4 position of early sterol precursors (e.g. lanosterol). Theoretically, the enzyme deficiency should lead to the accumulation of 4-methyl sterol precursors; however, the levels of these precursors in plasma of patients appear normal or only slightly increased. Also cholesterol levels are normal.

33.4.3 Genetics

CHILD syndrome is inherited as an X-linked dominant trait due to heterozygous mutations in the *NSDHL* gene encoding 3β-hydroxysteroid dehydrogenase and located on chromosome Xq28 [38, 39]. In one patient diagnosed with CHILD syndrome a heterozygous mutation was identified in the *EBP* gene [40]. So far some 10 female patients with CHILD syndrome have been analyzed at the molecular level.

33.4.4 Diagnostic Tests

As sterol analysis has been reported not to be diagnostic in this disorder, the only diagnostic test for CHILD syndrome is mutation analysis by sequencing the coding exons and flanking intronic sequences of the *NSDHL* gene [38, 39]. If no mutation is found in the *NSDHL* gene, one should consider also sequencing the *EBP* gene, as mutations in this gene also have been linked to CHILD syndrome [40].

33.4.5 Treatment and Prognosis

Since the clinical presentation of CHILD patients in general is far more severe than in CDPX2, the long-term outcome of patients is usually poor. Surgical corrections of skeletal abnormalities may be required.

33.5 Desmosterolosis (Desmosterol Reductase Deficiency)

33.5.1 Clinical Presentation

Only two patients with desmosterolosis have been reported. The first female infant died shortly after birth and suffered from multiple congenital malformations, including macrocephaly, hypoplastic nasal bridge, thick alveolar ridges, gingival nodules, cleft palate, total anomalous pulmonary venous drainage, ambiguous genitalia, short limbs and generalised osteosclerosis [41]. The second infant is a boy, who exhibited a far less severe phenotype. At three years of age, his clinical presentation included dysmorphic facial features, microcephaly, limb anomalies, and profound developmental delay [42]. Since the clinical presentation of the two patients is rather different, a further delineation of the clinical phenotype of desmosterolosis awaits the identification of additional patients.

33.5.2 Metabolic Derangement

Desmosterolosis is due to a deficiency of the enzyme sterol Δ^{24}-reductase (desmosterol reductase; enzyme 13 in

☐ Fig. 33.1), which catalyzes the reduction of the Δ^{24} double bond of sterol intermediates (including desmosterol) in cholesterol biosynthesis [43]. As a consequence, elevated levels of the cholesterol precursor desmosterol can be detected in plasma, tissue and cultured cells of patients with desmosterolosis [41–43].

33.5.3 Genetics

Desmosterolosis is an autosomal recessive disorder due to mutations in the *DHCR24* gene encoding 3β-hydroxysterol Δ^{24}-reductase and located on chromosome 1p31.1-p33. Sequence analysis of the *DHCR24* gene of the two patients revealed four different disease-causing missense mutations [43].

33.5.4 Diagnostic Tests

Laboratory diagnosis of desmosterolosis includes sterol analysis of plasma, tissues or cultured cells by GC-MS (detection of desmosterol) and mutation analysis by sequencing the coding exons and flanking intronic sequences of the *DHCR24* gene [43].

33.5.5 Treatment and Prognosis

No information on treatment and long-term outcome is available.

33.6 Lathosterolosis (Sterol Δ^5-Desaturase Deficiency)

33.6.1 Clinical Presentation

Only two patients with lathosterolosis have been reported. One female patient presented at birth with severe microcephaly, receding forehead, anteverted nares, micrognathia, prominent upper lip, high arched palate, postaxial hexadactyly of the left foot, and syndactyly between the second to fourth toes and between the fifth toe and the extra digit. From early infancy she suffered from cholestatic liver disease and, during infancy, severe psychomotor delay became apparent [44]. The second patient was a boy who presented at birth with SLOS-like features including growth failure, microcephaly, ptosis, cataracts, short nose, micrognathia, prominent alveolar ridges, ambiguous genitalia, bilateral syndactyly of the 2nd and 3rd toes, and bilateral postaxial hexadactyly of the feet. His clinical course was marked by failure to thrive, severe delay, increasing hepatosplenomegaly, increased gingival hypertrophy and death at the age of 18 weeks. Autopsy disclosed widespread storage of mucopolysaccharides and lipids within the macrophages and, to a lesser extent, parenchymal cells, of all organ systems and extensive demyelination of the cerebral white matter, and dystrophic calcification in the cerebrum, cerebellum, and brainstem [45].

33.6.2 Metabolic Derangement

Lathosterolosis is due to a deficiency of the enzyme sterol Δ^5-desaturase (enzyme 18 in ◘ Fig. 33.1), which introduces the C5-C6 double bond in lathosterol to produce 7-dehydrocholesterol, the ultimate precursor of cholesterol [41, 42]. As a consequence, elevated levels of lathosterol (and lowered cholesterol) can be detected in plasma, (tissue) and cultured cells of patients with lathosterolosis.

33.6.3 Genetics

Lathosterolosis is an autosomal recessive disorder due to mutations in the *SC5D* gene encoding 3β-hydroxysterol Δ^5-desaturase and located on chromosome 11q23.3. Sequence analysis of the *SC5D* gene of the two patients revealed three different disease-causing missense mutations [44, 45].

33.6.4 Diagnostic Tests

Laboratory diagnosis of lathosterolosis includes sterol analysis of plasma, tissues or cultured cells by GC-MS (detection of lathosterol) and mutation analysis by sequencing the coding exons and flanking intronic sequences of the *SC5D* gene [44, 45].

33.6.5 Treatment and Prognosis

No information on treatment and long-term outcome is available but it is possible that in some cases treatment for chronic cholestatic liver disease (e.g. fat-soluble vitamin supplementation) will be required.

33.7 Hydrops-Ectopic Calcification-Moth-Eaten (HEM) Skeletal Dysplasia or Greenberg Skeletal Dysplasia (Sterol Δ^{14}-Reductase Deficiency)

33.7.1 Clinical Presentation

HEM skeletal dysplasia, also known as Greenberg skeletal dysplasia, is a rare syndrome characterized by early *in utero* lethality. Affected fetuses typically present with severe foetal

hydrops, short-limb dwarfism, an unusual ›moth-eaten‹ appearance of the markedly shortened long bones, bizarre ectopic ossification centres and a marked disorganization of chondro-osseous histology and may present with polydactyly and additional nonskeletal malformations [35, 46, 47].

Genetically, HEM skeletal dysplasia appears allelic to Pelger-Huet anomaly [48], a rare benign autosomal dominant disorder of leukocyte development characterized by hypolobulated nuclei and abnormal chromatin structure in granulocytes of heterozygous individuals. Usually, these heterozygous individuals with Pelger-Huet anomaly do not show any evident clinical symptoms, but few (presumed) homozygotes for this defect with variable minor skeletal abnormalities and developmental delay have been reported.

33.7.2 Metabolic Derangement

HEM skeletal dysplasia is due to a deficiency of the enzyme sterol Δ^{14}-reductase (enzyme 15 in ◘ Fig. 33.1), which catalyzes the reduction of the Δ^{14} double bond in early sterol intermediates [49]. As a consequence, elevated levels of cholesta-8,14-dien-3β-ol (and minor levels of cholesta-8,14,24-trien-3β-ol) can be detected in tissues and cells of fetuses with HEM skeletal dysplasia. Heterozygous individuals with Pelger-Huet anomaly do not show aberrant sterol precursors.

33.7.3 Genetics

HEM skeletal dysplasia is an autosomal recessive disorder due to mutations in the *LBR* gene encoding lamin B receptor and located on chromosome 1q42 [46]. Lamin B receptor consists of an N-terminal lamin B/DNA-binding domain of ~200 amino acids followed by a C-terminal sterol reductase-like domain of ~450 amino acids, which exhibits the sterol Δ^{14}-reductase activity.

Disease-causing mutations have been detected in the *LBR* gene of 6 fetuses affected with HEM dysplasia, including missense and nonsense mutations and small deletions. In addition, several heterozygous splice-site, frame-shift and nonsense mutations have been detected in the *LBR* gene of individuals displaying Pelger-Huet anomaly [48]. The demonstration of Pelger-Huet anomaly in one of the parents of a foetus affected with HEM skeletal dysplasia confirms that Pelger-Huet anomaly represents the heterozygous state of 3β-hydroxysterol Δ^{14}-reductase deficiency [49].

33.7.4 Diagnostic Tests

Fetuses affected with HEM skeletal dysplasia are often detected by foetal ultrasound examination. Pelger-Huet anomaly can be diagnosed by microscopy of peripheral blood smears. Laboratory diagnosis of HEM skeletal dysplasia includes sterol analysis of tissues or cells by GC-MS (detection of cholesta-8,14-dien-3β-ol). Molecular analysis includes sequencing of the coding exons and flanking intronic sequences of the *LBR* gene [49].

33.7.5 Treatment and Prognosis

Most cases of Greenberg skeletal dysplasia terminate in early embryonic stages (10–20 weeks of gestation). One adult individual diagnosed with Pelger-Huet anomaly and homozygous for a splice-site mutation in the *LBR* gene has been described with developmental delay, macrocephaly and a ventricular septal defect. No information is available, however, on the effect of the mutation on cholesterol biosynthesis in this individual, if any.

33.8 Other Disorders

Accumulation of lanosterol has been described in some patients diagnosed with Antley-Bixler syndrome suggesting a defect of lanosterol C14-demethylase. However, no mutations in *CYP51*, the gene encoding lanosterol C14-demethylase have yet been described. Instead it appeared that a reduced activity of this enzyme (as well as enzymes of steroidogenesis) may occur as a result of mutations in the *POR* gene encoding cytochrome P450 oxidoreductase [50].

References

1. Goldstein JL, Brown MS (1990) Regulation of the mevalonate pathway. Nature 343:425-430
2. Hoffmann GF, Charpentier C, Mayatepek E et al (1993) Clinical and biochemical phenotype in 11 patients with mevalonic aciduria. Pediatrics 91:915-921
3. Drenth JPH, Haagsma CJ, van der Meer JWM et al (1994) Hyperimmunoglobulinemia D and periodic fever syndrome: the clinical spectrum in a series of 50 patients. Medicine 73:133-144
4. Frenkel J, Houten SM, Waterham HR et al (2000) Mevalonate kinase deficiency and Dutch type periodic fever. Clin Exp Rheumatol 18:525-532
5. Houten SM, Wanders RJ, Waterham HR (2000) Biochemical and genetic aspects of mevalonate kinase and its deficiency. Biochim Biophys Acta 1529:19-32
6. Houten SM, Frenkel J, Waterham HR (2003) Isoprenoid biosynthesis in hereditary periodic fever syndromes and inflammation. Cell Mol Life Sci 60:1118-1134
7. Houten SM, Kuis W, Duran M et al (1999) Mutations in MVK, encoding mevalonate kinase, cause hyperimmunoglobulinaemia D and periodic fever syndrome. Nat Genet 22:175-177
8. Drenth JP, Cuisset L, Grateau G et al (1999) Mutations in the gene encoding mevalonate kinase cause hyper-IgD and periodic fever syndrome. Nat Genet 22:178-181
9. Schafer BL, Bishop RW, Kratunis VJ et al (1992) Molecular cloning of human mevalonate kinase and identification of a missense mutation in the genetic disease mevalonic aciduria. J Biol Chem 267:13229-13238
10. Hoffmann GF, Sweetman L, Bremer HJ et al (1991) Facts and artefacts in mevalonic aciduria: development of a stable isotope dilution GCMS assay for mevalonic acid and its application to physiological fluids, tissue samples, prenatal diagnosis and carrier detection. Clin Chim Acta 198:209-227
11. Hoffmann GF, Brendel SU, Scharfschwerdt SR et al (1992) Mevalonate kinase assay using DEAE-cellulose column chromatography for first-trimester prenatal diagnosis and complementation analysis in mevalonic aciduria. J Inherit Metab Dis 15:738-746
12. Drenth JP, van der Meer JW (2001) Hereditary periodic fever. N Engl J Med 345:1748-1757
13. Simon A, Drewe E, van der Meer JW et al (2004) Simvastatin treatment for inflammatory attacks of the hyperimmunoglobulinemia D and periodic fever syndrome. Clin Pharmacol Ther 75:476-483
14. Takada K, Aksentijevich I, Mahadevan V et al (2003) Favorable preliminary experience with etanercept in two patients with the hyperimmunoglobulinemia D and periodic fever syndrome. Arthritis Rheum 48:2645-2651
15. Smith DW, Lemli L, Opitz JM (1964) A newly recognized syndrome of multiple congenital anomalies. J Pediatr 64:210-217
16. Langius FA, Waterham HR, Romeijn GJ et al (2003) Identification of three patients with a very mild form of Smith-Lemli-Opitz syndrome. Am J Med Genet 122A:24-29
17. Cunniff C, Kratz LE, Moser A et al (1997) Clinical and biochemical spectrum of patients with RSH/Smith-Lemli-Opitz syndrome and abnormal cholesterol metabolism. Am J Med Genet 68:263-269
18. Kelley RI, Hennekam RCM (2000) The Smith-Lemli-Opitz syndrome. J Med Genet 37:321-335
19. Tint GS, Irons M, Elias ER et al (1994) Defective cholesterol biosynthesis associated with the Smith-Lemli-Opitz syndrome. N Engl J Med 330:107-113
20. Waterham HR, Wanders RJA (2000) Biochemical and genetic aspects of 7-dehydrocholesterol reductase and Smith-Lemli-Opitz syndrome. Biochim Biophys Acta 1529:340-356
21. Witsch-Baumgartner M, Fitzky BU, Ogorelkova M et al (2000) Mutational spectrum in the delta7-sterol reductase gene and genotype-phenotype correlation in 84 patients with Smith-Lemli-Opitz syndrome. Am J Hum Genet 66:402-441
22. Witsch-Baumgartner M, Gruber M, Kraft HG et al (2004) Maternal apo E genotype is a modifier of the Smith-Lemli-Opitz syndrome. J Med Genet 41:577-584
23. Fitzky BU, Witsch-Baumgartner M, Erdel M et al (1998) Mutations in the delta7-sterol reductase gene in patients with the Smith-Lemli-Opitz syndrome. Proc Natl Acad Sci USA 95:8181-8186
24. Wassif CA, Maslen C, Kachilele-Linjewile S et al (1998) Mutations in the human sterol delta7-reductase gene at 11q12-13 cause Smith-Lemli-Opitz syndrome. Am J Hum Genet 63:55-62
25. Waterham HR, Wijburg FA, Hennekam RC et al (1998) Smith-Lemli-Opitz syndrome is caused by mutations in the 7-dehydrocholesterol reductase gene. Am J Hum Genet 63:329-338
26. Irons M, Elias ER, Abuelo D et al (1997) Treatment of Smith-Lemli-Opitz syndrome: results of a multicenter trial. Am J Med Genet 68:311-314
27. Sikora DM, Ruggiero M, Petit-Kekel K et al (2004) Cholesterol supplementation does not improve developmental progress in Smith-Lemli-Opitz syndrome. J Pediatr 144:783-791
28. Jira PE, Wevers RA, de Jong J et al (2000) Simvastatin. A new therapeutic approach for Smith-Lemli-Opitz syndrome. J Lipid Res 41:1339-1346
29. Spranger JW, Opitz JM, Bibber U (1971) Heterogeneity of chondrodysplasia punctata. Hum Genet 11:190-212
30. Happle R (1979) X-linked dominant chondrodysplasia punctata. Review of literature and report of a case. Hum Genet 53:65-73
31. Herman G (2000) X-Linked dominant disorders of cholesterol biosynthesis in man and mouse. Biochim Biophys Acta 1529:357-373

32. Braverman N, Lin P, Moebius FF et al (1999) Mutations in the gene encoding 3beta-hydroxysteroid-delta8-delta7 isomerase cause X-linked dominant Conradi-Hünermann syndrome. Nat Genet 22:291-294

33. Derry JM, Gormally E, Means GD et al (1999) Mutations in a delta8-delta7-sterol isomerase in the tattered mouse and X-linked dominant chondrodysplasia punctata. Nat Genet 22:286-290

34. Kelley RI, Wilcox WG, Smith M et al (1999) Abnormal sterol metabolism in patients with Conradi-Hünermann-Happle syndrome and sporadic chondrodysplasia punctata. Am J Med Genet 83:213-219

35. Offiah AC, Mansour S, Jeffrey I et al (2003) Greenberg dysplasia (HEM) and lethal X linked dominant Conradi-Hünermann chondrodysplasia punctata (CDPX2): presentation of two cases with overlapping phenotype. J Med Genet 40:e129

36. Mason DE, Sanders JO, MacKenzie WG et al (2002) Spinal deformity in chondrodysplasia punctata. Spine 27:1995-2002

37. Happle R, Koch H, Lenz W (1980) The CHILD syndrome. Congenital hemidysplasia with ichthyosiform erythroderma and limb defects. Eur J Pediatr 134:27-33

38. Liu XY, Dangel AW, Kelley RI et al (1999) The gene mutated in bare patches and striated mice encodes a novel 3beta-hydroxysteroid dehydrogenase. Nat Genet 22:182-187

39. Konig A, Happle R, Bornholdt D et al (2000) Mutations in the NSDHL gene, encoding a 3beta-hydroxysteroid dehydrogenase, cause CHILD syndrome. Am J Med Genet 90:339-346

40. Grange DK, Kratz LE, Braverman NE, Kelley RI (2000) CHILD syndrome caused by deficiency of 3beta-hydroxysteroid-delta8, delta7-isomerase. Am J Med Genet 90:328-335

41. FitzPatrick DR, Keeling JW, Evans MJ et al (1998) Clinical phenotype of desmosterolosis. Am J Med Genet 75:145-152

42. Andersson HC, Kratz L, Kelley R (2002) Desmosterolosis presenting with multiple congenital anomalies and profound developmental delay. Am J Med Genet 113:315-319

43. Waterham HR, Koster J, Romeijn GJ et al (2001) Mutations in the 3ß-hydroxysteroid Δ^{24}-reductase gene cause desmosterolosis, an autosomal recessive disorder of cholesterol biosynthesis. Am J Hum Genet 69:685-694

44. Brunetti-Pierri N, Corso G, Rossi M et al (2002) Lathosterolosis, a novel multiple-malformation/mental retardation syndrome due to deficiency of 3beta-hydroxysteroid-delta5-desaturase. Am J Hum Genet 71:952-958

45. Krakowiak PA, Wassif CA, Kratz L et al (2003) Lathosterolosis: an inborn error of human and murine cholesterol synthesis due to lathosterol 5-desaturase deficiency. Hum Mol Genet 12:1631-1641

46. Greenberg CR, Rimoin DL, Gruber HE et al (1988) A new autosomal recessive lethal chondrodystrophy with congenital hydrops. Am J Med Genet 29:623-632

47. Oosterwijk JC, Mansour S, van Noort G et al (2003) Congenital abnormalities reported in Pelger-Huet homozygosity as compared to Greenberg/HEM dysplasia: highly variable expression of allelic phenotypes. J Med Genet 40:937-941

48. Hoffmann K, Dreger CK, Olins AL et al (2002) Mutations in the gene encoding the lamin B receptor produce an altered nuclear morphology in granulocytes (Pelger-Huet anomaly). Nat Genet 31:410-414

49. Waterham HR, Koster J, Mooyer P et al (2003) Autosomal recessive HEM/Greenberg skeletal dysplasia is caused by 3 beta-hydroxysterol delta 14-reductase deficiency due to mutations in the lamin B receptor gene. Am J Hum Genet 72:1013-1017

50. Fluck CE, Tajima T, Pandey AV et al (2004) Mutant P450 oxidoreductase causes disordered steroidogenesis with and without Antley-Bixler syndrome. Nat Genet 36: 228-230

34 Disorders of Bile Acid Synthesis

Peter T. Clayton

Bile Acid Synthesis

Bile acids are biological detergents that are synthesised from cholesterol in the liver by modifications of the sterol nucleus and oxidation of the side chain. Synthesis of bile acids can occur by a number of pathways (◘ Fig. 34.1); the most important in adults starts with conversion of cholesterol to 7α-hydroxycholesterol. In infancy, other pathways are more important; one of these starts with the conversion of cholesterol to 27-hydroxycholesterol.

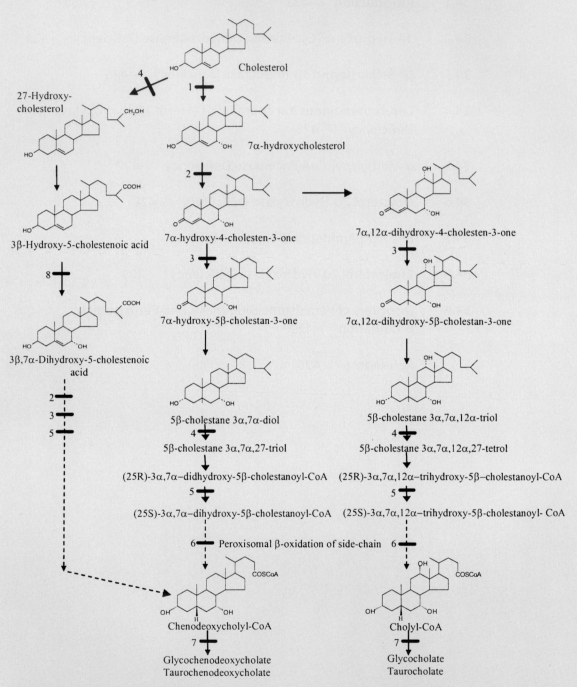

◘ **Fig. 34.1.** Major reactions involved in the synthesis of bile acids from cholesterol. **1**, cholesterol 7α-hydroxylase; **2**, 3β-hydroxy-Δ⁵-C₂₇-steroid dehydrogenase/isomerase; **3**, Δ⁴-3-oxosteroid-5β-reductase; **4**, sterol 27-hydroxylase; **5**, α-methylacyl-CoA race-mase; **6**, enzymes of peroxisomal biogenesis and β-oxidation; **7**, the bile acid amidating enzyme; **8**, oxysterol 7α-hydroxylase. Enzyme defects are depicted by *solid bars* across the arrows

Two inborn errors of metabolism affect the modifications of the cholesterol nucleus in both major pathways for bile acid synthesis: *3β-hydroxy- Δ⁵-C₂₇-steroiddehydrogenase (3 β-dehydrogenase) deficiency* and *Δ⁴-3-oxosteroid-5 β-reductase (5 β-reductase) deficiency*. These disorders produce cholestatic liver disease and malabsorption of fat and fat-soluble vitamins. Onset of symptoms is usually in the first year of life and, untreated, the liver disease can progress to cirrhosis and liver failure. Treatment with chenodeoxycholic acid and cholic acid can lead to dramatic improvement in the liver disease and the malabsorption. Neonatal cholestatic liver disease can also be the presenting feature of two disorders affecting oxidation of the cholesterol side chain – *sterol 27-hydroxylase deficiency [cerebrotendinous xanthomatosis (CTX)]* and *α-methylacyl-CoA racemase defic*iency. However, these disorders more commonly present later with neurological disease. CTX typically presents with cataracts and mental retardation in childhood, followed by motor dysfunction and tendon xanthomata in the second or third decade. Death may be caused by progressive motor dysfunction and dementia or by premature atherosclerosis. Chenodeoxycholic acid has been shown to halt or even reverse neurological dysfunction. α-Methylacyl-CoA racemase deficiency can produce a range of neurological problems in adult life including sensory motor neuropathy and pigmentary retinopathy. Other (so far rare) inborn errors of bile acid synthesis include *oxysterol 7α-hydroxylase deficiency* (rapidly progressive neonatal liver disease) a *bile acid amidation defect* (cholestatic liver disease and fat-soluble vitamin malabsorption), and *cholesterol 7α-hydroxylase deficiency* (adults with hyperlipidaemia and gall stones). In disorders of peroxisome biogenesis and peroxisomal β-oxidation, neurological disease usually predominates; these are considered in ▶ Chap. 40.

34.1   Introduction

Most of the known enzyme deficiencies of bile acid synthesis affect both the 27-hydroxycholesterol and the 7α-hydroxycholesterol pathways; the exceptions are cholesterol 7α-hydroxylase deficiency and oxysterol 7α-hydroxylase deficiency. Because of the broad specificity of many of the enzymes, the major metabolites are not those immediately proximal to the block. For instance, in 3β-hydroxy- Δ⁵-C₂₇-steroiddehydrogenase deficiency (enzyme 2 in ◘ Fig. 34.1), the major metabolite is not 7α–hydroxycholesterol but a series of unsaturated bile acids that have the normal bile acid side chain but persistence of the 3β, 7α-dihydroxy-Δ⁵ structure of the nucleus.

34.2   3β-Hydroxy-Δ⁵-C₂₇-Steroid Dehydrogenase Deficiency

34.2.1   Clinical Presentation

The first described patients with 3β-dehydrogenase deficiency presented with prolonged neonatal jaundice (conjugated bilirubin levels greater than 40 μM at the age of 2 months or older) associated with steatorrhoea. The stools were pale but not acholic. Rickets (due to malabsorption of vitamin D) was often apparent before the age of 6 months, and one patient developed a bleeding diathesis due to vitamin-K deficiency at the age of 9 months [1–3]. Routine investigations performed at age 2–6 months were not very helpful in distinguishing 3β-dehydrogenase deficiency from other causes of giant-cell hepatitis. The biochemical evidence of fat-soluble-vitamin malabsorption is perhaps more striking, e.g. plasma vitamin E concentration consistently less than 4 μM (normal age range = 11.5–35 μM) and the γ-glutamyl transpeptidase is usually normal or only minimally elevated [despite a significantly elevated aspartate aminotransferase (AST)]. The liver biopsy shows a periportal inflammatory infiltrate (often including eosinophils), giant cells, some hepatocellular necrosis and bridging fibrosis or even early cirrhosis. In untreated patients, pruritus often becomes apparent from the age of 6 months, and the problems of steatorrhoea and malabsorption of fat-soluble vitamins continue. In 1994, Jacquemin et al. described a group of patients with 3β-dehydrogenase deficiency who presented with jaundice, hepatosplenomegaly and steatorrhoea (a clinical picture resembling progressive familial intrahepatic cholestasis) between the ages of 4 months and 46 months [4]. Pruritus was absent in these children, in contrast to other children with severe cholestasis. The authors noted normal γ-glutamyl-transpeptidase activities in plasma, low serum cholesterol concentrations and low vitamin E concentrations. Presentation of 3β-dehydrogenase deficiency with chronic hepatitis in the second decade of life has also been described [5].

34.2.2   Metabolic Derangement

3β-Dehydrogenase catalyses the second reaction in the major pathway of synthesis of bile acids: the conversion of 7α-hydroxycholesterol to 7α-hydroxycholest-4-en-3-one. When the enzyme is deficient, the accumulating 7α-hydroxycholesterol can undergo side-chain oxidation with or without 12α-hydroxylation to produce 3β,7α-dihydroxy-5-cholenoic acid and 3β,7α,12α-trihydroxy-5-cholenoic acid, respectively. These unsaturated C₂₄ bile acids are sulphated in the C3 position; a proportion is conjugated to glycine, and they can be found in high concentrations in the urine. Concentrations of bile acids in the bile are low [6]. It is probable that the sulphated Δ⁵ bile acids cannot be secreted into the bile canaliculi and fuel bile flow in the same way as occurs with the normal bile acids. There are at least two possible ways in which this sequence of events might lead to damage to hepatocytes and, ultimately, to cirrhosis:

1. The abnormal metabolites produced from 7α-hydroxy-cholesterol may be hepatotoxic
2. Failure of bile acid-dependent bile flow may lead to hepatocyte damage, perhaps as a result of the accumulation of toxic compounds normally eliminated in the bile.

34.2.3   Genetics

3β-Dehydrogenase deficiency is an autosomal-recessive trait caused by mutations in the *HSD3B7* gene located on 16p11.2–12. In 2000, Schwarz et al. showed that the original patient described by Clayton et al. in 1987 was homozygous for a 2bp deletion in exon 6 of the gene (Δ1057–1058) [7]. In 2003, Cheng et al. reported mutations in 15 additional patients from 13 kindreds with 3β-dehydrogenase deficiency [8]. In patients with neonatal cholestasis, they identified deletions (310delC, 63delAG), a splice site mutation (340+1 G>T) and a missense mutation (E147K).

34.2.4   Diagnostic Tests

The diagnosis is established by demonstrating the presence of the characteristic Δ5 bile acids in plasma or urine. It is important to remember that bile acids with a Δ5 double bond and a 7-hydroxy group are acid labile. They may be destroyed by some of the methods that are used for solvolysis of sulphated bile acids prior to chromatographic analysis. Solvolysis is best performed using tetrahydrofuran/methanol/trifluoroacetic acid (900:100:1 volume ratio) [3,6]. Analysis by fast-atom-bombardment mass spectrometry (FAB-MS) overcomes the problem of lability [2, 9]. More recently, because electrospray ionisation tandem mass spectrometry (ESI-MS/MS) is in use in many laboratories, diagnostic bile acids have been detected using this methodology [10].

Plasma

If plasma bile acids are analysed using a gas chromatography (GC)-MS method that does not include a solvolysis step, the profile of non-sulphated bile acids that is obtained shows concentrations of cholic and chenodeoxycholic acid, which are extremely low for an infant with cholestasis. The concentration of 3β,7α-dihydroxy-5-cholestenoic acid is increased. Inclusion of a solvolysis step reveals the presence of high concentrations of 3β, 7α-dihydroxy-5-cholenoic acid (3-sulphate) and 3β,7α,12α-trihydroxy-5-cholenoic acid (3-sulphate). These can also be detected when plasma is analysed by FAB-MS or when a neonatal blood spot is analysed by ESI-MS/MS [10].

Urine

Urine analysed by negative ion FAB-MS shows the characteristic ions of the diagnostic unsaturated bile acids: mass/charge ratios (m/z) = 469, 485, 526 and 542. Using electrospray ionisation tandem mass spectrometry, the sulphated Δ5 bile acids (m/z 469 and 485) are detected as parents of m/z 97; glycine conjugates of sulphated Δ5 bile acids (m/z 526 and 542) are additionally detected as parents of m/z 74.

Fibroblasts

3β-Dehydrogenase can be assayed in cultured skin fibroblasts using tritiated 7α-hydroxycholesterol [11]. Patients show very low activity.

34.2.5   Treatment and Prognosis

Emergency treatment of coagulopathy with parenteral vitamin K may be required [3]. Vitamin D deficiency may be severe enough to require intravenous calcium as well as vitamin D therapy. However, long term treatment with fat-soluble vitamins is not required because bile acid replacement therapy corrects all the fat-soluble vitamin deficiencies. Untreated 3β-dehydrogenase deficiency has led to death from complications of cirrhosis before the age of 5 years; patients with milder forms of the disorder may survive with a chronic hepatitis into their second decade or beyond. The response to treatment depends upon the severity of the liver disease at the time of starting treatment. In patients with a bilirubin level less than 120 μM and an AST level less that 260 U/l, chenodeoxycholic acid therapy has led to a dramatic improvement in symptoms and in liver-function tests within 4 weeks and to an improvement in the liver biopsy appearances within 4 months. The dose of chenodeoxycholic acid that has been used is 12–18 mg/kg/day initially (for 2 months) followed by 9–12 mg/kg/day maintenance. In one infant with severe disease, chenodeoxycholic acid (15 mg/kg/day) led to a rise in bilirubin and AST [1]. Her treatment regime was changed to 7 mg chenodeoxycholic acid/kg/day plus 7 mg cholic acid/kg/day. Over the course of 15 months, her bilirubin and transaminases returned to normal, and a repeat liver biopsy showed a more normal parenchyma and less inflammation. The combination of cholic acid and chenodeoxycholic acid is probably the treatment of choice for patients with severe liver damage. Bile-acid-replacement therapy may work in one of two ways:

1. By fuelling bile-acid-dependent flow (hence directly relieving cholestasis)
2. By suppressing the activity of cholesterol 7α-hydroxylase (thereby reducing the accumulation of potentially toxic metabolites of 7α-hydroxycholesterol).

34.3   Δ⁴-3-Oxosteroid 5β-Reductase Deficiency

34.3.1   Clinical Presentation

Patients who excrete 3-oxo-Δ^4 bile acids as the major urinary bile acids can be divided into three groups – those who have proven mutations in *SRD5B1* (*AKR1D1*; the gene encoding the 5β-reductase enzyme) [12, 13], those in whom this has been excluded [14] and those in whom the results of gene analysis have not been published [15, 16]. In the latter two groups the cause of excretion of 3-oxo-Δ^4 bile acids remains uncertain and, since this pattern of urinary metabolite excretion can be a non-specific consequence of severe liver disease [17, 18], the description in this Chapter will focus on the five patients with proven 5β-reductase mutations.

In two of the four families described the parents were consanguineous [12, 13]. All five patients presented in the neonatal period with cholestatic jaundice with raised transaminases but normal γ-GT, low vitamin E, and prolonged clotting times which improved with parenteral vitamin K. Liver biopsies showed giant cell transformation, canalicular and hepatocellular cholestasis, portal inflammation, septal fibrosis, occasional necrotic foci and in some cases, increased extramedullary haemopoiesis. Without treatment, cholestasis persisted in all cases.

34.3.2   Metabolic Derangement

Mutations in *SRD5B1* lead to reduced activity of the hepatic enzyme that brings about the 5β(H) saturation of the C4 double bonds of bile acid precursors such as 7α-hydroxy-cholest-4-en-3-one and 7α,12α-dihydroxy-cholest-4-en-3-one. These intermediates can then undergo side-chain oxidation to produce the corresponding C24 bile acids. The mechanism of hepatocyte damage and cholestasis in 5β-reductase deficiency is unknown; as with 3β-dehydrogenase deficiency, toxicity of unsaturated intermediates and unsaturated bile acids and loss of bile-acid-dependent bile flow have been postulated.

34.3.3   Genetics

Primary 5β-reductase deficiency is an autosomal recessive disorder caused by mutations in *SRD5B1* (*AKR1D1*; the gene encoding the 5β-reductase enzyme) [12, 13]. The mutations that have been described are 385 C>T (missense), 467 C>G (missense), 511delT (frameshift, premature stop codon), 662C>T (missense) and 850 C>T (missense).

34.3.4   Diagnostic Tests

Plasma

GC-MS analysis of plasma bile acids reveals low or low normal concentrations of chenodeoxycholic acid (normal concentration 0.2–12.7 μM) and cholic acid (normal concentration 0.4–6.7 μM). In contrast, the plasma concentrations of 3-oxo-Δ^4 bile acids are markedly elevated, i.e. 7α-hydroxy-3-oxo-4-cholenoic acid > 1.5μ M and 7α,12 α-dihydroxy-3-oxo-4-cholenoic acid > 2.0 μM. Analysis of plasma bile acids by ESI-MS/MS shows taurine-conjugated (parents of m/z 80) and glycine conjugated (parents of m/z 74) 3-oxo-Δ^4 bile acids present at concentrations similar to those of their saturated analogues [12].

Urine

Analysis of urine by FAB-MS or ESI-MS/MS shows the presence of major ions attributable to the glycine conjugates of 7α-hydroxy-3-oxo-4-cholenoic acid and 7α,12α-dihydroxy-4-cholenoic acid (m/z = 444 and 460; parents of m/z 74), and their taurine conjugates (m/z = 494 and 510; parents of m/z 80) and sometimes the taurine conjugate of 7α,12α-dihydroxy-3-oxo-4-cholestenoic acid (m/z 552; parents of 80). The normal saturated bile acids (m/z 448, 464, 498, 514) are at background level.

The identities and relative amounts of urinary bile acids can be confirmed by GC-MS analysis following enzymatic deconjugation. In patients shown to have primary 5β-reductase deficiency, the 3-oxo-Δ^4 bile acids have comprised more than 90% of the total urinary bile acids; a lower percentage is found in most children whose excretion of 3-oxo-Δ^4 bile acids is secondary to liver damage of other aetiology.

34.3.5   Treatment and Prognosis

Emergency treatment for vitamin K deficiency may be required. Vitamin D may be needed for rickets. 5β-Reductase deficiency can progress rapidly to liver failure. However, treatment with bile acid replacement therapy can lead to normalisation of liver function and long term (at least 8 years) good health. Successful regimes include chenodeoxycholic acid plus cholic acid (8 mg/kg/day of each) and cholic acid alone. Thus, of the patients with proven 5β-reductase mutations, one infant progressed to liver failure, failed to respond to treatment with ursodeoxycholic acid or chenodeoxycholic acid and had a liver transplant at 19 weeks. One child failed to respond to ursodeoxycholic acid treatment but responded extremely well to treatment with chenodeoxycholic acid and cholic acid (started at 8 months) and was asymptomatic at the age of 10 years. One patient showed an initial response to chenodeoxycholic acid plus cholic acid but then deteriorated (possibly due to cytomegalovirus infection) and required transplantation. Twin patients

responded to treatment with cholic acid started at 8 months and they were well at 5 years.

34.4 Cerebrotendinous Xanthomatosis (Sterol 27-Hydroxylase Deficiency)

34.4.1 Clinical Presentation

It has recently become clear that cholestatic jaundice in infancy may be the first manifestation of cerebrotendinous xanthomatosis (CTX) [19]. However, it usually improves spontaneously. Chronologically, the next (or the first) symptom of CTX is often mental retardation detected during the first decade of life. Cataracts may also be present as early as 5 years of age. Wevers et al. [20] have documented four Dutch patients in whom persistent diarrhoea was present from early childhood. Motor dysfunction (spastic paresis, ataxia, expressive dysphasia) develops in approximately 60% of patients in the second or third decade of life. Tendon xanthomata may be detectable during the second decade of life but usually appear in the third or fourth decade. The Achilles tendon is the most common site; other sites include the tibial tuberosities and the extensor tendons of the fingers and the triceps. Premature atherosclerosis leading to death from myocardial infarction occurs in some patients. In others, death is caused by progression of the neurological disease with increasing spasticity, tremor and ataxia and pseudobulbar palsy. It is important to recognise that the neurological deterioration is very variable [21]. For example, some patients are normal intellectually but suffer from a neuropathy or mild spastic paresis; others have no neurological signs but present with psychiatric symptoms resembling schizophrenia. Magnetic-resonance imaging (MRI) of the brain may show diffuse cerebral atrophy and increased signal intensity in the cerebellar white matter on T2-weighted scans [22]. Osteoporosis is common in CTX and may produce pathological fractures; it is associated with low plasma concentrations of 25-hydroxy-vitamin D and 24,25-dihydroxy-vitamin D [23]. Patients with untreated CTX usually die from progressive neurological dysfunction or myocardial infarction between the ages of 30 years and 60 years.

34.4.2 Metabolic Derangement

CTX is caused by a defect in the gene for sterol 27-hydroxylase, the mitochondrial enzyme that catalyses the first step in the process of side-chain oxidation, which is required to convert a C27 sterol into a C24 bile acid [24]. 5β-Cholestane-3α,7α,12α-triol cannot be hydroxylated in the C27 position and accumulates in the liver. As a result, it is metabolised by an alternative pathway, starting with hydroxylation in the C25 position (in the endoplasmic reticulum). Further hydroxylations, e.g. in the C22 or C23 position, result in the synthesis of the characteristic bile alcohols that are found (as glucuronides) in the urine. Bile-acid precursors other than 5β-cholestane-3α,7α,12α-triol also accumulate. Some of these (e.g. 7α-hydroxy-cholest-4-en-3-one) are probably converted to cholestanol by a pathway involving 7α-dehydroxylation. Because patients with CTX have a reduced rate of bile-acid synthesis, the normal feedback inhibition of cholesterol 7α-hydroxylase by bile acids is disrupted. This further enhances the production of bile alcohols and cholestanol from bile-acid precursors. The major symptoms of CTX are produced by accumulation of cholestanol (and cholesterol) in almost every tissue of the body, particularly in the nervous system, atherosclerotic plaques and tendon xanthomata.

Sterol 27-hydoxylase is active in extrahepatic tissues, where it converts cholesterol into 27-hydroxycholesterol, which can be further metabolised and eliminated from cells. This pathway provides a route for the elimination of cholesterol; this route acts as an alternative to the high-density-lipoprotein-mediated reverse cholesterol transport [25]. Disruption of this pathway in CTX provides a further explanation for the accumulation of cholesterol in the tissues.

34.4.3 Genetics

CTX is inherited as an autosomal recessive trait. The cDNA encoding the 27-hydroxylase enzyme has been characterised, and the gene has been localised to chromosome 2q33-qter [24]. CTX can be caused by point mutations that lead to a production of an inactive enzyme (R362C and R446C) [26]. In Moroccan Jews, there appear to be two common mutations, both of which lead to failure of the production of sterol 27-hydroxylase mRNA. One is a frameshift mutation, the other is a splice-junction mutation [27]. Many other mutations have now been described.

34.4.4 Diagnostic Tests

Plasma

The concentration of cholestanol in plasma can be determined by GC or high-performance liquid chromatography (HPLC). Patients with CTX have plasma concentrations in the range of 30–400 μM (normal range = 2.6–16 μM). The plasma cholestanol-to-cholesterol ratio may be a better discriminant than the absolute cholestanol concentration. The following bile-acid precursors have been detected at increased concentrations in plasma: 7α-hydroxycholesterol, 7α-hydroxy-cholest-4-en-3-one, 7α,12α-dihydroxy-cholest-4-en-3-one. Plasma concentrations of bile acids are low; plasma concentrations of bile-alcohol glucuronides are elevated.

Urine

Negative ion FAB-MS or ESI-MS/MS indicate that major cholanoids in the urine are cholestanepentol glucuronides, giving rise to an ion of m/z ratio 627 [3, 28]. GC-MS analysis shows that the major alcohols are 3,7,12,23,25-pentols and 3,7,12,22,25-pentols in adults. Increased urinary bile-alcohol concentrations can be detected using an enzyme assay (7α-hydroxysteroid dehydrogenase) [29]. The urinary bile-alcohol excretion following cholestyramine administration has been used as a test for carriers of CTX [30].

Fibroblasts

27-Hydroxylation of C27 sterols can be measured in cultured skin fibroblasts, and the enzyme activity is virtually absent in fibroblasts from patients with CTX [31].

DNA

In certain populations in which one or two common mutations predominate, DNA analysis may prove to be a rapid method for diagnosis of both homozygotes and carriers of CTX (► above).

34.4.5   Treatment and Prognosis

The results of treatment with chenodeoxycholic acid were first reported in 1984 [32]. The rates of synthesis of cholestanol and cholesterol were reduced, and plasma cholestanol concentrations fell. A significant number of patients showed reversal of their neurological disability, with clearing of the dementia, improved orientation, a rise in intelligence quotient and enhanced strength and independence. The MRI appearances do not, however, show obvious improvement [33]. Urinary excretion of bile-alcohol glucuronides is markedly suppressed. Chenodeoxycholic acid almost certainly works by suppressing cholesterol 7α-hydroxylase activity; ursodeoxycholic acid, which does not inhibit the enzyme, is ineffective. Adults have usually been treated with a dose of 750 mg/day chenodeoxycholic acid. Other treatments that have been used in CTX include 3-hydroxy-3-methylglutaryl-coenzyme A (HMG-CoA) reductase inhibitors, (statins such as lovastatin) [34] and low-density lipoprotein apheresis [35]. There is insufficient information available to assess these forms of treatment at the present time. The osteoporosis seen in patients with CTX appears to be resistant to chenodeoxycholic acid therapy [36]. Cholestatic liver disease in infancy may be self-limiting but in those children in whom it is not, bile acid treatment has been successful; cholic acid is probably preferable to chenodoexycholic acid [19, 37].

34.5   α-Methylacyl-CoA Racemase Deficiency

34.5.1   Clinical Presentation

The first described patients with α-methylacyl-CoA racemase (AMACR) deficiency all had neurological disease [38]. Patient 1 presented in childhood with developmental delay and then had an acute encephalopathic illness at 18 that left him temporarily blind. On follow up into adult life he developed epilepsy, pigmentary retinopathy and an axonal sensory motor neuropathy affecting the legs more than the arms. Patient 2 developed a spastic paraparesis in her forties and she additionally had evidence of a demyelinating sensory motor polyneuropathy. Tremor and deep white matter changes on MRI have been described recently [39]. However, presentation with neonatal cholestatic liver disease has also been documented: In 2001, Van Veldhoven et al. described an infant with AMACR deficiency who presented with a coagulopathy due to vitamin K deficiency; a sibling had died of a major bleed with the same cause [40]. The infant had mild cholestatic jaundice with raised aspartate aminotransferase and, in contrast to 3β-dehydrogenase deficiency, 5β-reductase deficiency and CTX, a raised γ-GT. Liver biopsy showed a mild non-specific lymphocytic portal infiltrate and abundant giant cell transformation.

34.5.2   Metabolic Derangement

Side-chain oxidation of cholesterol produces the 25R isomer of 3α,7α,12α-trihydroxycholestanoyl-CoA [(25R)-THC-CoA], and α-oxidation of dietary phytanic acid produces (some) (2R)-pristanoyl-CoA. Before these substrates can undergo peroxisomal β-oxidation they need to be converted to the S isomers by AMACR. It is likely that decreased production of cholic acid and chenodeoxycholic acid contributes to cholestatic liver disease and fat-soluble vitamin malabsorption. The pathogenesis of the neurological disease is not understood.

34.5.3   Genetics

AMACR deficiency is caused by mutations in the *AMACR* gene on chromosome 5p13.2–5q11.1. Pathogenic mutations in the adults with neurological disease included S52P and L107P. The S52P mutation was also found in the siblings who presented with neonatal coagulopathy [41].

34.5.4   Diagnostic Tests

Analysis of plasma bile acids by GC-MS reveals increased concentrations of DHCA and THCA; HPLC-ESI-MS/MS can be used to show that it is the (25R) isomer of THCA that is accumulating. GC-MS analysis of fatty acids in plasma shows an elevated concentration of pristanic acid with mildly elevated/normal plasma phytanic acid concentration and normal very long chain fatty acids.

34.5.5 Treatment and Prognosis

Parenteral vitamin K may be life-saving. Cholic acid therapy was important in preventing continuing fat-soluble vitamin malabsorption in the cholestatic neonate described by Van Veldhoven et al. and Setchell et al. [40, 41]. Its role in improving the liver disease is less certain as, given that adults with the disorder do not show signs of liver disease, there may be spontaneous resolution (as in CTX). The role of a low phytanic acid diet is uncertain; it appeared to ameliorate neurological symptoms in at least one of the adults with neurological disease. The influence of bile acid therapy on the development and progression of neurological disease is also unknown at present.

34.6 Oxysterol 7α-Hydroxylase Deficiency

34.6.1 Clinical Presentation

Oxysterol 7α-hydroxylase has so far only been described in a single 10-week-old male infant with severe cholestasis, cirrhosis and liver synthetic failure [42].

34.6.2 Metabolic Derangement

This recessive disorder is due to mutations in the gene encoding microsomal oxysterol 7-hydroxylase, leading to inactivity of this enzyme and accumulation of 27-hydroxycholesterol, 3β-hydroxy-5-cholestenoic acid and 3β-hydroxy-5-cholenoic acid. The pathway of bile acid synthesis via 27-hydroxycholesterol (which is thought to be very important in infancy) is completely disrupted and the monohydroxy bile acids that accumulate are particularly hepatotoxic.

34.6.3 Genetics

The child was homozygous for an R388X mutation in the *CYP7B1* gene on chromosome 8q21.3.

34.6.4 Diagnostic Tests

Analysis of urine by FAB-MS revealed major peaks of m/z ratio 453 and 510 attributable to 3β-hydroxy-5-cholenoic acid 3-sulphate and its glycine conjugate. GC-MS analysis of plasma indicated that the main cholanoids were 3β-hydroxy-5-cholenoic acid and 3β-hydroxy-5-cholestenoic acid.

34.6.5 Treatment and Prognosis

The patient showed a deterioration with ursodeoxycholic acid and no improvement with cholic acid and required a liver transplant for hepatic failure at the age of 4 months.

34.7 Bile Acid Amidation Defect

34.7.1 Clinical Presentation

A 14-year-old boy was shown to excrete large amounts of unconjugated cholic acid in the urine; this was attributed to a defect in an enzyme involved in amidation (formation of glycine and taurine conjugates). He originally presented with cholestatic liver disease and went on to show evidence of vitamin-K deficiency and rickets. Two other patients (a five year old Saudi Arabian boy and his eight year old sister) had variable degrees of cholestasis. The boy had undergone a Kasai procedure in infancy for incorrectly diagnosed biliary atresia; the girl was asymptomatic at the time of diagnosis [43]. The disorder is also found amongst the Amish, in whom the presentation is with failure to thrive, with in some cases pruritus, and occasionally coagulopathy but without jaundice [44]. Two out of four affected patients suffered chronic upper respiratory infection. We have observed the characteristic pattern of bile acid metabolites of the amidation defect in a 3-month-old infant with cholestatic jaundice, vitamin D deficiency and a liver biopsy showing mild portal and focal lobular hepatitis.

34.7.2 Metabolic Derangement

Without the enzyme bile acid coenzyme A: amino acid N-acyl transferase, encoded by the *BAAT* gene, the CoA esters of chenodeoxycholic acid and cholic acid cannot be converted to their glycine and taurine conjugates. The unconjugated bile acids are much weaker at solubilising lipid in the gut. Hence the failure to thrive and fat-soluble vitamin malabsorption.

34.7.3 Genetics

Defective amidation of bile acids in the Amish is caused by homozygosity for a missense mutation (226 A>G; M76V) in the *BAAT* gene. The molecular basis of the disorder in other affected individuals has not been published.

34.7.4 Diagnostic Tests

Analysis of urine by negative ion FAB-MS or ESI-MS shows that the major urinary bile acid is an unconjugated trihydroxy-cholanoic acid (m/z 407); GC-MS shows that it is unconjugated cholic acid. Other bile acids that may be detected include sulphated dihydroxycholanoic acid(s) (m/z 471) and trihydroxycholanoic acids (m/z 487) and glucuronidated dihydroxycholanoic acid(s) and trihydroxycholanoic acid(s) (m/z 567 and 583).

34.7.5 Treatment and Prognosis

Treatment of vitamin K deficiency may be life saving, treatment of rickets may require 1α-hydroxycholecalciferol or 1,25-dihydroxycholecalciferol. The Amish patients probably had improvement in symptoms with ursodeoxycholic acid but it is important to note that familial hypercholanaemia of the Amish can be caused by defects in a gene responsible for integrity of tight junctions (*TJP2*) as well as by mutations in the *BAAT* gene.

34.8 Cholesterol 7α-Hydroxylase Deficiency

34.8.1 Clinical Presentation

Homozygous cholesterol 7α-hydroxylase deficiency has been detected in three adults with hypercholesterolaemia, hypertriglyceridaemia and premature gallstone disease [45]. One had premature coronary and peripheral vascular disease. Their LDL cholesterol levels were noticeably resistant to treatment with HMG-CoA reductase inhibitors (statins). A study of the kindred revealed that individuals heterozygous for the mutation were also hyperlipidaemic, indicating that this is a codominant disorder.

34.8.2 Metabolic Derangement

Cholesterol 7α-hydroxylase is the first step in the major pathway for bile acid synthesis (and therefore for cholesterol catabolism) in adults. Reduced activity of the enzyme leads to accumulation of cholesterol in the liver, leading to downregulation of LDL receptors and hypercholesterolaemia.

34.8.3 Genetics

Cholesterol 7α-hydroxylase deficiency is caused by mutations in the *CYP7A1* gene. The only mutation described to date is a frameshift mutation (L413fsX414) that results in loss of the active site and enzyme function.

34.8.4 Diagnostic Tests

In one homozygote the cholesterol content of a liver biopsy was shown to be increased. Faecal bile acid output was reduced and the ratio chenodeoxycholic acid derived faecal bile acids / cholic acid derived faecal bile acids was increased suggesting increased activity of the alternative 27-hydroxylase pathway for bile acid (predominantly chenodeoxycholic acid) synthesis.

34.8.5 Treatment and Prognosis

Treatment with a powerful HMG-CoA reductase inhibitor (atorvastatin) and niacin is required to bring plasma levels of cholesterol and triglycerides under control. The variability of the disorder and long-term prognosis is not known

34.9 Disorders of Peroxisome Biogenesis and Peroxisomal β-Oxidation

These are described in ► Chap. 40. Neurological disease usually dominates the clinical picture but some children with Zellweger syndrome or infantile Refsum's disease have quite marked cholestatic liver disease.

References

1. Clayton PT (1991) Inborn errors of bile acid metabolism. J Inherit Metab Dis 14:478-496
2. Clayton PT, Leonard JV, Lawson AM et al (1987) Familial giant cell hepatitis associated with synthesis of 3β,7α-dihydroxy- and 3β,7α,12α-trihydroxy-5-cholenoic acids. J Clin Invest 79:1031-1038
3. Horslen SP, Lawson AM, Malone M, Clayton PT (1992) 3β-Hydroxy-Δ⁵-C₂₇-steroid dehydrogenase deficiency; effect of chenodeoxycholic acid therapy on liver histology. J Inherit Metab Dis 15: 38-46
4. Jacquemin E, Setchell KDR, O'Connell NC et al (1994) A new cause of progressive intrahepatic cholestasis: 3β-hydroxy-Δ⁵-C₂₇-steroid dehydrogenase / isomerase deficiency. J Pediatr 125:379-384
5. Setchell KDR (1990) Disorders of bile acid synthesis. In: Walker WA, Durie PR, Hamilton JR et al (eds) Pediatric gastrointestinal disease. Pathophysiology, diagnosis and management, vol 2. Dekker, Philadelphia, pp 922-1013
6. Ichimiya H, Egestad B, Nazer H et al (1991) Bile acids and bile alcohols in a child with 3β-hydroxy-Δ⁵-C₂₇-steroid dehydrogenase deficiency: effects of chenodeoxycholic acid treatment. J Lipid Res 32:829-841
7. Schwarz M, Wright, AC, Davis DL et al (2000) The bile acid synthetic gene 3-beta-hydroxy-delta-5-C27-steroid oxidoreductase is mutated in progressive intrahepatic cholestasis. J Clin Invest 106:1175-1184
8. Cheng JB, Jacquemin E, Gerhardt M et al (2003) Molecular genetics of 3β-Hydroxy-Δ⁵-C27-steroid oxidoreductase deficiency in 16 patients with loss of bile acid synthesis and liver disease. J Clin Endocrinol Metab 88:1833-1841

9. Lawson AM, Madigan MJ, Shortland DB, Clayton PT (1986) Rapid diagnosis of Zellweger syndrome and infantile Refsum's disease by fast atom bombardment mass spectrometry of urine bile salts. Clin Chim Acta 161:221-231

10. Mills K, Mushtaq I, Johnson A et al (1998) A method for the quantitation of conjugated bile acids in dried blood spots using electrospray ionization mass spectrometry. Pediatr Res 43:361-368

11. Buchmann MS, Kvittingen EA, Nazer H et al (1990) Lack of 3β-hydroxy-Δ^5-C$_{27}$-steroid dehydrogenase/isomerase in fibroblasts from a child with urinary excretion of 3β-hydroxy-Δ^5-bile acids a new inborn error of metabolism. J Clin Invest 86:2034-2037

12. Lemonde HA, Custard EJ, Bouquet J et al (2003) Mutations in SRD5B1 (AKR1D1), the gene encoding Δ^4-3-oxosteroid 5β-reductase, in hepatitis and liver failure in infancy. Gut 52(10):1494-1499

13. Gonzales E, Cresteil D, Baussan C et al (2004) SRD5B1 (AKR1D1) gene analysis in delta(4)-3-oxosteroid 5beta-reductase deficiency: evidence for primary genetic defect. J Hepatol 40:716-718

14. Sumazaki R, Nakamura N, Shoda J et al (1997) Gene analysis in Δ^4-3-oxosteroid 5β-reductase deficiency. Lancet 349:329

15. Setchell KDR, Suchy FJ, Welsh MB et al (1988) Δ^4-3-Oxosteroid 5β-reductase deficiency described in identical twins with neonatal hepatitis. A new inborn error in bile acid synthesis. J Clin Invest 82:2148-2157

16. Schneider BL, Setchell KDR, Whittington PF et al (1994) Δ^4-3-Oxosteroid 5β-reductase deficiency causing neonatal liver failure and neonatal hemochromatosis. J Pediatr 124:234-238

17. Clayton PT, Patel E, Lawson AM et al (1988) 3-Oxo-Δ^4 bile acids in liver disease. Lancet i:1283-1284

18. Clayton PT (1994) Δ^4-3-Oxosteroid 5β-reductase deficiency and neonatal hemochromatosis (letter). J Pediatr 125:845 846

19. Clayton PT, Verrips A, Sistermans E et al (2002) Mutations in the sterol 27-hydroxylase gene (CYP27A) cause hepatitis of infancy as well as cerebrotendinous xanthomatosis. J Inherit Metab Dis 25:501-513

20. Wevers RA, Cruysberg JRM, Van Heijst AFJ et al (1992) Paediatric cerebrotendinous xanthomatosis. J Inherit Metab Dis 14:374-376

21. Kuriyama M, Fujiyama J, Yoshidome H et al (1991) Cerebrotendinous xanthomatosis: clinical features of eight patients and a review of the literature. J Neurol Sci 102:225-232

22. Bencze K, Polder DRV, Prockop LD (1990) Magnetic resonance imaging of the brain in CTX. J Neurol Neurosurg Psychiatry 53:166-167

23. Berginer VM, Shany S, Alkalay D et al (1993) Osteoporosis and increased bone fractures in cerebrotendinous xanthomatosis. Metabolism 42:69-74

24. Cali JJ, Russell DW (1991) Characterisation of human sterol 27-hydroxylase: a mitochondrial cytochrome P-450 that catalyses multiple oxidations in bile acid biosynthesis. J Biol Chem 266:7774-7778

25. Babiker A, Andersson O, Lund E et al (1997) Elimination of cholesterol in macrophages and endothelial cells by the sterol 27-hydroxylase mechanism. Comparison with high density lipoprotein-mediated reverse cholesterol transport. J Biol Chem 272:26253-26261

26. Cali JJ, Hsieh C-L, Francke U, Russell DW (1991) Mutations in the bile acid biosynthetic enzyme sterol 27-hydroxylase underlie cerebrotendinous xanthomatosis. J Biol Chem 266:7779-7783

27. Leitersdorf E, Reshef A, Meiner V et al (1993) Frameshift and splice-junction mutations in the sterol 27-hydroxylase gene cause cerebrotendinous xanthomatosis in Jews of Moroccan origin. J Clin Invest 91:2488-2496

28. Egestad B, Pettersson P, Skrede S, Sjövall J (1985) Fast atom bombardment mass spectrometry in the diagnosis of cerebrotendinous xanthomatosis. Scand J Clin Lab Invest 45:443-446

29. Koopman BJ, Molen JC, Wolthers BG, Waterreus RJ (1987) Screening for CTX by using an enzymatic method for 7α-hydroxylated steroids in urine. Clin Chem 33:142-143

30. Koopman BJ, Waterreus RJ, Brekel HWC, Wolthers BG (1986) Detection of carriers of CTX. Clin Chim Acta 158:179-186

31. Skrede S, Björkhem I, Kvittingen EA et al (1986) Demonstration of 26-hydroxylation of C27-steroids in human skin fibroblasts, and a deficiency of this activity in CTX. J Clin Invest 78:729-735

32. Berginer VM, Salen G, Shefer S (1984) Long-term treatment of CTX with chenodeoxycholic acid therapy. N Engl J Med 311:1649-1652

33. Berginer VM, Berginer J, Korczyn AD, Tadmor R (1994) Magnetic resonance imaging in cerebrotendinous xanthomatosis: a prospective clinical and neuroradiological study. J Neurol Sci 122:102-108

34. Lewis B, Mitchell WD, Marenah CB, Cortese C (1983) Cerebrotendinous xanthomatosis: biochemical response to inhibition of cholesterol synthesis. Br Med J 287:2122

35. Mimura Y, Kuriyama M, Tokimura Y et al(1993) Treatment of cerebrotendinous xanthomatosis with low density lipoprotein (LDL)-apheresis. J Neurol Sci 114:227-230

36. Chang WN, Lui CC (1997) Failure in the treatment of long-standing osteoporosis in cerebrotendinous xanthomatosis. J Formos Med Assoc 96:225-227

37. Clayton PT, Casteels M, Mieli-Vergani G, Lawson AM (1995) Familial giant cell hepatitis with low bile acid concentrations and increased urinary excretion of specific bile alcohols: a new inborn error of bile acid synthesis? Pediatr Res 37:424-431

38. Ferdinandusse S, Denis S, Clayton PT et al (2000) Mutations in the gene encoding peroxisomal alpha-methylacyl-CoA racemase cause adult-onset sensory motor neuropathy. Nat Genet 24:188-191

39. Clarke CE, Alger S, Preece MA et al (2004) Tremor and deep white matter changes in alpha-methylacyl-CoA racemase deficiency. Neurology 63:188-189

40. Van Veldhoven PP, Meyhi E, Squires RH et al (2001) Fibroblast studies documenting a case of peroxisomal 2-methylacyl-CoA racemase deficiency: possible link between racemase deficiency and malabsorption and vitamin K deficiency. Eur J Clin Invest 31:714-722

41. Setchell KD, Heubi JE, Bove KE et al (2003) Liver disease caused by failure to racemize trihydroxycholestanoic acid: gene mutation and effect of bile acid therapy. Gastroenterology 124:217-232

42. Setchell KDR, Schwarz M, O'Connell NC et al (1998) Identification of a new inborn error in bile acid synthesis: mutation of the oxysterol 7α-hydroxylase gene causes severe neonatal liver disease. J Clin Invest 102:1690-1703

43. Setchell KDR, O'Connell NC (2000) Disorders of bile acid synthesis and metabolism. In: Walker WA, Durie PR, Hamilton JR et al (eds) Paediatric gastrointestinal disease, 3rd edn. Decker, Hamilton, Ontario, pp 1138-1170

44. Carlton VE, Harris BZ, Puffenberger EG et al (2003) Complex inheritance of familial hypercholanemia with associated mutations in TJP2 and BAAT. Nat Genet 34:91-96

45. Pullinger CR, Eng C, Salen G et al (2002) Human cholesterol 7alpha-hydroxylase (CYP7A1) deficiency has a hypercholesterolemic phenotype. J Clin Invest 110:109-117

VIII Disorders of Nucleic Acid and Heme Metabolism

35 Disorders of Purine and Pyrimidine Metabolism

Georges van den Berghe, M.- Françoise Vincent, Sandrine Marie

Purine Metabolism

Purine nucleotides are essential cellular constituents which intervene in energy transfer, metabolic regulation, and synthesis of DNA and RNA. Purine metabolism can be divided into three pathways:

- The biosynthetic pathway, often termed *de novo*, starts with the formation of phosphoribosyl pyrophosphate (PRPP) and leads to the synthesis of inosine monophosphate (IMP). From IMP, adenosine monophosphate (AMP) and guanosine monophosphate (GMP) are formed. Further metabolism (not illustrated) leads to their di- and triphosphates, to their corresponding deoxyribonucleotides, and to RNA and DNA.

- The catabolic pathway starts from GMP, IMP and AMP, and produces uric acid, a poorly soluble compound, which tends to crystallize once its plasma concentration surpasses 6.5–7 mg/dl (0.38–0.47 mmol/l).

- The salvage pathway utilizes the purine bases, guanine, hypoxanthine and adenine, which are provided by food intake or the catabolic pathway, and reconverts them into, respectively, GMP, IMP and AMP. Salvage of the purine nucleosides, adenosine and guanosine, and their deoxy counterparts, catalyzed by kinases, also occurs.

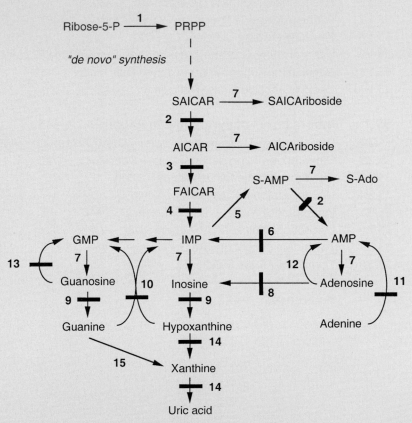

☐ **Fig. 35.1.** Pathways of purine metabolism. *AICAR*, aminoimidazolecarboxamide ribotide; *AMP*, adenosine monophosphate; *FAICAR*, formylaminoimidazolecarboxamide ribotide; *GMP*, guanosine monophosphate; *IMP*, inosine monophosphate; *P*, phosphate; *PRPP*, phosphoribosyl pyrophosphate, *S-Ado,* succinyladenosine; *SAICAR*, succinylaminoimidazolecarboxamide ribotide; *S-AMP*, adenylosuccinate, *XMP*, xanthosine monophosphate. **1**, PRPP synthetase; **2**, adenylosuccinase (adenylosuccinate lyase); **3**, AICAR transformylase; **4,** IMP cyclohydrolase (3 and 4 form ATIC); **5**, adenylosuccinate synthetase; **6**, AMP deaminase; **7**, 5′-nucleotidase(s), **8**, adenosine deaminase; **9**, purine nucleoside phosphorylase; **10**, hypoxanthine-guanine phosphoribosyltransferase; **11**, adenine phosphoribosyltransferase; **12**, adenosine kinase; **13**, guanosine kinase; **14**, xanthine oxidase (dehydrogenase). Enzyme defects are indicated by *solid bars* across the arrows

Inborn errors exist of the biosynthetic, catabolic, and salvage pathways of purine and pyrimidine metabolism, which are depicted in ◘ Fig. 35.1 and 35.3, respectively. The major presenting signs and laboratory findings in these inborn errors are listed in ◘ Table 35.1.

35.1 Inborn Errors of Purine Metabolism

Inborn errors of purine metabolism comprise errors of:
- *purine nucleotide synthesis*: phosphoribosylpyrophosphate (PRPP) synthetase superactivity, adenylosuccinase (ADSL) deficiency, AICA-ribosiduria caused by ATIC deficiency;
- *purine catabolism*: the deficiencies of muscle AMP deaminase (AMP-DA, also termed myoadenylate deaminase), adenosine deaminase (ADA), purine nucleoside phosphorylase (PNP) and xanthine oxidase;
- *purine salvage*: the deficiencies of hypoxanthine-guanine phosphoribosyltransferase (HGPRT) and adenine phosphoribosyltransferase (APRT). The deficiency of deoxyguanosine kinase causes mitochondrial DNA depletion (► also Chap. 15).

With the exception of muscle AMP-DA deficiency, all these enzyme defects are very rare.

35.1.1 Phosphoribosyl Pyrophosphate Synthetase Superactivity

Clinical Presentation

The disorder is mostly manifested by the appearance, in young adult males, of gouty arthritis and/or uric acid lithiasis, potentially leading to renal insufficiency [1, 2]. Uricemia can be very high, reaching 10–15 mg/dl (0.60–0.90 mmol/l) [normal adult values: 2.9–5.5 mg/dl (0.17–0.32 mmol/l)]. The urinary excretion of uric acid is also increased, reaching up to 2400 mg (14 mmol)/24 h, or 2.5 mmol/mmol creatinine [normal adult values: 500–800 mg (3-4.7 mmol)/24 h, or 02–0.3 mmol/mmol creatinine].

A few patients have been reported in which clinical signs of uric acid overproduction already appeared in infancy and were accompanied by neurologic abnormalities, mainly sensorineural deafness, particularly for high tones, but also hypotonia, locomotor delay, ataxia and autistic features [2].

Metabolic Derangement

The enzyme forms phosphoribosyl pyrophosphate (PRPP) from ribose-5-phosphate and ATP (◘ Fig. 35.1). PRPP is the first intermediate of the *de novo* synthesis of purine nucleotides (not shown in full detail in ◘ Fig. 35.1), which leads to the formation of inosine monophosphate (IMP), from which the other purine compounds are derived. PRPP synthetase is highly regulated. Various genetic regulatory and catalytic defects [1, 2] lead to superactivity, resulting in increased generation of PRPP. Because PRPP amidotransferase, the rate-limiting enzyme of the *de novo* pathway, is physiologically not saturated by PRPP, the synthesis of purine nucleotides increases, and hence the production of uric acid. PRPP synthetase superactivity is one of the few known examples of an hereditary anomaly of an enzyme which enhances its activity. The mechanism of the neurological symptoms is unresolved.

Genetics

The various forms of PRPP synthetase superactivity are inherited as X-linked traits. In the families in which the anomaly is associated with sensorineural deafness, heterozygous females have also been found with gout and/or hearing impairment [2]. Studies of the gene in six families revealed a different single base change in each of them [3].

Diagnostic Tests

Diagnosis requires extensive kinetic studies of the enzyme, which are performed on erythrocytes and cultured fibroblasts in a few laboratories in the world. The disorder should be differentiated from partial HGPRT deficiency, which gives similar clinical signs.

Treatment and Prognosis

Patients should be treated with allopurinol, which inhibits xanthine oxidase, the last enzyme of purine catabolism (◘ Fig. 35.1). This results in a decrease of the production of uric acid and in its replacement by hypoxanthine, which is about 10-fold more soluble, and xanthine, which is slightly more soluble than uric acid. Initial dosage of allopurinol is 10–20 mg/kg per day in children and 2–10 mg/kg per day in adults. It should be adjusted to the minimum required to maintain normal uric acid levels in plasma, and reduced in subjects with renal insufficiency. In rare patients with a considerable increase in *de novo* synthesis, xanthine calculi can be formed during allopurinol therapy [4]. Consequently, additional measures to prevent cristallization are recommended. These include a low purine diet (free of organ meats, fishes such as anchovy, herring, mackerel, salmon, sardines and tuna, dried beans and peas), high fluid intake and, since uric acid and xanthine are more soluble at alkaline than at acid pH, administration of sodium bicarbonate, potassium citrate or citrate mixtures to bring urinary pH to 6.0-6.5. Adequate control of the uricemia prevents gouty arthritis and urate nephropathy, but does not correct the neurological symptoms.

◘ **Table 35.1.** Main presenting clinical signs and laboratory data in inborn errors of purine and pyrimidine metabolism

Clinical signs	Diagnostic possibilities	Clinical signs	Diagnostic possibilities
Arthritis	PRPP synthetase superactivity	Muscle cramps	Muscle AMP deaminase deficiency
	HGPRT deficiency (partial)	Muscle wasting	Adenylosuccinase deficiency
Ataxia	PNP deficiency	Psychomotor delay	PRPP synthetase superactivity
	HGPRT deficiency (complete)		Adenylosuccinase deficiency
	Cytosolic 5'-nucleotidase superactivity		AICA-ribosiduria (ATIC deficiency)
Autistic features	PRPP synthetase superactivity		Combined xanthine and sulfite oxidase
	Adenylosuccinase deficiency		deficiency
	Dihydropyrimidine dehydrogenase		HGPRT deficiency (complete)
	deficiency		UMP synthase deficiency
	Cytosolic 5'-nucleotidase superactivity		Dihydropyrimidine dehydrogenase
Congenital blindness	AICA-ribosiduria (ATIC deficiency)		deficiency
Convulsions	Adenylosuccinase deficiency		Dihydropyrimidinase deficiency
	Combined xanthine and sulfite		Ureidopropionase deficiency
	oxidase deficiency		Cytosolic 5'-nucleotidase superactivity
	Dihydropyrimidine dehydrogenase	Recurrent infections	ADA deficiency
	deficiency		PNP deficiency
	Dihydropyrimidinase deficiency		Cytosolic 5'-nucleotidase superactivity
	Cytosolic 5'-nucleotidase superactivity	Renal insufficiency	PRPP synthetase superactivity
Deafness	PRPP synthetase superactivity		HGPRT deficiency (complete or partial)
Dysmorphic features	AICA-ribosiduria (ATIC deficiency)		APRT deficiency
Growth retardation	Adenylosuccinase deficiency	Self-mutilation	HGPRT deficiency (complete)
	ADA deficiency		
	UMP synthase deficiency	**Laboratory data**	**Diagnostic possibilities**
	Dihydropyrimidine dehydrogenase	Anemia	
	deficiency	Megaloblastic	UMP synthase deficiency
	Cytosolic 5'-nucleotidase superactivity	Hemolytic	ADA superactivity
Hypotonia	Adenylosuccinase deficiency		Pyrimidine 5'-nucleotidase deficiency
	Muscle AMP deaminase deficiency	Hyperuricemia	PRPP synthetase superactivity
	Ureidopropionase deficiency		HGPRT deficiency (complete or partial)
Kidney stones:		Hypouricemia	PNP deficiency
Uric acid	PRPP synthetase superactivity		Xanthine oxidase deficiency (isolated or
	HGPRT deficiency (complete or partial)		combined with sulfite oxidase deficiency)
Xanthine	Xanthine oxidase deficiency (isolated	Lymphopenia	
	or combined with sulfite oxidase	B and T-cells	ADA deficiency
	deficiency)	T-cells	PNP deficiency
2,8-Dihydroxyadenine	APRT deficiency	Orotic aciduria	UMP synthase deficiency
Orotic acid	UMP synthase deficiency		

ADA, adenosine deaminase; *APRT*, adenine phosphoriboysltransferase; *ATIC*, AICAR transformylase/IMP cyclohydrolase; *HGPRT*, hypoxanthine-guanine phosphoribosyltransferase; *PNP*, purine nucleoside phosphorylase; *PRPP*, phosphoribosyl pyrophosphate; *UMP*, uridine monophosphate.

35.1.2   Adenylosuccinase Deficiency

Clinical Picture

In the first reported presentation, often referred to as type I, patients display moderate to severe psychomotor retardation, frequently accompanied by epilepsy after the first years, and by autistic features (failure to make eye-to-eye contact, repetitive behavior, temper tantrums), seldom by severe growth retardation associated with muscular wasting [5, 6]. Rare patients, referred to as type II, are only mildly retarded [6], or display profound muscle hypotonia accompanied by slightly delayed motor development [7]. Other patients have been reported with convulsions starting within the first days to weeks of life [8, 9]. The marked clinical heterogeneity justifies systematic screening for the defi-

ciency in unexplained, profound as well as mild psychomotor retardation, and in neurological disease with convulsions and/or hypotonia.

Metabolic Derangement

Adenylosuccinase (ADSL, also named adenylosuccinate lyase), catalyzes two steps in purine synthesis (◘ Fig. 35.1): the conversion of succinylamino-imidazole carboxamide ribotide (SAICAR) into AICAR, along the *de novo* pathway, and that of adenylosuccinate (S-AMP) into AMP. Its deficiency results in accumulation in cerebrospinal fluid and urine of the succinylpurines, SAICA riboside and succinyladenosine (S-Ado), the products of the dephosphorylation, by 5'-nucleotidase(s), of the two substrates of the enzyme. Present evidence indicates that the more severe presentations of ADSL deficiency tend to be associated with S-Ado/SAICA riboside ratios around 1, whereas in milder clinical pictures these ratios are comprised between 2 and 4. This suggests that SAICA riboside is the offending compound, and that S-Ado could protect against its toxic effects. The ADSL defect is marked in liver and kidney, and variably expressed in erythrocytes, muscle, and fibroblasts [5, 6, 9]. The higher S-Ado/SAICA riboside ratios might be explained by a more profound loss of activity of the enzyme toward S-AMP than toward SAICAR, as compared with a parallel deficiency in severely affected patients [9]. The symptoms of the deficiency remain unexplained, but positron emission tomography reveals a marked decrease of the uptake of fluorodeoxyglucose in the cortical brain areas [10].

Genetics

The deficiency is transmitted as an autosomal recessive trait [5, 6]. Studies of the ADSL gene, localized on chromosome 22, have led to the identification of about 40 mutations [11-13] (ADSL mutations database home page, http://www.icp.ucl.ac.be/adsldb/). Most are missense mutations but a splicing error [12] and a mutation in the 5′UTR [14] have also been identified. Most frequently encountered, particularly in The Netherlands, and accounting for about one-third of the alleles investigated, is a R462H mutation. Most other mutations are found in single families, in which most patients are compound heterozygotes.

Diagnostic Tests

Diagnosis is based on the presence in cerebrospinal fluid and urine of SAICA riboside and S-Ado, which are normally undetectable. These can be recognized by various techniques. For systematic screening, a modified Bratton-Marshall test [15], performed on urine, appears most practical. False positive results are, however, recorded in patients who receive sulphonamides, for the measurement of which the test was initially devised. Several thin-layer chromatographic methods are also available [16]. Final diagnosis requires HPLC with UV detection [5]. Prenatal diagnosis of ADSL deficiency can be performed by mutation analysis on chorion villi [17].

Treatment and Prognosis

With the aim to replenish hypothetically decreased concentrations of adenine nucleotides in ADSL-deficient tissues, some patients have been treated for several months with oral supplements of adenine (10 mg/kg per day) and allopurinol (5-10 mg/kg per day). Adenine can be incorporated into the adenine nucleotides via adenine phosphoribosyltransferase (APRT, Fig. 35.1). Allopurinol is required to avoid conversion of adenine by xanthine oxidase, into minimally soluble 2,8-dihydroxyadenine, which forms kidney stones. No clinical or biochemical improvement was recorded, with the exception of weight gain and some acceleration of growth [6]. Oral administration of ribose (10

mmol/kg per day) has been reported to reduce seizure frequency in an ADSL-deficient girl [18]. Uridine (2 mmol/kg per day) also had a slight beneficial effect [19].

The prognosis for survival of ADSL-deficient patients is very variable. Mildly retarded patients have reached adult age, whereas several of those presenting with early epilepsy have died within the first months of life.

35.1.3 AICA-Ribosiduria

In a female infant [20] with profound mental retardation, marked dysmorphic features (prominent forehead and metopic suture, brachycephaly, wide mouth with thin upper lip, low-set ears, and prominent clitoris due to fused labia majora), and congenital blindness, a positive urinary Bratton-Marshall test led to the identification of a massive excretion of 5-amino-4-imidazolecarboxamide (AICA)-riboside, the dephosphorylated counterpart of AICAR (Fig. 35.1). Assay of ATIC, the bifunctional enzyme catalyzing the two last steps of *de novo* purine biosynthesis, revealed a profound deficiency of AICAR transformylase, and a partial deficiency of IMP cyclohydrolase. Sequencing of the ATIC gene showed a K426R change in the transformylase region in one allele, and a frameshift in the other. The discovery of this novel inborn error of purine synthesis reinforces the necessity to perform a Bratton-Marshall test [15] in all cases of unexplained mental retardation and/or neurological symptoms.

35.1.4 Muscle AMP Deaminase Deficiency

Clinical Picture

The deficiency of muscle AMP deaminase (AMP-DA, frequently referred to as *myoadenylate deaminase* in the clinical literature) is present in 1-2% of the Caucasian population. Most deficient individuals are asymptomatic. Nevertheless, some subjects, in whom the AMP-DA defect is termed primary, present with isolated muscular weakness, fatigue, cramps or myalgias following moderate to vigorous exercise, sometimes accompanied by an increase in serum creatine kinase and minor electromyographic abnormalities [21]. Muscular wasting or histological abnormalities are absent. Primary AMP-DA deficiency was initially detected in young adults, but later on wide variability was observed with respect to the age (1.5-70 years) of onset of the symptoms [22, 23]. Moreover, the enzyme defect has been detected in patients with hypotonia and/or cardiomyopathy, and in asymptomatic family members of subjects with the disorder. Secondary AMP-DA deficiency is found in association with several neuromuscular disorders amongst which amyotrophic lateral sclerosis, fascioscapulohumeral myopathy, Kugelberg-Welander syndrome, polyneuropathies, and Werdnig-Hoffmann disease [22, 23].

Metabolic Derangement

AMP-DA, adenylosuccinate synthetase and adenylosuccinase form the purine nucleotide cycle (◘ Fig. 35.2). Numerous functions have been proposed for this cycle in muscle (reviewed in [24]): (a) removal of AMP formed during exercise, in order to favor the formation of ATP from ADP by myokinase (adenylate kinase); (b) release of NH_3 and IMP, both stimulators of glycolysis and hence of energy production; (c) production of fumarate, an intermediate of the citric acid cycle, which also yields energy. It has therefore been proposed that the muscle dysfunction observed in primary AMP-DA deficiency is caused by impairment of energy production for muscle contraction. However, this does not tally with the vast number of asymptomatic AMP-DA-deficient individuals, and suggests that the deficiency might have a synergistic effect in association with other hitherto unidentified disorder(s).

It should be noted that muscle, liver and erythrocytes contain different isoforms of AMP-DA. A regulatory mutation of liver AMP-DA has been proposed as a cause of primary gout with overproduction of uric acid [25]. Individuals with a complete, although totally asymptomatic deficiency of erythrocyte AMP-DA have been detected in Japan, Korea and Taiwan [26].

Genetics

Primary AMP-DA deficiency is apparently transmitted as an autosomal recessive trait. *AMPD1*, the gene encoding muscle AMP-DA, is located on chromosome 1. In most individuals with the primary deficiency the defect is caused by a nonsense c.34C→T mutation resulting in a stop codon [27]. Population studies show that this mutant allele is found with a high frequency in Caucasians. This accords with the finding that about 2% of diagnostic muscle biopsies are AMP-DA deficient, and suggests that the mutation arose in a remote Western European ancestor. More recently, other more rare mutations of the *AMPD1* gene have been identified in AMP-DA deficient individuals. Interestingly, mutations of the *AMPD1* gene seem associated with improved outcome in heart diseases [28].

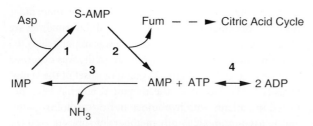

◘ **Fig. 35.2.** The purine nucleotide cycle. *IMP*, inosine monophosphate; *S-AMP*, adenylosuccinate; *AMP*, adenosine monophosphate; *ADP*, adenosine diphosphate; *ATP*, adenosine triphosphate; *Asp*, aspartate; *Fum*, fumarate. **1**, Adenylosuccinate synthetase; **2**, adenylosuccinase; **3**, AMP deaminase; **4**, also shown is myokinase (adenylate kinase)

Diagnostic Tests

Screening for the defect can be performed by an exercise test (▶ Chap. 3). A several-fold elevation of venous plasma ammonia, seen in normal subjects, is absent in AMP-DA deficiency. Final diagnosis is established by histochemical or biochemical assay in a muscle biopsy. In the primary defect, the activity of AMP-DA is below 2% of normal, and little or no immunoprecipitable enzyme is found. In the secondary defect, the activity is 2–15% of normal, and usually appreciable immunoreactivity is present [29]. In several large series of muscle biopsies for diagnostic purposes, low enzyme activities were found in about 2% of all specimens [22, 23].

Treatment and Prognosis

Patients may display a gradual progression of their symptoms, which may lead to the point that even dressing and walking a few steps lead to fatigue and myalgias. They should be advised to exercise with caution to prevent rhabdomyolysis and myoglobinuria. Administration of ribose (2–60 g per day orally in divided doses) has been reported to improve muscular strength and endurance [30].

35.1.5   Adenosine Deaminase Deficiency

Clinical Picture

The majority of patients display, within the first weeks or months after birth, a profound impairment of both humoral and cellular immunity, known as *severe combined immunodeficiency disease* (SCID). Multiple, recurrent infections rapidly become life-threatening [31, 32]. Cases with delayed infantile onset, later childhood onset, and even adult onset have, nevertheless, been reported. Caused by a broad variety of organisms, infections are mainly localized in the skin, the respiratory and the gastrointestinal tract. In the latter they often lead to intractable diarrhea, malnutrition and growth retardation. In affected children over 6 months of age, hypoplasia or apparent absence of lymphoid tissue is a suggestive sign. Bone abnormalities, clinically evident as prominence of the costochondral rib junctions, and radiologically as cupping and flaring thereof, are found in about half of the patients. In a few affected children neurological abnormalities are found, including spasticity, head lag, movement disorders, nystagmus and inability to focus. Hepatic dysfunction has also been reported [32, 33].

SCID can be confirmed by relatively simple laboratory tests: lymphopenia (usually less than 500 total lymphocytes per mm^3) involving both B and T cells, as well as hypogammaglobulinemia are almost invariably present. Whereas the IgM deficiency may be detected early, the IgG deficiency becomes manifest only after the age of 3 months, when the maternal supply has been exhausted. More elaborate tests show a deficiency of antibody formation following specific immunization and an absence or severe diminution of the lymphocyte proliferation induced by mitogens. The disease is progressive, since residual B- and T-cell function which may be found at birth, disappears later on.

Metabolic Derangement

The deficiency results in the accumulation in body fluids of adenosine, normally nearly undetectable (◘ Fig. 35.1), and deoxyadenosine (not shown in ◘ Fig. 35.1), another substrate of adenosine deaminase (ADA), derived from the catabolism of DNA. Inside lymphocytes, deoxyadenosine excess leads to accumulation of dATP which inhibits ribonucleotide reductase, an essential enzyme for the synthesis of DNA which has to proceed at a high rate during lymphocyte development and differentiation. More recently, dATP has also been reported to provoke thymic T-cell apoptosis [34]. Deoxyadenosine has moreover been shown to inactivate S-adenosylhomocysteine hydrolase [32], an enzyme which intervenes in methyl transfer, but how this affects lymphocyte function remains elusive.

Genetics

Approximately 1/3 of the cases of inherited SCID are X linked, whereas 2/3 are autosomal recessive. ADA deficiency is found only in the latter group, where it accounts for about 50% of the patients. The frequency of the deficiency is estimated at 1 per 100,000-500,000 births. Studies of the ADA gene, located on chromosome 20, have hitherto revealed over 70 mutations, the majority of which are single nucleotide changes, resulting in an either inactive or unstable enzyme [32]. Most patients carry two different mutations on each chromosome 20, but others, mainly from inbred communities, are homozygous for the mutation. Spontaneous in vivo reversion to normal of a mutation on one allele, as observed in tyrosinemia type I (► Chap. 18), has been reported [35].

Diagnostic Tests

The diagnosis is mostly performed on red blood cells. In general, severity of disease correlates with the loss of ADA activity: children with neonatal onset of SCID display 0–1% residual activity; in individuals with later onset, 1–5% of normal ADA activity are found [32]. It should be noted that only about 15% of the patients with the clinical and hematologic picture of inherited SCID are ADA-deficient. In the remaining patients, SCID is caused by other mechanisms. A few subjects have been described with ADA deficiency in red blood cells, but normal immunocompetence [32]. This is explained by the presence of residual ADA activity in their lymphocytes.

Treatment and Prognosis

Untreated, ADA deficiency as a rule invariably led to death, usually within the first year of life, unless drastic steps were taken, such as rearing in strictly sterile conditions from birth on. Treatment became possible with the advent of bone marrow transplantation. This remains the first choice provided an histocompatible donor is available, and gives a good chance for complete cure, both clinically and immunologically [36]. The graft provides stem cells, and hence T and B cells, which have sufficient ADA activity to prevent accumulation of adenosine and deoxyadenosine. Survival is, however, much lower with HLA-mismatched transplants.

If no histocompatible bone marrow donor is found, enzyme replacement therapy can be given. Repeated partial exchange transfusions with normal erythrocytes, irradiated before use to prevent graft-versus-host disease, result in marked clinical and immunological improvement in some patients, but in most response is poor or not sustained [36]. A much more effective enzyme replacement therapy is achieved with polyethylene glycol-modified ADA (PEG-ADA). Covalent attachment of PEG to bovine ADA results in marked extension of its half-life, and reduction of immunogenicity. Weekly to bi-weekly intramuscular injections of 15–30 units of PEG-ADA per kg result in mostly marked clinical improvement. In vitro immune function also significantly improves [37].

The first approved clinical trial of gene therapy was performed in 1990 in two girls with ADA deficiency [38]. Their peripheral blood T cells were collected, cultured with interleukin-2, corrected by insertion of the ADA gene by means of a retroviral vector, and reinfused. Because lymphocytes live only a few months, 11 or 12 infusions were given over two years to each patient. The number of T cells normalized, as did many cellular and humoral immune responses, no adverse events were observed and, remarkably, 10 years after the last cell infusion expression of the retroviral gene was still present [39]. Since as a precaution, patients continued to receive PEG-ADA although at reduced doses, benefits cannot be attributed unequivocally to gene therapy.

More recently, successful correction of ADA deficiency has been accomplished by gene therapy into hematopoietic stem cells which in theory have an unlimited life span, without concomitant PEG-ADA treatment, and with addition of a low-intensity, nonmyeloablative conditioning regimen [40]. It should be mentioned that gene therapy in X-linked, not ADA deficient SCID, although highly effective, as been placed on hold due to the development of leukemia in some patients [41].

35.1.6 Adenosine Deaminase Super-activity

A hereditary, approx. 50-fold elevation of red cell ADA, has been shown to cause non-spherocytic hemolytic anaemia [42]. The latter can be explained by an enhanced catabolism of the adenine nucleotides, including ATP, owing to the increased activity of ADA.

36.1.7 Purine Nucleoside Phosphorylase Deficiency

Clinical Picture

Recurrent infections are usually of later onset, starting from the end of the first year to up to 5-6 years of age, and are initially less severe than in ADA deficiency [43, 44]. A strikingly enhanced susceptibility to viral diseases, such as varicella, measles, cytomegalovirus and vaccinia has been reported, but severe candida and pyogenic infections also occur. One third of the patients have anemia, and two thirds display neurologic symptoms, including spastic tetra- or diplegia, ataxia and tremor. Immunological studies reveal an increasing deficiency of cellular immunity, reflected by a marked reduction in the number of T-cells. B-lymphocyte function is deficient in about one third of the patients.

Metabolic Derangement

The deficiency provokes an accumulation in body fluids of the 4 substrates of the enzyme which are normally nearly undetectable, namely guanosine, inosine (◘ Fig. 35.1), and their deoxycounterparts (not shown in ◘ Fig. 35.1), the latter derived from DNA breakdown. Formation of uric acid is thus severely hampered. The profound impairment of cellular immunity, characterizing the disorder, has been explained by an accumulation, particularly in T-cells, of excess dGTP. It is formed from deoxyguanosine, inhibits ribonucleotide reductase, and hence cell division.

Genetics

The deficiency is inherited in an autosomal recessive fashion. Studies of the PNP gene, located on chromosome 14, have revealed a number of molecular defects, among which a R234P mutation was most common [45].

Diagnostic Tests

Patients often display a striking decrease of the production of uric acid: plasma uric acid is usually below 1 mg/dl and may even be undetectable. However, in patients with residual PNP activity, uricemia may be at the borderline of normal. The urinary excretion of uric acid is usually also markedly diminished. Other causes of hypouricemia such as xanthine oxidase deficiency (▶ below), and drug administration (acetylsalicylic acid, thiazide diuretics), should be ruled out. Enzymatic diagnosis of PNP deficiency is usually performed on red blood cells.

Treatment and Prognosis

Until recently, most patients have died from overwhelming viral or bacterial infections, although at a later age than untreated ADA-deficient children. Treatments consisted of bone marrow transplantation and repeated transfusions of normal, irradiated erythrocytes [36, 44]. More recently, successful matched bone marrow transplantation has been reported [46]. Enzyme and gene therapy might become available in the near future.

35.1.8 Xanthine Oxidase Deficiency

Clinical Picture

Two deficiencies of xanthine oxidase (or dehydrogenase) are known: an isolated form [47], also termed hereditary *xanthinuria*, and a combined xanthine oxidase and *sulfite oxidase* deficiency [48]. Isolated xanthine oxidase deficiency can be completely asymptomatic, although in about one third of the cases kidney stones are formed. Most often not visible on X-ray, they may appear at any age. Myopathy may be present, associated with crystalline xanthine deposits. In the combined deficiency, the clinical picture of sulfite oxidase deficiency (which is also found as an isolated defect [49], ▶ Chap. 21) dominates that of the xanthine oxidase deficiency. The symptoms include neonatal feeding difficulties and intractable seizures, myoclonus, increased or decreased muscle tone, eye lens dislocation and severe mental retardation.

Metabolic Derangement

The deficiency results in the near total replacement of uric acid by hypoxanthine and xanthine as the end products of purine catabolism (◘ Fig. 35.1). Hereditary xanthinuria can result from a deficiency of xanthine oxidase (type I) or of both xanthine oxidase and aldehyde oxidase (type II). The latter is a closely related enzyme that metabolizes synthetic purine analogues such as allopurinol. In combined xanthine oxidase and sulfite oxidase deficiency there is in addition an accumulation of sulfite and of sulfur-containing metabolites, and a diminution of the production of inorganic sulfate. The combined defect is caused by the deficiency of a *molybdenum cofactor*, which is required for the activity of both xanthine oxidase and sulfite oxidase.

Genetics

The inheritance of both isolated xanthine oxidase deficiency and combined xanthine oxidase and sulfite oxidase deficiency is autosomal recessive. Studies of the xanthine oxidase gene, localized on chromosome 2, have led to the identification in hereditary xanthinuria type I of two mutations, resulting in a nonsense substitution and a termination codon, respectively [50]. Xanthinuria type II might be caused by mutation of a molybdenum cofactor sulferase gene [51]. More than 30 different mutations in three molybdenum cofactor biosynthetic genes have been identified in combined xanthine oxidase and sulfite oxidase deficiency [52].

Diagnostic Tests

Both in isolated and combined xanthine oxidase deficiency, plasma concentrations of uric acid below 1 mg/dl (0.06

mmol/L) are measured; they may decrease to virtually undetectable values on a low-purine diet. Urinary uric acid is reduced to a few percent of normal and replaced by hypoxanthine and xanthine. In the combined defect, these urinary changes are accompanied by an excessive excretion of sulfite and other sulfur-containing metabolites, such as S-sulfocysteine, thiosulfate and taurine. The enzymatic diagnosis requires liver or intestinal mucosa, the only human tissues which normally contain appreciable amounts of xanthine oxidase. Sulfite oxidase and the molybdenum cofactor can be assayed in liver and fibroblasts.

Treatment and Prognosis

Isolated xanthine oxidase deficiency is mostly benign but in order to prevent renal stones a low purine diet should be prescribed and fluid intake increased. The prognosis of combined xanthine oxidase and sulfite oxidase deficiency is very poor. So far, all therapeutic attempts, including low-sulfur diets, the administration of sulfate and molybdenum [48], and trials to bind sulfite with thiol-containing drugs, have been unsuccessful.

35.1.9 Hypoxanthine-Guanine Phosphoribosyltransferase Deficiency

Clinical Picture

The disorder can present under two forms. Patients with complete or near-complete deficiency of hypoxanthine-guanine phosphoribosyltransferase (HGPRT) display the Lesch-Nyhan syndrome [53]. Affected children generally appear normal during the first months of life. At 3 to 4 months of age, a neurological syndrome evolves, which includes delayed motor development, choreo-athetoid movements, and spasticity with hyperreflexia and scissoring. Over the years, the patients develop a striking, compulsive self-destructive behavior, involving biting of their fingers and lips, which leads to mutilating loss of tissue. Speech is hampered by athetoid dysarthria. Whereas most patients have IQ's around 50, some display normal intelligence. Approximately 50% of the patients have seizures. Soon or later they form uric acid stones. Mothers of Lesch-Nyhan patients have reported the finding of orange crystals on their affected son's diapers during the first few weeks after birth. Untreated, the uric acid nephrolithiasis progresses to obstructive uropathy and renal failure during the first decade of life. The latter clinical picture may, exceptionally, also be observed in early infancy.

Partial HGPRT deficiency is found in rare patients with gout. Most of them are normal on neurological examination, but occasionally spasticity, dysarthria and a spinocerebellar syndrome are found [54]. Whereas most patients with the Lesch-Nyhan syndrome do not develop gouty arthritis, this finding is common in partial HGPRT deficiency.

Metabolic Derangement

The considerable increase of the production of uric acid is explained as follows: PRPP, which is not utilized at the level of HGPRT (◘ Fig. 35.1), is available in increased quantities for the rate limiting, first enzyme of the *de novo* synthesis, PRPP amidotransferase (not shown in ◘ Fig. 35.1). Since the latter is normally not saturated with PRPP, its activity increases and the ensuing acceleration of the *de novo* synthesis results in the overproduction of uric acid.

The pathogenesis of the neurological symptoms is still not satisfactorily explained. A number of studies point to dopaminergic dysfunction, involving decreases of the concentration of dopamine and of the activity of the enzymes required for its synthesis, although dopaminergic drugs are not useful. Positron emission tomography of the brain with F-18 fluorodopa, an analogue of the dopamine precursor levodopa, has revealed a generalized decrease of the activity of dopa decarboxylase [55]. How the HGPRT defect leads to the deficit of the dopaminergic system, and how the latter results in the characteristic neuropsychiatric manifestations of the Lesch-Nyhan syndrome, remains to be clarified.

Genetics

Both the Lesch-Nyhan syndrome and the partial deficiencies of HGPRT are transmitted in a X-linked recessive manner. Studies of the HGPRT gene in large groups of unrelated patients have revealed a variety of defects, ranging from point mutations provoking single amino acid substitutions and henceforth enzymes with altered stability and/or kinetic properties, to extensive deletions resulting in suppression of enzyme synthesis [56]. These studies have contributed a great deal to the understanding of the clinical variation observed in human inherited disease, and provided support for the concept that, in X-linked disorders, new mutations constantly appear in the population. Presently, over 250 mutations of the HGPRT gene have been described, and molecular studies have led to precise prenatal diagnosis and efficient carrier testing of at-risk females [57].

Diagnostic Tests

Patients excrete excessive amounts of uric acid, ranging from 25 to 140 mg (0.15 to 0.85 mmol)/kg of body weight per 24 h, as compared to an upper limit of 18 mg (0.1 mmol)/kg per 24 h in normal children. Determination of the ratio of uric acid to creatinine (mg/mg) in morning samples of urine provides a screening test. This ratio is much higher in HGPRT deficiency than the normal upper limits of 2.5, 2.0, 1.0 and 0.6 for infants, 2 years, 10 years and adults, respectively [58]. Increased ratios are also found in other disorders with uric acid overproduction, such as PRPP synthetase superactivity, glycogenosis type I, lymphoproliferative diseases, and after fructose loading. The overproduction of uric acid is as a rule accompanied by an increase of serum urate, which may reach concentrations as high as 18 mg/dl (1 mmol/L). Occasionally, however, particularly before puberty, uricemia may be in the normal or high normal range.

Patients with the Lesch-Nyhan syndrome display nearly undetectable HGPRT activity in red blood cells [59]. In partial deficiencies, similar low or higher values may be found [60]. Rates of incorporation of hypoxanthine into the adenine nucleotides of intact fibroblasts correlate better with the clinical symptomatology than HGPRT activities in erythrocytes: patients with the complete Lesch-Nyhan syndrome incorporated less than 1.2% of normal, those with gout and neurological symptoms 1.2–10% of normal, and those with isolated gout, 10–55% of normal [60].

Treatment and Prognosis

Allopurinol, as detailed under *PRPP synthetase superactivity*, is indicated to prevent urate nephropathy. Allopurinol, even when given from birth, has, however, no effect on the neurological symptoms, which have sofar been resistant to all therapeutic attempts. Adenine has been administered, together with allopurinol, with the aim to correct a possible depletion of purine nucleotides. However, no or minimal changes in neurological behavior were recorded [61]. Patients should be made more comfortable by appropriate restraints, including elbow splints, lip guards and even tooth extraction, to diminish self-mutilation. Diazepam, haloperidol and barbiturates may sometimes improve choreoathetosis.

In a 22-year-old patient, bone marrow transplantation restored erythrocyte HGPRT activity to normal, but did not change neurological symptoms [62]. Recently, disappearance of self-mutilation was obtained by chronic stimulation of the globus pallidus [63].

35.1.10 Adenine Phosphoribosyltransferase Deficiency

Clinical Picture

The deficiency may become clinically manifest in childhood [64], even from birth [65], but also remain silent for several decades. Symptoms include urinary passage of gravel, small stones and crystals, frequently accompanied by abdominal colic, dysuria, hematuria and urinary tract infection. Some patients may even present with acute anuric renal failure [66]. The urinary precipitates are composed of 2,8-dihydroxyadenine, radiotranslucent, and undistinguishable from uric acid stones by routine chemical testing.

Metabolic Derangement

The deficiency results in suppression of the salvage of adenine (◘ Fig. 35.1), provided by food and by the polyamine pathway (not shown in ◘ Fig. 35.1). Consequently, adenine is oxidized by xanthine oxidase into 2,8-dihydroxyadenine, a very poorly soluble compound (solubility in urine, at pH 5 and 37°C, is about 0.3 mg/dl as compared to 15 mg/dl for uric acid).

The deficiency can be complete or partial. The partial deficiency is only found in the Japanese, among whom it is quite common [67]. Activities range from 10 to 30% of normal at supraphysiological concentrations of PRPP, but a 20- to 30-fold decrease in the affinity for PRPP results in near inactivity under physiological conditions.

Genetics

APRT deficiency is inherited as an autosomal recessive trait. All the type II Japanese patients carry the same c.2069T →C substitution in exon 5, resulting in a M136T change [67]. Approximately 80% are homogenous, with two other mutations accounting for nearly all the other cases. In Caucasians, approximately 30 mutations have been identified, some of which seem more common, also suggesting founder effects [68].

Diagnostic Tests

Identification of 2,8-dihydroxyadenine requires complex analyses, including UV and infrared spectrography, mass spectrometry and X-ray cristallography [64, 65]. It is therefore usually easier to measure APRT activity in red blood cells.

Treatment and Prognosis

In patients with symptoms, allopurinol should be given, as detailed under *PRPP synthetase superactivity*, to inhibit the formation of 2,8-dihydroxyadenine. Both in patients with stones and in those without symptoms, dietary purine restriction and high fluid intake are recommended. Alkalinization of the urine is, however, not advised: unlike that of uric acid, the solubility of 2,8-dihydroxyadenine does not increase up to pH 9 [64].

Ultimate prognosis depends on renal function at the time of diagnosis: late recognition may result in irreversible renal insufficiency requiring chronic dialysis, and early treatment in prevention of stones. Of note is that kidney transplantation has been reported to be followed by recurrence of microcrystalline deposits and subsequent loss of graft function [69].

35.1.11 Deoxyguanosine Kinase Deficiency

In several patients with the hepatocerebral form of mitochondrial DNA depletion syndrome (► also Chap. 15), characterised by early progressive liver failure, neurological abnormalities, hypoglycemia, and increased lactate, a deficiency of mitochondrial deoxyguanosine kinase was identified [70]. This enzyme phosphorylates the deoxycounterpart of guanosine (◼ Fig. 35.1) into deoxyGMP, and plays an essential role in the supply of precursors of mitochondrial DNA, particularly in liver and brain that lack a cytosolic form of the enzyme. A single nucleotide deletion in the mitochondrial deoxyguanosine kinase gene segregated with the disease in 19 patients in 3 kindreds [70]. Since then, othere mutations have been identified.

Pyrimidine Metabolism

Similarly to that of the purine nucleotides, the metabolism of the pyrimidine nucleotides can be divided into three pathways:

— The biosynthetic pathway starts with the formation of carbamoylphosphate by cytosolic carbamoylphosphate synthetase (CPS II), which is different from the mitochondrial CPS I which catalyzes the first step of ureogenesis (□ Fig. 20.1). This is followed by the synthesis of UMP, and hence of CMP and TMP.

— The catabolic pathway starts from CMP, UMP and TMP, and yields β-alanine and β-aminoisobutyrate which are converted into intermediates of the citric acid cycle.

— The salvage pathway, composed of kinases, converts the pyrimidine nucleosides, cytidine, uridine, and thymidine, into the corresponding nucleotides, CMP, UMP, and TMP.

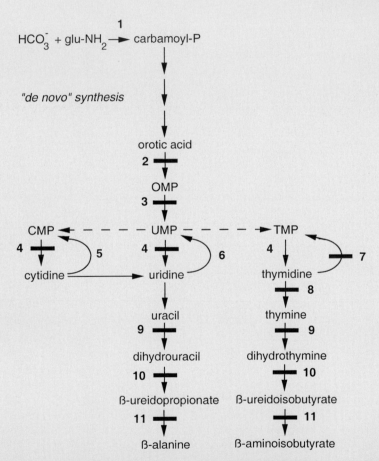

□ **Fig. 35.3.** Pathways of pyrimidine metabolism. *CMP*, cytidine monophosphate; *glu-NH$_2$*, glutamine; *OMP*, orotidine monophosphate; *PRPP*, phosphoribosylpyrophosphate; *TMP*, thymidine monophosphate; *UMP*, uridine monophosphate. **1**, carbamoylphosphate synthetase; **2**, orotate phosphoribosyltransferase; **3**, orotidine decarboxylase (**2** and **3** form UMP synthase); **4**, pyrimidine (cytosolic) 5'-nucleotidase; **5**, cytidine kinase; **6**, uridine kinase; **7**, thymidine kinase; **8**, thymidine phosphorylase; **9**, dihydropyrimidine dehydrogenase; **10**, dihydropyrimidinase; **11**, ureidopropionase. Enzyme deficiencies are indicated by *solid bars* across the arrows

35.2 Inborn Errors of Pyrimidine Metabolism

Inborn errors of pyrimidine metabolism comprise a defect of the *synthesis* of pyrimidine nucleotides (UMP synthase deficiency), and three inborn errors of pyrimidine *catabolism*: the deficiencies of dihydropyrimidine dehydrogenase (DPD) dihydropyrimidinase (DHP), and pyrimidine 5'-nucleotidase. More recently, superactivity of cytosolic 5'-nucleotidase, a fourth defect of pyrimidine catabolism, ureidopropionase deficiency, and deficiencies of thymidine phosphorylase and thymidine kinase, which cause mitochondrial diseases (▶ also Chap. 15), have been reported.

35.2.1 UMP Synthase Deficiency (Hereditary Orotic Aciduria)

Clinical Presentation

Megaloblastic anemia, which appears a few weeks or months after birth, is usually the first manifestation [71, 72]. Peripheral blood smears often show anisocytosis, poikilocytosis, and moderate hypochromia. Bone marrow examination reveals erythroid hyperplasia and numerous megaloblastic erythroid precursors. Characteristically, the anemia does not respond to iron, folic acid or vitamin B_{12}. Unrecognized, the disorder leads to failure to thrive and to retardation of growth and psychomotor development.

Metabolic Derangement

Uridine monophosphate (UMP) synthase is a bifunctional enzyme of the *de novo* synthesis of pyrimidines (◻ Fig. 35.3). A first reaction, orotate phosphoribosyltransferase (OPRT), converts orotic acid into OMP, and a second, orotidine decarboxylase (ODC), decarboxylates OMP into UMP. The defect provokes a massive overproduction of orotic acid and a deficiency of pyrimidine nucleotides [72]. The overproduction is attributed to the ensuing decrease of the feedback inhibition exerted by the pyrimidine nucleotides on the first enzyme of their *de novo* synthesis, cytosolic carbamoyl phosphate synthetase II (◻ Fig. 35.3). The deficiency of pyrimidine nucleotides leads to impairment of cell division, which results in megaloblastic anemia and in retardation of growth and development.

Genetics

Hereditary orotic aciduria is inherited as an autosomal recessive trait. The genetic lesion results in synthesis of an enzyme with reduced stability [73]. Three point mutations have been identified in two Japanese families [74].

Diagnostic Tests

Urinary analysis reveals a massive over excretion of orotic acid, reaching, in infants, 200- to 1000-fold the normal adult value of 1–1.5 mg per 24 h. Occasionally, orotic acid crystalluria is noted, particularly upon dehydration. Enzymatic diagnosis can be performed on red blood cells. In all patients reported hitherto, except one, both OPRT and ODC activities were deficient. This defect is termed type I. In a single patient, referred to as type II, only the activity of ODC was initially deficient, although that of OPRT also subsequently decreased [72].

Treatment and Prognosis

The enzyme defect can be by-passed by the administration of uridine, which is converted into UMP by uridine kinase (◻ Fig. 35.3). An initial dose of 100-150 mg/kg, divided over the day, induces prompt hematologic response and acceleration of growth. Further dosage should be adapted to obtain the lowest possible output of orotic acid. In some cases normal psychomotor development was achieved, but not in others, possibly owing to delayed onset of therapy.

35.2.2 Dihydropyrimidine Dehydrogenase Deficiency

Clinical Picture

Two forms occur. The first is found in children, most of whom display epilepsy, motor and mental retardation, often accompanied by generalized hypertonia, hyperreflexia, growth delay, dysmorphic features including microcephaly, and autistic features [75, 76]. In these patients, the deficiency of dihydropyrimidine dehydrogenase (DPD) is complete or near-complete. Nevertheless, the severity of the disorder is highly variable and even asymptomatic cases have been identified. The second clinical picture is found in adults who receive the pyrimidine analog, 5-fluorouracil, a classic treatment of various cancers including breast, ovary or colon [77, 78]. It is characterised by severe toxicity, manifested by profound neutropenia, stomatitis, diarrhea and neurologic symptoms, including ataxia, paralysis and stupor. In these patients, DPD deficiency is as a rule partial, and only revealed by 5-fluorouracil therapy.

Metabolic Derangement

The deficiency of DPD, which catalyzes the catabolism of uracil and thymine into dihydrouracil and dihydrothymine, respectively (◻ Fig. 35.3), leads to the accumulation of the former compounds [75]. Why a profound DPD deficiency becomes manifest in some pediatric patients, but not in others, is not known. How the defect leads to neurological symptoms also remains elusive, but reduction of the concentration of β-alanine, a neurotransmitter, may play a role. The marked potentiation of the action of the anticancer drug 5-fluorouracil, and henceforth of its toxicity, is explained by a block of the catabolism, via DPD, of this pyrimidine analog.

Genetics

The infantile form of DPD deficiency is inherited as an autosomal recessive trait. The DPD gene is localized on chromosome 1, and about 40 mutations have been identified. Most frequent is a splice site mutation (IVS14+1G>A), which results in skipping of a complete exon [76, 78, 79]. Strikingly, patients who carry the same mutation may display widely variable clinical symptoms. In the adult form of DPD deficiency, characterized by 5′-fluorouracil toxicity, approximately 25% of patients are heterozygotes for the IVS14+1G>A mutation [78].

Diagnostic Tests

Patients excrete high amounts of uracil (56–683 mmol/mol creatinine, as compared to 3–33 in control urine) and of thymine (7–439 mmol/mol creatinine, as compared to 0–4 in control urine). Elevations of uracil and thymine in plasma and cerebrospinal fluid are much less prominent [76]. Excretion of both compounds may also be less elevated in patients with high residual DPD activity. The pyrimidine catabolites can be detected by HPLC, GC-MS, and analysis of amino acids in urine before and after acid hydrolysis [80].

The enzyme defect can be demonstrated in the patients' fibroblasts, liver and blood cells, with the exception of erythrocytes [75, 76, 78]. In the pediatric patients, DPD deficiency is complete or near-complete; in the adult cancer patients experiencing acute 5-fluorouracil toxicity it is partial, with residual enzyme activities ranging from 3 to 30%.

Treatment and Prognosis

No treatment is available for pediatric patients. Symptoms usually remain the same, but death in early infancy of a more severely affected child has been reported. In the adult cancer patients, discontinuation of 5-fluorouracil results in slow resolution of the toxic symptoms [77, 78].

35.2.3   Dihydropyrimidinase Deficiency

Clinical Picture

This disorder was first reported in a single male baby of consanguineous parents, presenting with convulsions and metabolic acidosis [81]. Additional patients have been diagnosed since then [76]. As in DPD deficiency, the clinical picture varies from severe psychomotor retardation with epilepsy, dysmorphic features or microcephaly, to completely asymptomatic.

Metabolic Derangement

Dihydropyrimidinase (DHP) catalyzes the cleavage of dihydrouracil and dihydrothymine into, repectively, β-ureidopropionate and β-ureidoisobutyrate (❑ Fig. 35.3). Consequently, considerable quantities of dihydrouracil and dihydrothymine, which are normally found in small amounts,

are excreted in urine [76]. There is also a moderate elevation of uracil and thymine excretion. As in DPD deficiency, the reasons for the appearance and the mechanisms of the symptoms remain unexplained, and reduced concentrations of the neurotransmitter β-alanine may play a role. Increased sensitivity to 5-fluorouracil, leading to severe toxicity has also been reported [82].

Genetics

The defect is inherited as an autosomal recessive trait. Studies of the DHP gene, localized on chromosome 8, have led to the identification of one frameshift and five missense mutations in one symptomatic and five asymptomatic individuals [83]. Enzyme expression showed no significant difference in residual activity between the mutations of the symptomatic and the asymptomatic individuals.

Diagnostic Tests

Elevation of urinary dihydrouracil and dihydrothymine can be detected by the techniques used for measurement of uracil and thymine in DPD deficiency. Enzyme assay requires liver biopsy, since more accessible tissues do not possess DHP activity [81].

Treatment and Prognosis

There is no therapy and prognosis seems unpredictable. The first reported patient recovered completely and apparently displays normal physical and mental development [81]. In contrast, another patient had a progressive neurodegenerative clinical course [84].

35.2.4   Ureidopropionase Deficiency

In a female infant of consanguineous parents, presenting with muscle hypotonia, dystonic movements and severe developmental delay, in vitro H-NMR spectroscopy of urine revealed elevated ureidopropionic acid (also called N-carbamyl-β-alanine) and ureidoisobutyric acid (also called N-carbamyl-β-aminoisobutyric acid) [85]. These findings led to the identification of ureidopropionase deficiency (also termed β-alanine synthase) in the liver [86].

35.2.5   Pyrimidine 5′-Nucleotidase Deficiency

This defect, restricted to erythrocytes, leads to accumulation of pyrimidine nucleotides resulting in basophilic stippling and chronic hemolytic anemia [87]. The mechanism by which the increased pyrimidine nucleotides cause hemolysis remains unknown.

35.2.6 Cytosolic 5'-Nucleotidase Superactivity

Four unrelated children have been described with a syndrome including developmental delay, growth retardation, seizures, ataxia, recurrent infections, autistic features and hypouricosuria [88]. Studies in the patients' fibroblasts showed 6- to 20-fold elevations of the activity of cytosolic 5'-nucleotidase, measured either with a pyrimidine (UMP) or a purine (AMP) as substrate. Based on the possibility that this increased catabolism might cause a deficiency of pyrimidine nucleotides, the patients were treated with uridine at the dose of 1 g/kg per day. Remarkable developmental improvement, and a decrease in frequency of seizures and infections were recorded.

35.2.7 Thymidine Phosphorylase Deficiency

Patients with mitochondrial neurogastrointestinal encephalomyopathy (MNGIE), an autosomal recessive disease associated with multiple deletions of skeletal muscle mitochondrial DNA (▶ also Chap. 15), have been shown deficient in thymidine phosphorylase, owing to a variety of mutations [89]. The enzyme deficiency results in marked accumulation of thymidine, which most likely provokes imbalance of the mitochondrial nucleotides, and hence compromises the replication of mitochondrial DNA.

35.2.8 Thymidine Kinase Deficiency

In four independent patients with very severe, isolated myopathy, and depletion of muscular mitochondrial DNA (▶ also Chap. 15), two mutations of the gene encoding thymidine kinase-2, the mitochondrial form of the thymidine salvage enzyme, have been identified [90]. As in the deficiencies of deoxyguanosine kinase and thymidine phosphorylase, the defect likely produces imbalance of the mitochondrial nucleotides which disturbs the replication of mitochondrial DNA.

References

1. Sperling O, Boer P, Persky-Brosh S et al (1972) Altered kinetic property of erythrocyte phosphoribosylpyrophosphate synthetase in excessive purine production. Rev Eur Etud Clin Biol 17:703-706

2. Becker MA, Puig JG, Mateos FA et al (1988) Inherited superactivity of phosphoribosylpyrophosphate synthetase: association of uric acid overproduction and sensorineural deafness. Am J Med 85:383-390

3. Becker MA, Smith PR, Taylor W et al (1995) The genetic and functional basis of purine nucleotide feedback-resistant phosphoribosylpyrophosphate synthetase superactivity. J Clin Invest 96:2133-2141

4. Kranen S, Keough D, Gordon RB, Emmerson BT (1985) Xanthine-containing calculi during allopurinol therapy. J Urol 133:658-659

5. Jaeken J, Van den Berghe G (1984) An infantile autistic syndrome characterised by the presence of succinylpurines in body fluids. Lancet 2:1058-1061

6. Jaeken J, Wadman SK, Duran M et al (1988) Adenylosuccinase deficiency: an inborn error of purine nucleotide synthesis. Eur J Pediatr 148:126-131

7. Valik D, Miner PT, Jones JD (1997) First U.S. case of adenylosuccinate lyase deficiency with severe hypotonia. Pediatr Neurol 16:252-255

8. Van den Bergh FAJTM, Bosschaart AN, Hageman G et al (1998) Adenylosuccinase deficiency with neonatal onset severe epileptic seizures and sudden death. Neuropediatrics 29:51-53

9. Van den Berghe G, Vincent MF, Jaeken J (1997) Inborn errors of the purine nucleotide cycle: adenylosuccinase deficiency. J Inherit Metab Dis 20:193-202

10. De Volder AG, Jaeken J, Van den Berghe G et al (1988) Regional brain glucose utilization in adenylosuccinase-deficient patients measured by positron emission tomography. Pediatr Res 24:238-242

11. Stone RL, Aimi J, Barshop BA et al (1992) A mutation in adenylosuccinate lyase associated with mental retardation and autistic features. Nat Genet 1:59-63

12. Marie S, Cuppens H, Heuterspreute M et al (1999) Mutation analysis in adenylosuccinate lyase deficiency. Eight novel mutations in the re-evaluated full ADSL coding sequence. Hum Mutat 13:197-202

13. Kmoch S, Hartmannova H, Stiburkova B et al (2000) Human adenylosuccinate lyase (ADSL), cloning and characterization of full-length cDNA and its isoform, gene structure and molecular basis for ADSL deficiency in six patients. Hum Mol Genet 9:1501-1513

14. Marie S, Race V, Nassogne MC et al (2002) Mutation of a nuclear respiratory factor 2 binding site in the 5'untranslated region of the ADSL gene in three patients with adenylosuccinate lyase deficiency. Am J Hum Genet 71:14-21

15. Laikind PK, Seegmiller JE, Gruber HE (1986) Detection of 5'-phosphoribosyl-4-(N-succinylcarboxamide)-5-aminoimidazole in urine by use of the Bratton-Marshall reaction: identification of patients deficient in adenylosuccinate lyase activity. Anal Biochem 156:81-90

16. Sebesta I, Shobowale M, Krijt J, Simmonds HA (1995) Screening tests for adenylosuccinase deficiency. Screening 4:117-124

17. Marie S, Flipsen JWAM, Duran M et al (2000) Prenatal diagnosis in adenylosuccinate lyase deficiency. Prenat Diagn 20:33-36

18. Salerno C, D'Eufemia P, Finocchiaro R et al (1999) Effect of D-ribose on purine synthesis and neurological symptoms in a patient with adenylosuccinase deficiency. Biochim Biophys Acta 1453:135-140

19. Salerno C, Crifo C, Curatolo P, Ciardo F (2000) Effect of uridine administration to a patient with adenylosuccinate lyase deficiency. Adv Exp Biol Med 486:75-78

20. Marie S, Heron B, Bitoun P et al (2004) AICA-Ribosiduria: a novel, neurologically devastating inborn error of purine biosynthesis caused by mutation of ATIC. Am J Hum Genet 74:1276-1281

21. Fishbein WN, Armbrustmacher VW, Griffin JL (1978) Myoadenylate deaminase deficiency: a new disease of muscle. Science 200:545-548

22. Shumate JB, Katnik R, Ruiz M et al (1979) Myoadenylate deaminase deficiency. Muscle Nerve 2:213-216

23. Mercelis R, Martin JJ, de Barsy T, Van den Berghe G (1987) Myoadenylate deaminase deficiency: absence of correlation with exercise intolerance in 452 muscle biopsies. J Neurol 234:385-389

24. Van den Berghe G, Bontemps F, Vincent MF, Van den Bergh F (1992) The purine nucleotide cycle and its molecular defects. Progr Neurobiol 39:547-561

25. Hers HG, Van den Berghe G (1979) Enzyme defect in primary gout. Lancet 1:585-586

26. Ogasawara N, Goto H, Yamada Y et al (1987) Deficiency of AMP deaminase in erythrocytes. Hum Genet 75:15-18

27. Morisaki T, Gross M, Morisaki H et al (1992) Molecular basis of AMP deaminase deficiency in skeletal muscle. Proc Natl Acad Sci USA 89:6457-6461

28. Loh E, Rebbeck TR, Mahoney PD et al (1999) Common variant in AMPD1 gene predicts improved outcome in patients with heart failure. Circulation 23:1422-1425

29. Sabina RL, Fishbein WN, Pezeshkpour G et al (1992) Molecular analysis of the myoadenylate deaminase deficiencies. Neurology 42:170-179

30. Zöllner N, Reiter S, Gross M et al (1986) Myoadenylate deaminase deficiency: successful symptomatic therapy by high dose oral administration of ribose. Klin Wochenschr 64:1281-1290

31. Giblett ER, Anderson JE, Cohen F et al (1972) Adenosine-deaminase deficiency in two patients with severely impaired cellular immunity. Lancet 2:1067-1069

32. Hershfield MS, Arredondo-Vega FX, Santisteban I (1997) Clinical expression, genetics and therapy of adenosine deaminase (ADA) deficiency. J Inherit Metab Dis 20:179-185

33. Bollinger ME, Arredondo-Vega FX, Santisteban I et al (1996) Brief report: hepatic dysfunction as a complication of adenosine deaminase deficiency. N Engl J Med 334:1367-1371

34. Thompson LF , Vaughn JG, Laurent AB et al (2003) Mechanisms of apoptosis in developing thymocytes as revealed by adenosine deaminase-deficient fetal thymic organ cultures. Biochem Pharmacol 66:1595-1599

35. Hirschhorn R, Yang DR, Puck JM et al (1996). Spontaneous *in vivo* reversion to normal of an inherited mutation in a patient with adenosine deaminase deficiency. Nat Genet 13:290-295

36. Markert ML, Hershfield MS, Schiff RI, Buckley RH (1987) Adenosine deaminase and purine nucleoside phosphorylase deficiencies: evaluation of therapeutic interventions in eight patients. J Clin Immunol 7:389-399

37. Hershfield MS (1995) PEG-ADA replacement therapy for adenosine deaminase deficiency: an update after 8.5 years. Clin Immunol Immunopathol 76:S228-S232

38. Blaese RM, Culver KW, Miller AD et al (1995) T-lymphocyte-directed gene therapy for ADA-SCID: initial trial results after 4 years. Science 270:475-480

39. Muul LM, Tuschong LM, Soenen SL et al (2003) Persistence and expression of the adenosine deaminase gene for 12 years and immune reaction to gene transfer components: long-term results of the first clinical gene therapy trial. Blood 101:2563-2569

40. Aiuti A, Slavin S, Aker M et al (2002) Correction of ADA-SCID by stem cell gene therapy combined with nonmyeloablative conditioning Science 296:2410-2413

41. Cavazzana-Calvo M, Lagresle C, Hacein-Bey-Abina S, Fisher A (2005) Gene therapy for severe combined immunodeficiency. Annu Rev Med 56:585-602

42. Valentine WN, Paglia DE, Tartaglia AP, Gilsanz F (1977) Hereditary hemolytic anemia with increased red cell adenosine deaminase (45- to 70-fold) and decreased adenosine triphosphate. Science 195:783-785

43. Giblett ER, Ammann AJ, Wara DW et al (1975) Nucleoside phosphorylase deficiency in a child with severely defective T-cell immunity and normal B-cell immunity. Lancet 1:1010-1013

44. Markert ML (1991) Purine nucleoside phosphorylase deficiency. Immunodefic Rev 3:45-81

45. Markert ML, Finkel BD, McLaughlin TM et al (1997) Mutations in purine nucleoside phosphorylase deficiency. Hum Mutat 9:118-121

46. Carpenter PA, Ziegler JB, Vowels MR (1996) Late diagnosis and correction of purine nucleoside phosphorylase deficiency with allogeneic bone marrow transplantation. Bone Marrow Transplant 17:121-124

47. Dent CE, Philpot GR (1954) Xanthinuria, an inborn error (or deviation) of metabolism. Lancet 1:182-185

48. Wadman SK, Duran M, Beemer FA et al (1983) Absence of hepatic molybdenum cofactor : an inborn error of metabolism leading to a combined deficiency of sulphite oxidase and xanthine dehydrogenase. J Inherit Metab Dis 6[Suppl 1]:78-83

49. Shih VE, Abroms IF, Johnson JL et al (1977) Sulfite oxidase deficiency. Biochemical and clinical investigations of a hereditary metabolic disorder in sulfur metabolism. N Engl J Med 297:1022-1028

50. Ichida K, Amaya Y, Kamatani N et al (1997) Identification of two mutations in human xanthine dehydrogenase gene responsible for classical type I xanthinuria. J Clin Invest 99:2391-2397

51. Yamamoto T, Moriwaki Y, Takahashi S et al (2003) Identification of a new point mutation in the human molybdenum cofactor sulferase gene that is responsible for xanthinuria type II. Metabolism 52:1501-1504

52. Reiss J, Johnson JL (2003) Mutations in the molybdenum cofactor biosynthetic genes MOCS1, MOCS2, and GEPH. Hum Mutat 21:569-576

53. Lesch M, Nyhan WL (1964) A familial disorder of uric acid metabolism and central nervous system dysfuntion. Am J Med 36:561-570

54. Kelley WN, Greene ML, Rosenbloom FM et al (1969) Hypoxanthine-guanine phosphoribosyltransferase deficiency in gout. Ann Intern Med 70:155-206

55. Ernst M, Zametkin AJ, Matochik JA et al (1996) Presynaptic dopaminergic deficits in Lesch-Nyhan disease. N Engl J Med 334:1568-1572

56. Jinnah HA, De Gregorio L, Harris JC et al (2000) The spectrum of inherited mutations causing HPRT deficiency: 75 new cases and a review of 196 previously reported cases. Mutat Res 463:309-326

57. Alford RL, Redman JB, O'Brien WE, Caskey CT (1995) Lesch-Nyhan syndrome: carrier and prenatal diagnosis. Prenat Diagn 15:329-338

58. Kaufman JM, Greene ML, Seegmiller JE (1968) Urine uric acid to creatinine ratio - a screening test for inherited disorders of purine metabolism. Phosphoribosyltransferase (PRT) deficiency in X-linked cerebral palsy and in a variant of gout. J Pediatr 73:583-592

59. Seegmiller JE, Rosenbloom FM, Kelley WN (1967) Enzyme defect associated with a sex-linked human neurological disorder and excessive purine synthesis. Science 155:1682-1684

60. Page T, Bakay B, Nissinen E, Nyhan WL (1981) Hypoxanthine-guanine phosphoribosyltransferase variants: correlation of clinical phenotype with enzyme activity. J Inherit Metab Dis 4:203-206

61. Watts RWE, McKeran RO, Brown E et al (1974) Clinical and biochemical studies on treatment of Lesch-Nyhan syndrome. Arch Dis Child 49:693-702

62. Nyhan WL, Parkman R, Page T et al (1986) Bone marrow transplantation in Lesch-Nyhan disease. Adv Exp Med Biol 195A:167-170

63. Taira T, Kobayashi T, Hori T (2003) Disappearance of self-mutilating behavior in a patient with Lesch-Nyhan syndrome after bilateral chronic stimulation of the globus pallidus internus. Case report. J Neurosurg 98:414-416

64. Cartier P, Hamet M (1974) Une nouvelle maladie métabolique: le déficit complet en adénine-phosphoribosyltransférase avec lithiase de 2,8-dihydroxyadénine. C R Acad Sci Paris 279[série D]:883-886

65. Van Acker KJ, Simmonds HA, Potter C, Cameron JS (1977) Complete deficiency of adenine phosphoribosyltransferase. Report of a family. N Engl J Med 297:127-132

66. Greenwood MC, Dillon MJ, Simmonds HA et al (1982) Renal failure due to 2,8-dihydroxyadenine urolithiasis. Eur J Pediatr 138:346-349

67. Hidaka Y, Tarlé SA, Fujimori S et al (1988) Human adenine phosphoribosyltransferase deficiency. Demonstration of a single mutant allele common to the Japanese. J Clin Invest 81:945-950

68. Sahota A, Chen J, Stambrook PJ, Tischfield JA (1991) Mutational basis of adenine phosphoribosyltransferase deficiency. Adv Exp Med Biol 309B:73-76

69. Eller P, Rosenkranz AR, Mark W et al (2004) Four consecutive renal transplantations in a patient with adenine phosphoribosyltransferase deficiency. Clin Nephrol 61:217-221

70. Mandel H, Szargel R, Labay V et al (2001) The deoxyguanosine kinase gene is mutated in individuals with depleted hepatocerebral mitochondrial DNA. Nat Genet 29:337-341

71. Huguley CM, Bain JA, Rivers SL, Scoggins RB (1959) Refractory megaloblastic anemia associated with excretion of orotic acid. Blood 14:615-634

72. Smith LH (1973) Pyrimidine metabolism in man. N Engl J Med 288:764-771

73. Perry ME, Jones ME (1989) Orotic aciduria fibroblasts express a labile form of UMP synthase. J Biol Chem 264:15522-15528

74. Suchi M, Mizuno H, Kawai Y et al (1997) Molecular cloning of the human UMP synthase gene and characterization of point mutations in two hereditary orotic aciduria families. Am J Hum Genet 60:525-539

75. Berger R, Stoker-de Vries SA, Wadman SK et al (1984) Dihydropyrimidine dehydrogenase deficiency leading to thymine-uraciluria. An inborn error of pyrimidine metabolism. Clin Chim Acta 141:227-234

76. Van Gennip AH, Abeling NGGM, Vreken P, van Kuilenburg ABP (1997) Inborn errors of pyrimidine degradation: clinical, biochemical and molecular aspects. J Inherit Metab Dis 20:203-213

77. Tuchman M, Stoeckeler JS, Kiang DT et al (1985) Familial pyrimidinemia and pyrimidinuria associated with severe fluorouracil toxicity. N Engl J Med 313:245-249

78. Van Kuilenburg ABP (2004) Dihydropyrimidine dehydrogenase and the efficacy and toxicity of 5-fluorouracil. Eur J Cancer 40:939-950

79. Van Kuilenburg AB, Vreken P, Abeling NG et al (1999) Genotype and phenotype in patients with dihydropyrimidine dehydrogenase deficiency. Hum Genet 104:1-9

80. Van Gennip AH, Driedijk PC, Elzinga A, Abeling NGGM (1992) Screening for defects of dihydropyrimidine degradation by analysis of amino acids in urine before and after acid hydrolysis. J Inherit Metab Dis 15:413-415

81. Duran M, Rovers P, de Bree PK et al (1991) Dihydropyrimidinuria: a new inborn error of pyrimidine metabolism. J Inherit Metab Dis 14:367-370

82. Van Kuilenburg AB, Meinsma R, Zonnenberg BA et al (2003) Dihydropyrimidinase deficiency and severe 5-fluorouracil toxicity. Clin Cancer Res 9:4363-4367

83. Hamajima N, Kouwaki M, Vreken P et al (1998) Dihydropyrimidinase deficiency: structural organization, chromosomal localization, and mutation analysis of the human dihydropyrimidinase gene. Am J Hum Genet 63:717-726

84. Putman CW, Rotteveel JJ, Wevers RA et al (1997) Dihydropyrimidinase deficiency: a progressive neurological disorder ? Neuropediatrics 28:106-110

85. Assmann B, Göhlich-Ratmann G, Bräutigam C et al (1998) Presumptive ureidopropionase deficiency as a new defect in pyrimidine catabolism found with in vitro H-NMR spectroscopy. J Inherit Metab Dis 21[Suppl 2]:1

86. Van Kuilenburg AB, Meinsma R, Beke E et al (2004) Beta-ureidopropionase deficiency: an inborn error or pyrimidine degradation associated with neurological abnormalities. Hum Mol Genet 13:2793-2801

87. Valentine WN, Fink K, Paglia DE et al (1974) Hereditary hemolytic anemia with human erythrocyte pyrimidine 5'-nucleotidase deficiency. J Clin Invest 54:866-879

88. Page T, Yu A, Fontanesi J, Nyhan WL (1997) Developmental disorder associated with increased cellular nucleotidase activity. Proc Natl Acad Sci USA 94:11601-11606

89. Nishino I, Spinazzola A, Papadimitriou A et al (2000) MNGIE: an autosomal recessive disorder due to thymidine phosphorylase mutations. Ann Neurol 47:792-800

90. Saada A, Shaag A, Mandel H et al (2001) Mutant mitochondrial thymidine kinase in mitochondrial DNA depletion myopathy. Nat Genet 29:342-344

36 Disorders of Heme Biosynthesis

Norman G. Egger, Chul Lee, Karl E. Anderson

The Heme Biosynthetic Pathway

Heme (iron protoporphyrin), a metalloporphyrin with iron as the central metal atom, is the prosthetic group for many hemoproteins. It is produced mainly in the bone marrow (for hemoglobin), and in the liver (for cytochrome P450 enzymes). The pathway (Fig. 36.1) consists of eight enzymes; the first and last three are mitochondrial, the other four cytosolic.

The first enzyme of the pathway, 5-aminolevulinic acid synthase (ALAS), has a housekeeping form (termed ALAS1), and an erythroid form (termed ALAS2) encoded by a separate gene on the X chromosome. ALAS1 is especially active in liver, where it is subject to negative feedback by heme, and induced by a variety of drugs,

steroids and other chemicals that also induce cytochrome P450 enzymes [1, 2]. ALAS2 is induced by heme and erythropoietin but not by the factors that induce liver cytochrome P450 enzymes. This explains why such factors are important determinants of the clinical expression in hepatic porphyrias but not in erythropoietic porphyrias.

Mutations of ALAS2 are found in X-linked sideroblastic anemia. Mutations in genes for the other seven enzymes are found in the porphyrias. Deficiency of hepatic uroporphyrinogen decarboxylase, which occurs in porphyria cutanea tarda, can develop in the absence of a mutation of its gene.

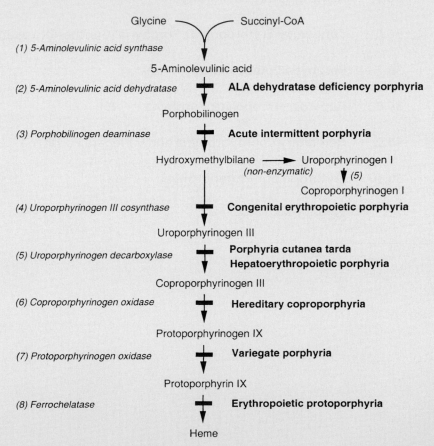

 Fig. 36.1. Pathway of heme biosynthesis. Intermediates and enzymes of the heme biosynthetic pathway are listed. *ALA,* 5-aminolevulinic acid; *CoA,* coenzyme A. The porphyrias caused by the various enzyme deficiencies (indicated by *solid bars* across the arrows) are given in *bold*

X-linked sideroblastic anemia is due to a deficiency of the erythroid form of the first enzyme in the heme biosynthetic pathway, 5-aminolevulinic acid synthase. Characteristics of the disease are variable, but typically include adult onset anemia, ineffective erythropoiesis with formation of ring sideroblasts, iron accumulation and pyridoxine responsiveness.

Porphyrias are metabolic disorders due to deficiencies of other enzymes of this pathway, and are associated with striking accumulations and excess excretion of heme pathway intermediates and their oxidized products. Symptoms and signs of the porphyrias are almost all due to effects on the nervous system or skin. The three most common porphyrias, *acute intermittent porphyria, porphyria cutanea tarda* and *erythropoietic protoporphyria*, differ considerably from each other. The first presents with acute neurovisceral symptoms and can be aggravated by some drugs, hormones and nutritional changes, and is treated with intravenous heme and carbohydrate loading. The skin is affected in the latter two although the lesions are usually distinct and treatment is different. Porphyrias are more often manifest in adults than are most metabolic diseases. All porphyrias are inherited, with the exception of porphyria cutanea tarda, which is due to an acquired enzyme deficiency in liver, although an inherited deficiency is a predisposing factor in some cases.

36.1 X-Linked Sideroblastic Anemia

36.1.1 Clinical Presentation

Sideroblastic anemia is a variable condition and can be either acquired or inherited. Its presence is suggested by hypochromic anemia in the presence of increases in serum iron concentration and transferrin saturation. The bone marrow contains nucleated erythrocyte precursors with iron-laden mitochondria surrounding the nucleus (ring sideroblasts). Progressive iron accumulation may occur as a result of ineffective erythropoiesis, leading to organ damage.

36.1.2 Metabolic Derangement

The inherited form is due to a deficiency of the erythroid form of 5-aminolevulinic acid synthase (ALAS2). Acquired forms have been attributed to alcohol, chemotherapy and to early stages of a myelodysplastic syndrome, which might affect one or more steps in heme synthesis. However, ALAS2 mutations have not been excluded in many of these cases.

36.1.3 Genetics

X-linked sideroblastic anemia is due to mutations of the ALAS2 gene. This disorder is heterogeneous, in that multiple mutations have been described [3, 4]. Phenotypic expression is variable [5]. Point mutations may occur in the pyridoxine binding site of the enzyme, and enzyme activity may be at least partially restored and anemia corrected by high doses of this vitamin.

36.1.4 Diagnostic Tests

Hypochromic anemia with evidence of iron overload suggests this diagnosis. Ring sideroblasts in the bone marrow and pyridoxine responsiveness is further evidence. Detection of an ALAS2 mutation and demonstration of its X-linked inheritance is important for a definite diagnosis. Screening for mutations of the gene associated with hemochromatosis *(HFE)* may identify patients at greater than expected risk for iron accumulation.

36.1.5 Treatment and Prognosis

Treatment consists of administration of pyridoxine and folic acid. The starting dose of pyridoxine is 100-300 mg/day followed by a maintenance dose of 100 mg/day. Phlebotomy to remove excess iron not only prevents organ damage, which is the primary cause of morbidity in this disease, but also may increase responsiveness to pyridoxine.

36.2 Classification of Porphyrias

These metabolic disorders are due to deficiencies of heme biosynthetic pathway enzymes and characterized by accumulation and excess excretion of pathway intermediates and their oxidized products. The photosensitizing effects of excess porphyrins cause cutaneous manifestations. Neurological effects are poorly explained, but are associated with increases in the porphyrin precursors, 5-aminolevulinic acid (also known as δ-aminolevulinic acid) and porphobilinogen.

5-Aminolevulinic acid and porphobilinogen are water-soluble and are excreted almost entirely in urine, as are porphyrins with a large number of carboxyl side chains (e.g. uroporphyrin, an octacarboxyl porphyrin). Protoporphyrin (a dicarboxyl porphyrin) is not soluble in water and is excreted entirely in bile and feces. Coproporphyrin (a tetracarboxyl porphyrin) is found in both urine and bile, and its urinary excretion increases when hepatobiliary function is impaired. Most of the porphyrin intermediates are porphyrinogens (reduced porphyrins) and these undergo autoxidation if they leave the intracellular environment and are then excreted primarily as the corresponding porphyrins. 5-Aminolevulinic acid, porphobilinogen and porphyrinogens are colorless and non-fluorescent, whereas oxidized porphyrins are reddish and fluoresce when exposed to ultraviolet light [6].

The porphyrias are classified with regard to the tissue where the metabolic defect is primarily expressed (hepatic and erythropoietic porphyrias), or the clinical presentation (acute neurovisceral or cutaneous porphyrias) (◨ Table 36.1).

Acute porphyrias (acute intermittent porphyria, variegate porphyria, hereditary coproporphyria and 5-aminolevulinic acid dehydratase porphyria) can cause acute attacks of potentially life-threatening neurovisceral symptoms (e.g. abdominal pain, neuropathy, and mental disturbances). All are associated with striking increases in 5-aminolevulinic acid, and three with increases in porphobilinogen.

Porphyrias accompanied by skin manifestations are termed *cutaneous porphyrias.* In these conditions, excitation of excess porphyrins in the skin by long-wave ultraviolet light (UV-A) leads to generation of singlet oxygen and cell damage. The two most common cutaneous porphyrias are porphyria cutanea tarda and erythropoietic protoporphyria. Variegate porphyria, and much less commonly hereditary coproporphyria, can also cause cutaneous symptoms.

Acute porphyria should be considered in patients with unexplained neurovisceral symptoms, such as abdominal pain. Diagnosis of active cases is based on measurement of porphyrin precursors and porphyrins in urine, blood and feces. Measurements of deficient enzymes and DNA methods are available for confirmation and for family studies.

36.3 Diagnosis of Porphyrias

In contrast to the nonspecific nature of symptoms, laboratory tests, if properly chosen and interpreted, can be both sensitive and specific [6]. The initial presentation determines the type of initial laboratory testing (◨ Table 36.2). In a severely ill patient with symptoms suggesting acute porphyria, it is very important to confirm or exclude this diagnosis promptly, because treatment is more successful if started soon after the onset of symptoms. Measurement of urinary porphyrin precursors (5-aminolevulinic acid and porphobilinogen) and total porphyrins is recommended when neurovisceral symptoms are suggestive of acute porphyria. Urinary porphobilinogen (and 5-aminolevulinic acid) is always markedly increased during attacks of acute intermittent porphyria but may be less increased in hereditary coproporphyria and variegate porphyria. 5-Aminolevulinic acid but not porphobilinogen is increased in 5-aminolevulinic acid dehydratase porphyria. The finding of normal levels of 5-aminolevulinic acid, por-

phobilinogen and total porphyrins effectively excludes all acute porphyrias as potential causes of current symptoms. Current recommendations are that all major medical centers should have capabilities for rapid screening of spot urine samples for excess porphobilinogen, and 5-aminolevulinic acid and total porphyrins be measured later on the same sample [7].

Total plasma porphyrins are increased in virtually all patients with blistering skin lesions due to porphyrias, and should be measured when a cutaneous porphyria is suspected [8, 9]. Plasma porphyrins may not be increased in all patients with the nonblistering photosensitivity found in erythropoietic protoporphyria, and measurement of erythrocyte protoporphyrin is more sensitive. Unfortunately, erythrocyte protoporphyrin is increased in many other erythrocytic disorders, and because this test lacks specificity, it does not alone confirm a diagnosis of erythropoietic protoporphyria.

Further laboratory evaluation is necessary if the initial tests are positive in order to distinguish between the different types of porphyria and establish a precise diagnosis. This is essential for management and genetic counseling.

36.4 5-Aminolevulinic Acid Dehydratase Porphyria

36.4.1 Clinical Presentation

This is the most recently described porphyria, and only 6 cases have been documented by molecular methods. Symptoms resemble those of acute intermittent porphyria, including abdominal pain and neuropathy. The disease may begin in childhood and in severe cases be accompanied by failure to thrive and anemia. Other causes of 5-aminolevulinic acid dehydratase deficiency and increased urinary 5-aminolevulinic acid need to be excluded, such as lead poisoning and hereditary tyrosinemia; these conditions can also present with symptoms resembling those in acute porphyrias.

36.4.2 Metabolic Derangement

This disorder is due to a homozygous or compound heterozygous deficiency of 5-aminolevulinic acid dehydratase, the second enzyme in the heme biosynthetic pathway (◨ Fig. 36.1). The enzyme is markedly reduced (<5% of normal) in affected individuals, and approximately half-normal in both parents, which is consistent with autosomal recessive inheritance (◨ Table 36.1). Lead poisoning can be distinguished by showing reversal of the inhibition of 5-aminolevulinic acid dehydratase in erythrocytes by the in-vitro addition of dithiothreitol. Hereditary tyrosinemia

Table 36.1. Enzyme deficiencies and classification of human porphyrias. Classifications are based on the major tissue site of overproduction of heme pathway intermediates (hepatic vs. erythropoietic) or the type of major symptoms (acute neurovisceral vs. cutaneous), but are not mutually exclusive

Disease	Enzyme	Porphyria classifications			
		Hepatic	Erythro-poietic	Acute	Cutaneous
5-Aminolevulinic acid dehydratase porphyria	*5-Aminolevulinic acid dehydratase*	? X		X	
Acute intermittent porphyria	*Porphobilinogen deaminase*[1]	X		X	
Congenital erythropoietic porphyria	*Uroporphyrinogen III cosynthase*		X		X
Porphyria cutanea tarda[2]	*Uroporphyrinogen decarboxylase*	X			X
Hepatoerythropoietic porphyria	*Uroporphyrinogen decarboxylase*	X	X		X
Hereditary coproporphyria	*Coproporphyrinogen oxidase*	X		X	X
Variegate porphyria	*Protoporphyrinogen oxidase*	X		X	X
Erythropoietic protoporphyria	*Ferrochelatase*		X		X

[1] This enzyme is also known as hydroxymethylbilane synthase, and formerly as uroporphyrinogen I synthase.
[2] Inherited deficiency of uroporphyrinogen decarboxylase is partially responsible for familial (type 2) porphyria cutanea tarda.

type 1 leads to accumulation of succinylacetone (2,3-dioxoheptanoic acid, a structural analog of 5-aminolevulinic acid and a potent inhibitor of the dehydratase, ▶ Chap. 18). Other heavy metals and styrene can also inhibit 5-aminolevulinic acid dehydratase.

36.4.3 Genetics

All well-documented cases were unrelated, and most had different mutations. Immunological studies to date have indicated that most mutant alleles produce a defective enzyme protein [10].

36.4.4 Diagnostic Tests

Characteristic findings include increases in urinary 5-aminolevulinic acid and coproporphyrin and erythrocyte zinc protoporphyrin, normal or slightly increased urinary porphobilinogen, and a marked decrease in erythrocyte 5-aminolevulinic acid dehydratase. Other causes of 5-aminolevulinic acid dehydratase deficiency must be excluded and the diagnosis confirmed by DNA studies [10]. The increase in urinary coproporphyrin (mostly isomer III) is probably due to metabolism of 5-aminolevulinic acid via the heme biosynthetic pathway in tissues other than the liver. Coproporphyrin III also increases in normal subjects after loading with exogenous 5-aminolevulinic acid [11]. Erythrocyte zinc protoporphyrin content is also increased, as in other homozygous cases of porphyria.

36.4.5 Treatment and Prognosis

There is little experience in treating this porphyria. In general, the approach is the same as in acute intermittent porphyria. Heme therapy was effective in most cases. It is prudent to avoid drugs that are harmful in other acute porphyrias.

36.5 Acute Intermittent Porphyria

36.5.1 Clinical Presentation

Symptoms appear during adult life and are more common in women than in men. Acute attacks of neurovisceral symptoms and signs are the most common presentation, although subacute and chronic manifestations can also occur. Attacks usually last for several days or longer, often require hospitalization, and are usually followed by complete recovery. Severe attacks may be much more prolonged and are sometimes fatal, especially if the diagnosis is delayed. Abdominal pain, the most common symptom, is usually steady and poorly localized, but is sometimes crampy. Tachycardia, hypertension, restlessness, fine tremors, and excess sweating suggest sympathetic overactivity. Nausea, vomiting, constipation, pain in the limbs, head, neck or chest, muscle weakness and sensory loss are also common. Dysuria, bladder dysfunction and ileus, with abdominal distention and decreased bowel sounds, may occur. However, increased bowel sounds and diarrhea are sometimes seen. Because the abdominal symptoms are neurological rather than inflammatory, tenderness, fever and leukocytosis are characteristically mild or absent. A peripheral neuropathy that is primarily motor can develop, and is manifested by muscle weakness that most often begins proximally in the upper extremities. It may progress to involve all extremities, respiratory muscles and even lead to bulbar paralysis. Tendon reflexes may be little affected or hyperactive in early stages, but are usually decreased or absent with advanced neuropathy. Muscle weakness is sometimes focal and asymmetric. Cranial and sensory nerves can be affected. Advanced motor neuropathy and death are rare unless porphyria is not recognized and appropriate treatment not instituted. Seizures may occur as an acute neurological manifestation of acute porphyrias, as a result of hyponatremia, or due to other causes unrelated to porphyria. Hyponatremia can be due to electrolyte depletion from vomiting or diarrhea, poor intake, renal sodium loss, or inappropriate antidiuretic hormone secretion. Persistent hypertension and impaired renal function may occur over the long term. Chronic abnormalities in liver function tests, particularly transaminases, are common, although few patients develop significant hepatic impairment. The risk of hepatocellular carcinoma is increased in this and other acute porphyrias, as well as in porphyria cutanea tarda [6, 12, 13].

36.5.2   Metabolic Derangement

Acute intermittent porphyria (AIP) is due to mutations that lead to loss of activity of porphobilinogen deaminase (also known as hydroxymethylbilane synthase and formerly as uroporphyrinogen I synthase), the third enzyme in the heme biosynthetic pathway (Fig. 36.1, Table 36.1). Inheritance is autosomal dominant, and the residual ~50% enzyme activity is mostly due to enzyme produced from the normal allele. Most heterozygotes remain asymptomatic with normal levels of urinary porphyrin precursors. When the disease is clinically expressed, accumulation of heme pathway intermediates in liver leads to increased excretion primarily in urine.

Apparently, the partial deficiency of porphobilinogen deaminase does not of itself greatly impair hepatic heme synthesis. However, when drugs, hormones, or nutritional factors increase the demand for hepatic heme, the deficient enzyme can become limiting. Induction of hepatic ALAS1 is then accentuated and 5-aminolevulinic acid and porphobilinogen accumulate. Excess porphyrins originate nonenzymatically from porphobilinogen, and perhaps enzymatically from 5-aminolevulinic acid transported to tissues other than the liver.

Most drugs that are harmful to patients with this and other acute hepatic porphyrias are known to have the capacity to induce the synthesis of cytochrome P450 enzymes and ALAS1 in the liver [2].

36.5.3   Genetics

More than 200 different mutations of the porphobilinogen deaminase gene have been identified in unrelated families [14]. The gene has two promoters, one of which is erythroid-specific. Erythroid-specific and housekeeping forms of this enzyme are derived from the same gene by alternative splicing of two primary transcripts. Most mutations in AIP lead to a deficiency of both isozymes. Mutations located in or near the first of the 15 exons in this gene can impair the synthesis of the housekeeping form but not the erythroid-specific form of porphobilinogen deaminase. Homozygous cases of acute intermittent porphyria are extremely rare, but should be suspected particularly if the disease is active early in childhood [15].

36.5.4   Diagnostic Tests

A substantial increase in urinary porphobilinogen is a sensitive and specific indication that a patient has either acute intermittent porphyria, hereditary coproporphyria or variegate porphyria (Table 36.2). A kit is available for the rapid detection of porphobilinogen at concentrations greater than 6 mg/l with a color chart for semiquantitative estimation of higher levels [16]; this enables major medical centers to provide for rapid in-house testing for these disorders [7]. Porphobilinogen remains increased between attacks of acute intermittent porphyria and becomes normal only after prolonged latency. Fecal total porphyrins are generally normal or minimally increased in acute intermittent porphyria, and markedly increased in the other two conditions. Total plasma porphyrins are charactistically increased in variegate porphyria, as discussed later, but are normal or only slightly increased in acute intermittent porphyria. Urinary porphyrins, and particularly coproporphyrin is generally more increased in hereditary coproporphyria and variegate porphyria. Urinary uroporphyrin can be increased in all of these disorders, especially when porphobilinogen is increased.

Decreased erythrocyte porphobilinogen deaminase helps to confirm a diagnosis of acute intermittent porphyria. However, falsely low activity may occur if there is a problem with processing or storing the sample. The erythrocyte enzyme is not deficient in all patients because some mutations of the porphobilinogen deaminase gene only reduce the housekeeping form of the enzyme. Furthermore, erythrocyte porphobilinogen deaminase has a wide normal range (up to 3-fold) that overlaps the range of patients with acute intermittent porphyria.

Measuring erythrocyte porphobilinogen deaminase is very useful for detecting asymptomatic carriers, if it is known that the propositus has a deficiency of the erythrocyte enzyme. Urinary porphobilinogen should also be measured when relatives are screened for this porphyria.

Table 36.2. First-line laboratory tests for screening for porphyrias and second-line tests for further evaluation when initial testing is positive

Testing	Symptoms suggesting porphyria	
	Acute neurovisceral symptoms	**Cutaneous photosensitivity**
First-line	Urinary 5-aminolevulinic acid, porphobilinogen and total porphyrins[1] (quantitative; random or 24 h urine).	*Blistering skin lesions*: Total plasma porphyrins[2] *Nonblistering*: Erythrocyte porphyrins[3]
Second-line	Total fecal porphyrins[1] Erythrocyte porphobilinogen deaminase Total plasma porphyrins[2]	Urinary 5-aminolevulinic acid, porphobilinogen and total porphyrins[1] Total fecal porphyrins[1]

[1] Fractionation of urinary and fecal porphyrins is usually not helpful unless the total is increased.
[2] The preferred method is by direct fluorescence spectrophotometry.
[3] Erythrocyte porphyrins are generally expressed as protoporphyrin, however the method detects other porphyrins as well. This test lacks specificity, because erythrocyte protoporphyrin is increased in many erythrocytic disorders.

Identification of the specific mutation in a known case enables the same mutation to be detected in relatives, most of whom are likely to be asymptomatic and can then be advised to take precautions to avoid exacerbating the disease.

36.5.5   Treatment and Prognosis

Intravenous hemin (heme arginate or hematin) is considered specific therapy for acute attacks because it represses hepatic ALAS1, and markedly reduces levels of 5-aminolevulinic acid and porphobilinogen. Severe attacks, with features such as nausea, vomiting, motor weakness and hyponatremia should be treated initially with hemin. Carbohydrate loading, usually accomplished by intravenous administration of 10% glucose, also has some repressive effect on ALAS1, but is much less effective. Glucose may be started initially until hemin is obtained. Heme arginate is the preferred form of hemin [17]. Degradation products of hematin (heme hydroxide) commonly cause phlebitis at the site of infusion and a transient anticoagulant effect. In countries where heme arginate is not available, hematin can be reconstituted with human albumin to stabilize the heme as heme albumin, which confers many of the advantages of heme arginate [18].

The standard regimen for hemin is 3–4 mg per kg body weight infused intravenously once daily for 4 days. Treatment of a newly diagnosed patient should be started only after a marked increase in urinary porphobilinogen is demonstrated using a rapid and reliable method. Recurrent attacks can be diagnosed on clinical grounds, since porphobilinogen remains elevated in most AIP patients between attacks, and the presenting signs and symptoms are often similar from one attack to the next. A longer course of treatment is seldom necessary if treatment is started early. Efficacy is reduced and recovery less rapid when treatment is delayed and neuronal damage is more advanced. Heme

therapy is not effective for chronic symptoms of acute porphyrias [19].

Most acute attacks are severe enough to require hospitalization for administration of intravenous hemin and observation for neurological complications and electrolyte imbalances. Narcotic analgesics are commonly required for abdominal, back or extremity pain, and small doses of a phenothiazine are useful for nausea, vomiting, anxiety, and restlessness. Chloral hydrate can be administered for insomnia. Diazepam in low doses is safe if a minor tranquilizer is required, although it needs to be kept in mind that benzodiazepines have some inducing effect on hepatic heme synthesis and may act in an additive fashion to other inducing influences. Bladder distention may require catheterization.

Carbohydrate loading can be tried instead of hemin for mild attacks. At least 300 g daily is recommended, and >500 g daily may be more effective. Carbohydrate can sometimes be given orally. However, nausea, vomiting and ileus usually prevent this approach. More complete parenteral nutrition should be considered for patients when oral intake is not possible for more than several days.

Abdominal pain may disappear within hours, and paresis begin to improve within days. Muscle weakness due to severe motor neuropathy may gradually resolve, but there may be some residual weakness.

Treatment of seizures is problematic, because almost all anticonvulsant drugs can exacerbate acute porphyrias. Bromides, gabapentin and probably vigabatrin can be given safely [20]. β–Adrenergic blocking agents may control tachycardia and hypertension in acute attacks of porphyria, but do not have a specific effect on the underlying pathophysiology [19].

An allogeneic liver transplant in a woman with severe, recurrent attacks of acute intermittent porphyria led to complete biochemical and clinical remission [21]. This experience supports the role of hepatic overproduction of porphyrin precursors in causing the neurological manifestations, but is not sufficient evidence for broad application of hepatic transplantation for acute porphyrias [7].

Identification and correction of precipitating factors such as certain drugs, inadequate nutrition, cyclic or exogenous hormones (particularly progesterone and progestins), and intercurrent infections can hasten recovery from an attack and prevent future attacks. Frequent cyclic attacks occurring in some women during the luteal phase of the cycle when progesterone levels are highest can be prevented by administration of a gonadotropin-releasing hormone analogue to prevent ovulation [22].

With prompt treatment of acute attacks and precautions to prevent further attacks, the outlook for patients with acute porphyrias is usually excellent. Fatal attacks have become much less common [12]. However, some patients continue to have attacks in the absence of identifiable precipitating factors. Some develop chronic pain and other symptoms, and may become addicted to narcotic analgesics. Such patients need to be followed closely because there is often coexistent depression and an increased risk of suicide.

36.6 Congenital Erythropoietic Porphyria (Gunther Disease)

36.6.1 Clinical Presentation

This is usually a severe disease with manifestations noted soon after birth, or even in utero. But clinical expression is variable and is determined in part by the degree of enzyme deficiency. Cutaneous features resemble those in porphyria cutanea tarda but in most cases are much more severe. Lesions include bullae and vesicles on sun-exposed skin, hypo- or hyperpigmented areas, hypertrichosis, and scarring. The teeth are reddish brown (erythrodontia) because of porphyrin deposition, and may fluoresce when exposed to long-wave ultraviolet light. Porphyrins are also deposited in bone. Hemolysis is almost invariably present and results from the markedly increased erythrocyte porphyrin levels, and is accompanied by splenomegaly. Life expectancy is often shortened by infections or hematological complications. There are no neurological manifestations.

Congenital erythropoietic porphyria can present in utero as nonimmune hydrops [23]. When this is recognized, intrauterine transfusion is possible, and after birth severe photosensitivity can be prevented by avoiding phototherapy for hyperbilirubinemia. Rarely, the disease develops in adults, and is associated with a myeloproliferative disorder.

36.6.2 Metabolic Derangement

This rare disorder is due to a severe deficiency of uroporphyrinogen III cosynthase, the fourth enzyme of the heme synthesis pathway (□ Fig. 36.1, □ Table 36.1). Hydroxymethylbilane (the substrate of the deficient enzyme) accumulates and is converted nonenzymatically to uroporphyrinogen I, a nonphysiological intermediate, which cannot be metabolized to heme. Therefore, uroporphyrin, coproporphyrin and other porphyrins accumulate in bone marrow, plasma, urine, and feces. Porphyrin accumulation in erythroid cells results in intramedullary and intravascular hemolysis, which leads to increased erythropoiesis. As a result, heme synthesis is actually increased in spite of the inherited enzyme deficiency, in order to compensate for porphyrin-induced hemolysis. Although the porphyrins that accumulate in this disease are primarily type I porphyrin isomers, type III isomers are also increased.

36.6.3 Genetics

Congenital erythropoietic porphyria is an autosomal recessive disorder. Patients have either homozygous or compound heterozygous mutations of the uroporphyrinogen III cosynthase gene. Like other porphyrias, this disease is genetically heterogeneous, and many different mutations have been identified [24]. Parents and other heterozygotes display intermediate deficiencies of the cosynthase. The disease can be diagnosed in utero by porphyrin measurements and DNA methods. Expansion of a clone of erythroid cells that carry a uroporphyrinogen III cosynthase mutation often accounts for adult-onset cases.

36.6.4 Diagnostic Tests

Erythrocyte and plasma porphyrins are markedly increased and usually consist mostly of uroporphyrin I. Coproporphyrin and even zinc protoporphyrin may be increased in erythrocytes. Porphyrins in urine are primarily uroporphyrin I and coproporphyrin I, and in feces mostly coproporphyrin I. Porphyrin precursors are not increased. The diagnosis should be confirmed by finding a marked deficiency in uroporphyrinogen III cosynthase activity and by mutation analysis.

36.6.5 Treatment and Prognosis

Protection of the skin from sunlight is essential. Minor trauma can lead to denudation of fragile skin. Bacterial infections should be treated promptly to prevent scarring and mutilation. Improvement in hemolysis has been reported after splenectomy. Oral charcoal may be helpful by increasing fecal excretion of porphyrins. High level blood transfusions and hydroxyurea may be effective by suppressing erythropoiesis and porphyrin synthesis [25, 26]. Bone marrow or stem cell transplantation is effective current therapy, and gene therapy may eventually be possible [27, 28].

36.7   Porphyria Cutanea Tarda

36.7.1   Clinical Presentation

This is the most common and readily treated form of porphyria and is manifested primarily by chronic, blistering skin lesions, especially on the backs of the hands, forearms, face and (in women) the dorsa of the feet. Neurological effects are not observed. Sun-exposed skin is also friable, and minor trauma may precede the formation of bullae or cause denudation of the skin. Small white plaques (milia) may precede or follow vesicle formation. Hypertrichosis and hyperpigmentation are also noted. Thickening, scarring and calcification of affected skin may be striking, and is referred to as pseudoscleroderma. Skin lesions are indistinguishable clinically from all other cutaneous porphyrias, except for erythropoietic protoporphyria (▶ later discussion). In pseudoporphyria, skin lesions resemble porphyria cutanea tarda but porphyrins are not significantly increased; presumably other photosensitizers are responsible.

Multiple susceptibility factors for porphyria cutanea tarda are commonly identified in an individual patient. A normal or increased amount of hepatic iron is a requirement for the disease. Others include moderate or heavy alcohol intake, hepatitis C infection, estrogen use and smoking. Infection with HIV is a less common association. There are geographic differences in the association with hepatitis C; in some locations more than 80% of patients are infected with this virus.

A large outbreak of this porphyria occurred in eastern Turkey in the 1950s from ingestion of wheat that was intended for planting, and had been previously treated with hexachlorobenzene as a fungicide. Porphyria cutanea tarda has been reported after exposure to other chemicals including di- and trichlorophenols and 2,3,7,8-tetrachlorodibenzo-p-dioxin (TCDD, dioxin). These halogenated polycyclic aromatic hydrocarbons induce an experimental porphyria in laboratory animals that biochemically closely resembles human porphyria cutanea tarda. Such toxic exposures are not evident in most human cases of sporadic porphyria cutanea tarda [29, 30].

36.7.2   Metabolic Derangement

This porphyria is caused by a profound deficiency of hepatic uroporphyrinogen decarboxylase, the fifth enzyme of the heme biosynthetic pathway (◘ Fig. 36.1, ◘ Table 36.1). Sporadic (type 1) and familial (types 2 and 3) forms of the disease have been described. These do not differ substantially in terms of clinical features or treatment. In all cases, a specific inhibitor of hepatic uroporphyrinogen decarboxylase, which has not yet been characterized, is generated from an intermediate of the heme biosynthetic pathway by an iron-dependent oxidative mechanism. Certain cytochrome P450 enzymes and low levels of ascorbic acid and carotenoids may contribute to this oxidative process within hepatocytes. The prevalence of HFE mutations is increased [30]. Individuals with type 2 disease from birth have half the normal enzyme activity and are therefore more susceptible to developing a more profound enzyme deficiency in the liver [29].

Patterns of excess porphyrins in this disease are complex and characteristic. Uroporphyrinogen, (an octacarboxyl porphyrinogen) undergoes a sequential, four-step decarboxylation to coproporphyrinogen (a tetracarboxyl porphyrinogen). Uroporphyrinogen and the hepta-, hexa-, and pentacarboxyl porphyrinogens accumulate. To complicate the porphyrin pattern further, pentacarboxyl porphyrinogen can be metabolized by coproporphyrinogen oxidase to a tetracarboxyl porphyrinogen termed isocoproporphyrinogen. These porphyrinogens accumulate first in liver, are mostly oxidized to the corresponding porphyrins, and then appear in plasma and are excreted in urine, bile and feces. Successful treatment may require some time before the massive porphyrin accumulations in liver are cleared.

36.7.3   Genetics

Porphyria cutanea tarda results from a liver-specific, apparently acquired deficiency of uroporphyrinogen decarboxylase. No mutations in this gene have been found in sporadic (type 1) porphyria cutanea tarda. The amount of hepatic uroporphyrinogen decarboxylase protein in type 1 disease, as measured immunochemically, is normal, as might be expected with an inhibitor of the enzyme.

An inherited partial deficiency of this enzyme contributes in type 2, which accounts for approximately 20% of patients with porphyria cutanea tarda. In these cases erythrocyte uroporphyrinogen decarboxylase is approximately 50% of normal in erythrocytes, and this feature is inherited as an autosomal dominant trait affecting all tissues. Type 2 becomes clinically manifest when hepatic uroporphyrinogen decarboxylase becomes profoundly inhibited, as in type 1. A number of mutations of the uroporphyrinogen decarboxylase gene have been identified in type 2 disease. Cases classified as type 3 disease, which are rare, have normal erythrocyte uroporphyrinogen decarboxylase activity but one or more relatives also have the disease. A genetic defect has not been clearly identified in type 3, and it is possible that these cases are not fundamentally different from type I [29].

36.7.4 Diagnostic Tests

Blistering skin lesions are found in all cutaneous porphyrias, except erythropoietic protoporphyria. Skin histopathology is not specific and does not establish a diagnosis of porphyria cutanea tarda or exclude pseudoporphyria. It is important to differentiate these conditions by laboratory testing before starting therapy.

Plasma porphyrins are increased in all patients with blistering skin lesions due to porphyria. The fluorescence spectrum of plasma porphyrins can readily distinguish variegate porphyria and erythropoietic protoporphyria from porphyria cutanea tarda (◘ Table 36.2). The diagnosis is best confirmed by increased total urinary porphyrins with a predominance of uroporphyrin and heptacarboxyl porphyrin. Total fecal porphyrins are usually less increased than in hereditary coproporphyria and variegate porphyria. In porphyria cutanea tarda, an increase in the proportion of fecal isocoproporphyrin, which can be expressed as a ratio to coproporphyrin, is distinctive.

VIII

36.7.5 Treatment and Prognosis

Repeated phlebotomy is standard treatment at most centers, although low-dose hydroxychloroquine (or chloroquine) is also effective. Patients are also advised to discontinue alcohol, estrogens, iron supplements, and other contributing factors. Phlebotomies remove iron and stimulate erythropoiesis, and utilization of storage iron for hemoglobin formation gradually reduces the serum ferritin to a target range of 15–20 ng/ml. This can usually be achieved by removal of only 5–6 units (450 ml each) of blood at 1–2 week intervals. Further iron depletion is of no additional benefit and may cause anemia and associated symptoms. Many more phlebotomies may be needed in patients who have marked iron overload, which is likely to be due to familial hemochromatosis. The plasma or serum porphyrin level falls somewhat more slowly than ferritin, and may not yet be normal when the target ferritin level is reached.

With treatment the activity of hepatic uroporphyrinogen decarboxylase gradually increases to normal. After remission, ferritin can return to pretreatment values without recurrence, in most cases. Postmenopausal women who have been treated for porphyria cutanea tarda can usually resume estrogen replacement without recurrence. Relapses seem to be more common in patients who resume alcohol intake, but will respond to further phlebotomies.

A low dose of hydroxychloroquine (100 mg twice weekly) or chloroquine (125 mg twice weekly) for several months gradually removes excess porphyrins from the liver. This is a suitable alternative when phlebotomy is contraindicated or difficult, and is preferred at some centers. Standard doses of these 4-aminoquinolines exacerbate photosensitivity and cause hepatocellular damage, and should not be used. Both may produce retinal damage, although this risk is very low, and may be lower with hydroxychloroquine than chloroquine. The mechanism by which these drugs remove porphyrins from the liver in this condition is not known [31]. This treatment is not effective in other porphyrias [19].

36.8 Hepatoerythropoietic Porphyria

36.8.1 Clinical Presentation

This rare disease is clinically similar to congenital erythropoietic porphyria and usually presents with red urine and blistering skin lesions shortly after birth. Mild cases may present later in life and more closely resemble porphyria cutanea tarda. Concurrent conditions, such as viral hepatitis, may accentuate porphyrin accumulation.

36.8.2 Metabolic Derangement

Hepatoerythropoietic porphyria is the homozygous form of familial (type 2) porphyria cutanea tarda, and is due to a substantial deficiency of uroporphyrinogen decarboxylase. Intermediate deficiencies of the enzyme are found in the parents, as expected for an autosomal recessive disorder (◘ Fig. 36.1, ◘ Table 36.1). The disease has features of both hepatic and erythropoietic porphyrias.

36.8.3 Genetics

This porphyria results from a homozygous or compound heterozygous state for mutations of the gene encoding uroporphyrinogen decarboxylase. The disease is genetically heterogeneous. Mutations found in this disease generally result in marked decreases in uroporphyrinogen decarboxylase activity, but some activity remains, so heme formation can occur [30].

36.8.4 Diagnostic Tests

The excess porphyrins found in urine, plasma and feces are similar to those in porphyria cutanea tarda. In addition, erythrocyte zinc protoporphyrin is increased, as in a number of other autosomal recessive porphyrias. This finding probably reflects an earlier accumulation of uroporphyrinogen in erythroblasts, which after completion of hemoglobin synthesis is metabolized to protoporphyrin.

Erythrocyte porphyrins in congenital erythropoietic porphyria are usually mostly uroporphyrin I and coproporphyrin I, but in some cases there is a predominance of zinc protoporphyrin. Hepatoerythropoietic porphyria is differentiated from congenital erythropoietic porphyria also by excess isocoproporphyrins in feces and urine, and by decreased erythrocyte uroporphyrinogen decarboxylase activity. It is important to document the diagnosis by molecular methods.

36.8.5 Treatment and Prognosis

Therapeutic options are essentially the same as in congenital erythropoietic porphyria.

36.9 Hereditary Coproporphyria and Variegate Porphyria

36.9.1 Clinical Presentation

These disorders can present with acute attacks that are identical to those in acute intermittent porphyria. However, unlike the latter disease, variegate porphyria and more rarely hereditary coproporphyria may cause blistering skin lesions that are indistinguishable from those of porphyria cutanea tarda. Symptoms are most common after puberty. Factors that exacerbate acute intermittent porphyria are important in both of these porphyrias. Variegate porphyria is particularly common in South Africa where most cases are descendants of a couple who emigrated from Holland and arrived in Cape Town in 1688 [32]. In rare homozygous cases of these porphyrias clinical manifestations begin in childhood.

36.9.2 Metabolic Derangement

Hereditary coproporphyria and variegate porphyria result from approximately 50% deficiencies of coproporphyrinogen oxidase and of protoporphyrinogen oxidase, respectively, which are the sixth and seventh enzyme of the heme biosynthetic pathway (Fig. 36.1, Table 36.1). In hereditary coproporphyria there is marked accumulation of coproporphyrin III (derived from autooxidation of coproporphyrinogen III), and urinary porphyrin precursors and uroporphyrin are increased particularly in association with acute attacks. Similar abnormalities are seen in variegate porphyria, but in addition protoporphyrin (derived from autooxidation of protoporphyrinogen) is increased in feces (and bile), and plasma porphyrins are increased. Protoporphyrinogen has been shown to inhibit porphobilinogen deaminase, which along with induction of hepatic ALAS1, may account for the increase in porphyrin

precursors during acute attacks, at least in variegate porphyria.

36.9.3 Genetics

Both of these porphyrias are autosomal dominant conditions. Homozygous cases are rare. Genetic heterogeneity is a feature of both. As expected, a single mutation (R59W) accounts for the many descendants with variegate porphyria in South Africa, which is an example of the founder effect [32].

36.9.4 Diagnostic Tests

Urinary 5-aminolevulinic acid and porphobilinogen are increased during acute attacks of these porphyrias, although the increases may be less and more transient than in acute intermittent porphyria. Urinary coproporphyrin increases may be more prominent and prolonged. However, coproporphyrinuria is a highly nonspecific finding. It can be seen in many medical conditions, especially when hepatic or bone marrow function is affected.

A marked, isolated increase in fecal coproporphyrin (especially isomer III) is distinctive for hereditary coproporphyria. Fecal coproporphyrin and protoporphyrin are about equally increased in variegate porphyria. An increase in fecal pseudo-pentacarboxyl porphyrin, which is a dicarboxyl porphyrin derived from protoporphyrin, is also diagnostically useful in variegate porphyria.

Increased plasma porphyrins and a fluorescence spectrum of plasma porphyrins (at neutral pH) is characteristic and very useful for rapidly distinguishing variegate porphyria from the other porphyrias. This is at least as sensitive as fecal porphyrin measurement for detecting variegate porphyria, although not as sensitive as a reliable assay for lymphocyte protoporphyrinogen oxidase or mutation analysis [33, 34].

Reliable assays for protoporphyrinogen oxidase and coproporphyrinogen oxidase in cultured fibroblasts or lymphocytes are available only in a few research laboratories. Erythrocytes cannot be used to measure these mitochondrial enzymes, because mature erythrocytes do not contain mitochondria. As in other porphyrias, identification of a mutation in an index case facilitates detection of relatives who carry the same mutation.

36.9.5 Treatment and Prognosis

Acute attacks are treated as in acute intermittent porphyria (▶ above). Cutaneous symptoms are more difficult to treat, and therapies that are effective for porphyria cutanea tarda (phlebotomy and low-dose hydroxychloroquine) are not effective in these conditions. Protection from sunlight is important.

36.10   Erythropoietic Protoporphyria

36.10.1   Clinical Presentation

Erythropoietic protoporphyria is the third most common porphyria. Cutaneous symptoms begin in childhood, and are generally much more prominent than objective changes by examination. Symptoms such as burning, itching, erythema, and swelling can occur within minutes of sun exposure, and the diffuse edema of sun-exposed areas may resemble angioneurotic edema. Other more chronic skin changes may include lichenification, leathery pseudo-vesicles, labial grooving, and nail changes. In contrast to other cutaneous porphyrias, blistering, milia, friability, and chronic skin changes such as scarring and hypertrichosis are not prominent. There is no fluorescence of the teeth and no neuropathic manifestations. Mild anemia with hypochromia and microcytosis is noted in some cases.

The severity of the symptoms is remarkably stable over time. Drugs that exacerbate hepatic porphyrias are not known to worsen this disease, although they are generally avoided as a precaution. Gallstones containing protoporphyrin may also develop. Some patients develop liver disease, which can progress rapidly to death from hepatic failure. This complication is accompanied by marked deposition of protoporphyrin in liver and increased levels in plasma and erythrocytes. A motor neuropathy may further complicate the course of liver decompensation in this disease, and is unexplained [35].

36.10.2   Metabolic Derangement

The inherited deficiency of ferrochelatase, the eighth and last enzyme in the heme biosynthetic pathway (◧ Fig. 36.1, ◧ Table 36.1) leads to increases in protoporphyrin in bone marrow, circulating erythrocytes, plasma, bile, and feces in this disease. Ferrochelatase is deficient in all tissues, but the deficient enzyme is rate-limiting for protoporphyrin metabolism primarily in bone marrow reticulocytes, which are the primary source of the excess protoporphyrin. Circulating erythrocytes and perhaps the liver contribute smaller amounts. Excess protoporphyrin is transported in plasma and excreted in bile and feces.

Erythrocyte protoporphyrin is mostly chelated with zinc in normal erythrocytes as well as in many other conditions where protoporphyrin in increased (e.g. lead poisoning, iron deficiency, and homozygous forms of porphyria). Formation of both heme and zinc protoporphyrin is catalyzed by ferrochelatase. Protoporphyrin accumulates mostly as free protoporphyrin in protoporphyria, because this enzyme is deficient. Free protoporphyrin diffuses more readily from erythrocytes into plasma than does zinc protoporphyrin, most of which remains in the erythrocyte for its full life span. Therefore, primarily reticulocytes and young circulating erythrocytes fluoresce when observed under long wave ultraviolet light.

Protoporphyrin is excreted in bile and may undergo enterohepatic circulation. Liver protoporphyrin content is not increased in uncomplicated protoporphyria. But large amounts of protoporphyrin derived primarily from the bone marrow can cause cholestasis and severe liver failure in some patients with protoporphyria.

36.10.3   Genetics

Many different mutations in the ferrochelatase gene have been identified in protoporphyria, and most express little or no ferrochelatase. The pattern of inheritance is best described as autosomal dominant, in that the primary inherited determinant of the disease in most families is a severe, disabling ferrochelatase mutation. As proposed in 1984, and supported by recent molecular evidence, most patients with clinically manifest disease have also inherited a normal, weakly expressed ferrochelatase allele [36-38]. This polymorphic allele, which expresses an aberrantly spliced mRNA that is subject to rapid degradation, is found in ~10% of normal Caucasians, and has no consequence in the absence of a mutant ferrochelatase allele that results in little or no enzyme activity [38]. Therefore, ferrochelatase activity is only 10-25% of normal in patients with manifest disease, rather than the expected ~50% for autosomal dominant inheritance, and many heterozygotes in a family have higher enzyme activity and no increase in erythrocyte protoporphyrin. Autosomal recessive inheritance, with two disabling mutations has been documented in a few families, where at least one of the two mutant ferrochelatase alleles expresses some enzyme activity [35].

36.10.4   Diagnostic Tests

The most sensitive screening test for this disorder is a determination of erythrocyte protoporphyrin, which under most circumstances is the predominant porphyrin in erythrocytes. This test lacks specificity because standard assays reflect all porphyrins that might be increased in many diseases, including free protoporphyrin (in protoporphyria), zinc protoporphyrin (in iron deficiency, lead poisoning, most homozygous cases of porphyria, and many other erythrocyte disorders), and very rarely uroporphyrin I and coproporphyrin I (in congenital erythropoietic porphyria). To gain specificity for protoporphyria, an increased erythrocyte protoporphyrin result is followed by a determination whether the protoporphyrin is free or complexed with zinc, using a simple ethanol extraction method.

The plasma porphyrin concentration is almost always increased, but less so than in other cutaneous porphyrias. Moreover, the excess protoporphyrin in plasma in this condition is particularly sensitive to light exposure, which may increase the chance of a falsely normal measurement. It is especially important to shield plasma samples from light if protoporphyria is suspected. The fluorescence spectrum of plasma porphyrins at neutral pH can distinguish erythropoietic protoporphyria from other porphyrias.

Total fecal porphyrins may be normal or increased in protoporphyria, with a predominance of protoporphyrin. Urinary porphyrins and porphyrin precursors are normal, unless the patient has liver impairment, in which case urinary porphyrins (especially coproporphyrin) may increase. Hepatic complications of the disease are often preceded by increasing levels of erythrocyte and plasma protoporphyrin, abnormal liver function tests, marked deposition of protoporphyrin in liver cells and bile canaliculi, and increased photosensitivity.

36.10.5 Treatment and Prognosis

Photosensitivity is managed by avoidance of sunlight. Oral β-carotene and cysteine improve tolerance to sunlight in some patients, perhaps by quenching singlet oxygen or free radicals. β-Carotene seems to be more effective in erythropoietic protoporphyria than in other cutaneous porphyrias. Cholestyramine may reduce protoporphyrin levels by interrupting its enterohepatic circulation. Iron deficiency, caloric restriction, and drugs or hormone preparations that impair hepatic excretory function should be avoided.

Treatment of liver complications is difficult. Transfusions or heme therapy may suppress erythroid and hepatic protoporphyrin production. Liver transplantation is sometimes required, but there is some risk that the new liver will also accumulate excess protoporphyrin and develop impaired function [39]. Operating room lights have produced severe skin and peritoneal burns in some patients with protoporphyria, liver failure, and marked increases in erythrocyte and plasma protoporphyrin concentrations. A patient with erythropoietic protoporphyria who underwent bone marrow transplantation for leukemia experienced complete remission of the porphyria [40]. Therefore, there is potential benefit from bone marrow replacement and gene therapy in this and other erythropoietic porphyrias [35].

References

1. Granick S (1966) The induction in vitro of the synthesis of δ-aminolevulinic acid synthetase in chemical porphyria: a response to certain drugs, sex hormones, and foreign chemicals. J Biol Chem 241:1359-1375
2. Anderson KE, Freddara U, Kappas A (1982) Induction of hepatic cytochrome P-450 by natural steroids: relationships to the induction of δ-aminolevulinate synthase and porphyrin accumulation in the avian embryo. Arch Biochem Biophys 217:597-608
3. Bekri S, May A, Cotter PD et al (2003) A promoter mutation in the erythroid-specific 5-aminolevulinate synthase (ALAS2) gene causes X-linked sideroblastic anemia. Blood 102:698-704
4. Cazzola M, May A, Bergamaschi G et al (2000) Familial-skewed X-chromosome inactivation as a predisposing factor for late-onset X-linked sideroblastic anemia in carrier females. Blood 96:4363-4365
5. Cazzola M, May A, Bergamaschi G et al (2002) Absent phenotypic expression of X-linked sideroblastic anemia in one of 2 brothers with a novel ALAS2 mutation. Blood 100:4236-4238
6. Anderson KE (2003) The porphyrias. In: Zakim D, Boyer T (eds) Hepatology. Saunders, Philadelphia, chap 11, pp 291-346
7. Anderson KE, Bloomer JE, Bonkovsky HL et al (2005) Recommendations for the diagnosis and treatment of the acute porphyrias. Ann Intern Med 142:439-450
8. Poh-Fitzpatrick MB, Lamola AA (1976) Direct spectrophotometry of diluted erythrocytes and plasma: a rapid diagnostic method in primary and secondary porphyrinemias. J Lab Clin Med 87:362-370
9. Poh-Fitzpatrick MB (1980) A plasma porphyrin fluorescence marker for variegate porphyria. Arch Dermatol 116:543-547
10. Sassa S (1998) ALAD porphyria. Semin Liver Dis 18:95-101
11. Shimizu Y, Ida S, Naruto H, Urata G (1978) Excretion of porphyrins in urine and bile after the administration of delta-aminolevulinic acid. J Lab Clin Med 92:795-802
12. Kauppinen R, Mustajoki P (1992) Prognosis of acute porphyria: occurrence of acute attacks, precipitating factors, and associated diseases. Medicine 71:1-13
13. Andant C, Puy H, Bogard C et al (2000) Hepatocellular carcinoma in patients with acute hepatic porphyria: frequency of occurrence and related factors. J Hepatol 32:933-939
14. Human Gene Mutation Database (www.hgmd.org).
15. Solis C, Martinez-Bermejo A, Naidich TP et al (2004) Acute intermittent porphyria: studies of the severe homozygous dominant disease provides insights into the neurologic attacks in acute porphyrias. Arch Neurol 61:1764-1770
16. Deacon AC, Peters TJ (1998) Identification of acute porphyria: evaluation of a commercial screening test for urinary porphobilinogen. Ann Clin Biochem 35:726-732
17. Tenhunen R, Mustajoki P (1998) Acute porphyria: treatment with heme. Semin Liver Dis 18:53-55
18. Bonkovsky HL, Healey BS, Lourie AN, Gerron GG (1991) Intravenous heme-albumin in acute intermittent porphyria: evidence for repletion of hepatic hemoproteins and regulatory heme pools. Am J Gastroenterol 86:1050-1056
19. Anderson KE (2003) Approaches to treatment and prevention of human porphyrias. In: Kadish KM, Smith K, Guilard R (eds) Porphyrin handbook, part II, vol 14. Academic Press, San Diego, chap 94, pp 247-284
20. Hahn M, Gildemeister OS, Krauss GL et al (1997) Effects of new anticonvulsant medications on porphyrin synthesis in cultured liver cells: potential implications for patients with acute porphyria. Neurology 49:97-106
21. Soonawalla ZF, Orug T, Badminton MN (2004) Liver transplantation as a cure for acute intermittent porphyria. Lancet 363:705-706
22. Anderson KE, Spitz IM, Bardin CW, Kappas A (1990) A GnRH analogue prevents cyclical attacks of porphyria. Arch Intern Med 150:1469-1474
23. Verstraeten L, Van Regemorter N, Pardou A et al (1993) Biochemical diagnosis of a fatal case of Gunther's disease in a newborn with hydrops-fetalis. Eur J Clin Chem Clin Biochem 31:121-128
24. Desnick RJ, Glass IA, Xu W et al (1998) Molecular genetics of congenital erythropoietic porphyria. Semin Liver Dis 18:77-84
25. Piomelli S, Poh-Fitzpatrick MB, Seaman C et al (1986) Complete suppression of the symptoms of congenital erythropoietic porphyria by long-term treatment with high-level transfusions. N Engl J Med 314:1029-1031

26. Guarini L, Piomelli S, Poh-Fitzpatrick MB (1994) Hydroxyurea in congenital erythropoietic porphyria (letter). N Engl J Med 330:1091-1092

27. Zix-Kieffer I, Langer B, Eyer D (1996) Successful cord blood stem cell transplantation for congenital erythropoietic porphyria (Gunther's disease). Bone Marrow Transplant 18:217-220

28. Fritsch C, Lang K, Bolsen K et al (1998) Congenital erythropoietic porphyria. Skin Pharmacol Appl Skin Physiol 11:347-357

29. Elder GH (2003) Porphyria cutanea tarda and related disorders. In: Kadish KM, Smith K, Guilard R (eds) Porphyrin handbook, part II, vol 14. Academic Press, San Diego, chap 88, pp 67-92

30. Egger NG, Goeger DE, Payne DA et al (2002) Porphyria cutanea tarda: multiplicity of risk factors including HFE mutations, hepatitis C, and inherited uroporphyrinogen decarboxylase deficiency. Dig Dis Sci 47:419-426

31. Egger NG, Goeger DE, Anderson KE (1996) Effects of chloroquine in hematoporphyrin-treated animals. Chem Biol Interact 102:69-78

32. Meissner P, Hift RJ, Corrigall A (2003) Variegate porphyria. In: Kadish KM, Smith K, Guilard R (eds) Porphyrin handbook, part II, vol 14. Academic Press, San Diego, chap 89, pp 93-120

33. Da Silva V, Simonin S, Deybach JC et al (1995) Variegate porphyria: diagnostic value of fluorometric scanning of plasma porphyrins. Clin Chim Acta 238:163-168

34. Long C, Smyth SJ, Woolf J et al (1993) Detection of latent variegate porphyria by fluorescence emission spectroscopy of plasma. Br J Dermatol 129:9-13

35. Cox TM (2003) Protoporphyria. In: Kadish KM, Smith K, Guilard R (eds) Porphyrin handbook, part II, vol 14. Academic Press, San Diego, chap 90, pp 121-149

36. Went LN, Klasen EC (1984) Genetic aspects of erythropoietic protoporphyria. Ann Hum Genet 48:105-117

37. Gouya L, Puy H, Robreau AM et al (2002) The penetrance of dominant erythropoietic protoporphyria is modulated by expression of wildtype FECH. Nat Genet 30:27-28

38. Bloomer J, Wang Y, Singhal A, Risheg H (2005) Molecular studies of liver disease in erythropoietic protoporphyria. J Clin Gastroenterol 39:S167-175

39. Do KD, Banner BF, Katz E (2002) Benefits of chronic plasmapheresis and intravenous heme-albumin in erythropoietic protoporphyria after orthotopic liver transplantation. Transplantation 73:469-472

40. Poh-Fitzpatrick MB, Wang X, Anderson KE et al (2002) Erythropoietic protoporphyria: altered phenotype after bone marrow transplantation for myelogenous leukemia in a patient heteroallelic for ferrochelatase gene mutations. J Am Acad Dermatol 46:861-866

VIII

IX Disorders of Metal Transport

37 Disorders in the Transport of Copper, Zinc and Magnesium

Roderick H.J. Houwen

Copper, Zinc and Magnesium

Copper is an essential component for a number of important metalloenzymes. Its absorption in the intestine, and excretion by the liver are tightly regulated to maintain adequate serum levels. This balance is disturbed in two inborn errors: Wilson disease and Menkes disease. *Wilson disease*, or hepatolenticular degeneration, is due to mutations in the *ATP7B* gene, encoding a copper-transport protein essential for the export of copper from the liver into bile. It is characterized by a gradual copper accumulation in the liver and, secondarily, in other organs, such as brain, kidney and cornea. Clinical symptoms result from copper accumulation in the liver and/or the brain. Early treatment with copper chelators or zinc is generally effective.

Menkes disease is a X-linked disorder due to mutations in the *ATP7A* gene, encoding a copper-transport protein required for the efflux of copper from cells. The disorder is characterized by a general copper deficiency. Patients manifest progressive neurodegeneration, which is usually fatal in infancy or childhood. Early therapy with copper histidine might have some benefits in selected patients.

Indian Childhood Cirrhosis (ICC), also known as *Idiopathic Copper Toxicosis* (ICT), is a rare copper storage disease seen in infants susceptible to high oral copper intake.

Zinc is a cofactor for over 100 enzymes and, as such, is involved in all major metabolic pathways. It is also essential for nucleic acid metabolism and protein synthesis and their regulation through so-called zinc-finger proteins. Zinc deficiency, either hereditary or acquired, has major detrimental effects, whereas high serum zinc has few, probably because of binding to albumin and α_2-macroglobulin.

Acrodermatitis enteropathica is due to mutations in the *SLC39A4* gene, encoding the major zinc importing carrier in the intestine. Symptoms typically start in infancy after the introduction of bottle feeding, and include periorificial and acral dermatitis, diarrhea, infections, and growth retardation. Therapy with zinc is extremely effective.

Zinc deficiency in breast fed babies presents with the same dermatological symptoms as acrodermatitis enteropathica, although the basic defect is probably different. Nevertheless, zinc therapy is equally effective.

Hyperzincemia with hypercalprotectinemia is characterized by extremely elevated levels of calprotectin thought to cause uncontrolled, harmful inflammatory reactions.

Autosomal dominant hyperzincemia without symptoms is most likely a non-disease.

Magnesium is the second most abundant intracellular cation and plays an essential role in many biochemical processes as well as neuromuscular excitability. Its homeostasis is regulated by the interplay between intestinal absorption and renal excretion.

Primary hypomagnesemia with secondary hypocalcemia generally presents in the first months of life with increased neuromuscular irritability or even frank convulsions. It is caused by mutations in the *TRPM6* gene, reducing uptake of magnesium from the gut. Magnesium suppletion is highly effective.

Hypomagnesemia with hypercalciuria and nephrocalcinosis provokes calcium deposition in the kidney, leading to renal failure, with few symptoms of hypomagnesemia. It is caused by mutations in the *CLDN16* gene, encoding a calcium and magnesium sensitive pore in the loop of Henle. Magnesium supplements do not prevent the development of end stage renal disease.

Isolated dominant hypomagnesemia provokes generalized convulsions and is caused by mutations in the *FXYD2* gene.

Isolated autosomal recessive hypomagnesemia has no other symptoms.

IX

37.1 Copper

37.1.1 Wilson Disease

Clinical Presentation

The overwhelming majority of cases display either hepatic or neurological symptoms, and the disease should be suspected in patients with liver disease without obvious cause or a movement disorder [1, 2]. In addition, the diagnosis is often made when siblings of a patient are screened. Occasionally, Wilson disease presents with isolated raised transaminases, Kayser-Fleischer rings or haemolysis.

Patients with hepatic symptoms generally present between 8 and 20 years of age, but may be as young as 3 or over 50. The presentation can be acute and severe with hepatitis, jaundice and impending liver failure. Transaminases, although raised, generally are much lower than in autoimmune or viral hepatitis [3]. While liver disease is rapidly progressive in some patients, in others jaundice can persist for months without progression to liver failure, or even subside. These patients ultimately develop liver cirrhosis and present several years later with neurological disease.

Neurological symptoms usually develop in the second or third decade, although patients may be as young as 8 years of age. Symptoms include dysarthria and diminished control of movements, accompanied in a later stage by tremors, rigidity and drooling in combination with swallowing problems. A frequent early sign is a deterioration in the quality of handwriting. In some patients psychiatric symptoms predominate, ranging from behavioural disturbances, often characterized by impulsivity and irritability, to frank psychosis.

Most patients have aminoaciduria in combination with excessive loss of bicarbonate, calcium and phosphate, and may develop renal stones or osteoporosis. Haemolytic anaemia, leading to gall-stones, may be present. Cardiomyopathy has also been described.

The greenish brown Kayser-Fleischer ring, located in the membrane of Descemet at the limbus of the cornea, can be seen with the naked eye in the majority of patients with full-blown neurological disease. Careful slit lamp examination will reveal this ring in almost all these patients. In contrast, in a substantial proportion of the patients presenting with liver disease and in most pre-symptomatic patients, the Kayser-Fleischer ring is absent. Conversely, a Kayser-Fleischer ring is occasionally found in patients with cholestatic liver disease. Its absence thus does not rule out Wilson disease, while its presence does not confirm the disorder.

Metabolic Derangement

Wilson disease is caused by reduced excretion of copper into bile, resulting in a gradual accumulation of copper in the liver and, secondarily, in the brain, kidneys and eye. A number of patients exhibit severe liver disease, while others redistribute copper to the brain, especially the basal ganglia, causing neurological disease. Copper excess exerts its hepatic toxicity by generating free radicals that oxidize the mitochondrial membranes, resulting in their swelling and loss of oxidative phosphorylation capacity. The characteristic Kayser-Fleischer ring is a deposit of copper and sulphur. The renal dysfunction is a consequence of copper accumulation in the renal tubules. The increased urinary copper excretion, characteristic for Wilson disease, is due to the loss of unbound, dialysable copper through the kidneys. This unbound copper can cause hemolysis in some patients.

The primary defect in Wilson disease is a lesion of a protein localized in the Golgi network, ATP7B, an adenosine triphosphatase (ATPase), which is responsible for the excretion of copper [4, 5] and for the incorporation of copper into ceruloplasmin. Owing to the reduced half-life of ceruloplasmin without copper, the concentration of serum ceruloplasmin is subnormal in Wilson disease. Rare patients, although unable to excrete copper into bile, can incorporate copper into ceruloplasmin and have normal serum ceruloplasmin [6].

Genetics

Wilson disease in an autosomal recessive condition caused by mutations in the *ATP7B* gene, localized on chromosome 13q14 [4, 5]. Its transcript, ATP7B, has six copper binding domains and is expressed predominantly in liver and kidney. ATP7B is highly homologous to APT7A, the protein defective in Menkes disease.

More than 200 mutations in the *ATP7B* gene have been described so far and are listed in the Wilson Disease Mutation Database (www.uofa-medical-genetics.org/wilson). The distribution of mutations within various racial groups is quite different, with the R778L mutation being common amongst Asian patients [7], the H1069Q mutation amongst European patients [8], and still other mutations being prevalent elsewhere. Most patients are compound heterozygotes. Mutations that completely destroy the function of the protein are generally found in patients who present early, while residual function is associated with late presentation. For example, patients homozygous for the non-functional R778L mutation tend to present earlier, with hepatic manifestations [7], whereas those homozygous for the H1069Q mutation present relatively late (i.e. around 21 years of age), with neurological symptoms, indicative of a relative slow build up of copper [8].

Diagnostic Tests

Wilson disease is characterized by low serum ceruloplasmin and serum copper, elevated urinary copper, and increased liver copper (◘ Table 37.1). These laboratory results should only be interpreted in combination, because each individual parameter can be abnormal in situations other than Wilson disease [9]. For example, liver copper is raised in liver cirrhosis, whereas serum ceruloplasmin is low in a substantial proportion of heterozygotes for Wilson disease, and in

◻ **Table 37.1.** Laboratory results in Wilson disease and controls

	Wilson disease	Normal
Serum ceruloplasmin (mg/l)	0–200	200–400
Serum copper (µmol/l)	<11	11–24
Urinary copper (µmol/24 h)	>1.6	<0.6
Liver copper (µg/g dry weight)	>250	<50

patients with hereditary aceruloplasminemia. Conversely, serum ceruloplasmin is normal in a small proportion of patients with Wilson disease.

Since over 90% of serum copper is normally bound to ceruloplasmin, it is generally low when serum ceruloplasmin is low, as is the case in Wilson disease. Characteristically the fraction of serum copper not bound to ceruloplasmin, called free serum copper, is raised. This sensitive parameter can be calculated with the knowledge that each mg of ceruloplasmin contains 3.4 µg of copper, provided the laboratory can reliably measure ceruloplasmin concentrations in the subnormal ranges, i.e. <200 mg/l.

Urinary copper excretion is determined in a 24 h collection, but is sensitive to contamination. Excretion is always increased in symptomatic patients, but may be normal or only borderline elevated in presymptomatic individuals. The diagnostic value of this parameter might be improved by administering a loading dose of penicillamine.

When Wilson disease is diagnosed in a family, siblings should be investigated. Analysis of mutations, or using closely linked markers, is more reliable than laboratory investigations of copper metabolism which cannot always distinguish between carriers and young patients who still have a low copper load.

Treatment and Prognosis

Prognosis is excellent for patients who start treatment before severe tissue damage has occurred, i.e. when presymptomatic or diagnosed at an early stage. Prognosis can still be good for those with more advanced disease, provided aggressive decoppering treatment is instituted immediately after diagnosis. Several therapeutic agents are available: penicillamine, trien and zinc. Tetrathiomolybdate is a relative new agent and experience is limited so far.

The first agent, penicillamine, has provided the largest experience. Penicillamine chelates copper by forming a stable complex that is subsequently excreted in urine. The initial dose for adults is 1–2 g/day, divided in four doses, together with 25 mg/day of pyridoxine. Approximately half of the patients with liver disease will recover, while the other half will need a transplant [10]. Of patients with neurological disease, approximately half will totally recover, 25% will recover but still have some residual disabilities, and 25% will either recover with severe remaining disabilities or die [11]. Of note, a significant proportion of patients with neurological disease will have an initial worsening of symptoms after starting penicillamine therapy. For these patients the chances of a total recovery are less. In addition, side effects and toxic reactions are seen in up to 20% of the patients treated with penicillamine and therapy has to be stopped in many. Given this suboptimal safety profile, alternatives for penicillamine have been sought, with trien (trientine) being the first to be introduced. This agent is also a copper chelator, with an efficacy that is approximately similar to penicillamine. However side effects seem less common [12].

Oral zinc has been used in the treatment of Wilson disease for more than 25 years. It induces metallothionein synthesis in the small intestinal epithelium. Since metallothionein binds copper preferentially over zinc, copper balance will become negative through faecal excretion, as villus cells are lost into the intestinal lumen. As compared to penicillamine, zinc does not have any serious side effects, although some patients experience gastric complaints on zinc sulphate. This can generally be solved by switching to zinc gluconate or zinc acetate. Given its favourable side effect profile, zinc seems the agent of choice in presymptomatic individuals. In patients with symptomatic disease (particularly with neurological symptoms) a small non-randomized, non-blinded trial showed similar outcomes for zinc and penicillamine [13]. Given the side effects of penicillamine and the frequency of initial deterioration in patients with neurological disease, zinc should be seriously considered in this group. In patients with hepatic disease, which can evolve rapidly, zinc seems less appropriate because it may have a slower effect on copper overload. Obviously, more trials are needed before final conclusions can be drawn. The initial dose of zinc sulphate for adults is 600 mg/day, divided in 3 doses; this dose can be doubled if insufficient effect is obtained. Urinary copper excretion should be followed: it should fall rapidly initially, and more slowly thereafter. A reasonable goal is to achieve an excretion below 2 µmol/day [1]. Copper depletion should be avoided: in the maintenance phase, 300 mg/day or even less can be sufficient.

Tetrathiomolybdate, a copper chelating agent with greater affinity for copper than penicillamine, has been used mainly for initial decoppering of patients with neurological symptoms [14]. The initial detioration, often seen in patients treated with penicillamine, appears to occur less frequently. Based on theoretical considerations and animal experiments, this agent could also have a place in the treatment of patients with liver disease, as current treatment modalities are suboptimal.

In patients presenting with severe liver disease, sufficient experience is only available for penicillamine. In this group, at least half will require a liver transplant [10]. Therefore other treatment modalities have been tried, such as the combination of zinc and penicillamine (or trien), tetrathiomolybdate, or addition of high dose vitamin E to the copper chelating therapy. Restoring normal plasma vitamin E levels seems to protect liver mitochondria against oxidative damage, and might be helpful in reversal of liver damage. However, none of these interventions have been investigated in a substantial number of patients.

37.1.2   Menkes Disease

Clinical Presentation

Symptoms generally appear at the age of 2 to 3 months, in males, when the neurodegeneration provoked by the disease becomes manifest with seizures and hypotonia [15]. Sometimes, non-specific signs can be present at birth, including prematurity, large cephalhematomas, skin laxicity and hypothermia, which are often not recognized as Menkes disease at that time. The hair, if present, can already exhibit the characteristic pili torti, which will appear later on in all. Patients loose earlier developmental milestones and, progressively, hypotonia is replaced by spasticity. A typical facial appearance, with sagging cheeks and frontal bossing, gradually becomes prominent. Feeding difficulties, vomiting and/or chronic diarrhea are common, and weight gain is generally insufficient; nevertheless, linear growth is relatively preserved. The loose skin, which is particularly prominent at the back of the neck and on the trunk, is a consequence of defective collagen crosslinking, as are the vascular tortuosity and bladder diverticula, which are present in virtually all patients. The latter are a frequent source of infection. Umbilical or inguinal hernias and/or a pectus excavatum are also commonly encountered.

Besides the more prevalent, severe Menkes phenotype, less severe forms occur in 10–15% of the patients, with the *occipital horn syndrome* being the mildest. This syndrome is characterized by connective tissue abnormalities with minimal effects on neurodevelopment [16]. Bone disease with demineralization, deformities and exostoses, particularly at the occipital insertion of the paraspinal muscles (hence its name), are characteristic. Furthermore, patients have urinary tract diverticuli, orthostatic hypotension and chronic diarrhea. Skin and joint laxity are common, but pili torti are rarely seen.

Metabolic Derangement

In Menkes disease, cellular copper uptake is normal, but copper cannot be exported from cells due to a defect of the ATP7A protein, a copper transporter localized in the Golgi network. When intracellular copper rises, the normal ATP7A protein is redistributed to a cytoplasmic vesicular compartment and the plasma membrane [17]. This renders copper available for excretion and incorporation into the enzymes that require copper. When ATP7A is defective, these pathways are blocked. Consequently, copper efflux from the intestinal cells is severely reduced, and insufficient copper will reach the circulation, pass the blood-brain barrier, and be incorporated into the cuproenzymes. (although specific mutations exist in which this latter function is spared [18]). Among the affected copper-requiring enzymes in the brain are dopamine β–hydroxylase, which is essential for catecholamine biosynthesis, peptidyl glycine monooxygenase, involved in the processing of neuropeptide precursors, and cytochrome-c-oxidase. Deficient activity of these enzymes is probably responsible for a significant part of the cerebral pathology in Menkes disease. Dysfunction of Cu/Zn superoxide dismutase seems to be compensated for by an increased activity of manganese superoxide dismutase, and as such probably does not contribute much to the neurodegeneration. Other enzymes influenced by copper deficiency are lysyloxidase, a critical enzyme in collagen cross-linking, and tyrosinase which is necessary for melanin formation.

Genetics

A rare condition with an incidence of approximately 1:250,000 [15], Menkes disease is inherited as an X-linked recessive trait. It is caused by mutations in the *ATP7A* gene, localized on chromosome Xq13.3, and expressed in all tissues, except liver. Its protein product, ATP7A, is highly homologous to APT7B, the protein defective in Wilson disease. The mutation spectrum in Menkes disease is wide, with lesions throughout the gene, without predominant mutations. Seven patients have been reported with chromosome abnormalities, mostly X-autosome translocations, visible on cytogenetic examination [19]. Gross deletions in the gene, encompassing one or more exons, or even almost the whole coding sequence, are found in approximately 15% of the cases [19]. Many single base pair changes or insertion/deletions of a few base pairs have been described. The vast majority of these mutations are predicted to introduce a premature stop codon, probably resulting in a truncated, non-functional protein. No straightforward genotype/phenotype correlations have been found so far, although most patients with the occipital horn syndrome have splice site mutations that potentially permit small amounts of *ATP7A* to be transcribed [20].

Diagnostic Tests

Reduced levels of serum copper (<11 µmol/l) and serum ceruloplasmin (<200 mg/l) support the diagnosis, but are not specific, since infants in the first months of life generally have low levels. An abnormal ratio between catecholamine metabolites in plasma and cerebrospinal fluid seems to be quite specific for Menkes disease [21], as is reduced urinary excretion of deoxypyridinoline, a metabolite formed in the cross-linking of collagen [22]. The copper retention, characteristic of Menkes disease, can be demonstrated by measuring the increased accumulation and reduced efflux of radiocopper in cultured fibroblasts [23]. Final diagnosis requires identification of the mutation.

Prenatal diagnosis is preferably done by mutation analysis. If the mutation is unknown DNA studies can still be informative by using intragenic microsatellite markers. Carrier detection too should be done by DNA analysis, especially as biochemical studies of copper accumulation in fibroblasts can give false negative results due to random inactivation of the X-chromosome.

Treatment and Prognosis

Classically, most patients die before three years of age due to infections or vascular complications, although with current medical care (improved feeding techniques) longer survival is not uncommon. Treatment is mainly symptomatic. Nevertheless, since symptoms can be attributed to insufficient copper for synthesis of cuproenzymes, a logical approach would be to administer parenteral copper to bypass the intestinal block, thereby making more copper available for incorporation into cuproenzymes. To this aim a number of inorganic copper salts have been used without clinical improvement. However, treatment with copper histidine, the physiological copper complex in humans, had significant clinical effects in four patients, resulting in near normal intellectual development, although the connective tissue abnormalities persisted [24]. Treatment in these patients was initiated in the first few months of life, which might have been a crucial factor, since copper treatment of brindled mice, a model for Menkes disease, prevented neurological damage, but only if started at day 7, while administration at day 10 was ineffective. Unfortunately however, early treatment with copper histidine in a larger series of 11 infants did not prevent death in 5 [15]. This therapy should nevertheless be considered for patients identified at an early age. When treatment is started after the onset of symptoms, meaningful neurological recovery seems impossible, although reduced irritability has been reported. Some evidence suggests that active ATP7A protein, albeit at a very low level, should be present for copper histidine therapy to work [18]. Still, 2 out of 4 patients succesfully treated by Christodoulou et al [24] had premature stop codons in ATP7A, reasonably preventing any functional protein to be synthesized.

37.1.3   Other Copper Storage Disorders

Indian Childhood Cirrhosis (ICC), is characterized by a normal serum ceruloplasmin and an extremely high liver copper (800–6500 µg/g dry weight) [25]. It is seen solely in young children. The usual outcome is liver failure, although this can be reverted by early decoppering therapy. The disorder is caused by an increased dietary copper intake in genetically susceptible individuals, due to the use of copper utensils when cooking milk. Eliminating this practice has virtually eradicated ICC. Although the disease is confined to India (hence its name) a similar disease has been seen in Tyrol (*Endemic Tyrolean Infantile Cirrhosis*, ETIC), which is also caused by using copper vessels when preparing milk [26]. Sporadic cases from all over Europe and Northern America have been described (generally labelled *Idiopathic Copper Toxicosis,* ICT), mostly associated with a high copper content of water in certain wells. Given the similarities in clinical and biochemical characteristics it seems possible that all three entities are in fact one and the same disease. Since many of the patients are from consanguineous families, it is probable that an autosomal recessive mutation is responsible. *MURR1*, the gene mutated in the copper toxicosis seen in Bedlington terriers, has been excluded as a candidate gene [27]. A human equivalent of the copper storage disease in Bedlington terriers has not yet been identified.

37.2   Zinc

37.2.1   Acrodermatitis Enteropathica

Clinical Presentation

Children with acrodermatitis enteropathica (AE) are healthy at birth, but develop symptoms some weeks after breast feeding has been stopped. The most striking clinical feature is a severe dermatitis, classically localized at the acral and periorificial sites [28, 29]. At onset, these skin lesions are erythematous, while after the first year of life pustular and hyperkeratotic changes become more prominent. Secondary infection with *Candida Albicans* and/or *Staphylococcus Aureus* is not uncommon. In addition to the skin lesions, seen in almost all patients, intermittent diarrhea can develop, which in more advanced stages can progress to intractable watery diarrhea and failure to thrive. If untreated, a significant fraction of the patients will have a gradual downhill course, although the majority seems to be able to survive without treatment into adulthood. Mood changes are an early sign of zinc deficiency, presenting as apathy and irritability in infancy and later on as depression. Infections are also frequent, and can be life threatening. Other clinical features include alopecia and nail deformities, as well as ophthalmological symptoms such as blepharitis, conjunctivitis and photophobia.

Metabolic Derangement

AE is caused by a partial block in the intestinal absorption of zinc, as demonstrated in vivo by oral application of ^{65}Zn [30]. Likewise, zinc absorption in intestinal biopsies of patients is reduced [31]. This defect is due to dysfunction of the protein involved in AE (ZIP4). The insufficient zinc absorption results in severe zinc deficiency with impairment of the function of many enzymes that have zinc as cofactor. Tissues with a high cellular turnover, such as skin, intestine, and lymphoid system are most severely affected.

Genetics

AE is an autosomal recessive disease caused by mutations in the *SLC39A4* gene localized on chromosome 8q24.3 [32, 33]. *SLC39A4* encodes a zinc transporter, ZIP4, with eight transmembrane domains, which probably form a zinc channel, and is expressed at the apical membrane of the enterocytes. Over 20 mutations have been identified so far, mainly in families from Europe, the Middle-East and North-Africa [34].

Diagnostic Tests

In most patients, serum zinc levels are lower (7.1±5.0 µmol/l) than normal (11.9–19.4 µmol/l) although values within the normal range are found in at least 15 % of patients [29]. Measurements of zinc in other tissues, such as hair and red or white blood cells, do not seem to improve diagnostic accuracy. In addition, several conditions, such as chronic diarrhea due to other causes, can present with low serum zinc. Therefore the diagnosis of AE can never be based on serum zinc. Other tests may contribute to a certain extent: low urinary zinc excretion (reflecting a low serum zinc level), low serum alkaline phosphatase activity, changes in the serum fatty acid profile, hypobetalipoproteinemia, reduction of serum vitamin A, and elevation of blood ammonia. In many patients, both humoral and cell-mediated immunity are depressed [35]. Small bowel biopsy generally shows partial to subtotal villous atrophy and Paneth cell inclusions on electron microscopy.

The defect in active zinc transport can be proven with radiolabeled zinc [30]. However, since this might not be available in most settings, a practical approach is to start zinc therapy when the clinical diagnosis is suspected, and await the response, which should occur within one week. When the clinical signs of acrodermatitis were equivocal one may consider to temporarily withdraw zinc therapy after some time to provoke a relapse, and in this way differentiate between true AE (which will relapse quickly) and acquired zinc deficiency.

Treatment and Prognosis

Before zinc supplementation was serendipitously found to correct the abnormalities in AE, patients were given breast milk and later on iodo-hydroxyquinolines. This generally resulted in partial or even total remission. Zinc therapy was introduced in 1975 [36], and is now used in all patient with AE. The usual dose is 150–400 mg zinc sulphate/day (equivalent to 35–90 mg elemental zinc/day), on which patients will start to show clinical improvement within days. Simultaneously, laboratory abnormalities such as serum zinc levels, urinary zinc excretion and alkaline phosphatase activity will normalize. Generally, the initial dose can be maintained throughout childhood, although some patients may need an increase during their growth spurt. After puberty, the requirements for zinc may be lower, but during pregnancy and lactation 400–500 mg zinc sulphate/day is needed. If the preparation causes gastric problems it may be encapsulated, or alternatively zinc gluconate or other zinc salts may be used. As zinc therapy will decopper patients it is necessary to monitor serum copper, and either reduce the dose of zinc or supplement copper if a deficiency is found. Prognosis is excellent since the introduction of zinc supplementation.

37.2.2   Zinc Deficiency in Breastfed Babies

Rarely, zinc deficiency with acrodermatitis can occur in breast-fed babies, especially in premature infants, as they have an increased zinc requirement in combination with a reduced capacity for zinc uptake in the gut [37]. Although this condition responds rapidly to oral zinc supplements, it is clearly different from AE, as it is seen exclusively during breast feeding and no impairment of intestinal zinc uptake can be found. The deficiency is caused by reduced levels of zinc in maternal milk, and its inheritance might be autosomal recessive [38].

37.2.3   Hyperzincemia with Hyper-calprotectinemia

Sampsom [39] described 5 patients with a new syndrome defined by high plasma zinc (77–200 µmol/l), recurrent infections, hepatosplenomegaly, arthritis, anemia and persistently raised concentrations of C-reactive protein. The majority of these patients also had severe growth retardation. Levels of serum calprotectin, the major zinc binding protein of phagocytes, were more than 1000 times the upper limit of normal. It is speculated that the very high concentration of this protein results in the uncontrolled and harmful inflammatory reactions which characterize this syndrome, while the hyperzincemia is caused by the zinc capturing properties of calprotectin. Inheritance of this syndrome is not clear yet.

37.2.4   Autosomal Dominant Hyper-zincemia Without Symptoms

Elevated serum zinc (40–70 µmol/l) was described by Smith et al in seven family members from one large pedigree. The condition seems to be inherited in an autosomal dominant fashion. Zinc concentrations in hair and erythrocytes were normal, as was serum albumin, to which most of the excess zinc seemed to be bound. There were no clinical symptoms, nor additional biochemical abnormalities, so this condition appears to be benign [40].

37.3   Magnesium

37.3.1   Primary Hypomagnesemia with Secondary Hypocalcemia

Clinical Presentation

Primary hypomagnesemia with secondary hypocalcemia (HSH) is a rare autosomal recessive disorder. It was first recognized in 1965 and since then more than 50 infants from all over the world have been described [41, 42]. Patients commonly present in the first months of life with generalized seizures or other symptoms of increased neuromuscular excitability such as irritability, poor sleeping, muscle spasms and/or tetany.

Metabolic Derangement

Primary hypomagnesemia is caused by impaired magnesium uptake from the gut [43]. A lowered renal threshold for magnesium may be a contributing factor [44]. The disease is caused by a defect of a protein, TRPM6, a member of the long transient receptor potential channel (TRPM) family, which complexes with its closest homolog, TRPM7, to form an ion-channel for magnesium at the cell surface. Genetic lesions of TPRM6 prevent assembly of this complex and hence impair magnesium transport [45].

Severe hypomagnesemia blocks synthesis and/or release of parathormone. In addition, when hypomagnesemia is present, the administration of parathormone (PTH) fails to induce a rise in serum calcium. The hypocalcemia in HSH is thus secondary to low parathormone levels in combination with some form of end organ resistance.

Genetics

Although a male/female ratio of 4 in the first reported patients led to the initial proposal of X-linked inheritance, further genetic investigations indicated autosomal recessive inheritance. This was clearly established when the gene was localized to a small interval on chromosome 9q22 by homozygosity mapping in three interrelated Bedouin kindreds. Within this interval, two groups subsequently identified mutations of the *TRPM6* gene [44, 46]. This gene is expressed in the small and large intestine as well as in the cells lining the distal tubules.

Diagnostic Tests

Primary hypomagnesemia is characterized by a very low serum magnesium (0.24 ± 0.11 mmol/l; normal 0.65–1.20 mmol/l) in combination with a low serum calcium (1.64 ± 0.41 mmol/l; normal 2.12–2.70 mmol/l). In the presence of serum hypomagnesemia, the urinary excretion of magnesium is reduced, and PTH levels are inappropriately low. No evidence for malabsorption of other nutrients is found, and renal function is not otherwise compromised.

Treatment and Prognosis

Untreated, the disorder will result in permanent neurological damage or death. However, magnesium supplementation corrects all clinical symptoms. Initially, magnesium should be given intravenously. The exact dose depends on the response of the patient, but is usually in the range of 0.5–1.5 ml/kg/day of a $MgSO_4$ 10% solution. After stabilization, therapy can be continued orally in an amount that must be adjusted to the clinical response. In a series of 15 patients the individual dosage varied between 0.7 and 3.5 mmol/kg/day of elemental magnesium. On this regimen, serum calcium normalized, but serum magnesium remained just below normal (0.53 ± 0.12 mmol/l) [42]. Dividing oral magnesium supplementation in three to five doses will reduce fluctuations of serum magnesium and will prevent the development of chronic diarrhea in many, but not all patients.

The prognosis of primary hypomagnesemia is good if the diagnosis is made early; with treatment growth and development is normal. However, patients who have frequent hypomagnesemia/hypocalcemia-induced convulsions, either before or after the diagnosis is made, are at risk for developing psychomotor retardation.

37.3.2   Hypomagnesemia with Hypercalciuria and Nephrocalcinosis

Clinical Presentation

Over 80 patients with familial hypomagnesemia with hypercalciuria and nephrocalcinosis (FHHNC) have been reported [47, 48]. Patients usually present during childhood with recurrent urinary tract infections, polyuria/polydipsia and/or hematuria. At presentation, renal stones are seen in 13–25% of patients, while nephrocalcinosis, rare at presentation, will ultimately develop in all. Clinical signs of hypomagnesemia such as seizures are less common, in line with only moderately depressed serum magnesium level. Ocular involvement, e.g. severe myopia and macular colobomata, is seen in a significant proportion of patients.

Metabolic Derangement

FHHNC is caused by a defect of paracellin-1, a protein localized in the thick ascending limb of Henle and the distal tubulus [49]. This is where magnesium and calcium are passively reabsorbed through the paracellular pathway. Paracellin-1, as part of the tight junction, is thought to contribute to the formation of a calcium and magnesium sensitive pore, through which this reabsorption takes place. Disturbance of this process leads to renal loss of magnesium and calcium, with secondary development of nephrocalcinosis and ultimately renal failure.

Genetics

The gene encoding paracellin-1, *CLDN16* (formerly *PCLN-1*), belongs to the claudin multigene family [49] and is localized on chromosome 3q27-q29. So far, over 20 distinct mutations have been identified, all single base pair changes. First degree family members of patients with FHHNC have a tendency towards mild hypomagnesemia, hypercalciuria and renal stone formation, indicating that heterozygosity for *CLDN16* mutations also predisposes to a mildly disturbed renal handling of magnesium and calcium. Interestingly, *CLDN16* is also expressed in the cornea and retinal epithelium, thereby providing a link between defects in paracellin-1 and the ocular pathology observed in some patients.

Diagnostic Tests

Serum magnesium is low (mean 0.40 mmol/l, range 0.23–0.61 mmol/l) [48], but less so than in primary hypomagnesemia. Median calcium excretion is 10.0 mg/kg/24 h (normal 4–6 mg/kg/24 h). Serum calcium is somewhat below the lower level of normal in about half of the patients. Other biochemical abnormalities include hypocitraturia and mild hyperuricemia. At diagnosis, glomerular filtration rate is already reduced in the majority of patients, and subsequently deteriorates further. Renal sonography shows nephrocalcinosis, with its characteristic medullary distribution, early in the course of the disease.

Treatment and Prognosis

Oral magnesium salts are used to supplement renal loss, while thiazide diuretics are given to reduce calcium excretion rates in an effort to prevent the progression of nephrocalcinosis, which correlates with development of renal failure. However, these strategies do not seem to significantly influence the progression of renal failure. In a recent series of 33 patients, all showed a deterioration in glomerular filtration rate, and one third developed end stage renal disease during adolescence [48]. The median age at end stage renal disease in this group was 14.5 years (range 5.5–37.5 years).

37.3.3 Isolated Dominant Hypomagnesemia

This disorder was first described by Geven et al in two Dutch families [50]. The index cases presented with generalized convulsions, which led to the detection of the hypomagnesemia (0.40 mmol/l; normal 0.65–1.20 mmol/l). Subsequent evaluation showed a reduced tubular threshold for magnesium in combination with lowered calcium excretion. Autosomal dominant inheritance was evidenced by investigation of the families of the two probands: the same combination of hypomagnesemia and hypocalciuria was found in 22 out of 47 family members. Interestingly, none of them had any clinical symptom of magnesium deficiency.

In the two families, a locus for this disorder was mapped to chromosome 11q23, revealing a similar haplotype for all cases in both pedigrees, which suggests a common ancestor. Within the *FXYD2* gene, residing in this interval, a hetero-

zygous G123A mutation was identified [51]. This gene encodes the γ-subunit of a Na^+K^+-ATPase, which is expressed in the distal tubules, the main site of renal magnesium reabsorption. Obviously normal function of the Na^+K^+-ATPase is necessary for adequate renal magnesium handling, and the mutation identified in the γ-subunit specifically impairs its activity, accounting for the dominant negative effect of the mutation seen in these families. The exact pathophysiologic mechanism leading to the low serum magnesium and the associated low urinary calcium excretion is not yet clear. The disorder seems genetically heterogeneous since an American family with a similar phenotype has been described that does not map to the 11q23 locus [52].

37.3.4 Isolated Autosomal Recessive Hypomagnesemia

Isolated autosomal recessive hypomagnesemia has been described in two children from a consanguineous family [53]. Apart from the hypomagnesemia due to increased urinary magnesium excretion, no biochemical abnormality was reported. This disorder can be distinguished from autosomal dominant hypomagnesemia by the normal calcium excretion in the urine.

37.3.5 Other Metals

Aceruloplasminemia is an autosomal recessive disorder characterized by accumulation of iron in liver, spleen, pancreas, retina and basal ganglia by the fourth or fifth decade of life [54, 55]. Clinically the disease consists of the triad of adult-onset neurological disease (chorea, cerebellar ataxia, dystonia, Parkinsonism and psychiatric signs), retinal degeneration and diabetes mellitus. The elevated iron concentration is associated with increased lipid peroxidation suggesting that increased oxidative stress is involved in neuronal cell death. More than 30 aceruloplasminemia-causing mutations in the ceruloplasmin gene have been identified. Desferrioxamine, a high-affinity iron chelator, reduces body iron stores and may therefore ameliorate diabetes as well as hepatic and neurological symptoms [56].

Manganese-related disease (prolidase deficiency) is discussed in ► Chap. 30; molybdenum-related disease (combined deficiency of sulfite oxidase and xanthine oxidase) is discussed in ► Chap. 35.

References

1. Brewer GJ, Yuzbasiyan-Gurkan V (1992) Wilson disease. Medicine 71:139-164
2. Houwen RHJ, van Hattum J, Hoogenraad TU (1993) Wilson disease. Neth J Med 43:26-37
3. Strand S, Hofmann WJ, Grambihler A et al (1998) Hepatic failure and liver cell damage in acute Wilson's disease involve CD95 (APO-1/Fas) mediated apoptosis. Nat Med 4:588-593

4. Bull PC, Thomas GR, Rommens JM et al (1993) The Wilson disease gene is a putative copper transporting P-type ATPase similar to the Menkes gene. Nat Genet 5:327-337

5. Tanzi RE, Petrukhin K, Chernov I, et al (1993) The Wilson disease gene is a copper transporting ATPase with homology to the Menkes disease gene. Nat Genet 5:344-350

6. Forbes JR, Cox DW (2000) Copper-dependent trafficking of Wilson disease mutant ATP7B proteins. Hum Mol Genet 9:1927-1935

7. Liu XQ, Zhang YF, Liu TT et al (2004) Correlation of ATP7B genotype with phenotype in Chinese patients with Wilson disease. World J Gastroenterol 10:590-593

8. Stapelbroek JM, Bollen CW, Ploos van Amstel JK, et al (2004) The H1069Q mutation in ATP7B is associated with late and neurologic presenttaion in Wilson disease: results of a meta-analysis. J Hepatol 41:758-763

9. Ferenci P, Caca K, Loudianos G et al (2003) Diagnosis and phenotypic classification of Wilson disease. Liver International 23:139-142

10. Nazer H, Ede RJ, Mowat AP, Williams R (1986) Wilson's disease: clinical presentation and use of prognostic index. Gut 27:1377-1381

11. Walshe JM, Yealland M (1993) Chelation treatment of neurological Wilson's disease. Q J Med 86:197-204

12. Dahlman T, Hartvig P, Löfholm M et al (1995) Long-term treatment of Wilson's disease with triethylene tetramine dihydrochloride (trientine). Q J Med 88:609-616

13. Czlonkowska A, Gajda J, Rodo M (1996) Effects of long-term treatment in Wilson's disease with D-penicillamine and zinc sulphate. Neurol 243:269-273

14. Brewer GJ, Hedera P, Kluin KJ et al (2003) Treatment of Wilson disease with Ammonium Tetrathiomolybdate. III. Initial therapy in a total of 55 neurologically affected patients and follow-up with zinc therapy. Arch Neurol 60:379-385

15. Kaler SG (1998) Diagnosis and therapy of Menkes syndrome, a genetic form of copper deficiency. Am J Clin Nutr 67:1029S-1034S

16. Tsukahara M, Imaizumi K, Kawai S, Kajii T (1994) Occipital horn syndrome: report of a patient and review of the literature. Clin Genet 45:32-35

17. Petris MJ, Mercer JFB (1999) The Menkes protein (ATP7A;MNK) cycles via the plasma membrane both in basal and elevated extracellular copper using a C-terminal di-leucine endocytic signal. Hum Mol Genet 8:2107-2115

18. Kim BE, Smith K, Petris MJ (2003) A copper treatable Menkes disease mutation associated with defective trafficking of a functional Menkes copper ATPase. J Med Genet 40:290-295

19. Tümer Z, Møller LB, Horn N (2003) Screening of 383 unrelated patients affected with Menkes disease and finding of 57 gross deletions in ATP7A. Hum Mutat 22:457-464

20. Møller LB, Tümer Z, Lund C et al (2000) Similar splice-site mutations of the ATP7A gene lead to different phenotypes: classical Menkes disease or occipital horn syndrome. Am J Hum Genet 66:1211-1220

21. Kaler SG, Goldstein DS, Holmes C et al (1993) Plasma and cerebrospinal fluid neurochemical pattern in Menkes disease. Ann Neurol 33:171-175

22. Kodoma H, Sato E, Yanagawa Y et al (2003) Biochemical indicator for evaluation of connective tissue abnormalities in Menkes' disease. J Pediatr 142:726-728

23. Tümer Z, Horn N (1998) Menkes disease: Underlying genetic defect and new diagnostic possibilities. J Inherit Metab Dis 21:604-612

24. Christodoulou J, Danks DM, Sarkar B et al (1998) Early treatment of Menkes disease with parenteral cooper-histidine: long-term follow-up of four treated patients. Am J Med Genet 76:154-164

25. Tanner MS (1998) Role of copper in Indian childhood cirrhosis. Am J Clin Nutr 67:1074S-1081S

26. Müller T, Feichtinger H, Berger H, Müller W (1996) Endemic Tyrolean infantile cirrhosis: an ecogenetic disorder. Lancet 347:877-880

27. Müller T, van de Sluis B, Zhernakova A et al (2003) The canine copper toxicosis gene MURR1 does not cause non-Wilsonian hepatic copper toxicosis. J Hepatol 38:164-168

28. Aggett PJ (1983) Acrodermatitis enteropathica. J Inherit Metab Dis 6:39S-43S

29. Van Wouwe JP (1989) Clinical and laboratory diagnosis of acrodermatitis enteropathica. Eur J Pediatr 149:2-8

30. Lombeck I, Schnippering HG, Ritzl F et al (1975) Absorption of zinc in acrodermatitis enteropathica. Lancet i:855

31. Atherton DJ, Muller DPR, Aggett PJ, Harries JT (1979) A defect in zinc uptake by jejunal biopsies in acrodermatitis enteropathica. Clin Sci 56:505-507

32. Küry S, Dréno B, Bézieau S et al (2002) Identification of SLC39A4, a gene involved in acrodermatitis enteropathica. Nat Genet 31:239-240

33. Wang K, Zhou B, Kuo YM et al (2002) A novel member of a zinc transporter family is defective in acrodermatitis enteropathica. Am J Hum Genet 71:66-73

34. Küry S, Kharfi M, Kamoun R et al (2003) Mutation spectrum of human SLC39A4 in a panel of patients with Acrodermatitis Enteropathica. Hum Mutat 22:337-338

35. Antilla PH, Von Willebrand E, Simell O (1986) Abnormal immune responses during hypozincaemia in acrodermatitis enteropathica. Acta Paediatr Scand 75:988-992

36. Neldner KH, Hambidge KM (1975) Zinc therapy of acrodermatitis enteropathica. N Engl J Med 292:879-882

37. Stevens J, Lubitz L (1998) Symptomatic zinc deficiency in breast-fed term and premature infants. J Paed Child Health 34:97-100

38. Sharma NL, Sharma RC, Gupta KR, Sharma RP (1988) Self-limiting acrodermatitis enteropathica. A follow-up study of three interrelated families. Int J Dermatol 27:485-486

39. Sampsom B, Fagerhol MK, Sunderkötter C et al (2002) Hyperzincaemia and hypercalprotectinaemia: a new disorder of zinc metabolism. Lancet 360:1742-1745

40. Smith JC, Zeller JA, Brown ED, Ong SC (1976) Elevated plasma zinc: a heritable anomaly. Science 193:496-498

41. Dudin KI, Teebi AS (1987) Primary hypomagnesaemia. A case report and literature review. Eur J Pediatr 146:303-305

42. Shalev H, Phillip M, Galil A et al (1998) Clinical presentation and outcome in primary familial hypomagnesaemia. Arch Dis Child 78:127-130

43. Milla PJ, Aggett PJ, Wolff OH, Harries JT (1979) Studies in primary hypomagnesaemia: evidence for defective carrier-mediated small intestinal transport of magnesium. Gut 20:1028-1033

44. Walder RY, Landau D, Meyer P et al (2002) Mutation of TRPM6 causes familial hypomagnesemia with secondary hypocalcemia. Nat Genet 31:171-174

45. Chubanov V, Waldegger S, Schnitzler MM et al (2004) Disruption of TRPM6/TRPM7 complex formation by a mutation in the TRPM6 gene causes hypomagnesemia with secondary hypocalcemia. Proc Natl Acad Sci USA 101:2894-2899

46. Schlingmann KP, Weber S, Peters M et al (2002). Hypomagnesemia with secondary hypocalcemia is caused by mutations in TRPM6, a new member of the TRPM family. Nat Genet 31:166-170

47. Benigno V, Canonica CS, Bettinelli A et al (2000) Hypomagnesaemia-hypercalciuria-nephrocalcinosis: a report of nine cases and a review. Nephrol Dial Transplan 15:605-610

48. Weber S, Schneider L, Peters M et al (2001) Novel paracellin-1 mutations in 25 families with familial hypomagnesemia with hypercalciuria and nephrocalcinosis. J Am Soc Nephrol 12:1872-1881

49. Simon DB, Lu Y, Choate KA et al (1999) Paracellin-1 a renal tight junction protein required for paracellular Mg2+ resorption. Science 285:103-106

50. Geven WB, Monnens LA, Willems HL et al (1987) Renal magnesium wasting in two families with autosomal dominant inheritance. Kidney Int 31:1140-1144

51. Meij IC, Koenderink JB, van Bokhoven H et al (2000) Dominant isolated renal magnesium loss is caused by misrouting of the Na^+K^+-ATP-ase γ-subunit. Nat Genet 26:265-266

52. Kantorovich V, Adams JS, Gaines JE et al (2002) Genetic heterogeneity in familial renal magnesium wasting. J Clin Endocrinol Metab 87:612-617

53. Geven WB, Monnens LAH, Willems JL et al (1987) Isolated autosomal recessive renal magnesium loss in two sisters. Clin Genet 32:398-402

54. Miyajima H, Nishimura Y, Mizoguchi K et al (1987) Familial apoceruloplasmin deficiency associated with blepharospasm and retinal degeneration. Neurology 37:761-767

55. Kono S, Miyajima H (2006) Molecular and pathological basis of aceruloplasminemia. Biol Res 39:15-23

56. Miyajima H, Takahashi Y, Kamata T (1997) Use of desferrioxamine in the treatment of aceruloplasminemia. Ann Neurol 41:404-407

X Organelle-Related Disorders: Lysosomes, Peroxisomes, and Golgi and Pre-Golgi Systems

38 Disorders of Sphingolipid Metabolism

Marie-Thérèse Vanier

Sphingolipid Structure and Metabolism

The common lipophilic moiety of all sphingolipids is ceramide (◘ Fig. 38.1), in which a long-chain sphingoid base (such as sphingosine) is attached by an amide linkage to a long-chain or very long-chain fatty acid. In glycosphingolipids, the carbohydrate moiety is linked to the primary alcohol group of the sphingoid base. Neuraminic acid-containing glycosphingolipids are named gangliosides. The hydrophilic portion of sphingomyelin is phosphorylcholine. Synthesis and degradation of sphingolipids take place in different cellular compartments. Glycosphingolipids are formed in the Golgi apparatus by sequential addition of monosaccharides to ceramide, catalysed by specific glycosyltransferases. Sphingolipids are then transported and inserted in the plasma membrane where they play a structural and functional role. For degradation (◘ Fig. 38.1), after transport by the endosomal pathway to the lysosome, they are hydrolysed stepwise by specific sphingohydrolases, some of which may need co-factors called sphingolipid activator proteins for their *in vivo* action.

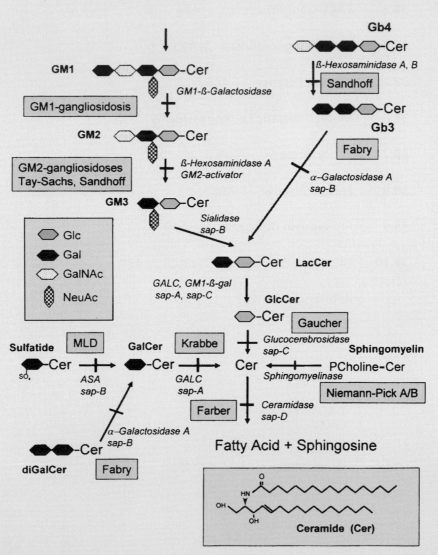

◘ Fig. 38.1. Sphingolipid structure and degradation. *Glc*, glucose; Gal, galactose; *GalNAc*, galactosamine; *NeuAc*, N-acetylneuraminic acid; *LacCer*, lactosylceramide; *GlcCer*, glucosylceramide; *GalCer*, galactosylceramide; *Gb3*, globotriaosylceramide; *Gb4*, globotetraosylceramide (globoside); *diGalCer*, galabiosyl-ceramide; *GM1*, GM1 ganglioside; *GM2*, GM2 ganglioside; *ASA*, arylsulfatase A; *GALC*, galactocerebrosidase; *MLD*, metachromatic leukodystrophy; *sap*, saposin. Enzyme defects are indicated by *solid bars* across the arrows

Sphingolipidoses are a subgroup of lysosomal storage disorders in which sphingolipids accumulate in one or several organs as the result of a primary deficiency in enzymes or activator proteins involved in their degradative pathway. Traditionally, this subgroup also includes Niemann-Pick disease type C, characterized by impaired cellular trafficking of several lipids. With the exception of Fabry disease, which is X-linked recessive, sphingolipidoses have an autosomal recessive inheritance. The clinical presentation and course of the classical forms of the various diseases are often characteristic. With the help of relevant procedures (imaging, neurophysiology, ophthalmologic examination…), careful examination of the patient and perusal of the disease history (especially age and type of first symptom) should lead to a provisional diagnosis and oriented biochemical tests. Late-onset forms are often more difficult to recognize, and foetal presentations have also been overlooked in the past. No overall screening procedure is yet available to date. In most sphingolipidoses, the diagnosis is made by demonstration of the enzymatic defect, generally expressed in most cells, organs or even serum (leukocytes represent the most widely used enzyme source). In specific diseases, more complex biochemical tests or/and a molecular genetics assessment may be necessary. The past 15 years have seen the era of specific therapies for non-neuronopathic Gaucher disease and Fabry disease. But in spite of active research on animal models, knowledge on pathophysiology and progress toward therapy of the neurological forms in human patients remain to date limited.

38.1   Gaucher Disease

38.1.1   Clinical Presentation

Historically, three clinical phenotypes are recognized, but the full disease spectrum is actually a continuum. All types are panethnic but type 1 has a particularly high prevalence in the Ashkenazi Jewish population (carrier frequency 1:13).

Type I, defined by the lack of neurological symptoms, constitutes 80-90% of all cases [1]. Most commonly diagnosed in adults, it can present at any age [2]. There is a wide variability in the pattern and severity of the symptoms, from extremely handicapping forms to asymptomatic individuals, with most symptomatic patients having visceral, haematological and (more frequently in adults) skeletal disease [3]. Children often show severe splenomegaly, generally associated with hepatomegaly, but the degree of visceromegaly is highly variable, both in children and adults. This may lead to anemia, thrombocytopenia, and thus a bleeding tendency.

Leukopenia is less frequent. Children may show delayed growth and menarche. Subcapsular splenic infarctions may cause attacks of acute abdominal pain and medullary infarction of long bones, excruciating pain often referred to as bone crises. Essentially in adult patients, bone involvement represents a major cause of morbidity. Aseptic necrosis of the femoral head and spontaneous fractures due to osteopenia are other common complications. Lung involvement with diffuse infiltration may occur. In adults, pulmonary hypertension has been described in rare, usually splenectomized, patients. An association of Gaucher disease with parkinsonism has been reported [4].

Type II (Acute neuronopathic Gaucher disease). Classically, patients present early in infancy with brainstem dysfunction (horizontal gaze palsy, convergent squint, dysphagia), visceromegaly, retroflexion of the neck, pronounced spasticity, failure to thrive and cachexia. The downhill clinical course is rapid and few of these patients survive beyond the age of 2 years. This form has been subcategorized as type IIB [5]. Type IIA refers to patients with strabismus, paucity of facial movements but less or no sign of pyramidal involvement, irritability or cognitive impairment and a slower course (they may survive up to 5 years), who fill the gap between type II and type III.

Foetal and lethal neonatal variants are reminiscent of the phenotype seen in the knock-out Gaucher mouse. Many of these cases are associated with hydrops foetalis, some have been described as »collodion babies« [6, 7].

Type III (Subacute or chronic neuronopathic Gaucher disease) is heterogeneous. The most common form consists of severe systemic involvement and supranuclear saccadic horizontal gaze palsy, with or without developmental delay, hearing impairment and other brainstem deficits [8]. The second most common phenotype shows a relatively mild systemic disease but progressive myoclonic encephalopathy, with seizures, dementia and death. There are also patients with severe systemic involvement and supranuclear gaze palsy who develop a progressive myoclonic encephalopathy. A particular presentation with cardiac involvement (heart valve and aortic calcification), supranuclear gaze palsy, mild hepatosplenomegaly and bone disease, has been associated with homozygosity for the D409H mutation. In neurological Gaucher disease, extrapyramidal involvement has also been observed. Distinction between type II and type III, sometimes difficult, is however critical due to differences in prognosis and treatment.

38.1.2   Metabolic Derangement

The primary metabolic defect resides in a block of the lysosomal degradation of glucosylceramide (glucocerebroside) and glucosylsphingosine. In the vast majority of cases this is due to a deficiency of acid β-glucosidase (glucocerebrosidase, glucosylceramidase). Exceedingly rare cases, presenting as type III [9] are due to a deficiency of the saposin (SAP) *sap*-C, required for the *in vivo* hydrolysis of glucosylceramide. Glucosylceramide (glucocerebroside) accumulates massively in liver and spleen of patients in all types. Although elevated in cerebral gray matter of type II and type III patients, its concentration in brain remains low. Pathophysiology of the disease is poorly understood [10]. Glucosylsphingosine, a highly cytotoxic compound, Ca^{2+}, and inflammatory responses seem to be involved.

38.1.3 Genetics

The disease (except for *sap*-C deficiency) is caused by mutations of the *GBA* (acid β-glucosidase) gene (1q21). N370S, the most common mutation in Ashkenazim, is also very frequent in Caucasian populations. Even as a genetic compound with another mutant allele, N370S is always associated with a non-neuronopathic phenotype. The severity can vary widely in Gaucher patients with the same genotype, including N370S homozygotes [11-13]. The second most frequent mutation, L444P, first described in Norbottnian type III, is more frequently associated with types II and III. Complex alleles due to genetic rearrangements are more often associated with severe forms, including perinatal lethal forms [7].

38.1.4 Diagnostic Tests

Bone marrow examination may reveal Gaucher cells, often multinucleated reticuloendothelial cells with a vacuolated cytoplasm with »wrinkled tissue paper« appearance. In serum, levels of chitotriosidase, angiotensin-converting enzyme, tartrate resistant acid phosphatase and the chemokine CCL18/PARK are typically very elevated. These markers are used to monitor treated patients (▶ below). The demonstration of a deficient glucocerebrosidase activity in lymphocytes (preferably) or leukocytes can be done using an artificial fluorogenic substrate. Cultured cells have a much higher activity. Studies of lipids in liver and/or spleen might allow a retrospective diagnosis in autopsy material. Frozen tissue is optimal, but sphingolipid analysis is possible on formalin-fixed tissues. In *sap*-C deficiency, glucocerebrosidase activity is normal, but the findings of Gaucher cells on a bone marrow smear and strikingly elevated chitotriosidase levels should lead to molecular analysis of the *PSAP* gene. Study of the lipid profile in a liver biopsy would also demonstrate pathognomonic glucosylceramide storage.

38.1.5 Treatment and Prognosis

Two approaches are currently available for the specific treatment of type I (and to some extent type 3) patients: enzyme replacement therapy (ERT) and substrate reduction therapy (SRT). Splenectomy enhances the risk of progression of the disease at other sites, especially bone and lung and can generally be avoided by institution of ERT. Pregnancy is not contraindicated in untreated patients, although bleeding may become critical before and after birth. There is now a good experience of ERT throughout pregnancy [14]. Enzyme therapy is conducted with slow infusions of a recombinant enzyme modified to expose mannose groups (imiglucerase), ensuring optimal uptake by macrophages. During the past 10 years more than 3000 patients have been treated worldwide, and this form of therapy has largely proven its safety and effectiveness. Guidelines have been published [15–17]. The natural history of type I can be dramatically improved. ERT prevents progressive manifestations, and ameliorates Gaucher disease-associated anemia, thrombocytopenia, organomegaly, bone pain, and bone crises. However, the enzyme does not cross the blood-brain barrier, and this treatment has no effect on the neurological manifestations of type II [5]. While ERT aims at restoring the degradation rate of the accumulated substrate, SRT tends to reduce the cell burden by slowing down the rate of synthesis of the substrate to a level where it can be slowly cleared by a deficient enzyme with some residual activity. The iminosugar N-butyldeoxynojirimycin (NB-DNJ), that partially inhibits glucosylceramide synthase, has recently been approved as an oral medication (miglustat) for treatment of mild forms of Gaucher type 1. Sustained therapeutic effects have been observed [18]. More experience is still needed to precisely define its optimal use. Contraception is mandatory in treated patients (males and females). Because miglustat partially crosses the blood brain barrier, trials are currently conducted in less severe neurological forms of the disease.

38.2 Niemann-Pick Disease Type A and B

Since the early 80's, the heterogenous group of Niemann-Pick disease has been divided in two clearly separate entities, based on their metabolic defect: sphingomyelinase deficiencies, including the historical types A and B, and lipid trafficking defects, corresponding to Niemann-Pick disease type C (neither caused by the deficiency of a lysosomal enzyme nor its cofactor, ▶ below).

38.2.1 Clinical Presentation

Sphingomyelinase deficiencies have historically been categorized into a severe, acute neuronopathic form, or type A, and a non-neuronopathic form, or type B, but there appears to be a continuum ranging from mild to severe type B, and then from late-onset neurological forms toward severe classical type A. Type A has its highest prevalence in Ashkenazim and is rare in other ethnic groups. Type B does not have an Ashkenazi Jewish predilection and appears more frequent in Southern Europe, North Africa, Turkey and the Arabian peninsula than in Northern Europe.

Classical type A patients have a quite uniform presentation [19]. First symptoms are often vomiting, diarrhea, or both, and failure to thrive, often appearing in the first weeks of life. Neonatal cholestatic icterus is very rare. Prominent and progressive hepatosplenomegaly and lymphadenopathy occur in most cases before 3 to 4 months of age, and sometimes in the neonatal period. Hypotonia and muscle weakness are common. Psychomotor retardation becomes evident around 6 months of age. Initial axial hypotonia is later combined with bilateral pyramidal signs. Slowed nerve conduction velocity is generally present. A cherry-red spot in the retina is detected in about half of the patients. Severe cachexia is common. Loss of motor function and intellectual deterioration continues to the point where patients become spastic and rigid. Seizures are rare. Brownish-yellow discoloration and xanthomas may be detected in the skin. Death usually occurs between 1.5 to 3 years. Cases with a milder systemic involvement, slightly protracted onset of neurological symptoms and slower course are also seen.

A growing number of *intermediate cases* are being described, especially from Germany and the Czech Republic, of late infantile or juvenile neurological onset, and of adults with neurological disease [20, 21].

Type B is a chronic disease. Most typically, the presenting sign is splenomegaly or hepatosplenomegaly in late infancy or childhood, but the age of discovery may occur from birth until late adulthood. The severity of hepatosplenomegaly and associated signs is highly variable. Splenectomy is seldom necessary. Only rare cases have been described with severe liver disease. The most constant associated signs are radiographic abnormalities of the lung (diffuse, reticulonodular infiltrations) and interstitial lung disease with variable impairment of pulmonary function [22]. In cases presenting in infancy or childhood, growth restriction is common in late childhood and adolescence [23], with delay in skeletal age and puberty. Abnormal lipid profiles, mildly elevated liver transaminases, low platelet count are other common findings In adult patients, pulmonary reticular fibrosis may be the initial sign. Dyspnea on exertion is the most common pulmonary symptom, but some patients may eventually become oxygen-dependent. True type B patients do not have neurological involvement and are intellectually intact, although ophthalmoscopic examination may reveal a retinal macular halo or cherry red maculae [24]. In a recent longitudinal study [25] the disease was characterized by hepatosplenomegaly with progressive hypersplenism, worsening atherogenic lipid profile, gradual deterioration in pulmonary function and stable liver dysfunction. Although there are severe forms, the most frequent clinical phenotype is that of a moderately serious disorder compatible with an essentially normal life span.

38.2.2   Metabolic Derangement

A primary deficiency of the lysosomal (or acid) sphingomyelinase resulting from mutations on the *SMPD1* gene leads to the progressive accumulation of sphingomyelin in systemic organs in all types of the disease, and in brain in the neuronopathic forms. Sphingomyelin storage is massive in liver and spleen in type A and slightly less in type B. A significant increase of unesterified cholesterol occurs secondarily. By *in vitro* measurements using the natural substrate, a marked sphingomyelinase deficiency is observed in all patients. *In situ* hydrolysis of labelled sphingomyelin by living cultured fibroblasts demonstrates a significant level of residual activity in many type B patients, suggesting that the mutated enzyme has retained enough catalytic activity to limit accumulation and protect the brain. Sphingosylphosphorylcholine (increased in type A brain) may participate in the pathogenesis of the brain dysfunction.

38.2.3   Genetics

More than 100 disease-causing mutations of the *SMPD1* (acid sphingomyelinase) gene (11p15) are known [21, 26]. In Ashkenazi Jewish type A patients, 3 mutations (R496L, L302P, fsP330) account for > 90% of alleles. R608del, highly prevalent in North African patients, is the most common type B mutation (20–30% of alleles) in many countries. So far, it has always been correlated with a type B phenotype whatever the nature of the second mutated allele. Q292K is associated with late-onset neurological involvement. Recently, it has been demonstrated that the *SMPD1* gene is paternally imprinted, and that some heterozygous carriers present with signs of the disease because of preferential expression of the maternal allele owing to methylation of the paternal allele [26a].

38.2.4   Diagnostic Tests

Bone marrow usually reveals the presence of (non-specific) foamy histiocytes or sea-blue histiocytes. Chitotriosidase is moderately elevated. The diagnosis is made by demonstration of a deficiency in sphingomyelinase activity in leukocytes (or lymphocytes) or in cultured cells (which have a much higher level of activity). The choice of a specific substrate is critical. Sphingomyelin radioactively labelled on the choline moiety is the gold standard but a promising fluorogenic substrate has recently been made available.

Problems have been reported using synthetic substrates [20]. The *in vitro* assay does not reliably distinguish A from B phenotypes. The loading test in living fibroblasts is more informative but has limitations [21].

38.2.5   Treatment and Prognosis

No specific therapy is yet available. Experience of bone marrow transplantation (BMT) is limited but did not appear to improve symptoms in type A patients. In type B, splenectomy may have a deleterious effect on the lung disease. Most type B females enjoy uncomplicated pregnancies, although careful monitoring for bleeding is advisable. Preclinical trials using the human recombinant enzyme have been conducted in a knock-out mouse model, that led to correction of the storage process in liver, spleen and lung (but, as expected, not in brain), providing the proof of principle for enzyme replacement therapy for type B [27]. Clinical trials in type B patients are planned for the near future. In the mouse model, several other experimental therapeutic measures have been tried with partial results, including intracerebral gene therapy using an AAV vector [28].

38.3   GM1-Gangliosidosis

38.3.1   Clinical Presentation

First descriptions of infantile GM1 gangliosidosis by B.H. Landing and J.S. O'Brien emphasized its characteristics of a neurovisceral lipidosis sharing features with both Tay-Sachs disease and Hurler disease. Forms with an almost exclusive neuronal storage were recognized later.

In the *typical early infantile form* (or type 1), children are often hypotonic in the first days or weeks of life, with poor head control. The arrest in neurological development is observed at 3 to 6 months of age. Feeding difficulties and failure to thrive are common. Many infants have facial and peripheral oedema. In typical cases, dysmorphic features may be present very early or develop with time, with a puffy face, moderate macroglossia, hypertrophic gums, depressed nasal bridge, chipmunk face, but an increasing number of infantile patients have presented without dysmorphic expressions. Hepatomegaly and later splenomegaly are almost always present. Dorsolumbar kyphoscoliosis is common. After a few months, signs of visual failure appear, with often a pendular nystagmus. A macular cherry red spot is found in about 50% of cases, but seldom before 6 months of age. As time passes, hypotonia gives way to spasticity. Rapid neurological regression is usual after the first year of life, with generalized seizures, swallowing disorder, decerebrate posturing and death, often before age 2. Radiological signs in the long bones and spine are constant in clinically severe patients, but can be minimal in cases with only psychomotor deteriora-

tion. Subperiosteal bone formation can be present at birth. Widening of the diaphyses and tapering of the extremities appear later. At the age of 6 months, striking Hurler-like bone changes are seen with vertebral beaking in the thoracolumbar zone, broadening of the shafts of the long bones with distal tapering, and widening of the metacarpal shafts with proximal pinching of the four lateral metacarpals.

A severe *neonatal form* with cardiomyopathy has been described. GM1-gangliosidosis is also a cause of non-immune *fetal hydrops*.

The *late infantile variant* (or type 2) usually begins between 12 and 18 months, with unsteadiness in sitting or standing, or difficulty in walking [19]. Regression is rapid and severe, a spastic quadriparesis develops, associated with pseudobulbar signs. Seizures are frequent and may become a major problem. The patients are not dysmorphic and hepatosplenomegaly is not present. Vision is generally normal. Radiography of the spine reveals moderate but constant changes with a mild anterosuperior hypoplasia of the vertebral bodies at the thoracolumbar junction.

The term, adult form has been employed to designate the *chronic late-onset form* of GM1 gangliosidosis, with onset in late childhood, adolescence or adulthood. Dysarthria and extrapyramidal signs, especially dystonia, are the most common signs [29]. Cognitive impairment is absent to moderate, and there are no ocular abnormalities. Bone changes are inconstant. The course of the disease is very slow. Re-evaluation of Japanese adults with a spinocerebellar ataxia-like syndrome, progressive dementia and low intracellular β-galactosidase indicate that they belong to another, new disease entity, with secondary deficiency of β-galactosidase.

38.3.2   Metabolic Derangement

GM1-gangliosidosis is due to a deficiency of lysosomal acid β-galactosidase, which cleaves glycoconjugates containing a terminal β-galactosidic linkage and is necessary for the degradation not only of GM1 ganglioside and other glycosphingolipids, but also of galactose-containing oligosaccharides and keratan sulfates. As a consequence, the most severe forms of the disease combine features of a neuronal lipidosis, a mucopolysaccharidosis and an oligosaccharidosis. Acid β-galactosidase functions in a multienzyme lysosomal complex with neuraminidase, the protective protein/cathepsin A (PPCA) and N-acetyl-galactosamine-6-sulfate sulfatase [30]. This explains the quite similar clinical phenotype of *galactosialidosis*, a distinct condition due to the deficiency of PPCA, that causes a combined secondary deficiency of acid β-galactosidase and acid sialidase (neuraminidase). Finally, β-galactosidase deficiency can be associated with two clinically different diseases, GM1 gangliosidosis, with prominent features of a sphingolipidosis, and Morquio B disease (mucopolysaccharidosis type IVB), in which abnormalities of mucopolysaccharide metabolism prevail. In tissues from patients with GM1 gangliosidosis, three major groups of accumulated compounds have been been identified: the sphingolipid GM1 ganglioside, glycoprotein-derived oligosacccharides, and keratan sulfate. A massive storage of GM1 occurs in brain tissue. Increased levels of its lysocompound, potentially of pathogenetic significance, have been reported. Galactose-containing oligosaccharides have been found in liver and urine. Keratan sulfate and other mucopolysaccharides accumulate in liver and spleen. The amount of keratan sulfate excretion in urine is less in GM1 gangliosidosis than in Morquio B disease.

38.3.3   Genetics

About 50 mutations of the gene for acid β-galactosidase, *GLB1* (3p21.33) have been described. Neither the type nor location of the mutation correlates well with a specific phenotype.

38.3.4   Diagnostic Tests

Vacuolated lymphocytes may be found in peripheral blood, and foamy histiocytes in the bone marrow. Radiographic bone examination showing Hurler-like abnormalities (▶ above) may suggest the diagnosis. In the infantile form, cranial computerized tomography (CT) and magnetic resonance imaging (MRI) usually give non-specific results, with diffuse atrophy of the central nervous system (CNS) and features of myelin loss in the cerebral white matter. Lesions in the basal ganglia may be present in the adult form. Analysis of urinary oligosaccharides is a good orientation test. In the classical early infantile form, excretion is massive, with a pathognomonic profile. Oligosaccharide excretion can, however, be much less in forms with predominant neurodegenerative disease. Mucopolysaccharide analysis in urine usually shows increased levels of keratan sulfate. The diagnosis is established by demonstration of a deficient activity of acid β-galactosidase, which can be measured on leukocytes using an artificial chromogenic or fluorogenic substrate. A subsequent study of neuraminidase (in leukocytes or cultured fibroblasts) should be performed systematically in every β-galactosidase deficient patient to exclude galactosialidosis.

38.3.5   Treatment and Prognosis

No specific treatment is available to date. Substrate reduction therapy using miglustat is a potential approach for clinical trials in late onset forms.

38.4   GM2-Gangliosidoses

GM2-gangliosidoses consist of 3 different genetic and biochemical subtypes: Tay-Sachs disease (or B variant), Sandhoff disease (or 0 variant), and GM2 activator deficiency (AB variant). All are characterized by impaired lysosomal catabolism of ganglioside GM2 which requires 3 gene products: the β-hexosaminidase α- and β-subunits and the GM2 activator protein. Tay-Sachs disease corresponds to a deficiency of the α-subunit and thus of hexosaminidase A (αβ heterodimer), Sandhoff disease to a deficiency of the β-subunit and thus of both hexosaminidase A and B (ββ homodimer). Classical Tay-Sachs disease has a much higher incidence in the Ashkenazi Jewish population than in other ethnic groups. Infantile forms are by far the most common, but juvenile and adult forms are also recognized. A particular enzymatic variant of Tay-Sachs disease (the B1 variant) has a high incidence in northern Portugal [31] and is globally more frequent in southern Europe. Variant AB is exceedingly rare (<10 reported cases).

38.4.1   Clinical Presentation

The infantile forms of the 3 subtypes have a very similar presentation. Around 4 to 6 months of age, motor weakness and hypotonia is the usual earliest sign, almost constantly associated with a typical startle response to sounds with extension of the arms (hyperacusis). Hypotonia progresses, with loss of acquired milestones. Loss of visual attentiveness is also seen early, and ophthalmoscopic examination almost invariably reveals a typical macular cherry red spot in the retina. Blindness follows and spasticity, disordered swallowing and seizures develop. Macrocephaly begins by 18 months of age. By year 3 the child is demented and decerebrate. Death often occurs due to aspiration pneumonia. In Sandhoff disease, in spite of an additional accumulation of glycolipids and oligosaccharides in visceral organs, organomegaly and bony abnormalities are rarely observed.

Late infantile and juvenile forms are mostly due to a deficiency of hexosaminidase A (often B1 variant). The onset of symptoms is usually between 2 to 10 years of age with ataxia, incoordination and dysarthria, followed by progressive psychomotor deterioration, spasticity and seizures. Myoclonus can be prominent. Cherry red spots are inconstant.

Chronic or adult forms may show variable presentations, with pyramidal and extrapyramidal signs, movement disorders (dystonia, athetosis, ataxia), psychosis (reported in 30-50% of adult onset patients), and a syndrome of lower motor neuron and spinocerebellar dysfunction with supranuclear ophthalmoplegia. Some patients may show autonomic dysfunction. Adult onset Sandhoff disease may present as spinocerebellar degeneration [32].

38.4.2   Metabolic Derangement

The normal catabolism of GM2 ganglioside requires the GM2 activator protein to first bind to and extract GM2 from the plasma membrane before presenting it for cleavage to hexosaminidase A (the $\alpha\beta$ heterodimer). Hexosaminidase B, the $\beta\beta$ homodimer, hydrolyses other substrates with a terminal hexosamine (glycoproteins and glycolipids), but not ganglioside GM2. In Tay-Sachs disease (affecting the α subunit), hexosaminidase A only is deficient. In Sandhoff disease (affecting the β-subunit) both hexosaminidases are inactivated. In GM2-activator deficiency, the substrate is not made available to the otherwise normally functioning enzyme. All types are characterized by storage of GM2 ganglioside in neurons. This results in meganeurites with aberrant neurite formation that may play a role in the pathophysiological mechanisms. GM2 storage is very pronounced in infantile forms, less in juvenile forms, and even less in adult forms. Increased levels of lyso-GM2 have also been reported in infantile forms. In Sandhoff disease, asialo-GM2 also accumulates in brain, while other compounds, such as globoside and oligosaccharides, accumulate in liver and other visceral organs. Apoptotic neuronal death and macrophage/microglial-mediated inflammation [33] have been suggested as possible mechanisms of the neurodegeneration process.

38.4.3   Genetics

More than 100 mutations of the *HEXA* gene (on chromosome 15) have been identified. Three mutations - an insertion (+TATC 1278) and a splicing defect (IVS12 1421+1G>C) in infantile cases, a missense (G269S) in adult forms – account for 95% of the Ashkenazi Jewish alleles. Mutations at codon 178 altering the 3-dimensional structure of the enzyme, such as R178H, result in a particular enzymatic variant or B1 variant presenting as a juvenile form in the homozygous state. A relatively good genotype-phenotype correlation has been reported. More than 20 mutations of the *HEXB* gene and 5 of the GM2 activator gene (both on chromosome 5) have been described.

38.4.4   Diagnostic Tests

In Tay-Sachs and Sandhoff disease, confirmation of the clinical diagnosis is easy by appropriate enzyme testing on leukocytes or cultured fibroblasts. The assay for total hexosaminidases (A+B) using a synthetic fluorogenic substrate is straightforward and allows the diagnosis of Sandhoff disease. Differential assay of hexosaminidase A using heat or acid inactivation does not identify patients with the B1 variant. To diagnose hexosaminidase A deficiencies, the direct assay using the sulfated synthetic substrate (4-MU-6-sulfo-β-glucosaminide) specific for the α-subunit, is the method of choice. A high residual activity is found in Sandhoff disease, due to excess of hexosaminidase S ($\alpha\alpha$ dimer). In GM2 activator deficiency, hexosaminidase A activity measured *in vitro* is normal. Electron microscopic examination of a skin or conjunctival biopsy may provide strong evidence in favour of the diagnosis by demonstrating concentric lamellated bodies in nerve endings. Loading studies in living fibroblasts using radiolabeled gangliosides is a possible biochemical approach. The definitive diagnosis requires gene sequencing.

38.4.5   Treatment

Seizures are generally responsive to standard treatment. No effective curative treatment is currently available. Following encouraging results of experimental therapy in the mouse models for Tay-Sachs and Sandhoff disease using NB-DNJ (miglustat) [34], a clinical trial has been initiated with this drug in patients with a late-onset form of Tay-Sachs disease. Interim results are awaited in the near future. Studies in animals with NSAIDs suggested that inflammation may also contribute to disease progression. Gene therapy is still at an early experimental stage in mouse models.

38.5   Krabbe Disease

38.5.1   Clinical Presentation

Krabbe disease (or globoid cell leukodystrophy) leads to demyelination of the central and peripheral nervous system. Its estimated overall incidence is between 1 in 100 000 to 200 000 live births. It is more frequent in Scandinavia (but not Finland). The classic infantile form accounts for >85% of cases. Later onset cases appear to be more common in southern Europe, especially Italy and Sicily. The incidence of adult-onset cases is probably underestimated.

In the *infantile form* [35], the disease usually starts within the first 6 months (sometimes before the age of 3 months). Initial symptoms include increasing irritability, crying, vomiting and other feeding problems, hyperesthesia, tonic spasms on light or noise stimulation, and signs of peripheral neuropathy. Episodic unexplained fever is also common. This stage with hypertonic episodes is followed by permanent opisthotonic posturing with characteristic flexed upper extremities and extended lower extremities. Seizures may appear. Hyperpyrexia and hypersalivation are frequent. As the disease progresses, blindness occurs, followed by loss of bulbar functions and hypotonia. Death occurs from hyperpyrexia, respiratory complications or aspiration, classically before the age of 2 years, but in current practice, not so rarely later.

Clinical diagnosis of late-onset Krabbe disease is much more difficult. The *late infantile and juvenile forms* start between the ages of 15 months and 10 years (most cases before the age of 5 years). The first signs are often gait disturbances (spastic paraparesis or ataxia or both, sometimes spastic hemiplegia) in a previously normal or mildly retarded child. Visual failure with optic atrophy is also a common symptom, especially in the late infantile form [36]. At variance with the infantile form, peripheral neuropathy is only present in approximately half of the cases. Time of onset and severity of mental deterioration is variable. Seizures are reported as infrequent, but when present, can be a major therapeutic problem. The course of the disease is quite variable and unpredictable, even in siblings. Many patients show initial rapid deterioration followed by gradual progression lasting for years.

Adult patients [37, 38] often present with spastic paraplegia, with or without peripheral neuropathy, and usually do not show mental deterioration.

38.5.2   Metabolic Derangement

Krabbe disease results from galactosylceramidase (or galactocerebrosidase, cerebroside β-galactosidase) deficiency, a lysosomal enzyme that catabolizes galactosylceramide, a major lipid component of myelin, and also lactosylceramide and galactosylsphingosine. *In vivo*, galactosylceramide degradation further requires the saposin (SAP) *sap*-A. A single case has been reported due to *sap*-A deficiency [39]. Galactosylceramidase deficiency leads to an accumulation of galactosylceramide in the pathognomonic »globoid cells« (multinuclear macrophages) seen in the demyelinating lesions of the white matter, and of a toxic metabolite galactosylsphingosine (psychosine) in the oligodendrocytes and the Schwann cells. Psychosine, a highly apoptotic compound increased in brain of infantile patients, is thought to play a major role in the pathogenesis of the disease and, more specially, to underlie the early destruction of oligodendrocytes characteristic of the infantile form, and thus an arrest of myelin formation [40].

38.5.3   Genetics

The galactosylceramidase *(GALC)* gene is located on 14q31. The 502T/del complex mutant allele that associates a large deletion and a polymorphism seems to originate from Sweden and makes up close to 50% of the mutant alleles in most European countries and in the USA. T513M and Y551S are also frequent. G270D is common among late-onset patients [41]. Some common polymorphisms (especially 1637G>C and 502C>T) influence enzyme activity and may be responsible for a pseudodeficiency state, particularly when in compound heterozygocity with a disease-causing allele. One infantile case was assigned to a mutation in the sap-A domain of the *PSAP* gene.

38.5.4   Diagnostic Tests

Motor nerve conduction velocity is consistently low in infantile and juvenile cases, but may be normal in adult patients. MRI shows areas of hyperintensity on T2 weighted images that correlate well with areas of demyelination and globoid cell accumulation [42]. In late onset cases, T2 images may show more localized areas of hyperintensity or may in a few cases appear normal [43, 44]. In some adult cases, high signal intensity was localized only in the corticospinal tract. In typical infantile cases, CT shows diffuse cerebral atrophy with hypodensity of the white matter. Calcifications may be observed in the thalamus, basal ganglia and periventricular white matter. Brain stem evoked potentials have also been studied [45]. Protein in cerebrospinal fluid (CSF) is usually elevated in infantile cases, but inconstantly in late-onset cases. The ultimate diagnosis is made by studying galactosylceramidase activity in leukocytes or cultured fibroblasts, which is best done using the natural radiolabeled substrate. An alternative fluorogenic substrate has recently been proposed, for which experience is still limited. Prior to any prenatal diagnosis, an enzymatic study of both parents is mandatory, as pseudodeficiencies due to polymorphisms are quite common. Screening for 502T/del and T513M in newly diagnosed patients and for the 2 common polymorphisms is useful. In the patient with *sap*-A deficiency, galactosylceramidase activity was deficient in leukocytes, but not in cultured fibroblasts (*sap*-A possibly stabilizes galactosylceramidase).

38.5.5   Treatment

In advanced disease, supportive analgesic treatment of the often severe pain that may result from radiculopathy is important, as well as treatment of spasticity. Allogenic BMT or cord-blood transplantation may be effective in preventing onset or halting progression of the disease in late-onset cases [46, 47, 47a]. On the other hand, the general experience of BMT undertaken in symptomatic infantile cases has been extremely poor. But in classic infantile forms that had been diagnosed prenatally, umbilical-cord blood transplantation in presymptomatic newborns (12–44 days) favourably altered the natural history of the disease [48].

38.6   Metachromatic Leukodystrophy

38.6.1   Clinical Presentation

Metachromatic leukodystrophy (MLD) is panethnic, with reported incidences ranging between 1:40 000 (Sweden) and 1:170 000 (Germany), apart from specific ethnic groups with higher frequency.

The *late infantile form* [49] is most common. First symptoms appear between the age of 1 to 2 years. Most children have begun to walk, although about 15% never do so independently. Around 14-16 months of age, the child develops progressive difficulties in locomotion, weak lower limbs and falls. Examination usually shows hypotonia, and reduced or absent tendon reflexes due to peripheral neuropathy with extensor plantar responses. Walking and then standing soon becomes impossible. The child shows spastic quadriplegia, together with speech deterioration, gradual mental regression, optic atrophy leading to blindness, followed by a vegetative state and death.

The onset of the *juvenile form* [19, 50] ranges between 3 and 14 years. Failure in school, behavioural problems or disturbance of cognitive function may precede motor abnormalities, especially in patients with a later onset (>6 years). Progressive difficulties in walking with pyramidal signs and peripheral neuropathy together with cerebellar ataxia constitute the most common presentation, but various other symptoms can occur such as hemiplegia, dystonia, and choreoathetosis. Seizures may develop.

Two distinct types of *adult MLD* have been identified [51]. In the first group, patients have a predominant motor disease, with pyramidal and cerebellar signs, dystonia and peripheral neuropathy, or isolated peripheral neuropathy [52]. In the second group, behavioural and psychiatric problems (often confused with schizophrenia) are the presenting symptoms, followed by dementia and spastic paresis [53].

38.6.2   Metabolic Derangement

The primary metabolic defect is a block in lysosomal degradation of sulfatide (or galactosylceramide-sulfate) and other sulfated glycolipids. *In vivo*, the sulfatide is presented to the enzyme arylsulfatase A (ASA) as a 1:1 complex with the SAP *sap*-B. A deficiency of either ASA or *sap*-B can cause MLD. Few (<15) cases with *sap*-B deficiency have been documented, most with a late infantile form. Sulfatide is a prominent lipid component of the myelin sheath. Its ratio to galactocerebroside plays a role in stability and physiological properties of this membrane. Progressive accumulation of sulfatides (and possibly lysosulfatide) in the central and peripheral nervous system will soon lead to disruption of the newly formed myelin and intense demyelination. In MLD, sulfatide also accumulates in the kidney, which is reflected by a very abnormal excretion of sulfatide in urine sediment.

38.6.3   Genetics

More than 80 different mutations of the arylsulfatase A *ARSA* gene (22q13) are known. The 3 more frequent alleles

among European patients are 459+1G>A (severe phenotype), P426L (mild phenotype) and I179S (mild phenotype). Two very frequent polymorphisms of the *ARSA* gene, one leading to the loss of an N-glycosylation site, and the second to the loss of a polyadenylation signal, result in a reduction of the amount of enzyme, and constitute the molecular basis of ASA pseudodeficiency. They often occur jointly, but can also be found independently. In some countries, as many as 15% of the general population carry one such pseudodeficiency (*pd*) allele [54–56].

38.6.4   Diagnostic Tests

In most patients, motor nerve conduction velocities of peripheral nerves are decreased and sensory nerve action potentials have a diminished amplitude with a prolonged peak latency [57]. Decreased nerve conduction is not always present in adult MLD. MRI shows bilateral symmetrical changes of the central white matter, with diffuse hyperintensity in T2-weighted images and slight hypointensity in T1-weighted images. The changes often first appear in the periventricular white matter and progress with the disease. Cerebellar atrophy is also common. Abnormalities are also described by diffusion MRI and proton magnetic resonance spectroscopy (MRS) [58, 59]. The CSF protein content is usually elevated in late infantile patients (although not at an early stage), inconstantly in the juvenile form and rarely in the adult form.

Determination of arylsulfatase A in leukocytes (or cultured fibroblasts) using p-nitrocatechol-sulfate as a substrate constitutes the first biochemical test. Pseudodeficiency is a major pitfall. Individuals homozygous for a *pd* allele (1–2% of the European population) have about 5–15% of normal ASA activity but no detectable clinical abnormality or pathology. The same applies to subjects compound heterozygotes for a disease-causing *mld* and a *pd* allele. A deficient ASA activity is therefore not enough to conclude to the diagnosis of MLD, even in a patient with a leukodystrophy. Molecular screening for the *pd* allele is useful, but the same allele may carry a *pd* polymorphism and an *mld* mutation [56]. The study of sulfatides in the urinary sediment circumvents the problem. MLD patients excrete massive (late infantile and juvenile patients) or significant (adult-onset type) amounts of sulfatides, while subjects with an ASA pseudodeficiency are within or slightly above the normal range. ASA pseudodeficiency also poses problems in prenatal diagnosis. It is mandatory in a newly diagnosed family to study enzyme activity and to screen for the *pd* allele in the index case and both parents.

In MLD patients with *sap*-B deficiency, the regular *in vitro* ASA assay will not show a deficiency, and studies of glycolipid excretion in urine are essential. The profile is pathognomonic, showing a combined MLD pattern and Fabry pattern. The ultimate diagnosis will require molecular analysis of the *PSAP* gene. Another complex biochemical test – loading of labeled sulfatides in living fibroblasts – also detects these patients and can differentiate pseudodeficient individuals from MLD patients [60].

Finally, an arylsulfatase A deficiency may be due to a primary deficiency in the formylglycine-generating enzyme (FGE) encoded by the *SUMF1* gene, which leads to the disease called *multiple sulfatase deficiency or Austin disease* (also known as mucosulfatidosis). Usually these patients display not only a leukodystrophy, but also a Hurler-like aspect and/or ichthyosis. The correct diagnosis will be reached by assaying other sulfatases (arylsulfatase B, iduronate-2-sulfatase…) and demonstrating a combined deficiency.

38.6.5   Treatment and Prognosis

Symptomatic treatment of spasticity and of pain from radiculopathy is important. Currently, there is no satisfactory treatment of any form of MLD. Allogenic BMT has been performed in a number of cases [47, 61–63]. It is generally considered that adult-onset and juvenile-onset patients beneficiate from BMT, with slowing of the disease progression and improvement of cognitive functions. The indication remains controversial in the late infantile form. While it is generally accepted that clearly symptomatic patients are not candidates, a number of yet asymptomatic affected siblings have been transplanted. In the latter, a significant difference in survival and in involvement of the CNS has been observed compared to the untransplanted sibling. Unfortunately, BMT has no effect on the peripheral neuropathy, which becomes a significant problem. Recent experimental studies in the *ARSA* knock-out mouse model suggest that achieving a high level of enzyme activity might overcome this limitation [64, 65].

38.7   Fabry Disease

38.7.1   Clinical Presentation

Fabry disease, the only X-linked sphingolipidosis, is associated with severe multiorgan dysfunction. Its incidence has been estimated from 1 in 40 000 to 60 000 live births for males. Heterozygous females can be symptomatic [66, 67]. Although clinical onset occurs in childhood, disease presentation may be subtle, leading to retarded diagnosis or misdiagnosis. Males with the classic form have a disease onset during the first decade, typically with crises of severe pain in the extremities (acroparesthesia) provoked by exertion or temperature changes. Episodic Fabry crises of acute pain may last hours to days. Unexplained bouts of fever and hypohidrosis, heat, cold and exercise intolerance, gastrointestinal problems, corneal dystrophy (cornea verti-

cillata) that does not affect vision, are other manifestations. At this stage, renal function, urinary protein excretion and cardiac function and structure are generally still normal [68]. Characteristic skin lesions, angiokeratomas, appear on the lower part of the abdomen, buttocks and scrotum in 80% of patients. Progressive renal involvement, that may result in end-stage renal disease and require dialysis or transplantation, occurs in adulthood. Cardiac manifestations include left ventricular hypertrophy, valvular disease (mitral insufficiency), ascending aortic dilatation, coronary artery disease and conduction abnormalities leading to congestive heart failure, arythmias and myocardial infarction. Cerebrovascular manifestations include early stroke, transient ischemic attacks, white matter lesions, hemiparesis, vertigo or dizziness, and complications of vascular disease, in particular hearing loss. Clinical manifestations in heterozygous females range from asymptomatic to full-blown disease as severe as in affected males, but with globally a later onset and slower progression. [69, 70]. Atypical variants with a milder, later onset phenotype have been described. The cardiac variant presents with cardiomegaly and mild proteinuria usually after 40 years of age.

38.7.2   Metabolic Derangement

The primary defect is a deficiency of the lysosomal enzyme α-galactosidase A which releases galactose from ceramide trihexoside (globotriasylceramide, Gb3) and related glycosphingolipids (especially galabiosylceramide, Gb2), due to mutations of the *GLA* gene. This results in progressive accumulation of Gb3 in vascular endothelial cells, perithelial and smooth muscle cells, leading to ischemia and infarction especially in the kidney, heart and brain. Early and substantial deposition of Gb3 occurs in podocytes, leading to proteinuria, and with age, in cardiomyocytes, causing cardiac hypertrophy and conduction abnormalities. Small-fibre polyneuropathy is the cause of pain and anhidrosis.

38.7.3   Genetics

Fabry disease has an X-linked recessive transmission. More than 300 mutations of the *GLA* gene have been described. There are also numerous reported polymorphisms. Many mutations are private; a number are recurrent in specific countries. The mutation N215S seems to be associated with the cardiac variant. *De novo* mutations are rare.

38.7.4   Diagnostic Tests

In affected males with the classic or variant phenotype, the disease is readily diagnosed by showing an α-galactosidase

A deficiency in leukocytes. Plasma or dried blood spots have also been advocated as better suited to large-scale screening, but subsequent confirmation in leukocytes is advisable. In contrast, heterozygous females may have normal to low levels of activity, and thus enzyme assay is not reliable for carrier detection. If the subject is related to a patient, molecular analysis is of course the test of choice, but if not, systematic sequencing incurs the risk of difficult interpretation due to the numerous polymorphisms. Studies of urinary glycolipids are useful. Large amounts of Gb3 and Gb2 are excreted by untreated male hemizygotes (except patients with a renal graft and those with a cardiac variant), and smaller but still significant amounts by heterozygote females, symptomatic or not. At present urinary Gb3 appears as the most reliable biochemical test for carrier screening [71]. Plasma Gb3 is elevated in male patients, but not in female carriers. Measurement of Gb3 is used to monitor treatment [72].

38.7.5 Treatment and Prognosis

The disease results in a significant reduction in life expectancy [70]. There is also the psychosocial burden of a rare, chronic and progressive disease. It is essential that patients are diagnosed early, that the family receives genetic counselling, and that proper treatment is given. Guidelines have been published [69]. Alleviation of pain and treatment of the renal and cardiac disease are important issues. In recent years, significant progress has occurred; ERT with recombinant α-galactosidase A has been shown to decrease pain and to stabilize renal function [73–75]. In Europe, two essentially similar products, agalsidase alpha and agalsidase beta, have been approved.

38.8 Farber Disease

38.8.1 Clinical Presentation

Farber lipogranulomatosis, a very rare disease, is clinically heterogeneous. Onset can be during infancy or much later, causing death within the first year or in some cases at an adult age. Foetal forms are known [76]. The most frequent signs are periarticular subcutaneous nodules and joint swelling, contractures, hoarseness due to laryngeal involvement. Hepatomegaly can be present as well as a macular cherry red spot. Neurological manifestations are variable, and may include severe psychomotor deterioration with seizures, or mild neurological involvement. Juvenile onset patients may show neurological involvement only. The clinical description of later-onset cases is poorly documented.

38.8.2 Metabolic Derangement and Genetics

The deficiency of acid ceramidase leading to the storage of ceramide in various organs is due to mutations of the ceramidase gene (8p21.3–22).

38.8.3 Diagnostic Tests

Electron microscopy of excised nodule or a skin biopsy may reveal numerous inclusions with typical curvilinear bodies in histiocytes, and »banana bodies« in Schwann cells. In vitro measurement of ceramidase activity requires a specific substrate available in very few laboratories. Loading test in cultured fibroblasts using a precursor of ceramide (sphingomyelin of sulfatide) and subsequent study of ceramide turnover is another suitable approach [60]. The latter test also gives abnormal results in prosaposin deficiency.

38.8.4 Treatment and Prognosis

Currently there is no specific therapy. Symptomatic treatment is based on analgesics, corticotherapy, and plastic surgery. Good results of BMT have been reported only in patients without central nervous system involvement [77, 78].

38.9 Prosaposin Deficiency

38.9.1 Clinical Presentation

The 5 known cases have shown almost the same course with severe neurovisceral storage disease manifesting immediately after birth with rapidly fatal course and death between 4 and 17 weeks of age. The patients have hepatosplenomegaly, hypotonia, massive myoclonic bursts, abnormal ocular movements, dystonia and seizures [79].

38.9.2 Metabolic Derangement and Genetics

Sphingolipid activator proteins are small glycoproteins required as cofactors for the lysosomal degradation of sphingoglycolipids with short hydrophilic head groups. They act either by solubilizing the substrate or by mediating enzyme binding to the membrane or modifying the enzyme conformation. The *PSAP* gene (10q21) encodes the prosaposin protein which is transported to the lysosome where it is processed to 4 homologous proteins (*sap*-A to D). *sap*-A is a cofactor for degradation of galactosyl and lactosylceramide; its deficiency causes a Krabbe disease variant (1 case published); *sap*-B is involved in the *in vivo* degradation of sulfatides and Gb3, and its deficiency causes an MLD variant (~ 15 cases known); *sap*-C is necessary for hydrolysis of glucosylceramide and its deficiency causes a Gaucher disease variant (3 cases known). Although no patient has been described with sap-D deficiency, this factor is implicated in ceramide degradation. Prosaposin deficiency is due to the combined lack of all 4 *sap*- factors, explaining a tissue storage of all the lipids cited above. The disorder is autosomal recessive. The 3 mutations identified in patients explain abolished production of the prosaposin precursor and thus of all 4 factors.

38.9.3   Diagnostic Tests

Gaucher-like cells are found in bone marrow. Study of glycolipids in urine sediment shows a pattern close to that described for *sap* B deficiency. Galactocerebrosidase activity was reported deficient in leukocytes and fibroblasts. The loading test in living fibroblasts described for Farber disease will show a severe block in ceramide hydrolysis. Lipid studies in liver tissue reveal a combined increase of glucosylceramide, lactosylceramide and ceramide. Final assessment requires molecular analysis of the *PSAP* gene.

38.10   Niemann-Pick Disease Type C

Niemann-Pick type C disease (NPC) is characterized by a complex defect in cellular lipid trafficking. The last decade has seen recognition that mutations in two different genes, *NPC1* or *NPC2*, can cause the disease, identification of the genes, and increasing knowledge on the gene products.

38.10.1   Clinical Presentation

NPC is panethnic, with an estimated incidence of 1 in 120 000 to 150 000. It includes former type D (a Nova Scotia *NPC1* isolate). The clinical course is very heterogeneous [80, 81]. The age of presentation may vary from the perinatal period to adulthood. The systemic (liver, spleen and lung) involvement and neurological disease follow an independent course.

Systemic involvement: *Foetal hydrops or foetal ascites* can occur. Liver involvement is often present in early life. Nearly half of the patients have a neonatal cholestatic icterus with hepatosplenomegaly. A few of them worsen and die from hepatic failure before 6 months, defining a *neonatal, cholestatic rapidly fatal form*. In most patients, the icterus resolves spontaneously, and only hepatosplenomegaly remains. Some infants develop a severe respiratory insufficiency. In children without a history of cholestatic icterus,

hepatosplenomegaly is often the presenting symptom. This may stay the only sign for many years, until onset of neurological symptoms. Hepatosplenomegaly, usually mild to moderate, tends to diminish with time, and is not always present in patients with a juvenile or adult onset neurological disease. On the other hand, there are 3 well documented older adults with splenomegaly only. Apart from these exceptional cases and infants dying early, all NPC patients develop neurological symptoms.

Neurological involvement: The patients are generally classified according to the age of onset of the neurological symptoms, which correlates with the following course and life span.

In the *severe infantile neurological form*, infants with hepatosplenomegaly show a delay in motor milestones that becomes evident between the age of 1–2 years, and hypotonia. Many never learn to walk. Mental status is less affected. The disease progresses towards pyramidal tract involvement, pronounced spasticity and mental regression. Signs of white matter involvement are present. Survival rarely exceeds 5 years.

Classic NPC (60–70% of the cases) includes patients with *late infantile* and *juvenile neurological onset.* In the late infantile form, hepatosplenomegaly is almost invariably present. The child often presents with gait problems and clumsiness between 3 and 5 years, due to ataxia. Language delay is frequent. The motor problems worsen, and cognitive dysfunction appears. In the juvenile form, neurological symptoms appear between 6 and 15 years, onset is more insidious and variable. Splenomegaly can be absent. School problems with difficulty in writing, impaired attention are common and may lead to misdiagnosis. The child becomes clumsier, shows more learning disabilities, and ataxia become obvious. In both forms, vertical supranuclear gaze palsy, with initial slowing of vertical saccadic velocity, is almost constant and a characteristic sign. Gelastic cataplexy occurs in about 20% of patients and can be the presenting symptom. As ataxia progresses, dysphagia, dysarthria, and dementia develop. Action dystonia is also frequent. About half of the patients with the classic form develop seizures, that may become difficult to treat. In a later stage, the patients develop pyramidal signs and spasticity, and swallowing problems. Most require gastrostomy. Death usually occurs between 7 and 12 years in late infantile onset patients, and is very variable in the juvenile form, some patients being still alive by age 30 or more.

In *adult onset patients*, presentation (as late as 60 years) is even more insidious. Some patients show severe ataxia, dystonia and dysarthria with a variable cognitive dysfunction, while psychiatric symptoms and dementia are dominating in others [82]. Epilepsy is rare in adult NPC. These patients may not show splenomegaly nor vertical gaze palsy.

38.10.2   Metabolic Derangement

When either the NPC1 or the NPC2 protein is non-functional, the cellular trafficking of endocytosed LDL-derived cholesterol is impaired, resulting in accumulation of unesterified cholesterol in the endosomal/lysosomal system and delay in homeostatic reactions. This specific abnormality constitutes the basis for biological diagnosis of NPC. Recent studies [83] demonstrated non-redundant functional cooperativity of NPC1 and NPC2, supporting the view that the two proteins function in tandem or in sequence in the same metabolic pathway. However, their precise function and relationship remain unclear (▸ [84] for reviews). The pattern of accumulating lipids is different in brain and in non neural organs. In extraneural organs sphingolipid accumulation is likely secondary to the cholesterol storage. On the other hand, neurons store only minimal amount of cholesterol but significant amount of glycolipids, including GM3 and GM2, and the question of which accumulation is primary in brain remains controversial. Excess GM2 ganglioside in neurons is said to correlate with ectopic dendritogenesis and meganeurite formation. The mechanisms for early loss of Purkinje cells or neurofibrillary tangles formation are unknown [85].

38.10.3   Genetics

Approximately 95% of patients have mutations in the *NPC1* gene (18q11) which encodes a large membrane glycoprotein with late endosomal localization. The remainder have mutations in the *NPC2* gene (14q24.3) which encodes a small soluble lysosomal protein that binds cholesterol with high affinity. More than 200 disease-causing mutations of the *NPC1* gene are known, as well as 60 polymorphisms. I1061T is the most frequent allele in patients of Western European descent and P1007A the second most frequent one. Only 21 families are known with mutations in the *NPC2* gene [80].

38.10.4   Diagnostic Tests

Neuroimaging is generally not contributive. Chitotriosidase in serum is moderately elevated. The foamy histiocytes that may be found in bone marrow aspirates stain positive with filipin. The definitive diagnosis requires cultured fibroblasts. The pathognomonic accumulation of cholesterol in lysosomes is visualized by fluorescence microscopy after staining with filipin. Typical (but not variant) patients also show a very decreased rate of LDL-induced cholesteryl ester formation. Filipin staining will give unequivocal results in about 80% of patients, but in the remaining, described as variant, the level of cholesterol is much less, and the test requires conditioning of the cells in lipoprotein-deficient serum followed by a 24 h challenge with LDL-containing culture medium [86]. Even though, interpretation of results in variant cells can be very difficult [87].

38.10.5   Treatment and Prognosis

No specific treatment is available [88]. Cataplectic attacks can be treated by clomipramine or modafilin. Management of epilepsy, when present, is essential. Cholesterol-lowering drugs did not show evidence of neurological benefit. Treatment of NPC1 mutant mice with NB-DNJ (miglustat), an inhibitor of glycosphingolipid synthesis, led to a delay in onset of neurological symptoms and a prolonged life span [89]. Controlled clinical trials of miglustat in patients are ongoing and interim results are eagerly awaited. The recent finding that neurosteroidogenesis is compromised in the NPC1 mutant mouse may also have a significant bearing since the mice responded favourably to allopregnanolone injections early in life [90]. Significant problems for therapeutic trials are the wide spectrum of clinical phenotypes and the lack of good disease markers.

References

1. Charrow J, Andersson HC, Kaplan P et al (2000) The Gaucher registry: demographics and disease characteristics of 1698 patients with Gaucher disease. Arch Intern Med 160:2835-2843
2. Grabowski GA, Andria G, Baldellou A et al (2004) Pediatric non-neuronopathic Gaucher disease: presentation, diagnosis and assessment. Consensus statements. Eur J Pediatr 163:58-66
3. Wenstrup RJ, Roca-Espiau M, Weinreb NJ, Bembi B (2002) Skeletal aspects of Gaucher disease: a review. Br J Radiol. 75[Suppl 1]:A2-12
4. Varkonyi J, Rosenbaum H, Baumann N et al (2003) Gaucher disease associated with parkinsonism: four further case reports. Am J Med Genet A 116:348-351
5. Vellodi A, Bembi B, de Villemeur TB et al (2001) Management of neuronopathic Gaucher disease: a European consensus. J Inherit Metab Dis 24:319-327
6. Stone DL, Sidransky (1999) Hydrops fetalis: lysosomal storage disorders in extremis. Adv Pediatr 46:409-440
7. Mignot C, Gelot A, Bessieres B et al (2003) Perinatal-lethal Gaucher disease. Am J Med Genet A 120:338-344
8. Dreborg S, Erikson A, Hagberg B (1980) Gaucher disease – Norrbottnian type. I. General clinical description. Eur J Pediatr 133:107-118
9. Pampols T, Pineda M, Giros ML et al (1999) Neuronopathic juvenile glucosylceramidosis due to sap-C deficiency: clinical course, neuropathology and brain lipid composition in this Gaucher disease variant. Acta Neuropathol (Berl) 97:91-97
10. Vellodi A (2005) Lysosomal storage disorders. Br J Haematol 128:413-431
11. Sidransky E (2004) Gaucher disease: complexity in a «simple» disorder. Mol Genet Metab 83:6-15
12. Amato D, Stachiw T, Clarke JT, Rivard GE (2004) Gaucher disease: variability in phenotype among siblings. J Inherit Metab Dis 27:659-669
13. Lachmann RH, Grant IR, Halsall D, Cox TM (2004) Twin pairs showing discordance of phenotype in adult Gaucher's disease. QJM 97:199-204

14. Elstein Y, Eisenberg V, Granovsky-Grisaru S et al (2004) Pregnancies in Gaucher disease: a 5-year study. Am J Obstet Gynecol 190:435-441

15. Weinreb NJ, Charrow J, Andersson HC et al (2002) Effectiveness of enzyme replacement therapy in 1028 patients with type 1 Gaucher disease after 2 to 5 years of treatment: a report from the Gaucher Registry. Am J Med 113:112-119

16. Weinreb NJ, Aggio MC, Andersson HC et al (2004) Gaucher disease type 1: revised recommendations on evaluations and monitoring for adult patients. Semin Hematol 41:15-22

17. Baldellou A, Andria G, Campbell PE et al (2004) Paediatric non-neuronopathic Gaucher disease: recommendations for treatment and monitoring. Eur J Pediatr 163:67-75

18. Elstein D, Hollak C, Aerts JM et al (2004) Sustained therapeutic effects of oral miglustat (Zavesca, N-butyldeoxynojirimycin, OGT 918) in type I Gaucher disease. J Inherit Metab Dis 27:757-766

19. Lyon G, Adams RD, Kolodny EH (1996) Neurology of hereditary metabolic diseases of children. McGraw Hill, New York

20. Harzer K, Rolfs A, Bauer P et al (2003) Niemann-Pick disease type A and B are clinically but also enzymatically heterogeneous: pitfall in the laboratory diagnosis of sphingomyelinase deficiency associated with the mutation Q292 K. Neuropediatrics 34:301-306

21. Pavlù-Pereira H, Asfaw B, Poupetová H et al (2005) Acid sphingomyelinase deficiency. Phenotype variability with prevalence of intermediate phenotype in a series of 25 Czech and Slovak patients. A multi-approach study. J Inherit Metab Dis 28:203-227

22. Mendelson DS, Wasserstein MP, Desnick RJ et al (2006) Chest radiograph, high-resolution CT, and pulmonary function findings in Niemann-Pick disease type B. Radiology 238:339-345

23. Wasserstein MP, Larkin AE, Glass RB et al (2003) Growth restriction in children with type B Niemann-Pick disease. J Pediatr 142:424-428

24. McGovern MM, Wasserstein MP, Aron A et al (2004) Ocular manifestations of Niemann-Pick disease type B. Ophthalmology 111:1424-1427

25. Wasserstein MP, Desnick RJ, Schuchman EH et al (2004) The natural history of type B Niemann-Pick disease: results from a 10-year longitudinal study. Pediatrics 114:e672-e677

26. Simonaro CM, Desnick RJ, McGovern MM et al (2002) The demographics and distribution of type B Niemann-Pick disease: novel mutations lead to new genotype/phenotype correlations. Am J Hum Genet 71:1413-1419

26a. Simonaro CM, Park J-H, Eliyahu E et al (2006) Imprinting at the SMPD1 locus: Implications for acid sphingomyelinase-deficient Niemann-Pick Disease. Am J Hum Genet 78:865-870

27. Miranda SR, He X, Simonaro CM et al (2000) Infusion of recombinant human acid sphingomyelinase into Niemann-Pick disease mice leads to visceral, but not neurological, correction of the pathophysiology. FASEB J 14:1988-1995

28. Shihabuddin LS, Numan S, Huff MR et al (2004) Intracerebral transplantation of adult mouse neural progenitor cells into the Niemann-Pick-A mouse leads to a marked decrease in lysosomal storage pathology. J Neurosci 24:10642-10651

29. Muthane U, Chickabasaviah Y, Kaneski C et al (2004) Clinical features of adult GM1 gangliosidosis: report of three Indian patients and review of 40 cases. Mov Disord 19:1334-1341

30. Pshezhetsky AV, Ashmarina M (2001) Lysosomal multienzyme complex: biochemistry, genetics, and molecular pathophysiology. Prog Nucleic Acid Res Mol Biol 69:81-114

31. Pinto R, Caseiro C, Lemos M et al (2004) Prevalence of lysosomal storage diseases in Portugal. Eur J Hum Genet 12:87-92

32. Bolhuis PA, Oonk JG, Kamp PE et al (1987) Ganglioside storage, hexosaminidase lability, and urinary oligosaccharides in adult Sandhoff's disease. Neurology 37:75-81

33. Myerowitz R, Lawson D, Mizukami H et al (2002) Molecular pathophysiology in Tay-Sachs and Sandhoff diseases as revealed by gene expression profiling. Hum Mol Genet 11:1343-1350

34. Jeyakumar M, Butters TD, Cortina-Borja M et al (1999) Delayed symptom onset and increased life expectancy in Sandhoff disease mice treated with N-butyldeoxynojirimycin. Proc Natl Acad Sci USA 96:6388-6393

35. Hagberg B, Kollberg H, Sourander P, Akesson HO (1969) Infantile globoid cell leucodystrophy (Krabbe's disease). A clinical and genetic study of 32 Swedish cases 1953-1967. Neuropädiatrie 1:74-88

36. Lyon G, Hagberg B, Evrard P et al (1991) Symptomatology of late onset Krabbe's leukodystrophy: the European experience. Dev Neurosci 13:240-244

37. Kolodny EH, Raghavan S, Krivit W (1991) Late-onset Krabbe disease (globoid cell leukodystrophy): clinical and biochemical features of 15 cases. Dev Neurosci 13:232-239

38. Henderson RD, MacMillan JC, Bradfield JM (2003) Adult onset Krabbe disease may mimic motor neurone disease. J Clin Neurosci 10:638-639

39. Spiegel R, Bach G, Sury V et al (2005) A mutation in the saposin A coding region of the prosaposin gene in an infant presenting as Krabbe disease: first report of saposin A deficiency in humans. Mol Genet Metab 84:160-166

40. Suzuki K (1998) Twenty five years of the »psychosine hypothesis«: a personal perspective of its history and present status. Neurochem Res 23:251-259

41. Wenger DA, Rafi MA, Luzi P (1997) Molecular genetics of Krabbe disease (globoid cell leukodystrophy): diagnostic and clinical implications. Hum Mutat 10:268-279

42. Husain AM, Altuwaijri M, Aldosari M (2004) Krabbe disease: neurophysiologic studies and MRI correlations. Neurology 63:617-620

43. Barone R, Bruhl K, Stoeter P et al (1996) Clinical and neuroradiological findings in classic infantile and late-onset globoid-cell leukodystrophy (Krabbe disease). Am J Med Genet 63:209-217

44. Loes DJ, Peters C, Krivit W (1999) Globoid cell leukodystrophy: distinguishing early-onset from late-onset disease using a brain MR imaging scoring method. AJNR Am J Neuroradiol 20:316-323

45. Aldosari M, Altuwaijri M, Husain AM (2004) Brain-stem auditory and visual evoked potentials in children with Krabbe disease. Clin Neurophysiol 115:1653-1656

46. Krivit W, Shapiro EG, Peters C et al (1998) Hematopoietic stem-cell transplantation in globoid-cell leukodystrophy. N Engl J Med 338:1119-1126

47. Peters C, Steward CG (2003) Hematopoietic cell transplantation for inherited metabolic diseases: an overview of outcomes and practice guidelines. Bone Marrow Transplant 31:229-239

47a. Boelens JJ (2006) Trends in haematopietic cell transplantation for inborn errors of metabolism. J Inherit Metab Dis 29:413-420

48. Escolar ML, Poe MD, Provenzale JM et al (2005). Transplantation of umbilical-cord blood in babies with infantile Krabbe's disease. N Engl J Med 352:2069-2081

49. Hagberg B (1963) Clinical symptoms, signs and tests in metachromatic leukodystrophy. In: Folch-Pi J, Bauer H (eds) Brain lipids and lipoproteins and the leukodystrophies. Elsevier, Amsterdam, pp 134-146

50. Haltia T, Palo J, Haltia M, Icen A (1980) Juvenile metachromatic leukodystrophy. Clinical, biochemical, and neuropathologic studies in nine new cases. Arch Neurol 37:42-46

51. Baumann N, Turpin JC, Lefevre M, Colsch B (2002) Motor and psychocognitive clinical types in adult metachromatic leukodystrophy: genotype/phenotype relationships? J Physiol Paris 96:301-306

52. Comabella M, Waye JS, Raguer N et al (2001) Late-onset metachromatic leukodystrophy clinically presenting as isolated peripheral neuropathy: compound heterozygosity for the IVS2+1G-->A mutation and a newly identified missense mutation (Thr408Ile) in a Spanish family. Ann Neurol 50:108-112

53. Shapiro EG, Lockman LA, Knopman D, Krivit W (1994) Characteristics of the dementia in late-onset metachromatic leukodystrophy. Neurology 44:662-665

54. Gieselmann V, Zlotogora J, Harris A et al (1994) Molecular genetics of metachromatic leukodystrophy. Hum Mutat 4:233-242

55. Berger J, Loschl B, Bernheimer H et al (1997) Occurrence, distribution, and phenotype of arylsulfatase A mutations in patients with metachromatic leukodystrophy. Am J Med Genet 69:335-340

56. Rafi MA, Coppola S, Liu SL et al (2003) Disease-causing mutations in cis with the common arylsulfatase A pseudodeficiency allele compound the difficulties in accurately identifying patients and carriers of metachromatic leukodystrophy. Mol Genet Metab 79:83-90

57. Cameron CL, Kang PB, Burns TM et al (2004) Multifocal slowing of nerve conduction in metachromatic leukodystrophy. Muscle Nerve 29:531-536

58. Sener RN (2003) Metachromatic leukodystrophy. Diffusion MR imaging and proton MR spectroscopy. Acta Radiol 44:440-443

59. Oguz KK, Anlar B, Senbil N, Cila A (2004) Diffusion-weighted imaging findings in juvenile metachromatic leukodystrophy. Neuropediatrics 35:279-282

60. Kudoh T, Wenger DA (1982) Diagnosis of metachromatic leukodystrophy, Krabbe disease, and Farber disease after uptake of fatty acid-labeled cerebroside sulfate into cultured skin fibroblasts. J Clin Invest 70:89-97

61. Krivit W, Shapiro E, Kennedy W et al (1990) Treatment of late infantile metachromatic leukodystrophy by bone marrow transplantation. N Engl J Med 322:28-32

62. Malm G, Ringden O, Winiarski J et al (1996) Clinical outcome in four children with metachromatic leukodystrophy treated by bone marrow transplantation. Bone Marrow Transplant 17:1003-1008

63. Kidd D, Nelson J, Jones F et al (1998) Long-term stabilization after bone marrow transplantation in juvenile metachromatic leukodystrophy. Arch Neurol 55:98-99

64. Biffi A, De Palma M, Quattrini A et al (2004) Correction of metachromatic leukodystrophy in the mouse model by transplantation of genetically modified hematopoietic stem cells. J Clin Invest 113:1118-1129

65. Matzner U, Herbst E, Hedayati KK et al (2005) Enzyme replacement improves nervous system pathology and function in a mouse model for metachromatic leukodystrophy. Hum Mol Genet 14:1139-1152

66. Whybra C, Kampmann C, Willers I et al (2001) Anderson-Fabry disease: clinical manifestations of disease in female heterozygotes. J Inherit Metab Dis 24:715-724

67. Guffon N (2003) Clinical presentation in female patients with Fabry disease. J Med Genet 40:e38

68. Ries M, Gupta S, Moore DF et al (2005) Pediatric Fabry disease. Pediatrics 115: e344-e355

69. Desnick RJ, Brady R, Barranger J et al (2003) Fabry disease, an under-recognized multisystemic disorder: expert recommendations for diagnosis, management, and enzyme replacement therapy. Ann Intern Med 138:338-346

70. Mehta A, Ricci R, Widmer U et al (2004) Fabry disease defined: baseline clinical manifestations of 366 patients in the Fabry Outcome Survey. Eur J Clin Invest 34:236-242

71. Mills K, Morris P, Lee P et al (2005) Measurement of urinary CDH and CTH by tandem mass spectrometry in patients hemizygous and heterozygous for Fabry disease. J Inherit Metab Dis 28:35-48

72. Whitfield PD, Calvin J, Hogg S et al (2005) Monitoring enzyme replacement therapy in Fabry disease--role of urine globotriaosylceramide. J Inherit Metab Dis 28:21-33

73. Beck M, Ricci R, Widmer U, Dehout F et al (2004) Fabry disease: overall effects of agalsidase alfa treatment. Eur J Clin Invest 34:838-844

74. Wilcox WR, Banikazemi M, Guffon N et al (2004) Long-term safety and efficacy of enzyme replacement therapy for Fabry disease. Am J Hum Genet 75:65-74

75. Warnock DG (2005) Fabry disease: diagnosis and management, with emphasis on the renal manifestations. Curr Opin Nephrol Hypertens 14:87-95

76. Kattner E, Schafer A, Harzer K (1997) Hydrops fetalis: manifestation in lysosomal storage diseases including Farber disease. Eur J Pediatr 156:292-295

77. Vormoor J, Ehlert K, Groll AH et al (2004) Successful hematopoietic stem cell transplantation in Farber disease. J Pediatr 144: 132-134

78. Yeager AM, Uhas KA, Coles CD, Davis PC et al (2000) Bone marrow transplantation for infantile ceramidase deficiency (Farber disease). Bone Marrow Transplant 26:357-363

79. Harzer K, Paton BC, Poulos A et al (1989) Sphingolipid activator protein deficiency in a 16-week-old atypical Gaucher disease patient and his fetal sibling: biochemical signs of combined sphingolipidoses. Eur J Pediatr 149:31-39

80. Vanier MT, Millat G (2003) Niemann-Pick disease type C. Clin Genet 64:269-281

81. Patterson M., Vanier MT (2004) Niemann-Pick disease type C. In: Zimran A(ed) Glycolipid storage disorders. Adis Communications, Abingdon, pp 79-89

82. Lossos A, Schlesinger I, Okon E et al (1997) Adult-onset Niemann-Pick type C disease. Clinical, biochemical, and genetic study. Arch Neurol 54:1536-1541

83. Sleat DE, Wiseman JA, El Banna M et al (2004) Genetic evidence for nonredundant functional cooperativity between NPC1 and NPC2 in lipid transport. Proc Natl Acad Sci USA 101:5886-5891

84. Niemann-Pick disease type C, Liscum L, Sturley SL (eds) (2004) Biochim Biophys Acta 1685:1-90

85. Walkley SU, Suzuki K (2004) Consequences of NPC1 and NPC2 loss of function in mammalian neurons. Biochim Biophys Acta 1685:48-62

86. Vanier MT, Rodriguez-Lafrasse C, Rousson R et al (1991) Type C Niemann-Pick disease: spectrum of phenotypic variation in disruption of intracellular LDL-derived cholesterol processing. Biochim Biophys Acta 1096:328-337

87. Vanier MT, Suzuki K (1998) Recent advances in elucidating Niemann-Pick C disease. Brain Pathol 8:163-174

88. Patterson MC, Platt F (2004) Therapy of Niemann-Pick disease, type C. Biochim Biophys Acta 1685:77-82

89. Zervas M, Somers KL, Thrall MA, Walkley SU (2001) Critical role for glycosphingolipids in Niemann-Pick disease type C. Curr Biol 11:1283-1287

90. Griffin LD, Gong W, Verot L, Mellon SH (2004) Niemann-Pick type C disease involves disrupted neurosteroidogenesis and responds to allopregnanolone. Nat Med 10:704-711

39 Mucopolysaccharidoses and Oligosaccharidoses

J. Ed Wraith

Mucopolysaccharides

Mucopolysaccharides (now preferentially termed gly-cosaminoglycans, GAGs) are essential constituents of connective tissue, including cartilage and vessel walls. They are composed of long sugar chains, containing highly sulfated, alternating uronic acid and hexosamine residues, assembled into repeating units. The poly-saccharide chains are bound to specific core proteins within complex macromolecules called proteoglycans.

Depending on the composition of the repeating units, several mucopolysaccharides are known (◘ Fig. 39.1). Their degradation takes place inside the lysosomes and requires several acid hydrolases. Deficiencies of specific degradative enzymes have been found to be the cause of a variety of eponymous disorders, collectively termed mucopolysaccharidoses.

Dermatan sulfate

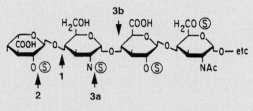

Heparan sulfate

Keratan sulfate

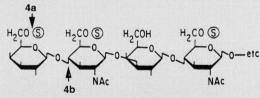

◘ **Fig. 39.1.** Main repeating units in mucopolysaccharides and location of the enzyme defects in the mucopolysaccharidoses (MPS). *NAc*, *N*-acetyl; *S*, sulfate. **1**, α-iduronidase (MPS I: Hurler and Scheie disease); **2**, iduronate sulfatase (MPS II: Hunter disease); **3a**, heparan *N*-sulfatase (MPS IIIa: Sanfilippo A disease); **3b**, α-*N*-acetylglucosaminidase (MPS IIIb: Sanfilippo B disease) **4a**, N-acetylgalactosamine-6-sulfatase (MPS IVa: Morquio A disease); **4b**, β-galactosidase (MPS IVb: Morquio B disease); **6**, NAc-galacto-samine 4-sulfatase (MPS VI: Maroteaux-Lamy disease); **7**, β-glucu-ronidase (MPS VII: Sly disease)

Genetic defects in enzymes that are involved in the lysosomal degradation of the mucopolysaccharides (glycosaminoglycans, GAGs) (■ Fig. 39.1) and the oligosaccharide chains of glycoproteins (■ Fig. 39.8) lead to chronic and progressive storage disorders that share many clinical features. These vary from facial dysmorphism, bone dysplasia (dysostosis multiplex), hepatosplenomegaly, neurological abnormalities, developmental regression and a reduced life expectancy at the severe end of the clinical spectrum, to an almost normal clinical phenotype and life span in patients with more attenuated disease. Mucopolysaccharidoses (MPS) and oligosaccharidoses are transmitted in an autosomal recessive manner, except for the X-linked MPS II (*Hunter syndrome*). Diagnosis of these disorders is initially by detecting partially degraded GAG or oligosaccharide in urine and confirmed by specific enzyme assays in serum, leukocytes or skin fibroblasts.

For the majority of disorders treatment is palliative but there have been important advances in the use of specific enzyme replacement therapy strategies for some MPS and this is an area of very rapid development. In addition, haematopoietic stem cell transplantation (HSCT) can improve outcome in carefully selected patients with MPS (especially MPS IH, *Hurler syndrome*), but this procedure is associated with significant morbidity and mortality.

Gene augmentation/transfer using a variety of vectors has been successful in cultured cells and animal models but has not yet been successfully performed in a human patient with one of these disorders.

It is important to remember that prenatal diagnosis is possible for all the MPS and oligosaccharidoses.

39.1 Clinical Presentation

39.1.1 Mucopolysaccharidoses

Mucopolysaccharidoses (MPS), like all lysosomal storage diseases, are chronic, progressive multisystem disorders. Affected infants are usually normal at birth and the disease is only diagnosed as the phenotype evolves with time. Infants with an MPS-like phenotype present at birth are most likely to have mucolipidosis type II (*I-Cell disease*). There is very wide clinical heterogeneity within this group of disorders.

In general the MPS disorders present in one of three ways:

1. as a dysmorphic syndrome e.g. MPS IH (*Hurler*), MPS II (*Hunter*), MPS VI (*Maroteaux-Lamy*);
2. with learning difficulties, behavioral disturbance and dementia e.g. MPS III (*Sanfilippo*);
3. as a severe bone dysplasia e.g. MPS IV (*Morquio*).

Patients with MPS often have a facial appearance, which is characteristically labelled coarse, although most parents find the term objectionable. A combination of subcutaneous storage and involvement of the facial bones in the dysostosis produces the typical appearance, seen in its most developed form in MPS IH (*Hurler syndrome*, ■ Fig. 39.2). Underdevelopment of the mid-facial skeleton and the firm puffiness associated with subcutaneous storage results in a flat nasal bridge and a blurring of the facial features. The lips and tongue are thickened and the hair is often abundant and dull. The persistent nasal discharge detracts further from the child's general appearance. A dark synophyris is a characteristic finding and affected children are often hirsute. The facial phenotype is much less obvious in patients with MPS III and absent in patients with MPS IV. A brief clinical summary for all of the disorders discussed is given in ■ Table 39.1.

Hurler Syndrome (MPS IH) and Scheie Disease (MPS IS)

Patients with MPS I have deficiency of the enzyme α-L-iduronidase (■ Fig. 39.1) and accumulate the glycosaminoglycans (GAGs) dermatan and heparan sulfate (DS, HS). Infants with severe disease (MPS 1H, *Hurler syndrome*) are usually diagnosed in the first year of life [1]. Upper airway obstruction and frequent ear, nose and throat infections dominate the clinical picture at an early stage and then the full clinical picture of short stature, hepatosplenomegaly, increasing facial dysmorphism, cardiac disease, progressive learning difficulties and corneal clouding evolve over the second and third years of life. The bone dysplasia that occurs in MPS and other lysosomal storage disorders, dysostosis multiplex, is most florid in MPS IH and in-

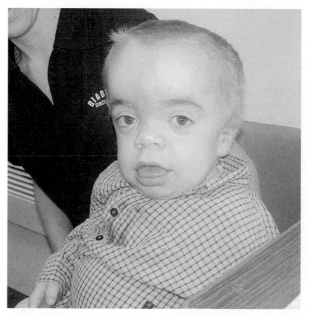

■ **Fig. 39.2.** Facial features of Hurler syndrome (MPS IH)

◻ **Table 39.1.** Mucopolysaccharidoses (MPS), mucolipidoses (ML) and oligosaccharidoses (OS) – diagnostic data

Disease	Enzyme deficiency	Storage material	Chromosome location	Gene mutations	Screening test	Diagnostic test	Prenatal diagnosis	Main clinical features
Mucopolysaccharidoses								
MPS I (Hurler, Scheie, Hurler/Scheie)	Iduronidase	DS, HS	4p16.3	W402X, Q70X plus many others	Urine GAGs	WBC enzyme assay	CVB[1]	HSM, CNS, SD, DYS, OPH, CAR
MPS II (Hunter)	Iduronate-2-sulfatase	DS, HS	Xq27-28	No common mutations	Urine GAGs	Plasma enzyme assay	CVB[2]	HSM, CNS, SD, DYS, OPH, CAR, SK
MPS III (Sanfilippo)								
IIIA	Heparan-N-sulfatase	HS	17q25.3	R245H,R74C and many others	Urine GAGs	WBC enzyme assay	CVB	CNS, SD (+/−), DYS (+/−)
IIIB	N-acetyl-glucos-aminidase	HS	17q21.1	No common mutations	Urine GAGs	Plasma enzyme assay	CVB	CNS, SD (+/−), DYS (+/−)
IIIC	Acetyl CoA glucosamine N-acetyl transferase	HS	Uncertain	Unknown	Urine GAGs	WBC enzyme assay	CVB	CNS, SD (+/−), DYS (+/−)
IIID	N-acetyl-glucosamine-6-sulfatase	HS	12q14	Very few patients studied	Urine GAGs	WBC enzyme assay	CVB	CNS, SD (+/−), DYS (+/−)

Storage material: *DS*, dermatan sulphate; *HA*, hyaluronic acid; *HS*, heparan sulphate; *KS*, keratan sulphate.
Screening test: *GAGs*, glycosaminoglycans; *oligos*, oligosaccharides.
Diagnostic test: *WBC*, white blood cells.
Prenatal diagnosis: *AF*, amniotic fluid; *CVB*, chorion villus biopsy.
Main clinical features: *CAR*, cardiac disease; *CNS*, central nervous system regression; *CRS*, cherry red spot; *DYS*, dysmorphic appearance; *HF*, hydrops fetalis; *HSM*, hepatosplenomegaly; *OPH*, eye signs: corneal clouding or opthalmoplegia; *SD*, dysostosis multiplex; *SK*, dermatological signs; *SKA*, angiokeratoma. (+/−), sign not always present or mild.

[1] Low activity in CVB – caution re: contamination with maternal decidua.
[2] Always do fetal sexing as some unaffected female fetuses will have very low enzyme results.
[3] Difficult because of cross-reactivity from other sulphatases.
[4] Lysosomal UDP-N-acetylglucosamine-l-phosphotransferase.

◻ Table 39.1 (continued)

Disease	Enzyme deficiency	Storage material	Chromosome location	Gene mutations	Screening test	Diagnostic test	Prenatal diagnosis	Main clinical features
MPS IV (Morquio)								
IVA	N-Acetylgalactosamine-6-sulfatase	KS	16q24	I113F (UK and Ireland)	Urine GAGs	WBC enzyme assay	CVB	SD, CAR, OPH (+/–)
IVB	β-Galactosidase	KS	3p21-pter	No common mutations	Urine GAGs	WBC enzyme assay	CVB	SD, CAR
MPS VI (Maroteaux-Lamy)	N-Acetylgalactosamine-4-sulfatase	DS	5q13-q14	No common mutations	Urine GAGs	WBC enzyme assay	CVB[3]	HSM, SD, DYS, OPH, CAR
MPS VII (Sly)	β-Glucuronidase	HS, DS	7q21.1-q22	Very few patients studied	Urine GAGs	WBC enzyme assay	CVB	HF, HSM, CNS, SD, DYS, OPH, CAR
MPS IX	Hyaluronidase	HA	3p21.3	Very few patients studied	None	Cultured cells	Unknown	Unknown
Mucolipidoses								
ML I (Sialidosis I)	Neuraminidase	Sialic acid	10pter-q23	No common mutations	Urine sialic acid	Cultured cells	Cultured cells	CNS, CRS, SD (+/–)
ML II (I Cell)	Transferase[4]	Many	4q21-q23	No common mutations	Urine oligos	Plasma enzyme assay	Cultured cells or AF	HSM, CNS, SD, DYS, OPH, CAR
ML III (pseudo-Hurler)	Transferase[4]							
IIIA	As ML II	Many	12q23.3	Very few patients studied	Urine oligos	Plasma enzyme assay	Cultured cells or AF	HSM (+/–), CNS (+/–), SD, DYS (+/–), CAR
IIIC	Transferase - δ-subunit	Many	16p13.3	Very few patients studied	Urine oligos	Plasma enzyme assay	Cultured cells or AF	as ML III A
ML IV	Unknown	Unknown	19p13.2-13.3	R750W (20%)	None	Histology	Histology of CVB	CNS, OPH

Table 39.1 (continued)

Disease	Enzyme deficiency	Storage material	Chromosome location	Gene mutations	Screening test	Diagnostic test	Prenatal diagnosis	Main clinical features
OligosaccharidosesS								
α-Mannosidosis	α-Mannosidase	α-mannosides	19p13.2-q12	R750W (20%)	Urine oligos	WBC enzyme assay	CVB	HSM, SD, DYS, CAR, CNS (+/−)
β-Mannosidosis	β-Mannosidase	β-mannosides	4p	Very few patients studied	Urine oligos	WBC enzyme assay	CVB	CNS, HSM (+/−)
Fucosidosis	Fucosidase	Fucosides glycolipids	1p24	No common mutations	Urine oligos	WBC enzyme assay	CVB	CNS, SKA
Aspartylglucos-aminuria	Aspartylglucosamini-dase	Aspartyl-glucos-amine	4q32-33	C163S (90% of Finnish patients)	Urine oligos	WBC enzyme assay	CVB	CNS, DYS, SD (+/−)
Schindler disease	α-N-Acetylgalactos-aminidase	N-Acetylgalactos-amineglycolipids	22q13.1-13.2	Very few patients studied	Urine oligos	WBC enzyme assay	CVB	CNS, SD, SKA
ISSD (infantile sialic acid storage disease)	Sialic acid transporter	Sialic acid	6q14-q15	R39C/other mutation	Urine oligos	Cultured cells	AF	HF, DYS, HSM, CNS, SD
Salla disease	As ISSD	As ISSSD	As ISSD	R39C homozygotes	As ISSD	As ISSD	As ISSD	CNS (+/−)
Galactosialidosis	Neuraminidase and β-galactosidase protective protein	Oligosaccharides, sialic acid	20q13.3	Very few patients studied	Urine oligos	Cultured cells	Cultured cells	HF, CNS, SD, HSM

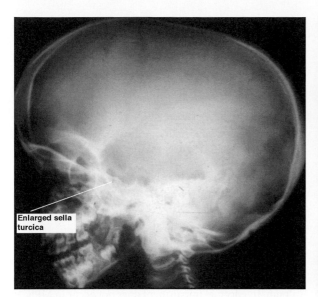

□ Fig. 39.3. Dysostosis multiplex in Hurler syndrome (MPS 1H): scaphiocephalic skull with expanded sella turcica

Enlarged sella turcica

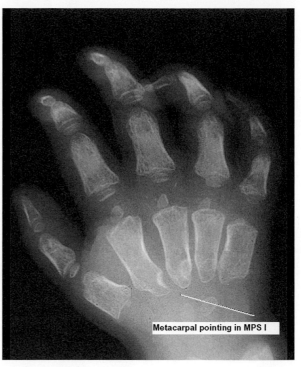

Metacarpal pointing in MPS I

□ Fig. 39.5. Dysostosis multiplex in Hurler syndrome (MPS IH): proximal pointing of the metacarpals

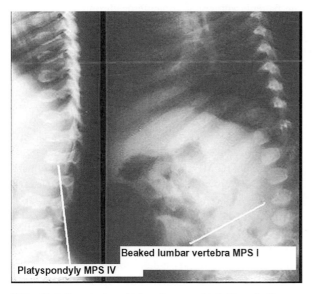

Beaked lumbar vertebra MPS I

Platyspondyly MPS IV

□ Fig. 39.4. Dysostosis multiplex: platospondyly in Morquio disease (MPS IV) and beaked lumbal vertebra in Hurler syndrome (MPS IH)

cludes abnormalities in the skull (enlarged sella turcica, scaphiocephaly, □ Fig. 39.3), broad ribs, hook shaped vertebrae (□ Fig. 39.4), prominent pointing of the metacarpals (□ Fig. 39.5) and underdevelopment of the pelvic bones. Patients with severe MPS I usually die before the age of 10 years as a result of cardiorespiratory disease.

At the other end of this clinical spectrum patients with *Scheie disease* (MPS IS) are intellectually normal, often of normal height and can live a normal life span although many patients become disabled as a result of degenerative joint disease, corneal opacity and cardiac valve lesions.

The symptoms of patients between these two extremes can be extremely variable (*Hurler-Scheie*, MPS I II-S) and can include short stature, coarse facies, corneal clouding, joint stiffness, deafness, and valvular heart disease. The onset of symptoms in MPS IH-S is observed between ages three and eight years and there is usually little or no intellectual dysfunction. Death usually occurs during the second or third decade of life from cardiac disease.

Hunter Syndrome (MPS II)

MPS II (iduronate-2-sulfatase deficiency, *Hunter syndrome*) differs from other MPS in that its inheritance is X-linked recessive. Unlike Fabry disease, manifesting female heterozygotes are exceptionally rare. Like MPS I this disorder is a spectrum with severely affected patients sharing many of the clinical signs and symptoms of patients with the severe form of MPS I, with the exception that the cornea remains clear in MPS II. Prominent Mongolian blue spots and a characteristic papular rash are other features prominent in severe MPS II (□ Fig. 39.6). Patients with the more attenuated form of MPS II can live well into adult life and a number have gone on to have their own families. Attenuated patients with MPS I, II and VI are at risk of developing cervical myelopathy due to dural thickening (pachymeningitis cervicalis) and thickening of the transverse ligaments even if the craniocervical junction has a stable arrangement (□ Fig. 39.7). This often presents insidiously with loss of endurance before more obvious signs of ascending paralysis become apparent.

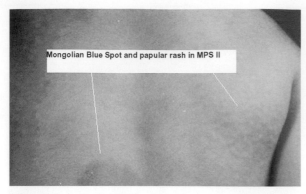

◻ Fig. 39.6. Papular rash and Mongolian blue spot in Hunter syndrome (MPS II)

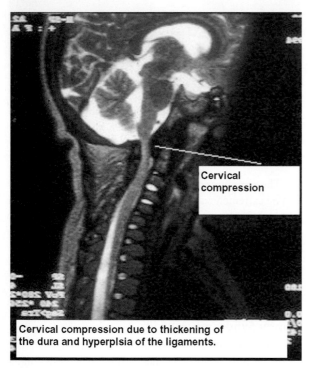

◻ Fig. 39.7. MRI scan of the craniocervical junction demonstrating cervical cord compression in Hunter syndrome (MPS II)

Sanfilippo Syndrome (MPS III)

Patients with all four subtypes of MPS III (A, B, C and D, *Sanfilippo syndrome*) have a defect in the catabolism of heparan sulfate. This results in a disorder which primarily affects the central nervous system whereas somatic abnormalities are usually mild. This often leads to a considerable delay in diagnosis. The condition has three phases [2]. The first phase, usually before diagnosis, consists of developmental delay alone, often primarily affecting speech. Some patients have ear disease and will fail hearing tests which is the usual reason given, initially, for the speech delay. In the second phase (age 3–10 years) the illness is dominated by a severe behavioral disturbance, characterized by hyperactiv-

ity, challenging behavior and profound sleep disturbance. The third phase of the illness (usually after the first decade) is associated with continuing loss of skills and slow deterioration into a vegetative state, death usually occurring in the early in the third decade. Like in all the other MPS there is considerable heterogeneity and not all patients will follow this pattern of deterioration. Patients with MPS III may be of normal height, have mild dysostosis multiplex and often develop seizures in the second or third phase of their illness.

Morquio Disease (MPS IV)

MPS IV (*Morquio disease*) is caused by a defect in the degradation of keratan sulfate. In classical *Morquio type A* (MPS IVA, galactose-6-sulfatase deficiency) the patients are affected by a very severe skeletal dysplasia characterized by vertebral platyspondyly (◻ Fig. 39.4), hip dysplasia and genu valgum. Intellectual impairment does not occur in MPS IV, but height prognosis is very poor with affected adults ranging from 95-105 cm fully grown. Odontoid dysplasia is often associated with atlanto-axial subluxation and renders the patients vulnerable to acute or chronic cervical cord compression. In *Morquio B* (MPS IVB, β-galactosidase deficiency, GM_1-gangliosidosis) the skeletal involvement is not as pronounced, but the patients have prominent central nervous system disease and a slowly progressive neurodegenerative course.

Maroteaux-Lamy Syndrome (MPS VI)

Patients with MPS VI (N-acetylgalactosamine-4-sulfatase deficiency, *Maroteaux-Lamy syndrome*) have somatic features resembling MPS I, but without neurological impairment. This disorder is rare but important to recognize early as enzyme replacement therapy (ERT) is available to treat affected patients [3].

Sly Syndrome (MPS VII)

MPS VII (β-glucuronidase deficiency, *Sly syndrome*) is a very variable disorder which has hydrops fetalis as its most common presentation. Patients that survive pregnancy have a clinical disease similar to MPS I including the same degree of clinical heterogeneity.

Natowicz Syndrome (MPS IX)

The newest MPS disorder, MPS IX (hyaluronidase deficiency, *Natowicz syndrome*), has been fully characterized in one patient in whom the phenotype was mild with short stature and periarticular soft tissues masses the dominant clinical findings [4, 5].

Oligosaccharides/Glycoproteins

Almost all the secreted and membrane-associated proteins of the body are glycosylated, as well as numerous intracellular proteins, including the lysosomal acid hydrolases. A great variety of oligosaccharide chains are attached to the protein backbone via the hydroxyl group of serine or threonine (O-linked), or via the amide group of asparagine (N-linked), to form tree-like structures (◘ Fig. 39.8). The chains usually have a core composed of N-acetylglucosamine and mannose, often contain galactose, fucose and N-acetylgalactosamine, and frequently possess terminal sialic acids (N-acetylneu-

raminic acid). Oligosaccharide chains with a terminal mannose-6-phosphate are involved in the targeting of lysosomal enzymes to lysosomes. This recognition marker is synthesized in two steps from UDP-N-acetyl-glucosamine (◘ Fig. 39.9). Deficiencies of the enzymes required for the degradation of the oligosaccharide chains cause oligosaccharidoses (glycoprotein storage diseases). Defects of the synthesis of the mannose-6-phosphate recognition marker result in the mislocalization of lysosomal enymes. Defects of the synthesis of the oligosaccharide chains are discussed in ▶ Chap. 41.

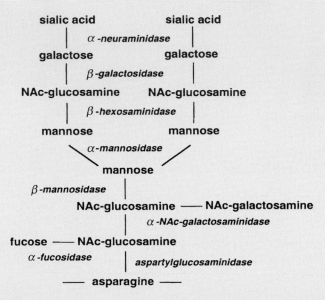

◘ **Fig. 39.8.** General composite example of a glycoprotein oligosaccharide chain. *NAc, N*-acetyl. Degradative enzymes are listed in *italics*

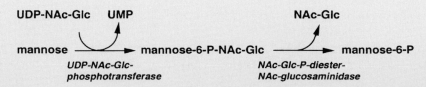

◘ **Fig. 39.9.** Synthesis of the mannose-6-phosphate recognition marker. *NAc-Glc, N*-acetylglucosamine; *UDP*, uridine diphosphate; *UMP*, uridine monophosphate. Enzymes are listed in *italics*

39.1.2   Oligosaccharidoses

Oligosaccharidoses or glycoprotein storage disorders share many features in common with MPS disorders but the urine GAG screen is normal or only shows non-specific abnormalities. For convenience, the mucolipidoses (ML), disorders that combine clinical features of MPS and sphingolipidoses, are also considered here. These include *sialidosis I* (ML I) caused by α–neuraminidase deficiency, and *mucolipidosis II* (ML II), and its milder allelic variant *mucolipidosis III,* (ML III), both caused by the deficiency of UDP-N-acetylglucosamine-1-phosphotransferase, an enzyme not involved in lysosomal degradation but in the synthesis of a recognition marker.

Mannosidosis

A deficiency of α-mannosidase gives rise to the extremely variable disorder *a-mannosidosis*. A mild Hurler phenotype associated with variable learning difficulties, hepatosplenomegaly, deafness and skeletal dysplasia, is complicated by an immune deficiency which can dominate the clinical progression of the disease [6]. *β-Mannosidosis* due to a deficiency of β-mannosidase is much less prevalent than α-mannosidosis and is very variable but severe learning difficulties, challenging behavior, deafness and frequent infections are relatively common [7].

Fucosidosis

Patients with *fucosidosis* lack the typical facial dysmorphism seen in the other disorders described in this chapter. Deficiency of α-fucosidase activity leads to a severe neurodegenerative disorder often with seizures and a mild dysostosis. Affected patients often exhibit prominent, widespread angiokeratomas and like all oligosaccharidoses the clinical course of the patients can be variable [8].

Galactosialidosis

Galactosialidosis is caused by combined deficiency of the lysosomal enzymes β-galactosidase and α-neuraminidase. The combined deficiency has been found to result from a defect in protective protein/cathepsin A (PPCA), an intralysosomal protein which protects these enzymes from premature proteolytic processing. The clinical features of affected patients includes hydrops fetalis as well as a more slowly progressive disorder associated with learning difficulties, dysostosis multiplex and corneal opacity.

Transport Defects

The allelic disorders *Salla disease* and *infantile free sialic acid storage disease (ISSD)* result from mutations within the gene coding for sialin, a lysosomal membrane protein that transports sialic acid out of lysosomes. *ISSD* has a severe phenotype with infantile onset (including severe visceral involvement, cardiomyopathy, skeletal dysplasia and learning difficulties), while the Finnish variant, *Salla disease*, has a milder phenotype with later onset. Both disorders cause learning difficulties but *ISSD* is generally fatal in early childhood whilst patients with *Salla* disease survive into middle-age [9].

Aspartylglucosaminuria

This disorder, due to a deficiency of aspartylglucosaminidase, has a high prevalence in Finland and is rare in other countries. A characteristic facial dysmorphism has been described (hypertelorism, a short and broad nose, simple ears with small or missing lobule and thick lips [10]) and a slowly progressive psychomotor retardation occurs with death in middle age [11].

Schindler Disease

This disease, due to α-N-acetylgalactosaminidase deficiency, is a rare, clinically heterogeneous disorder with a wide spectrum including an early onset neuroaxonal dystrophy and a late onset form characterised by abundant angiokeratoma. There are discrepancies between genotype and phenotype in the disease and it has been suggested that other factors (apart from the enzyme deficiency) may contribute to the severe neurological findings in early-onset patients [12].

Sialidosis (ML I)

This disorder is characterized by the progressive lysosomal storage of sialic acid-rich glycopeptides and oligosaccharides caused by a deficiency of the enzyme neuraminidase. The *sialidoses* are distinct from the *sialurias* (infantile sialic acid storage disease, *ISSD* and *Salla disease*) in which there is storage and excretion of free sialic acid, rather than bound sialic acid. The clinical spectrum in sialidosis ranges widely from a presentation with hydrops fetalis to the comparatively slowly progressive cherry-red spot myoclonus syndrome.

I-Cell Disease (ML II) and Pseudo-Hurler (ML III)

Patients with ML II have a Hurler-like phenotype often presenting in the newborn period. There is often a very severe skeletal dysplasia and patients often have a small head circumference due to premature sutural synostosis. Cardiomyopathy and severe coronary artery disease can be present and are often the most important indicators of prognosis. Most patients die in the first decade of life. ML III is extremely variable and many patients survive into adult life with little or no learning problems. Skeletal dysplasia including an erosive arthropathy affecting ball and socket joints (shoulder and hips) can be extremely disabling in adults with ML III.

Pycnodysostosis

Cathepsin K is a recently identified lysosomal cysteine proteinase, abundant in osteoclasts, where it is felt to play a vital role in the resorption and remodelling of bone. A deficiency of this enzyme was shown to be associated with the skeletal dysplasia, pycnodysostosis, the disorder thought to be the cause of Toulouse-Lautrec's disability [13]. In addition to short stature (150–160 cm), affected individuals have a generalised increase in bone density; wormian bones of the skull with open fontanelles, partial absence of the distal phalanges and bone fragility. Dental abnormalities are also common.

39.2 Metabolic Derangements

A summary of the metabolic derangements (enzyme deficiency and storage material) in the MPS, oligosaccharidoses and mucolipidoses is given in ◘ Table 39.1.

The vast majority of disorders are defects of single enzymes, involved in the degradation of mucopolysaccharides (◘ Fig. 39.1) or oligosaccharides (◘ Fig. 39.8). The exceptions are *ML II* (and III), the transport defects (*ISSD and Salla disease*), and *galactosialidosis. ML II and III* share the same post-translational modification defect due to the absence of UDP-N-acetylglucosamine-1-phosphotransferase (◘ Fig. 39.9), the enzyme necessary to synthesize the recognition marker required to target various newly formed lysosomal enzymes to the lysosomes. As a result the enzymes are secreted into the extracellular space, where high activity is found; inside the cells the enzyme levels are considerably reduced. In the transport defects the gene encoding the lysosomal membrane protein sialin is defective. Urinary excretion of sialic acid is considerably elevated in these conditions. The combined defect of neuraminidase and β-galactosidase (*galactosialidosis*) is caused by a lack of the protective protein cathepsin A (PPCA) responsible for stabilization of the enzyme complex within the lysosome and protecting them for rapid proteolytic degradation. PPCA may also have a role in the protection of elastin binding protein (EBP) at the cell surface [14].

What is not clear is how the metabolic derangement leads to the clinical and functional defects seen in the patients especially those affecting the central nervous system. The clinical phenotype partially depends on the type and amount of storage substance but the pathogenic cascades leading to disease in the brain remain poorly understood [15].

39.3 Genetics

With the exception of MPS II, *Hunter syndrome*, this group of disorders are inherited in an autosomal recessive manner. ◘ Table 39.1 gives details of the genes involved and also common mutations where present. All families should be referred for genetic counselling, carrier detection where available and necessary, and also given information on the availability of prenatal diagnosis.

39.4 Diagnostic Tests

The diagnosis of MPS and oligosaccharidoses is based on clinical suspicion, supported by appropriate clinical and radiological examinations followed by urinary examination for glycosaminoglycan (GAG) and oligosaccharide excretion and then specific enzyme assay usually on white blood cells. The diagnostics tests including screening, prenatal and postnatal diagnosis are given in ◘ Table 39.1. Many different methods have been used in the initial biochemical screening of these patients. As most patients excrete (except MPS IX) increased amounts of GAG or oligosaccharides, analysis of urine for the presence of these substances is the most common first step in the diagnostic process. Spot tests are inexpensive and can be performed rapidly, but unfortunately they can be unreliable and may miss patients with MPS III and MPS IV. For quantitative measurement of urine GAG, several tests are available (e.g. determination of uronic acid based on the carbazol method or the spectrophotometric assay using the dye dimethylmethylene blue). Two dimensional electrophoresis of GAG extracted from urine allows the clearest discrimination between different GAG species. Thin-layer chromatography is used to detect abnormal urinary excretion of oligosaccharides and sialic acid. All urinary screening tests can give false negative results, especially in older patients with an attenuated clinical course. Patients with ML II and III are often missed and therefore definitive diagnosis should be confirmed by the appropriate enzyme assays (◘ Table 39.1).

39.5 Treatment and Prognosis

Palliative care remains an important aspect of the holistic management of patients with these disorders. For some conditions it is the only available therapy. Multidisciplinary management is essential and patients are best managed in specialist centres with access to a comprehensive range of clinical and supporting services. The careful use of medication to treat sleep disturbance (e.g. melatonin) and challenging behaviour (e.g. risperidone) is important in those conditions primarily affecting the brain (e.g. MPS III). Ear, nose and throat surgery, orthopaedic review and neurosurgical intervention may all be indicated at some stage in affected patients. The anaesthetic considerations must not be forgotten as the facial dysmorphism, skeletal dysplasia and upper airways obstruction present in many of these patients can be a challenge.

Attempts at curative treatment for this group of disorders currently includes haematopoietic stem cell transplantation (HSCT) and enzyme replacement therapy (ERT). Substrate reduction therapy (SRT) has not been attempted in humans with this class of disorders and gene augmentation (gene therapy) has not been successfully applied in affected humans although results in some animal studies are promising.

The largest experience of the use of HSCT is in MPS I (severe variant, *Hurler syndrome*). A number of cell sources have been used including umbilical cord blood [16]. Results are variable and although developmental progress can be preserved the patients often have severe residual orthopaedic problems [17]. In addition the success of this therapy is limited by high morbidity and mortality. Experience in other MPS disorders is more limited and efficacy has not clearly been demonstrated in any of the other MPS conditions with the exception of MPS VI.

ERT is a safer approach to treatment for these disorders but is limited by an inability of the enzyme to cross the blood brain barrier when given intravenously. Aldurazyme (Biomarin/Genzyme) is now available for MPS I to treat the non-neurological manifestations of the disease. This follows a successful clinical trial where patients with MPS I demonstrated improvements in endurance and respiratory function following treatment with Aldurazyme [18]. Naglazyme (Biomarin) is also available to treat MPS VI and therapy is likely to be available for MPS II in the near future.

Successful gene augmentation trials have not yet been demonstrated in patients with this group of disorders. Problems with sustained gene expression have hindered progress, but this method of treatment may be the most likely to be successful in disorders with prominent central nervous system components. Animal models are proving a valuable resource to evaluate early studies of gene augmentation [19, 20] and it is only a matter of time before this is revisited in the human.

References

1. Cleary MA, Wraith JE (1995) The presenting features of mucopolysaccharidosis type IH (Hurler syndrome). Acta Paediatr 84:337-339
2. Cleary MA, Wraith JE (1993) Management of mucopolysaccharidosis type III. Arch Dis Child 69:403-406
3. Harmatz P, Whitley CB, Waber L et al (2004) Enzyme replacement therapy in mucopolysaccharidosis VI (Maroteaux-Lamy syndrome). J Pediatr 144:574-580
4. Natowicz MR, Short MP, Wang Y et al (1996) Clinical and biochemical manifestations of hyaluronidase deficiency. N Engl J Med 335:1029-1033
5. Triggs-Raine B, Salo TJ, Zhang H et al (1999) Mutations in HYAL1, a member of a tandemly distributed multigene family encoding disparate hyaluronidase activities, cause a newly described lysosomal disorder, mucopolysaccharidosis IX. Proc Natl Acad Sci USA 96:6296-6300
6. Malm D, Halvorsen DS, Tranebjaerg L et al (2000) Immunodeficiency in alpha-mannosidosis: a matched case-control study on immunoglobulins, complement factors, receptor density, phagocytosis and intracellular killing in leucocytes. Eur J Pediatr 159:699-703
7. Bedilu R, Nummy KA, Cooper A et al (2002) Variable clinical presentation of lysosomal beta-mannosidosis in patients with null mutations. Mol Genet Metab 77:282-290
8. Willems PJ, Gatti R, Darby JK et al (1991) Fucosidosis revisited: a review of 77 patients. Am J Med Genet 38:111-131
9. Kleta R, Morse RP, Orvisky E et al (2004) Clinical, biochemical, and molecular diagnosis of a free sialic acid storage disease patient of moderate severity. Mol Genet Metab 82:137-143
10. Arvio MA, Peippo MM, Arvio PJ et al (2004) Dysmorphic facial features in aspartylglucosaminuria patients and carriers. Clin Dysmorphol 13:11-15
11. Arvio M (1993) Follow-up in patients with aspartylglucosaminuria. Part I. The course of intellectual functions. Acta Paediatr 82:469-471
12. Sakuraba H, Matsuzawa F, Aikawa S et al (2004) Structural and immunocytochemical studies on alpha-N-acetylgalactosaminidase deficiency (Schindler/Kanzaki disease). J Hum Genet 49:1-8
13. Gelb BD, Shi GP, Chapman HA, Desnick RJ (1996) Pycnodysostosis, a lysosomal disease caused by cathepsin K deficiency. Science 273:1236-1238
14. Malvagia S, Morrone A, Caciotti A et al (2004) New mutations in the PPBG gene lead to loss of PPCA protein which affects the level of the beta-galactosidase/neuraminidase complex and the EBP-receptor. Mol Genet Metab 82:48-55
15. Walkley SU (2004) Pathogenic cascades and brain dysfunction. In: Platt FM, Walkley SU (eds) Lysosomal disorders of the brain. Oxford University Press, Oxford, pp 290-316
16. Staba SL, Escolar ML, Poe M et al (2004) Cord-blood transplants from unrelated donors in patients with Hurler's syndrome. N Engl J Med 350:1960-1969
17. Weisstein JS, Delgado E, Steinbach LS et al (2004) Musculoskeletal manifestations of Hurler syndrome: long-term follow-up after bone marrow transplantation. J Pediatr Orthop 24:97-101
18. Wraith JE, Clarke LA, Beck M et al (2004) Enzyme replacement therapy for mucopolysaccharidosis I: a randomized, double-blinded, placebo-controlled, multinational study of recombinant human alpha-L-iduronidase (laronidase). J Pediatr 144:581-588
19. Ellinwood NM, Vite CH, Haskins ME (2004) Gene therapy for lysosomal storage diseases: the lessons and promise of animal models. J Gene Med 6:481-506
20. Mango RL, Xu L, Sands MS et al (2004) Neonatal retroviral vector-mediated hepatic gene therapy reduces bone, joint, and cartilage disease in mucopolysaccharidosis VII mice and dogs. Mol Genet Metab 82:4-19

40 Peroxisomal Disorders

Bwee Tien Poll-The, Patrick Aubourg, Ronald J.A. Wanders

Peroxisomal Functions

Peroxisomes are cell organelles which derive their name from the presence of catalase, an enzyme that converts hydrogen peroxide into oxygen and water. As with lysosomes, they are found in all human cells except erythrocytes; however, unlike lysosomes, they possess anabolic besides catabolic functions (◘ Fig. 40.1). Peroxisomes are mainly involved in lipid metabolism: they synthesize ether-phospholipids, called plasmalogens, which are important constituents of cell membranes, and myelin, and β-oxidize very long chain fatty acids. Peroxisomes are also involved in the oxidation of phytanic acid, a chlorophyll derivative, and in the formation of bile acids from mevalonate via cholesterol. In addition they are concerned with the catabolism of lysine via pipecolic acid and glutaric acid (◘ Fig. 23.1), and of glyoxylate (◘ Fig. 43.1).

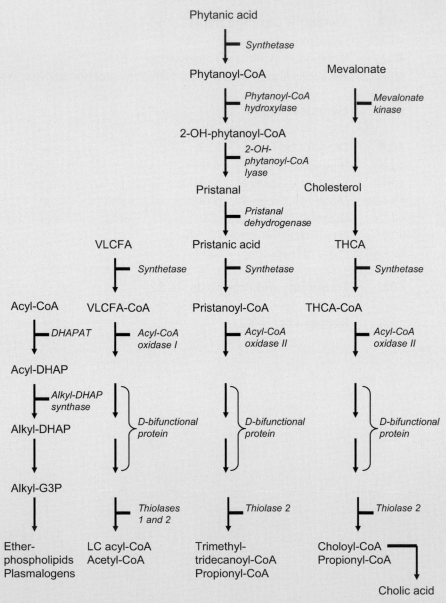

◘ **Fig. 40.1.** Schematic representation of the main peroxisomal functions. *CoA*, coenzyme A; *DHAP*, dihydroxyacetone phosphate; *DHAPAT*, dihydroxyacetone phosphate acyltransferase; G3P, glycerol-3-phosphate; *LC*, long-chain; *THCA*, trihydroxycholestanoic acid; *VLCFA*, very-long-chain fatty acid

Peroxisomal disorders can be recognized by the presence of dysmorphic features, neurological abnormalities and hepatic and gastrointestinal dysfunction. Widely different features that can occur include the following:

- Craniofacial dysmorphism, skeletal abnormalities, shortened proximal limbs, calcific stippling of the epiphyses
- Encephalopathy, seizures, peripheral neuropathy, abnormal gait, hypotonia
- Ocular abnormalities such as retinopathy, cataracts
- Liver disease with hyperbilirubinaemia, hepatomegaly and cholestasis
- Failure to thrive

Possibilities for (dietary) treatment are limited.

40.1 Clinical Presentation

At least 16 clinically and biochemically marked heterogeneous disorders linked to peroxisomal dysfunction have been identified [1–4]. In general, the onset of symptoms is not accompanied by an acute event or by abnormal routine laboratory tests indicating metabolic derangement. Most often, the presentation is associated either with variable neurodevelopment delay beginning in infancy or early childhood or with progressive neurological manifestations starting in the school-age period. Given the diversity of the clinical and biochemical abnormalities manifest in peroxisomal disorders, it is easier to regard the clinical diagnosis as a function of both the age of the patient (◘ Table 40.1) and the presence of one or more of the following predominant features: dysmorphism, neurological dysfunction and liver disease [5].

40.1.1 Dysmorphism

Dysmorphic features, including large fontanelles, a high forehead, epicanthic folds and abnormal ears, may be mistaken for chromosomal disorders such as Down syndrome. Other abnormalities include: rhizomelic shortening of limbs in rhizomelic chondrodysplasia punctata (RCDP), calcific stippling of epiphyses in RCDP and classical Zellweger syndrome (CZ), or renal cysts in CZ. These congenital manifestations are indicative of dysmorphogenesis during the prenatal period [6].

◘ **Table 40.1.** Clinical symptoms of peroxisomal disorders related to age

Symptoms	Disorder
Neonatal period Hypotonia, poor reactivity, seizures Craniofacial dysmorphism Skeletal abnormalities Conjugated hyperbilirubinemia	 CZ, ZeS Acyl-CoA oxidase deficiency D-BP-deficiency RCDP (typical/atypical)
First 6 months Failure to thrive Hepatomegaly, prolonged jaundice Gastro-intestinal problems, Hypocholesterolemia Vitamin-E deficiency Visual abnormalities	 ZeS D-BP-deficiency Milder forms of RCDP
Six months to 4 years Failure to thrive Neurologic presentation Psychomotor retardation Visual and hearing impairment (ERG, VEP, BAEP) Osteoprosis	 ZeS Late onset white matter disease Milder forms of RCDP
Beyond 4 years Behavior changes, cognitive deterioration, White-matter demyelination; spastic paraparesis Visual and hearing impairment Peripheral neuropathy, gait abnormality	 X-linked ALD/AMN Classical Refsum disease Racemase deficiency

ALD, adrenoleukodystrophy; *AMN*, adrenomyeloneuropathy; *BAEP*, brain auditory-evoked potentials; *ERG*, electroretinogram; *RCDP*, rhizomelic chondrodysplasia punctata; *VEP*, visual-evoked potentials; ZeS, Zellweger spectrum; *CZ*, classical Zellweger; *D-BP*, D-bifunctional protein.

40.1.2   Neurological Dysfunction

In the neonatal period, the predominant symptoms are often severe hypotonia with loss of reactivity (which can be mistaken for a neuromuscular disease). Abnormalities in neuronal migration (neocortical dysplasia) and cerebral/cerebellar white matter disease may be present in CZ, neonatal adrenoleukodystrophy (NALD) and D-bifunctional protein (D-BP) deficiency. All disorders with an abnormal peroxisomal lipid metabolism present with neurological dysfunction of different severity. An increasing number of inborn errors of metabolism without evident biochemical abnormalities on routine laboratory screening should also be considered in the diagnosis [6].

40.1.3   Hepatic and Gastrointestinal Disease

The predominant manifestations may be hepatomegaly, cholestasis, hyperbilirubinemia, prolonged jaundice, osteoporosis and failure to thrive, especially in disorders with a generalized deficiency of peroxisomal enzymes.

The Neonatal Period

In order to facilitate the recognition of peroxisomal disorders, their clinical presentation is categorized according to age. Two prototypes of peroxisomal disorders with a neonatal presentation are CZ syndrome, the most severe condition, and RCDP. Their phenotypes are distinct from the other peroxisomal disorders and should not cause difficulties in the differential diagnosis. Patients with NALD, D-BP deficiency and mevalonic aciduria may also have prominent manifestations in the neonatal period.

Classical Zellweger Syndrome

It is characterized by:
- Errors of morphogenesis
- Neuronal migration disorder, germinolytic cysts, delay of myelination
- Severe hypotonia and weakness from birth; seizures
- Sensorineural deafness
- Ocular abnormalities
- Degenerative changes
- Liver disease
- Failure to thrive
- Absence of recognizable hepatic peroxisomes (presence of peroxisomal ghosts)
- Death in the first year

Infants with CZ display such characteristic physical features that the diagnosis can usually be made on examination alone (◘ Fig. 40.2a). Typically they show profound hypotonia and weakness, a severe swallowing disorder, very large fontanelles, eye abnormalities (mostly pigmentary retinopathy, cataracts), seizures, and hepatomegaly [3].

Classical Rhizomelic Chondrodysplasia Punctata

Classical RCDP is characterized clinically by the presence of severe proximal shortening of limbs, typical facial dysmorphism, cataracts, calcific stippling of epiphyses, coronal clefts of vertebral bodies, multiple joint contractures, and severe growth and psychomotor delays [3].

The calcific stippling may disappear after the age of 2 years. The chondrodysplasia punctata is more widespread than in CZ and may involve extra-skeletal tissues. Some patients have ichthyosis. The classical RCDP phenotype is genetically heterogeneous involving not only the *PEX7*-gene (RCDP type 1) but also two other genes of plasmalo-

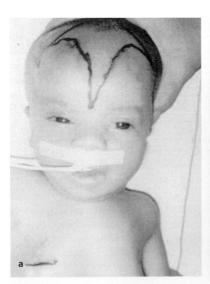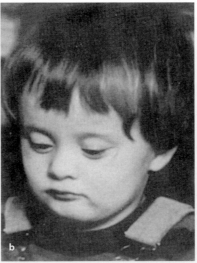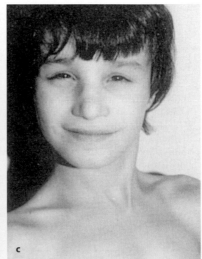

◘ **Fig. 40.2a–c.** Three patients with multiple enzyme defects and defective peroxisome assembly. **a** Classical Zellweger syndrome at 2 weeks of age. **b** Infantile Refsum disease at 2 years of age; note facial dysmorphism resembling Down syndrome. **c** Infantile Refsum disease at 11 years of age gen synthesis enzymes (RCDP type 2 and type 3). These three types are clinically indistinguishable from each other. Conversely, patients may have a *PEX7*-gene deficiency and present a milder clinical phenotype without significant shortening of the long bones, much milder developmental delay, ability to walk, and prolonged survival [7–10]. Classical RCDP and its milder variants must be distinguished from other forms of chondrodysplasia punctata, such as the autosomal dominant Conradi-Hünermann form with normal intellect (Chap. 33), the autosomal recessive rhizomelic form (*McKusick 215100*), the X-linked recessive form (*McKusick 302950*), and the X-linked dominant form (lethal in males).

Neonatal Adrenoleukodystrophy

Patients with NALD are somewhat less severely affected than those with CZ [3]; dysmorphic features are less striking and may even be absent. The large majority present with neonatal or early infantile onset hypotonia and seizures. They have progressive white matter disease, may have pachypolymicrogyria, and usually die in late infancy. Calcific stippling of epiphyses and renal cysts are absent. Some patients can mimick a neuromuscular disorder resembling spinal muscular atrophy (Werdnig-Hoffman disease) [11]. Patients with a single peroxisomal β-oxidation enzyme defect but with clinical manifestations resembling those of NALD, straight-chain fatty acyl-coenzyme A deficiency [12, 13] have been described. Liver peroxisomes are normal or appear to be enlarged in size, whereas in CZ and NALD, they are morphologically absent or severely decreased in number.

D-Bifunctional Protein Deficiency

D-BP deficiency is the most frequently diagnosed single enzyme defect of peroxisomal fatty acid β-oxidation that mimics the clinical manifestations of CZ or NALD. Patients show severe nervous system involvement with hypotonia, seizures, absence of significant development and peripheral neuropathy [14–16]. The majority of patients with D-BP deficiency has evidence of a neuronal migration defect and progressive white matter disease [17].

Dysmorphic features may be present: prominent forehead, flat nasal bridge and micrognathia. Most of the patients died between the ages of 6 months and 2 years.

First Six Months of Life

During this period of life, the predominant symptoms may be hepatomegaly associated (or not associated) with prolonged jaundice, liver failure and nonspecific digestive problems (anorexia, vomiting, diarrhea) leading to failure to thrive and osteoporosis. Hypocholesterolemia, hypolipoproteinemia and decreased levels of fat-soluble vitamins, symptoms that resemble a malabsorption syndrome, are frequently present (◘ Table 40.1). These patients can be erroneously diagnosed as having a congenital defect of glycosylation or a defect of the mitochondrial respiratory chain. Most CZ patients develop hepatomegaly and seizures and do not survive beyond this period. There is clinical, biochemical, and genetic overlap among the three phenotypes, CZ, NALD and infantile Refsum disease (IRD) [3, 18] and the name Zellweger spectrum has been proposed for the whole group.

Infantile Refsum Disease

IRD is the least severe phenotype within the group of peroxisome biogenesis disorders with generalized peroxisomal dysfunction. IRD patients have a later onset of variable initial symptoms, no neuronal migration disorder, no progressive white matter disease, and their survival is variable. They may have external features reminiscent of CZ or little or no facial dysmorphism (◘ Fig. 40.2b, c). Their cognitive and motor development varies between severe global handicap and moderate learning disorder with deafness and visual impairment due to retinopathy. Many patients show growth failure and hyperoxaluria [19]. Cerebral findings on magnetic resonance imaging (MRI) are variable and differ from CZ in predominance of regressive over developmental changes [20]. Most patients with IRD survive infancy and some even reach adulthood.

Between Six Months and Four Years

During this period of life, cognitive and motor dysfunction becomes evident (◘ Table 40.1). Sensorineural hearing impairment is associated with abnormal brain-stem auditory evoked responses. Various ocular abnormalities can be observed including retinopathy, cataracts, optic nerve atrophy, glaucoma and brushfield spots. The electroretinogram and visual evoked responses are frequently disturbed, and this may precede the fundoscopie abnormalities. Retinopathy associated with hearing loss, developmental delay and dysmorphism may be mistaken for other diseases, including malformative syndromes [21]. In this respect, the boundaries between malformative syndromes and inborn errors are not well delineated. Most CZ and NALD patients do not survive this period.

Late Onset White Matter Disease

Late-onset cerebral white matter disease/leukodystrophy has been described in patients with a Zellweger spectrum phenotype, either following IRD or following normal early development and in the absence of distinct external features [22]. Two patients without facial dysmorphism had normal neurodevelopmental milestones during their first year, followed by rapid deterioration including severe hypotonia, seizures, retinopathy and deafness. A third patient initially diagnosed with IRD developed cerebral white matter degeneration in the third year of life. MRI in all three patients revealed cerebral demyelination with sparing of subcortical fibers and pronounced central cerebellar demyelination.

Beyond Four Years of Age
X-linked Adrenoleukodystrophy/Adrenomyelo-neuropathy

This is the most common peroxisomal disorder (1/20,000 males and females). Considerable clinical variability exists within the same family [23]. The phenotype neither correlates with genotype nor the biochemical abnormality in plasma or fibroblasts (accumulation of saturated VLCFA). The childhood cerebral form is the most severe form (40% of all ALD phenotypes), with onset of neurologic symptoms between 5 and 12 years of age, leading to a vegetative state and death in a few years. Affected males may present with school failure, an attention-deficit disorder or behavioural changes (due to visuo-spatial deficits and/or central hearing loss) as first manifestations, followed by severe visual and hearing impairment, quadriplegia and cerebellar ataxia; seizures or signs of intracranial hypertension are not uncommon. Hypoglycemic and/or salt losing episodes with increased skin pigmentation reflect adrenal insufficiency which may precede, coincide with or follow the onset of neurological involvement. Most childhood patients show characteristic symmetric cerebral lesions on computerized tomography (CT) or MRI involving the white matter in the parietal and occipital lobes. Following intravenous injection of contrast, a garland-like contrast enhancement adjacent to demyelinating lesions reflects an inflammatory reaction that coincides with rapid neurological deterioration. The initial topography of demyelinating lesions markedly influences the progression of the disease. The occipital forms progress much more rapidly than the frontal forms which are more frequently observed in adolescents.

Adrenomyeloneuropathy (AMN) affects 60% of adult ALD male (20-50 years) and 60% of heterozygous women (>40 years) [24]. The presentation in both sexes is with progressive spastic paraparesis. 35% of men with AMN subsequently develop cerebral demyelination, which, although initially progresses more slowly than in boys, has ultimately the same fatal prognosis. Women with AMN do not develop cerebral demyelination.

The percentage of de novo ALD gene mutations is less than 6% and consequently it is important to undertake genetic counselling and screening in ALD families in order to detect individuals at risk including heterozygous women, boys with adrenal insufficiency, and those asymptomatic with normal neuroimaging.

Classical Refsum Disease

CRD is another peroxisomal disorder with clinical onset during the school-age or adolescence period. Retinopathy, peripheral neuropathy, cerebellar ataxia and elevated cerebrospinal fluid protein level without an increased number of cells are the main features. Less constant findings are sensorineural hearing loss, anosmia, skin changes, and skeletal and cardiac abnormalities [25, 26]. Mental retardation, liver dysfunction and dysmorphism are absent. The onset of symptoms is typically in late childhood or adolescence, but may be as late as the fifth decade. Without treatment patients show a gradually progressive deterioration.

Recently, patients with atypical Refsum disease have been reported with a defect in the peroxisomal fatty acid β-oxidation pathway, α-methyl-CoA racemase deficiency [27]. The clinical manifestations of racemase deficiency are those of an adult-onset sensory motor neuropathy and retinopathy.

Some patients with peroxisomal assembly defects can also present with an isolated motor and sensory peripheral neuropathy mimicking at first Charcot-Marie-Tooth disease [5].

40.2   Metabolic Derangements

The metabolic abnormalities observed in the different peroxisomal disorders (PDs) are listed in ◘ Table 40.2 and ◘ Table 40.3 and follow logically from the nature of the different genetic defects and the subsequent loss of function of the (enzyme) proteins involved. Meaningful discussion of the metabolic abnormalities observed in the various PDs first requires discussion of the principle features of peroxisome biogenesis and the metabolic functions of peroxisomes (▸ [28] for review).

40.2.1   Peroxisome Biogenesis

Peroxisome biogenesis resembles the biogenesis of mitochondria in many respects. As with mitochondria, peroxisomes multiply by division of pre-existing peroxisomes. However, in contrast, peroxisomes lack their own DNA, all peroxisomal proteins being encoded by the nuclear genome. Peroxisomal proteins are synthesized on free polyribosomes, and are specifically directed to peroxisomes via distinct peroxisomal targeting signals (PTS). This is true for both peroxisomal membrane and matrix proteins. Matrix proteins such as catalase, acyl-CoA oxidase, and dihydroxy-acetonephosphate acyltransferase (DHAPAT) contain one of two targeting signals (PTS1 and PTS2), which are small conformational units made up of a short stretch of amino acids. In the case of PTS1, the targeting information is contained in the last 3 carboxy terminal amino acids of a protein, whereas the PTS2 is determined by a stretch of nine amino acids. The PTS1 and PTS2 signals are recognized in the cytosol by specific receptors, the so-called PTS1- and PTS2-receptors, as encoded by the *PEX5* and *PEX7* gene, respectively. These two receptors carry the proteins specifically to the peroxisomal membrane after which the peroxisomal protein import machinery transports the proteins across the peroxisomal membrane where they catalyze their specific metabolic function. The peroxisomal protein im-

□ **Table 40.2.** Classification of peroxisomal disorders

No.	Disorder	Abbreviation	Protein involved	Gene	Chromosome
Disorders of Peroxisome Biogenesis					
1.	Zellweger syndrome	ZS	Different peroxins including: Pex1p,2p, 3p,5p,6p,10p,12p, 13p,14p,16p,19p,26p	*PEX1,2, 3,5,6,10,12,13, 14,16,19,26*	Multiple loci
2.	Neonatal adrenoleukodystrophy	NALD			
3.	Infantile Refsum disease	IRD			
4.	Rhizomelic chondrodysplasia punctata Type 1	RCDP Type 1	Pex7p	*PEX7*	6q22-924
Single Peroxisomal Enzyme Deficiencies					
5.	X-linked adrenoleukodystrophy	X-ALD	ALDP	*ABCD1*	Xq28
6.	Acyl-CoA oxidase deficiency (pseudo-ne-onatal ALD)	ACOX1-deficiency	Straight-chain acyl-CoA oxidase (SCOX/ACOX1)	*ACOX1*	17q25.1
7.	D-bifunctional protein deficiency	D-BP deficiency	D-BP/MF2/MFEII/D-PBE	*HSD17B4*	5q2
8.	2-Methylacyl-CoA racemase deficiency	Racemase deficiency	AMACR	*AMACR*	5q13.2-p.12
9.	Rhizomelic chondrodysplasia punctata Type 2 (DHAPAT deficiency)	RCDP Type 2	DHAPAT	*GNPAT*	1q42.1-42.3
10.	Rhizomelic chondrodysplasia punctata Type 3 (alkyl-DHAP synthase deficiency)	RCDP Type 3	ADHAPS	*AGPS*	2q33
11.	Refsum disease (phytanoyl-CoA hydroxy-lase deficiency)	ARD	Phytanoyl-CoA hydroxylase (PhyH)	*PHYH / PAHX*	10p15-p14
12.	Hyperoxaluria Type 1	PH1	Alanine glyoxylate aminotransferase (AGT)	*AGXT*	2q37.3
13.	Glutaric acidemia Type 3	GA3	?	?	?
14.	Acatalasaemia	–	Catalase	*CAT*	11p13
15.	Mulibrey nanism	MUL	Trim37p	*TRIM*	17q22-23

port system is a multi-protein complex and mutations in any gene coding for a component of this multi-protein complex leads to a defect in peroxisome biogenesis. This implies that peroxisome biogenesis involves the correct expression of multiple so-called *PEX* genes of which 15 have been identified in humans [28]. The corresponding proteins are called peroxins. These data explain the marked genetic heterogeneity among the peroxisome biogenesis disorders as discussed in detail later.

40.2.2 Metabolic Functions of Peroxisomes

Peroxisomes catalyze a number of important metabolic functions (□ Fig. 40.1). From the point of view of peroxisomal disorders the main peroxisomal functions in humans are: (1) fatty acid β-oxidation; (2) etherphospholipid biosynthesis; (3) fatty acid α-oxidation, and (4) glyoxylate detoxification.

1. *Fatty acid β-oxidation*: although most dietary fatty acids (FAs) are degraded in mitochondria, the peroxisomal β-oxidation system is of a crucial importance since a number of FAs can only be oxidized by the peroxisomal system. These FAs include: (1) very-long-chain FAs (VLCFAs); (2) pristanic acid, and (3) di- and trihydroxycholestanoic acid (DHCA and THCA). The latter compounds are intermediates in the biosynthesis of bile acids, are formed from cholesterol in the liver, and are β-oxidized within peroxisomes to produce chenodeoxycholic acid and cholic acid, respectively. In order to oxidize these different FAs, peroxisomes contain the enzymatic machinery required to catalyze four consecutive steps of β-oxidation. The first step of peroxisomal β-oxidation is catalyzed by one of two acyl-CoA oxidases of which one is specific for straight-chain FAs, whereas the other is specific for branched-chain substrates. The second and third step of peroxisomal β-oxidation are catalyzed by a single protein with both hydratase and 3-hydroxyacyl-CoA dehydrogenase ac-

Table 40.3. Biochemical characteristics of different peroxisomal disorders

	Diagnostic group								
	1			2			3	4	
	ZSDs (ZS,NALD,IRD)	AOXD	DBPD	RCDP Type 1	RCDP Type 2	RCDP Type 3	X-ALD	RD	AMACRD
Plasma									
▪ Very-long chain fatty acids	↑	↑	↑	N	N	N	↑	N	N
▪ Di- and trihydroxy-cholestanoic acid	↑	N	↑a	N	N	N	N	N	↑
▪ Phytanic acid	N-↑b	N	N-↑d	N-↑b	N	N	N	↑	N-↑
▪ Pristanic acid	N-↑c	N	N-↑c	N	N	N	N	N	↑
Erythrocytes									
▪ Plasmalogen level	↓	N	N	↓	↓	↓	N	N	N
Liver									
▪ Peroxisomes	deficient	present, but abnormal	present, but abnormal	present	present	present	present	N	NK

ZSDs, Zellweger spectrum disorders; *ZS*, Zellweger syndrome; *NALD*, neonatal adrenoleukodystrophy; *IRD*, infantile Refsum disease; *AOXD*, acyl-CoA oxidase 1 deficiency; *DBPD*, D-bifunctional protein deficiency; *RCDP*, rhizomelic chondrodysplasia punctata; *X-ALD*, X-linked adrenoleukodystrophy; *RD*, Refsum disease; *AMACRD*, 2-methylacyl-CoA racemase deficiency; *N*, normal; *NK*, not known.

a Di- and trihydroxycholestanoic acid are not elevated in all DBPD-patients (▶ [15]).

b Phytanic acid is derived from dietary sources only and may therefore vary from normal to elevated in patients in whom phytanic acid α-oxidation is deficient.

c Pristanic acid is derived from dietary sources only either directly or indirectly from phytanic acid via α-oxidation and may therefore vary from normal to elevated if pristanic acid β-oxidation is deficient.

d Phytanic acid is often elevated if pristanic acid β-oxidation is impaired even if phytanic acid β-oxidation per se is normal.

☐ Table 40.3 (continued)

	Diagnostic group								
	1			**2**			**3**	**4**	
	ZSDs (ZS,NALD,IRD)	AOXD	DBPD	RCDP Type 1	RCDP Type 2	RCDP Type 3	X-ALD	RD	AMACRD
Fibroblasts									
▪ Plasmalogen synthesis	→	N	N	→	→	→	N	N	N
▪ DHAPAT	→	N	N	→	→	→[d]	N	N	N
▪ Alkyl DHAP synthase	→	N	N	→	N	→	N	N	N
▪ C26:0 β-oxidation	→	→	→	N	N	N	→	N	N
▪ Pristanic acid β-oxidation	→	N	→	N	N	N	N	N	→
▪ Acyl-CoA oxidase 1	→	→	N	N	N	N	N	N	N
▪ D-Bifunctional protein	→	N	→	N	N	N	N	N	N
▪ Phytanic acid α-oxidation	→	N	N	→	N	N	N	→	N
▪ Phytanoyl CoA hydroxylase	→	N	N	→	N	N	N	→	N
▪ Peroxisomes	deficient	present, but abnormal	present, but abnormal	present	present	present	present	present	present
▪ Mutant gene	*PEX1,2,3,5,6,10, 12,13,14,16,19,26*	*ACOX1*	*HSD17B4*	*PEX7*	*GNPAT*	*AGPS*	*ABCD1*	*PAHX*	*AMACR*

tivities, named D-bifunctional protein (D-BP). Finally, two thiolases are involved in the fourth and last step of peroxisomal β-oxidation, one specific for straight-chain substrates, and the other reactive with both straight-chain and branched-chain substrates. Degradation of some FAs requires the active participation of auxiliary enzymes of which 2-methylacyl-CoA racemase (AMACR) plays an indispensable role in the degradation of pristanic acid and DHCA and THCA.

2. *Etherphospholipid biosynthesis:* peroxisomes play an essential role in etherphospholipid biosynthesis since the first two steps of the pathway, i.e. the formation of acyl-dihydroxyacetone phosphate (DHAP) and its subsequent conversion to alkyl-DHAP, are localized in peroxisomes. The enzymes involved are called dihydroxyacetonephosphate acyltransferase (DHAPAT) and alkyl-dihydroxyacetonephosphate synthase (alkyl-DHAP synthase or ADHAPS). The alkyl-DHAP, as synthesized in peroxisomes, is subsequently transported to the endoplasmic reticulum where conversion into plasmalogens takes place.

3. *Phytanic acid α-oxidation:* in contrast to straight-chain FAs and 2-methyl branched-chain FAs, 3-methyl branched-chain FAs, such as phytanic acid, cannot undergo β-oxidation. Nature has solved this problem by creating a mechanism called α-oxidation, which allows oxidative removal of the terminal carboxyl group as CO_2 by 2-OH-phytanoyl-CoA lyase (Fig. 40.1). This mechanism ensures that 3-methyl branched-chain FAs are chain-shortened into the 2-methyl FA, pristanal, which can then undergo regular β-oxidation [29].

4. Glyoxylate detoxification: glyoxylate is a toxic compound since it is readily converted into oxalate, which precipitates in the presence of calcium ions. For this reason glyoxylate needs to be detoxified rapidly as catalyzed by the peroxisomal enzyme alanine glyoxylate aminotransferase (AGT, ▶ also Chap. 43) [30].

40.2.3 Metabolic Abnormalities in the Different Peroxisomal Disorders

Peroxisome Biogenesis Defects

Within the group of peroxisome biogenesis defects distinction must be made between the Zellweger spectrum disorders (ZSDs) on the one hand, and RCDP type 1 on the other hand. In ZSDs, peroxisome biogenesis is fully defective which results in a generalized loss of peroxisomal functions as reflected in the following abnormalities: (1) impaired *de novo* plasmalogen biosynthesis, resulting in a generalized deficiency of plasmalogens in tissues including erythrocytes, (2) impaired peroxisomal β-oxidation, resulting in the accumulation of VLCFAs, pristanic acid, DHCA and THCA in plasma, and (3) deficient α-oxidation of phytanic acid, resulting in the accumulation of phytanic acid in plasma.

Because phytanic acid and pristanic acid are derived from exogenous sources only, the accumulation of these compounds in ZSDs is both diet- and age-dependent. In RCDP type 1 there is a selective defect in peroxisome biogenesis, affecting the PTS2 pathway only, whereas the PTS1 route is unimpaired. This implies that in RCDP most peroxisomal proteins are correctly targeted to peroxisomes with the exception of the three PTS2 proteins, which include: (1) phytanoyl-CoA hydroxylase, (2) alkyl-DHAP synthase, and (3) peroxisomal thiolase, with the ultimate consequence that plasmalogen biosynthesis and phytanic acid α-oxidation are deficient in RCDP type 1. This results in a deficiency of tissue plasmalogens and the accumulation of phytanic acid, again in an age- and diet-dependent manner. Peroxisomal β-oxidation, however, is completely normal in RCDP, which explains the normal plasma levels of VLCFAs, pristanic acid, and DHCA plus THCA (Table 40.3).

Single Peroxisomal Enzyme Deficiencies

The single peroxisomal enzyme deficiencies can be subdivided into separate groups, based on the metabolic pathway affected. The disorders of peroxisomal β-oxidation include X-linked adrenoleukodystrophy (X-ALD), acyl-CoA oxidase deficiency, D-bifunctional protein deficiency, and 2-methylacyl-CoA racemase deficiency. In X-ALD there is a selective defect in the peroxisomal β-oxidation of VLCFAs but normal oxidation of the branched-chain fatty acids, pristanic acid, and DHCA and THCA. This is based on the notion that the protein defective in X-ALD, called ALDP, is involved in the uptake of VLCFAs across the peroxisomal membrane, but is not involved in the uptake of pristanic acid and DHCA and THCA. In D-bifunctional protein (D-BP) deficiency oxidation of all peroxisomal fatty acids is impaired, which follows logically from the fact that D-BP is the single enzyme protein involved in the oxidation of all peroxisomal fatty acids. This explains the accumulation of VLCFAs, pristanic acid and DHCA and THCA in patients deficient in D-BP. It should be noted that 3 different subgroups of D-BP deficiency have been identified. In acyl-CoA oxidase deficiency the oxidation of VLCFA is affected with normal oxidation of pristanic acid and DHCA and THCA. Finally, in 2-methylacyl-CoA racemase (AMACR) deficiency, only the peroxisomal β-oxidation of the branched-chain fatty acids (pristanic acid, and DHCA plus THCA) is impaired with normal oxidation of VLCFAs, thus explaining the accumulation of pristanic acid and DHCA and THCA in patients with this disorder.

The group of plasmalogen biosynthesis defects includes (1) DHAPAT deficiency, and (2) alkyl-DHAP synthase deficiency. In both cases plasmalogen biosynthesis is deficient. In contrast to RCDP type 1, which is due to mutations in the *PEX7* gene, in RCDP type 2 and type 3, where there are isolated deficiencies of plasmalogen biosynthesis, phytanic acid α-oxidation is unaffected and as a consequence, plasma phytanic acid levels are normal. Refsum disease is so far

519 **40**

the only isolated disorder of phytanic acid α-oxidation. As a consequence of the deficient activity of the enzyme phytanoyl-CoA hydroxylase, phytanic acid accumulates in a time and diet dependent manner.

In hyperoxaluria type 1, where alanine glyoxylate aminotransferase is dysfunctional, there is an accumulation of glyoxylate, which is subsequently converted into glycolate and oxalate (► Chap. 43). Finally, glutaric aciduria type 3 and acatalasemia are rare disorders described in a few patients only.

40.3 Genetics

With the exception of X-ALD the pattern of inheritance of peroxisomal disorders is autosomal recessive. Zellweger spectrum disorders show a large degree of genetic heterogeneity (◘ Table 40.2) with 11 different genes currently implicated, including the following *PEX* genes: *PEX1*, *PEX2*, *PEX3*, *PEX5*, *PEX6*, *PEX10*, *PEX12*, *PEX13*, *PEX16*, and *PEX26* [28]. Mutations in one particular *PEX* gene, i.e. *PEX7*, are associated with a different clinical phenotype, RCDP type 1 [31, 32]. As discussed above, RCDP is itself also genetically heterogeneous since RCDP type 2 and 3 are caused by mutations in the *GNPAT* and *AGPS* genes coding for the first (DHAPAT) and second (alkyl-DHAP synthase) enzyme of plasmalogen biosynthesis, respectively. The other peroxisomal disorders do not show genetic heterogeneity, although it should be emphasized that mild mutations in the *PEX7* gene can give rise to a phenotype resembling Refsum disease rather than RCDP. Finally, Mulibrey nanism is caused by mutations in the gene *TRIM*, which codes for a peroxisomal protein Trim37p, with no known function in peroxisome biogenesis.

40.4 Diagnostic Tests

Firstly, it is important to emphasize that there is no single laboratory test capable of identifying all peroxisomal disorders in a single analysis. Selection of the appropriate investigation(s) should be based on the clinical presentation. We have introduced the concept of diagnostic groups in order to develop logical guidelines for the laboratory diagnosis of the various disorders. These diagnostic groups include:

1. Zellweger spectrum disorders (ZS, NALD, IRD), acyl-CoA oxidase deficiency and D-bifunctional protein deficiency.
2. Rhizomelic chondrodysplasia punctata spectrum disorders. This group includes all peroxisomal forms of RCDP including type 1, 2, and 3.
3. X-Linked adrenoleukodystrophy complex. This includes all types including childhood cerebral ALD (CCALD) and adrenomyeloneuropathy (AMN).

4. The remaining peroxisomal disorders. This includes Refsum's disease, 2-methylacyl-CoA racemase (AMACR) deficiency, hyperoxaluria type 1, glutaryl-CoA oxidase deficiency, acatalasaemia and Mulibrey nanism.

40.4.1 Diagnostic Group 1

VLCFAs are abnormal in all disorders belonging to diagnostic group 1 [28] (◘ Table 40.3); normal levels exclude a ZSD or peroxisomal β-oxidation defect. If VLCFAs are abnormal, additional tests should then be performed to allow further discrimination. These include the analysis of plasmalogens in erythrocytes and bile acid intermediates, pristanic acid, and phytanic acids in plasma. If erythrocyte plasmalogens levels are deficient, the patient definitely suffers from a ZSD. Normal plasmalogen levels usually, but not always, point to a peroxisomal β-oxidation defect. These analyses should be followed by detailed studies in fibroblasts to establish whether there is a disorder of peroxisome biogenesis or an isolated β-oxidation disorder. In case of ZSD, complementation studies must be performed with the ultimate aim to identify the gene defective in the patient (► [28] for more details).

Although definitely not a first line diagnostic test, pipecolic acid analysis has been found to be helpful in the identification of patients, especially since amino acid analysis is often done as part of a general screening program for inborn errors of metabolism [33].

If fibroblast studies have shown an isolated peroxisomal β-oxidation defect, the activity of the different enzymes involved are measured to establish whether the diagnosis is acyl-CoA oxidase deficiency or D-bifunctional protein deficiency; this is then followed by subsequent molecular analysis.

40.4.2 Diagnostic Group 2

The clinical similarities of the 3 peroxisomal forms of RCDP warrant their inclusion in a single diagnostic group. Erythrocyte plasmalogens are always deficient, even in milder cases, making this a reliable initial laboratory test [8]. Abnormal results should be followed by detailed studies in fibroblasts to further discriminate between type 1, 2, and 3 (► [28] for full details).

40.4.3 Diagnostic Group 3

Analysis of plasma VLCFAs is a reliable initial test to verify whether a patient is affected by X-ALD [34]. If abnormal, one may proceed with fibroblast studies followed by molecular analysis. It should be noted, however, that fibroblast studies are not absolutely obligatory and direct molecular studies in blood cells can be performed too. Heterozygote detection is not as straightforward. Plasma VLCFAs have been found to be normal in about 5–15% of obligate heterozygotes. We advocate performing molecular studies and omitting VLCFA analysis in families in which the molecular defect has been established in the index patient. Where there is no family history plasma VLCFA analysis can be performed, followed, where necessary, by fibroblast studies including immunofluorescence using antibodies against the ALD-protein. This method may be especially rewarding since the product of the mutant X-ALD allele often produces an unstable protein, so that in heterozygotes a mosaic pattern is observed [35].

40.4.4   Diagnostic Group 4

Refsum Disease

Plasma phytanic acid levels, although varying widely in patients with Refsum disease, are always abnormal and therefore a reliable test for the disease. Definite diagnosis requires measurements of phytanoyl-CoA hydroxylase in fibroblasts followed by molecular analyses. In a subset of patients no mutations are found in the hydroxylase gene. Studies have shown that in these patients there is a mutation in *PEX7* [36].

Primary Hyperoxaluria Type 1 (PH1)

In most patients with PH1 there is increased urinary excretion of glyoxylate, oxalate and glycolate. Until recently, definitive diagnosis of PH1 was thought to require assessment of AGT activity in liver. However, AGT activity may not be deficient in a large group of patients in which the enzyme is mistargeted to the mitochondria. Consequently the current consensus is that molecular analysis of the *AGT* gene is the preferred method to definitively establish a diagnosis.

Peroxisomal 2-Methylacyl-CoA Racemase (AMACR) Deficiency

Patients with a deficiency of AMACR are unable to degrade pristanic acid and the bile acid intermediates. Plasma studies should, therefore, include analysis of pristanic acid by GC-MS and bile acid intermediates, preferably by tandem-MS. If abnormal, fibroblast studies, including measurements of racemase activity, and molecular studies should be performed [27].

40.4.5   Histological Detection

The abundance, size and structure of liver peroxisomes can be studied using the diaminobenzidine (DAB) procedure (which reacts with the peroxisomal marker enzyme catalase) and immunocytochemical techniques with antibodies against matrix and membrane peroxisomal proteins [37]. When peroxisomes are lacking, virtually all of the catalase is present in the cytosolic fraction instead of the particulate fraction. The term »membrane ghosts« refers to empty-looking vesicles surrounded by a membrane. These vesicles represent abortive peroxisomes, which are unable to import their matrix enzyme proteins. In some patients with a Zellweger spectrum, a mosaic distribution of peroxisomes in the liver and in fibroblasts can be observed [38].

40.4.6   Prenatal Diagnosis

Prenatal diagnosis of all peroxisomal disorders is now feasible, usually using chorionic villus biopsy material and there is a shift from biochemical to molecular methods. In X-ALD, prenatal diagnosis relies first on sex determination using the detection of Y chromosome in maternal blood. ALD male fetuses are then identified using chorionic-villus samples by direct analysis of the ALD protein and/or ALD gene mutation. When the ALD gene mutation is not yet known in the index case, VLCFA can be measured on cultured chorionic-villus samples or amniocytes.

40.5   Treatment and Prognosis

In classical Refsum disease reduction of phytanic acid levels by a low-phytanate diet (especially prohibition of ruminant meats and fats), with or without plasmapheresis, has been successful in arresting the progress of the peripheral neuropathy. However, when the diet is too strict (less than 10 mg phytanate/day), it may lead to a reduction in the energy intake, weight loss and a paradoxical rise in plasma phytanic acid levels followed by clinical deterioration. This is due to the mobilization of phytanate from lipids stored in adipose tissue.

In X-linked ALD patients, hematopoietic cell transplantation (HCT) can stabilize or even reverse cerebral demyelination when the procedure is performed at a very early stage [39], i.e. when patients have no evident neurological or neuropsychological symptoms. The indication of HCT relies in each case on careful clinical and neuroimaging analysis of disease progression and severity. In practice, less than 50% of patients who are candidate for HCT can be transplanted during the therapeutic window in which the procedure is efficacious. There is no treatment for the inflammatory cerebral form of ALD. Lorenzo's oil (a mixture of oleic and erucic acid) allows normalisation of plasma VLCFA but unfortunately has no curative or preventive effects. Follow-up with brain MRI every 6 months should be undertaken in all asymptomatic boys from 4 to 12 years in order to detect early signs of cerebral demyelination. This will allow any early changes to be identified so that HCT can be undertaken, provided an HLA-matched donor is available.

The ALD mouse model develops a late-onset AMN-like phenotype, and overexpression of the ALD related gene prevents these neurological abnormalities. A trial is currently ongoing to test whether drugs that upregulate this gene are neuroprotective in humans with AMN.

For patients with abnormal peroxisomal assembly and defects that originate before birth, the possibilities for treatment are very limited. Attempts in these patients have included dietary measures such as phytanic acid restriction, treatment of deficiencies of fat soluble vitamins, especially vitamins E and K, decreasing the abnormal bile acids using cholic and deoxycholic acid, and the correction of docosahexaenoic acid and alkylglycerol deficiency by oral supplements [40-42]. In view of its lack of efficacy treatment remains primarily supportive.

References

1. Fournier B, Smeitink JAM, Dorland L et al (1994) Peroxisomal disorders: a review. J Inherit Metab Dis 17:470-486
2. Moser A, Rasmussen M, Naidu S et al (1995) Phenotype of patients with peroxisomal disorders subdivided into sixteen complementation groups. J Pediatr 127:13-22
3. Gould JS, Raymond GV, Valle D (2001) The peroxisome biogenesis disorders. In: Scriver C, Baudet AL, Valle D et al (eds) The metabolic and molecular bases of inherited disease, 8th edn. McGraw-Hill, New York, pp 3181-3217
4. Wanders RJA, Barth PG, Heymans HSA (2002) Peroxisomal disorders. In: Rimoin DL, Conner JM, Pyeritz RE, Korf BR (eds) Principles and practice of medical genetics, 4th edn. Churchill Livingstone, London, pp 2752-2787
5. Baumgartner MR, Poll-The BT, Verhoeven NM et al (1998) Clinical approach to inherited peroxisomal disorders. A series of 27 patients. Ann Neurol 44:720-730
6. Poll-The BT, Saudubray JM, Ogier H et al (1987) Clinical approach to inherited peroxisomal disorders. In: Vogel F, Sperling K (eds) Human genetics. Springer, Berlin Heidelberg New York, pp 345-351
7. Poll-The BT, Maroteaux P, Narcy C et al (1991) A new type of chondrodysplasia punctata associated with peroxisomal dysfunction. J Inherit Metab Dis 14:361-363
8. Smeitink JAM, Beemer FA, Espeel M et al (1992) Bone dysplasia associated with phytanic acid accumulation and deficient plasmalogen synthesis: a peroxisomal entity amenable to plasmapheresis. J Inherit Metab Dis 15:377-380
9. Barth PG, Wanders RJA, Schutgens RBH, Staalman C (1996) Variant RCDP with normal phytanic acid: clinico-biochemical delineation of a subtype and complementation studies. Am J Med Genet 62:164-168
10. White AL, Modaff P, Holland-Morris F, Pauli RM (2003) Natural history of rhizomelic chondrodysplasia punctata. Am J Med Genet 118A:332-342
11. Baumgartner MR, Verhoeven NM, Jacobs C et al (1998) Defective peroxisome biogenesis with a neuromuscular disorder resembling Werdnig-Hoffman disease. Neurology 51:1427-1432
12. Poll-The BT, Roels F, Ogier H et al (1988) A new peroxisomal disorder with enlarged peroxisomes and a specific deficiency of acyl-CoA oxidase (pseudo-neonatal adrenoleukodystrophy). Am J Hum Genet 42:422-434
13. Suzuki Y, Shimozawa N, Yajima S et al (1994) Novel subtype of peroxisomal acyl-CoA oxidase deficiency and bifunctional enzyme deficiency with detectable enzyme protein: identification by means of complementation analysis. Am J Hum Genet 54: 36-43
14. Watkins PA, Chen WN, Harris CJ et al (1989) Peroxisomal bifunctional enzyme deficiency. J Clin Invest 83:771-777
15. Van Grunsven EG, van Berkel E, Mooijer PAW et al (1999) Peroxisomal bifunctional protein deficiency revisited: resolution of its true enzymatic and molecular basis. Am J Hum Genet 64:99-107
16. Wanders RJA, Barth PG, Heymans HSA (2001) Single peroxisomal enzyme deficiencies. In: Scriver C, Baudet AL, Valle D et al (eds) The metabolic and molecular bases of inherited disease, 8th edn. McGraw-Hill, New York, pp 3219-3256
17. Kaufmann WE, Theda C, Naidu S et al (1996) Neuronal migration abnormality in peroxisomal bifunctional enzyme defects. Ann Neurol 39:268-271
18. Poll-The BT, Saudubray JM, Ogier H et al (1987) Infantile Refsum disease: an inherited peroxisomal disorder. Comparison with Zellweger syndrome and neonatal adrenoleukodystrophy. Eur J Pediatr 146:477-483
19. Poll-The BT, Gootjes J, Duran M et al (2004) Peroxisome biogenesis disorders with prolonged survival: phenotypic expression in a cohort of 31 patients. Am J Med Genet 126A:333-338
20. Barth PG, Majoie CBLM, Gootjes J et al (2004) Neuroimaging of peroxisome biogenesis disorders (Zellweger spectrum) with prolonged survival. Neurology 62:439-444
21. Poll-The BT, Maillette de Buy Wenniger-Prick LJ, Barth PG, Duran M (2003) The eye as a window to inborn errors of metabolism. J Inherit Metab Dis 26:229-244
22. Barth PG, Gootjes J, Bode H et al (2001) Late onset white matter disease in peroxisome biogenesis disorder. Neurology 57:1949-1955
23. Dubois-Dalcq M, Feigenbaum V, Aubourg P (1999) The neurobiology of X-linked adrenoleukodystrophy, a demyelinating peroxisomal disorder. Trends Neurosci 22:4-12
24. van Geel BM, Bezman L, Loes DJ et al (2001) Evolution of phenotypes in adult male patients with X-linked adrenoleukodystrophy. Ann Neurol 49:186-194
25. Skjeldal OH, Stokke O, Refsum S et al (1987) Clinical and biochemical heterogeneity in condition with phytanic acid accumulation. J Neurol Sci 77:87-96
26. Plant GR, Hansell DM, Gibberd FB, Sidey MC (1990) Skeletal abnormalities in Refsum's disease (heredopathia atactica polyneuritiformis). Br J Radiol 63:537-541
27. Ferdinandusse S, Denis S, Clayton PT et al (2000) Mutations in the gene encoding peroxisomal α-methylacyl-CoA racemase cause adult-onset sensory motor neuropathy. Nat Genet 24:188-191
28. Wanders RJA (2004) Metabolic and molecular basis of peroxisomal disorders: A review. Am J Med Genet 126A:355-375
29. Wanders RJA, Jansen GA, Lloyd MD (2003) Phytanic acid alpha-oxidation, new insights into an old problem: a review. Biochim Biophys Acta 1631:119-135
30. Danpure CJ (2004) Molecular aetiology of primary hyperoxaluria type 1. Nephron Exp Nephrol 98:e39-44
31. Braverman N, Chen L, Lin O et al (2002) Mutation analysis of PEX7 in 60 probands with rhizomelic chondrodysplasia punctata and functional correlations of genotype with phenotype. Hum Mutat 20:284-297
32. Motley AM, Brites P, Gerez L et al (2002) Mutational spectrum in the PEX7 gene and functional analysis of mutant alleles in 78 patients with rhizomelic chondrodysplasia punctata type 1. Am J Hum Genet 70:612-624
33. Peduto A, Baumgartner MR, Verhoeven NM et al (2004) Hyperpipecolic acidemia: a diagnostic tool for peroxisomal disorders. Mol Genet Metab 82:224-230

34. Bezman L, Moser AB, Raymond GV et al (2001) Adrenoleukodystrophy: incidence, new mutation rate, and results of extended family screening. Ann Neurol 49:512-517
35. Kemp S, Pujol A, Waterham HR et al (2001) ABCD1 mutations and the X-linked adrenoleukodystrophy mutation database: role in diagnosis and clinical correlations. Hum Mutat 18:499-515
36. Van den Brink DM, Brites P, Haasjes J et al (2003) Identification of PEX7 as the second gene involved in Refsum disease. Am J Hum Genet 72:471-477
37. Roels F, Espeel M, De Craemer D (1991) Liver pathology and immunocytochemistry in congenital peroxisomal disease: a review. J Inherit Metab Dis 14:853-875
38. Depreter M, Espeel M, Roels F (2003) Human peroxisomal disorders. Microsc Res Tech 61:203-223
39. Shapiro E, Krivit W, Lockman L et al (2000) Long-term beneficial effect of bone marrow transplantation for childhood onset cerebral X-linked adrenoleukodystrophy. Lancet 356:713-718
40. Robertson EF, Poulos A, Sharp P et al (1988) Treatment of infantile phytanic acid storage disease: clinical, biochemical and ultrastructural findings in two children treated for 2 years. Eur J Pediatr 147:133-142
41. Setchell KD, Bragetti P, Zimmer-Nechemias L et al (1992) Oral bile acid treatment and the patient with Zellweger syndrome. Hepatology 15:198-207
42. Martinez M, Vazquez E, Garcia Silva MT et al (2000) Therapeutical effects of docosahexaenoic acid in patients with generalized peroxisomal disorders. Am J Clin Nutr 71:376S-385S

41 Congenital Disorders of Glycosylation

Jaak Jaeken

Synthesis of N-Glycans

This complex synthesis proceeds in three stages, schematically represented in ■ Fig. 41.1.

(1) Formation in the cytosol of *nucleotide-linked sugars*, mainly guanosine diphosphate-mannose (GDP-Man), also uridine diphosphate glucose (UDP-Glc) and UDP-N-acetylglucosamine (UDP-GlcNAc), followed by attachment of GlcNAc and Man units to dolichol phosphate, and flipping (indicated by circular arrows) of the nascent oligosaccharide structure into the endoplasmic reticulum (ER).

(2) Stepwise *assembly* in the ER, by further addition of Man and Glc of the 14-unit oligosaccharide precursor, dolichol pyrophosphate-N-acetylglucosamine$_2$-mannose$_9$-glucose$_3$ (indicated by an asterix in the lower left part of ■ Fig. 41.1).

(3) Transfer of this precursor onto the nascent protein (depicted in the left part of ■ Fig. 41.1), followed by final *processing* of the glycan in the Golgi apparatus by trimming and attachment of various sugar units.

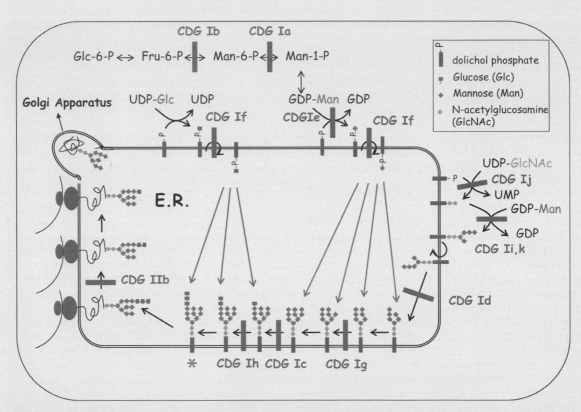

■ **Fig. 41.1.** Schematic representation of the synthesis of N-glycans. *ER*, endoplasmic reticulum; *Fru*, fructose; *GDP*, guanosine diphosphate; *Glc*, glucose; *GlcNAc*, N-acetylglucosamine; *Man*, mannose; *P*, phosphate; *UDP*, uridine diphosphate; *, dolichol pyrophosphate-N-acetylglucosamine$_2$-mannose$_9$-glucose$_3$. Defects are indicated by *solid bars* across the arrows. Modified after Matthijs et al [7]

41.1   Introduction

Numerous proteins are glycosylated with tree or antenna-like oligosaccharide structures (■ Fig. 41.1), also termed glycans, attached to the polypeptide chain. Most extracellular proteins, such as serum proteins (transferrin, clotting factors), most membrane proteins, and several intracellular proteins (such as lysosomal enzymes), are glycoproteins. The glycans are defined by their linkage to the protein: *N-glycans* are linked to the amide group of asparagine, and *O-glycans* are linked to the hydroxyl group of serine or threonine. Synthesis of *N*-glycans, schematically represented in ■ Fig. 41.1, proceeds in three stages: formation of nucleotide-linked sugars, asssembly, and processing. Synthesis of *O*-glycans involves assembly but no processing, and occurs mainly in the Golgi apparatus. It forms a diversity of structures such as *O*-xylosylglycans, *O*-mannosylglycans, and *O-N*-acetylgalactosaminylglycans.

In recent years, a series of defects of the synthesis of the oligosaccharide chains of glycoproteins, named Congenital Disorders of Glycosylation (CDG), have been identified. Twelve disorders of the assembly of *N*-glycans (CDG-I group) are currently known, designated CDG-Ia to CDG-Il. They are listed in ■ Table 41.1, together with the main clinically affected organs and systems, defective proteins, and genes. The location of each enzyme defect is shown in

■ Fig. 41.1. Six disorders of the processing of *N*-glycans (CDG-II group) are known and designated CDG-IIa to CDG-IIf. They are also listed in ■ Table 41.1, and the location of the CDG-IIb defect is shown in ■ Fig. 41.1. Five disorders of *O*-glycosylation have been identified and are listed in ■ Table 41.2. The defect of oligosaccharide-chain processing that leads to deficiency of the mannose-6-phosphate recognition marker of the lysosomal enzymes and causes mucolipidosis II or III is discussed in ▶ Chap. 39. The deficiencies of the lysosomal enzymes that degrade the oligosaccharide side chains, and cause oligosaccharidoses, are also discussed in that chapter.

Patients with CDG form a rapidly growing group, with a very broad spectrum of clinical manifestations. Moreover, their discovery has opened new avenues in the field. In 2004, a combined *N*- and *O*-glycosylation defect was reported. It is due to the deficiency of a subunit, COG7, in a protein complex involved in trafficking and function of the glycosylation machinery. Also in 2004, the first defect in lipid glycosylation was reported, namely the deficiency of GM3 synthase, an enzyme involved in the synthesis of gangliosides.

Because of the large number of CDGs, only the most frequently reported types will be considered in detail in this chapter: CDG-Ia, CDG-Ib, and CDG-Ic. The hereditary multiple exostoses syndrome, Walker-Warburg syndrome, muscle-eye-brain disease, and the newly discovered dis-

■ Table 41.1. Genetic N-glycosylation disorders

Name	Main clinically affected organs and systems	Defective protein	Defective gene
CDG-I			
CDG-Ia	Nervous system, fat tissue, other organs[1]	Phosphomannomutase 2	*PMM2*
CDG-Ib	Intestine, liver	Phosphomannose isomerase	*MPI*
CDG-Ic	Nervous system	Glucosyltransferase I	*hALG6*
CDG-Id	Nervous system	Mannosyltransferase VI	*hALG3*
CDG-Ie	Nervous system	Dolichol-P-Man synthase I	*DPM1*
CDG-If	Nervous system, skin	Lec35	*Lec35*
CDG-Ig	Nervous system	Mannosyltransferase VIII	*hALG12*
CDG-Ih	Intestine, liver	Glucosyltransferase II	*hALG8*
CDG-Ii	Nervous system, eyes, liver	Mannosyltransferase II	*hALG2*
CDG-Ij	Nervous system	UDP-GlcNAc: dolichol phosphate N-acetylglucosamine 1-phosphate transferase	*DPAGT1*
CDG-Ik	Nervous system, liver	Mannosyltransferase I	*hALG1*
CDG-Il	Nervous system, liver	Mannosyltransferase VII/IX	*hALG9*
CDG-II			
CDG-IIa	Nervous system, skeleton, intestine, immune system, dysmorphism	N-acetylglucosaminyltransferase II	*MGAT2*
CDG-IIb	Nervous system, dysmorphism	Glucosidase I	*GLS1*
CDG-IIc	Nervous system, immune system, dysmorphism	GDP-fucose transporter 1	*FUCT1*
CDG-IId	Nervous system, skeletal muscles	β-1,4-galactosyltransferase 1	*B4GALT1*
CDG-IIe[2]	Nervous system, liver, skeleton	Conserved oligomeric Golgi complex, subunit 7	*COG7*
CDG-IIf	Megathrombocytopenia, neutropenia	CMP-sialic acid transporter	*SLC35A1*

[1] Eyes, heart, liver, kidneys, skeleton, gonads, immune system.

[2] Preliminary assignment.

◘ Table 41.2. Genetic O-glycosylation disorders

Name	Main clinically affected organs and systems	Defective protein	Defective gene
Defects in O-xylosylglycan synthesis			
Multiple exostoses syndrome	Cartilage	Glucuronyltransferase/N-acetyl-D-hexosaminyltransferase	*EXT1/EXT2*
Progeroid variant of Ehlers-Danlos syndrome	Generalized rapid aging	β-1,4-Galactosyltransferase 7	*B4GALT7*
Defects in O-mannosylglycan synthesis			
Walker-Warburg syndrome		O-mannosyltransferase 1	*POMT1*
Muscle-eye-brain disease	Brain, eyes, skeletal muscles	O-mannosyl-β-1,2-N-acetylglucosaminyltransferase 1	*POMGnT1*
Defect in O-N-acetylgalactosaminylglycan synthesis			
Familial tumoral calcinosis	Skin, subcutaneous tissues, kidneys	ppGaNTase-T3[1]	*GALNT3*

[1] UDP-N-acetyl-α-D-galactosamine: polypeptide N-acetylgalactosaminyltransferase 3.

orders will also be briefly discussed. For recent reviews ▶ [1–7], and for recent reports not covered by these reviews ▶ [8–10].

CDG should be considered a possible diagnosis in any unexplained clinical condition. The rationale for this recommendation is twofold: Firstly, the extremely broad clinical spectrum covered by the some 25 known CDGs and, secondly, since about 1% of the human genome is involved in glycosylation, it is more than probable that the majority of CDGs have still to be discovered. We predict that these will also include diseases due to defects in organ-specific glycosylation (brain-CDG, kidney-CDG, etc). Also, as has already been shown for hereditary multiple exostoses, Walker-Warburg syndrome and others, there is no doubt that known diseases with unknown etiology will continue to be identified as CDGs.

CDG should be considered particularly in multi-organ disease with neurological involvement. Isoelectrofocusing of serum transferrin is still the screening method of choice but it is important to realize that it is able to detect only a limited number of CDGs, namely N-glycosylation disorders associated with sialic acid deficiency. The (partial) deficiency of sialic acid in these forms of CDG causes one of two main types of cathodal shift (◘ Fig. 41.2 and Sect. 41.2.1). A type 1 pattern indicates an assembly disorder and CDG-Ia and CDG-Ib should be considered first. If these are excluded the next step is dolichol-linked glycan analysis which will usually locate the site of the defect. A type 2 pattern indicates a disorder of processing; protein-linked glycan analysis should then be performed in an attempt to identify the defective step.

41.2   Congenital Disorders of Protein N-Glycosylation

41.2.1   Phosphomannomutase 2 Deficiency (CDG-Ia)

Clinical Presentation

CDG-Ia is by far the most frequent CDG with at least 550 patients known worldwide. The symptomatology can be recognized shortly after birth. The nervous system is affected in all patients, and most other organs are involved in a variable way. The neurological picture comprises alternating internal strabismus and other abnormal eye movements, axial hypotonia, psychomotor retardation (IQ typically between 40 and 60), ataxia and hyporeflexia. After infancy,

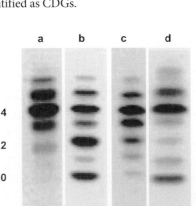

◘ Fig. 41.2a–d. Serum transferrin isoelectrofocusing patterns. **a** normal pattern; **b** type 1 pattern; **c** and **d** type 2 patterns; 0, 2, 4 indicate the number of sialic acid residues

symptoms include retinitis pigmentosa, often stroke-like episodes, and sometimes epilepsy. As a rule there is no regression. During the first year(s) of life, there are variable feeding problems (anorexia, vomiting, diarrhea) that can result in severe failure to thrive. Other features are a variable dysmorphism, which may include large, hypoplastic/dysplastic ears, abnormal subcutaneous adipose-tissue distribution (fat pads, orange peel skin), inverted nipples, and mild to moderate hepatomegaly, skeletal abnormalities and hypogonadism. Some infants develop a pericardial effusion and/or cardiomyopathy. At the other end of the clinical spectrum are patients with a very mild phenotype (no dysmorphic features, slight psychomotor retardation). Patients often have an extraverted and happy appearance. Neurological investigations reveal (olivoponto) cerebellar hypoplasia, variable cerebral hypoplasia and peripheral neuropathy. Liver pathology is characterized by fibrosis and steatosis, and electron microscopy shows myelin-like lysosomal inclusions in hepatocytes but not in Kupffer cells [11–16].

Metabolic Derangement

Phosphomannomutase (PMM) catalyzes the second committed step in the synthesis of guanosine diphosphate (GDP) mannose, namely the conversion of mannose-6-phosphate into mannose-1-phosphate, which occurs in the cytosol (◘ Fig. 41.1). CDG-Ia is due to the deficiency of PMM2, the principal isozyme of PMM. Since GDP-mannose is the donor of the mannose units used in the ER to assemble the dolichol-pyrophosphate oligosaccharide precursor, the defect causes hypoglycosylation, and hence deficiency and/or dysfunction of numerous glycoproteins, including serum proteins (such as thyroxin-binding globulin, haptoglobin, clotting factor XI, antithrombin, cholinesterase etc.), lysosomal enzymes and membranous glycoproteins.

Genetics

PMM deficiency is inherited as an autosomal-recessive trait due to mutations of the *PMM2* gene on chromosome 16p13. At least 60 mutations (mainly missense) have been identified. The most frequent mutation causes a R141H substitution which, remarkably, has not yet been found in the homozygous state, pointing to a lethal condition [17]. The frequency of this mutation in the normal Belgian population is as high as 1/50. The incidence of PMM deficiency is not known; in Sweden it has been estimated at 1:40,000.

Prenatal testing should only be offered in families with a documented PMM deficiency and mutations in the *PMM2* gene. It cannot be performed by any assay that determines the glycosylation of proteins since this has been found to be normal in the foetus [7].

Diagnostic Tests

The diagnosis of congenital disorders of *N*-glycosylation in general (and of PMM deficiency in particular) is usually made by isoelectrofocusing and immunofixation of serum transferrin [18] (◘ Fig. 41.2). Normal serum transferrin is mainly composed of tetrasialotransferrin and small amounts of mono-, di-, tri-, penta- and hexasialotransferrins. The partial deficiency of sialic acid (a negatively charged and end-standing sugar) in CDG causes a cathodal shift. Two main types of cathodal shift can be recognized. Type 1 is characterized by an increase of both disialo- and asialotransferrin and a decrease of tetra-, penta- and hexasialotransferrins; in type 2 there is also an increase of the tri- and/or monosialotransferrin bands. In PMM2 deficiency, a type 1 pattern is found. A type 1 pattern is also seen in the secondary glycosylation disorders, chronic alcoholism and hereditary fructose intolerance. A shift due to a transferrin protein variant has first to be excluded (by isoelectrofocusing after neuraminidase treatment, studying another glycoprotein and investigating the parents). The carbohydrate-deficient transferrin (CDT) assay is also useful for the diagnosis of sialic acid-deficient CDG. It quantifies the total sialic acid-deficient serum transferrin. A draw-back is a non-negligible number of false-positive results. Recently, capillary zone electrophoresis of total serum has been introduced for the diagnosis of CDG [19].

In addition to the above-mentioned serum glycoprotein abnormalities, laboratory findings include elevation of serum transaminase levels, hypoalbuminemia, hypocholesterolemia, and tubular proteinuria. To confirm the diagnosis, the activity of PMM should be measured in leukocytes or fibroblasts.

Treatment and Prognosis

No effective treatment is available. The promising finding that mannose is able to correct glycosylation in fibroblasts with PMM deficiency [20] could not be substantiated in patients [21]. There is a substantially increased mortality (~20%) in the first years of life due to severe infection or vital organ involvement (liver, cardiac or renal insufficiency).

41.2.2   Phosphomannose-Isomerase Deficiency (CDG-Ib)

Clinical Presentation

Three groups independently reported this CDG first in 1998 [22–24]. Some 20 patients have been described. Most have presented with hepatic-intestinal disease without notable dysmorphism, and with or without only minor neurological involvement. Symptoms started between the ages of 1 and 11 months. One patient had recurrent vomiting and liver disease that disappeared after the introduction of solid food at the age of 3 months. A healthy adult has been reported who had transient feeding problems in childhood. In the other patients, symptoms persisted and consisted of various combinations of recurrent vomiting, abdominal pain, protein-losing enteropathy, recurrent thromboses, gastrointestinal bleeding, liver disease and symptoms of (hyperinsulinemic or normoinsulinemic) hypoglycemia. In 1985, four infants from Quebec were reported with a similar syndrome who retrospectively were shown most probably to have the same disease [25].

Metabolic Derangement

Phosphomannose-isomerase (PMI) catalyzes the first committed step in the synthesis of GDP-mannose, namely the conversion of fructose-6-phosphate into mannose-6-phosphate (■ Fig. 41.1). Hence the blood biochemical abnormalities are indistinguishable from those found in PMM2 deficiency. Since the substrate of PMI, fructose-6-phosphate, is efficiently metabolized in the glycolytic pathway, it does not accumulate intracellularly.

Genetics

Inheritance of PMI deficiency is autosomal recessive. The gene has been localised to chromosome 15q22. Several mutations have been identified. Prenatal diagnosis is only possible if the molecular defect is known in the proband [7].

Diagnostic Tests

Serum transferrin isoelectrofocusing shows a type 1 pattern. The diagnosis is confirmed by finding a decreased activity of PMI in leukocytes or fibroblasts and/or (a) mutation(s) in the corresponding gene.

Treatment and Prognosis

PMI deficiency is the most rewarding CDG to diagnose because, so far, it is the only one known that can be efficiently treated. Mannose is the therapeutic agent [24]. Hexokinases phosphorylate mannose to mannose 6-phosphate, thus bypassing the defect. An oral dose of 1 g mannose/kg body weight per day (divided in 5 doses) is used. The clinical symptoms usually disappear rapidly but it takes several months before the transferrin isoelectrofocusing pattern improves significantly. Nevertheless, several patients with proven PMI deficiency, including one receiving mannose treatment, have died.

41.2.3 Glucosyltransferase I Deficiency (CDG-Ic)

Clinical Presentation

CDG-Ic is the second most common N-glycosylation disorder with more than 30 patients identified since its description in 1998 [26]. Clinical features in common with CDG-Ia are hypotonia, strabismus and seizures, but psychomotor retardation is milder, there is less dysmorphism, and usually no retinitis pigmentosa or cerebellar hypoplasia. A few patients have had protein-losing enteropathy, a consistent feature in CDG-Ib and CDG-Ih [27].

Metabolic Derangement

Glucosyltransferase I deficiency is a defect in the attachment in the ER of the first of three glucose molecules to the dolichol-linked mannose$_9$-N-acetylglucosamine$_2$ intermediate (■ Fig. 41.1). It causes hypoglycosylation of serum glycoproteins, because non-glucosylated oligosaccharides are a sub-optimal substrate for the oligosaccharyltransferase and are, therefore, transferred to proteins with a reduced efficiency. For an unknown reason, the blood glycoproteins are unusually low (particularly factor XI and coagulation inhibitors, such as antithrombin and protein C). The reason the clinical picture in these patients is much milder that that of PMM deficient patients may be because a deficiency in glucosylation of the dolichol-linked oligosaccharides does not affect the biosynthesis of GDP-mannose and, hence, does not affect the biosynthesis of compounds such as GDP-fucose or the biosynthesis of glycosylphosphatidylinositol-anchored glycoproteins.

Genetics

Inheritance of this glucosyltransferase deficiency is autosomal recessive. The gene maps to chromosome 1p22.3. A333V is a common mutation. Prenatal diagnosis is only reliable if the molecular defect is known in the proband [7].

Diagnostic Tests

This disease illustrates that, even in cases of mild psychomotor retardation without any specific dysmorphic features, isoelectrofocusing of serum sialotransferrins should be performed. When a type 1 pattern is found, PMM and PMI deficiency must be considered first. If these enzymes show normal activities, the next step is the analysis of the dolichol-linked oligosaccharides in fibroblasts. If the major fraction of these oligosaccharides consists of nine mannose and two N-acetylglucosamine residues without the three glucose residues that are normally present, this specific glucosyltransferase activity should be measured in fibroblasts (which is only undertaken by very few laboratories). If deficient activity is found then this should be followed by mutation analysis.

Treatment and Prognosis

No efficient treatment is available. The long-term outcome is unknown since all reported patients have been children.

41.3 Congenital Disorders of Protein O-Glycosylation

41.3.1 Hereditary Multiple Exostoses

Hereditary multiple exostoses is an autosomal dominant disease with a prevalence of 1/50 000, and characterized by the formation on the ends of long bones of cartilage-capped tumors, known as osteochondromas [28]. These are often present at birth but usually not diagnosed before early childhood. Their growth slows at adolescence and stops in adulthood. A small percentage of these lesions are subject to malignant degeneration. Complications may arise from compression of peripheral nerves and blood vessels.

The basic defect resides in a Golgi-localised protein complex, termed exostosin-1/exostosin-2 (EXT1/EXT2), which adds D-glucuronic and N-acetylglucosamine units in the synthesis of heparan sulfate (◘ Fig. 39.1). It has been hypothesized that mutations in these glycosyltransferases impair the synthesis of a glycosaminoglycan that exerts a tumor-suppression function. This would explain the higher risk of affected individuals to develop chondrosarcomas and osteosarcomas.

Fifty mutations in the *EXT1* gene, localized on chromosome 8q24.1, and 25 mutations in *EXT2,* on chromosome 11p11-p12, have been identified [29]. Mutations in the two genes are responsible for over 70% of the cases of hereditary multiple exostoses. Prenatal diagnosis can be performed by mutation analysis.

41.3.2 Walker-Warburg Syndrome

Walker-Warburg Syndrome is one of some 25 neuronal migration disorders known in humans. It is characterized by brain and eye dysgenesis associated with congenital muscular dystrophy. Male patients often have testicular defects. Psychomotor development is absent. The brain lesions consist of »cobblestone« lissencephaly, agenesis of the corpus callosum, cerebellar hypoplasia, hydrocephaly and sometimes encephalocoele [30]. The disease usually runs a fatal course before the age of one year, and only symptomatic treatment is available.

The metabolic derangement is an aberrant glycosylation of α-dystroglycan, an external membrane protein expressed in muscle, brain and other tissues [31]. Most glycans of this heavily glycosylated protein seem to be *O*-linked via mannose, and they control the interaction with extracellular matrix proteins. Disrupted glycosylation of α-dystroglycan (and probably other glycoproteins) results in loss of this interaction and hence in progressive muscle degeneration and abnormal neuronal migration (overmigration) in the brain. In about 20% of the patients this disrupted glycosylation is due to a defective *O*-mannosyltransferase-1, which catalyzes the first step in the synthesis of the *O*-mannosylglycan core. It is caused by mutations in the gene *POMT1*, located on chromosome 9q34.1 [32].

41.3.3 Muscle-Eye-Brain Disease

Muscle-eye-brain disease is a neuronal migration/congenital muscular dystrophy syndrome similar to Walker-War-

burg syndrome but less severe, and with longer survival [30]. The defect is in protein *O*-mannosyl-β-1,2-N-acetyl-glucosaminyltransferase 1, catalyzing the second step in the synthesis of the *O*-mannosylglycan core [33]. The disease is autosomal recessive and due to mutations in *POMGnT1*, located on chromosome 1p34-p33.

41.4 Newly Discovered Disorders

41.4.1 COG7 Deficiency

COG7 deficiency was identified in two siblings born small for gestational age and with perinatal asphyxia. They had dysmorphic features, particularly of the face, encephalopathy, and cholestatic liver disease. Both died, at 5 and 10 weeks, respectively. There was a mild to moderate increase of lysosomal enzyme activities in plasma. There was a type 2 pattern on serum transferrin isoelectrofocusing [34].

Studies of fibroblast glycoproteins showed a partial *N*- and *O*-glycosylation defect caused by a decreased transport of CMP-sialic acid and UDP-galactose into the Golgi, and a reduced activity of two glycosyltransferases involved in the galactosylation and sialylation of *O*-glycans. The localization of the 8-subunit conserved oligomeric Golgi (COG) complex, involved in trafficking and function of the glycosylation machinery, was found to be abnormal: on indirect immunofluorescence there was diffuse cytoplasmic staining instead of Golgi staining of COG5, COG6 and COG8. The basic defect was eventually localised to the COG7 subunit, and a homozygous intronic mutation was found in *COG7*, located on chromosome 16p [10].

41.4.2 GM3 Synthase Deficiency

This first identified glycolipid glycosylation disorder was detected in an Old Order Amish pedigree [9]. The first symptoms consisted of poor feeding and irritability appearing between 2 weeks and 3 months of age. Epilepsy developed within the first year (grand mal and other presentations) and was difficult to control. Moreover, the patients showed profound developmental stagnation with regression. On brain magnetic resonance imaging there was diffuse atrophy at an older age.

The metabolic derangement was identified as a defect of lactosylceramide α-2,3 sialyltransferase (also called GM3 synthase) which can be measured in plasma or fibroblasts. The defect causes accumulation of lactosylceramide associated with a decrease of the gangliosides of the GM3 and GD3 series. A homozygous mutation was found in *SIAT9*, localized on chromosome 2p11.2.

References

1. Aebi M, Hennet T (2001) Congenital disorders of glycosylation: genetic model systems lead the way. Trends Cell Biol 11:136-141
2. Freeze HH (2002) Human disorders in N-glycosylation and animal models. Biochim Biophys Acta 1573:388-393
3. Jaeken J (2003) Komrower lecture: Congenital disorders of glycosylation (CDG): it's all in it! J Inherit Metab Dis 26:99-118
4. Lowe JB, Marth JD (2003) A genetic approach to mammalian glycan function. Annu Rev Biochem 72:643-691
5. Marquardt T, Denecke J (2003) Congenital disorders of glycosylation: review of their molecular bases, clinical presentations and specific therapies. Eur J Pediatr 162:359-379
6. Jaeken J, Carchon H (2004) Congenital disorders of glycosylation: a booming chapter of pediatrics. Curr Opin Pediatr 16:434-439
7. Matthijs G, Schollen E, Van Schaftingen E (2004) The prenatal diagnosis of congenital disorders of glycosylation (CDG). Prenat Diagn 24:114-116
8. Frank CG, Grubenmann CE, Eyaid W, Berger EG, Aebi M, Hennet T (2004) Identification and functional analysis of a defect in the human ALG9 gene: definition of congenital disorders of glycosylation type IL. Am J Hum Genet 75:146-150
9. Simpson MA, Cross H, Proukakis C et al (2004) Infantile onset symptomatic epilepsy syndrome caused by a homozygous loss-of-function mutation of GM3 synthase. Nat Genet 36:1225-1229
10. Wu X, Steet RA, Bohorov O et al (2004) Mutation of the COG complex subunit gene *COG7* causes a lethal congenital disorder. Nat Med 10:518-523
11. Jaeken J, Vanderschueren-Lodeweyckx M, Casaer P et al (1980) Familial psychomotor retardation with markedly fluctuating serum proteins, FSH and GH levels, partial TBG-deficiency, increased serum arylsulphatase A and increased CSF protein: a new syndrome? Pediatr Res 14:179
12. Jaeken J, Stibler H, Hagberg B (1991) The carbohydrate-deficient glycoprotein syndrome: a new inherited multisystemic disease with severe nervous system involvement. Acta Paediatr Scand [Suppl] 375:1-71
13. Van Schaftingen E, Jaeken J (1995) Phosphomannomutase deficiency is a cause of carbohydrate-deficient glycoprotein syndrome type I. FEBS Lett 377:318-320
14. Jaeken J, Besley G, Buist N et al (1996) Phosphomannomutase deficiency is the major cause of carbohydrate-deficient glycoprotein syndrome type I. J Inherit Metab Dis 19[Suppl 1]:6
15. de Zegher F, Jaeken J (1995) Endocrinology of the carbohydrate-deficient glycoprotein syndrome type 1 from birth through adolescence. Pediatr Res 37:395-401
16. Van Geet C, Jaeken J (1993) A unique pattern of coagulation abnormalities in carbohydrate-deficient glycoprotein syndrome. Pediatr Res 33:540-541
17. Matthijs G, Schollen E, Van Schaftingen E, Cassiman J-J, Jaeken J (1998) Lack of homozygotes for the most frequent disease allele in carbohydrate-deficient glycoprotein syndrome type 1 A. Am J Hum Genet 62:542-550
18. Jaeken J, van Eijk HG, van der Heul C et al (1984) Sialic acid-deficient serum and cerebrospinal fluid transferrin in a newly recognized genetic syndrome. Clin Chim Acta 144:245-247
19. Carchon H, Chevigné R, Falmagne JB, Jaeken J (2004) Diagnosis of congenital disorders of glycosylation by capillary zone electrophoresis of serum transferrin. Clin Chem 50:101-111
20. Panneerselvam K, Freeze HH (1996) Mannose corrects altered N-glycosylation in carbohydrate-deficient glycoprotein syndrome fibroblasts. J Clin Invest 97:1478-1487
21. Kjaergaard S, Kristiansson B, Stibler H et al (1998) Failure of short-term mannose therapy of patients with carbohydrate-deficient glycoprotein syndrome type I A. Acta Paediatr 87:884-888
22. de Koning TJ, Dorland L, van Diggelen OP et al (1998) A novel disorder of N-glycosylation due to phosphomannose isomerase deficiency. Biochem Biophys Res Commun 245:38-42
23. Jaeken J, Matthijs G, Saudubray J-M et al (1998) Phosphomannose isomerase deficiency: a carbohydrate-deficient glycoprotein syndrome with hepatic-intestinal presentation. Am J Hum Genet 62:1535-1539
24. Niehues R, Hasilik M, Alton G et al (1998) Carbohydrate-deficient glycoprotein syndrome type Ib: phosphomannose isomerase deficiency and mannose therapy. J Clin Invest 101:1414-1420
25. Pelletier VA, Galeano N, Brochu P et al (1985) Secretory diarrhea with protein-losing enteropathy, enterocolitis cystica superficialis, intestinal lymphangiectasia and congenital hepatic fibrosis: a new syndrome. J Pediatr 108:61-65
26. Burda P, Borsig L, de Rijk-van Andel J et al (1998) A novel carbohydrate-deficient glycoprotein syndrome characterized by a deficiency in glucosylation in the dolichol-linked oligosaccharide. J Clin Invest 102:647-652
27. Grünewald S, Imbach T, Huijben K et al (2000) Clinical and biochemical characteristics of congenital disorder of glycosylation type Ic, the first recognized endoplasmic reticulum defect in N-glycan synthesis. Ann Neurol 47:776-781
28. Wicklund CL, Pauli RM, Johnston D, Hecht JT (1995) Natural history study of hereditary multiple exostoses. Am J Med Genet 55:43-46
29. Wuyts W, Van Hul W (2000) Molecular basis of multiple exostoses: mutations in the EXT1 and EXT2 genes. Hum Mutat 15:220-227
30. Cormand B, Pikho H, Bayes M et al (2001) Clinical and genetic distinction between Walker-Warburg syndrome and muscle-eye-brain disease. Neurology 56:1059-1069
31. Muntoni F, Brockington M, Blake DJ, Torelli S, Brown SC (2002) Defective glycosylation in muscular dystrophy. Lancet 360:1419-1421
32. Beltrán-Valero de Bernabé D, Currier S, Steinbrecher A et al (2002) Mutations in the O-mannosyltransferase gene *POMT1* give rise to the severe neuronal migration disorder Walker-Warburg syndrome. Am J Hum Genet 71:1033-1043
33. Yoshida A, Kobayashi K, Manya H et al (2001) Muscular dystrophy and neuronal migration disorder caused by mutations in a glycosyltransferase, POMGnT1. Dev Cell 1:717-724
34. Spaapen LJM, Bakker JA, van der Meer SB et al (2005) Fatal outcome in sibs with a newly recognized multiple glycosylation disorder involving O-and N-glycosylation. J Inherit Metab Dis 28:707-714

42 Cystinosis

Michel Broyer

Lysosomal Porters for Cystine and Related Compounds

Intralysosomal cystine is formed by protein catabolism in the organelle, and is normally exported by a cystine porter (◘ Fig. 42.1) which contains a membrane protein, *cystinosin*. Defects of this protein cause lysosomal accumulation of cystine. Cysteamine can enter into the lysosome and combine with cystine. This results in the formation of cysteine (which can be exported by the cysteine porter), and of the mixed disulfide, cysteine-cysteamine (which, due to its structural analogy, can be exported by the lysine porter).

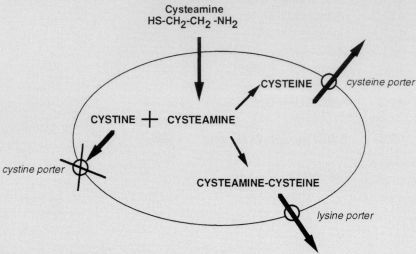

◘ **Fig. 42.1.** Lysosomal export of cystine and related compounds

Cystinosis is a generalized lysosomal storage disease classified into three clinical phenotypes, of which the nephropathic or infantile form is by far the most frequent. The first symptoms start at about 6 months of age with anorexia, polyuria, failure to thrive and are manifestations of a Fanconi proximal renal tubulopathy. The natural history is that of end stage renal disease between 6 and 12 years. Survival beyond this age is associated with the development of extrarenal complications in eyes, thyroid, gonads, endocrine pancreas, muscle and central nervous system. An intermediate or juvenile onset form, and a benign or adult form limited to the eyes, are caused by lesions of the same gene. The lysosomal cystine accumulation leads to cellular dysfunction of many organs without a clear mechanism. The disease is caused by mutations in the *CTNS* gene coding for *cystinosin*, a lysosomal carrier protein. The diagnosis is ascertained by measurement of cystine in leukocytes. Treatment is both supportive and specific, the latter based on cysteamine, which effectively decreases cystine accumulation.

42.1   Infantile Cystinosis

42.1.1   Clinical Presentation

First Stage

The first 3–6 months of life are usually symptom-free. The first symptoms mostly develop before the age of 1 year [1]. They include feeding difficulties, anorexia, vomiting, polyuria, constipation and failure to thrive. If the diagnosis is delayed, severe rickets develops after 10-18 months in spite of correct vitamin D supplementation. Due to a severe concentrating defect a polyuria of 2-5 liters/day develops rapidly. Urine from cystinotic patients is characteristic, being pale and cloudy with a peculiar odour, probably due to aminoaciduria, and the diagnosis can be immediately suspected if both glucose and protein are found. When the disease has become symptomatic, the full expression of the Fanconi syndrome is generally present at first examination. It includes normoglycemic glycosuria, generalized aminoaciduria, tubular proteinuria (with massive excretion of β2-microglobulin and lysozyme), decreased reabsorption of phosphate with hypophosphatemia, excessive losses of potassium and sodium bicarbonate leading to hypokalemia, hyponatremia and metabolic acidosis. Hypercalciuria is also massive and hypouricemia is constant. Tubular loss of carnitine may cause carnitine depletion. Kidney biopsy shows tubular abnormalities and some cystine crystals; plurinucleated glomerular podocytes are also characteristic.

The general reabsorptive defect of the proximal tubule explains the severe hydroelectrolyte imbalance which may be life-threatening. Episodes of fever, probably related to dehydration, are also commonly noted. Lithiasis has been reported in rare cases, related to the high urinary excretion of urate, calcium and organic acids, and nephrocalcinosis may be observed [2]. Blond hair and a fair complexion with difficult tanning after exposure to the sun are often noted in white Caucasian cystinotic children.

Involvement of the eye is a primary symptom of cystinosis, starting with photophobia, which usually appears at 2 or 3 years of age. Ophthalmological examination with a slit lamp and a biomicroscope reveals cystine crystal deposits. There are also fundal abnormalities with typical retinopathy and subsequent alterations of the retinogram which usually appear later.

End Stage Renal Failure

The natural history of the disease includes severe stunting of growth and a progressive decrease of the glomerular filtration rate, leading to end stage renal failure (ESRF) between 6 and 12 years of age. The pathological basis is a dramatic atrophy of the kidneys with glomerular sclerosis and tubulo-interstitial fibrosis. Early onset of renal failure without marked Fanconi syndrome has been reported [3]. This evolution may be delayed by cysteamine treatment especially when started in the first months of life. This treatment also improves growth velocity. During the course of renal deterioration, the decrease in glomerular filtration is reflected by an improvement of urinary losses and a spurious regression of the Fanconi syndrome. In ESRF severe renal hypertension may develop. Repeat nasal bleeding is sometimes observed in cystinotic patients on dialysis [4]. After kidney transplantation there is no recurrence of the Fanconi syndrome even if cystine crystals are seen in the graft, where they are located inside macrophages or leukocytes. When found, tubular symptoms in grafted patients are due to a rejection reaction.

Late Symptoms

The advent of renal replacement therapy and transplantation has uncovered the continued cystine accumulation in extrarenal organs and has emphasized the multisystemic nature of cystinosis, which may additionally involve the eyes, thyroid, liver, spleen, pancreas muscle and central nervous system (CNS) [4–7].

Late Ocular Complications

The severity of eye involvement differs from one patient to another [8]. Corneal deposits accumulate progressively in the stroma of the cornea and iris in all patients and on the surface of the anterior lens and retina in some. Photophobia, watering and blepharospasm may become disabling; these symptoms are often related to erosion of corneal epithelium, leading eventually to keratopathy. Photophobia may be prevented and even completely cured by cysteamine eyedrops [9]. Sight may be progressively reduced, leading to blindness in a few patients who already had major ocular symptoms at an early age and a severe retinopathy. Cataract has been reported [10].

Endocrine Disturbances

Hypothyroidism. Thyroid dysfunction usually appears between 8 and 12 years of age, but it may be earlier or later. It is rarely overt with clinical symptoms, but rather discovered by systematic assessment of thyroid function [4], and it may be partly responsible for the growth impairment. Cysteamine was reported to delay or prevent thyroid dysfunction [11].

Gonadal Function. Abnormalities in the pituitary testicular axis with a low plasma testosterone and high FSH/LH level [12, 13] seem common in male patients with cystinosis. They may preclude full pubertal development. Female patients exhibit pubertal delay but seem to have more normal gonadal functions and there are several reports of successful pregnancies.

Endocrine Pancreas. Postoperative hyperglycemia and permanent insulin-dependent diabetes have been reported in several series of cystinotic patients after kidney transplantation. In patients not treated by cysteamine, 50% had diabetes according to the WHO definition [14]. The exocrine pancreas is usually not affected except in one reported case with steatorrea [15].

Liver and Spleen Involvement

Hepatomegaly and splenomegaly occur in one-third to one-half of the cases after 15 years of age who did not receive cysteamine [4]. Hepatomegaly is related to enlarged Kupffer's cells that transform into large foam cells containing cystine crystals. This enlargement may be the cause of portal hypertension with gastroesophageal varices. Splenomegaly is also related to the development of foam cells in the red pulp. Hematological symptoms of hypersplenism may be noted. A recent study has shown that cysteamine prevents this type of complications [16].

Muscle

A distinctive myopathy, potentially leading to a severe handicap, has been reported in some patients with generalised muscle atrophy and weakness, mainly of distal muscles of all limbs but with more severe involvement of the interosseous muscles and those of the thenar eminence [17, 18]. Pharyngeal and oral dysfunction, which may also cause voice changes, is often observed and has been attributed to muscle dysfunction [19, 20]. Pulmonary dysfunction with an extraparenchymal pattern of restrictive lung disease was recently reported in a series of adult nephropathic cystinotics up to 40 years of age; it was directly correlated with the severity of myopathy [21]. It is not clear if a case of cardiomyopathy reported in a patient was directly related to cardiac cystine accumulation [22].

Central Nervous System

Cystinosis does not affect general intellectual performances. Nevertheless, several kinds of neurologic complications have been reported in cystinosis. Convulsions may occur at any age, but it is difficult to evaluate the direct responsibility of cystinosis in this complication aside uremia, electrolyte desequilibrium, drug toxicity etc. A subtle and specific visuoperceptual defect and lower cognitive performances with sometimes subtle impairment of visual memory and tactile recognition have been reported [23, 24]. More severe CNS abnormalities with various defects have also been described [7, 25]. The clinical symptoms include hypotonia, swallowing and speech difficulties, development of bilateral pyramidal signs and walking difficulties, cerebellar symptoms and a progressive intellectual deterioration leading to a pseudo-bulbar syndrome. In other cases, acute ischemic episodes may occur with hemiplegia or aphasia. This cystinotic encephalopathy was only observed above 19 years of age, and at present it is difficult to know its actual incidence. The effectiveness of cysteamine treatment for the prevention of CNS involvement is not known either. Cysteamine treatment was associated in some cases with an improvement of neurologic symptoms [25]. Brain imaging in cystinosis may show several types of abnormalities. Brain atrophy, calcifications and abnormal features of white matter on magnetic resonance imaging (MRI) examination are commonly observed after 15 to 20 years of age [25, 26].

42.1.2   Metabolic Derangement

Movement of cystine out of cystinotic lysosomes is significantly decreased in comparison with that of normal lysosomes [27, 28]. Consequently, cystine accumulates in many tissues including kidney, bone marrow, conjunctiva, thyroid, muscle, choroid plexus, brain parenchyma and lymph nodes. This abnormality is related to a molecular defect of cystinosin, the protein that transports cystine across the lysosomal membrane. The function of this carrier molecule was demonstrated in a cell model after deletion of the lysosomal targeting signal directing cystinosin to the plasma membrane [29, 30]. Cystine transport out of the lysosomes is H^+ driven. Why lysosomal cystine accumulation leads to cellular dysfunction is not clear. It has been shown that cystine loading of proximal tubular cells in vitro was associated with ATP depletion [31] and inhibition of Na^+ dependent transporters [32]. Cellular cystine excess may also inhibit pyruvate kinase [33], deplete the glutathione cell pool, thereby favouring oxidative stress [34] and apoptosis. This knowledge has led to the use of cysteamine which increases the transport of cystine out of the lysosome. A knock-out mouse model lacking cystinosin recently reported will certainly be helpfull for the understanding of the metabolic derangement [35].

42.1.3   Genetics

Nephropathic cystinosis is an autosomal recessive disorder. The gene, mapped to chromosome 17 and named *CTNS*, encodes a protein of 367 amino acids which has the structure of an integral membrane protein with 7 membrane spanning domains and two lysosomal targeting signals [36]. More than 50 mutations in the first 10 exons and in the promotor of the gene have been identified in association with cystinosis. The most common is a 57 kb deletion found in 76% of patients of European descent. In the other cases, point mutations or shorter deletions are found on both alleles, some of them clustering in certain ethnic and/or geographical areas [37].

Adolescent and adult forms have the same mode of inheritance with mutations that do not disrupt the open reading frame and are generally found in the intertransmembrane loops or in the N-terminal region.

42.1.4   Diagnostic Tests

The reported incidence of the disease is about 1:180,000 live births [38]. Cystinosis is ascertained by measurement of the free cystine content, usually in leukocytes, which in patients with nephropathic cystinosis is about 10-50 times normal [39]. The assay uses a protein-binding technique on white blood cells, is very sensitive, and can be carried out on small, 3 ml blood samples. In cystinosis, the level is usually 5 to 15 nmol of 1/2 cystine/mg protein. The technique enables detection of heterozygous carriers with levels of 0.5 to 1.4 nmol 1/2 cystine/mg protein. In control subjects, cystine is undetectable or <0.4. Ion exchange column chromatography may also be used but is less sensitive. The results obtained on polymorphonuclear leukocytes are approximately twice those obtained on mixed leukocytes and this must be taken in consideration when comparing data. The measurements may also be carried out on fibroblasts, conjunctiva and muscle. S-labeled cystine incorporation in cultured skin fibroblasts, amniotic cells, or chorionic villi enables a prenatal diagnosis in the first trimester to be made [40]. The diagnosis can also be made by mutation analysis if a deletion is found in locus D17 S829, or another mutation in the gene. The use of markers close to this locus when a first child is affected also allows prenatal diagnosis. At birth, diagnosis is possible on placenta or cord blood white cells.

42.1.5   Treatment

The therapy of nephropathic cystinosis is both supportive and specific.

Supportive Treatment of the Tubular Losses
Several abnormalities have to be corrected:

Water. The water intake must be adjusted to diuresis, short-term weight variation and, if necessary, plasma protein concentration. Fluid requirement increases with external temperature and with fever. It is also increased by the required mineral supplements.

Acid Base Equilibrium. Sodium and potassium bicarbonate, which have a better gastric tolerance than citrate, must be given in order to obtain a plasma bicarbonate level between 21 and 24 mmol/l. This is sometimes difficult and may require large amounts of buffer, up to 10–15 mmol/kg.

Sodium. Sodium losses sometimes remain uncompensated after achieving acid base equilibrium. This is recognizable by a persistent hyponatremia with failure to thrive.

Potassium. Hypokalemia requires potassium supplements in order to maintain serum potassium above 3 mmol/l. Four to 10 mmol/kg are usually necessary to achieve this goal. Prescription of Amiloride at a dose of 2-5 mg/day may help in some cases.

Phosphorus. Hypophosphatemia must be corrected with a supplement of sodium/potassium phosphate at a dose of 0.3–1 g/day. The aim is to obtain a plasma phosphate just above 1.0 to 1.2 mmol/l. This poorly tolerated supplement may be gradually withdrawn after some months or years. Excessive phosphorus prescription may lead to nephrocalcinosis.

Vitamin D Supplementation. Since tubular 1α -hydroxylation is diminished in this disease, it is justified to give 1α- or 1α-25-OHD$_3$ (0.10–0.50 µg/day), especially in cases of symptomatic rickets. These prescriptions must be carefully adjusted by regular follow-up of serum calcium.

Carnitine supplementation at a dose of 100 mg/kg per day in four divided doses has been proposed in order to correct muscle carnitine depletion [41].

All these supplements need to be given regularly in order to replace the losses which are permanent. A good way to achieve this goal is to prepare in advance all the supplements, except vitamin D, in a bottle containing the usual amount of water for the day. Losses of water, potassium and sodium may be drastically reduced by the prescription of indomethacin at a dose of 1.5–3 mg/kg in two separate doses [42]. It has been shown that the angiotensin converting enzyme (ACE) inhibitor, enalapril, diminishes albuminuria and possibly slows down the degradation of renal function [43]. When renal degradation progresses, and the glomerular filtration rate decreases, indomethacin must be stopped; at this time tubular losses also decrease and the mineral supplements must be adjusted and progressively tapered off in order to avoid overload, especially with sodium and potassium. At the dialysis stage, mineral supplements are usually no longer necessary.

Feeding problems may require tube or gastric button feeding and in some cases continuous or intermittent total parenteral nutrition [44].

Renal Replacement Therapy

There is no specific requirement for cystinotic children for this procedure at this stage. Hemodialysis or continuous ambulatory/cyclic peritoneal dialysis (CAPD/CPD) are both effective and applied according to the circumstances. As for any child with ESRF, kidney transplantation is considered the best approach. Results of kidney transplantation in the European Dialysis and Transplant Association (EDTA) pediatric registry were better than for any other primary renal disease in children [45].

Supportive Treatment of Extrarenal Complications

Hypothyroidism, even if asymptomatic, should be treated with L-thyroxine supplementation. Growth failure, one of the most striking complications of nephropathic cystinosis, was reported to be improved by administration of recombinant growth hormone at a dose of 1 U/kg/week [46]. Portal hypertension may lead to ascites and bleeding esophageal varices, rendering a portal bypass necessary. Hypersplenism with permanent leukopenia and/or thrombocytopenia may be an indication for splenectomy. Photophobia and watering may be improved by local symptomatic therapy such as vitamin A eye drops, artificial tears, topical lubricants, and thin bandage soft contact lenses. It has been shown that eye drops containing 0.5% cysteamine are able to prevent corneal deposits [9], and may decrease and even suppress the deposits already present. Corneal graft has been rarely performed, with variable results.

Specific Therapy

Several attempts have been made to suppress lysosomal cystine storage. Dietary restriction of sulfur amino acids has no effect. Only one drug, cysteamine ($HS-CH_2-NH_2$), has been employed in cystinosis with apparent benefit, as shown in a prospective study [47]. Cysteamine is now commercially available as cysteamine bitartrate (Cystagon). The dose is progressively increased from 10 to 50 mg/kg of cysteamine base per day. Cysteamine is rapidly absorbed and its maximum effect, assessed by cystine assay in leukocytes, occurs after 1-2 h, and lasts no longer than 6 h [48]. Consequently, it has to be given in 4 separate doses – one every 6 h – in order to obtain the best prevention of cystine accumulation. Careful monitoring of polymorphonuclear leukocyte cystine content is essential since the response to cysteamine is variable. Polymorphonuclear leukocyte cystine content should be determined 6 hours after the last dose and just prior the next: the aim is to keep cystine content under 2, or better, under 1 nmol of 1/2 cystine per mg

of protein. The drug should be started as soon as the diagnosis is confirmed [49, 50]. The good results obtained in the treatment of nephropathy have encouraged the use of cysteamine in patients who are at risk of developing extrarenal complications. Side effects of the drug, which are related to increased acid output, include nausea and vomiting and can be managed with omeprazole [51]. Less commonly, allergic rashes, seizures and neutropenia are seen. In addition, cysteamine is responsible for an unpleasant breath odour so that compliance with 4 doses per day is difficult to maintain in the long term, especially in adolescents [52].

42.2   Adolescent Cystinosis

This is a very rare, milder form of the disease, with later clinical onset and delayed evolution to ESRF. It represents less than 2 or 3% of the cases in a recent epidemiological study [38]. The first symptoms usually appear after 6–8 years of age. Proteinuria may be misleading because its severity is sometimes in the nephrotic range. Fanconi syndrome may be absent [53] or mild, and tubular losses are less important than in infantile cystinosis. The same is true for extrarenal symptoms. ESRF develops around 15 years of age in most patients.

The diagnosis is ascertained by the assessment of the cystine content of leukocytes, which has been found similar to that of infantile cases.

42.3   Adult Benign Cystinosis

Adult or benign cystinosis was first reported by Cogan et al in 1957 [54]. This exceptional disorder [55] is characterized by the presence of cystine crystals in the eye and the bone marrow. Crystals in the cornea are usually found by chance examination. The level of cystine in leukocytes is intermediate between that of heterozygotes and homozygotes for nephropathic cystinosis. All systemic manifestations of the other forms of cystinosis are lacking. The mutations in the *CTNS* gene that have been found in these patients encode a protein that allows sufficient residual cystine transport.

References

1. Broyer M, Guillot M, Gubler MC, Habib R (1981) Infantile cystinosis: a reappraisal of early and late symptoms. Adv Nephrol 10:137-166
2. Theodoropoulos DS, Shawker TH, Heinrichs C, Gahl WA (1995). Medullary nephrocalcinosis in nephropathic cystinosis. Pediatr Nephrol 9:412-418
3. Van't Hoff WG, Ledermann SE, Waldron M, Trompeter RS (1995). Early-onset chronic renal failure as a presentation of infantile nephropathic cystinosis. Pediatr Nephrol 9:483-484

4. Broyer M, Tete MJ, Gubler MC (1987) Late symptoms in infantile cystinosis. Pediatr Nephrol 1:519-524

5. Gahl WA, Kaiser-Kupfer MI (1987) Complications of nephropathic cystinosis after renal failure. Pediat Nephrol 1:260-268

6. Geelen JM, Monnens LA, Levtchenko EN (2002) Follow-up and treatment of adults with cystinosis in the Netherlands. Nephrol Dial Transplant 17:1766-1770

7. Theodoropoulos DS, Krasnewich D, Kaiser-Kupfer MI, Gahl WA (1993). Classic nephropathic cystinosis as an adult disease. JAMA 270:2200-2204

8. Dureau P, Broyer M, Dufier JL (2003) Evolution of ocular manifestations in nephropathic cystinosis: a long-term study of a population treated with cysteamine. J Pediatr Ophthalmol Strabismus 40:142-146

9. Gahl WA, Kuehl EM, Iwata F et al (2000) Corneal crystals in nephropathic cystinosis: natural history and treatment with cysteamine eyedrops. Mol Genet Metab 71:100-120

10. Fahey DK, Fenton S, Mohamed Q, Logan P (2001) Cystinosis, cataract surgery, and corneal erosions. J Cataract Refract Surg 27:2041-2043

11. Kimonis VE, Troendle J, Rose SR et al (1995) Effects of early cysteamine therapy on thyroid function and growth in nephropathic cystinosis. J Clin Endocrinol Metab 80:3257-3261

12. Chik C L, Friedman A, Merriam G R, Gahl WA (1993) Pituitary-testicular function in nephropathic cystinosis. Ann Intern Med 119:568-575

13. Tete MJ, Broyer M (1999) Thyroid and gonads involvement in cystinosis. In: Broyer M (ed) Cystinosis. Elsevier, Paris, pp 70-74

14. Robert JJ, Tête MJ, Guest G et al (1999) Diabetes mellitus in patients with infantile cystinosis after renal transplantation. Pediatr Nephrol 13:524-529

15. Fivusch B, Flick J, Gahl WA (1988) Pancreatic exocrine insufficiency in a patient with nephropathic cystinosis. J Pediatr 112:49-51

16. Gagnadoux MF, Tête MJ, Guest G et al (1999) Hepatosplenic disorders in nephropathic cystinosis. In: Broyer M (ed) Cystinosis. Elsevier, Paris, pp 70-74

17. Gahl WA, Dalakas M, Charnas L, Chen K (1988). Myopathy and cystine storage in muscles in a patient with nephropathic cystinosis. N Engl J Med 319:1461-1464

18. Vester U, Schubert M, Offner G, Brodehl J (2000) Distal myopathy in nephropathic cystinosis. Pediatr Nephrol 14:36-38

19. Sonies BC, Ekman EF, Andersson HC et al (1990) Swallowing dysfunction in nephropathic cystinosis. N Engl J Med 323:565-570

20. Trauner DA, Fahmy RF, Mishler DA (2001) Oral motor dysfunction and feeding difficulties in nephropathic cystinosis. Pediatr Neurol 24(5):365-368

21. Anikster Y, Lacbawan F, Brantly M et al (2001) Pulmonary dysfunction in adults with nephropathic cystinosis. Chest 119:394-401

22. Dixit MP, Greifer I (2002) Nephropathic cystinosis associated with cardiomyopathy: a 27-year clinical follow-up. BMC Nephrol 3:8

23. Colah S, Trauner DA (1997) Tactile recognition in infantile nephropathic cystinosis. Dev Med Child Neurol 39:409-413

24. Ballantyne AO, Trauner DA (2000) Neurobehavioral consequences of a genetic metabolic disorder: visual processing deficits in infantile nephropathic cystinosis. Neuropsychiatry Neuropsychol Behav Neurol 13:254-263

25. Broyer M, Tete MJ, Guest G et al (1996) Clinical polymorphism of cystinosis encephalopathy. Results of treatment with cysteamine. J Inherit Metab Dis 19:65-75

26. Nichols S, Press G, Schneider J, Trauner D (1990). Cortical atrophy and cognitive performance in infantile nephropathic cystinosis. Pediatr Neurol 6:379-381

27. Gahl WA, Bashan N, Tietze F (1982) Cystine transport is defective in isolated leukocyle Lysosomes from patients with cystinosis. Science 217:1263-1265

28. Gahl WA, Tietze F, Bashan N (1983) Characteristics of cystine countertransport in normal and cystinotic Lysosome-rich granular fractions. Biochem J 216:393-400

29. Kalatzis V, Cherqui S, Antignac C, Gasnier B (2001) Cystinosin, the protein defective in cystinosis, is a H(+)-driven lysosomal cystine transporter. EMBO J 20:5940-5949

30. Kalatzis V, Nevo N, Cherqui S et al (2004) Molecular pathogenesis of cystinosis: effect of CTNS mutations on the transport activity and subcellular localization of cystinosin. Hum Mol Genet 13:1361-1371

31. Foreman JW, Bowring MA, Lee J et al (1987) Effect of cystine dimethyl ester on renal solute handling and isolated renal tubule transport in the rat: a new model of the Fanconi syndrome. Metabolism 36:1185-1191

32. Cetinkaya I, Schlatter E, Hirsch Jr et al (2002) Inhibition of Na(+)-dependent transporters in cystine-loaded human renal cells: electrophysiological studies on the Fanconi syndrome of cystinosis. J Am Soc Nephrol 13:2085-2093

33. Feksa LR, Cornelio A, Dutra-Filho CS et al (2004) Inhibition of pyruvate kinase activity by cystine in brain cortex of rats. Brain Res 1012:93-100

34. Chol M, Nevo N, Cherqui S et al (2004) Glutathione precursors replenish decreased glutathione pool in cystinotic cell lines. Biochem Biophys Res Commun 324:231-235

35. Cherqui S, Sevin C, Hamard G et al (2002) Intralysosomal cystine accumulation in mice lacking cystinosin, the protein defective in cystinosis. Mol Cell Biol 22:7622-7632

36. Town M, Jean G, Cherqui S et al (1998) A novel gene encoding an integral membrane protein is mutated in nephropathic cystinosis. Nat Genet 18:319-324

37. Kalatzis V, Antignac C (2003) New aspects of the pathogenesis of cystinosis. Pediatr Nephrol 18:207-215

38. Cochat P, Cordier, Lacôte C, Saïd MH (1999) Cystinosis: epidemiology in France. In: Broyer M (ed) Cystinosis. Elsevier, Paris, pp 28-35

39. Schneider JA, Wong V, Bradley K, Seegmiler JE (1968) Biochemical comparisons of the adult and childhood forms of cystinosis. N Engl J Med 279:1253-1257

40. Patrick AD, Young EP, Mossman J et al (1987) First trimester diagnosis of cystinosis using intact chorionic villi. Prenat Diagn 7:71-74

41. Gahl WA, Bernardini IM, Dalakas MC et al (1993) Muscle carnitine repletion by long-term carnitine supplementation in nephropathic cystinosis. Pediatr Res 34:115-119

42. Haycock GB, Al-dahhan J, Mak RHK, Chantler C (1982) Effect of indomethacin on clinical progress and renal function in cystinosis. Arch Dis Child 57:934-939

43. Levtchenko E, Blom H, Wilmer M et al (2003) ACE inhibitor enalapril diminishes albuminuria in patients with cystinosis. Clin Nephrol 60:386-389

44. Elenberg E, Norling LL, Kleinman RE, Ingelfinger JR (1998) Feeding problems in cystinosis. Pediatr Nephrol 12:365-370

45. Broyer M on behalf of the EDTA registry committee (1989) Kidney transplantation in children, data from the EDTA registry. Transplant Proc 21:1985-1988

46. Wühl E, Haffner D, Offner G et al (2001) Long-term treatment with growth hormone in short children with nephropathic cystinosis. J Pediatr 138:880-887

47. Gahl WA, Reed G, Thoene JG et al (1987) Cysteamine therapy for children with nephropathic cystinosis. N Engl J Med 316:971-977

48. Belldina EB, Huang MY, Schneider JA et al (2003) Steady-state pharmacokinetics and pharmacodynamics of cysteamine bitartrate in paediatric nephropathic cystinosis patients. Br J Clin Pharmacol. 56:520-525

49. Da Silva VA, Zurbrug RP, Lavanchu P et al (1985) Long term treatment of infantile nephropathic cystinosis with cysteamine. N Engl J Med 313:1460-1463

50. Kleta R, Bernardini I, Ueda M et al (2004) Long-term follow-up of well-treated nephropathic cystinosis patients. J Pediatr 145:555-560

51. Dohil R, Newbury RO, Sellers ZM et al (2003) The evaluation and treatment of gastrointestinal disease in children with cystinosis receiving cysteamine. J Pediatr 143:224-230

52. Schneider JA (2004) Treatment of cystinosis: Simple in principle, difficult in practice. Pediatrics 145:436-438

53. Hory B, Billerey C, Royer J, Saint Hillier Y (1994) Glomerular lesions in juvenile cystinosis: report of 2 cases. Clin Nephrol 42:327-330

54. Cogan DG, Kuwabara T, Kinoshita J et al (1957) Cystinosis in a adult. JAMA 164:394

55. Anikster Y, Lucero C, Guo J et al (2000) Ocular nonnephropathic cystinosis: clinical, biochemical, and molecular correlations. Pediatr Res 47:17-23

X

43 Primary Hyperoxalurias

Pierre Cochat, Marie-Odile Rolland

Oxalate Metabolism

Oxalate is a poorly soluble end-product of the metabolism of a number of amino acids, particularly glycine, and of other compounds such as sugars and ascorbic acid. The immediate precursors of oxalate are glyoxylate and glycolate (◘ Fig. 43.1). The main site of synthesis of glyoxylate and oxalate is the liver peroxisome, which can also detoxify glyoxylate by reconversion into glycine, catalyzed by alanine: glyoxylate aminotransferase (AGT). In the cytosol, glyoxylate can be converted into oxalate by lactic acid dehydrogenase (LDH); it can also be converted into glycolate by glyoxylate reductase (GR) and into glycine by glutamate: glyoxylate aminotransferase (GGT). Glycolate can also be formed from hydroxypyruvate, a catabolite of glucose and fructose. Hydroxypyruvate can be converted into L-glycerate by LDH and into D-glycerate by hydroxypyruvate reductase (HPR), which also has a GR activity.

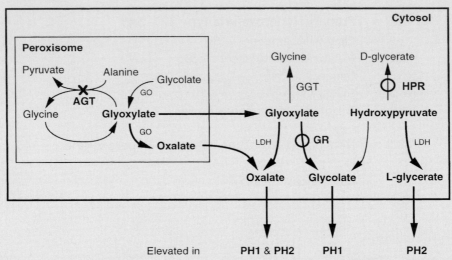

◘ **Fig. 43.1.** Major reactions involved in oxalate, glyoxylate and glycolate metabolism in the human hepatocyte. *AGT*, Alanine: glyoxylate aminotransferase; *GGT*, glutamate: glyoxylate aminotransferase; *GO*, glycolate oxidase; *GR*, glyoxylate reductase; *HPR*, hydroxypyruvate reductase; *LDH*, lactate dehydrogenase; *X*, metabolic block in primary hyperoxaluria type 1 (PH1), *O*, metabolic block in primary hyperoxaluria type 2 (PH2)

◘ **Fig. 43.2.** Bone histology in primary hyperoxaluria type 1: calcium oxalate deposition shown by polarized light microscopy

Primary hyperoxalurias (PH) are rare diseases which are characterized by overproduction and accumulation of oxalate in tissues.

PH1 caused by deficiency or mistargeting of alanine: glyoxylate aminotransferase (AGT) in liver peroxisomes is the most frequent and most severe form. Deposits of calcium oxalate crystals in the kidney lead to stones, nephrocalcinosis and deteriorating kidney function, while bone disease is the most severe extrarenal involvement. Careful conservative treatment (high fluid intake, calcium-oxalate crystallization inhibitors, and pyridoxine) should be started early as it may prolong kidney survival. Liver and kidney transplantation are the final current options. Hyperoxaluria and hyperglycoluria are indicative of PH1.

PH2, caused by glyoxylate-reductase (GR) deficiency in the liver and other tissues, is less frequent and less severe, and treatment less demanding. Hyperoxaluria without hyperglycoluria and increased urinary excretion of L-glycerate differentiate it from PH1.

In addition, there are isolated reports of PH without either AGT or GR deficiency, so that it is likely that there is at least another form of PH (PH3) yet to be explained.

Primary hyperoxaluria (PH) results from endogenous overproduction of oxalic acid and accumulation of oxalate within the body. The main target organ is the kidney since oxalate is excreted in the urine leading to nephrocalcinosis, recurrent urolithiasis and subsequent renal impairment. Primary hyperoxaluria is associated with increased urinary excretion of glycolate in PH1, and of L-glycerate in PH2 (◘ Fig. 43.1). Secondary hyperoxaluria also occurs and is attributed to increased intestinal absorption or excessive intake of oxalate.

43.1 Primary Hyperoxaluria Type 1

43.1.1 Clinical Presentation

PH1, is the most common form of PH. Five different presentations are described: i) a rare infantile form with early nephrocalcinosis and kidney failure; ii) a late-onset form with occasional stone passage in late adulthood; iii) the most common form with recurrent urolithiasis and progressive renal failure leading to a diagnosis of PH1 in childhood or adolescence; iv) a rare condition where the diagnosis is first made following recurrence in a transplanted kidney; and v) pre-symptomatic subjects in whom PH1 is discovered from family history [1].

Renal Involvement

PH1 presents with symptoms referable to the urinary tract in more than 90% of the cases: loin pain, hematuria, urinary tract infection, passage of stones, nephrocalcinosis, uremia, metabolic acidosis, growth delay, and anemia. Oxalate exerts a toxic effect on mitochondrial function of renal epithelial cells and therefore leads to direct tubular damage [2]. However, the most common presentation is stone disease. Calculi – multiple, bilateral and radio-opaque - are made of calcium oxalate. Nephrocalcinosis, best demonstrated by ultrasound, is present on plain abdomen x-ray at an advanced stage.

The median age at initial symptoms is 5 to 6 years, ranging from birth to the 6th decade. End-stage renal disease (ESRD) is reached by the age of 25 years in half of patients [1].

The infantile form often presents as a life-threatening condition because of rapid progression to ESRD due to both early oxalate load and immature glomerular filtration rate (GFR): one-half of the patients experience ESRD at the time of diagnosis and 80 % develop ESRD by the age of 3 years [3, 4].

Extrarenal Involvement

When GFR falls to below 30 to 50 ml/min per 1.73 m^2, continued overproduction of oxalate by the liver along with reduced oxalate excretion by the kidneys leads to a critical saturation point for plasma oxalate (Pox >30 to 50 μmol/l) so that oxalate deposition occurs in many organs [5].

Bone is the major compartment of the insoluble oxalate pool and the bone oxalate content is higher (15 to 910 μmol oxalate/g bony tissue) than among ESRD patients without PH1 (2 to 9 μmol/g). Calcium oxalate crystals accumulate first in the metaphyseal area and form dense suprametaphyseal bands on x-ray. Later on, oxalate osteopathy (◘ Fig. 43.2) leads to pain, erythropoietin-resistant anemia, and spontaneous fractures.

Along with the skeleton, systemic involvement includes many organs because of progressive vascular lesions: heart (cardiomyopathy, arrhythmias, and heart block), nerves (polyradiculoneuropathy), joints (synovitis, chondrocalcinosis), skin (calcium oxalate nodules, livedo reticularis), soft tissues (peripheral gangrene), retina (flecked retinopathy) and other visceral lesions (e.g. intestinal infarction, hypothyroidism) [1].

Systemic involvement – named oxalosis – is responsible for poor quality of life leading to both disability and severe complications. Indeed PH1 is one of the most life-threatening hereditary renal diseases, mainly in developing countries where the mortality rate may reach 100% in the absence of adequate treatment [1].

43.1.2   Metabolic Derangement

PH1 is due to a deficiency or to a mistargeting to the mitochondria of the liver-specific pyridoxal-phosphate-dependent peroxisomal enzyme AGT [6]. The resulting decreased transamination of glyoxylate into glycine leads to subsequent increase in its oxidation to oxalate, a poorly soluble end-product. In patients with a presumptive diagnosis of PH, 10 to 30% are identified as non-PH1 because AGT activity and immunoreactivity are normal [7]. Among PH1 patients, 75% have undetectable enzyme activity (*enz-*) and the majority of these also have no immunoreactive protein (cross reacting material, *crm-*). In the rare *enz-/crm+* patients, a catalytically inactive but immunoreactive AGT is found within the peroxisomes. The remaining PH1 patients have AGT activity in the range of 5 to 50% of the mean normal activity (*enz+*), and the level of immunoreactive protein parallels the level of enzyme activity (*crm+*). In *enz+/crm+* patients, the disease is caused by a mistargeting of AGT: about 90% of the immunoreactive AGT is localized in the mitochondria instead of in the peroxisomes, where only 10% of the activity is found; almost all patients who are pyridoxine-responsive are in this group [8]. Interestingly, human hepatocyte AGT, which is normally exclusively localized within the peroxisomes, is unable to function when diverted to the mitochondria. However patients with a primary peroxisomal disorder – e.g. Zellweger syndrome – do not exhibit hyperoxaluria.

43.1.3   Genetics

PH1 is the most common form of PH (1:60,000 to 1:120,000 live births per year in Europe) [1, 9]. Due to autosomal recessive inheritance, it is much more frequent when parental consanguinity is present. Indeed it is responsible for less than 0.5% of ESRD in children in Europe *versus* 10.4% in Kuwait [10].

The AGT gene (*AGXT*) is a single copy gene located at the telomeric end of chromosome 2q36-q37; the 43 kDa protein contains 392 amino acids. Polymorphisms have been identified: the most common in Europe and North America is P11L, which introduces a weak mitochondrial mistargeting sequence at the N-terminal end of the protein.

More than 50 mutations, some of them more frequent, have been identified so far. They affect either the enzyme or its targeting: G82Q is associated with loss of AGT activity; I244T with increased AGT degradation; G41R with intraperoxisomal AGT aggregation; G170R with peroxisome-to-mitochondrion mistargeting, sometimes with significant AGT catalytic activity; G170R and F152I with pyridoxine responsiveness; homozygous patients for the 33insC mutation with ESRD during infancy [11-17]. I244T and G170R mutations are common in European and North-American patients (~40% of mutant alleles) and interact with the P11L

polymorphism which plays an important role in phenotype determination [13]. DNA analysis among different ethnic groups has revealed the presence of specific mutations, founder effects and phenotype-genotype correlations among North-African, Japanese, Turkish and Pakistani populations [18].

43.1.4   Diagnosis

The diagnosis of PH1 is still being often delayed for years following initial symptoms. The combination of both clinical and radiological signs is a strong argument for PH1, i.e. the association of renal calculi, nephrocalcinosis and renal impairment; family history may bring additional information. Crystalluria and infrared spectroscopy are of major interest for identification and quantitative analysis of crystals and stones, showing calcium oxalate monohydrate crystals (type Ic whewellite) with a crystal number >200/mm^3 in case of heavy hyperoxaluria [19]. Such crystals can also be identified in urine or tissues by polarized light microscopy or infrared spectroscopy. Fundoscopy may show flecked retina.

In patients with normal or significant residual GFR, concomitant hyperoxaluria and hyperglycoluria are indicative of PH1, but some patients do not present with hyperglycoluria. In dialysed patients, plasma oxalate (± glycolate): creatinine ratio and oxalate (± glycolate) measurement in dialysate may be contributive [20].

A definitive diagnosis of PH1 requires the measurement of AGT activity in a liver biopsy. Despite controversial information about the relationship between AGT activity and the severity of the disease [11], liver biopsy assessment is mandatory if a liver transplantation is being considered.

Mutation analysis is based on the sequencing of the 11 exons from the index case and gene segregation is checked in both parents. In the presence of a typical presentation, mutation analysis of the most frequent mutations according to local background may provide a useful first line test without liver biopsy [16].

Prenatal diagnosis can be performed from DNA obtained from crude chorionic villi or amniocytes, on the basis of a restricted analysis of exons including the familial mutation. Such a procedure allows the identification of normal, affected and carrier fetuses.

43.1.5   Treatment and Prognosis

Supportive Treatment

Conservative measures should be started as soon as the diagnosis has been made and even suspected. The aims are to decrease oxalate production and to increase the urinary solubility of calcium oxalate. The risk of stone formation is increased when urine oxalate exceeds 0.4–0.6 mmol/l, especially if urine calcium exceeds 4 mmol/l; therefore supportive therapy should be adapted to keep concentrations of oxalate and calcium below these limits. This should be attempted by giving high fluid intake (>2 l/m^2 per day) and supported by calcium-oxalate crystallization inhibitors, such as citrate (potassium or sodium), 100 to 150 mg/kg per day in 3 to 4 divided doses [9, 21]. When it is not available, crystallization inhibition may be obtained by using sodium bicarbonate, magnesium or orthophosphate [9]. Diuretics require careful management: furosemide will maintain a high urine output with the risk of an increased calciuria whereas the diuretic effect of hydrochlorothiazide is less marked but is associated with an appreciable decrease of calcium excretion. Restriction of dietary oxalate intake (beet, root, strawberries, rhubarb, spinach, coffee, tea, nuts) has limited influence on the disease as oxalate of dietary origin contributes very little to hyperoxaluria in PH [9]. Calcium restriction is not recommended, because less calcium would then bind oxalate and form insoluble calcium-oxalate complexes in the gut. Ascorbic acid supplementation is not recommended as it is a precursor of oxalate.

The absence of intestinal oxalate-degrading bacterium *Oxalobacter formigenes* has been found to be associated with hyperoxaluria, so that increased amounts of such a microorganism in the gut might decrease disposable oxalate.

The main purpose of therapy is to lower both Pox and plasma calcium-oxalate saturation. The effects of conservative measures can be assessed by serial determinations of crystalluria score and calcium oxalate supersaturation software [21, 22].

Pyridoxine (cofactor of AGT) sensitivity (i.e. >30% reduction of urinary oxalate excretion) is found in 10–30% of patients, so that it must be tested early at a daily dose of 2–5 mg/kg with stepwise increase up to 10–20 mg/kg as megadoses of pyridoxine may induce sensory neuropathy [9]. Response to pyridoxine, best detected by oxalate and glycolate measurement, may delay the progression to ESRD [3, 21]; the patients most likely to respond are those with homozygous G170R or F152I mutation, who also experience preserved renal function over time under adequate treatment [17].

The treatment of stones should avoid open and percutaneous surgery because further renal lesions will alter GFR. The use of extra-corporeal shock wave lithotripsy may be an available option in selected patients but the presence of nephrocalcinosis may be responsible for parenchymal damage.

Bilateral nephrectomy is recommended in most patients on renal replacement therapy in order to limit the risk of infection, obstruction and passage of stones.

Renal Replacement Therapy

Dialysis. Conventional dialysis is unsuitable for patients who have reached ESRD because it cannot clear sufficient amounts of oxalate [8]. In such patients, Pox ranges between 70 and 150 µmol/l (control values <7.4 µmol/l) (◻ Table 43.1). Therefore daily hemodialysis (6-8 h per session) using high-flux membranes would be required but such a strategy cannot be routinely used [6]. The challenge is to keep predialysis Pox below 50 µmol/l in order to limit the progression of systemic oxalosis. Conventional long-term hemodialysis is generally contraindicated because it only prolongs a miserable existence.

The benefit of (pre-) post-transplantation hemodialysis is still debated and should be limited to patients with either oliguria or severe systemic burden and subsequent long lasting oxalate release from skeleton.

Kidney transplantation allows significant removal of soluble Pox. However, because the biochemical defect is in the liver, overproduction of oxalate and subsequent deposition in tissues continues unabated. The high rate of urinary oxalate excretion originates from both ongoing oxalate production from the native liver and oxalate deposits in tissues. Due to oxalate accumulation in the graft, isolated kidney transplantation is no longer recommended, because of a 100% recurrence rate leading to poor graft survival and patient quality of life [23]. Indeed, renal transplantation does not prevent the progression of skeletal and vascular complications. The chances of a successful transplantation are unrelated to residual AGT activity but success is improved if it is performed preemptively. Good results have been reported in selected patients after early renal transplantation with

◻ **Table 43.1.** Plasma and urine concentrations of oxalate, glycolate and L-glycerate: control values [8, 9]

Urine	Oxalate per day		<0.50 mmol/1.73 m^2
	Oxalate: creatinine	age <1 year 1–4 years 5–12 years adult	<0.20 mmol/mmol <0.13 mmol/mmol <0.08 mmol/mmol <0.07 mmol/mmol
	Glycolate per day	child adult	<0.55 mmol/1.73 m^2 <0.26 mmol/1.73 m^2
	Glycolate: creatinine	age <1 year 1–4 years 5–12 years adult	<0.07 mmol/mmol <0.09 mmol/mmol <0.05 mmol/mmol <0.04 mmol/mmol
	L-Glycerate: creatinine		<0.03 mmol/mmol
Plasma	Oxalate	child adult	<7.4 µmol/l <5.4 µmol/l
	Oxalate: creatinine	child adult	<0.19 µmol/µmol <0.06 µmol/µmol

Conversion factors: oxalate (COOH-COOH) 1 mmol = 90 mg; glycolate (COOH-CH$_2$OH) 1 mmol = 76 mg.

vigorous pre- and postoperative dialysis [24]; however, living donors should be avoided because the overall results are poor [23, 25]. Isolated kidney transplantation may be regarded as a temporary solution in some countries before managing the patient in a specialized center for further combined liver-kidney procedure.

Independent of initial pyridoxine response, it is recommended to check it again following isolated renal transplantation [24].

Enzyme Replacement Therapy

Ideally, any kind of transplantation should precede advanced systemic oxalate storage [1, 26]. Further assessment of the oxalate burden needs therefore to be predicted by monitoring sequential GFR, Pox (❏ Table 43.1), calcium oxalate saturation and systemic involvement (bone mineral density, bone histology) [1, 27, 28].

Rationale for Liver Transplantation. Since the liver is the only organ responsible for glyoxylate detoxification by AGT, the excessive production of oxalate will continue as long as the native liver is left in place. Therefore any form of enzyme replacement will succeed only when the deficient host liver has been removed. Liver transplantation will supply the missing enzyme in the correct organ (liver), cell (hepatocyte) and intracellular compartment (peroxisome). The ultimate goal of organ replacement is to change a positive whole-body accretion rate into a negative one by reducing endogenous oxalate synthesis and providing good oxalate clearance via either native or transplanted kidney.

Combined Liver-Kidney Transplantation. In Europe, 8 to 10 combined liver-kidney transplantations per year have been reported; the results are encouraging, as patient survival approximates 80% at 5 years and 65–70% at 10 years [25]. Comparable results have been reported from the United States Renal Data System, with a 76% death-censored graft survival at 8 years post transplantation [23]. Such a strategy can be successfully proposed to infants with PH1 [4]. In addition, despite the potential risks for the grafted kidney due to oxalate release from the body stores, kidney survival is about 95% three years post-transplantation and the GFR ranges between 40 and 60 ml/min per 1.73 m^2 after 5 to 10 years [25, 26].

Isolated liver transplantation might be the first-choice treatment in selected patients before advanced chronic renal failure has occurred, i.e. at a GFR between 60 and 40 ml/min/1.73 m^2 [29]. Such a strategy has a strong rationale but raises ethical controversies. Around 25 patients have received an isolated liver transplant without uniformly accepted guidelines, since the course of the disease is unpredictable and a sustained improvement can follow a phase of rapid decrease in GFR [26, 29].

Post-transplantation Reversal of Renal and Extra-renal Involvement. Deposits of calcium oxalate in tissues can be remobilized according to the accessibility of oxalate burden to the blood stream. After combined transplantation, Pox returns to normal before urine oxalate does, and oxaluria can remain elevated as long as several months [8, 24, 25]. Therefore there is still a risk of recurrent nephrocalcinosis or renal calculi that might jeopardize graft function. Glycolate, which is soluble and does not accumulate, is excreted in normal amounts immediately after liver transplantation.

Thus, independent of the transplantation strategy, the kidney must be protected against the damage that can be induced by heavy oxalate load suddenly released from tissues. Forced fluid intake (3-5 l/1.73 m^2 per day) supported by diuretics and the use of crystallization inhibitors is the most important approach. Pox, crystalluria and calcium oxalate saturation are helpful tools in renal management after combined liver-kidney transplantation [22, 28, 30]. The benefit of daily high-efficiency (pre-) post-transplant hemodialysis is still debated; it will provide a rapid drop in Pox but also an increased risk of urine calcium-oxalate supersaturation and therefore should be limited to patients with significant systemic involvement [1, 28, 30].

Combined transplantation should be planned when the GFR ranges between 20 and 40 ml/min per 1.73 m^2 because, at this level, oxalate retention increases rapidly [26, 29]. In patients with ESRD, vigorous hemodialysis should be started and urgent liver-kidney transplantation should be performed. Even at these late stages, damaged organs, such as the skeleton or the heart, do benefit from enzyme replacement [30], which results in an appreciable improvement in quality of life.

Donors for Combined Liver-Kidney Transplantation. The type of donor -cadaver or living- depends mainly on the physician and the country where the patient is treated [3, 31, 32]. According to the timing of transplantation, a living donor may be considered because of the restricted number of potential biorgan deceased donors. A living donor can be proposed in a preemptive procedure using either isolated liver or synchronous liver-kidney transplantation. In patients with ESRD and systemic involvement, a metachronous transplantation procedure might be an option since first-step liver transplantation will then allow oxalate clearance by vigorous hemodialysis before considering further kidney transplantation from the same (living) donor.

Management of Pregnancy

There is limited information about outcomes and complications of pregnancy in women with PH. In a recent paper including 26 pregnancies in 11 patients with PH1 [33], outcomes were: 19 term infants, 2 preterm infants, 3 miscarriages, 1 stillbirth and 1 abortion. No maternal complications were reported in half of the pregnancies; in the remaining ones, the most common problems were hypertension, urinary tract infection and stone-associated symptoms. Only one patient experienced a loss of renal function. Most infants had no complications.

Future Trends

Although gene therapy has been advocated, many years of research will be required before considering its potential use [8]. Different AGT crystal forms have been recently obtained for some polymorphic variants, and aminoacid changes found in these crystals may affect AGT stability [34]. A better understanding of such changes will allow designing pharmacological agents that will stabilize AGT such as chemical chaperones without the need for organ transplantation [Danpure, personal communication].

43.2   Primary Hyperoxaluria Type 2

43.2.1   Clinical Presentation

PH2 has been documented in less than 50 patients but there are some unreported cases. Median age at initial symptoms is 1 to 2 years, and the classical presentation is urolithiasis but stone-forming activity is lower than in PH1 [21]. GFR is usually maintained during childhood and systemic involvement is exceptional.

43.2.2   Metabolic Derangement

PH2 is characterized by the absence of an enzyme with glyoxylate reductase (GR), hydroxypyruvate reductase (HPR), and D-glycerate-dehydrogenase activities (◘ Fig. 43.1) [16]. GR plays a role in the reduction of cytosolic glyoxylate and has a predominant hepatic distribution [16].

43.2.3   Genetics

There is evidence for autosomal recessive transmission and the gene encoding the enzyme GR/HPR (*GRHPR)* has been located on chromosome 9cen [35]. Several missense, nonsense, and deletion mutations have been identified [16].

43.2.4   Diagnosis

The biochemical hallmark is the increased urinary excretion of L-glycerate (◘ Table 43.1) but the definitive diagnosis requires measurement of GR activity in a liver biopsy as some PH2 patients have normal L-glycericaciduria [36]. However, in the presence of hyperoxaluria without hyperglycoluria, a diagnosis of PH2 should be considered and screening of the most frequent mutation (c.103delG) may be a first line molecular genetic approach [16].

43.2.5   Treatment and Prognosis

The overall long-term prognosis is better than for PH1. ESRD occurs in 12% of patients, between 23 and 50 years of age. As in PH1, supportive treatment includes high fluid intake, crystallization inhibitors and prevention of complications. Kidney transplantation has been performed in some patients, often leading to recurrence including hyperoxaluria and L-glycerate excretion [37]. Liver transplantation has therefore been suggested, but more data are needed concerning the tissue distribution of the deficient enzyme and the biochemical impact of hepatic GR/HPR deficiency before such a strategy can be recommended.

Fourteen pregnancies have been reported in 5 patients with PH2 [33]: 11 led to term infants, 1 to a preterm infant, 1 to stillbirth and 1 to abortion. There were no maternal complications in 6 pregnancies but 2 had hypertension, 2 experienced urinary tract infections, and 2 had stone problems. Ten infants had no significant complications; others suffered from pulmonary complications (1), cerebral palsy (1) or developmental delay (1).

43.3   Non-Type 1 Non-Type 2 Primary Hyperoxaluria

There are some isolated reports of PH without either AGT or GR/HPR deficiency and of PH with hyperglycoluria in the absence of AGT deficiency [14]. It is therefore likely that there is at least another form of PH yet to be explained.

References

1. Cochat P, Collard LBDE (2004) Primary hyperoxalurias. In: Avner ED, Harmon WE, Niaudet P (eds) Pediatric nephrology, 5th edn. Lippincott Williams & Wilkins, Baltimore, pp 807-816
2. Cao LC, Honeyman TW, Cooney R et al (2004) Mitochondrial dysfunction is a primary event in renal cell oxalate toxicity. Kidney Int 66:1890-1900
3. Cochat P, Koch Nogueira PC, Mahmoud AM et al (1999) Primary hyperoxaluria in infants: medical, ethical and economic issues. J Pediatr 135:746-750
4. Millan MT, Berquist WE, So SK et al (2003) One hundred percent patient and kidney allograft survival with simultaneous liver and kidney transplantation in infants with primary hyperoxaluria: a single-center experience. Transplantation 76:1458-1463
5. Monico CG, Wilson DM, Bergert JH, Milliner DS (1999) Renal oxalate handling and plasma oxalate concentration in patients with chronic renal insufficiency and in patients with primary hyperoxaluria. J Am Soc Nephrol 10:82A
6. Diaz C, Catalinas FD, de Alvaro F et al (2004) Long daily hemodialysis sessions correct systemic complications of oxalosis prior to combined liver-kidney transplantation : case report. Ther Apher Dial 8:52-55
7. Fargue S, Chevalier-Prost F, Rolland MO, Cochat P (2002) Diagnosis of primary hyperoxaluria type 1: a one-centre experience. Pediatr Nephrol 17:C52

8. Barratt TM, Danpure CJ (1999) Hyperoxaluria. In: Barratt TM, Avner ED, Harmon WE (eds) Pediatric nephrology, 4th edn. Williams & Wilkins, Baltimore, pp 609-619

9. Leumann E, Hoppe B (2001) The primary hyperoxalurias. J Am Soc Nephrol 12:1986-1993

10. Al-Eisa AA, Samhan M, Naseef M (2004) End-stage renal disease in Kuwaiti children: an 8-year experience. Transplant Proc 36:1788-1791

11. Amoroso A, Pirulli D, Florian F et al (2001) AGXT gene mutations and their influence on clinical heterogeneity of type 1 primary hyperoxaluria. J Am Soc Nephrol 12:2072-2079

12. Coulter-Mackie MB, Applegarth D, Toone JR, Henderson H (2004) The major allele of the alanine: glyoxylate aminotransferase gene: seven novel mutations causing primary hyperoxaluria. Mol Genet Metab 82:64-68

13. Lumb MJ, Danpure CJ (2000) Functional synergism between the most common polymorphism in human alanine: glyoxylate aminotransferase and four of the most common disease-causing mutations. J Biol Chem 275:36415-36422

14. Monico CG, Persson M, Ford GC et al (2002) Potential mechanisms of marked hyperoxaluria not due to primary hyperoxaluria I or II. Kidney Int 62:392-400

15. Nogueira PC, Vuong TS, Bouton O et al (2000) Partial deletion of the AGXT gene (EX1_EX7del): a new genotype in hyperoxaluria type 1. Hum Mutat 15:384-385

16. Rumsby G, Sharma A, Cregeen DP, Solomon LR (2001) Primary hyperoxaluria type 2 without L-glyceraciduria: is the disease underestimated? Nephrol Dial Transplant 16:1697-1699

17. van Woerden CS, Groothoff JW, Wijburg FA et al (2004) Clinical implications of mutation analysis in primary hyperoxaluria type 1. Kidney Int 66:746-752

18. Basmaison O, Rolland MO, Cochat P, Bozon D (2000) Identification of 5 novel mutations in the AGXT gene. Hum Mutat 15:577

19. Daudon M, Jungers P, Lacour B (2004) Intérêt clinique de l'étude de la cristallurie. Ann Biol Clin 62:379-393

20. Wong PN, Law ELK, Tong GMW et al (2003) Diagnosis of primary hyperoxaluria type 1 by determination of peritoneal dialysate glycolic acid using standard organic-acids analysis method. Perit Dial Int 23:S210-213

21. Milliner DS, Wilson DM, Smith LH (2001) Phenotypic expression of primary hyperoxaluria: comparative features of types I and II. Kidney Int 59:31-36

22. Jouvet P, Priquelier L, Gagnadoux MF et al (1998) Crystalluria: a clinically useful investigation in children with primary hyperoxaluria post-transplantation. Kidney Int 53:1412-1416

23. Cibrik DM, Kaplan B, Arndorfer JA, Meier-Kriesche HU (2002) Renal allograft survival in patients with oxalosis. Transplantation 74:707-710

24. Monico CG, Milliner DS (2001) Combined liver-kidney and kidney-alone transplantation in primary hyperoxaluria. Liver Transplant 11:954-963

25. Jamieson NV (1998) The results of combined liver/kidney transplantation for primary hyperoxaluria 1984-1997. The European PH1 transplant registry report. European PH1 Transplantation Study Group. J Nephrol 11[Suppl 1]:36-41

26. Ellis SR, Hulton SA, McKiernan PJ et al (2001) Combined liver-kidney transplantation for primary hyperoxaluria in young children. Nephrol Dial Transplant 16:348-354

27. Behnke B, Kemper MJ, Kruse HP, Muller-Wiefel DE (2001) Bone mineral density in children with primary hyperoxaluria type 1. Nephrol Dial Transplant 16:2236-2239

28. Hoppe B, Kemper MJ, Bokenkamp A et al (1998) Plasma calcium oxalate supersaturation in children with primary hyperoxaluria and end-stage renal failure. Kidney Int 56:268-274

29. Shapiro R, Weismann I, Mandel H et al (2001) Primary hyperoxaluria type 1: improved outcome with timely liver transplantation: a single center report of 36 children. Transplantation 72:428-432

30. Gagnadoux MF, Lacaille F, Niaudet P et al (2001) Long term results of liver-kidney transplantation in children with primary hyperoxaluria. Pediatr Nephrol 16:946-950

31. Nakamura M, Fuchinoue S, Nakajima I et al (2001) Three cases of sequential liver-kidney transplantation from living related donors. Nephrol Dial Transplant 16:166-168

32. Nolkemper D, Kemper MJ, Burdelski M et al (2000) Long-term results of pre-emptive liver transplantation in primary hyperoxaluria type 1. Pediatr Transplant 4:177-181

33. Norby SM, Milliner DS (2004) Outcomes and complications of pregnancy in women with primary hyperoxaluria. Am J Kidney Dis 43:277-285

34. Zhang X, Roe SM, Pearl LH, Danpure CJ (2001) Crystallization and preliminary crystallographic analysis of human alanine: glyoxylate aminotransferase and its polymorphic variants. Acta Crystallogr 57:1936-1937

35. Cramer SD, Ferree PM, Lin K et al (1999) The gene encoding hydroxypyruvate reductase (GRHPR) is mutated in patients with primary hyperoxaluria type II. Hum Mol Genet 8:2063-2069

36. Cramer SD, Ferree PM, Lin K et al (1999) The gene encoding hydroxypyruvate reductase (GRHPR) is mutated in patients with primary hyperoxaluria type II. Hum Mol Genet 8:2063-2069

37. Rumsby G, Williams E, Coulter-Mackie M (2004) Evaluation of mutation screening as a first line test for the diagnosis of the primary hyperoxaluria. Kidney Int 66:959-963

38. Johnson SA, Rumsby G, Cregeen D, Hulton SA (2002) Primary hyperoxaluria type 2 in children. Pediatr Nephrol 17:597-601

Foundations dealing with hyperoxaluria:

http://www.ohf.org	The Oxalosis and Hyperoxaluria Foundation
http://www.airg-france.org	Association pour l'Information et la Recherche sur les Maladies Rénales Génétiques

X

Subject Index

D

G

Printing and Binding: Stürtz GmbH, Würzburg